Natural Language Semantics

Formation and Valuation

Brendan S. Gillon

The MIT Press
Cambridge, Massachusetts
London, England

© 2019 Massachusetts Institute of Technology

All rights reserved. No part of this book may be reproduced in any form by any electronic or mechanical means (including photocopying, recording, or information storage and retrieval) without permission in writing from the publisher.

This book was set in Nimbus Roman by Westchester Publishing Services. Printed and bound in the United States of America.

Library of Congress Cataloging-in-Publication Data

Names: Gillon, Brendan S., author.
Title: Natural language semantics : formation and valuation / Brendan S. Gillon.
Description: Cambridge, MA : The MIT Press, [2019] | Includes bibliographical references and index.
Identifiers: LCCN 2018017619 | ISBN 9780262039208 (hardcover ; alk. paper)
Subjects: LCSH: Semantics—Textbooks. | English language—Semantics—Textbooks. | LCGFT: Textbooks.
Classification: LCC P325 .G55 2019 | DDC 401/.43—dc23 LC record available at https://lccn.loc.gov/2018017619

10 9 8 7 6 5 4 3 2 1

To my children, Evan and Colleen, who have enriched my life so very much

Contents

	Preface	ix
	Table of Greek Letters	xv
1	**Language, Linguistics, Semantics: An Introduction**	1
2	**Basic Set Theory**	51
3	**Basic English Grammar**	105
4	**Language and Context**	175
5	**Language and Belief: Implicatures and Presuppositions**	209
6	**Classical Propositional Logic: Notation and Semantics**	237
7	**Classical Propositional Logic: Deduction**	275
8	**English Connectors**	321
9	**Classical Predicate Logic**	375
10	**Grammatical Predicates and Minimal Clauses in English**	397
11	**Classical Quantificational Logic**	457
12	**Enrichments of Classical Quantificational Logic**	521
13	**The Lambek Calculus and the Lambda Calculus**	551
14	**Noun Phrases in English**	615
15	**Conclusion**	673
	References	683
	List of Symbols	693
	Index	701

Preface

This book started, as many—even most—such books, as course notes for an introductory course in natural language semantics both for undergraduates students and for first-year graduate students. Even so, it nonetheless contains some of my own research.

The book's central question is the central question of natural language semantics: namely, how is one's understanding of a complex expression of natural language determined by one's understanding of the expressions making it up? This question is investigated with respect to English. And while English is the principal target language, other languages are occasionally discussed.

To answer the central question requires a characterization of how complex expressions are formed, on the one hand, and a way of characterizing one's understanding of the expressions, on the other. The characterization of how complex expressions of a natural language are formed is its syntax, and there are well-understood logical techniques to formalize natural language syntax. Characterizing how an expression is understood is another matter entirely. Traditionally, one speaks of what one understands when one understands an expression of natural language as the expression's meaning. But the notion of meaning is an elusive one and its nature continues to remain controversial. However, there is an ersatz for meaning, namely, set theoretic entities, which, when used as values assigned to expressions, provide at least a partial answer to the central question. The idea is to assign set theoretic entities to all expressions and to do so in such a way that the set theoretic entity assigned to a complex expression is determined by the set theoretic entities assigned to the simpler expressions making it up. The situation is analogous to the situation in logic. The notation of a logic has complex expressions that are made up of simpler expressions. Set theoretic entities are assigned to the expressions by assigning entities to the simpler expressions and using the formation rules, sometimes known as valuation rules, to state how an entity is assigned to the complex expressions made up from the simpler ones. It turns out that, in some cases, the techniques of logic can be applied to natural language expressions, off the shelf as it were; in many cases, however, the techniques of logic have to be adapted to the vagaries of natural language syntax. The subject matter of the book comprises, then on the one hand, the study of the patterns of the formation of complex expressions from simpler ones and the adoption or adaption from logic of rules, or valuations, whereby values are assigned to the complex expressions on the basis of the values assigned to the simpler ones.

This conception of natural language semantics is shared by many, if not most, approaches to the contemporary study of natural language semantics. However, the treatment in this book differs in several important ways from treatments found in other textbooks. Let me explain how.

The patterns in the expressions of a natural language semantics are rich and varied, and the theoretical approaches to them based on logic are numerous and demand a fair degree of logical knowledge. Students coming to the study of semantics, however, rarely have studied any logic and usually have studied no mathematics since secondary school. Most authors have therefore decided to teach one theoretical approach, their own, giving only a minimum explanation of the relevant logical techniques, and choosing only examples that suit the theory being taught. Thus, the facts presented are those that the theory handles best and the logic taught is just the logic used by that theory.

While such textbooks are often very good at introducing students to a particular theory of natural language semantics, they have, in my view, several drawbacks. First, having studied one of these textbooks, their readers are not in a position to read and understand research based on other theories of natural language semantics. This has the lamentable effect of balkanizing the study of natural language semantics: students trained in one theory shrink from reading research within other theories and therefore do not consider treatments of the same area of research done within alternative theories. Second, since the phenomena presented are just those that illustrate the theory in question, readers have no real idea either of the complexity of the relevant data or of the limits of the logical techniques they have been taught. Third, these textbooks, with the exception of Gamut's *Logic, Language and Meaning* (1991), underestimate the training that students with no postsecondary training in mathematics require in order to learn how to use the logical techniques appropriately and creatively. It is unrealistic in the extreme to expect that students who, for the most part, have done no mathematics since secondary school, can pick up the logic that was pioneered by mathematicians and logicians as recently as just the middle of the last century. Indeed, the perennial complaint I hear from students and colleagues who have been taught natural language semantics in this way and who have had no undergraduate level training in mathematics is that they use the logical techniques they have been taught by rote, imitating what they have seen without much, if any, understanding of them.

This book aspires to enable readers with little or no postsecondary mathematical training to avoid these pitfalls. Rather than aiming to take readers on the shortest route possible to the point where they can read the literature of one particular theory, often losing all except those with an antecedent knack for mathematics, I take a longer route, which, though slower, takes diligent readers, regardless of mathematical or logical talent, to the point where they can read the literature in a variety of theories. First, I concentrate on those ideas of logic essential to the various natural language semantic theories using logical techniques and take the time necessary to ensure that these ideas are understood and mastered. Thus, where other texts relegate the explanation of the notions of set theory necessary to the exposition of the logical techniques to an appendix of one or two dozen pages or to the same number of pages scattered here and there along the way, I gather all the notions into one fifty-page chapter, right after the introductory chapter, and provide extensive explanations and exercises. Second, I strive to ensure that readers have a good grasp of the range of patterns in English expressions and make clear to them which patterns have been successfully treated, which, though not treated, are amenable to a successful treatment, and which seem out of reach of any known treatment. The book, therefore, devotes much space to presenting the patterns of English that are to be treated, drawing on the best available contemporary descriptive grammars of English. Finally, most students who come to linguistics are students in the faculty of arts. They have little or no exposure to what empirical inquiry is, beyond what they may have gleaned from science classes in secondary school. In addition, many naive or outdated ideas about the nature of empirical inquiry circulate in linguistics. True, few, if any, linguists appeal to behaviorist and verificationist ideas of scientific method, popular among linguists in the first half of the last century, but many appeal to the ideas that emerged in the middle of the last century, such as Karl Popper's falsificationism, and which, though serving as a healthy corrective to excesses of behaviorism and verificationism, themselves commit excesses that more recent work in the philosophy of science has corrected. These ideas, explained in the first chapter, are used and illustrated in subsequent chapters.

The hope is that the book will provide readers with a mastery of the basic logical tools so that either they can read the more accessible research in natural language semantics, regardless of the theory assumed by the author of the research, or should they require further background, they can at least read introductory expositions of the theory easily and quickly. In addition, it is also hoped that readers will acquire an appreciation not only of the breadth of the field but also of how reasoning in empirical research is done.

Though the material in the book began as material used in a one-semester course in natural language semantics, the material in it now is used in a sequence of two courses, each one semester long. The first semester covers chapters 1 through 6 and 8 through 10. This course centers on constituency grammars, the kind of grammar developed at the inception of what was known as transformational grammar and other subsequently developed grammars such as lexical functional grammar and head-driven phrase structure grammar. The limitation of the first course is that students get no exposure to the kind of grammar that grew from categorial grammar. The second semester presents type logical grammar, or Lambek-typed grammar as I prefer to call it, and discusses the similarities and differences between the two kinds of grammars. This material is found in chapters 7 and 11 through 15.

The material taught in the course is entirely self-contained. Though many students have taken both an introductory course in linguistics and an introductory course in logic, neither course is necessary. Indeed, every year I have some students who have taken neither introductory course take my course and still perform well in it. It is only students who have little confidence in their mathematical abilities that I encourage to take an introductory logic course before taking my course. The question is not whether students can learn the logic, the question is only how much time they require to learn it. Those who require a little more time are encouraged to take a logic course either before or concurrently with my course. My experience has been that students who know how to study, even if they have no penchant for mathematics, have no trouble learning the logic taught in the course.

Despite its length, this is an introductory book and many relevant topics have been omitted. To begin with, little has been said about languages other than English, though much of what applies to English applies to many other languages. Moreover, many aspects of English relevant to semantics have been omitted. Also, many relevant forms of logic have been omitted, including modal logic, multivalued logic, and dynamic logic, all with important applications in the study of natural language semantics. (An overview of the omissions is found in chapter 15, section 3.)

Although over the years my notes grew into detailed written chapters, I did not think to publish the results until very recently. I am most grateful to Marc Lowenthal of MIT Press for the interest he has taken in this book and his encouragement. I am also very appreciative of the many improvements brought to my original manuscript by the editing skills of Ms. Deborah Grahame-Smith and other members of her team.

During the many years I have been writing this book, I accumulated a great debt to the many students, colleagues, and friends who contributed to this work in a variety of ways. I apologize to those whose names have slipped my mind with the passage of time.

Let me first express my appreciation to the two decades of students who took my class, especially to the assiduous and curious ones whose questions made me revisit my explanations time and again. A number of these students—including Robin Anderson, Gabriel Gaudreault, Isabelle Oke, Symon Stevens Guille, and Daniel Zackon—helped with editing and formatting various versions of these chapters. I would also like to express my appreciation to Hisako Noguchi, Stephan Hurtubise, and Isabelle Oke for their assistance with the formidable task of helping me proofread and index the book. I have also many times had the good fortune to be assisted by very bright and dedicated teaching assistants who were instrumental in improving the exposition of this work and the exercises: September Cowley, Daniel Goodhue, and Walter Pedersen. My sincere thanks to each of them. I am indebted to a number of friends and colleagues with whom I discussed this work or who were kind enough to read and comment on early versions of the chapters. They gave freely of their time, providing the explanations and advice that I sought and their thoughts and comments proved invaluable. They include Nigel Duffield, Colleen Gillon, Wilfried Hodges, Michael Makkai, Janet Martin Nielsen, David Nicolas, Jeff Oaks, Paul Pupier, Peter Scharf, and Ben Shaer. I am also very grateful to those who engaged in a detailed reading and commenting on many of the chapters: Alan Bale, Brian van den

Broek, Benjamin Caplan, Steven Davis, Jeff Pelletier, Dirk Schlimm, Robert Seeley, and Ken Turner.

Finally, I would like to take this opportunity to acknowledge my personal and intellectual debt to my late colleague Jim Lambek, who sought me out when I first started at McGill University, became a friend and always encouraged me in my work. I admired Jim not only for his remarkable mathematical achievements, only a few of which I have the mathematical background to appreciate, but particularly for his exceptionally broad intellectual interests, spanning mathematics and its history to philosophy and linguistics.

Table of Greek Letters

Greek Alphabet

Lowercase	Uppercase	Name
α	A	alpha
β	B	beta
γ	Γ	gamma
δ	Δ	delta
ϵ	E	epsilon
ζ	Z	zeta
η	H	eta
θ	Θ	theta
ι	I	iota
κ	K	kappa
λ	Λ	lambda
μ	M	mu
ν	N	nu
ξ	Ξ	xi
o	O	omicron
π	Π	pi
ρ	P	rho
σ	Σ	sigma
τ	T	tau
υ	Υ	upsilon
ϕ	Φ	phi
χ	X	chi
ψ	Ψ	psi
ω	Ω	omega

1 Language, Linguistics, Semantics: An Introduction

1   The Study of Language before the Twentieth Century

The field of linguistics has its origins in the study of grammar. The study of grammar arises when one attempts to describe the patterns of a language. Many civilizations undertook to describe their classical languages. The Hellenistic civilization undertook to describe Classical Greek. The civilization of classical India undertook to describe its language, Classical Sanskrit. The tradition initiated in Europe by the Greeks was carried on by the Romans, and later by the Medieval Europeans. When, at the end of the Middle Ages, Latin lost its grip as the intellectual lingua franca of Europe and the various European languages acquired the status of languages of culture, these languages came to be studied grammatically—always through the lens of Classical Latin. At the same time, the voyages of discovery brought Europeans into contact with the various languages of the world, and the need to teach those languages to other Europeans led in many cases to the formulation of grammars, based, not surprisingly, on the model of the grammars of Classical Latin. Initially, many authors writing such grammars sought to impose some uniformity on language usage through the prescription of rules they themselves invented. Later, many authors sought simply to describe the language usage they observed.

Indeed, nearly down to the present day, the grammar taught in school has had only tangential connections with the studies pursued by professional linguists; for most people prescriptive grammar has become synonymous with "grammar," and most educated people continue to regard grammar as an item of folk knowledge open to speculation by all, and in no way a discipline requiring special preparation such as is assumed for chemistry or law.

Another important development leading to the creation of the field of linguistics was the rise of philology.[1] At the end of the eighteenth century, Europeans came into extensive

---

1. Philology is the "study of literature that includes or may include grammar, criticism, literary history, language history, systems of writing, and anything else that is relevant to literature or to language as used in literature." "Philology," Webster's Third New International Dictionary, Unabridged, Merriam-Webster, 2002, accessed June 26, 2010, http://unabridged.merriam-webster.com.

contact with India. Early on, it was observed that classical European languages bore a remarkable resemblance to Classical Sanskrit. This observation, reenforced by the well-recognized common origin of Latin for the Romance languages, as they have come to be known, led to the hypothesis that languages had a common origin. Thus, historical linguistics was born: the undertaking to reconstruct the language underlying a family of languages.

Even at the beginning of the nineteenth century, many fallacies about language still plagued its study. First, the older the state of a language, the better it is. Second, the principal goal in the study of language is to establish, or reestablish, its pristine form. Third, the categories of language are rational categories and are, to a greater or lesser extent, those enshrined in Latin or Greek.

2 The Birth of Linguistics

By the end of the nineteenth century, these assumptions had been fully abandoned by those who pursued the scholarly study of language. The discovery and study of non-Indo-European languages, especially those of the indigenous peoples of North America, had the immediate effect of showing the inapplicability of many of the grammatical concepts derived from Greek and Latin grammars that were used to treat the languages of Western Europe. Those interested in studying North American indigenous languages sought to devise new grammatical concepts whereby to describe these languages. This, in turn, had the effect of breaking the commonly made, but fallacious, link between grammar and reason, for one came to recognize that different languages achieve the same expressive ends in different ways and that these differences are arbitrary and hence not to be appraised as more or less reasonable.

This also had the effect of breaking the grip of prescriptivism over the study of language. Prescriptivism espouses the view that the purpose of grammar is to establish and to defend so-called correct usage. Prescriptivists usually maintain that what they prescribe and proscribe accord with the canons of reason. However, differences in grammatical patterns have nothing to do with what is reasonable. For the most part, the favoring of one choice over another is nothing more than a prejudice reflecting social class or a resistance to regularization. Thus, linguistics, securing for itself its own conceptual apparatus to be applied to the resolution of factual questions, became a fully autonomous, empirical discipline.

In the nineteenth century, fascination with the origin and evolution of language was preeminent. However, at the beginning of the twentieth century, Ferdinand de Saussure (1857–1913) observed in his *Cours de linguistique générale* (Bally and Sechehaye 1972) that the study of language could be undertaken from two distinct points of view: from the point of view of the structure of a language at any given moment in time (*synchronic*) and from the point of view of how a language's structure changes over time (*diachronic*). During the twentieth century, two other important developments took place: the link between the study of language and the study of psychology, on the one hand, and the application of logic to the study of the structure of language and meaning, on the other. We turn to these developments now.

2.1 Linguistics and Psychology

The link between the study of language and psychology goes back to the beginning of modern psychology, when Wilhelm Wundt (1832–1920) established psychology as an empirical science, independent of philosophy (Wundt 1904). It was the American linguist, Leonard Bloomfield (1887–1949), who first viewed the study of language, or linguistics as it had come to be known, as a special branch of psychology.

At that time, psychology had come under the influence of behaviorism, a movement initiated by the American psychologist John Broadus Watson (1878–1958), which had a deep and lasting impact on a host of areas that investigate organisms and their behavior, including all of the social sciences (Watson 1925). Up to that time, animal behavior was explained in terms of instinct, while human behavior was explained in terms of the mind, its states and its activities. The primary method of investigation in human psychology was introspection. Behaviorism rejected all appeal to any unobservable entities such as either instincts, in the case of nonhuman animals, or the mind, its states and its activities, in the case of humans. Behaviorism admitted only observable and measurable data as objects worthy of scientific study. Behaviorists, accordingly, confined themselves to the study of an organism's observable and measurable physical *stimulus,* its observable and measurable *response* elicited by the stimulus and any biological processes relating stimulus to response.

Bloomfield (1933, chap. 2) sought to make linguistics scientific by recasting it in terms of behaviorist psychology. While most linguists today still regard linguistics as a special domain within psychology, their view about what kind of theory a theory of psychology should be is very different from that of Bloomfield.

By the end of the Second World War, behaviorism was on the decline and the notion of instinct was being rehabilitated as a scientifically respectable one. Unlike earlier thinkers, who would ascribe instincts to animals with little or no experimental confirmation, the new advocates of instinct, or innate endowment, sought empirical confirmation of their ascriptions. Thus, early ethologists,[2] whose work began to appear right after the war, sought to determine whether certain forms of behavior exhibited by a species of animal were innate (instinctive) or acquired (learned). Ethologists are especially interested in the kinds of behavior characteristic of an animal where a determinate sequence of actions, once

---

2. Ethology is the study of animal behavior that grew out of the work of the zoologists Nikolaas Tinbergen (1907–1988), Konrad Lorenz (1903–1989), and Kurt von Frisch (1886–1982). It investigates animal behavior using a combination of laboratory and field sciences, with strong ties to certain other disciplines such as neuroanatomy, ecology, and evolutionary biology. The three shared the 1973 Nobel Prize in Physiology or Medicine.

initiated, goes to completion. They call such a sequence of actions a *fixed action pattern*. The question arises: is such behavior innate or is it acquired? And if it is acquired, how is it acquired? In some cases, the behavior is innate, that is, the animal can perform the behavior in question once it has the required organs, for example, the collection of food. In other cases, the behavior is acquired. In cases of acquired behavior, the further question arises: does the young animal acquire the fixed action pattern from repeated exposure to the same behavior in the adults of its species or is the young animal innately predisposed to exhibit the behavior once it is triggered by an appropriate stimulus from its environment? A particularly good way to address such questions is the *deprivation,* or *isolation,* experiment.

In such experiments, one seeks to determine whether a young animal raised isolated from its conspecifics, or animals of the same species as it, thereby being deprived of exposure to their behavior, either fails altogether to be able to perform the behavior or manages only to perform it poorly. If it performs the behavior correctly, then one can reasonably conclude that the behavior is innate and not acquired. However, should it fail to produce the behavior, one can then try to figure out what experience the animal requires for the behavior to emerge. It may be that brief exposure to the behavior in its conspecifics is sufficient for the behavior to emerge, or it may be that both exposure and practice are required.

Though some behavior in animals is innate, most behavior emerges as a result of the animal's innate endowment and its experience. The latter is especially evident in the case of behavior that is characteristic of a mature organism but of which the immature organism is incapable. On the one hand, experience alone is not sufficient for the behavior to emerge, since an organism that is not disposed to develop the behavior never will. Cats, for example, cannot build nests, and birds cannot swim upstream to spawn eggs. On the other hand, innate endowment alone is also not sufficient for the behavior to emerge, since ex hypothesi the organism cannot at inception perform the behavior. The question then becomes: what balance between innate endowment and experience is required to bring about the organism's ability to exercise the capacity that it eventually acquires? To ascertain what the innate endowment is, one must fix what the capacity is that the organism comes to exercise and which experiences are necessary for it to acquire the capacity. This leads to the following suggestive picture:

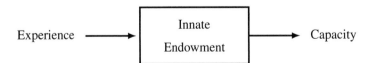

In addition, it has been demonstrated repeatedly, for the capacity that underpins a certain form of behavior to develop in an organism, the organism must experience a certain form of stimulus, and it must do so during a limited period of time known as the *critical period*. This is well illustrated by experimental work done by C. Blakemore and G. Cooper (1970), reported in Frisby (1980, 95). Mature cats have the capacity to detect horizontal and vertical lines. Blakemore and Cooper undertook the following: Newly born kittens were segregated into two groups. One group was placed in an environment set up in such a way that the kittens saw only vertical lines, while the other group was placed in an environment set up in such a way they saw only horizontal lines. After some time, neurophysiological recordings from their brains showed that the kittens raised in an environment with only vertical stripes possessed only vertically tuned striate neurons, while those raised in an environment with only horizontal stripes possessed only horizontally tuned striate neurons.[3] Kittens, then, pass through a period as a result of which exposure to vertical and horizontal stripes permits them to come to be able to detect such stripes. Clearly, then, kittens have an innate endowment that, when properly stimulated, enables them to acquire a capacity to detect vertical and horizontal stripes.

2.1.1 Language acquisition

Linguistic behavior is a form of intraspecial communicative behavior, that is, behavior whereby one member of a species signals information to another member of the same species. However, not all intraspecial communicative behavior is alike. For some species, its repertoire of communicative behavior comprises a finite, indeed very small, set of discrete, or digital, signals. For example, vervet monkeys have three vocal signals, one used on the sighting of a leopard, another on the sighting of a python, and a third on the sighting of an eagle. Honey bees, in contrast, have an infinite repertoire of continuous, or analog, signals to communicate the location of nectar. A honey bee returning to the hive conducts a so-called *waggle dance,* the axis of which indicates the direction of the nectar and the rate indicates the distance (von Frisch 1950, 1974). The repertoire of human linguistic expressions, however, is unlike either. It is unique among the repertoires of all animals: it is both discrete and infinite.

Linguistic behavior is both unique to humans and common among humans. Moreover, it is a form of behavior that humans are incapable of engaging in at birth but which they are capable of engaging in later. It is natural to conceive of this development along the lines indicated earlier:

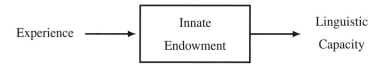

However, the linguistic capacity is not the only capacity thought to be unique to humans. As Aristotle pointed out centuries ago, the capacity to reason is unique to humans. It might be thought, then, that our linguistic capacity results somehow from our general ability to

---

3. Striate neurons are neurons in the striate cortex, which is one of the centers of the brain responsible for visual awareness.

reason. This is certainly a widely held layman's view of language. However, it has been argued that this is not the case, rather the human linguistic capacity arises from a peculiar part of the innate endowment.

Like his predecessor, Leonard Bloomfield, Noam Chomsky (born 1928) has also emphasized the link between linguistics and psychology. However, unlike Bloomfield, Chomsky has been a vociferous critic of behaviorism, particularly in the study of language. Indeed, his celebrated review of *Verbal Behavior* by the well-known behaviorist psychologist Burrhus Frederic Skinner (1904–1990) was extremely influential both inside of and outside of linguistics (Chomsky 1959b). In many, many of his publications, Chomsky developed a very different view of the nature of language. In particular, he has argued that linguistic behavior is underpinned by a linguistic capacity, which, he maintains, emerges from several capacities. These include the capacity to reason, the capacity to remember, the capacity to focus one's attention on various things, as well as, among other capacities, the capacity to form and recognize grammatical sentences. This capacity, which Chomsky once dubbed *grammatical competence* and now calls the *I-language,* together with other capacities pertinent to humans using and understanding language, results in the human linguistic capacity. He has also argued that humans have an innate endowment unique to them and specific to the development of the linguistic capacity. Chomsky has variously called this innate endowment *language acquisition device, universal grammar,* and *language faculty*. In early work, Chomsky depicted his view with the following adaptation of the earlier diagrams:

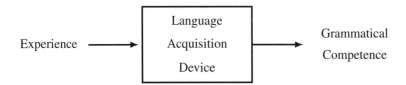

The argument adduced by Chomsky again and again to support this hypothesis is the so-called *poverty of the stimulus argument* (see Chomsky [1960] 1962, 528–530; 1965, 57–58; 1967, 4–6). The argument is based on a number of observations, which, when taken together, furnish presumptive evidence in favor of a working hypothesis to the effect that humans have an innate predisposition to acquire grammatical competence.

The first observation is that the structure of a language, over which a child gains mastery, is both complex and abstract from its acoustic signal. In particular, Chomsky has pointed out that the basic expressions making up a complex expression of a natural language have a structure that is over and above the linear order of the successive basic expressions making it up and that this additional structure is not contained in the acoustic signal conveying the complex expression.

Language, Linguistics, Semantics

The second observation is that, while the structure of linguistic expressions is both abstract from its acoustic signal and complex, the grammatical competence whereby a human produces and understands sentences is acquired by a child in a short span of time. Third, it is observed, this competence is acquired, even though the child has little exposure to signals carrying examples of the relevant structure and even though many of the utterances exhibit these structures in a defective way—being interrupted or otherwise unfinished sentences.

Fourth, it is observed that, in spite of important differences in the sample of utterances to which children of the same linguistic community are exposed, nonetheless they do converge on the same grammatical competence, as reflected in the convergence of the expressions they find acceptable.

The fifth observation is that the rules that are characteristic of grammatical competence are not taught. Consider, for example, an anglophone child's mastery of the rule for plural noun formation in English. All that anglophone children are taught in school is that the letter -*s* is added to a noun (with the exception of such words as *man*, *foot*, and so on). But this is not the rule that anglophone children learn when they are learning to speak English, for the simple reason that it is not the rule of English plural formation. The actual rule is more complex. And no adult, native English speaker can state the rule, unless he or she has been linguistically trained. To be sure, there is a suffix. But, it is pronounced differently, depending on what the sound immediately preceding it is. Thus, the plural suffix, when attached to the word *cat,* yields one sound, namely [s]; when attached to the word *dog,* it yields another, namely [z]; and when attached to the word *bush,* it yields still another, namely, [iz]. Children discern and master the difference without instruction.

Another example of anglophone speakers mastering an aspect of English without any instruction is this. Consider this pair of sentences:

(1.1)   John promised Mary to leave.

(1.2)   John persuaded Mary to leave.

Every native speaker of English knows that the agent of the leaving expressed in sentence (1.1) is the agent of the promising, that is, the person denoted by the subject of the main verb, whereas the agent of the leaving expressed in sentence (1.2) is the patient of the persuading, that is, the person denoted by the direct object. It cannot be due to any explicit instruction, for dictionary entries for the relevant verbs do not provide that kind of information about the verbs. And it cannot be due to the left-to-right order of the words, for the two sentences differ from one another only in the choice of verb.

Similarly, every native speaker of English knows when the third-person personal pronoun *it* can be used and when it cannot be used, as illustrated next, yet no speaker can state the rule governing its distribution.

(2.1)   John threw out the magazine without reading it.
        *John threw out the magazine without reading ___.
(2.2)   *Which magazine did John throw out ___ without reading it.
        Which magazine did John throw out ___ without reading ___.
(2.3)   *I never saw the magazine which John threw out ___ without reading it.
        I never saw the magazine which John threw out ___ without reading ___.

(Underlining is used to indicate which expression is the antecedent of which pronoun. An asterisk prefixed to an expression indicates that native speakers judge it as unacceptable.)

Sixth, it is generally acknowledged that a child's acquisition of his grammatical competence is independent of his intelligence, motivation, and emotional makeup. And finally, it is believed that no child is predisposed to learn one language rather than another. A child born to unilingual Korean speakers, if raised from birth by unilingual French speakers, will learn French as easily as any other child learning French born to unilingual French speakers, just as a child born to unilingual French speakers, if raised from birth by unilingual Korean speakers, will learn Korean as easily as any other child learning Korean born to unilingual Korean speakers.

These seven observations give rise to the following limits on any hypothesis about the acquisition language, namely, that it cannot be so rich as to predispose a child to acquire competence in the grammar of one language over that of another, for, as noted, no child is more disposed to learn one language over another. At the same time, the innate endowment cannot be so poor as to fail to account for a child's rapid acquisiton of grammatical competence, in light of the abstract yet uniform nature of the competence; the quality of his exposure; the poverty of his exposure; and the independence of his acquisition from his intelligence, motivation, and emotional makeup (Chomsky 1967, 3). In short, this innate endowment cannot be so rich as to preclude the acquisition of some attested language but it must be rich enough to ensure that one can acquire any attested language within the limits of time and access to data (Chomsky 1967, 2).

This argument, though widely accepted by linguists, was initially greeted with skepticism by empirically minded philosophers such as Hilary Putnam (1967), who, disputing some of the observations herein, argued for the conclusion that human linguistic competence is the result of the general human ability to learn. More recently, connectionists have presented computational models suggesting that it is indeed possible to abstract constituency structure from the acoustic signal (Ellman et al. 1996). Notice, however, as stressed by Fodor (1981), the debate is not about whether there is an innate endowment to account for language learning—this no one truly disputes—rather, the debate is about the nature of the innate endowment, in particular, about whether or not the necessary innate endowment is specific to language learning.

Another kind of argument to support the hypothesis that humans have a special aptitude to learn language is based on the claim that humans pass through a period critical to the acquisition of the linguistic capacity (Lenneberg 1967, 142–153). Basic morality excludes performing a deprivation experiment in which one would deprive an infant of exposure to language during its childhood to see whether the adult could later learn a language. Nevertheless, it has been claimed that such deprived children have been found. One case was that of a so-called wild child of Aveyron, found in the wilds in France in the nineteenth century, unable to speak, and reported never to have acquired a command of French, despite rigorous training (Lane 1976). Yet, too many facts about this boy's situation remain unclear for experts to see in him confirmation or disconfirmation of the hypothesis that humans have an innate capacity specific to acquiring language. More recently, in the latter half of the twentieth century, in the United States, more specifically in Los Angeles, a young girl named Genie was discovered who had been locked up in a room by herself from infancy. She too never acquired normal English fluency, again despite extensive training (Curtiss 1988). However, in spite of careful documentation of her history, confounding factors preclude experts from arriving at any consensus as to whether or not her case provides evidence for or against the hypothesis.

Still another argument to support the hypothesis is based on the claim that children are much more successful than adults in learning a second language. While researchers have conducted studies to support this claim, the interpretation of the studies has been disputed by other authors and other studies have failed to replicate them.

2.1.2 Grammatical competence

The human linguistic capacity is not something that can be directly observed. Rather, it must be figured out from behavior. The relevant behavior is the set of expressions composing the language used. Thus, the first step in characterizing the capacity is to characterize the expressions of the language.

Important steps in this direction were taken, amazingly enough, over two thousand five hundred years ago by unknown thinkers of those Indo-European tribes that moved into what is today Pakistan and northwest India. These tribes, known as Indo-Aryans, had a keen interest in their language, Sanskrit. So advanced was their knowledge of Sanskrit that by the fifth century BCE, they had formulated what today we call a *generative grammar* of Sanskrit. The monument that testifies to this astonishing achievement is the *Aṣṭādhyāyī* (fourth century BCE), the world's earliest extant grammar. This grammar, either written or compiled by Pāṇini, a speaker of the language, of whom we know almost nothing, is neither a list of observations about the language, nor is it a descriptive grammar of the kind compiled by modern field linguists; rather, it comprises a finite set of rules and a finite set of minimal expressions from which each and every proper expression of Sanskrit can be derived in a finite number of steps. Such grammars, unknown elsewhere in the world until the middle of the last century, are today known as *generative grammars*. It is this conception of grammar that is the basis of all mathematically rigorous treatments of natural language.

Pāṇini's grammar embodies a number of insights. One of them is made explicit by the great Sanskrit grammarian, Patañjali (second century BCE), in his *Great Commentary,* or *Mahābhāṣya,* a commentary on Pāṇini's *Aṣṭādhyāyī*. In it, Patañjali observes that there is no finite upper bound on the set of possible correct Sanskrit expressions, so that the learning of the language requires the learning of its vocabulary and its rules. He writes:

the recitation of each particular word is not a means for understanding of grammatical expressions. Bṛhaspati addressed Indra[4] during a thousand divine years going over the grammatical expressions by speaking each particular word, and still he did not attain the end. With Bṛhaspati as the instructor, Indra as the student and a thousand divine years as the period of study, the end could not be attained, so what of the present day when he who lives a life in full lives at most a hundred years?... Therefore the recitation of each particular word is not a means for the understanding of grammatical expressions. But then how are grammatical expressions understood? Some work containing general and particular rules has to be composed. (Kielhorn 1880, vol. 1, 5–6; translated by Staal 1969, 501–502)

The insight is that the number of grammatical expressions in Sanskrit is so large that it would be impossible to learn them one by one; instead, one must learn a finite set of rules that can be applied to a finite set of basic expressions.[5] Indeed, the *Aṣṭādhyāyī* and its appendices comprise just that: a finite list of basic expressions and a finite set of rules that together generate all and only the grammatical expressions of Sanskrit.

Another insight embodied in the grammar is that a complex expression is understood on the basis of an understanding of the basic expressions making it up. This insight is also found in the works of Medieval European logicians such as Peter Abelard (1079–1142) and John Buridan (fourteenth century) and apparently independently in the works of modern European logicians such as Gottlob Frege (1848–1925) and Rudolf Carnap (1891–1979). The *Aṣṭādhyāyī* embodies this insight by pairing each sentence of Sanskrit that it generates with a situation whose parts are associated with the minimal expressions making up the sentence.[6] The empirical basis for this insight arose from the following kind of observation. Consider these two sentences uttered in precisely the same circumstances:

(3.1)   A **cow** is a mammal.

(3.2)   A **rock** is a mammal.

These sentences form what linguists call a *minimal pair*: a pair of linguistic items that are alike in all relevant respects except two. The two sentences in (3) are alike in all relevant respects except that, where the first sentence has the word *cow*, the second sentence has the word *rock*, and where native speakers of English judge the first sentence true, they

---

4. In Vedic mythology, Bṛhaspati is the preceptor of the gods, the master of sacred wisdom, charms, hymns, and rites, and Indra is one of the chief gods.

5. This anticipates by two thousand five hundred years claims to the same effect about human languages in general made by Noam Chomsky and by Donald Davidson.

6. The relevant discussion occurs in connection with rule A 1.2.45, that is, rule 45 of chapter 2 of book I of his *Aṣṭādhyāyī*. See Bronkhorst 1998 for a contemporary treatment.

judge the second false. The obvious explanation of why native speakers of English assign different truth values to the sentences in (3) is that they understand the words *cow* and *rock* differently.[7]

But what precisely is the relation that determines how complex expressions are made out of simpler expressions? Here we turn to the idea of *immediate constituency analysis,* a key idea of the American structuralist linguists, who were working in North America in the first half of the twentieth century. Though its three essential ingredients are embodied in the rules of Pāṇini's *Aṣṭādhyāyī,* it was Leonard Bloomfield, the founder of the movement, himself a student of the Indian grammatical tradition, who gave explicit formulation to the ideas. They were further elaborated and applied by his successors: in particular, by Bernard Bloch (1946), Zellig Harris (1946), Eugene Nida (1948), Rulon Wells (1947), and Charles Hockett (1954). Immediate constituency analysis has three essential ingredients. The first is that each complex expression can be analyzed into immediate subexpressions—typically two, which themselves can be analyzed, in turn, into their immediate constituents and that this analysis can be continued until minimal constituents are reached. The second is that each expression can be put into a set of expressions that can be substituted one for the other in a more complex expression without compromising its acceptability. And the third is that each of these sets of expressions can be assigned a syntactic category.

Using immediate constituency analysis, one can show that sentence (4.0) can be analyzed in two different ways, each having a different meaning.

(4.0)   Galileo saw a patrician with a telescope.

(4.1)   Galileo saw [NP a patrician [PP with a telescope]].
        A patrician with a telescope was seen by Galileo.

(4.2)   Galileo saw [NP a patrician] [PP with a telescope].
        The patrician was seen with a telescope by Galileo.

To see this, consider the following circumstances. Galileo is looking out the window of his apartment in Venice through his telescope at a patrician walking empty-handed through Saint Mark's Square. Sentence (4.0) is judged false, should it be understood according to the sentence's annotation in (4.1), where the prepositional phrase (PP) *with a telescope* is taken as a modifier of the noun phrase (NP) *a patrician,* and it is judged true, should it be understood according to the sentence's annotation in (4.2), where the prepositional phrase *with a telescope* is taken as a modifier of the verb *saw.*

These kinds of facts suggest that immediate constituency is crucial in the determination of the meaning of a complex expression by the meanings of its subexpressions. After all,

---

7. Indian grammarians in particular, and Indian thinkers in general, called this method of using minimal pairs the method of *anvaya* (*concomitance* or *agreement*) and *vyatireka* (*difference* or *exclusion*). Some readers might recognize that this way of proceeding is what is known to Europeans and Americans as Mill's method of agreement and difference. Mill's method dates back to at least the time of Medieval English philosopher Robert Grosseteste (ca. 1175–1253). These ideas are briefly discussed in section 3.2 of this chapter.

what else but the grouping due to the immediate constituent analysis could explain how it is that the very same sentence can be judged both true and false with respect to one and the same circumstances?

Immediate constituency analysis also throws light on the absence of any finite bound on the expressions of a natural language. As we shall see in greater detail in later chapters, a constituent of a certain type can have as a constituent another constituent of the same type. Thus, for example, a prepositional phrase may contain another prepositional phrase that in turn contains still another prepositional phrase.

(5.1)   Bill sat [PP behind the first chair].

(5.2)   Bill sat [PP behind the chair [PP behind the first chair]].

(5.3)   Bill sat [PP behind the chair [PP behind the chair [PP behind the first chair]]].

⋮ ⋮

Indeed, it is easy to show that many kinds of constituents in English, besides prepositional phrases, have this property. Coordinated independent clauses do, for example. Consider the English connector *and*. It can be put between two independent clauses (say (6.1) and (6.2)) to form an independent, compound clause—(6.3), which itself can be joined to either of the initial clauses to form still another independent compound clause (6.4).

(6.1)   It is raining.

(6.2)   It is snowing.

(6.3)   It is raining and it is snowing.

(6.4)   It is raining and it is snowing and it is raining.

⋮ ⋮

The same thing can be done with the connector *or*. Relative clauses furnish still another example of a constituent having this property.

(7.1)   Bill saw a man.

(7.2)   Bill saw a man who saw a man.

(7.3)   Bill saw a man who saw a man who saw a man.

(7.4)   Bill saw a man who saw a man who saw a man who saw a man.

⋮ ⋮

Indeed, this property, whereby a constituent of a certain type can contain as a constituent another of the same type, seems to be a property of every known human language. And it is by dint of this property, known to mathematicians as *recursion,* that each human language comprises an infinite number of expressions. Put another way, the infinite set of expressions that composes a language can be obtained, or *generated,* by applying a finite set of recursive rules to a finite set of basic expressions, or minimal elements. These and other rules are said to compose the language's grammar and are thereby said to characterize the corresponding grammatical competence.

The hypothesis of grammatical competence involves idealization. The idealization is simple enough. For example, it is clear that human beings have the capacity to do arithmetic. The mastery of this capacity does not require that the person exercising the capacity exercise it flawlessly. No doubt every person, no matter how arithmetically competent, makes arithmetic errors. The simple fact of making arithmetic errors in no way impugns the ascription of the arithmetic competence. What would warrant the withdrawal of the ascription would be either persistent error or an inability to recognize errors once pointed out. Indeed, without an ascription of arithmetic competence, it would make no sense even to speak of errors and to try to characterize them in any way that might yield insight into human psychology. In other words, one can study errors only against a standard.

As a result, it is useful to distinguish human grammatical competence from performance, discounting all slips of the tongue, mispronunciations, hesitation pauses, stammering, stuttering—in short, discounting anything attributable to irrelevant factors such as memory limitations, distractions, shifts of attention and interest, and the malfunctioning of the physiological and neurological mechanisms involved in language behavior (Lyons 1977, 586). Such an idealization, or regularization, of the data is essential to a proper understanding not only of grammatical competence but also of linguistic performance. Derogations from something can be determined once what the derogations are derogations from has been correctly characterized. It should be emphasized that these errors of performance are not set aside to be ignored, but to be eventually the object of more careful study.

2.1.3 Autonomy

Just as the exercise of the human linguistic capacity is thought to result from the exercise of several capacities together—including the grammatical competence; so the exercise of grammatical competence is thought to result from the exercise of component competences, whose joint exercise is required for the exercise of the grammatical competence. Moreover, just as it is held that the various capacities making up the linguistic capacity are distinguishable and not reducible one to the other, so it is held that the various components of grammatical competence are distinguishable and not reducible one to the other. Another way to put this is to say that the capacities making up the linguistic capacity and the competences making up grammatical competence are autonomous from one another.

To understand better what is meant by *autonomy*[8] applied to the various component competences of grammatical competence, let us consider how various facts are

---

8. The idea of *autonomy,* when Chomsky first used the term, met, surprisingly, with much resistance. Like many of Chomsky's coinages, it was the use of a term for a familiar concept, which in fact accurately characterized much of the linguistic practice of then as well as now, namely, that the various customary areas of study, though related, do not reduce, one to the other.

explained. To begin with, consider the following expressions. The unacceptability of these expressions is to be explained, not by the rules of English phonology, but by the rules of English syntax.

(8.1) *Butted to when in did sorry he town.

(8.2) *Bill called up me.

The first fails to have any syntactic structure, while the second violates the syntactic rule whereby a particle forming a constituent with a verb must succeed a pronoun, which is the constituent's direct object.

Similarly, the unacceptability of the next pair of expressions is to be explained, in the first case, by the rule of English that the noun of the subject noun phrase and the verb of the verb phrase agree in grammatical number and, in the second case, by the fact the noun *mouse* is an exception to the rule of plural formation.

(9.1) *The boys is here.

(9.2) *The mouses have escaped.

Notice that the sentences in (9) are not uninterpretable. If uttered either by an adult for whom English is not a native language or by a child, they would be interpreted, in the first case, as saying the same thing either as the sentence *The boys are here* or as the sentence *The boy is here* and, in the second case, as saying the same thing as *The mice have escaped*. Consequently, no linguist would suggest that such sentences are unacceptable for reasons of semantics.

Consider finally another set of unacceptable expressions. They do not violate any recognized rules of English phonology or morphology. Moreover, each corresponds in syntactic structure to a perfectly acceptable English sentence.

(10.1) The girl loves the boy.
        *The stone loves the boy.

(10.2) Carla drinks water.
        *Quadruplicity drinks procrastination.
        (Russell 1940, 275)

(10.3) Fred harvested a large bail of hay.
        *Fred harvested a magnetic puff of amnesia.
        (Cruse 1986, 2)

Such expressions, if interpretable at all, are interpretable only by reconstruing the meaning of one or another of its words. Such sentences are explained as unacceptable, not for reasons of phonology, morphology, or syntax, but for reasons of semantics, in particular, because of a failure of the meanings of various expressions to cohere, as it were.

Language, Linguistics, Semantics

Thus, we see how expressions, all judged to be unacceptable, have their unacceptability explained by different explanatory resources: some are explained by phonological rules, some by morphological rules, others by syntactic rules, and still others by semantic rules. Thus, phonology, syntax, morphology, and semantics are taken to be autonomous, though related, components of a theory of grammar.

It is important to note that the mutual autonomy of these various components is not drawn into question by the fact that they may have overlapping domains of application. Here is one elementary example of how morphology and syntax on the one hand and phonology on the other can have overlapping domains of application. In French, for example, the choice of the form of a word may be determined in some cases by purely morphosyntactic considerations and in other cases by phonological considerations.

(11.1) Son logement est beau.
His dwelling is beautiful.

(11.2) Sa demeure est belle.
His dwelling is beautiful.

(12.1) Sa femme est belle.
His wife is beautiful.

(12.2) Son épouse est belle.
His wife is beautiful.

Thus, in (11), the choice between the possessive adjectives *sa* and *son* is determined by the gender of the following noun: *logement* is masculine, so the form of the possessive adjective is the masculine form *son,* while *demeure* is feminine, so the form of the possessive adjective is the feminine form *sa*. In (12.2), however, even though *épouse* is feminine, the appropriate form of the possessive adjective is the masculine form *son,* which is required by the fact that the immediately following word begins with a vowel.[9]

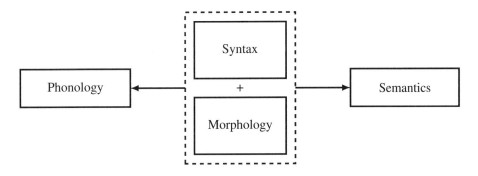

---

9. A similar alternation is found in English. Compare *a rubber band* with *an elastic band*.

Not only have linguists held that the various components of grammar are autonomous with respect to one another, they have also held that grammatical competence and world knowledge are also autonomous from one another. Thus, the unacceptability of the next pair of sentences is to be ascribed to different sources: the first to a violation of a grammatical rule of syntax and the second to a conflict with our beliefs about the world.

(13.1) *I called up him.

(13.2) *The man surrounded the town.

Expressions such as those in (13.2), which seem to be in every respect like declarative sentences and yet seem to make no sense, are sometimes said to be *semantically anomalous* expressions. They are said to be anomalous, instead of false, since they do not seem to be expressions that are easily liable to being judged as either true or false. The English philosopher Gilbert Ryle (1900–1976) said that such sentences contain category mistakes (Ryle 1949, 16). They contrast with the sentences in (14), which are judged inexorably false.

(14.1) I dislike everything I like. (Leech 1974, 6)

(14.2) My uncle always sleeps awake. (Leech 1974, 7)

(14.3) Achilles killed Hector, but Hector did not die.

Corresponding to sentences that are judged inexorably false are those that are judged inexorably true.

(15.1) I like everything I like.

(15.2) My brother always sleeps asleep.

(15.3) If Achilles killed Hector, then Hector died.

The inexorability of such judgments is thought to arise, not from a speaker's knowledge of the world, but from his or her knowledge of English. This inexorability contrasts with the lack of inexorable falsity of either of these sentences.

(16.1) Some man is immortal.

(16.2) My uncle always sleeps standing on one toe. (Leech 1974, 7)

After all, the truth of these sentences is conceivable, as is the falsity of the next two sentences.

(17.1) All men are mortal.

(17.2) No cat is a robot.

However, it is not always clear whether or not a given sentence is inexorably true or inexorably false. Consider the next sentence.

(18) This man is not pregnant.

Is this sentence necessarily true? And if so, is it necessarily true because of the meanings of the words in it? As Lyons (1995, 122) notes, it is not inconceivable that biotechnology will some day permit a fetus-bearing womb to be implanted in a man who can later deliver the child by cesarean section.

There are sentences that are true, no matter what, by dint of the grammar of English, and there are others that are false, no matter what, by dint of the grammar of English. There are still others that are true or false, depending on what the world is like. However, as we shall see later, drawing lines between these various classes of sentences is not always easy.

2.2 Linguistics and Logic

If the link between linguistics and psychology is strong, the link with logic is no less so. Logic, initially, sought to distinguish good arguments from bad ones. More particularly, it sought to identify which argument forms preserve truth and which do not. Since arguments are communicated and, to that extent, expressed in a language, it is natural to use the forms of language to identify the forms of arguments. It is not surprising, then, that those interested in logic have been interested in language and have made, in their pursuit of logic, interesting observations about language and have developed important insights into language.

The intertwining of logical and linguistic concerns is evident from the beginning of the study of logic in Europe. In the course of developing his syllogistic, Aristotle introduces the distinction between subject and predicate, a distinction that has been with us ever since as both a grammatical and a logical one. Nor were these the only linguistic distinctions drawn by Aristotle: Aristotle seems to have been the first European to have identified conjunctions as a lexical class as well as to have identified tense as a feature of verbs, among many other things. The Stoics too, a philosophical movement of the Hellenistic period, also had an interest in logical and linguistic matters, among other things. They identified the truth-functionality of *and, or,* and *if* and distinguished verbal aspect from verbal tense, again, among many other things. This mixture of logical and linguistic concerns appears again in the *logica moderna* of the Middle Ages, especially with its treatment of *syncategoremata*, those expressions like the English expressions *all* (*omnis*), *both* (*uterque*), *no* (*nullus*), *unless* (*nisi*), *only* (*tantum*), *alone* (*solus*), *infinitely many* (*infinita in pluralia*), numerals (*dictiones numerales*), and so on.

The next major development to stimulate still further the intertwining of logical and linguistic concerns is the development of the formalization of mathematics. In particular, classical quantificational logic was developed as a means to represent mathematical reasoning. It does this by providing a notation in terms of which, it is generally agreed, all mathematical arguments can be framed. By focusing on mathematical arguments, or proofs, logic turned its attention to how all parts of a mathematical proof could be put into notation and how that notation could be rigorously specified. The study of how the notation could be rigorously specified led to recursion theory. The study of how the notation was to be interpreted led to model theory. Both of these developments have had a fundamental impact on linguistics: the first provided the basis for the formalization of grammatical rules, and the second brought to the attention of philosophers, and later of linguists, the central question of semantics: how do the meanings of constituent expressions contribute to the meaning of the expression of which they are constituents?

In the remainder of this section, using a few extremely simple examples for illustration, we will learn about these two ideas, which are fundamental to the study of both logic and natural language. While the examples may seem utterly contrived, the basic idea is not. Indeed, the basic idea is familiar to anyone who has studied secondary school mathematics, though not the choice of basic symbols, made with a view to helping to avoid distracting, extraneous detail. We can use these simple examples to maintain our bearing as we wade more deeply into the complexity that inevitably accompanies exposure to a wider and wider range of natural language phenomena.

2.2.1 Formation rules

What is recursion? We shall not try to give a rigorous mathematical definition here. However, we shall give a perfectly rigorous illustration. Consider the set SL, whose members are the sequences of one or more occurrences of instances of the letters *A, B, C,* or *D.* The set SL, then, includes not only the letters *A, B, C,* and *D,* but also sequences of these letters such as *AB, BD, DC, AAA, DCBAADD,* and many, many others. It includes only such sequences. Thus, it does not include *AEC, FBE,* and so on. As readers can easily see, SL has an infinite number of members; after all, given any sequence of these letters, one can obtain a different sequence of these letters by adding any one of the four letters to the sequence given.

The foregoing characterization of SL is not a recursive specification. The following is, however. SL comprises the elements *A, B, C,* and *D* as well as any expression that can be obtained from an expression already in SL, as it were, by suffixing to it either *A, B, C,* or *D.* With L as the set of elements *A, B, C,* and *D,* we can give a formal recursive definition of SL.

(19)   FRs: SUFFIXATION FORMATION RULE for SL

(19.1)   If x is an expression of L, then x is an expression of SL.

(19.2)   If y is an expression of SL and z is an expression of L, then yz is an expression of SL.

(19.3)   Nothing else is a member of SL.

We shall call definitions of this kind *formation rules.* We shall refer to this particular formation rule as the *suffixation formation rule* (for SL), which we shall abbreviate as FRs.

Language, Linguistics, Semantics

Such a definition is said *to generate* the members of SL on the basis of L. Let us see how FRs generates a specific expression of SL, say the expression $BACD$. A, B, C, and D, being expressions of L, are, by (19.1), also expressions of SL. Since B is an expression of L, then by (19.1), B is an expression of SL. This is represented in the following diagram by the line that has $B \in L$ above and $B \in SL$ below. For the time being, we will regard $\in$ as shorthand for "is an expression of." Since B is an expression of SL, or equivalently $B \in SL$, and A is an expression of L, or equivalently $A \in L$, it follows by (19.2) that BA is an expression of SL. This is represented in the diagram by two lines, one with $B \in SL$ above, another with $A \in L$ above, and the two converging to $BA \in SL$ below them. Another application of the rule in (19.2) yields the expression BAC. A fourth application yields the expression $BACD$. In short, each transition from one level of the diagram to the level immediately beneath it corresponds to the application of one of the rules in (19), where what is on the upper level corresponds to the *if*-clause of the rule and what is on the lower level corresponds to the *then*-clause of the same rule.

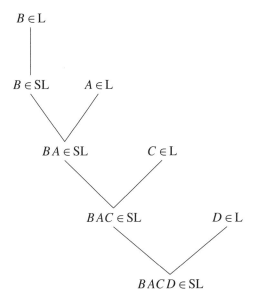

This diagram illustrates two facts. First, it illustrates that simpler expressions are combined, two at a time, into a more complex expression. We shall refer to two expressions that combine to form a more complex expression as the (immediate) subexpressions of the more complex expression. For example here, BA and C are the immediate subexpressions of BAC. Second, it depicts that, in a finite number of steps, any expression of SL can be obtained from the simplest expressions of SL, namely those in L, or equivalently, that, in a finite number of steps, any expression of SL can be decomposed into the simplest expressions of SL, namely those of L.

A little reflection on (19) shows that all and only the expressions of SL can be obtained on the basis of (19). In particular, it should be obvious that (19.1) and (19.2) guarantee that all the sequences of one or more occurrences of members of L are included, and that (19.3) guarantees that no other sequences are included.

Suffixation is not the only way in which the expressions in SL can be generated. They can also be generated by prefixation. Thus, the members of SL can be recursively specified as comprising the elements of L as well as any expression that can be obtained by prefixing an element of L to an expression already in SL.

(20)   FRp: PREFIXATION FORMATION RULE for SL

(20.1) If x is an expression of L, then x is an expression of SL.

(20.2) If y is an expression of L and z is an expression of SL, then yz is an expression of SL.

(20.3) Nothing else is a member of SL.

As we shall now show, $BACD$ is also an expression of SL according to the recursive definition in (20). Since D is an expression of L, it is also, by (20.1), an expression of SL. Now C is also an expression of L. So by (20.2), CD is an expression of SL; as is ACD, again by (20.2), since A is an expression of L. Finally, B, being an expression of L, is prefixed by (20.2) to ACD, thereby yielding $BACD$ as an expression of SL. We depict these steps next.

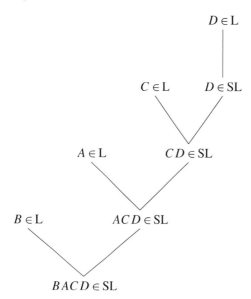

Thus, as we just showed, the expression $BACD$ is a member of SL, both by the recursive specification in (19) and by the recursive specification in (20). However, as the two diagrams make clear, each recursive specification generates the same expression in different ways. Thus, while *BAC* is a subexpression of *BACD,* according to the recursive specification in (19), it is not a subexpression of *BACD* according to the recursive specification in (20).

The form of recursive specification illustrated by (19) and (20) is far simpler than the form of recursive specification used by linguists to generate the expressions of a natural language. Two types of recursive specification are usually employed by linguists: that of a constituency grammar and that of a categorial grammar. The first arose out of the study of language itself, the other out of the study of logic. The first has its origins in Pāṇini's *Aṣṭādhyāyī,* for a great number of its phonological and morphological rules are, to a very close approximation, instances of what are also called today *context sensitive rules* (Staal 1965). Such rules were not used again in grammars for another two thousand five hundred years, when the American structuralist linguist Leonard Bloomfield (1933), who had closely studied Pāṇini's grammar (Rogers 1987), rediscovered their utility, applying them not only in morphology and phonology, as Pāṇini had done, but in syntax as well. Their use was subsequently greatly developed by Zellig Harris (1946; 1951), Bernard Bloch (1946), Rulon Wells (1947), Eugene Nida (1948), and Charles Hockett (1954), among others. Shortly thereafter, Noam Chomsky, who had studied mathematical systems known as *rewriting*, or *semi-Thue*, *systems* (Chomsky 1956; 1959a; 1963), suggested that such systems be used to formalize immediate constituency analysis (Chomsky 1957).

Another way to specify recursively the expressions of natural language is to use what is known as the *Lambek calculus* (Lambek 1958), named after its discoverer, Joachim Lambek (1922–2014). The Lambek calculus is a generalization of *categorial grammar,* which Kazimierz Ajdukiewicz (1890–1963), drawing on ideas found in the *Fourth Logical Investigation* of Edmund Husserl (1859–1938), had devised to give a mathematical characterization of the notation of classical quantificational logic. Ajdukiewicz (1935) explains some aspects of the mathematics of the notation by pointing out, in an anecdotal way, some of its possible applications in the study of natural language structure. But Ajdukiewicz's mention of natural language is completely opportunistic, for his concerns were entirely with logic. Yehoshua Bar-Hillel (1915–1975) was the first person to explore seriously how categorial grammar might be applied to the study of natural language (Bar-Hillel 1953).

We shall learn in later chapters how each of these two types of recursive specification can be applied in the analysis of the syntax of the expressions of natural language and how they relate to one another.

2.2.2 Valuation rules

Having given an idea of what recursion is and how it pertains to the analysis of the syntax of natural language, let us turn to what model theory is and how it pertains to the way in which the meanings of constituent expressions contribute to the meaning of the expression of which they are constituents. Linguists and philosophers refer to this property as *compositionality*.

Model theory is concerned, among other things, with the relation of symbols of logical notation to mathematical objects. The basic idea can be illustrated with the expressions of SL. Let us consider the question: are the expressions of SL sufficient to permit one to name each and every natural number, that is, to name 0, 1, 2, … ? The answer is yes. It can be done by riding piggyback on the recursive specification of SL to assign values recursively to each expression of SL. Here is one way to do it. First, we assign 0 to *A,* 1 to *B,* 2 to *C,* and 3 to *D.* Next, we assign a value to a complex expression following its recursive specification. We shall call such rules *valuation rules*. As we shall see, each valuation rule is defined in light of an antecedently given formation rule.

To get a better idea of what valuation rules involve, let us consider an example. Recall the formation rule FRs defined in (19). Each complex expression comprises two immediate constituent expressions, a left-hand one and a right-hand one. The value assigned to the complex expression is the one obtained by multiplying the value assigned to the left-hand immediate constituent expression by 4 and adding to it the value of the right-hand immediate constituent expression. Let us call this assignment i. It is defined in two clauses. The first stipulates which values are to be assigned to the expressions in L: the expressions *A, B, C,* and *D* are assigned the values 0, 1, 2, and 3, respectively. The second states how, on the basis of values assigned to the parts of an expression, a value is to be assigned to the expression itself: if an expression consists in expression y of SL and expression z of L, then the value assigned to the complex expression yz is the sum of the value assigned to z and four times the value assigned to y. Using the notation $i(x) = y$ to mean i assigns y to x, we can state the definition of i as follows:

(21)   VRi: VALUATION RULE i for FRs

(21.1) $i(A) = 0, i(B) = 1, i(C) = 2,$ and $i(D) = 3$.

(21.2) If y is an expression of SL and z is an expression of L, then $i(yz) = 4 \cdot i(y) + i(z)$.

To see how VRi works, let us notice some of its features. First, just as (19.1) specifies the elements whereby the expressions of SL are generated, so (21.1) specifies their values. Next, just as (19.2) specifies how a complex expression is constituted by two immediate constituent expressions, so (21.2) specifies how the value of a complex expression is determined by the values of its immediate constituent expressions. In brief, VRi (21) works in tandem with FRs (19).

Let us see how. Recall how $BACD$ is generated by FRs (19). First, (19.1) states that the elements of L are expressions of SL. Thus, as we saw, B is an expression of SL. By (19.2), BA is an expression of SL. At the same time, (21.1) assigns 0 to *A* (that is, $i(A) = 0$) and 1 to *B* (that is, $i(B) = 1$). According to (21.2), the value assigned to BA is four times the value assigned to *B* plus the value of *A,* that is, $4 \cdot i(B) + i(A)$, or $4 \cdot 1 + 0$, or 4. Since BA

is an expression of SL and C is an element of L, according to (19.2), BAC is an expression of SL. According to (21.2), since BA is assigned the value 4 and C is assigned the value 2, BAC is assigned the value $4 \cdot 4 + 2$, or 18. Finally, since BAC is an expression of SL and D is an element of L, $BACD$ is an expression of SL and its value is $4 \cdot 18 + 3$, or 75. All of this is nicely displayed in the following diagram.

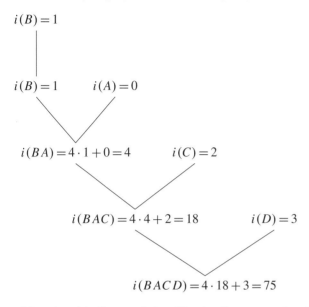

Note that this diagram is just like the diagram used earlier to depict how FRs (19) generates the expression *BACD*, except the node labels indicating the categories of the various subexpressions have been replaced with labels indicating the values assigned to the various subexpressions.

The reader should be able to see that each expression of SL formed by FRs and assigned a value by VRi denotes a natural number and that each natural number is denoted by some expression of SL formed from the rule of FRs and assigned a value by the rule VRi.

VRi is not the only way the expressions of SL formed by FRs can be assigned natural numbers as values in such a way that each expression denotes a natural number and each natural number is denoted by some expression. To see this, let us retain the recursive specification of the expressions of SL given by FRs (19), but we will replace the recursive specification of values given by VRi (21) with the one in (22). The specification in (22) assigns the very same values to the elements of L, but instead of multiplying the value of the left-hand immediate constituent expression by 4, it multiplies the value of the right-hand immediate constituent by 4 and then adds that value to the value of the left-hand immediate constituent expression.

(22) VRj: VALUATION RULE j for FRs

(22.1) $j(A) = 0$, $j(B) = 1$, $j(C) = 2$, and $j(D) = 3$.

(22.2) If y is an expression of SL and z is an expression of L, then $j(yz) = j(y) + 4 \cdot j(z)$.

Let us see what value VRj (22) assigns to *BACD* when it is generated by FRs (19). As before, (19.1) states that the elements of L are expressions of SL. So, *B* is an expression of SL. By (19.2), *BA* is an expression of SL too. As with (21.1), (22.1) assigns 0 to *A* (that is, $j(A) = 0$) and 1 to *B* (that is, $j(B) = 1$). According to (22.2), the value assigned to *BA* is the value assigned to *B* plus four times the value assigned to *A*, that is, $j(B) + 4 \cdot j(A)$, or $1 + 4 \cdot 0$, or 1. Since *BA* is an expression of SL and *C* is an element of L, again, according to (19.2), *BAC* is an expression of SL. Since (22.2) has assigned 1 to *BA*, it assigns to *BAC* the sum of the value assigned to *BA* plus four times the value assigned to *C*, that is, $1 + 4 \cdot 2$, or 9. Finally, (22.2) assigns to *BACD* the sum of the value assigned to *BAC* plus four times the value assigned to *D*, that is, $9 + 4 \cdot 3$, or 21. Again, we display these calculations in the diagram.

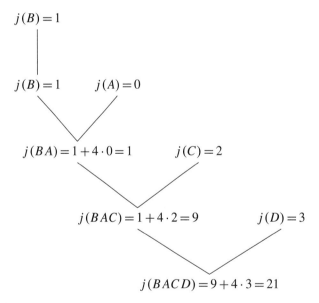

Since FRs (19) is used to generate the expression *BACD*, the diagram used previously to depict the generation of *BACD* by FRs (19) is used again here. However, this diagram differs from the immediately preceding one for (21), in which values are assigned to the expressions making up the expression *BACD*. This is because the assignment of values displayed in the previous diagram is determined by VRi (21), whereas the assignment of values displayed here is determined by VRj (22).

Language, Linguistics, Semantics

It is probably not surprising that, even though the expressions of SL are generated in the same way, that is by the recursive specification in FRs (19), and even though the elements of L are assigned the very same values, the values assigned to the complex expressions nonetheless differ, depending, as we saw, on which valuation rule is applied. What might be surprising is that the values assigned to the complex expressions differ when the formation rules differ, even though the very same values are assigned to the elements of L and the very same way of assigning values to a complex expression on the basis of the values assigned to its constituent expressions is used.

To see how this is so, let us define the rule VRk. This valuation rule assigns the very same values as VRi to the expressions of L, and, like VRi, VRk assigns a value to the complex expression that results from multiplying the left-hand immediate constituent expression by 4 and adding to it the value of the immediate right-hand expression. However, unlike VRi, which works in tandem with FRs, VRk works in tandem with FRp.

(23)   VRk: VALUATION RULE k for FRp

(23.1) $k(A) = 0$, $k(B) = 1$, $k(C) = 2$, and $k(D) = 3$.

(23.2) If y is an expression of L and z is an expression of SL, then $k(yz) = 4 \cdot k(y) + k(z)$.

What value does VRk (23) assign to $BACD$ when it is generated by FRp (20)? It assigns the value 11 to CD, the value 11 to ACD and the value 15 to $BACD$. Thus, (19) working in tandem with (21) assigns $BACD$ the value 75, while (23), working in tandem with (20), assigns $BACD$ the value 15.

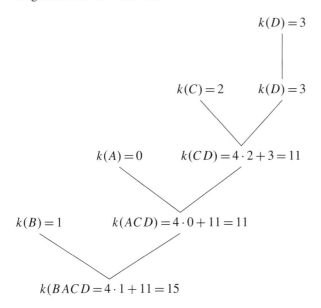

Just as both the pair of FRs and VRi and the pair FRs and VRj bring it about that each expression of SL denotes some natural number and that each natural number be denoted by some expressions, so too does the pair of FRp and VRk.

Although formation rules can be stated independently of valuation rules, valuation rules cannot be stated independently of formation rules. Moreover, what value is assigned to a complex expression depends both on the valuation rule used and the formation rule with which it is used in tandem.

2.2.3 Grammar: Formation rules and valuation rules

Bearing in mind the foregoing discussion of formation rules and valuation rules, let us elaborate on the similarity between the expressions of SL and those of natural language. As we saw, expressions of a natural language such as English have a recursive structure. A value can be associated with complex expressions of English. For example, simple declarative sentences such as *cows are animals* have associated with them the value of being true. Using the recursive specification of the expressions of SL given by FRs (19) and assigning them values in accordance with VRi (21), one can associate a value with the expression *CAD,* namely, the value 35.

Now, the value of 35, associated with the complex expression *CAD* and the value of truth associated with the declarative sentence *Cows are animals* are determined, at least in part, by the values of their simplest subexpressions. In particular, a change in a simple subexpression of a more complex expression can lead to a change in the value associated with the more complex expression. Thus, just as the replacement of *C* in *CAD* by *B* leads to a change in the value expressed, since *CAD* expresses 35 and *BAD* expresses 19, so the replacement of *cows* in *cows are animals* by *rocks* leads to a change in the truth, since *cows are animals* is true, while *rocks are animals* is false.

Nor is the value to be associated with a complex expression, be it one of SL or one of natural language, simply the aggregate, as it were, of the values associated with its generators. For, if that were so, then the value to be associated with the expression *BAD* would be the same as the value to be associated with the expression *DAB,* since, after all, they have the same smallest subexpressions. Similarly, the value to be associated with the expression *Alice saw Bill* would be the same as the value to be associated with the expression *Bill saw Alice,* since again the two expressions have the same smallest constituents. But that too need not be so, for one could be true while the other is false.

From the foregoing discussion emerges the following conception of a grammar for a natural language. A grammar of a natural language comprises a finite set of (syntactic) rules whereby, from an inventory of basic expressions, or words, known as a *lexicon,* are generated all and only the acceptable expressions of the language. The finite set of syntactic rules are paired with semantic rules to which, on the basis of a suitable assignment of values to the words of the language, values are assigned to the complex expressions they make up. An important corollary of this conception of grammar is that just as there are no valuation rules without formation rules, there are no semantic rules without syntactic rules.

This is precisely the conception of grammar implicit in the *Aṣṭādhyāyī*, Pāṇini's grammar of Classical Sanskrit. The *Aṣṭādhyāyī* itself comprises a little more than four thousand aphorisms (*sūtra*), for the most part, rules (*vidhi*).[10] It is supplemented with two appendices, one an annotated list of basic verbal roots, called the *dhātu-pāṭha,* and another annotated list of basic nominal stems and suffixes, called the *gaṇa-pāṭha*. The rules in the grammar state how to pair basic morphemes with the things they express, and, once an initial string of basic morphemes is established, there are other rules that apply so that once the morphemes constituting a sentence are properly interpreted with respect to a situation, possible or real, in the world, the sentence they constitute correctly expresses the situation in question. Here, however, the grammar falls short. Though the grammar has explicit recursive rules, for example in compound formation, it does not have explicit recursive rules for many of the other recursive structures of Classical Sanskrit. Naturally, then, it also fails to have the requisite semantic rules to assign values to structures recursively.

The remedy to such shortcomings did not appear until the second quarter of the twentieth century, when the one discipline to concern itself with making precise how values associated with subexpressions determine the value of the expressions they compose appeared, namely the subdiscipline of logic known as model theory. Its founder, Alfred Tarski (1901–1983), recognized the pertinence of model theory to the study of how complex expressions in a natural language acquire their meaning from the expressions that make them up, though he himself (1932, sec. 1; 1944, 347) doubted that a satisfactory formal account of this property of natural language expressions could be worked out. However, his student Richard Montague (1932–1971) pursued such an undertaking in the early 1960s (1970a; 1970b; 1973), at which point the same challenge was taken up by David Lewis (1972), Renate Bartsch and Theo Vennemann (1972), and Max Cresswell (1973). They too used categorial grammars to characterize the recursive structure of natural language expressions.

Semantics, then, addresses two central questions: what meanings are to be associated with the basic expressions of a language? and, how does the meaning of simpler expressions contribute to the meaning of the complex expressions the simpler ones make up? The utility of model theory is to enlighten us on how to proceed with answering the latter question. If a satisfactory answer is obtained, then, it will explain not only how changes in our understanding of complex expressions can arise from changes in their constituents, but also how it is that humans are able to understand completely novel complex expressions. After all, one's understanding of complex expressions cannot be accounted for by appeal to memorization of a language's expressions, as explicitly noted by Patañjali

---

10. The aphorisms also include definitions of technical terms (*samjñā*) as well as meta-rules (*paribhāṣā*), rules that govern rules.

two thousand three hundred years ago, any more than an appeal to memorization can explain how it is that humans knowing elementary arithmetic can understand previously unencountered numerals. Rather, one is able to understand novel complex expressions since they are combinations of simple ones that are not novel and which are antecedently understood.

However, model theory will not enlighten us about how to answer the first question: what values are associated with the various minimal expressions of a language? Answers to this question are surrounded with controversy. Views can be divided into those that hold the values to be real entities and those that hold them to be ideal entities. For some thinkers, these real entities are mental, states of the mind or concepts. Such views are found in Aristotle (see *De Interpretatione,* chaps. 1–3 passim), the speculative grammarians such as Roger Bacon (fl. thirteenth century) and Thomas of Erfurt (fl. fourteenth century), Enlightenment philosophers, most notably John Locke (1632–1704), nineteenth-century psychologists such as Wilhelm Wundt (1832–1920), and contemporary philosophers such as Jerry Fodor (1971). For other thinkers, the entities are real entities, many of which are found in the world. Views of this kind have been advocated by such diverse thinkers as William of Ockham (ca. 1285–1349) and Ludwig Wittgenstein (1889–1951), as well as by behaviorist linguists such as Leonard Bloomfield. The remaining view is that meaning is abstract or ideal, which, like numbers, are to be found neither in the mind nor in the world. Such a view could be attributed to the Stoics (cf. their notion of *lekton*); it can certainly be attributed to Gottfried Leibniz (1646–1716), Gottlob Frege (1848–1925), and Edmund Husserl (1859–1938).

It might be thought that an answer to the question of how values of simpler expressions combine to yield values for the complex expressions they make up presupposes a satisfactory answer to the philosophically vexed question of what values are the values to be associated with minimal expressions of natural language. The problem can be finessed, provided that suitable ersatz values are found. Here again, model theory furnishes help, for, as a first approximation, the very same kinds of values used in model theory can be used in semantics, as we shall see.

Exercises: Formation rules and valuation rules

These exercises are intended to familiarize the reader with the two fundamental ideas introduced in this section, namely, the idea of formation rules and the idea of valuation rules. These exercises fall into four groups. The first group pertains to the simplest numeral system, where each numeral is just a string of tallies. The second pertains to the the formation and valuation rules we have discussed. The third and fourth groups ask the reader to apply these ideas, respectively, to various numeral systems that humans have devised for counting and to the counting numerals found in various actual human languages. We shall return to natural language counting numerals in the exercises in chapter 3.

Language, Linguistics, Semantics

1. The simplest numeral system is the tally numeral system. It uses a single cipher, |. A tally numeral is just a string of such a cipher. Let TN be the set of all finite sequences of tallies: $\{|, ||, |||, ||||, \ldots\}$.

(a) Define TN using a prefixation formation rule. Define an accompanying valuation rule so that each tally numeral is assigned some positive integer and each positive integer is assigned to some tally numeral. Define another accompanying valuation rule so that each tally numeral is assigned some natural number and each natural number is assigned to some tally numeral. (The natural numbers comprise the positive integers and 0.)

(b) Define TN using a suffixation formation rule. As in the last exercise, define two valuation rules: one for the positive integers, and one for the natural numbers.

(c) Define TN using a formation rule that is different from either of the rules devised for the previous two exercises. Define two valuation rules: one for the positive integers, and one for the natural numbers.

2. Formation and valuation rules for SL

(a) Use the rule pair FRs (19) and VRi (21), first, to diagram the generation of each of these expressions of SL: *BD, AB, BCD, CDA, DCA, CABD,* and *BCAD;* and second, to find an SL expression for each of these numbers: 5, 11, 16, 23, 37, 64, and 76.

Do the same thing for the rule pair FRs (19) and VRj (22) and for the rule pair FRp (20) and VRk (23).

(b) Formulate a valuation rule, to be called VRh, which, like VRk, is to be paired with FRp (20) and which, while agreeing with the values VRk assigns to the members of L, disagrees with it on how it assigns values to the complex expressions of SL, multiplying instead the value assigned to the right-hand subexpression by 4 and adding to it the value of the left-hand subexpression.

Now use the rule pair FRp (20) and VRh, first, to diagram the generation of each of these expressions of SL: *BD, AB, BCD, CDA, DCA, CABD,* and *BCAD;* and second, to find an SL expression for each of these numbers: 5, 11, 16, 23, 37, 64, and 75.

(c) The rule pair FRs (19) and VRi (21) permits SL to have more than one expression for the same number. First, find some examples of several expressions for the same number and then state a general rule for finding expressions of SL that are expressions of the same number. Second, revise the second clause in FRs (19) so that each number has exactly one expression for it.

Carry out a similar exercise for the rule pair FRp and VRh.

(d) Consider the following formation rule valuation rule pair:

(24.1)   If $x \in$ L, then $x \in$ SL.

(24.2)   If $y \in$ SL, $y \neq A$ and $z \in$ L, then $yz \in$ SL.

(24.3)   Nothing else is a member of SL.

(25.1) $h(A) = 0, h(B) = 1, h(C) = 2$, and $h(D) = 3$.

(25.2)   If $y \in$ SL, $y \neq A$ and $z \in$ L, then $h(yz) = a \cdot h(y) + h(z)$.

Are any natural numbers left unexpressed or are any natural numbers expressed by more than one expression in SL, when a in (25.2) is 3? If so, identify the unexpressed natural numbers or the natural numbers for which more than one expression in SL exists. Answer the same questions, when a in (25.2) is 5.

(e) Consider the formation rule FRb defined as follows:

(26.1)   If $x \in$ L, then $x \in$ SL.

(26.2)   If $y \in$ SL and $z \in$ SL, then $yz \in$ SL.

How does it differ from FRs (19) and FRp (20) in its generation of, say, BAD? Does it generate all and only the expressions of SL? Try to write a valuation rule for it. Discuss the challenges that writing such a rule presents.

3. For each of the following numeral systems, state the formation rule to generate the set of its numerals and then state a valuation rule for it that will assign the intended value to each expression generated by the formation rule. The formation rules should be stated with no reference to how the expressions are to be interpreted. Moreover, what corresponds to the second condition of the formation rules for expressions in SL may in fact comprise several conditions and each condition, which must be written in an *if-then* form, may have rather elaborate *if*-clauses.

(a) The Minoans, who were a non-Indo-European people, flourished (ca. 3000–1100 BCE) on the island of Crete during the Bronze Age. Their numeral system was based on five ciphers. For ease of typography, let us use the following letters from the Latin alphabet—U, D, H, T, and M—as their ciphers. Each numeral comprises a sequence of these ciphers. No numeral contains more than nine occurrences of any one cipher. Thus, the longest numeral contains 45 ciphers and it denotes the integer 99,999. Moreover, no two numerals denote the same positive integer. In other words, each positive integer has a unique numeral denoting it. Finally, while such expressions as DUU and MTHHU are examples of Minoan numerals, UDU, TMHHU, and UHMT are not (see Ifrah 1994, vol. 1, chap. 15, 433–438 for more details).

(b) The set of roman numerals can be expressed in terms of the ciphers I, V, X, L, C, D, and M alone. The ciphers I, X, C, and M never occur more than four times in a numeral, while the ciphers V, L, and D never occur more than once. In other words, they include expressions such as IIII, VIIII, XXXX, LXXXX, CCCC, DCCCC, and MMMMDCCCC, but exclude the expressions such as DD, LL, VV as well as expressions such as IIIII, XXXXX, CCCCC, and MMMMM. The form of roman numerals entertained for this exercise also exclude the expressions: IV, IX, XL, XC, CD, and CM. (Indeed, these expressions were a

Language, Linguistics, Semantics

later innovation in the roman numeral notation.) Like Minoan numerals, no two numerals denote the same positive integer.

(c) The ancient Greeks used the twenty-four letters of their alphabet together with three letters from the Phoenician alphabet and a kind of comma to express positive integers. For convenience, we shall use the letters of the Roman alphabet. Since the Roman alphabet has only twenty-six letters, we shall use the twenty-six uppercase letters and the lowercase letter z, in other words, A, ..., Z, and z. To avoid confusion with the use of the comma in set theoretical notation, we shall use the asterisk instead. The alphabetical symbols have the following values:

CIPHER	VALUE	CIPHER	VALUE	CIPHER	VALUE
A	1	J	10	S	100
B	2	K	20	T	200
C	3	L	30	U	300
D	4	M	40	V	400
E	5	N	50	W	500
F	6	O	60	X	600
G	7	P	70	Y	700
H	8	Q	80	Z	800
I	9	R	90	z	900

Letters are strung together with no letter occurring more than once and with the letters denoting larger values occurring to the left of ones denoting smaller values. Thus, KE is a numeral but EK is not. Here are three numerals and their values: KE denotes 25, LG denotes 37, and VQG denotes 487. Multiples of one thousand up to nine thousand are denoted by one of the first nine letters with an asterisk in front of it. Thus, *BVQG denotes 2487. Finally, no two numerals denote the same positive integer.

4. Now, we shall consider three sets of counting numerals for the positive integers from 1 to 99,999. These sets progressively approximate the counting numerals of a few well-behaved, real human languages.

For each set,

- describe in your own words the pattern of these counting numerals;
- devise a formation rule that will generate all and only the counting numerals belonging to the set; and

(NB: the formation rules should be stated with no reference to how the expressions are to be interpreted.)

- state a valuation rule for the formation rule you have devised so that each counting numeral of the set, as generated by the formation rule you have devised, receives its intended value.

(a) The set of idealized counting numerals, or IN:

The following are the basic expressions:

one, two, three, four, five, six, seven, eight, nine, ones, ten, hundred, thousand, myriad.

Here is a sample of the counting numerals for the cardinal numbers from 1 to 99,999 comprising IN:

one ones, two ones, three ones, four ones, five ones, six ones, seven ones, eight ones, nine ones, one ten,

one ten one ones, ... , one ten nine ones,

two ten, two ten one ones, ... , nine ten nine ones,

one hundred, one hundred one ones, ... , one hundred nine ones, one hundred one ten, one hundred one ten one ones, ... , nine hundred nine ten nine ones,

one thousand, one thousand one ones, ... , nine thousand nine hundred nine ten nine ones,

one myriad, one myriad one ones, ... , nine myriad nine thousand nine hundred nine ten nine ones.

(b) The set of numerals WN:

No human language contains an equivalent of the expression *ones* used in (a). We now consider the set of counting numerals for the cardinal numbers from 1 to 99,999, which is just like the set IN, except that *ones* is not among the basic expressions.

Here is a sample of the counting numerals composing WN for the cardinal numbers from 1 to 99,999:

one, two, three, four, five, six, seven, eight, nine, one ten,

one ten one, ... , one ten nine,

two ten, two ten one, ... , nine ten nine,

one hundred, one hundred one, ... , one hundred nine, one hundred one ten, one hundred one ten one, ... , nine hundred nine ten nine,

one thousand, one thousand one, ... , nine thousand nine hundred nine ten nine,

one myriad, one myriad one, ... , nine myriad nine thousand nine hundred nine ten nine.

(c) The set of numerals ZN:

Several languages, including Chinese, Japanese, Korean, Mongolian, Tibetan, and ancient Turkish, have counting numerals for the cardinal numbers from 1 to 99,999 very similar to the counting numerals in WN. They differ from the expressions in WN in a number of small, but important ways. For example, the word corresponding to the word *one* never precedes the word corresponding to the word *ten*, unless the expression corresponding to *one ten* is part of a larger counting numeral. Thus, the counting numeral for 11

Language, Linguistics, Semantics

corresponds to *ten one,* but the counting numeral for 511 corresponds to *five hundred one ten one*.

Here is a sample of the counting numerals for the numbers from 1 to 99,999 in an approximation of Mandarin Chinese:[11]

yī (*one,*) èr (*two,*) sān (*three,*) sì (*four,*) wǔ (*five,*) liù (*six,*) qī (*seven,*) bā (*eight,*) jiǔ (*nine,*) shí (*ten,*)

shí yī (*eleven,*) ... , shí jiǔ (*nineteen,*)

èr shí (*twenty,*) èr shí yī (*twenty-one,*) ... , jiǔ shí jiǔ (*ninety-nine,*)

yì bǎi (*one hundred,*) yì bǎi yī (*one hundred one,*) ... , yì bǎi yì shí (*one hundred ten,*) ... , jiú bái jiǔ shí jiǔ (*nine hundred ninety-nine,*)

yí qiān (*one thousand,*) yí qiān yī (*one thousand one,*) ... , jiú qiān jiú bái jiǔ shí jiǔ (*nine thousand nine hundred ninety nine,*)

yí wàn (*ten thousand,*) yì wàn yì (*ten thousand one,*) ... jiǔ wàn jiǔ qiān jiǔ bǎi jiǔ shí jiǔ (*nine myriad nine thousand nine hundred ninety nine.*)

3 Conclusion

In this chapter, we have situated linguistics, the formal study of language, with respect to both its historical antecedents and other branches of knowledge, in particular, psychology on the one hand and logic on the other. Moreover, we have situated semantics with respect to the other subdisciplines of linguistics—namely, phonetics, phonology, morphology and syntax—and aspects of human thought, which, though not part of the study of language proper, are nonetheless pertinent—namely, human beliefs about the world. Bearing this in mind, we shall now indicate what topics are covered in this book, how the topics are related to one another, and in what way they are presented.

3.1 Topics Covered and Their Interconnections

The fundamental question to be explored in this book is the one set out in the last section of this chapter: how is the value of a complex expression of natural language, and of English in particular, determined by the values of the expressions making it up? To address the question, we must understand how expressions in a natural language, and in English in particular, are made up. As we saw, the makeup of natural language expressions is characterized by their constituent structure. Since our principal language of study here is

---

[11]. In the sample, *yì bǎi yī* means, in fact, one hundred ten, *yì qiān yī* means one thousand one hundred, and *yì wàn yī* means eleven thousand. To express one hundred one, one thousand one, ten thousand one, and similar positive integers requires another basic expression, namely, *líng* (*zero*), as well as additional rules and additional complications.

English, chapter 3 provides a brief review of basic, traditional English grammar as well as an introduction to constituency grammars, the grammars that grew out of the immediate constituent analysis of the American structuralist linguists.

The discipline concerned with making explicit how the values of complex expressions are determined by the values of the expressions making them up is logic. A number of chapters of this book then, are devoted to setting out the basics of logic. Presented in this book are the fundamentals of classical propositional logic (chapters 6 and 7), of first-order predicate logic (part of which, called here classical predicate logic, is presented in chapter 9 and part of which, called here classical quantificational logic, is presented in chapter 11). Readers who have studied logic should review these chapters both to be aware of the notation used here (not all books on logic use the same notation) and to familiarize themselves with some facets of propositional logic and first-order predicate logic not usually found in introductory textbooks on the subject. In addition, this book includes a chapter on various enrichments of classical quantification logic (chapter 12), in particular, generalized quantificational logic, and on what are called the Lambek calculus and the Lambda calculus (chapter 13).

The presentation of logic in this book is completely self-contained: no knowledge of logic is presupposed. All the concepts of logic used in this book are presented in all the detail necessary for the beginner. Moreover, since the study of logic, as the study of any area of modern mathematics, requires familiarity with some concepts of set theory, the basics of set theory required here are given in chapter 2. Almost all of these basics are familiar from the study of high school algebra.

The chapters on those parts of English grammar that we shall be examining are preceded by the relevant chapters on logic. The chapter of classical propositional logic precedes the chapter on the grammar of coordination and subordination of English clauses (chapter 7). The chapter on classical predicate logic precedes the chapter on the grammar of minimal, independent English clauses (chapter 10). The chapter on classical quantificational logic (chapter 11), the chapter on generalized quantificational logic (chapter 12), and the chapter on the Lambek calculus and Lambda calculus (chapter 13) precede the chapter on the grammar of English noun phrases (chapter 14).

In addition to the chapters on basic English grammar and basic set theory, there are two other chapters, one on meaning and context and the other on meaning and communicative presumptions. In particular, chapter 4 makes clear that one's understanding of the meaning of an expression depends in part on an appropriate understanding of the context in which it is used. Chapter 5 shows how some of the presumptions underlying successful linguistic communication shape one's understanding of natural language expressions.

In short, the essential ideas of natural language semantics are presented in chapters 6 through 14 and the necessary preliminaries are presented in chapters 2 through 5. In addition, the conclusion (chapter 15) shows how the material taught in this book relates to approaches to natural language semantics advocated by different scholars in the field.

3.2 Some Remarks on Empirical Inquiry

Now, let me explain a little about the nature of linguistic inquiry. As indicated, all linguists agree that the discipline of linguistics is an empirical one. All are engaged in arriving at a finite set of rules designed to characterize all and only the regularities, or patterns, in the expressions of human language; these rules are formulated in light of which expressions native speakers accept and which they reject. Though judgments of acceptability by native speakers have customarily been the primary source of data for linguistic theory, they are by no means the only source. Cross-linguistic comparison also helps the linguist to identify patterns. Also important are data from studies of language impairment arising from congenital developmental disorders or from damage to the brain, as well as studies of first and second language acquisition. While these latter sources of data are very important, their study involves their own specialized knowledge, which cannot be adequately incorporated into a book devoted to the basic concepts of natural language semantics.

Although there are different approaches to gathering data for the study of linguistics, they are all instances of systematic, empirical inquiry. Such inquiry involves individual observations, observational regularities, and, very often also, theoretical hypotheses. Observational regularities are arrived at in one of two ways: either they are extrapolated from individual observations or they are shown to follow from a theoretical hypothesis. Strictly speaking, one cannot observe directly an observational regularity, though one can observe instances of it. Nor can one directly observe a theoretical hypothesis, though if it is to be contentful, it must have logical consequences that are observable. In the remainder of this section, we shall elaborate on these ideas, since they will be repeatedly applied in several of the chapters that follow. The topic of systematic, empirical inquiry is both complex and subtle; the exposition here will be brief and, therefore, greatly simplified.

3.2.1 Extrapolation of an observational regularity

Much systematic, empirical inquiry begins with an observational regularity extrapolated from commonsense observations. These observations could be of phenomena that arise on their own or of phenomena provoked by experimental intervention. A regularity arising from the observation of such phenomena is a pattern that can be expressed by a universal statement.[12] In natural language, universal statements have several equivalent forms. In English, sentences of the following forms express universal statements:

(27.1)　　All As are C.

(27.2)　　If something is A, it is C.

(27.3)　　Being A is a sufficient condition for being C.

(27.4)　　Being C is a necessary condition for being A.

---

12. There are also what are called *statistical regularities,* but since they play no role in this book, they are not discussed.

The next four sentences are expressions of the same universal statement.

(28.1) All whales have lungs.

(28.2) If something is a whale, it has lungs.

(28.3) Being a whale is a sufficient condition for having lungs.

(28.4) Having lungs is a necessary condition for being a whale.

The A conditions and the C conditions in a universal statement, however, may each include several properties.

(29.1) All pieces of copper expand when heated.

(29.2) All ice floats on water.

(29.3) If the temperature of a gas increases and its pressure remains the same, its volume increases.

(29.4) If a body falls from rest in a vacuum on earth at sea level, the distance it covers in time t is $t^2 \cdot 4.9$ meters.

Indeed, observational regularities may form an interconnected system whose entirety can be expressed only by a very large set of universal statements of a particular form: for example, the observational regularities exhibited by the taxonomies of plants and animals, initiated by Aristotle and greatly expanded and refined by the Swedish botanist and zoologist Carl Linnaeus (1701–1778); or the observational regularities of the proportions in which substances combine to form other substances exhibited in the periodic table of elements, arrived at by the Russian chemist Dimitri Mendeleev (1834–1907), building on extensive and meticulous observations of experimentally obtained phenomena of many who went before him.

A universal statement expressing an observational regularity is a universal statement that expresses a regularity and each of whose instances are in principle open to commonsense observation. Here is an example:

(30) All ravens are black.

Not all universal statements express regularities. For example, the statement in (31) does not express an observational regularity, though the statement is universal and instances of it are in principle open to commonsense observation.

(31) All the screws in this lawn mower are rusty.

What distinguishes universal statements expressing regularities from those that do not? This question does not have a simple answer. Suffice it to say here that universal statements expressing regularities support subjunctive and counterfactual conditionals, whereas universal statements such as the one in (31) do not.[13] For example, the statement in (30) supports the subjunctive conditional that, should a bird be a raven, then it would be black, as well as the counterfactual conditional that, if something were a raven, then it would be black. But the statement in (31) supports neither kind of conditional. Even if the statement in (31) is true, neither the subjunctive conditional that should some screw be in the lawn mower in question, it would be rusty, nor the counterfactual conditional that if some brass screw had been in the lawn mower in question, it would be rusty, is true. For if they were true, it would be impossible for a screw to be in the lawn mower and not be rusty. But surely it is possible to change any rusty screw in the lawn mower for one that is not rusty.

Very often observational regularities emerge only after many, painstaking observations. It is far from evident to commonsense observation, for example, that whenever a ray of light is reflected at a plane surface, the angle of incidence is the same as the angle of reflection. Some observational regularities are so commonsensical that one is unaware of making any extrapolation. For example, it seems unlikely that anyone consciously extrapolated that all ravens are black from careful examination of one raven after another. At the same time, history is filled with examples of commonsensical extrapolations that turn out not to be regularities. Anyone living in Europe before the seventeenth century undoubtedly believed, on the basis of very few observations, that all swans are white. Anyone who at that time might have seriously suggested otherwise would have been dismissed as an arrant fool. Yet, when Europeans arrived in Australia, they soon discovered that swans in Australia are black.

As stated, observations can be of phenomena that arise on their own or of phenomena provoked by experimental intervention. A particularly clear and simple example of an observational regularity involving experimental intervention is found in *Elements of Chemical Philosophy,* an early textbook on chemistry written by one of the founders of modern chemistry, Humphry Davy (1778–1829).

Whoever will consider with attention the slender green vegetable filaments (*conferva rivularis*) which in the summer exist in almost all streams, lakes, or pools, under the different circumstances of shade and sunshine, will discover globules of air upon the filaments that are shaded. He will find that the effect is owing to the presence of light. This is an *observation;* but it gives no information respecting the nature of the air. Let a wine glass filled with water be inverted over the conferva, the air will collect in the upper part of the glass, and when the glass is filled with air, it may be closed by the hand, placed in its usual position and an inflamed taper introduced into it; the taper will burn with more brilliancy than in the atmosphere. This is an *experiment*. If the phenomena are reasoned upon, and the question is put, whether all vegetables of this kind, in fresh or in salt water, do not produce such air under like circumstance, the enquirer is guided by analogy: and when this is determined to be the case by new trials, a general scientific truth is establish—that all confervae in the sunshine produce a species of air that supports flame in a superior degree; which has been shown to be the case by various minute investigations. (1812, 1–2, cited in Hacking 1983, 152)

---

13. The grammar of such conditionals is touched on in section 6 of chapter 8.

(Humphry Davy uses the word "observation" for phenomena that are observed on their own; I use the word to denote both phenomena observed on their own and phenomena obtained from experimental intervention.)

We have said that observational regularities are expressed by universal statements. A universal statement is false if there is even one counter example. In other words, even if all the observed instances of a universal statement are true, it still might turn out to be false, since there is always the possibility of a falsifying instance, either overlooked in the present or past or lurking somewhere in the future. Thus, while a universal statement can be proved false by just one counterexample, it can never be proved true, though it can, as the number of its observed instances without any counterexamples increases, become more and more strongly confirmed.

One might think that, if one comes across a counterexample to a universal statement with otherwise substantial confirmation purporting to express an observational regularity, all the work that went into extrapolating the purported observational regularity was for naught. But this need not be true. Here is an instructive example. Even to this day, only a finite number of observations has been made of the night sky, though human beings have been observing it for tens of thousands of years and studying it for thousands. Obviously, then, the stargazers of ancient China, ancient Egypt, and ancient Mesopotamia made only a finite number of observations of the stars. Yet we can be certain that, from the finite number of those observations, they made the following commonsensical extrapolations: the stars do not move with respect to one another, but they do move across the sky from east to west. In particular, each star appearing on the eastern horizon shortly after sunset rises steadily in the sky, passes overhead, and then descends toward the western horizon, disappearing with the rise of the sun in the east.

At some point in historical ancient times, stargazers in Mesopotamia and, later and independently, stargazers in China, noticed, not one, but five star-like celestial bodies that do not remain at a fixed distance from the stars nor from one another. Rather, these five star-like celestial bodies appear to move among the stars of the zodiac. We have no idea of whether the first stargazers to notice these five bodies, which appear to common sense just like stars, regarded them as stars. But had they so regarded them, then the observation of their motion refutes the universal statement expressing one of the observational regularities just noted, namely, that no star moves with respect to any other. What we do know is that the Classical Greek astronomers called these five celestial bodies wanderers, or *planets,* regarding them, not as counterexamples to a universal statement purporting to express an observational regularity, but as exceptional celestial bodies. Moreover, as the Greek and Chinese astronomers knew, each planet evinces its own regularity of motion through the stars of the zodiac. Each returns to its same position relative to the stars: the moon once every 30 days roughly, Mercury once every 88 days, Venus once every 225 days, the sun once every year, Mars once every 2 years, Jupiter once every 12 years, and Saturn once every 29.5 years. In fact, the Classical Greek astronomers, instead of grouping the planets with what we today call the stars, grouped them with the sun and the moon, which they also called planets, because they too wander among the stars of the zodiac, returning regularly to the same position relative to the stars, the moon every twenty-seven days and the sun every year.

As a matter of fact, the problem of finding a suitable explanation for the observational regularities of the planets, the sun, and the moon was the driving force behind astronomy from the time of Plato (ca. 428–348 BCE), who proposed this problem to his student, Eudoxus of Cnidus (ca. 395–337 BCE), to the Renaissance, when the correct account finally emerged over a period of a century and a half. It started with Nicolas Copernicus (1473–1543) showing in detail how planets, including the earth, orbit the sun and it was perfected by Johannes Kepler (1571–1630), who used the extensive and meticulous astronomical observations of Tycho Brahe (1546–1601) to show that the orbits of the planets are not circles, but ellipses.

One important lesson to be drawn from this case is that counterexamples to a universal statement purporting to express an observational regularity may, when suitably restricted, not only be vindicated but may lead to the discovery of other observational regularities. We shall have many occasions in this book to see counterexamples to universal statements purporting to express patterns in linguistic expressions and to see that these counterexamples themselves evince patterns.

3.2.2 Evidence for or against a theoretical hypothesis

Observational regularities invite explanation. This usually takes the form of a theoretical hypothesis in which something not open to commonsense observation is posited with a view to explaining one or more accepted observational regularities. Let us consider a simple example in which a hypothesis of something not open to commonsense observation is posited to explain an observational regularity. As undoubtedly noticed by any seafaring people, the ancient Greeks had observed again and again that, as ships go out to sea, they not only grow smaller in appearance but they also disappear from bottom to top. Though such observations have obviously been made only a finite number of times, no doubt people have instinctively extrapolated from such observations to the observational regularity that whenever a ship goes out to sea, it not only shrinks in apparent size but also disappears from bottom to top. What accounts for this observational regularity? The Classical Greeks hypothesized that the earth is a sphere.

Whereas commonsense observation by itself could neither prove nor disprove this hypothesis, still commonsense observation can provide evidence either for or against it. We consider first how commonsense observation can provide evidence for the hypothesis.

Surely, at the very least, for a hypothesis to be accepted as an explanation of an observational regularity, the observational regularity must be shown to follow from the hypothesis. If it is so shown, then we have evidence for the hypothesis. Moreover, should other observational regularities come to light and be shown to follow from the hypothesis as well, we would have even more evidence for it.

Let us illustrate both of these points with our example of the hypothesis that the earth is a sphere and the observational regularity that whenever a ship goes out to sea, it not only shrinks in apparent size but also disappears from bottom to top. The statement that the earth is a sphere is not open to commonsense observation before the last third of the twentieth century. Moreover, the statement entails the observational regularity that whenever a ship goes out to sea, it disappears from bottom to top. After all, if one were to stand on an utterly flat surface, as the surface of a body of water appears to be on a completely calm day, then one's line of sight with respect to the flat surface would extend as far as the flat surface does. If, however, one were to stand on a perfectly spherical surface, then one's line of sight with respect to the spherical surface would extend up to, but not beyond, those points on the surface picked out by a straight line passing through one's eyes and tangent to the surface of the sphere. Anything moving directly away from the observer beyond such a point would sink below that line, and when it does, it would disappear from bottom to top, as though the object were sinking below the horizon. Evidence for the hypothesis that the earth is a sphere is the fact that it entails the observational regularity that, whenever a ship goes to sea, it disappears from bottom to top.

At the same time, other observational regularities follow from the hypothesis, and with time, these observational regularities received commonsense observational support. For example, the hypothesis that the earth is a sphere entails the following observational regularity: travelers traveling on the surface of the earth in an apparently straight line, which is in fact a great circle, inevitably and eventually return to their points of departure. This does not follow from the observational regularity of ships going out to sea disappearing from bottom to top. Another remarkable consequence was one worked out by Eratosthenes of Cyrene (270–180 BCE): he deduced from the hypothesis, using some elementary Euclidean geometry, the circumference of the earth. This too does not follow from the observational regularity of ships going out to sea disappearing from bottom to top, even if it is supplemented by all the mathematics one likes. In fact, it was a conviction in the truth of the hypothesis of the earth's sphericity and the two observational consequences just mentioned that led the Italian explorer Christopher Columbus (1451–1506) to set off for the Far East, sailing west from the Iberian Peninsula[14] and that led the Portugese explorer Ferdinand Magellan (1480–1521) to set off in 1519 on an expedition to circumnavigate the earth.

In a similar fashion, but with much greater complexity and on a vastly larger scale, Darwin's theory of natural selection, a hypothesis not open to confirmation by immediate commonsensical observation, is now accepted as a correct theory that accounts for the

---

14. As it happens, the value Christopher Columbus relied on for the circumference of the earth was substantially smaller than the real value. Fortunately for him and his crews, the American continents intervened between Spain and China.

observational regularities embodied in the botanist's taxonomy of plants and the zoologist's taxonomy of animals, and the theory of the chemical bond, also a hypothesis not open to confirmation by immediate commonsensical observation, is accepted as a correct theory that accounts for the observational regularities found in the periodic table.

Now let us turn to the question of what counts as evidence against a theoretical hypothesis. One might think that any observation that is logically inconsistent with a theoretical hypothesis refutes it. However, though such observations are evidence against the hypothesis, they rarely are taken to refute the hypothesis. To begin with, if a wide range of observational regularities follow from a theoretical hypothesis, even if observations are made that are inconsistent with the hypothesis, one might call into question the observation, instead of the hypothesis. Moreover, even if one accepts the observation that is inconsistent with the hypothesis, one need not completely abandon the hypothesis; after all, it might be possible to modify the hypothesis to make it consistent with the new observations. In addition, as we shall see, the observational consequences of a hypothesis rarely, if ever, follow from a hypothesis alone; they almost always require other assumptions, sometimes ones that have been made explicit, sometimes ones that remain implicit and of which its proponents are unaware. It might be possible to keep the hypothesis and to modify one of the other assumptions. Each of these points will be illustrated in the chapters that follow.

So far, we have been considering examples of one hypothesis put forth to explain an observational regularity. However, it can happen that two or more hypotheses can account for the very same observational regularities. The question then arises: which hypothesis should one adopt? In such cases, one seeks to make an observation that is consistent with one hypothesis but not with the other. However, it is often not easy to come by such decisive observations.

Let us consider an illustration of this point. Recall the observational regularities noted that at night the stars do not move with respect to one another but they do move across the sky from east to west and that each appears on the eastern horizon shortly after sunset, rising steadily in the sky, passing overhead, and then descending toward the western horizon, finally disappearing when the sun rises in the east. At a later time, ancient astronomers greatly refined their description of this observational regularity, observing that, if a star is, say, directly overhead at midnight one night, it will appear at that position the next night at four minutes to midnight.

To account for these regularities, the Classical Greek astronomers adopted the hypothesis that the stars are situated on a celestial sphere that rotates on an axis going through the star Polaris once every twenty-three hours and fifty-six minutes. This hypothesis, like the hypothesis of the earth's sphericity, was not open to confirmation by immediate commonsense observation. Nonetheless, the observational regularities just described follow from the hypothesis. Moreover, other observational regularities also follow from the hypothesis. For example, it follows that each star maintains its relative position both to the other stars and to the observer even on overcast nights when an observer cannot see the stars. Indeed, it follows that each star maintains its relative position both to the other stars and to the observer even during the day when sunlight renders observation of the stars impossible.

Though the hypothesis of the celestial sphere was the dominant hypothesis to account for the motion of the stars, another hypothesis was put forth by Aristarchus of Samos (ca. 310–230 BCE). He hypothesized that the earth, a sphere, rotates around its axis from west to east once every twenty-three hours and fifty-six minutes. The very same observational regularities follow from both hypotheses. As a matter of fact, no naked eye observation of the fixed stars is inconsistent with either hypothesis. However, the hypothesis that the earth rotates on its axis once every twenty-three hours and fifty-six minutes seemed utterly absurd to anyone living before the rise of modern science, for it is utterly inconsistent with commonsense observations about bodies on rotating surfaces. Bodies that are not fixed to a rotating wheel are thrown off, once the wheel starts to rotate rapidly. The Classical Greeks knew to a good first approximation the circumference of the earth, so that they could calculate that if the earth were rotating, it would be rotating at 424 meters a second, a speed virtually unimaginable to anyone then living. To their minds, anything rotating that quickly would throw off anything not secured to it.

3.2.3 Testing observational regularities

As we saw, observational regularities can be arrived at in two ways: by an extrapolation from various individual observations and by showing it to follow from a hypothesis. Observational regularities, then, play a crucial role in systematic, empirical inquiry. It should come as no surprise, then, that the question of how to test observational regularities is one that scholars have reflected on for a long time, indeed, all the way back through Medieval Christian, Islamic, and Jewish philosophers to the Classical Greeks and independently by Classical Indian grammarians. While the idea had been explicitly formulated by such medieval intellectual giants as Robert Grosseteste (ca. 1175–1253), it is most widely associated with John Stuart Mill (1806–1873), who, in the nineteenth century, elaborated the methods at length and illustrated their application with examples from the science of his day. While Mill's formulation of these methods has been rightly criticized, there is no doubt that his statements point to essential features of systematic, empirical investigation. We shall now state and illustrate the three that find application in the pages that follow.

The first method is his method of agreement (Mill [1843] 1881, bk. 3, chap. 8, sec. 1). One way to see it is as the well-known practice of finding confirmatory instances of the universal statement expressing an observational regularity. This consists in finding new instances of the A conditions of the universal statement and observing whether they are also instances of the C conditions. (See the sentences in (27).) Let us consider an example. Edward Jenner (1749–1823), a doctor living in Berkeley, Gloucestershire, learned that many people believed that dairymaids who contracted cowpox, a relatively benign disease, did not contract smallpox, a then often fatal disease. He had noted in his diary that

Sarah Portlock, Mary Barge, Elizabeth Synne, Simon Nichols, Joseph Merret, and William Rodway had contracted cowpox, but did not contract smallpox, even when exposed to it. He extrapolated from these observations the observational regularity that those who contract cowpox do not contract smallpox. To test this, in May 1796, he took pus from Sarah Nelmes, a dairymaid sick with cowpox, and put it into the arm of a healthy eight-year-old boy, James Phipps, who had never had smallpox. The boy contracted cowpox. Forty-eight days later, Jenner put smallpox pus into Phipps's arm, but he never fell ill. Based on these cases, Jenner advocated the use of inoculation with cowpox as a prophylactic against the contraction of smallpox.

The second method is Mill's method of difference (Mill [1843] 1881, bk. 3, chap. 8, sec. 2). It can be viewed as a way to determine whether a condition that is one of several A conditions is required for the C conditions or is superfluous. An example will help make the idea clear. Until the seventeenth century, it was almost universally believed in Europe that simple living organisms could come into being by spontaneous generation. In particular, it was thought that maggots would be spontaneously generated from putrifying meat. The observational regularity was thought to be that putrifying meat produces maggots. In 1668, Francesco Redi (1626–1697) conducted an experiment to refute this view, only part of which we shall describe here. He placed meat into a sealed jar preventing any contact between the meat and flies. The meat in the jar putrified, but at no point did maggots appear on the meat. He concluded that putrifying meat did not spontaneously produce maggots. In particular, what he showed is that, in the absence of flies alighting on the meat, the putrifying meat failed to produce maggots.

Finally, we turn to the joint method of agreement and difference (Mill [1843] 1881, bk. 3, chap. 8, sec. 4). As the name suggests, this method combines the two previous methods. It seeks both to find confirmatory instances of the observational regularity and to confirm that one or another of the A conditions is required for the C conditions. Again, we turn to an example to make clear what the method is. Anthrax is a serious, infectious disease, caused by a bacterium, which afflicts grazing animals, such as cows, horses, sheep, and donkeys, as well as humans. Louis Pasteur (1822–1895), having identified the bacterium that he thought caused the disease, developed a vaccine against it. On the basis of this hypothesis, Pasteur inferred the observational regularity that a grazing animal inoculated with the vaccine he developed would not succumb to anthrax. In the spring of 1881, then, under the auspices of the Agricultural Society of Melun, at a farm in Pouilly-le-Fort, he vaccinated one group of twenty-four sheep, a goat, and several cattle, while leaving unvaccinated another group of twenty-four sheep, a goat, and several other cattle. The first group was vaccinated twice, two weeks apart. Fifteen days after the second vaccination of the first group, he injected into all the animals of both groups live anthrax bacteria. All the animals in the unvaccinated group died within three days, whereas all the vaccinated animals were unaffected. In short, the experiment provided confirmatory evidence that vaccination is, among various A conditions for the prevention of anthrax, a requisite one. In light of the devastating effects of an outbreak of anthrax, it is hardly surprising that the report of the demonstration created an international sensation. This example of the joint method of agreement and difference is today described as a controlled experiment, where the group of unvaccinated animals constitutes the *control group.*

Though the illustrations given herein all involve causation, and though Mill himself clearly thought of them as methods pertaining to observational causal regularities, they have wider use. As it happens, these methods have long been known to linguists. Indeed, these methods were employed by the ancient Indian grammarians, who also called them, coincidentally, the methods of agreement (*anvaya*) and difference (*vyatireka*). The joint method is familiar to linguists today as the method of *minimal pairs,* a technique most commonly used by phonologists, but in fact used in all areas of linguistic investigation. We shall be using such evidence extensively in the chapters that follow. Indeed, such evidence has already been adduced several times in earlier sections of this chapter.

In our investigation of the semantics of English, we shall be pursuing similar objectives. We shall learn about observational regularities of English usage, many recorded in comprehensive, descriptive grammars of English, and others not. We shall consider various hypotheses, which are not themselves open to confirmation by immediate commonsensical observation, to account for the observational regularities. In some cases, we shall see that the observational regularities have exceptions. At that point, we shall grapple with the question of whether the exceptions constitute counterexamples to the regularities so that the regularities should be abandoned or the exceptions themselves evince observational regularities so that fresh hypotheses should be sought to account for the latter regularities. In other cases, we shall find that the very same observational regularity can be accounted for by different hypotheses, and we shall explore ways in which such competing hypotheses might be distinguished by observation so that we can retain one and discard the other.

Exercises: Some remarks on empirical inquiry

1. For each of the experiments described, state the observational regularity being tested and explain which of the three methods set out in section 3 has been applied to the observational regularity. If the observational regularity is obtained from a theoretical hypothesis, state what it is.

(a) Walter Reed (1851–1902), James Carroll (1854–1907), and Jesse W. Lazear (1866–1900) conducted the following experiment to establish a suggestion by Carlos Juan Findlay (1833–1915) that the mosquito *Aedes aegypti* is the vector of yellow fever.

In November of 1900, they had erected a small building into which no mosquito could penetrate. A wire mosquito screen divided the room into two spaces. In one of these spaces fifteen mosquitoes, which had fed on yellow fever patients, were set free. A non-immune volunteer entered the room with the mosquitoes and was bitten by seven mosquitoes. Four days later, he suffered an attack of yellow fever. (Copi 1953, 446–447)

(b) Farmers in Europe had long known that anthrax afflicts grazing animals and humans, but not chickens. Louis Pasteur noted that chickens have a body temperature between 43° and 44° Celsius, whereas grazing animals and humans have a body temperature around 37° Celsius. Pasteur suspected that this difference in body temperature of the animals accounted for the difference in susceptibility of the animals to the disease. To confirm his suspicion, he conducted the following experiment.

He injected a number of chickens with anthrax bacilli and placed them in a cold bath, lowering their body temperature to 37° Celsius. Once the chickens showed symptoms of the disease, he removed some of them from the cold bath and warmed those to the point where their body temperature returned to normal. The ones left in the cold bath died, whereas those removed from it and warmed up to their normal body temperature did not.

(a) Reider F. Sognnaes reports the following experiment that he conducted.

We have recently obtained conclusive experimental evidence that there can be no tooth decay without bacteria and a food supply for them. In germ-free laboratories at the University of Notre Dame and the University of Chicago, animals innocent of oral micro-organisms do not develop cavities. Where animals in normal circumstances average more than four cavities each, the germ-free rats shown no signs of caries. At the Harvard School of Dental Medicine we have demonstrated the other side of the coin, that food debris also must be present. Rats that have plenty of bacteria in their mouths but are fed by tube directly to the stomach do not develop cavities. In a pair of rats joined by surgery so that they share a common blood circulation, the one fed by mouth develops tooth decay, and one fed by tube does not. (Sognnaes 1957, 112–113)

2. Strato of Lampsacus was, from 286 to 268 BCE, head of the Lyceum, the school in Athens that Aristotle had founded in 335 BCE. He is said to have made the following argument.

If one drops a stone or any other weight from a height of about a finger's breadth, the impact made on the ground will not be perceptible, but if one drops the object from a height of a hundred feet or more, the impact on the ground will be a powerful one. Now there is no other cause for this powerful impact. For the weight of the object does not increase, the object itself had not become greater, it does not strike a greater space of ground, nor is it impelled by a greater [external force]: rather it moves more quickly. (Lloyd 1973, 16)

What is Strato of Lampsacus's theoretical hypothesis? What is the observational regularity? And how does he think the observational regularity follows from the theoretical hypothesis?

3. For each of the two experiments presented, identify the hypothesis and, without appealing to any facts beyond common sense, state the observational regularity being tested, state the theoretical hypothesis from which it follows, and explain which of the three methods set out in section 3 has been applied. In addition, critically discuss the experiment in relation to the hypothesis.

(a) A horse doctor in the Jura Mountains in the east of France, Louvrier, claimed that he had a cure for anthrax. It comprised rubbing a sick cow so as to make it as warm as possible, to then cut gashes in the skin in which was poured turpentine and then to apply a thick layer of manure mixed with hot vinegar. Pasteur conducted the following experiment.

In the presence of Louvrier and a commission of farmers, Pasteur injected four healthy cows with live anthrax. The next day, when all four cows were seen to be suffering from anthrax, Pasteur requested Louvrier to treat two cows, while the other two were left untreated. In the end, one cow treated by Louvrier died and the other recovered, and one untreated cow died and the other recovered.

(b) Jan Baptista van Helmont (1579–1644) conducted the following experiment to substantiate his view that a tree is made primarily of water. Here is what he reported.

That all plants immediately and substantially stem from the element water alone I have learnt from the following experiment. I took an earthen vessel in which I placed two hundred pounds of earth dried in an oven, and watered with rain water. I planted in it the stem of a willow tree weighing five pounds. Five years later it had developed a tree weighing one hundred and sixty-nine pounds and about three ounces. Nothing but rain (or distilled water) had been added. The large vessel was placed in earth and covered by an iron lid with a tin-surface that was pierced with many holes. I have not weighed the leaves that came off in the four autumn seasons. Finally I dried the earth in the vessel again and found the same two hundred pounds of it diminished by about two ounces. Hence one hundred and sixty-four pounds of wood, bark and roots had come up from water alone. (Howe 1965, 408–409)

4. Find a report of a deprivation experiment, summarize the report, and explain how the experiment works in terms of Mill's methods.

5. The following two sentences are judged true and false, respectively. State a hypothesis, choose one of Mill's methods whereby these facts can be said to support the hypothesis and explain how the method you have chosen, applied to these facts, supports the hypothesis.

(32.1)   A cow is a mammal.

(32.2)   A rock is a mammal.

SOLUTIONS TO SOME OF THE EXERCISES

2.2.3   Grammar: Formation rules and valuation rules

1. Tally numeral system

Formation rule:

(TF1)   If $x \in \{|\}$, then $x \in$ TN.

(TF2)   If $y \in$ TN and $z \in \{|\}$, then $yz \in$ TN.

(TF3)   Nothing else is a member of TN.

Valuation rule:

(TV1) $v(|) = 1$.

(TV2) If $y \in \text{TN}$ and $z \in \{|\}$, then $h(yz) = v(y) + v(z)$.

TF2 is a suffixation rule. Notice that, even if TF2 had been formulated as a prefixation rule, the valuation rule could remain unchanged.

2. SL

(a) $i(BD) = 7, i(BCD) = 27, i(DCA) = 56, i(BCAD) = 99$

$5 = i(BB), 16 = i(BAA), 37 = i(CBB), 76 = i(BADA)$

$j(BD) = 13, j(BCD) = 21, j(DCA) = 11, j(CABD) = 18$

$5 = j(BB), 16 = j(ABD), 23 = j(DCD), 64 = j(ABDDDDD)$

$k(BD) = 7, k(BCD) = 15, k(DCA) = 20, k(CABD) = 15$

$5 = k(BB), 16 = k(DBA), 37 = k(DDDB), 64 = k(DDDDDBA)$

(b) VRh:

(24.1) $h(A) = 0, h(B) = 1, h(C) = 2$, and $h(D) = 3$.

(24.2) If $y \in L$ and $z \in \text{SL}$, then $h(yz) = h(y) + 4 \cdot h(z)$.

$h(BD) = 13, h(BCD) = 57, h(DCA) = 11, h(BCAD) = 201$
$5 = h(BB), 16 = h(AAB), 37 = h(BBC), 64 = h(AAAB)$

(c) The rule pair FRs (19) and VRi (21) has expressions such as AB and AAB that each denote 1. To avoid such redundant expressions, one need add only one simple condition has to the clause (19.2) of FRs.

(d) When $a = 4$, all the natural numbers are expressed and no natural number is expressed more than once.

When $a = 3$, all natural numbers are expressed and some are expressed more than once.

When $a = 5$, some natural numbers are not expressed.

(e) This question is worth devoting some time to.

3. Numeral systems

(a) Minoan numerals

We first define a formation rule for MN, the set of Minoan numerals which can be expressed in terms of the ciphers U, D, H, T, and M. To do so, we let N be the set of the five ciphers {U, D, H, T, M}. In addition, we order the ciphers in N: M outranks T, H, D, and U; T outranks H, D, and U; H outranks D and U; D outranks only U; and U outranks nothing. This ordering is similar to the familiar ordering of the letters of the Roman alphabet whereby A outranks all Roman letters subsequent to A, B outranks all Roman letters subsequent to B, and so on.

Formation rule for MN

(MF1) If $x \in$ N, then $x \in$ MN;

(MF2) If (1) $y \in$ MN and $z \in$ MN, (2) no symbol in z outranks any symbol in y and (3) in y and z taken together, no member of N occurs more than nine times, then $yz \in$ MN;

(MF3) Nothing else is a member of MN.

We now define v, a valuation rule, which assigns the intended positive integers to each expression in MN.

Valuation rule for MN

(MV1) $v(\text{U}) = 1$, $v(\text{D}) = 10$, $v(\text{H}) = 100$, $v(\text{T}) = 1000$ and $v(\text{M}) = 1000$;

(MV2) If (1) $y \in$ MN and $z \in$ MN, (2) no symbol in z outranks any symbol in y and (3) in y and z taken together, no member of N occurs more than nine times, then $v(yz) = v(y) + v(z)$.

4. Natural language counting numerals

(a) Idealized counting numerals

We first define a formation rule for IN, the set of idealized counting numerals.

To do so, we divide up the basic expressions into two sets, U and B. Let U have the nine basic expressions: *one, two, three, four, five, six, seven, eight,* and *nine*. Let B have the remaining basic expressions: *ones, ten, hundred, thousand,* and *myriad* Next, we observe that each expression in IN has a form. To describe the form, we partition the set B into five sets, O, D, H, T, and M. Each of these sets has just one member: O has *ones*, D has *ten*, H has *hundred*, T has *thousand*, and M has *myriad*. Each expression in IN can be seen to have the form:

(UM)(UT)(UH)(UD)(UO)

where no subexpression corresponding to a pair of parentheses need occur, but it cannot be that every pair of parentheses has no corresponding subexpression.

Notice that the basic expressions in each complex expression in IN respects an order. We order the basic expressions in B as follows: *myriad* outranks *thousand, hundred,* and *ten*; *thousand* outranks *hundred* and *ten*; and *hundred* outranks *ten*.

Formation rule for IN

(IF1) If $x \in$ U and $y \in$ B, then $xy \in$ IN;

(IF2) If $x, y \in$ IN and each expression of B in x outranks each expression of B in y, then $xy \in$ IN;

(IF3) Nothing else is a member of IN.

We now define v, a valuation rule, which assigns the intended positive integers to each expression in IN.

Valuation rule for IN

(IV0) $v(one) = 1$, $v(two) = 2$, $v(three) = 3$, $v(four) = 4$, $v(five) = 5$, $v(six) = 6$, $v(seven) = 7$, $v(eight) = 8$, $v(nine) = 9$, $v(ones) = 1$, $v(ten) = 10$, $v(hundred) = 100$, $v(thousand) = 1000$; and $v(myriad) = 10000$;

(IV1)   If $x \in U$ and $y \in B$, then $v(xy) = v(x) \bullet v(y)$;

(IV2)   If $x, y \in IN$ and each B symbol in x outranks each B symbol in y, then $v(xy) = v(x) + v(y)$.

SOME REMARKS:

- Note that no expression either of U or of B is an expression of IN. However, each expression of IN has expressions of U and of B as subexpressions. Thus, we require a clause to assign values to the expressions of U and of B. Clause (IV0) does this.
- Clause (IV1) assigns values to those expressions arise from clause (IF1).
- Clause (IV2) assigns values to those expressions which arise from clause (1F2).

3.2 Some remarks on empirical inquiry

1(a) Let A be the nonimmune volunteer who entered the room and was bitten by seven mosquitoes that had fed on yellow fever patients.

  Theoretical hypothesis:
Whatever causes yellow fever is transmitted by mosquitoes from those who have the illness to those who do not.

  Observational regularity:
Whoever is not immune to yellow fever and is bitten by a mosquito that has fed on someone who has yellow fever contracts yellow fever.

  Method of agreement:
The fact that A is not immune to yellow fever and is bitten by a mosquito that has fed on someone who has yellow fever and contracts yellow fever is evidence for the observational regularity.

2 Basic Set Theory

1 Introduction

In this chapter, we shall learn about the principal notions of set theory used in the study of logic and natural language. We shall learn about *sets* and their *membership*. We shall learn about certain basic relations between sets, such as one set being a *subset* of another and one set being *disjoint* from another. We shall also learn about operations on sets: namely, the operations of *union, intersection, difference,* and *complementation*. In addition to sets, we shall be introduced to *sequences,* which are *ordered sets*. Ordered sets differ from sets in that the members of the former are in an order relative to one another, whereas those of the latter are not. We shall also learn about sets of sets and various operations on them. Finally, we shall study *relations* and *functions*.

2 Sets and Their Members

Entities, be they abstract or concrete, can form sets, and the entities that form a set are members of the set they form. Thus, for example, all chairs with exactly four legs form a set, and any chair with exactly four legs is a member of that set.

Sets, like numbers, have no spatial or temporal location; hence, like numbers, they are abstract entities. As a result, even if every member of a set is a concrete entity, the set formed from those entities is itself, nevertheless, an abstract entity. Thus, every chair with four legs is a concrete entity; however, the set of all such chairs is an abstract entity.

As mentioned, sets may also be formed from abstract entities. The *natural numbers,* for example, are abstract entities, and they form the set whose members are: 0, 1, 2, 3, ... Indeed, sets themselves can form sets.

Finally, a set can be formed from both concrete and abstract entities. For example, all chairs with exactly four legs and all natural numbers together form a set. Any chair with exactly four legs, as well as any natural number, is a member of this set. This set should not be confused with the two-membered set whose only members are the set of natural numbers and the set of chairs with exactly four legs. No chair nor any natural number is a member of this latter set.

The most fundamental relation in set theory is the relation that obtains between a member of a set and the set itself. This relation is called the *set membership* relation, and it is denoted by the symbol $\in$, a variant of the lowercase Greek letter epsilon ϵ. Membership in a set is expressed by inserting the sign for set membership between a name for a member of a set and a name for the set. Thus, if A is a set and b is one of its members, then this fact can be expressed as follows: $b \in A$. Besides saying that b is a member of A, one may also say: b belongs to A, b is an element of A, or simply, b is in A.

A set is determined by its members and nothing else. This fact leads to two ways of naming sets: *list notation* and *abstraction notation*. In list notation, one encloses a list of the set's members within curly brackets. The expression $\{1, 2, 3\}$ denotes the set whose members are the numbers 1, 2, and 3. In abstraction notation, also known as *set builder notation*, one writes within curly brackets, a variable, followed by either a colon or a vertical bar, and a description, usually containing the intial variable, which is satisfied by all and only the members of the set. Thus, the following expression in abstraction notation denotes the set of four-legged chairs:

$\{x : x$ is a four-legged chair$\}$.

There are two features about list notation that one should bear in mind. First, the order in which elements in the list within the curly brackets are given is irrelevant. In other words, no matter how the names of the members in the list are rearranged, the set denoted by the list within curly brackets is always the same. Thus, the very same set is denoted both by $\{a, b, c\}$ and by $\{b, a, c\}$. Second, a redundant item in a list in list notation is irrelevant; that is to say, if two lists differ only in the repetition of names in the lists, the sets denoted by the two lists within curly brackets are the same. So, the very same set is denoted both by $\{a, b, c, a\}$ and by $\{c, a, b, b\}$. In short, two names for the same set, both in list notation, can differ only in the order or the degree of redundancy of the names within the curly brackets.

Not only does list notation permit the very same set to be named in two different ways, but so does abstraction notation. Thus, the set of even (natural) numbers,

$\{x : x$ is an even natural number$\}$,

is the same as the set of natural numbers divisible by 2,

$\{x : x$ is a natural number and x is divisible by 2$\}$.

Finally, a name of a set in list notation and a name of a set in abstraction notation may denote one and the same set. For example, $\{1, 2, 3\}$ can also be denoted by:

$\{x : x$ is a natural number greater than 0 and less than 4$\}$.

Here are some other conventions governing the notation of set theory that will be observed in this book. Uppercase letters of the Roman alphabet pertain exclusively to sets: uppercase letters from the beginning of the alphabet (e.g., $A, B, C, \ldots$) are parameters, standing for fixed sets, while uppercase letters from the end of the alphabet (e.g., $\ldots X, Y, Z$) are variables ranging over sets. Lowercase letters may pertain both to nonsets—hence, to entities, concrete or abstract, from which sets may be formed—or to sets. Thus, presented with an uppercase letter, one may be sure that one is dealing with a set; presented with a lowercase letter, however, one cannot be sure whether the relevant entity is a set or some other entity different from a set. Lowercase letters from the beginning of the alphabet (e.g., $a, b, c, \ldots$) are parameters, standing for fixed entities that may or may not be sets, while lowercase letters from the end of the alphabet (e.g., $\ldots, x, y, z$) are variables ranging over entities that may or may not be sets. Sometimes we shall have occasion to use letters in calligraphic font, which, if from the beginning of the alphabet, are parameters and stand for fixed sets of sets, and if from the end of the alphabet, are variables and range over sets of sets.

2.1 Some Important Sets

Several sets commonly referred to in mathematics have standard names. One is the set of natural numbers, usually denoted by $\mathbb{N}$. The natural numbers have two important properties: first, the addition of two natural numbers always yields a natural number, and second, the multiplication of two natural numbers always yields a natural number. However, the subtraction of one natural number from another does not always yield a natural number. Thus, while the natural number 15 subtracted from the natural number 17 yields the natural number 2, the natural number 17 subtracted from the natural number 15 yields no natural number.

In order to obtain a set of numbers large enough to guarantee that the subtraction of any two numbers always yields a member of that set, the natural numbers must be increased to include a negative counterpart for every natural number other than 0. This new set of numbers is known as the *integers,* and the symbol used to denote this set is $\mathbb{Z}$.

Within the set of integers, it is usual to distinguish the set of positive integers $\{+1, +2, +3, \ldots\}$ from the negative integers $\{\ldots, -3, -2, -1\}$. Positive integers are usually written without the plus sign. The symbol used to denote the set of positive integers is $\mathbb{Z}^+$, and the one used to denote the set of negative integers is $\mathbb{Z}^-$. Clearly, the integers include the natural numbers, since the natural numbers are just the positive integers together with 0. Finally, it is convenient to have symbols denoting the set of the first n positive integers. For example, we shall refer to the set $\{1, 2, 3\}$ with the complex symbol $\mathbb{Z}_3^+$, to the set $\{1, 2, 3, 4\}$ with the complex symbol $\mathbb{Z}_4^+$, and more generally, to the set $\{1, 2, 3, \ldots, n\}$ with the complex symbol $\mathbb{Z}_n^+$.

2.2 The Size of a Set

The size, or *cardinality*, of a set is the number of distinct members in it. A set may be either finite or infinite, depending on whether the number of distinct members it has is either finite or infinite, respectively. The size, or cardinality, of the set $\{2, 3, 5, 7\}$ is four. The official way to write this is to put the name of the set between a pair of vertical lines and to write the resulting expression as equal to the appropriate number. Thus,

$|\{2, 3, 5, 7\}| = 4.$

The same statement about the cardinality of this set can also be made using abstraction notation:

$|\{\, x : x \text{ is a prime number and } 1 < x < 10 \,\}| = 4.$

Some kinds of finite sets have special names. Sets with exactly two members are called *doubleton sets*. Sets with exactly one member are called *singleton sets*. There is exactly one special set with no members and it is called the *empty set* or the *null set*. One very suggestive, though uncommon, symbol for the empty set is $\{\,\}$. The symbol used here is the commonly used symbol $\emptyset$. The empty set, or $\emptyset$, is the set theoretic counterpart to the number zero, or 0. Zero and the empty set are not the same. They must never be confused: 0, or zero, is a number and $\emptyset$, or the empty set, is a set.

Here are some examples to illustrate the various concepts we have introduced so far:

$|\{2, 4, 6\}| = 3, \text{ but } |\{\{2, 4, 6\}\}| = 1;$

$|\{\mathbb{N}\}| = 1, \text{ but } |\mathbb{N}| \text{ is infinite};$

$|\emptyset| = 0, \text{ but } |\{\emptyset\}| = 1.$

There is one more special set known either as the *universal set* or the *universe of discourse*. This set is often denoted by U (mnemonic for *universe*); here, however, we shall denote it by V to avoid confusion with another symbol to be introduced later. The idea, as suggested by the second term, is that, in any conversation, some background set of entities is presupposed as the things potentially to be talked about. These entities comprise not all the entities that one might possibly think of, but only some small portion of them. Thus, for example, if, in casual conversation, someone says that *Everyone enjoyed the party*, it is clear that the speaker means not every human being on the planet, but rather every person in attendance at the party referred to. If the topic of discussion is plane Euclidean geometry, the universal set is the set of points in a plane; if the topic is arithmetic, the universal set is the set of natural numbers. The presupposition of a universe of discourse that is less than the totality of all possible entities is a practical necessity for most topics of conversation, be they mathematical or otherwise; such a presupposition turns out to be of logical necessity if the topic of discussion is set theory.

2.3 Relations between Sets

In this section, we shall be concerned with three relations a set can bear to a set: the subset relation (symbolized by $\subseteq$), the proper subset relation (symbolized by $\subset$), and the disjointness relation (symbolized by $\perp$). We begin with the subset relation.

Definition 1 Subset relation
$X \subseteq Y$ iff[1] every member of X is a member of Y.

The set $\{7, 13, 23\}$, for example, is a subset of the set $\{7, 13, 23, 31\}$, that is,

$$\{7, 13, 23\} \subseteq \{7, 13, 23, 31\},$$

since, as one can easily verify, every member of the first set is a member of the second. At the same time, $\{1, 5, 9\}$ is not a subset of $\{5, 9, 10\}$, that is,

$$\{1, 5, 9\} \nsubseteq \{5, 9, 10\},\text{[2]}$$

since 1 is a member of the first set but not of the second. From the discussion earlier, it should be evident that $\mathbb{N}$ is a subset of $\mathbb{Z}$, that is,

$$\mathbb{N} \subseteq \mathbb{Z}.$$

An intuitively clear way to depict this relation, provided by Gottfried Leibniz (1646–1716) and popularized by Leonhard Euler (1707–1783), is this: to represent a set A as a subset of a set B, one encloses a circle representing A with a circle representing B, as shown in figure 2.1.

In an Euler diagram of the subset relation (fig. 2.1), the relation between the sets is depicted by the positioning, with respect to one another, of the circles representing the related sets. Indeed, the use of relative positioning within a rectangle of closed figures representing sets to depict the relevant relation between sets is characteristic of Euler diagrams. While immediately intuitive, this form of diagram turns out to be neither as flexible nor as general as one might want.

John Venn (1834–1923) devised a way to improve on the flexibility and generality of Euler diagrams. In a Venn diagram, the position of closed figures representing sets does not depict anything about the relation between the sets. Rather, in a Venn diagram, two closed figures are drawn overlapping, as shown in figure 2.2. Further notation is added to depict the relation obtained between the sets represented by the closed figures.

Notice that in figure 2, the rectangle is partitioned into four distinct regions, which are called *cells*. One cell, labeled 1, lies outside of both circles; another, labeled 2, is the crescent-shaped region that includes the left-hand circle but excludes the right-hand circle;

---

1. The term 'iff' is short for "if and only if." Such statements are called *biconditionals*.
2. A slash through a mathematical symbol turns that part of the statement involving the symbol with the slash into the negation of the very same part without the slash. For example, $1 \notin \{2, 3\}$ is short for *it is not the case that* $1 \in \{2, 3\}$.

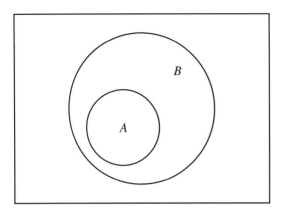

Figure 2.1
Euler diagram of $A \subseteq B$.

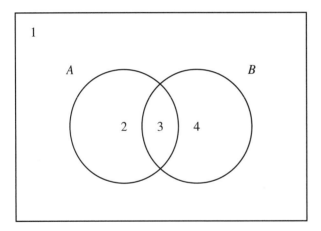

Figure 2.2
Venn diagram schema for relating sets A and B.

another, labeled 4, is the crescent-shaped region that includes the right-hand circle and excludes the left-hand one; and the remaining one, labeled 3, is the lens-shaped region that is the overlap between the two circles.

To depict a relation between two sets in a Venn diagram, one annotates one or more of the cells with an appropriate symbol. The placement of a star ($\star$) within a region indicates that the set represented by the region is nonempty, and the shading of a region indicates that the set represented by the region is empty. To depict the set A as a subset of B, one shades in the left-hand crescent-shaped cell, as shown in figure 2.3.

Readers should study the Euler and Venn diagrams of the subset relation to see that indeed the very same state of affairs is depicted in both diagrams.

Basic Set Theory

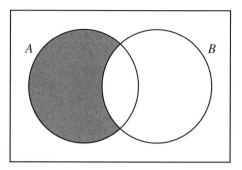

Figure 2.3
Venn diagram for $A \subseteq B$.

Three facts are important about the subset relation: First, if one set is a subset of a second and the second of a third, then the first set is a subset of the third. Second, if one set is a subset of a second and the second of the first, then the sets are identical. Third, every set is a subset of itself. We record these facts in notation:

Fact 1 Facts about the subset relation

(1) If $X \subseteq Y$ and $Y \subseteq Z$, then $X \subseteq Z$.

(2) If $X \subseteq Y$ and $Y \subseteq X$, then $X = Y$.

(3) $X \subseteq X$.

While the third fact is not very clear from the diagrams just introduced, the first and second are. We begin with the first fact. That set A is a subset of set B is depicted by enclosing a circle representing A within a circle representing B. And that B is a subset of C is depicted by enclosing the circle representing B within a circle representing C. Once this is done, the circle representing C will have enclosed the circle representing A, thereby depicting A as a subset of C. This property can be depicted equally well by a Venn diagram.

That set A is a subset of set B is depicted by shading in the crescent-shaped region of the circle representing A excluded from the circle representing B. That B is a subset of C is depicted by shading in the crescent-shaped region of the circle representing B excluded from the circle representing C. Once this is done, the crescent-shaped region of the circle representing A excluded from the circle representing C will have been shaded in, thereby depicting A as a subset of C.

It is important to stress that these diagrams are *not* proofs; they are simply aids to help one *see*, as it were, the fact that requires proof. An Euler diagram is of no help in depicting the second fact, but a Venn diagram is. To depict A and B, as the very same set, one must shade in the crescent-shaped region of the circle representing the set A, excluded from the circle representing the set B, as well as shade in the crescent-shaped region of the circle

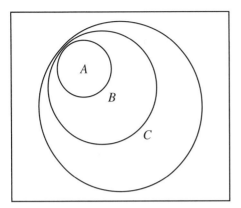

Figure 2.4
Euler diagram for $A \subseteq B \subseteq C$.

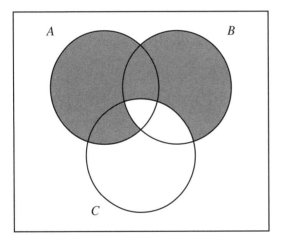

Figure 2.5
Venn diagram for $A \subseteq B \subseteq C$.

representing the set B, excluded from the circle representing the set A. That is, each of these crescent-shaped regions must be shaded in. Such a diagram not only depicts A as a subset of B but also B as a subset of A. Conversely, a diagram that depicts A as a subset of B and B as a subset of A, does so by shading in the crescent-shaped region of the circle representing the set A, excluded from the circle representing the set B, as well as the crescent-shaped region of the circle representing the set B, excluded from the circle representing the set A. Of the region confined by the two circles depicting the sets A and B, only the lens-shaped region of their overlap is left unshaded, thereby depicting A and B as one and the same set.

We now turn to the second relation, the proper subset relation.

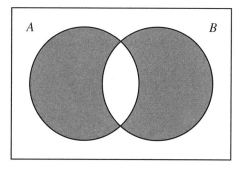

Figure 2.6
Venn diagram for $A = B$.

Definition 2 Proper subset relation
$X \subset Y$ iff every member of X is a member of Y and some member of Y is not a member of X.

Notice that the definition of the proper subset relation has two conditions. The first is that each member of X is a member of Y, and the second is that Y has a member that X does not have. Putting these two conditions together means that Y has all of X's members and then some. Thus, for example, the set {7, 13, 23, 31} has all the members of {7, 13, 23}, and then one, namely, 31.

{7, 13, 23} $\subset$ {7, 13, 23, 31}.

Similarly, $\mathbb{Z}$ has all the members of $\mathbb{N}$ and then some. Indeed, $\mathbb{Z}$ has, in addition to all the natural numbers, all the negative integers.

$\mathbb{N} \subset \mathbb{Z}$.

Clearly, no set is a proper subset of itself, for while any set contains all of its members, it does not contain any others.

The proper subset relation is depicted, in a Venn diagram, in the same way as the subset relation, except that, in addition to the shading in of one crescent-shaped region (an annotation of the diagram corresponding to the first part of the definition), a star ($\star$) is placed in the other crescent-shaped area (an annotation corresponding to the other part of the definition).

A little reflection should convince the reader that the Euler diagram in figure 2.1 to depict the subset relation is really a diagram depicting the proper subset relation. In fact, there is no way, without introducing further annotation, to depict in an Euler diagram an instance of the subset relation that is not an instance of the proper subset relation.

Here are three important facts about the proper subset relation. Like the subset relation, if one set is a proper subset of a second and the second of a third, then the first set is a

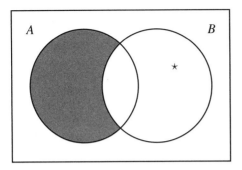

Figure 2.7
Venn diagram for $A \subset B$.

proper subset of the third. Second, if one set is a proper subset of another, then the other is not a proper subset of the one. Third, no set is a proper subset of itself.

Fact 2 Facts about the proper subset relation

(1) If $X \subset Y$ and $Y \subset Z$, then $X \subset Z$.

(2) If $X \subset Y$, then $Y \not\subset X$.

(3) $X \not\subset X$.

We close this exposition of the relations of a set to a set with the third relation, the *disjointness* relation.

Definition 3 Disjointness relation
$X \perp Y$ iff no member of X is a member of Y.

The sets $\{1, 2\}$ and $\{3, 4\}$ are disjoint from one another, that is,

$\{1, 2\} \perp \{3, 4\}$.

In contrast, the sets $\{1, 2, 3\}$ and $\{3, 4, 5\}$ are not disjoint from one another, that is,

$\{1, 2, 3\} \not\perp \{3, 4, 5\}$

since both have 3 as a member. Two infinite sets that are disjoint from one another are the sets $\mathbb{Z}^+$ and $\mathbb{Z}^-$; there is no integer that is both positive and negative.

In an Euler diagram, the disjointness relation is depicted by two nonoverlapping circles, as shown in figure 2.8. In a Venn diagram, it is depicted by the shading in of the lens-shaped region of two overlapping circles.

As one can see from the diagram, if one set is disjoint from a second, then the second is disjoint from the first. We restate this fact in notation as follows:

Basic Set Theory

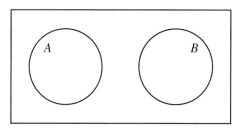

Figure 2.8
Euler diagram for $A \perp B$.

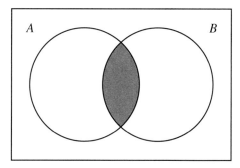

Figure 2.9
Venn diagram for $A \perp B$.

Fact 3 Fact about the disjointness relation
If $X \perp Y$, then $Y \perp X$.

Exercises: Set membership and relations between sets

1. Using set abstraction notation, rename the following sets:
(a) $\{10, 11, 12, 13, \ldots, 97, 98, 99\}$
(b) $\{2, 4, 6, 8, 10, \ldots\}$
2. Rename each of the following sets using list notation and state the cardinality of each:
(a) $\{x : x$ is a two-digit numeral whose units digit is the same as its tens digit$\}$
(b) $\{x : x$ is a two-digit numeral whose units digit is twice its tens digit$\}$
3. Let $A = \{1, 4\}$, $B = \{4\}$, $C = \{1, 2, 3\}$, and $D = \{1, 2\}$. Determine the truth or falsity of the following:
(a) $1 \in A$ (g) $3 \in C$
(b) $B \subseteq D$ (h) $B \subset C$

(c) $2 \in B$
(d) $B \perp C$
(e) $B \perp D$
(f) $C \subseteq D$
(i) $4 \in D$
(j) $A \subset B$
(k) $C \perp D$
(l) $A \perp B$

4. Let $A = \{1, 3, 5\}$, $B = \{2, 4\}$, $C = \{1, 2, 3\}$, and $D = \{3, 4\}$. Determine the truth or falsity of the following:

(a) $C \subset A$ and $A \perp B$.

(b) $C \subseteq D$ or $B \perp C$.

(c) If $A \perp C$, then $D \subseteq B$.

(d) $C \perp A$ and $D \subset A$.

5. Let $A = \{a, c, e\}$, $B = \{b, c\}$, $C = \{a, b, c\}$, $D = \{d, e, \emptyset\}$, and $E = \{A, b, C, d\}$, where a, b, c, d, and e are distinct objects. Determine the truth or falsity of the following:

(a) $A \in E$
(b) $B \subseteq A$
(c) $C \subset B$
(d) $D \perp C$
(e) $D \perp B$
(f) $c \subset C$
(g) $B \in e$
(h) $E \perp A$

6. Determine the truth or falsity of the following:

(a) $\emptyset \in \emptyset$
(b) $\emptyset \subseteq \emptyset$
(c) $\emptyset \in \{\emptyset\}$
(d) $\emptyset \subseteq \{\emptyset\}$
(e) $\{\emptyset\} \in \emptyset$
(f) $\{\emptyset\} \subseteq \emptyset$

7. Determine the truth or falsity of the following:

(a) $\mathbb{Z}^- \subset \mathbb{N}$
(b) $\mathbb{N} \subset \mathbb{Z}$
(c) $\mathbb{N} = \mathbb{Z}^+$
(d) $\mathbb{N} \perp \mathbb{Z}^-$
(e) $\mathbb{N} \perp \mathbb{Z}^+$
(f) $\mathbb{N} \subset \mathbb{Z}^-$
(g) $\mathbb{Z}^+ \subseteq \mathbb{N}$
(h) $\mathbb{Z}^+ \subseteq \mathbb{Z}^-$

8. Which of the following sets are infinite and which are finite? If the set is finite, state its cardinality (if possible).

(a) The set of letters of the Roman alphabet.

(b) The set of even integers.

(c) A proper subset of a finite set.

(d) The set of even prime numbers.

3 Operations on Sets

We shall learn about four operations on sets: union, intersection, difference, and complementation. We begin with the binary operation of *union*, symbolized by $\cup$.

Definition 4 Union
$x \in X \cup Y$ iff $x \in X$ or $x \in Y$.

The definition stipulates that the set formed from the union of two sets comprises precisely those elements that are found in *either* (and possibly both). Here are some examples:

$\{1, 2, 5\} \cup \{3, 6\} = \{1, 2, 3, 5, 6\}$;

$\{1, 2, 5\} \cup \{1, 3, 4\} = \{1, 2, 3, 4, 5\}$;

$\mathbb{N} \cup \mathbb{Z}^- = \mathbb{Z}$.

The same idea can be conveyed by using the Venn diagram schema in figure 2.2. In that diagram, the area comprising the cells 2, 3, and 4 represents $A \cup B$.

Inspection of the Venn diagram reveals that union has several interesting facts, which we state now and which we recapitulate in notation in Fact 4. First, as stated in (1), the order in which one takes the union of two sets makes no difference. Second, as stated in (2) and (3), if one takes the union of two sets, each set taken to form the union is a subset of their union. Third, as stated in (4), has two parts: if two sets are subsets of the same set, then the union of the two sets is also a subset of it, and if the union of two sets is a subset of a set, then each of the sets from which the union has been formed is a subset of the set. Fourth, as stated in (5), also has two parts: if one set is a subset of another, then their union is the second set, and if the union of two sets yields one of the sets, then the set not yielded is a subset of the other set. Fifth, as stated in (6), the union of a set with itself is just the set itself. And sixth, as stated in (7), the union of any set with the empty set is just the set itself.

Fact 4 Facts about union

(1) $X \cup Y = Y \cup X$.

(2) $X \subseteq X \cup Y$.

(3) $Y \subseteq X \cup Y$.

(4) $X \subseteq Z$ and $Y \subseteq Z$ iff $X \cup Y \subseteq Z$.

(5) $X \subseteq Y$ iff $Y = X \cup Y$.

(6) $X \cup X = X$.

(7) $X \cup \emptyset = X$.

We now turn to the binary operation, *intersection,* symbolized by $\cap$.

Definition 5 Intersection

$x \in X \cap Y$ iff $x \in X$ and $x \in Y$.

The definition stipulates that the set formed from the intersection of two sets comprises precisely those elements that are found in *both*. Here are a few examples:

$\{1, 2, 6\} \cap \{2, 3, 6\} = \{2, 6\}$;

$\{1, 2, 5\} \cap \{1, 2, 4\} = \{1, 2\}$;

$\{1, 2, 3\} \cap \{6, 5, 4\} = \emptyset$.

The Venn diagram in figure 2.2 also helps to convey how the intersection works: the area of cell 3 represents $A \cap B$.

Facts analogous to those that hold of union hold for intersection. They can be easily seen from inspection of the Venn diagram. First, as stated in (1) in Fact 5, the order in which one takes the intersection of two sets makes no difference. Second, as stated in (2) and (3), if one takes the intersection of two sets, the intersection is a subset of each set taken to form the intersection. Third, as stated in (4), has two parts: if some set is a subset of two sets, then it is also a subset of their intersection, and if a set is a subset of the intersection of two sets, it is a subset of each of the two sets forming the intersection. Fourth, as stated in (5), also has two parts: if one set is a subset of another, then their intersection is the first set, and if the intersection of two sets yields one of the sets, then the set yielded is a subset of the other set. Fifth, as stated in (6), the intersection of a set with itself is just the set itself. And sixth, as stated in (7), the intersection of any set with the universal set is just the set itself. Here are these same facts stated in notation.

Fact 5 Facts about intersection

(1) $X \cap Y = Y \cap X$.

(2) $X \cap Y \subseteq X$.

(3) $X \cap Y \subseteq Y$.

(4) $Z \subseteq X$ and $Z \subseteq Y$ iff $Z \subseteq X \cap Y$.

(5) $Y \subseteq X$ iff $Y = X \cap Y$.

(6) $X \cap X = X$.

(7) $X \cap \emptyset = \emptyset$.

The third and last binary operation is *difference*. The difference of one set from another is not to be confused with the subtraction of one number from another. The difference of one set from another involves the removal from the first set of any its members that are also in the second, for example:

$\{1, 3, 5\} - \{2, 4, 5\} = \{1, 3\}$.

Since 5 is the only element in common to the two sets, it is removed from the first set to yield {1, 3}.

Here is the definition of (set) difference.

Definition 6 Difference
$x \in X - Y$ iff $x \in X$ and $x \notin Y$.

The Venn diagram (fig. 2.2) once again sheds light on how set difference works. The area of cell 2 represents the result of subtracting B from A (that is, $A - B$).

We now come to the last operation, *complementation,* a unary operation.

Definition 7 Complementation
$x \in -X$ iff $x \in V$ and $x \notin X$.

The definition stipulates that the complement of a set is formed from all the elements in the universe that are *not* in the set whose complement is being formed. Suppose that the universe $V = \{1, 2, 3, 4, 5\}$. Then,

$-\{1, 2, 5\} = \{3, 4\}$,

$-\{5, 4, 1\} = \{3, 2\}$.

Again, referring to the Venn diagram (fig. 2.2), one sees that the area that cells 1 and 4 compose represents $-A$, while that which cells 1 and 2 compose represents $-B$.

Several facts that hold for complementation are easily seen from considering a suitable Venn diagram. First, as stated in (1) in Fact 6, a set and its complement have nothing in common; in other words, their intersection is the empty set. Second, as stated in (2), the universe of discourse is exhausted by the union of a set with its complement. A third fact, stated in (3), is that a set and its complement's complement are the same. The next two facts, known together as DeMorgan's laws, connect complementation with union and intersection: the union of the complement of two sets is the same as the complement of their intersection, and the intersection of the complement of two sets is the same as the complement of their union. The sixth fact, which is particularly clear when illustrated with an Euler diagram, is that complementation reverses the subset relation. Finally, the set that results from the intersection of a set with the complement of a set is the same as the one that results from removing elements of the second set from the first.

Fact 6 Facts about complementation

(1) $X \cap -X = \emptyset$.

(2) $X \cup -X = V$.

(3) $--X = X$.

(4) $-X \cup -Y = -(X \cap Y)$.

(5) $-X \cap -Y = -(X \cup Y)$.
(6) $X \subseteq Y$ iff $-Y \subseteq -X$.
(7) $X - Y = X \cap -Y$.

Exercises: Operations on sets

1. Let the universal set V be $\{1, 2, 3, 4, 5, 6\}$. Let $A = \{1, 3, 5\}$, $B = \{2, 4\}$, $C = \{1, 2, 3\}$, $D = \{4, 5, 6\}$, and $E = \{3\}$. Rename each of the following sets using list notation:
   (a) $A \cap (A \cup (B \cap -E))$
   (b) $A \cup (C \cap -E)$
   (c) $(A \cap -D) \cap (B \cup C)$
   (d) $A \cup (A \cap ((D \cap -A) \cap (-C \cup B)))$
   (e) $(A \cup D) \cap -(B \cap C)$
   (f) $(C \cap -A) \cap (B \cup -A)$
   (g) $-(B \cap C) \cap -B$
   (h) $C \cap -(C \cap -D)$
   (i) $C \cap (B \cup D)$

2. Let the universal set be $\mathbb{Z}$. Simplify the following expressions wherever possible.
   (a) $\mathbb{N} \cap \mathbb{Z}^+$
   (b) $\mathbb{Z}^+ \cup \mathbb{N}$
   (c) $\mathbb{Z}^+ \cup \mathbb{Z}^-$
   (d) $\mathbb{Z}^- \cap \mathbb{Z}^+$
   (e) $\mathbb{Z}^+ \cap -\mathbb{N}$
   (f) $(\mathbb{N} - \mathbb{Z}^+) \cup \mathbb{Z}^+$
   (g) $\mathbb{N} \cap -(\mathbb{Z}^+ - \{0\})$

3. Determine which of the following claims are true and which are false. If a claim is false, provide a counterexample.
   (a) $X \cap -Y \subseteq X$
   (b) $X \cap Y \perp X \cap -Y$
   (c) If $X \subseteq Y$ and $Y \subseteq Z$, then $-Z \subseteq -X$.
   (d) If $X \subseteq Z$ and $Y \subseteq Z$, then $Y \cap -X \subseteq Z$.

4 Sequences

We now turn from sets to sequences. Sequences, which are a special kind of ordered set, differ from sets precisely because their members are ordered, whereas the members of a set are not ordered. To explain the notion of a sequence, let us begin by discussing the special case of a sequence with just two members, also known as an ordered pair.

Basic Set Theory

4.1 Ordered Pairs

An ordered pair is similar to, but quite distinct from, a doubleton. They are similar insofar as they are often named in similar ways. One usually names a doubleton by putting down two names, separated by a comma, and enclosing the whole thing in a pair of braces. For example, the doubleton comprising just 1 and 2 is denoted as follows: $\{1, 2\}$. One usually names an ordered pair by putting down two names, separated by a comma, and enclosing the whole thing in a pair of angle brackets, for example, $\langle 1, 2 \rangle$.[3]

As stated, sets are unordered. Hence, the order in which the elements of a doubleton are listed does not matter. So, for example, $\{1, 2\} = \{2, 1\}$. An ordered pair, however, is ordered. Hence, the order in which its members are listed does matter. Thus, $\langle 1, 2 \rangle \neq \langle 2, 1 \rangle$. Since the order of the members of an ordered pair matters, each position is given a name. The first member is said to be the *first coordinate* and the second the *second coordinate*.

Another difference between an ordered pair and a set is that redundancy in an order pair matters, whereas redundancy in a set does not. Thus, while $\langle 1, 1 \rangle$ is an ordered pair, $\{1, 1\}$ is not a doubleton, since $\{1, 1\} = \{1\}$ but $\langle 1, 1 \rangle \neq \langle 1 \rangle$. Moreover, $\langle a, b \rangle$ must be an ordered pair, regardless of whether or not a is identical with b; whereas $\{a, b\}$ may be a singleton, not a doubleton, depending on whether or not a is identical with b. Finally, every doubleton is a set, but no ordered pair is a set.

Two ordered pairs are identical just in the case where their first coordinates are identical as well as their second coordinates. That is,

Definition 8 Equality of ordered pairs
$\langle x, y \rangle = \langle w, z \rangle$ iff $x = w$ and $y = z$.

Just as doubletons are a special kind of set, so an ordered pair is a special kind of sequence. And just as every natural number is the cardinality of some set, so too every natural number is the length of some sequence. Thus, singletons have a cardinality of one, doubletons a cardinality of two, and so on.

Ordered pairs are sequences of length two, and there are sequences of length three, four, five, and so on, and even of length one. Since order matters in an ordered set, one may refer to a member of a sequence set by its position in the list. If the sequence has length n, it is often referred to as an n-tuple and the member in the ith position is said to be the ith coordinate. Two sequences are, in fact, the very same sequence just in the case they have the same length and they are coordinate by coordinate identical. We restate this more formally as follows:

Definition 9 Equality of sequences
$\langle x_1, \ldots, x_i, \ldots, x_n \rangle = \langle y_1, \ldots, y_i, \ldots, y_n \rangle$ iff $x_i = y_i$, for each $i \in \mathbb{Z}_n^+$.

---

3. While logicians and set theorists typically use angle brackets to form names for sequences, it is customary in much of mathematics, and especially in algebra, to use parentheses instead. We shall follow the practice of logicians and set theorists.

Thus, for example, $\langle 1, 2, 3 \rangle$ is not equal to $\langle 1, 3, 2 \rangle$, for even though they have the same members, some of the coordinates of one have members different from the corresponding coordinates of the other. Specifically, while the first coordinates of each of the triples are equal, the second coordinates are not, and neither are the third coordinates.

4.2 Cartesian Product

We now come to an important binary operation, the *Cartesian product,* which is denoted by the symbol ×. This operation, also known as the *direct product,* creates from two sets a set of ordered pairs where the first and second coordinates of each pair in the resulting set are taken from the first set and the second set, respectively. In spite of the symbol used, the Cartesian product should not be confused with the arithmetical operation of multiplication. The arithmetic operation of multiplication takes a pair of numbers and yields a number; the set theoretic operation of Cartesian product takes a pair of sets and yields a set. However, the two operations are connected, as we shall see, and it is the existence of this connection that justifies the choice of the very same symbol for the two distinct operations.

Definition 10 Cartesian product
Let X and Y be sets. $\langle x, y \rangle \in X \times Y$ iff $x \in X$ and $y \in Y$.

Logically equivalent to this definition is this equality:

$X \times Y = \{\langle x, y \rangle : x \in X \text{ and } y \in Y\}$.

Here are some examples to clarify the definition:

$\{1, 2\} \times \{3, 4\} = \{\langle 1, 3 \rangle, \langle 1, 4 \rangle, \langle 2, 3 \rangle, \langle 2, 4 \rangle\}$;

$\{1, 2\} \times \{2, 3, 4\} = \{\langle 1, 2 \rangle, \langle 1, 3 \rangle, \langle 1, 4 \rangle, \langle 2, 2 \rangle, \langle 2, 3 \rangle, \langle 2, 4 \rangle\}$.

There are two important points to bear in mind concerning this definition. First, the Cartesian product of two sets is a set, not an ordered pair. Second, each member of a Cartesian product of two sets is an ordered pair.

As explained, the operations of Cartesian product and arithmetic multiplication are distinct. Moreover, these operations are not trivially analogous, since arithmetic multiplication is commutative and associative, whereas the Cartesian product is neither. In other words, neither $X \times Y = Y \times X$, nor $X \times (Y \times Z) = (X \times Y) \times Z$ holds. Readers are invited to find specific cases to bear out the claim that the Cartesian product is not commutative. The following example shows that it is not associative. Let X be $\{1, 2\}$, Y be $\{3, 4\}$, and Z be $\{5, 6\}$. Now,

$Y \times Z = \{\langle 3, 5 \rangle, \langle 3, 6 \rangle, \langle 4, 5 \rangle, \langle 4, 6 \rangle\}$

and, as we saw earlier,

$X \times Y = \{\langle 1, 3 \rangle, \langle 1, 4 \rangle, \langle 2, 3 \rangle, \langle 2, 4 \rangle\}$.

Basic Set Theory

Thus,

$$(X \times Y) \times Z = \{\langle\langle 1, 3\rangle, 5\rangle, \langle\langle 1, 4\rangle, 5\rangle, \langle\langle 2, 3\rangle, 5\rangle, \langle\langle 2, 4\rangle, 5\rangle,$$
$$\langle\langle 1, 3\rangle, 6\rangle, \langle\langle 1, 4\rangle, 6\rangle, \langle\langle 2, 3\rangle, 6\rangle, \langle\langle 2, 4\rangle, 6\rangle\}$$

and

$$X \times (Y \times Z) = \{\langle 1, \langle 3, 5\rangle\rangle, \langle 1, \langle 3, 6\rangle\rangle, \langle 1, \langle 4, 5\rangle\rangle, \langle 1, \langle 4, 6\rangle\rangle,$$
$$\langle 2, \langle 3, 5\rangle\rangle, \langle 2, \langle 3, 6\rangle\rangle, \langle 2, \langle 4, 5\rangle\rangle, \langle 2, \langle 4, 6\rangle\rangle\}.$$

As readers can verify, $(X \times Y) \times Z$ and $X \times (Y \times Z)$ have different members and hence are distinct sets. In particular, $\langle\langle 1, 3\rangle, 5\rangle$ is a member of $(X \times Y) \times Z$, but it is not a member of $X \times (Y \times Z)$. Relevant here is the fact that $\langle\langle 1, 3\rangle, 5\rangle$ and $\langle 1, \langle 3, 5\rangle\rangle$ are not equal. The first coordinate of the ordered pair $\langle\langle 1, 3\rangle, 5\rangle$ is itself an ordered pair, namely, $\langle 1, 3\rangle$, while the first coordinate of $\langle 1, \langle 3, 5\rangle\rangle$ is 1, which is not an ordered pair.

To know the size of the set that is the Cartesian product of two sets, one can multiply the cardinalities of the two sets forming the product, that is,

Fact 7 Fact about Cartesian product

$|X \times Y| = |X| \times |Y|$.

This is the reason why mathematicians use the same symbol both for arithmetic multiplication (used on the right side of the equation) and for the Cartesian product (used on the left side of the equation).

Naturally, it is possible to take the Cartesian product of a set with itself. The result is a set of ordered pairs in which the first and second coordinates are drawn from the same set, for example:

$\{1, 2\} \times \{1, 2\} = \{\langle 1, 1\rangle, \langle 1, 2\rangle, \langle 2, 1\rangle, \langle 2, 2\rangle\}.$

It is common to use exponent notation here. Thus, if A is a set, the Cartesian product of it with itself—that is, $A \times A$—is also expressed as A^2. This is in keeping with the use of the same sign for both multiplication of numbers and for the Cartesian product of sets.

Exercises: Cartesian product

1. Let $A = \{2, 4, 6\}$, $B = \{1, 3\}$, and $C = \{3, 5\}$. Calculate the following:
   (a) $A \times B$
   (b) C^2
   (c) $B^2 \times A$
   (d) $(A \times B) \cap (A \times C)$
   (e) $C^2 \cup B^2$
   (f) $(B \cap C) \times A$
   (g) $(A \cap C) \times A^2$
   (h) $(B \cup C) \times B$

2. Determine which of the following claims are true and which are false. If a claim is false, provide a counterexample.

(a) If $X = \emptyset$, then $X \times Y = Y \times X$.

(b) If $X \times Y = Y \times X$, then $X = Y$.

(c) If $X \subseteq Y$, then $X \times Z \subseteq Y \times Z$.

(d) If $X \subset Y$, then $X \times Z \subset Y \times Z$.

(e) $X \times (Y \cap Z) = (X \cap Y) \times (X \cap Z)$.

(f) $X \cup (Y \times Z) = (X \cup Y) \times (X \cup Z)$.

(g) $X \cap (Y \times Z) = (X \cap Y) \times (X \cap Z)$.

5 Families of Sets

We noted earlier that a set can have sets as members. A set where all members are sets is often said to be a *family* of sets. Thus, for example, $\{\{1, 2\}, \{2, 3\}, \{1, 3\}\}$ is a set all of whose members are sets. It is, then, a family of sets. In this section, we shall learn about an operation that creates a family of sets from a set and about two operations that create a set from a family of sets.

5.1 The Power Set Operation

There is an operation that operates on a set to yield a family of sets. It is known as the *power set operation,* and it will be denoted by the expression Pow. In particular, the power set operation collects into one set all the subsets of a given set. Consider, for example, the set $\{1, 2\}$. Its power set is the set of all its subsets, namely:

$\{\emptyset, \{1\}, \{2\}, \{1, 2\}\}$.

Definition 11 Power set
$X \in \text{Pow}(Y)$ iff $X \subseteq Y$.

The definition of the power set operation is in the form of a biconditional. Logically equivalent to this definition is the following equality:

$\text{Pow}(Y) = \{X : X \subseteq Y\}$.

Here are two more examples of applications of the power set operation:

$\text{Pow}(\{1\}) = \{\emptyset, \{1\}\}$;

$\text{Pow}(\{1, 2, 3\}) = \{\emptyset, \{1\}, \{2\}, \{3\}, \{1, 2\}, \{2, 3\}, \{1, 3\}, \{1, 2, 3\}\}$.

The first two facts of Fact 8 identify permanent members of any power set: the set from which the power set is formed and the empty set. The third fact says that a set's cardinality determines its power set's cardinality. Since each subset of a set results from suppressing,

Basic Set Theory

or not, one or more of the set's members, if the set has n members, then there are 2^n ways of forming subsets of the set.

Fact 8 Facts about the power set

(1) $X \in \text{Pow}(X)$.

(2) $\emptyset \in \text{Pow}(X)$.

(3) $|\text{Pow}(X)| = 2^{|X|}$.

Exercises: The power set operation

1. Let $V = \{a, b, c, d\}$ (where a, b, c, d are all distinct from one another). Calculate the following:

(a) $\{ X \subseteq V : a \in X \}$

(b) $\{ X \subseteq V : \{b\} \subseteq X \}$

(c) $\{ X \subseteq V : a \in X \text{ and } b \in X \}$

(d) $\{ X \subseteq V : \{a, b\} \perp X \}$

(e) $\{ X \subseteq V : \emptyset \subset X \}$

2. Let $V = \{a, b, c, d\}$ (where a, b, c, d are all distinct from one another). Calculate the following:

(a) $\{ X \subseteq V : |X| = 2 \}$

(b) $\{ X \subseteq V : |X| = 3 \}$

(c) $\{ X \subseteq V : |X| = 4 \}$

(d) $\{ X \subseteq V : |X| = 5 \}$

(e) $\{ X \subseteq V : |X| < 1 \}$

(f) $\{ X \subseteq V : |X| < 2 \}$

3. Let $V = \{a, b, c, d\}$ and $A = \{a, b\}$. (where a, b, c, d are all distinct from one another). Calculate the following:

(a) $\{ X \subseteq V : X \cap A = A \}$

(b) $\{ X \subseteq V : X \cap -A = A \}$

(c) $\{ X \subseteq V : -X \cup A = X \}$

(d) $\{ X \subseteq V : X \cup A = X \}$

(e) $\{ X \subseteq V : X \cup A = \emptyset \}$

(f) $\{ X \subseteq V : X \cap A = V \}$

4. Let $V = \{a, b, c, d\}$ and $A = \{a, b\}$ (where a, b, c, d are all distinct from one another). Calculate the following:

(a) $\{ X \subseteq V : |X \cap A| = 1 \}$

(b) $\{ X \subseteq V : |X \cap A| = 2 \}$

(c) $\{ X \subseteq V : |X \cap A| = 3 \}$

5. Determine which of the following claims are true and which are false. If a claim is false, provide a counterexample.

(a) If $X \subset Y$, then $\text{Pow}(X) \subset \text{Pow}(Y)$.

(b) If $X \perp Y$, then $\text{Pow}(X) \perp \text{Pow}(Y)$.

(c) $\text{Pow}(X \cup Y) \subseteq \text{Pow}(X) \cup \text{Pow}(Y)$.

(d) $\text{Pow}(X \times Y) = \text{Pow}(X) \times \text{Pow}(Y)$.

5.2 Operations on Families of Sets

In this section, three operations on families of sets are introduced. They are each generalizations of three binary operations introduced earlier: union, intersection, and Cartesian product. The binary operations take pairs of sets and yield a set. Their generalizations take not just a pair of sets but a family of sets and yield a set. We begin with *generalized union*.

Definition 12 Generalized union
$x \in \bigcup \mathcal{Z}$ iff, for some $Y \in \mathcal{Z}, x \in Y$.

This definition leads to the following equality:

$\bigcup \mathcal{Z} = \{x : x \in Y, \text{ for some } Y \in \mathcal{Z}\}$.

Definition 13 Generalized intersection
$x \in \bigcap \mathcal{Z}$ iff, for each $Y \in \mathcal{Z}, x \in Y$.

This definition leads to the following equality:

$\bigcap \mathcal{Z} = \{x : x \in Y, \text{ for each } Y \in \mathcal{Z}\}$.

To see how these operations apply, consider this family of sets:

$\mathcal{A} = \{\{1, 2, 3\}, \{2, 3, 4\}, \{3, 4, 5\}\}$.

$\bigcup \mathcal{A} = \{1, 2, 3, 4, 5\}$.

$\bigcap \mathcal{A} = \{3\}$.

Now, in all cases where the cardinality of the family of sets is finite, generalized union and generalized intersection reduce to a finite iteration of the binary operations of union and intersection, respectively. In other words, if $\mathcal{A} = \{A_1, \ldots, A_n\}$, then:

$\bigcup \mathcal{A} = A_1 \cup \ldots \cup A_n$;

and

$\bigcap \mathcal{A} = A_1 \cap \ldots \cap A_n$.

We now come to the third generalized operation, *generalized Cartesian product*, also known as the *generalized direct product*. Just as the Cartesian product applies to a pair of sets to yield the set of all pairings of members of the first set with members of the second, so the generalized Cartesian product applies to any number of sets to yield the set of all sequences where the member in the first coordinate is taken from the first set, the member in the second coordinate is taken from the second set, and so on.

Consider, for example, the family of sets $\mathcal{F}$, whose members—$\{a, b, e\}$, $\{b, d, e\}$, $\{d, e\}$, $\{a, c, e\}$—have been indexed as A_1, A_2, A_3, and A_4, respectively. Its generalized Cartesian product is a set of quadruples, the first coordinate of each quadruple of which is a member

Basic Set Theory

of A_1, the second coordinate of each quadruple of which is a member of A_2, the third coordinate of each quadruple of which is a member of A_3, and the fourth coordinate of each quadruple of which is a member of A_4. Thus, $\langle a, b, e, c \rangle$ is a member of the generalized cartesian product of $\mathcal{F}$ (as indexed), since $a \in A_1$, $b \in A_2$, $e \in A_3$, and $c \in A_4$. However, $\langle a, b, c, e \rangle$ is not a member, since its third coordinate c is not a member of A_3.

Notice that the indexing of the sets is crucial, for the indexing indicates which set supplies members for which coordinate. To begin with, had the sets in $\mathcal{F}$ been indexed differently, the membership in its generalized Cartesian product could be different. For example, should A_3 be the set $\{a, c, e\}$ and A_4 be the set $\{d, e\}$, then $\langle a, b, e, c \rangle$ would not be a member of the generalized Cartesian product of $\mathcal{F}$ (under this new indexation), since while $a \in A_1$, $b \in A_2$, and $e \in A_3$, $c \notin A_4$, which, under the new indexation, names $\{d, e\}$.

Also, notice that, as the family of sets is indexed by $\{1, 2, 3, 4\}$, the sequences, or n-tuples, in its generalized Cartesian product are sequences of length four, that is, quadruples. The generalized Cartesian product of a family of sets indexed by $\mathbb{Z}_3^+$ contains sequences of length three, that is, triples. More generally, the generalized Cartesian product of a family of sets indexed by $\mathbb{Z}_n^+$ contains only n-tuples. It is also possible to use the set of positive integers $\mathbb{Z}^+$ to index a family of sets. In such a case, the generalized Cartesian product of that family has sequences of an infinite length.

Though it is possible to define generalized Cartesian product in a much more general way, we shall restrict our definition to those cases where the indexing set is either the set of the first n positive integers or the set of positive integers. In keeping with the analogy of using a larger version of the symbol for binary union for generalized union and using a larger version of the symbol for binary intersection for generalized intersection, we shall use a larger version of the symbol for Cartesian product for generalized Cartesian product.[4]

Definition 14 Generalized Cartesian product

(1) Finite generalized Cartesian product:

$\langle x_1, \ldots, x_n \rangle \in \bigtimes \{X_1, \ldots, X_n\}$ iff, for each $i \in \mathbb{Z}_n^+$, $x_i \in X_i$.

(2) Infinite generalized Cartesian product:

$\langle x_1, \ldots, x_n, \ldots \rangle \in \bigtimes \{X_1, \ldots, X_n, \ldots\}$ iff, for each $i \in \mathbb{Z}^+$, $x_i \in X_i$.

This definition has two corresponding equalities:

$\bigtimes \{X_1, \ldots, X_n\} = \{ \langle x_1, \ldots, x_n \rangle : \text{for each } i \in \mathbb{Z}_n^+, x_i \in X_i \}$

and

$\bigtimes \{X_1, \ldots, X_n, \ldots\} = \{ \langle x_1, \ldots, x_n, \ldots \rangle : \text{for each } i \in \mathbb{Z}^+, x_i \in X_i \}$.

---

4. A more common symbol for $\bigtimes$ is the upper case Greek letter for pi, Π, mnemonic for *product*.

Before turning to some details pertaining to the operation, let us take note of a common notational simplification. Recall the set $\mathcal{F}$, introduced at the beginning of this section. Its generalized union, expressed as $\bigcup \mathcal{F}$, can also be expressed as $\bigcup \{A_1, A_2, A_3, A_4\}$, since $\mathcal{F} = \{A_1, A_2, A_3, A_4\}$. However, this latter expression is somewhat cumbersome. Using indexation, one could write the slightly shorter expression $\bigcup_{i \in \mathbb{Z}_4^+} \{A_i\}$. Now, it is customary in writing mathematical expressions to omit parentheses, brackets, and braces wherever they are not essential. It is therefore standard to shorten the last expression still further to the expression: $\bigcup_{i \in \mathbb{Z}_4^+} A_i$.

It is important to point out that nothing prevents the indexed family from containing but one set. In that case, the one set provides the members for each of the coordinates. For example, let the family $\{\{a, b\}\}$ be indexed by $\mathbb{Z}_3^+$. Its generalized Cartesian product is the set of all triples each of whose coordinates is drawn from the set $\{a, b\}$. The cardinality of this set of triples is eight. Indeed, whenever the family is a singleton set, say $\{A\}$ is indexed by $\mathbb{Z}_n^+$, then mathematicians write A^n, for the same reason that they write B^2, instead of $B \times B$.

Finally, a useful fact about generalized Cartesian product pertains to the cardinality of the sets it produces: the generalized Cartesian product of an indexed family of finite sets is the multiplicative product of the cardinalities of each of indexed sets. Thus, in the previous example, the cardinality of $\times_{i \in \mathbb{Z}_4^+} \mathcal{F}$ is the multiplicative product of the cardinalities of A_1, A_2, A_3, and A_4: that is, $| \times_{i \in \mathbb{Z}_4^+} \mathcal{F}| = |A_1| \times |A_2| \times |A_3| \times |A_4|$, or $3 \times 3 \times 2 \times 3$, or fifty-four.

Exercises: Operations on families of sets

1. Let V be $\{1, 2, 3, 4\}$. Moreover, let $A_1 = \{1, 2, 3\}$, $A_2 = \{3, 4\}$, $A_3 = \{1\}$, $A_4 = \emptyset$, and $A_5 = \{3, 4\}$. Calculate the following:

   (a) $\bigcup \{A_1, A_2, A_3, A_4, A_5\}$
   (b) $\bigcap \{A_1, A_2, A_3, A_4, A_5\}$
   (c) $\bigcap \{-A_3\}$
   (d) $\bigcup \{-A_5\}$
   (e) $\bigcup -A_4$
   (f) $\bigcap (A_2 \cup A_5)$
   (g) $\bigcup \{-A_1, -A_2, -A_3\}$
   (h) $\bigcap \{-A_2, -A_3, -A_4\}$
   (i) $\times_{i \in \mathbb{Z}_2^+} A_i$
   (j) $\times_{i \in \mathbb{Z}_3^+} A_i$
   (k) $\times_{i \in \mathbb{Z}_4^+} A_i$

2. Let V be $\mathbb{N}$. And let $B_1 = \{0, 1, 2\}$, $B_2 = \{1, 2, 3\}$, $B_3 = \{2, 3, 4\}$, ..., $B_n = \{n-1, n, n+1\}$. Calculate the following:

   (a) $\bigcup \{B_1, \ldots, B_5\}$
   (b) $\bigcup \{B_1, B_4, B_7\}$
   (c) $\bigcap \{B_1, B_2, B_3\}$
   (d) $\bigcap \{-B_2, -B_4, \ldots, -B_{2n}, \ldots\}$
   (e) $\bigcup \{-B_2, -B_4, \ldots, -B_{2n}, \ldots\}$
   (f) $\bigcup \{B_3, B_6, \ldots, B_{3n}, \ldots\}$

6 Relations

A *relation* is something that connects a number of entities. It is important to distinguish a relation from the instances that instantiate it. While this distinction may sound a bit arcane, it is not. After all, one easily distinguishes the color red from the instances that instantiate it. That is, one easily distinguishes the color red, on the one hand, from red fire trucks, red pencils, red hats, and so forth, on the other. The distinction between a relation and the instances that instantiate it is perfectly parallel. That is, the relation of being a father is distinct from any given pair comprising a father and the person of whom he is the father.

From a mathematical point of view, a relation comprises a set of sequences and an instance of the relation is any one of these sequences. In particular, a *binary relation* comprises a set of ordered pairs and each of its instances is an ordered pair in this set. Usually, the order matters: a pair consisting of a father and his son is an instance of the fatherhood relation, while a pair consisting of a son and his father is an instance of the relation of being a son. A relation that has triples as instances is a *ternary relation*. And one which has quadruples as instances is a *quaternary relation*. In general, a relation that has n-tuples as instances is an n-ary relation.

Often the instances of a relation are thought of as being drawn from the members of some background set. Such a background set is called the relation's *domain*. To be more specific, consider the binary relation of being a sibling. It is often convenient to see this relation against some background set, for example, the people in some village. While a village will typically have many pairs of people who are siblings, it might also have people whose parents had only one child. Such people will not appear in any ordered pair. A relation seen against a background set is said to be *a relation on a set*. If the relation is a binary relation, then it is said to be *a binary relation on a set*. The set is called the relation's domain.

It is also sometimes convenient to view a relation, not against one background set from which the members of each n-tuple are drawn, but against an n-tuple of sets, where an instance of the relation is a sequence drawn from the corresponding sets. To grasp more easily what is meant here, let us consider another binary relation, the relation of being the national capital of a country. Rather than seeing it against the background set comprising all the cities and countries in the world, we could view it as a relation from one set, the set of cities in the world, to another set, the set of countries in the world. A binary relation viewed in this way comprises three ingredients. The first ingredient is called the binary relation's *domain*. It is the set of those things that could be related by the relation to something, but need not be related to anything. The second ingredient is the binary relation's *codomain*. It is the set of things to which something may be related by the relation but to which nothing need be related. The third ingredient is the binary relation's *graph*. It specifies which things of the domain are in fact related to which things of the codomain. In our example of binary relation, the domain is the set of all cities in the world. But not every city is a capital of some country. Thus, the domain includes cities that are not capitals and hence not related by the relation. At the same time, the codomain is all the countries of the world. Every country does, in fact, have a capital. Thus, in this example, everything in the codomain is involved in the relation. But this is an accident of our example. A codomain, like a domain, may include items that are not involved in the relation. The relation's graph specifies which city is the capital of which country.

The remainder of this section has three parts. The first is devoted to the topic of binary relations on a set. As we shall see in chapter 3, natural language expressions embody a very important binary relation, namely the relation of constituency. To prepare the way for understanding this relation, we shall familiarize ourselves with a number of important properties of binary relations on sets, namely, the properties of *reflexivity, irreflexivity, symmetry, asymmetry, antisymmetry, transitivity, intransitivity*, and *connectedness*. The second part is devoted to the topic of binary relations from a set to a set. Understanding binary relations from this point of view permits one to understand easily one of the most fundamental concepts used in the study of the semantics of natural language, the concept of a function. This is the topic of the third part. We learned about the concept of a function in high school, where it was introduced as part of the study of algebra. In this last part, we shall see, the concept of a function in a much more general setting that will ease the task of seeing how it applies to the study of natural language.

6.1 Binary Relations on a Set

A binary relation on a set comprises a set, called the binary relation's domain, and its binary graph, where the members of the ordered pairs of the graph are drawn from the domain. This can be restated more formally as follows:

Definition 15 Binary relation on a set
$\langle D, G \rangle$ is a binary relation on the set D iff $G \subseteq D^2$.

We now turn to two different, convenient and intuitive ways to depict binary relations on a set. As was said, a binary graph is simply a set of ordered pairs. The first is by means of a matrix and the second by means of a directed graph.

A binary relation can be viewed as a matrix that is depicted by means of a table. The members of the domain are listed in an irredundant and exhaustive fashion from top to bottom along the left side of the table, and they are listed again in the same order from left to right along the top of the table. The first member of an ordered pair is located in the list in the form of a column at the left of the table and the second member of the pair is located in the list in the form of the row at the top of the table. Membership in the binary relation's graph is signified by a plus in the appropriate column and row, while nonmembership is signified by a minus. Consider the binary relation R_1:

$\langle \{1, 2, 3, 4\}, \{\langle 1, 2 \rangle, \langle 1, 3 \rangle, \langle 2, 2 \rangle, \langle 2, 3 \rangle, \langle 2, 4 \rangle, \langle 3, 4 \rangle\} \rangle.$

It can be viewed as a matrix, which is depicted in the following table.

R_1	1	2	3	4
1	−	+	+	−
2	−	+	+	+
3	−	−	−	+
4	−	−	−	−

Another way to see a binary relation is by a directed graph. As we are about to see, the word *graph* is used in a number of different ways by mathematicians. Earlier, we introduced the word *graph* as denoting a set of *n*-tuples. In this sense, a binary graph is a set of ordered pairs. The reason a set of ordered pairs is called a *graph* comes from elementary algebra. Readers will recall that a straight line can be used to represent the real numbers. Such a representation is called the *real number line*. Two such lines, placed at right angles to one another and intersecting at the points corresponding to 0, determine a plane. The two real number lines are used to identify uniquely each point in the plane: each point is assigned a pair of real numbers, known as the Cartesian coordinates of that point, and each pair of real numbers determines exactly one point in the plane. For this reason, the points in a plane can be thought of as a set of ordered pairs. A graph is a set of points in a plane. Hence, a graph is a set of ordered pairs. Thus, the word *graph* applied to points in a plane is applied to the set of ordered pairs of real numbers and, by extension, to any set of ordered pairs.

But the graphs of elementary algebra are not the only ways of depicting ordered pairs. Graph theory, a branch of mathematics different from elementary algebra, provides a different way of thinking about ordered pairs. The notion is that of a *directed graph*. A directed graph comprises a domain, whose members are called *nodes* and which are depicted as points in a plane, and a set of ordered pairs, each member of which is called a *directed edge* or an *arc* and which is depicted by arrows; the directed edge goes from the point depicting the node that is the first coordinate of the ordered pair to the point depicting the node that is the second coordinate of the same ordered pair.

We can see a binary relation as a directed graph, and when we do so, it is depicted into the following way. Any set can accommodate many binary relations. Indeed, for any fixed set of points representing nodes, there are as many binary relations on those nodes as there are different combinations of arrows linking the points. The largest set of ordered pairs on a domain D is $D \times D$, or D^2. This is known as the domain's *universal* binary relation. Every other binary relation on D has a graph that is a proper subset of D^2. Indeed, each binary relation on D corresponds to exactly one of the subsets of D^2. Thus, the number of binary relations on D is $2^{(|D|^2)}$. As we just noted, one distinguished binary relation on D is the universal relation; another is the *empty* or *null* binary relation on D. Its graph is the empty set. A third distinguished binary relation on a set is the *identity* relation. Its graph

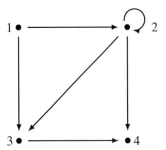

Figure 2.10
Directed graph for R_1.

is the set of ordered pairs where each member of the structure's universe is paired with itself and itself alone. For example, if the domain comprises four elements, call them *a, b, c,* and *d*, then the graph of the identity relation on this set is: $\{\langle a, a\rangle, \langle b, b\rangle, \langle c, c\rangle, \langle d, d\rangle\}$. In this book, should we wish to refer to the identity relation on a set D, we shall use the notation I_D.

Before turning to some of the special properties of binary relations on sets, we should introduce a convenient notational convention. Recall that a binary graph is a set of ordered pairs. Each ordered pair in a binary relation's graph specifies that the first member of the ordered pair is related by the binary relation to the second member. It is convenient, then, to write, instead of the lengthier expression $\langle a, b\rangle \in G_R$, where G_R is the graph of the binary relation R, the shorter expression aRb.

Binary relations on a set have a number of special properties. They include, but are not limited to, those of *reflexivity, irreflexivity, symmetry, asymmetry, antisymmetry,* and *transitivity*. Let us consider each of these properties in turn.

A binary relation on a set is reflexive only in the case where every member of the set bears the relation to itself. This can be put more formally as follows:

Definition 16 Reflexivity of a binary relation
Let R be a binary relation on the set D. R is reflexive iff, for each $x \in D$, xRx.

Any person is as tall as himself and any person is at least as old as himself. So both the relations of *being as tall as* and *being at least as old as* are reflexive on the domain of people. Moreover, any natural number is less than or equal to itself, so the relation of being less than or equal to on $\mathbb{N}$ is reflexive. Finally, any positive integer evenly divides itself. Thus, the relation of evenly dividing on $\mathbb{Z}^+$ is also reflexive.

The binary relation R_2 on the set *{a, b, c, d }*, whose graph is

$$G_{R_2} = \{\langle a, a\rangle, \langle b, b\rangle, \langle b, d\rangle, \langle c, b\rangle, \langle c, c\rangle, \langle d, a\rangle, \langle d, d\rangle\}$$

is reflexive. The table and diagram of its graph follow.

R_2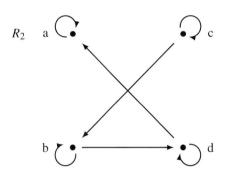

R_2	a	b	c	d
a	+	−	−	−
b	−	+	−	+
c	−	+	+	−
d	+	−	−	+

A reflexive relation, when depicted by a diagram for a directed graph, is one each of whose points has an arrow that both emanates from and returns to itself, that is, each point has one arrow whose nock and tip coincide with it. We shall call such arrows *circular arrows* or *loops*. The table for a reflexive relation has a plus in each cell along its principal diagonal, that is the cells that go from the top left to the bottom right.

Definition 17 Irreflexivity of a binary relation
Let R be a binary relation on the set D. R is irreflexive iff, for no $x \in D$, xRx.

Many binary relations are irreflexive. They include being a child of, since no one is his own child; being a cousin of, since no one is his or her own cousin; being taller than, since no one is taller than himself or herself; and being older than, since no one is older than himself or herself.

An irreflexive relation on the set $\{a, b, c, d\}$ is R_3, whose graph is specified as follows:

$G_{R_3} = \{\langle a, b\rangle, \langle a, d\rangle, \langle b, a\rangle, \langle b, c\rangle, \langle b, d\rangle, \langle c, b\rangle, \langle c, d\rangle, \langle d, c\rangle\}$.

Notice that the table for the graph of this irreflexive relation has a principal diagonal with only minuses and the diagram contains no circular arrows. Such is the case for any irreflexive relation.

R_3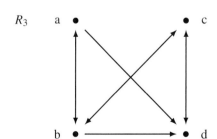

R_3	a	b	c	d
a	−	+	−	+
b	+	−	+	+
c	−	+	−	+
d	−	−	+	−

There are binary relations that are neither reflexive nor irreflexive. Having confidence in is one such relation: some people have confidence in themselves and some do not. The relation R_4 is another example of a relation that is neither reflexive nor irreflexive.

Definition 18 Symmetry of a binary relation
Let R be a binary relation on the set D. R is symmetric iff, for each $x, y \in D$, if xRy, then yRx.

Being a sibling of and being as tall as are two symmetric relations. After all, if one person is a sibling of another, then the other is a sibling of the one. Many relations fail to be symmetric. Being at least as tall as is not symmetric, since any person is at least as tall as anyone shorter than he, but no person shorter than someone is ever as tall as he. Alas, neither admiring nor loving is symmetric. Finally, being a sister of is a symmetric relation only on domains containing nothing but human females.

The diagram for a directed graph of a symmetric relation is one in which every arrow emanating from a point terminates at a point from which emanates an arrow returning to the point of emanation of the first arrow.

R_4 is a symmetric relation on the set $\{a, b, c, d\}$.

$G_{R_4} = \{\langle a, b\rangle, \langle a, d\rangle, \langle b, a\rangle, \langle b, b\rangle, \langle b, c\rangle, \langle c, b\rangle, \langle c, d\rangle, \langle d, a\rangle, \langle d, c\rangle\}$

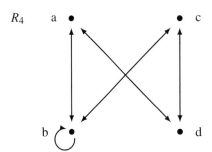

R_4	a	b	c	d
a	−	+	−	+
b	+	+	+	−
c	−	+	−	+
d	+	−	+	−

Notice that the graph diagram has the following characteristic: if an arrow emanates from a node to a node, then an arrow terminates into the first node from the second.[5] The counterpart of this pattern in the table for this relation's graph is the symmetry about its principal diagonal: if one folds the table along the principal diagonal, each cell to the west or south of the diagonal will coincide with exactly one cell to the east or north of the diagonal and the coinciding cells will match with respect to whether or not they contain a plus or a minus. This symmetry about the principal diagonal is a general fact about symmetric binary relations.

---

5. If a pair of points is connected by two arrows, each going in a direction opposite to the other, then a simplification of the diagram is to replace the pair of arrows with one double-headed one.

Basic Set Theory

Definition 19 Asymmetry of a binary relation
Let R be a binary relation on the set D. R is asymmetric iff, for each $x, y \in D$, if xRy, then it is not the case that yRx.

Many common sense binary relations are asymmetric. In addition to those expressed by comparative adjectives, natural relations such as being a parent of and being a child of are asymmetric. If one person is the parent of a second, the second is not a parent of the first.

The binary relation R_5 on the set $\{a, b, c, d\}$ with the graph

$$\{\langle a, b\rangle, \langle a, c\rangle, \langle c, b\rangle, \langle d, a\rangle, \langle d, c\rangle\}$$

is asymmetric. The following graph appears both in the form of a graph diagram and in the form of a table.

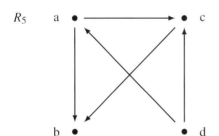

R_5	a	b	c	d
a	−	+	+	−
b	−	−	−	−
c	−	+	−	−
d	+	−	+	−

Notice that the graph diagram has the following characteristic: if an arrow emanates from a node to a node, no arrow terminates into the first node from the second. The counterpart of this pattern in the table for this relation's graph is an asymmetry about its principal diagonal: if one folds the table along the principal diagonal, each cell with a plus is paired with a cell with a minus. Notice that this is compatible with a cell with a minus being paired with a cell with a minus. In other words, any transposition of a member in a relation's graph yields a member that is not in the graph.

Another kind of binary relation is an antisymmetric one. While words denoting symmetric relations and words denoting asymmetric relations are plentiful in natural language, words denoting antisymmetric ones are rare, being confined to words used in mathematics. Two mathematical examples of antisymmetric relations are that of being divisible and that of being less than or equal to. If, for example, one number divides another and the other the one, then they are, in fact, the same number.

Definition 20 Antisymmetry of a binary relation
Let R be a binary relation on the set D. R is antisymmetric iff, for each $x, y \in D$, if xRy and yRx, then $x = y$.

An antisymmetric relation R_6 on the set $\{a, b, c, d\}$ is specified by the following graph.

$$\{\langle a, b\rangle, \langle a, c\rangle, \langle b, b\rangle, \langle c, b\rangle, \langle c, c\rangle, \langle d, a\rangle, \langle d, d\rangle\}$$

The diagram and table of the graph follow.

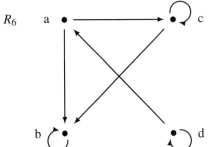

R_6	a	b	c	d
a	−	+	+	−
b	−	+	−	−
c	−	+	+	−
d	+	−	−	+

The directed graph for an antisymmetric relation is one in which no double-headed arrows occur, though loops may. An antisymmetric relation is just like an asymmetric relation, except that while the graph of an asymmetric relation may not have any ordered pairs whose first and second coordinates are the same, the graph of an antisymmetric relation may. In other words, if one folds over the table for the graph of an antisymmetric relation along the principal diagonal, each plus in a nondiagonal cell to the west or south of the diagonal coincides with a minus in a cell to the east or north of the diagonal. This means that every asymmetric relation is antisymmetric.

Definition 21 Transitivity of a binary relation
Let R be a binary relation on the set D. R is transitive iff, for each $x, y, z \in D$, if xRy and yRz, then xRz.

Dividing evenly, being smarter than, being at least as tall as, and being at least as old as are four transitive relations. For example, if one person is at least as tall as a second and the second is at least as tall as a third, then the first is at least as tall as the third.

Relation R_7, whose graph follows, specifies a transitive relation on the set $\{a, b, c, d\}$.

$$\{\langle a, b\rangle, \langle a, c\rangle, \langle b, b\rangle, \langle b, c\rangle, \langle c, c\rangle, \langle d, a\rangle, \langle d, b\rangle, \langle d, c\rangle, \langle d, d\rangle\} \tag{2.1}$$

Its matrix and its directed graph appear as follows.

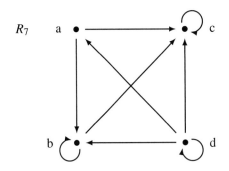

R_7	a	b	c	d
a	−	+	+	−
b	−	+	+	−
c	−	−	+	−
d	+	+	+	+

Basic Set Theory

Transitive relations have no simple characterization in terms of their tables, but the diagrams of their directed graphs are distinctive: there is an arrow that directly connects any two points that are connected by a sequence of two arrows.

Definition 22 Intransitivity of a binary relation
Let R be a binary relation on the set D. R is intransitive iff, for each $x, y, z \in D$, if xRy and yRz, then it is not the case that xRz.

Many binary relations are typically intransitive, for example, being a father of and being a son of. Thus, if Alan is the father of Boris and Boris is the father of Carl, Alan is not the father of Carl. Other binary relations that are intransitive are those of being immediately before.

Relation R_8 on the set $\{a, b, c, d\}$, whose graph is

$$\{\langle a, c \rangle, \langle c, b \rangle, \langle d, a \rangle\} \tag{2.2}$$

is intransitive. Its matrix and its directed graph follow.

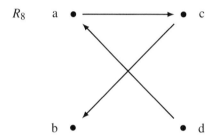

R_8	a	b	c	d
a	−	−	+	−
b	−	−	−	−
c	−	+	−	−
d	+	−	−	−

Intransitive relations also have no simple characterization in terms of their tables. The diagrams of their directed graphs never have an arrow that directly connects any two points that are connected by a sequence of two arrows.

Definition 23 Connectedness of a binary relation
Let R be a binary relation on the set D. R is connected iff, for each $x, y \in D$ that are distinct, either xRy or yRx.

An example of a connected relation is the relation of being less than, as it applies to the set of natural numbers; after all, pick any two distinct natural numbers—say 19 and 87—either $19 < 87$ or $87 < 19$.

The diagram for a directed graph depicting a connected relation is one in which any two distinct points have an arrow connecting them. A connected relation R_9, on the set $\{a, b, c, d\}$, is specified by this graph:

$$\{\langle a, d \rangle, \langle b, a \rangle, \langle b, b \rangle, \langle b, c \rangle, \langle b, d \rangle, \langle c, a \rangle, \langle d, c \rangle\} \tag{2.3}$$

R_9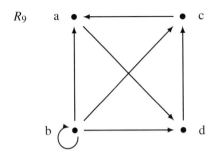

R_9	a	b	c	d
a	−	−	−	+
b	+	+	+	+
c	+	−	−	−
d	−	−	+	−

The pattern in the table for a connected relation is that at least one of the two corresponding cells opposite each other with respect to the principal diagonal must have a plus.

The different properties of binary relations on sets defined in this discussion are not the only properties. But these are the most important, and they include kinds of binary relations on sets that are important to the study of the mathematical stucture of natural language.

Many binary relations on sets customarily identified in mathematics combine some of these properties. We shall identify two that play a central role in this book, namely, a *strict order* and a *partial order*.

Definition 24 Strict order

Let R be a binary relation on the set D. R is a strict order on D iff R is both asymmetric and transitive on D.

Here is one example of a strict order on the set $\{a, b, c, d\}$. The reader can verify that it is indeed asymmetric and transitive.

$G_{S_1} = \{\langle a, c \rangle, \langle a, d \rangle, \langle b, a \rangle, \langle b, c \rangle, \langle b, d \rangle, \langle c, d \rangle\}$

S_1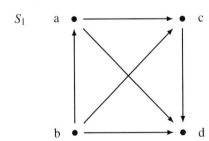

S_1	a	b	c	d
a	−	−	+	+
b	+	−	+	+
c	−	−	−	+
d	−	−	−	−

Other, more intuitive, examples of strict orders are the relation of being less than, over say the natural numbers, and the relation of being poorer than, over a group of people.

Definition 25 Partial order

Let R be a binary relation on the set D. R is a partial order on D iff R is reflexive, antisymmetric, and transitive on D.

Basic Set Theory

The arithmetical relations of less than or equal to and greater than or equal to are partial orders. The following is also an example of a partial order.

$G_{P_1} = \{\langle a, a\rangle, \langle a, c\rangle, \langle a, d\rangle, \langle b, a\rangle, \langle b, b\rangle, \langle b, c\rangle, \langle b, d\rangle, \langle c, c\rangle, \langle c, d\rangle, \langle d, d\rangle\}$

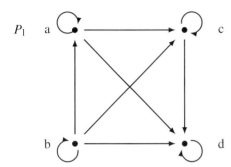

P_1	a	b	c	d
a	+	−	+	+
b	+	+	+	+
c	−	−	+	+
d	−	−	−	+

Exercises: Binary relations on a set

1. Which of the following are binary relations on a set?
(a) $\langle\{a, b, c\}, \{\langle a, b\rangle, \langle b, b\rangle, \langle b, a\rangle, \langle a, a\rangle\}\rangle$
(b) $\langle\{a, b, c\}, \emptyset\rangle$
(c) $\langle\{a, b, c\}, \{\langle a, b\rangle, \langle b, c\rangle, \langle c, d\rangle, \langle a, c\rangle, \langle c, a\rangle\}\rangle$
(d) $\langle\emptyset, \emptyset\rangle$
(e) $\langle A, A \times A\rangle$

2. Consider the seven binary relations listed here. For each relation, determine which of the following properties it has: reflexivity, irreflexivity, symmetry, asymmetry, antisymmetry, or transitivity. (Assume that a, b, c, and d are all distinct.)

R_1: $\langle\{a, b, c\}, \{\langle a, a\rangle, \langle b, b\rangle\}\rangle$
R_2: $\langle\{a, b, c\}, \emptyset\rangle$
R_3: $\langle\{a, b, c\}, \{\langle a, a\rangle, \langle b, c\rangle, \langle c, b\rangle\}\rangle$
R_4: $\langle\{a, b, c\}, \{\langle a, b\rangle, \langle a, a\rangle, \langle c, c\rangle, \langle b, c\rangle, \langle b, b\rangle, \langle c, a\rangle\}\rangle$
R_5: $\langle\{a, b, c\}, \{\langle a, a\rangle, \langle b, b\rangle, \langle c, c\rangle\}\rangle$
R_6: $\langle\{a, b, c\}, \{\langle a, b\rangle, \langle b, a\rangle, \langle c, c\rangle\}\rangle$
R_7: $\langle\{a, b, c, d\}, \{\langle a, d\rangle, \langle b, d\rangle, \langle c, d\rangle, \langle d, d\rangle\}\rangle$

3. Which of the properties of binary relations on a set does the identity relation on a set have?

6.2 Binary Relations from a Set to a Set

We now turn to a second way to see a binary relation, namely, as a specification comprising the set of things that may be related, the set of things that may be related to, and the actual pairing. Here is the formal definition of a binary relation.

Definition 26 Binary relation of a set to a set
Let R be $\langle X, Y, G \rangle$. R is a binary relation iff X and Y are sets and $G \subseteq X \times Y$.

The diagrams for bipartite directed graphs are particularly apt for depicting binary relations from a set to a set. In a diagram for a bipartite directed graph, there are two columns of points. The first column depicts the nodes in the domain and the second column the nodes in the codomain. The nodes are listed in an irredundant and exhaustive manner. Arrows connecting a point in the left-hand column with a point in the right-hand column corresponds to the ordered pairs in the graph of the binary relation.

Notice that a binary relation from a set to a set does not require that the sets be disjoint. Indeed, it permits the domain and the codomain to be identical. A binary relation from the domain $\{1, 2, 3, 4, 6\}$ to the codomain $\{2, 3, 5\}$ is depicted.

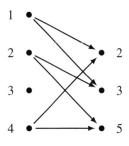

Suppose that one has two sets A and B. How many binary relations are there from A to B? The answer is that there are as many binary relations from A to B as there are distinct sets of ordered pairs whose first coordinate is a member of A and whose second coordinate is a member of B. Thus, the number of binary relations from A to B is $2^{|A \times B|}$, or $2^{|A| \times |B|}$. As before, when the graph of the binary relation is $A \times B$, the relation is the universal relation from A to B, and when it is the empty set, it is the null or empty binary relation from A to B.

Binary relations from a set to a set also have special properties: *left totality*, *right totality*, *left monogamy*, and *right monogamy*. Each of these properties plays a role in determining functions and various kinds of functions.

A binary relation is left total precisely when each member of its domain bears the relation to some member of its codomain.

Definition 27 Left totality

Let $R = \langle X, Y, G \rangle$ be a binary relation. R is left total iff, for each $x \in X$, there is a $y \in Y$, such that xRy.

The relation R_1,

$R_1 = \langle A, B, G \rangle$ where $A = \{a, b, c\}$
$B = \{2, 4, 6, 8\}$
$G = \{\langle a, 2 \rangle, \langle a, 6 \rangle, \langle b, 4 \rangle, \langle b, 8 \rangle, \langle c, 8 \rangle\}$

is left total, while the relation R_2,

$R_2 = \langle C, D, H \rangle$ where $C = \{d, e, f\}$
$D = \{1, 3, 5, 7\}$
$H = \{\langle d, 1 \rangle, \langle d, 5 \rangle, \langle f, 5 \rangle, \langle f, 7 \rangle\}$

is not.

Seen in terms of a diagram for a bipartite directed graph, a left total relation has at least one arrow emanating from each point in its left-hand column. R_1 satisfies this characterization, as one can see from an inspection of its bipartite directed graph diagram; R_2 does not, since no arrow emanates from e.

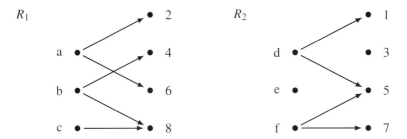

A binary relation can also be depicted with a table: the left-most column lists the members of the relation's domain, the top-most row lists the members of its codomain, and each spot in the appropriate row and column is marked with a plus or minus to signify membership in the relation's graph. Like the diagrams for bipartite directed graphs, these diagrams, too, reflect whether or not a binary relation is left total. Consider the tables for R_1 and R_2.

R_1	2	4	6	8
a	+	−	+	−
b	−	+	−	+
c	−	−	−	+

R_2	1	3	5	7
d	+	−	+	−
e	−	−	−	−
f	−	−	+	+

A binary relation is left total only in the case where its table has a plus in each of its rows. R_1's table satisfies this condition, whereas R_2's table does not.

A binary relation is right total precisely when each member of its codomain has the relation borne to it by some member of its domain.

Definition 28 Right totality
Let $R = \langle X, Y, G \rangle$ be a binary relation. R is right total iff, for each $y \in Y$, there is a $x \in X$, such that xRy.

The relation R_1 is right total, but the relation R_2 is not. Characterizations fully parallel to those given for left totality apply here. Looking at the bipartite directed graph diagram for R_2, one sees that there is one element in its codomain into which no arrow terminates. Hence, R_2 is not right total. Looking at the diagram for R_1, however, one sees that each element in its codomain has at least one arrow terminating into it. For that reason, R_1 is right total. In general, a binary relation is right total only in the case where its bipartite directed graph diagram has the following property: each node in the codomain has at least one arrow terminating into it.

Right totality can also be read off the table for a binary relation: a binary relation is right total only in the case where its table has a plus in each of its columns. R_1's table satisfies this condition, whereas R_2's table does not.

In the examples given, the very same relation is both left total and right total. This is a co-incidence. Consider S_1, which is just like R_1, except that c is not related to 8. S_1 is right total, but not left total. Moreover, consider S_2, which is just like R_2, except that e is related to 5. S_2 is left total, but not right total.

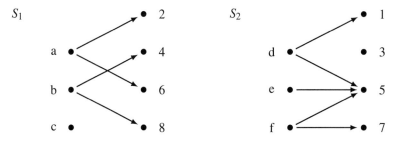

A binary relation is left monogamous precisely when no member of its domain bears the relation to more than one member of its codomain. Neither R_1 nor R_2 are left monogamous, since, on the one hand, $aR_1 2$ and $aR_1 6$, and, on the other hand, $fR_2 5$ and $fR_2 7$. The relation T_1 is, however, left monogamous.

$$T_1 = \langle A, D, I \rangle \quad \text{where} \quad A = \{a, b, c\}$$
$$D = \{1, 3, 5, 7\}$$
$$I = \{\langle a, 1 \rangle, \langle b, 7 \rangle, \langle c, 7 \rangle\}$$

Basic Set Theory

Notice that, in the bipartite directed graph diagram of T_1, at most one arrow emanates from any point in its left-hand column. This is a defining characteristic of left monogamous relations.

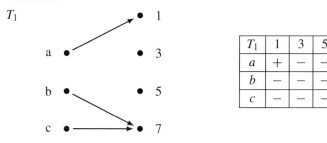

Left monogamy is manifested in a binary relation's table by the fact that no row has more than one plus. Readers can verify that T_1 is left monogamous by inspecting its table. The formal definition of left monogamy is given as follows.

Definition 29 Left monogamy[6]
Let $R = \langle X, Y, G \rangle$ be a binary relation. R is left monogamous iff, for each $x \in X$ and each $y, z \in Y$, if xRy and xRz, then $y = z$.

Paired with the notion of left monogamy is the notion of right monogamy. A binary relation is right monogamous precisely when no member of its codomain has the relation borne to it by more than one member of its domain.

Definition 30 Right monogamy
Let $R = \langle X, Y, G \rangle$ be a binary relation. R is right monogamous iff, for each $x, z \in X$ and for each $y \in Y$, if xRy and zRy, then $x = z$.

None of the relations R_1, R_2, or T_1 is right monogamous. T_1, for example, is not right monogamous, since both bT_17 and cT_17. However, T_2 is right monogamous.

$T_2 = \langle C, B, J \rangle$ where $C = \{d, e, f\}$
$B = \{2, 4, 6, 8\}$
$J = \{\langle d, 2 \rangle, \langle d, 4 \rangle, \langle f, 6 \rangle\}$

Like left monogamy, right monogamy can be read off its bipartite directed graph diagram or its table. In the bipartite directed graph of a right monogamous binary relation, at most one arrow terminates into any point in its right-hand column. In the table for such a binary relation, at most one plus occurs in any of its columns. This is illustrated as follows.

---

6. Left monogamy is also known as being single-valued.

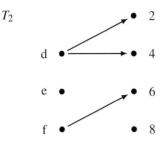

T_2	2	4	6	8
d	+	+	−	−
e	−	−	−	−
f	−	−	+	−

Exercises: Binary relations from a set to a set

1. Write out all binary relations from $\{1, 2\}$ to $\{a, b\}$.

2. For each relation given, determine whether it is left total, left monogamous, right total, or right monogamous.

(a) $\langle \{a, b, c, e\}, \{1, 3, 5\}, \{\langle a, 1\rangle, \langle c, 3\rangle, \langle b, 5\rangle\}\rangle$

(b) $\langle \{a, b, c\}, \{1, 3, 4, 5, 6, 7\}, \{\langle a, 3\rangle, \langle c, 1\rangle, \langle c, 4\rangle, \langle a, 6\rangle, \langle b, 5\rangle\}\rangle$

(c) $\langle \{a, b, c, d, e\}, \{4, 6, 8\}, \{\langle a, 4\rangle, \langle a, 6\rangle, \langle c, 6\rangle, \langle d, 8\rangle, \langle e, 8\rangle\}\rangle$

(d) $\langle \{a, b, c, d\}, \{1, 4, 5, 6\}, \{\langle a, 1\rangle, \langle c, 1\rangle, \langle c, 4\rangle, \langle a, 6\rangle, \langle b, 5\rangle\}\rangle$

(e) $\langle \{a, c, d\}, \{1, 4, 6, 8\}, \{\langle a, 1\rangle, \langle a, 6\rangle, \langle c, 4\rangle, \langle d, 1\rangle, \langle d, 8\rangle\}\rangle$

(f) $\langle \{a, b, c, e\}, \{1, 3, 5\}, \{\langle a, 1\rangle, \langle c, 1\rangle, \langle b, 5\rangle, \langle e, 1\rangle\}\rangle$

(g) $\langle \{a, b, c\}, \{1, 3, 4, 5, 6, 7\}, \{\langle c, 3\rangle, \langle a, 1\rangle, \langle a, 4\rangle, \langle c, 6\rangle, \langle b, 5\rangle, \langle a, 5\rangle\}\rangle$

(h) $\langle \{b, c, d, e\}, \{1, 3, 5\}, \{\langle e, 5\rangle, \langle d, 5\rangle, \langle d, 1\rangle, \langle b, 3\rangle, \langle c, 1\rangle\}\rangle$

(i) $\langle \{a, b, c, e\}, \{1, 3, 4, 5\}, \{\langle a, 5\rangle, \langle b, 4\rangle, \langle c, 3\rangle, \langle e, 1\rangle\}\rangle$

(j) $\langle \{a, b, c, e\}, \{1, 3, 5\}, \{\langle a, 1\rangle, \langle c, 1\rangle, \langle b, 1\rangle, \langle e, 1\rangle\}\rangle$

6.3 Function

In high school algebra, everyone studies functions. Such functions are typically rather complex, hiding their rather simple nature. *Functions* involve three things: a domain, codomain, and a graph. The domain is a set of entities—entities of any sort, though in high school algebra, it is usually the set of real numbers. The codomain is also a set of entities—again, entities of any sort, though in high school algebra, it is usually the set of real numbers. To understand the graph of a function, imagine the domain and the codomain to have all of their elements exhaustively listed in their own respective lists, each list comprising a column of names. The graph of the function, then, is a pairing of members on the first list with members of the second list represented by an arrow going from **each** member on the first list to **exactly one** member on the second list. In other words, a function is a binary relation that is left total and left monogamous.

To see better what is involved, consider R and S. R is a function, while S is not. Observe that R comprises two lists, the list on the left, $\{a, b, c,\}$ being the domain and the list on the right, $\{2, 4, 6, 8\}$, being the codomain. Further, exactly one arrow connects each member of the domain with some member of the codomain. S fails to be a function for two reasons: first, there is no arrow connecting e to any member of the codomain; second, both d and f have more than one arrow connecting them to members of the codomain.

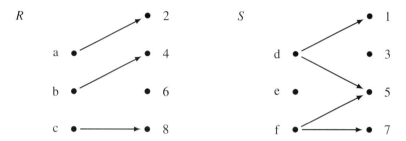

Since a function is a special kind of binary relation, the terminology and notation that apply to binary relations also apply to functions. However, since the mathematics of functions preceded the mathematics of binary relations, much of the mathematics of functions has notation and terminology peculiar to it. Thus, it is customary to denote a function, not as

$$f = \langle X, Y, G \rangle,$$

but as

$$f: X \to Y,$$

specifying its graph separately. The graph's specification is not in the usual notation for a set of ordered pairs, but in the form of a vertical, irredundant list of pairs, where a butted arrow ($\mapsto$) between the elements of each pair replaces the angle brackets that would otherwise enclose them. For example, the following functional relation,

$$f = \langle \{1, 2, 3, 4\}, \{a, b, c\}, \{\langle 1, a \rangle, \langle 2, a \rangle, \langle 3, c \rangle, \langle 4, b \rangle\} \rangle, \tag{2.4}$$

whose graph can be expressed by either the following bipartite graph diagram or table,

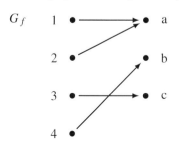

G_f	a	b	c
1	+	−	−
2	+	−	−
3	−	−	+
4	−	+	−

is displayed as follows:

$$f: \{1, 2, 3, 4\} \to \{a, b, c\} \quad \text{where} \quad \begin{aligned} 1 &\mapsto a \\ 2 &\mapsto a \\ 3 &\mapsto c \\ 4 &\mapsto b \end{aligned}$$

This way of displaying a function's graph is in a form similar to that of a bipartite graph, except here, all arrows are parallel, and some members of the codomain may not appear, whereas in a proper bipartite graph, all arrows need not be parallel, and not only must each member from the domain appear exactly once, but also each member from the codomain must appear exactly once.

Thus, functions with finite graphs may be displayed in three different forms: in the form of a bipartite graph, in the form of a table, and in the form of a vertical list. It might be helpful at this point to rehearse how the defining properties of a function, namely left totality and left monogamy, are reflected in these displays. In the bipartite graph display, each element in the left-hand column is connected by an arrow to exactly one element in the right-hand column: it is connected by at least one arrow, to satisfy the condition of left totality, and it is connected by at most one arrow, to satisfy the condition of left monogamy. In the table display, there is exactly one plus in each row of the table: there is at least one plus to satisfy the condition of left totality and there is at most one plus to satisfy the condition of left monogamy. And finally, in the vertical list display, each member of the domain appears exactly once: it appears at least once, to satisfy the condition of left totality, and it appears at most once, to satisfy the condition of left monogamy.

For most mathematical applications, the domain and codomain of a function are infinite, and so the function's graph cannot be enumerated. In such cases, it is customary to write down a rule from which any ordered pair in the function's graph can be calculated. We shall call such a rule *a rule of association*. For example, the function, part of whose graph is set out in the following,

$$g: \mathbb{Z}^+ \to \mathbb{Z}^- \quad \text{where} \quad \begin{aligned} 1 &\mapsto -2 \\ 2 &\mapsto -4 \\ 3 &\mapsto -6 \\ \vdots &\mapsto \vdots \end{aligned}$$

can be also be specified in either of the two ways that follow.

$g: \mathbb{Z}^+ \to \mathbb{Z}^-$ *where* $x \mapsto -2x$;

$g: \mathbb{Z}^+ \to \mathbb{Z}^-$ *where* $x \mapsto y$ *such that* $-2x = y$;

Sometimes, the rule of association must be more complex: the domain of the function is partitioned and a distinct rule is given for the elements of each set in the partition. For example,

Basic Set Theory

$$h: \mathbb{Z} \to \mathbb{Z} \text{ where } \begin{cases} x \mapsto x+1 & \text{for } x < -1 \\ x \mapsto 2x & \text{for } -1 \leq x \leq 3 \\ x \mapsto 4x-5 & \text{for } 3 < x \end{cases}$$

Thus, the enumeration of the elements of a function's graph is usually replaced by a rule of association. One cannot stipulate arbitrarily such rules of association and be assured that one has obtained a function, that is, a pairing that is both left total and left monogamous. This lack of assurance is a departure from what was encountered with sets and arbitrarily chosen binary relations. Within a fixed universe of discourse, one can use a property to define a set: the property may not have any possessors, in which case the set defined is empty, but there is a set nonetheless, the empty set. Similarly, one can use a rule of association to define a binary relation: the rule may not manage to pair anything in the domain with anything in the codomain, in which case the graph of the relation is empty, but there is a graph nonetheless. But one cannot set down any arbitrary rule of association and expect to obtain a relation that is both left total and left monogamous. Thus, to define a function, one defines the underlying relation and then establishes that the relation in question is both left total and left monogamous. It is this proof that guarantees that the definition given is truly a definition of a function and not of some nonfunctional, possibly empty, relation.

Mathematicians refer to the establishment of left totality as the establishment of *existence*. This is because left totality is defined as follows: for each member of the domain, there *exists* a member of the codomain such that the former is related to the latter. Mathematicians refer to the establishment of left monogamy as the establishment of *uniqueness*. This is because left monogamy requires that any member of the codomain associated with a member of the domain be *uniquely* associated with it.

The fact that functions are left total and left monogamous means that each member of the function's domain is related by the functions's graph to a unique member of its codomain. This unique connection permits one to refer unambiguously to a member of the function's range by any member of its domain connected with it. In other words, if one knows which function one is talking about, once one specifies a member of the domain, one in effect uniquely specifies a member of its codomain.

In (2.4), 2 is uniquely related to a by f. Insofar as 2 is related to a by f, one writes in infix notation $2fa$. However, in light of the fact that 2 is uniquely related to a by f, one can more perspicuously write in this traditional notation $f(2) = a$.

Functions have their own special vocabulary. Let us familiarize ourselves with it. Functions are often referred to as *maps* or *mappings,* on the one hand, or as *assignments,* on the other. Thus, one speaks of a map or mapping *from* a set, its domain, *to* a set, its codomain. A function, then, maps a member of its domain to a member of its codomain. In the previous example, f maps 2 to a. One also speaks of a function as an assignment of members of its codomain to members of its domain. A function, then, assigns a member *of* its codomain *to* a member of its domain. When members of a function's domain and codomain are paired by the function, they are referred to as the function's *argument* and *value,* respectively. Thus, a is said to be the value of the argument 2 *under* the function f. It is also said to be the value of the function f *at* the argument 2. An argument and a value of a function are also referred to as its *preimage* and its *image,* respectively, in which case a is said to be the image of the 2, its preimage, under f.

$$f\ (\_) = \ \underline{}$$

$$\uparrow \uparrow$$

argument;   value;
preimage   image

This new notation permits us to recast the definitions of the functions g and h given earlier into a more customary format:

$g \colon \mathbb{Z}^+ \to \mathbb{Z}^-$ where $g(x) = -2x$.

$h \colon \mathbb{Z} \to \mathbb{Z}$ where $h(x) = \begin{cases} x+1 & \text{if } x < -1 \\ 2x & \text{if } -1 \leq x \leq 3 \\ 4x - 5 & \text{if } 3 < x \end{cases}$

Mathematicians have identified a number of different kinds of functions. Functions that are right total are said to be *surjections*. In other words, a surjection is a function each element of whose codomain has a preimage. Functions that are right monogamous are said to be *injections*. In other words, an injection is a function where distinct preimages have distinct images. Functions that are both right total and right monogamous are *bijections*. Thus, bijections are functions that are both injections and surjections. In addition, mathematicians speak of *constant* functions. Constant functions are ones where the very same element of the codomain is the image of every element in the domain. The following function is a constant function; every element of its domain is mapped to a:

$\langle \{1, 2, 3, 4\},\ \{a, b, c\},\ \{\langle 1, a \rangle, \langle 2, a \rangle, \langle 3, a \rangle, \langle 4, a \rangle\} \rangle$.

There are also so-called *partial* functions. The expression is slightly unfortunate, since partial functions need not be functions. Recall that a function is, by definition, a binary relation that is both left total and left monogamous. A partial function is a binary relation that is left monogamous. It need not, however, be left total. For example, the following binary relation is a partial function:

$\langle \{1, 2, 3, 4\},\ \{a, b, c\},\ \{\langle 1, c \rangle, \langle 3, a \rangle, \langle 4, a \rangle\} \rangle$.

It is not, however, a function, since it is not left total. Nonetheless, since every function is both left total and left monogamous, it is left monogamous and hence a partial function.

Basic Set Theory

The idea behind the term *partial function* is that a partial function behaves like a function only with respect to part of its domain.

We now come to six notions pertaining to functions that will play an important role in later chapters of the book. The first two pertain to relations between functions, either where one restricts a function to a subset of its domain or where one extends a function by adding elements to its domain. The third pertains to functions obtained from functions called *near variant functions*. The fourth pertains to a special kind of function known as a *characteristic function*. These are functions that have a domain with exactly two particular elements, one associated with true and the other with false. The fifth notion pertains to functions that work in tandem, or the *product* of two or more functions. And the last notion is one that establishes a systematic connection between between binary relations, on the one hand, and functions from the domain of the binary relation to the power set of its codomain.

6.3.1 Restrictions and extensions of functions

One function is said to *extend* a second function, or equivalently, the second function is said to be an *extension* of the first function, just in case the domain, codomain, and graph of the extension are supersets, respectively, of the domain, codomain, and graph of the function that has the extension.

Definition 31 Extension
Let f and g be functions. g is an extension of f iff $D_f \subseteq D_g$, $C_f \subseteq C_g$, and $G_f \subseteq G_g$.

Thus, for example, the function m is an extension of the function l.

$l = \langle \{1, 2, 3\}, \{b, c\}, \{\langle 1, c\rangle, \langle 2, b\rangle, \langle 3, b\rangle\}\rangle,$
$m = \langle \{1, 2, 3, 4\}, \{a, b, c\}, \{\langle 1, c\rangle, \langle 2, b\rangle, \langle 3, b\rangle, \langle 4, c\rangle\}\rangle,$

Notice that, even if the domain to which a function will be extended is specified, there is usually no unique extension. Thus, l can be extended to the domain $\{1, 2, 3, 4\}$ in three different ways, depending on which member of the codomain is assigned to 4.

The converse, so to speak, of an extension is a restriction. A function restricted to a subset of its domain is the function whose domain is the intersection of the initial function's domain and the restricting set and whose graph is the set of ordered pairs in the initial function whose first coordinates are in the intersection. Thus, the function m restricted to the set $\{1, 2, 3\}$ is the function l. Or to put it another way, the function l is the restriction of the function m to the set $\{1, 2, 3\}$. Notice, by the way, that the function l is equally the restriction of m to the set $\{1, 2, 3, 5\}$, since the intersection of the set $\{1, 2, 3, 5\}$ with the domain of m is the set $\{1, 2, 3\}$. Here is the formal definition:

Definition 32 Restriction
Let f, or $\langle D, C, G\rangle$, be a function. Let X be a set. Then, the restriction of f to X is the function $\langle X \cap D, C, G \cap X \times C\rangle$.

6.3.2 Near variant functions

In chapter 11 and following, we will have occasion to talk about functions that are exactly alike, except for what value they assign to at most one member of their domain. Consider the functions j and k:

$$j = \langle \{1,2,3,4\}, \{a,b,c\}, \{\langle 1,c\rangle, \langle 2,b\rangle, \langle 3,a\rangle, \langle 4,b\rangle\}\rangle;$$
$$k = \langle \{1,2,3,4\}, \{a,b,c\}\{\langle 1,c\rangle, \langle 2,b\rangle, \langle 3,a\rangle, \langle 4,a\rangle\}\rangle.$$

They have the same domain and the same codomain. Moreover, their graphs are exactly the same, except for the ordered pair that has 4 as its first coordinate, that is, $\langle 4,b\rangle \in G_j$ but $\langle 4,b\rangle \notin G_k$ and $\langle 4,a\rangle \notin G_j$ but $\langle 4,a\rangle \in G_k$. In other words, for each member x of the $\{1,2,3,4\}$ except 4, $j(x) = k(x)$. When $x = 4$, $j(x) \neq k(x)$, since $j(x) = b$ and $k(x) = a$. The first is a near variant of the second.

Definition 33 Near variant functions
Let f and g be functions. f is a near variant of g iff $D_f = D = D_g$, $C_f = C = C_g$, and D has at most one element x such that $f(x) \neq g(x)$.

One immediate consequence of this definition is that each function is a near variant of itself.

It is convenient, when one function is a near variant of another, to name one function in terms of the other. Thus, for example, j from the previous example could be named by $k_{4\mapsto b}$. The function $k_{4\mapsto b}$ is just like the function k, except that it maps 4 to b. But this is just the function j. Equally, $j_{4\mapsto a}$ is a function just like j except it maps 4 to a. But this is just the function k. Note finally that $j_{4\mapsto b}$ is just j and $k_{4\mapsto a}$ is just k.

6.3.3 Characteristic functions

A characteristic function is a function that characterizes, as it were, a set. In particular, it specifies for a set which entities are members of the set and which are not. It does so by assigning members of the set some distinguished value, customarily 1, and assigning nonmembers another distinguished value, customarily 0. The set of entities that are in the domain of a characteristic function, if not explicitly stated, is usually evident from context. We define such a function as follows:

Definition 34 Characteristic function
Let X be a set. f is a characteristic function for Y, a subset of X, iff

(1) f is a function from X into $\{0,1\}$, and
(2) for each $x \in X$, $f(x) = 1$, if $x \in Y$, and $f(x) = 0$, if $x \notin Y$.

Earlier, we introduced the idea of a set's power set, which is the set of all subsets of a given set. Now we introduce the idea of a set of characteristic functions for a set, which is the set of all functions from the set into $\{0,1\}$.

Definition 35 Characteristic functions of a set
Let X be a set. Then, $\text{Chr}(X) = \text{Fnc}(X, \{0, 1\})$.

As we are about to illustrate, but not prove, for any set X, being a characteristic function of is a bijection between $\text{Pow}(X)$ and $\text{Chr}(X)$. Consider the set A, which has precisely three members, a, b, and c. It has eight subsets, since the number of subsets of a set is 2 raised to the cardinality of the set. Listed in the following table on the left-hand side are the subsets of A, indexed by the members of $\mathbb{Z}_8^+$ and listed on the right-hand side are the graphs of the functions from A into $\{0, 1\}$, also indexed by the members of $\mathbb{Z}_8^+$. (Notice that A, a subset of A, is indexed as A_1.)

$\text{Pow}(\{a, b, c\})$ $\text{Chr}(\{a, b, c\})$
$A_1 = \{a, b, c\}$ $G_{f_1} = \{a \mapsto 1, b \mapsto 1, c \mapsto 1\}$
$A_2 = \{a, b\}$ $G_{f_2} = \{a \mapsto 1, b \mapsto 1, c \mapsto 0\}$
$A_3 = \{a, c\}$ $G_{f_3} = \{a \mapsto 1, b \mapsto 0, c \mapsto 1\}$
$A_4 = \{a\}$ $G_{f_4} = \{a \mapsto 1, b \mapsto 0, c \mapsto 0\}$
$A_5 = \{b, c\}$ $G_{f_5} = \{a \mapsto 0, b \mapsto 1, c \mapsto 1\}$
$A_6 = \{b\}$ $G_{f_6} = \{a \mapsto 0, b \mapsto 1, c \mapsto 0\}$
$A_7 = \{c\}$ $G_{f_7} = \{a \mapsto 0, b \mapsto 0, c \mapsto 1\}$
$A_8 = \emptyset$ $G_{f_8} = \{a \mapsto 0, b \mapsto 0, c \mapsto 0\}$

As the table makes clear, there is an obvious association of each set with each function. A_1 obviously associates with f_1, A_2 with f_2, and more generally, A_i with f_i, for each $i \in \mathbb{Z}_8^+$. Associated in that way, we can think of each function as indicating, for its associated set, whether some member of A is in that set; in other words, it assigns 1 to a member of A, if the member of A is in the associated subset, and it assigns 0 to a member of A, if it is not. Under this view, f_6, for example, indicates that b is in A_6, but neither a nor c is. Thus, f_6 can be said to be the characteristic function of A_6. And more generally, f_i is the characteristic function of A_i, for each $i \in \mathbb{Z}_8^+$.

Characeristic functions will play an important role in chapter 12 and following.

6.3.4 Product functions

Here, we are going to learn about the *product functions*, an idea that will have an important role in chapter 15. A product function is one that makes two or more functions work in tandem. Consider, for example, two functions, one on the natural numbers and one on the letters of the Roman alphabet. The first function is the doubling function, called here δ. It assigns to each natural number its double: that is, for each $x \in \mathbb{N}$, $x \mapsto 2x$. The second function is a succedent function, called here σ, which assigns to each letter of the Roman alphabet but the last the letter that succeeds it in the alphabet's customary ordering and assigns to the last letter, z, the first letter, a. In other words, σ is a function on $\mathbb{A}$, the letters of the Roman alphabet, where $a \mapsto b, \ldots, y \mapsto z$, and $z \mapsto a$.

We shall now define the product of these two functions, which we call $\langle \delta, \sigma \rangle$. This function is one in which δ and σ work in tandem. For these functions to work in tandem means that they must apply in tandem to an ordered pair of elements, the first of which is from $\mathbb{N}$, the domain of δ, and the second of which is from $\mathbb{A}$, the domain of σ. Thus, the application of $\langle \delta, \sigma \rangle$ to the ordered pair $\langle 5, b \rangle$ means that δ applies to 5 and σ applies to b to yield an ordered pair $\langle \delta(5), \sigma(b) \rangle$, or $\langle 10, c \rangle$. In short, $\langle \delta, \sigma \rangle(\langle 5, b \rangle) = \langle 10, c \rangle$. Similarly, $\langle \delta, \sigma \rangle(\langle 23, y \rangle) = \langle \delta(23), \sigma(y) \rangle = \langle 46, z \rangle$ and $\langle \delta, \sigma \rangle(\langle 1, z \rangle) = \langle \delta(1), \sigma(z) \rangle = \langle 2, a \rangle$.

Generalizing on this example, we state the definition of the product of a pair of unary functions.

Definition 36 Product of unary functions

Let f and g be functions from X into W and Y into Z, respectively. Then, $\langle f, g \rangle$ is a function from $X \times Y$ into $W \times Z$, where, for each $x \in X$ and $y \in Y$, $\langle f, g \rangle(\langle x, y \rangle) = \langle f(x), g(y) \rangle$.

Taking the product of two unary functions is not commutative. That is, in general, it is not true that $\langle f, g \rangle$ is the same as $\langle g, f \rangle$. For example, $\langle \delta, \sigma \rangle(\langle 5, b \rangle) = \langle \delta(5), \sigma(b) \rangle = \langle 10, c \rangle$ but $\langle \sigma, \delta \rangle(\langle b, 5 \rangle) = \langle \sigma(b), \delta(5) \rangle = \langle c, 10 \rangle$ and $\langle c, 10 \rangle \ne \langle 10, c \rangle$.

In the preceding examples, we defined the product of a pair of unary functions. We can just as well define the product of a pair of binary functions. We now illustrate this possibility. Let the first function be the addition function on the set of natural numbers, which we shall designate with the usual plus sign. Let us now define another binary function, one on strings of lowercase letters of the Roman alphabet. By the set of strings of lowercase letters over the Roman alphabet, we mean: $\{a, b, \ldots, z, aa, ab, ac, \ldots, az, ba, bb, \ldots\}$. Let us call this set $\mathbb{A}^*$. Concatenation, which takes a pair of strings of letters and concatenates the second to the end of the first, is a binary function on this set. Concatenation applies to the pair of strings, *zba* and *wwqrst*, taken from $\mathbb{A}^*$, to yield *zbawwqrst*, in $\mathbb{A}^*$. In other words, letting · designate concatenation, we put into notation what we just said as follows: *zba* · *wwqrst* = *zbawwqrst*.

The product of addition and concatenation, or $\langle +, \cdot \rangle$, is a function from pairs of pairs of members of $\mathbb{N} \times \mathbb{A}^*$ into $\mathbb{N} \times \mathbb{A}^*$; in other words, $\langle +, \cdot \rangle \colon (\mathbb{N} \times \mathbb{A}^*) \times (\mathbb{N} \times \mathbb{A}^*) \to \mathbb{N} \times \mathbb{A}^*$. Thus, $\langle +, \cdot \rangle(\langle 15, bca \rangle, \langle 29, df \rangle) = \langle 44, bcadf \rangle$.

6.3.5 Binary relations and their associated functions

Let us turn to the sixth and last notion pertaining to functions, which we shall see again in chapter 12. There is a systematic connection between binary relations, on the one hand, and functions from the domain of the binary relation to the power set of its codomain, which results from this rule: take as the function's domain the domain of the binary relation, take as the function's codomain the power set of the relation's codomain, and take as the graph the set of ordered pairs where the first member is from the domain and the second member is the set of all those members of the relation's codomain to which the first member is

Basic Set Theory

related by the relation. Let us apply this rule to the binary relation T_3, whose domain is the set $\{a, b, c,\}$ whose codomain is the set $\{1, 3, 5\}$, and whose graph comprises the ordered pairs $\{\langle a, 1\rangle, \langle a, 5\rangle, \langle b, 1\rangle, \langle b, 3\rangle\}$. The rule says that the domain of the function corresponding to T_3 is T_3's domain, or the set $\{a, b, c\}$, the codomain of the function is Pow($\{1, 3, 5\}$), which is the power set of T_3's codomain, and the graph of the function is the set $\{\langle a, \{1, 5\}\rangle, \langle b, \{1, 3\}\rangle, \langle c, \emptyset\rangle\}$.

Now, let us consider the inverse. Given a function from a set to a set of sets, one can define a binary relation, where the domain of the relation is the same as the domain of the function, where the codomain of the relation is the generalized union of the codomain of the function, and where the graph of the relation consists of the pairs where the first member is from the domain and the second member is in one or other of the sets in the set of sets, which is the codomain of the function. Let us rehearse these instructions with an example. Consider the function whose domain is the set $\{a, b, c, d\}$, whose codomain is the set $\{\{1, 2\}, \{2, 3\}\{1, 3, 4\}, \{2, 4\}\}$, and whose graph is $\{\langle a, \{1, 2\}\rangle, \langle b, \{1, 3, 4\}\rangle, \langle c, \{2, 3\}\rangle, \langle d, \{2, 4\}\rangle\}$. The binary relation corresponding to this function has the same domain, namely $\{a, b, c, d\}$. Its codomain is the generalized union of the codomain of the function, that is, $\bigcup \{\{1, 2\}, \{2, 3\}\{1, 3, 4\}, \{2, 4\}\}$, or $\{1, 2, 3, 4\}$. The graph is $\{\langle a, 1\rangle, \langle a, 2\rangle, \langle b, 1\rangle, \langle b, 3\rangle, \langle b, 4\rangle, \langle c, 2\rangle, \langle c, 3\rangle, \langle d, 2\rangle, \langle d, 4\rangle\}$.

Definition 37 Binary relation's associated function
Let R be a binary relation from D to C. Then, its associated function, im_R, from D into Pow(C), assigns to x, a member of D, the set $\{y: Rxy\}$.

If we were to be rigorous, we would have to prove that the relation just defined, which assigns to a binary relation a function of the kind described, is both left total and left monogamous. In fact, it is also right total and right monogamous. Having these four properties, the relation is a bijection. This fact we shall use in chapter 12.

Exercises: Functions

1. Consider the function $f: \mathbb{R} \to \mathbb{R}^7$ whose graph is determined by the following rule: $x \mapsto 2x^2 + 3$, that is, $f(x) = 2x^2 + 3$. Determine, for each of the following cases, the value.

(a) $f(2)$

(b) $f(-3)$

---

7. The set of real numbers, denoted $\mathbb{R}$, is defined as the set of all numbers that can be written in decimal notation, even ones where the decimal expansion does not end. It includes both the rational numbers (numbers that can be written as p/q, where p and q are integers) and the irrational numbers (numbers that cannot be written as p/q). Examples of rational numbers are $2 (= 2/1)$, $1/3$, $-23.44 (= -586/25)$. Examples of irrational numbers are π $(= 3.14159\ldots)$, $\sqrt{2} (= 1.41421\ldots)$, $\log_{10} 3 (= 0.47712\ldots)$.

(c) $f(0)$

(d) $f(0.5)$

(e) $f(-0.25)$

(f) $f(-2)$

2. Which of the following binary relations are functions? If one is not a function, explain why it is not.

(a) $\langle \{a, b, c\}, \{1, 2\}, \{\langle a,1\rangle, \langle b,2\rangle, \langle c,1\rangle\}\rangle$

(b) $\langle \{1, 2, 3\}, \{b, c\}, \{\langle 1, b\rangle, \langle 2, b\rangle, \langle 3, c\rangle, \langle 1, c\rangle\}\rangle$

(c) $\langle \{a, b, c\}, \{2, 3\}, \{\langle a, 1\rangle, \langle b, 2\rangle\}\rangle$

(d) $f: \mathbb{Z} \to \mathbb{Z}$ where $x \mapsto y$ such that $x^2 = y$

(e) $g: \mathbb{Z} \to \mathbb{Z}$ where $x \mapsto y$ such that $x = y^2$

(f) $h: \mathbb{Z} \to \mathbb{Z}$ where $\begin{cases} x \mapsto 3x & \text{for} \quad x < -1 \\ x \mapsto 4x & \text{for} \quad -1 < x < 3 \\ x \mapsto 5x & \text{for} \quad 3 < x \end{cases}$

(g) $k: \mathbb{Z} \to \mathbb{Z}$ where $\begin{cases} x \mapsto x+5 & \text{for} \quad x \leq 0 \\ x \mapsto x+3 & \text{for} \quad 0 \leq x \leq 2 \\ x \mapsto x+1 & \text{for} \quad 2 \leq x \end{cases}$

(h) $l: \mathbb{Z} \to \mathbb{Z}$ where $\begin{cases} x \mapsto x & \text{for} \quad x < 0 \\ x \mapsto 2x & \text{for} \quad 0 \leq x \leq 2 \\ x \mapsto x^2 & \text{for} \quad 2 < x \end{cases}$

(i) $\langle \{1, 2\}, \{a, b\}, \emptyset \rangle$

(j) $\langle \emptyset, \{a, b\}, \emptyset \rangle$

(k) $m: \mathbb{R} \to \mathbb{R}$ where $\begin{cases} x \mapsto 1 & \text{if } x \text{ is rational} \\ x \mapsto -1 & \text{if } x \text{ is irrational} \end{cases}$

3. Let P be the points in a (Euclidean) plane π and let S be the straight lines in π. Which of the following are functions?

(a) $P \times S \to S$ where $\langle p, s_i \rangle$ is assigned s_j and s_j is perpendicular to s_i and p lies on s_j.

(b) $P \times P \to S$ where $\langle p_i, p_j \rangle$ is assigned s and s lies on both p_i and p_j.

(c) $S \times S \to P$ where $\langle s_i, s_j \rangle$ is assigned p and p lies on both s_i and s_j.

4. Consider the function f, or $\langle \{1, 2, 3\}, \{3, 6, 9\} \{\langle 1, 6\rangle, \langle 2, 3\rangle, \langle 3, 9\rangle\}\rangle$.

(a) List all extensions of f that result from enlarging its domain to include 4.

(b) List all extensions of f that result from enlarging its codomain to include 8.

(c) List all extensions of f that result from enlarging its domain to include 5 and its codomain to include 11.

Basic Set Theory 101

5. Consider the function g, or $\langle\{1, 2, 3, 4\}, \{5, 6, 7, 8\}, \{\langle 1, 8\rangle, \langle 2, 6\rangle, \langle 3, 5\rangle, \langle 4, 5\rangle\}\rangle$. Identify the functions that result from the restriction of g to the following sets.

(a) $\{1, 2, 3\}$ (d) $\emptyset$ (g) $\{1, 2, 6, 9\}$
(b) $\{3, 4\}$ (e) $\{1, 2, 3, 4\}$ (h) $\{9\}$
(c) $\{1\}$ (f) $\mathbb{N}$ (i) $\mathbb{Z}^-$

6. How many distinct restrictions are there of a function whose domain has cardinality n, a natural number?

7. Consider the function h, or $\langle\{a, b, c\}, \{2, 3, 5, 7\}\{\langle a, 3\rangle, \langle b, 3\rangle, \langle c, 7\rangle\}\rangle$. How many near variants does h have? How many near variants does h have distinct from h? Explain how you arrived at your answer.

8. Let A be the $\{a, b, c, d\}$, whose cardinality is 4. Identify the graphs for the characteristic functions for the following subsets of A:

(a) $\{a, b, c\}$ (c) $\emptyset$ (e) $\{a\}$
(b) $\{b, c\}$ (d) $\{a, b, d\}$ (f) $\{d\}$

9. Consider the following product functions.

(a) Let f be a function that adds 2 to each natural number; let g be a function that appends the letter a to each member of $\mathbb{A}^*$, the set of strings of lowercase letters over the Roman alphabet. Indicate the value $\langle f, g\rangle$ assigns to each of the following arguments.

(i) $\langle 3, b\rangle$ (iii) $\langle 19, cca\rangle$ (v) $\langle 101, fdage\rangle$
(ii) $\langle 9, dc\rangle$ (iv) $\langle 25, aaa\rangle$ (vi) $\langle 256, cccaaac\rangle$

(b) Let $\times$ be the multiplication on $\mathbb{N}$; let $\cdot$ be the concatenation on $\mathbb{A}^*$. Indicate the value $\langle\times, \cdot\rangle$ assigns to each pair of the following arguments.

(i) $\langle\langle 3, b\rangle, \langle 2, dbb\rangle\rangle$ (iv) $\langle\langle 2, dbb\rangle, \langle 2, b\rangle\rangle$
(ii) $\langle\langle 9, abc\rangle, \langle 5, def\rangle\rangle$ (v) $\langle\langle 1, b\rangle, \langle 2, a\rangle\rangle$
(iii) $\langle\langle 11, dzx\rangle, \langle 7, azbxcw\rangle\rangle$ (vi) $\langle\langle 31, dbdbdb\rangle, \langle 2, ee\rangle\rangle$

10. Consider the following binary relation:
$\langle\{a, b, c\}, \{1, 2, 3, 4\}, \{\langle a, 1\rangle, \langle a, 3\rangle, \langle b, 1\rangle, \langle b, 3\rangle, \langle b, 4\rangle, \langle c, 3\rangle, \langle c, 2\rangle\}\rangle$.
State the graph of the corresponding function from $\{a, b, c\}$ into Pow($\{1, 2, 3, 4\}$).

11. Consider the following function relation:
$\langle\{a, b, c\}, \{1, 2, 3, 4\}, \{\langle a, \{1, 3\}\rangle, \langle b, \{3\}\rangle, \langle c, \{2, 4\}\rangle,)\}\rangle$.
State the graph of the corresponding relation from $\{a, b, c\}$ to $\{1, 2, 3, 4\}$.

SOLUTIONS TO SOME OF THE EXERCISES

2 Sets and their members

1. (a) $\{x : x$ is expressed by a two-digit numeral$\}$ or
   $\{x : x \in \mathbb{N}$ and $10 \leq x \leq 99\}$
2. (a) $\{11, 22, 33, 44, 55, 66, 77, 88, 99\}$
3. True: a, d, e, g; false: b, c, f, h, i, j, k, l.
4. True: none.
5. True: a, d, e, h; false: b, c, f, g.
6. True: b, c, d; false: a, e, f.
7. True: b, d, g; false: a, c, e, f, h.
8. (a) finite (26 members)   (c) finite

3 Operations on sets

1. (a) $\{1, 3, 5\}$ (c) $\{1, 3\}$ (e) $\{1, 3, 4, 5, 6\}$
   (g) $\{1, 3, 5, 6\}$ (i) $\{2\}$
2. (a) $\mathbb{Z}^+$ (c) $\mathbb{Z} - \{0\}$ (e) $\emptyset$ (g) $\{0\}$
3. True: all.

4.2 Cartesian product

1. (a) $\{\langle 2, 1\rangle, \langle 2, 3\rangle, \langle 4, 1\rangle, \langle 4, 3\rangle, \langle 6, 1\rangle, \langle 6, 3\rangle\}$
   (b) $\{\langle 3, 3\rangle, \langle 3, 5\rangle, \langle 5, 5\rangle, \langle 5, 3\rangle\}$
   (c) $\{\langle\langle 1, 1\rangle, 2\rangle, \langle\langle 1, 1\rangle, 4\rangle, \langle\langle 1, 1\rangle, 6\rangle, \langle\langle 1, 3\rangle, 2\rangle, \langle\langle 1, 3\rangle, 4\rangle, \langle\langle 1, 3\rangle, 6\rangle,$
       $\langle\langle 3, 1\rangle, 2\rangle, \langle\langle 3, 1\rangle, 4\rangle, \langle\langle 3, 1\rangle, 6\rangle, \langle\langle 3, 3\rangle, 2\rangle, \langle\langle 3, 3\rangle, 4\rangle, \langle\langle 3, 3\rangle, 6\rangle\}$
   (d) $\{\langle 2, 3\rangle, \langle 4, 3\rangle, \langle 6, 3\rangle\}$
   (e) $\{\langle 1, 1\rangle, \langle 1, 3\rangle, \langle 3, 1\rangle, \langle 3, 3\rangle, \langle 3, 5\rangle, \langle 5, 3\rangle, \langle 5, 5\rangle\}$
   (f) $\{\langle 3, 2\rangle, \langle 3, 4\rangle, \langle 3, 6\rangle\}$
   (g) $\emptyset$
   (h) $\{\langle 1, 1\rangle, \langle 3, 1\rangle, \langle 1, 3\rangle, \langle 3, 3\rangle, \langle 5, 1\rangle, \langle 5, 3\rangle\}$
2. True: a, c; false: b, d, e, f, g.

5.1 The power set operation

1. (a) $\{\{a\}, \{a, b\}, \{a, c\}, \{a, d\}, \{a, b, c\}, \{a, c, d\}, \{a, b, d\}, \{a, b, c, d\}\}$

Basic Set Theory

   (c) $\{\{a,b\},\{a,b,c\},\{a,b,d\},\{a,b,c,d\}\}$

   (e) $\{\{a\},\{b\},\{c\},\{d\},\ldots,\{a,b,c,d\}\}$ (i.e., $\text{Pow}(V) - \{\emptyset\}$)

2. (a) $\{\{a,b\},\{a,c\},\{a,d\},\{b,c\},\{b,d\},\{c,d\}\}$

   (c) $\{\{a,b,c,d\}\}$

   (e) $\{\emptyset\}$

3. (a) $\{\{a,b\},\{a,b,c\},\{a,b,d\},\{a,b,c,d\}\}$

   (c) $\emptyset$

   (e) $\emptyset$

4. (a) $\{\{a\},\{b\},\{a,c\},\{b,c\},\{a,d\},\{b,d\},\{a,c,d\},\{b,c,d\}\}$

   (c) $\emptyset$

5. True: a; false: b, c, d.

5.2 Operations on families of sets

1. (a) $\{1,2,3,4\}$ (c) $\{2,3,4\}$ (e) $\emptyset$

   (g) $\{1,2,3,4\}$ (i) $\{\langle 1,3\rangle,\langle 1,4\rangle,\langle 2,3\rangle,\langle 2,4\rangle,\langle 3,3\rangle,\langle 3,4\rangle\}$ (k) $\emptyset$

2. (a) $\{0,1,2,3,4,5,6\}$ (c) $\{2\}$ (e) $\mathbb{N}$

6.1 Binary relations on a set

1. (a) yes   (c) no   (e) yes

2. R_1: Symmetric, antisymmetric, and transitive.

   R_3: Symmetric.

   R_5: Reflexive, symmetric, transitive, and antisymmetric.

   R_7: Antisymmetric and transitive.

6.2 Binary relations from a set to a set

1. Hint: there are sixteen such relations.

2.

	Left monogamy	Right monogamy	Left totality	Right totality
(a)	True	True	False	True
(c)	False	False	False	True
(e)	False	False	True	True
(g)	False	False	True	False
(i)	True	True	True	True

3 Basic English Grammar

1 Introduction

Ever since humans have thought about their own language, they have noticed patterns as well as anomalies within these patterns. One of the first patterns to be discovered was that similar sounds occur again and again in the utterances of a language. A hypothesis to account for this pattern is that a language is based on a finite number of minimal sounds, known as *phones* or *segments,* into which all complex utterances in the language can be segmented or decomposed. An alphabet is a notation that presupposes this hypothesis;[1] the ultimate refinement of this notation is the *International Phonetic Alphabet*, or IPA.

Another crucial pattern recognized in all literate societies is that utterances comprise similar sequences of phones, or *words*. It is words and the patterns of their variation in form that have been the principal object of study for traditional grammar. Traditional grammars are a form of systematic, empirical inquiry, though their empirical nature was often not recognized as such by those who composed them.

In this chapter, we begin the study of the structure of acceptable English expressions. Before doing so, however, we shall provide a brief overview of fundamental terms used in the study of English grammar, an understanding of which will be presupposed in subsequent sections of this chapter as well as in subsequent chapters. These technical terms include those for the various parts of speech and other technical terms such as *gender, case, number, coordination, subordination, tense, aspect, mood, voice, infinitive, participle, gerund, subject, direct object, indirect object, complement, predicate, modifier, auxiliary verb, copular verb,* and so on. These terms, used in traditional English grammar, will be explained in an informal and intuitive way.

Readers who are unfamiliar with the fundamentals of English grammar should read this section carefully; those who are familiar with the fundamentals should nonetheless read

---

1. Having an alphabet is not a necessary condition for being able to posit the pattern and formulate the hypothesis. Sanskrit grammarians seem to have worked entirely without written notation.

it cursorily, since this section serves a second purpose, namely, to prepare the readers to appreciate the contribution of contemporary, formal methods in the study of natural language. As we shall see in section 2, English grammar, presented in the style of traditional English grammar, is not able to address the central questions of the study of natural language raised in chapter 1. The point of this discussion is to enhance the readers' appreciation of the insights that result from the application of empirical and mathematical methods to the study of natural language set out in brief in the first chapter. In section 3, we shall set out a grammar that we shall call a *constituency grammar*. Unlike traditional grammars, this grammar has a mathematical structure, which we shall discuss in some detail. We shall also show how it provides answers to the central questions of semantics raised in chapter 1. This grammar, which has its modern origins in the work of American Structuralist linguists, is far from complete. As we shall see, the facts of English require that it be enriched. Some of the enrichment will be carried out in this chapter and more will be carried out in subsequent chapters. Section 4 will take stock of where we stand with respect both to the questions raised in chapter 1 and the remarks made there on what rigorous empirical inquiry into the study of language is like.

2 Traditional Grammar of English

Traditional grammar in Europe, dating back to classical Greece, distinguishes two units of grammatical analysis: the sentence (*logos*), which is defined as the unit that expresses a complete thought, and the word (*lexis*), which is the minimal unit of meaning. Traditional grammars develop categories of each kind of unit. We shall review each in turn, beginning with words, or lexical units.

2.1 Parts of Speech

In Europe, the categorization of words goes back to before Plato. However, the earliest complete set of lexical categories was provided by Aristarchus of Samothrace (217–145 BCE) and set out by his pupil Dionysius of Thrax (ca. 100 BCE) in his *Téchnē Grammatikē*. The eight lexical categories, or parts of speech (*mérē lógou*), comprised the noun (*ónoma*), the verb (*rhēma*), the participle (*metochē*), the article (*árthon*), the pronoun (*antónymiā*), the preposition (*próthesis*), the adverb (*epirrhēma*), and the conjunction (*sýndesmos*). Subsequent Greek grammarians such as Apollonius Dyscolus (ca. 100 CE) made modifications to these categories, as well as the Latin grammarians Donatus (ca. 400 CE) and Priscian (ca. 500 CE).[2] The parts of speech, or lexical categories, used today for Western European languages, including English, are set out here, together with the criteria whereby the categories are assigned.

---

2. See Robins (1966) for a brief history; for a more complete history, see Michael (1970).

- *Noun*   A noun is a word or a group of words that names a person, a place, an idea, or a thing (where things include objects, activities, qualities, and conditions).
- *Pronoun*   A pronoun is a word that stands for a noun. That for which a pronoun stands is its antecedent.
- *Verb*   A verb is a word or a group of that expresses an action, a condition, or a state.
- *Adjective*   An adjective is a word or group of words expressing a quality.
- *Adverb*   An adverb is a word or group of words that modifies a verb, an adjective, or another adverb.
- *Preposition*   A preposition is a word that shows a relation between its object and another word in the sentence.
- *Conjunction*   A conjunction is a word that connects two sentences or two parts of a sentence.
- *Interjection*   An interjection is a word or group of words that expresses some feeling in an exclamatory fashion.

Having set out this categorization of the words of English, the so-called lexical categories, let us see in more detail what traditional grammar has to say about each of these categories. We shall discuss each category in turn, pointing out what subcategories, if any, there are, as well as the prominent properties and uses of each category.

2.1.1   Nouns

Typical English nouns include *Julius Caesar, Lake Meech, capitalism, table, soccer, beauty,* and *fatigue*. English nouns, like nouns in other languages, have often been divided into subcategories. These subcategories of nouns have included proper nouns, common nouns, abstract nouns, concrete nouns, collective nouns, and compound nouns. The bases for such subcategorization are diverse.

It has been customary since the Stoics to distinguish between proper nouns and common nouns. The idea is that some entities are uniquely named by a noun. A proper noun is one that is proper to, unique to, a particular entity. *Julius Caesar* is a noun that is proper to the Roman consul of the first century BCE. A common noun is one that a number of entities have in common. *Table,* for example, is a noun that all tables have in common. While common nouns are single words, proper nouns sometimes comprise several words, for example, *Age of Reason* and *The Black Sea*.

Common nouns are quite diverse in what they denote. They may denote physical entities (*table*), mental entities (*concept*), states (*fatigue*), events (*picnic*), and conditions (*disease*). Indeed, it seems that there is nothing they cannot denote. Since the Middle Ages, however, common nouns have often been divided into *concrete* nouns (that is, nouns denoting concrete entities, for example, *rock, leaf, person*) and *abstract* nouns (that is, nouns denoting abstract entities, for example, *justice, truth, scorn*).

Still another kind of common noun often identified in traditional grammar are *collective* nouns (that is, nouns that denote collections). They include such nouns as *platoon, flock,* and *group* (see Quirk et al 1985, chap. 5.108).

And lastly, there are so-called *compound* nouns. These nouns are identified, not by what they denote, but by how they are formed. They are nouns that are put together, or compounded, from nouns. The compound noun *ticket office* is compounded from *ticket* and *office,* while the compound noun *sister-in-law* is put together from two nouns and a preposition.

English nouns exhibit three features: grammatical number, case, and gender. Grammatical number is a feature of many English nouns. Gender is a feature of very few. And case is a feature of even fewer.

English nouns show a distinction between *plural* and *singular*. In most cases, the plural is indicated by the suffix *-s* and the singular by its absence. Thus, *flowers* is a plural noun, while *flower* is a singular one. Many English nouns have irregular plural forms. They include not only such words as *mouse, man*, and *tooth,* but also such borrowed words as *phenomenon* (*phenomena*) and *concerto* (*concerti*). The distinction between plural and singular, as marked in most cases by the presence and absence of the suffix *-s,* is known as grammatical number.

Many languages—for example, French, German, Latin, Sanskrit, and Swahili—have what is called gender, or more specifically, grammatical gender. It is a classification of nouns into categories or kinds, which has little or nothing to do with natural gender, or sex, which is a feature of animals. Each noun in Latin, for example, is assigned to one of three different classes: masculine, feminine, or neuter. The Latin noun *dolor* (*pain*), for example, is masculine, while *fāma* (*reputation*) is feminine, and *mare* (*sea*) is neuter. Although in languages that have gender, words denoting animals or humans that are female have feminine grammatical gender, this is not invariably so. In German, for example, the word for a young girl, *mädchen,* denotes a human that is female, but the word belongs to the grammatical category of neuter. English used to have grammatical gender, but it no longer does. It does have, however, words whose usage requires they denote male humans, such as the pronouns *he* and *him,* or words whose usage requires that they denote female humans, such as the pronouns *she* and *her*. Here are some other examples of English words with natural gender.

Masculine	man, boy, father, gander, Henry
Feminine	woman, girl, mother, mare, Alice
Neuter	flower, water, stone, city

Case is a feature of many Indo-European languages, including German, Latin, Russian, Czech, and Sanskrit, to mention but a few. It used to be a feature of English, but now its only vestiges are found among English personal pronouns. Traditional English grammars identify three cases: nominative, objective, and possessive. The following three words are said to be the nominative, objective, and possessive case forms of *he, him,* and *his,* respectively. With the exception of a few personal pronouns, as just illustrated, there is no difference of form between English nouns in the nominative and objective cases.

	SINGULAR	PLURAL	SINGULAR	PLURAL
NOMINATIVE	*boy*	*boys*	*man*	*men*
OBJECTIVE	*boy*	*boys*	*man*	*men*
POSSESSIVE	*boy's*	*boys'*	*man's*	*men's*

A noun may bear a number of relations to verbs. A noun may be a verb's subject, its object or its indirect object, as exemplified by the next three sentences.

(1.1) *Bill* runs five miles every day.

(1.2) The guard spotted *Bill*.

(1.3) The instructor gave *Bill* a book.

Bill is also the indirect object in the next sentence, which is synonymous with the one in (1.3).

(2) The instructor gave a book to *Bill*.

As we shall see shortly, there are special verbs, such as the verb *to be* and *to become,* known as *copular* verbs. Nouns that follow such verbs are known as *predicate nominatives.*

(3) Soccer is *a sport*.

Nouns are sometimes in a special relation to other nouns, further specifying them. When one does so, it is said to be in *apposition* to the noun it is further specifying. Thus, in the next sentence, *my brother* is in apposition to the noun *Paul*.

(4) The instructor gave a book to Paul, *my brother*.

One noun being in apposition to another should be distinguished from one noun being an *objective complement* of another.

(5) We consider them *our friends*.

Thus, in the last sentence, *our friends* is an objective complement; it is not in apposition to *them*.

A noun may also be an *object* of a preposition.

(6.1) The child wrote on *the sofa*.

(6.2) Amy went to *the store*.

Finally, nouns, especially proper nouns, can be used for direct address.

(7)   *Helen,* please leave the room.

2.1.2   Pronouns

A pronoun is a word that stands for a noun. That for which a pronoun stands is its *antecedent*. Traditional grammar identifies many kinds of English pronouns. They include *personal* pronouns (for example, *I, we, you*), *relative* pronouns (for example, *who, which, that*), *interrogative* pronouns (for example, *what* and *who*), *demonstrative* pronouns (for example, *this, that, these, those*), *indefinite* pronouns (for example, *anyone, somebody*), *reflexive* pronouns (those suffixed with *-self*), and *reciprocal* pronouns (*each other* and *one another*).

English pronouns exhibit differences in grammatical number. They also, in some instances, exhibit differences in gender (for example, *he* versus *she*). The personal pronouns also exhibit differences in what is called grammatical *person*, a difference also found among verbs. Thus, personal pronouns are classified into *first person* (pronouns that include the speaker in its reference, for example, *I* and *we*), *second person* (pronouns that include the person or persons one is speaking to as its reference, for example, *you*), and *third person* (pronouns that refer neither to the speaker or to the person or persons being spoken to, for example, *he, she,* and *they*). (The technical term *person* is explained and illustrated in detail in chapter 4, section 2.1.)

SINGULAR	FIRST	SECOND	THIRD		
			MAS	FEM	NEU
NOMINATIVE	I	you	he	she	it
OBJECTIVE	me	you	him	her	it
POSSESSIVE	my/mine	your/yours	his	her/hers	its

English pronouns have many of the same uses as nouns. However, in addition, some have a special use from which their name derives: they stand in for other nouns, called their *antecedents*. However, not all pronouns have antecedents; only relative pronouns, demonstrative pronouns, and the third-person personal pronouns do. (This topic will be discussed in detail in chapter 4, section 3.)

2.1.3   Verbs

A verb is a word or a group of words that expresses an action, a condition, or a state. English verbs fall into four major subcategories: *intransitive, transitive, copular* (also known as *linking verbs*), and *auxiliary*. Intransitive verbs are ones that can form a sentence without any nouns following them.

(8.1)   Birds *fly*.

(8.2)   The woman *worked* rapidly.

In contrast, transitive verbs require a noun to follow them.

(9.1)   We *saw* the fire.

(9.2)   The man *threw* the ball.

Some transitive verbs require two nouns to follow them.

(10.1)   He *made* the boy a kite.

(10.2)   He *made* a kite for the boy.

The second noun to follow is the *direct object* and the first is the *indirect object*. Sentences such as the one in (10.1) alternate with those in which the direct object follows the verb and the indirect object follows a preposition that itself occurs after the direct object, as shown in (10.2). Finally, some transitive verbs require an *objective complement*.

(11.1)   The class *elected* Mary president.

(11.2)   We *consider* John a fool.

Some verbs can be both transitive and intransitive.

(12.1)   They *broke* the glass.

(12.2)   The glass *broke*.

(13.1)   They *dropped* the curtain.

(13.2)   The curtain *dropped* suddenly.

Copular verbs, or linking verbs, differ from intransitive and transitive verbs. Intransitive verbs do not require any noun to occur after them, whereas transitive verbs do. *Copular verbs* are verbs that can be followed by an adjective instead of a noun.

(14.1)   The table *is* sturdy.

(14.2)   Every child *became* sad.

(14.3)   The actress *looks* weary.

(14.4)   He *seems* tall.

Some copular verbs may also be followed by a noun.

(15.1)   A table *is* a piece of furniture.

(15.2)   The woman *became* an actress.

Finally, we come to *auxiliary,* or *helping,* verbs. They include, for example, such verbs as *to be, to have, may, can, must, will, shall.*

(16.1) I *must* leave.

(16.2) The soldier *will* be brave.

(16.3) You *should have been* sleeping.

(16.4) I *do* not know.

English verbs, like English pronouns, are distinguished both by person and number. They also exhibit *tense* and *aspect.*

Person and number are most clearly exemplified by the verb *to be.*

to be	SINGULAR	PLURAL
FIRST:	(I) *am*	(we) *are*
SECOND:	(you) *are*	(you) *are*
THIRD:	(he) *is*	(they) *are*

Verbs often express actions. The form of a verb indicates a relation between the time of the action expressed and the time the verb is uttered. Thus, the following three sentences differ insofar as the action expressed precedes, is simultaneous with, and succeeds the time of the sentence's utterance.

TENSE
PAST   John *heard* the bell.
PRESENT   John *hears* the bell.
FUTURE   John *will hear* the bell.

The verbs of these sentences are said to be in the *past, present* and *future* tenses, respectively.

Aspect pertains to whether or not the action expressed by the verb is ongoing or not. English distinguishes between *progressive,* or *continuous,* and *perfect.*

PAST PROGRESSIVE   John *was listening* to the bell.
PRESENT PROGRESSIVE   John *is listening* to the bell.
FUTURE PROGRESSIVE   John *will be listening* to the bell.

PAST PERFECT   John *had heard* the bell.
PRESENT PERFECT   John *has heard* the bell.
FUTURE PERFECT   John *will have heard* the bell.

PRESENT PERFECT PROGRESSIVE   John *has been laughing.*
PAST PERFECT PROGRESSIVE   John *had been laughing.*
FUTURE PERFECT PROGRESSIVE   John *will have been listening.*

Some of the verbs in these examples occur in sentences as single words, for example, *hears* in the sentence *John hears the bell,* while others occur with auxiliary verbs, for example, *will hear* in the sentence *John will hear the bell*. Those verb forms that include auxiliary verbs are said to be *compound,* or *periphrastic,* verb forms, while those without are *simple* verb forms.

English verbs are also distinguished according to what is called *voice*. English has two voices: *active* and *passive*.

(17.1)   *John* wrote the letter.

(17.2)   *They* saw the prime minister.

(17.3)   *The explosion* wounded the soldier.

The verbs in these sentences are in the active voice. Here are the same sentences, but put in the passive voice.

(18.1)   The letter was written by *John*.

(18.2)   The prime minister was seen by *them*.

(18.3)   The soldier was wounded by *the explosion*.

In English, verbs in the passive voice require the auxiliary verb *to be*.

Verb tense forms are classified by *mood:* the indicative mood, the imperative mood, and the subjunctive mood. These classes are roughly correlated with the force of the clause in which they occur. It is said that when the tense of the verb of a clause is in the indicative mood, the clause purports to express a factual state of affairs.

(19.1)   Bill *runs* five miles every day.

(19.2)   The guard *spotted* Bill.

(19.3)   The instructor *will give* Bill a book.

When the tense of the verb of a clause is in the imperative mood, the clause expresses a command. The form of verbs in the imperative mood is simple: with the exception of the imperative of the verb *to be,* the form is just that of the first-person singular, present, active. The imperative of the verb *to be* is *be*.

(20.1)   Please *come* here and *sit* down.

(20.2)   *Be* punctual.

(20.3)   *Have* another piece of pie.

Finally, when the tense of the verb of a clause is in the subjunctive mood, the clause purports to express a nonfactual state of affairs, including states of affairs that are wished for. The form of the present subjunctive is the same as that of the imperative.

(21.1) It is essential that the mission *succeed*.

(21.2) The director insists that you *be* present at the meeting.

(21.3) If I *were* you, I would not go.

The form of verbs discussed so far are known as *finite* forms. They are the forms that verbs in simple sentences must have. However, verbs are also liable to *nonfinite* forms. These include *infinitives, participial,* and *gerundial* forms. English verbs have six infinitival forms, distinguished from one another by the features of tense, aspect, and voice. The six infinitival forms of the verb *to stop* follow.

ACTIVE	PRESENT	PERFECT
SIMPLE	*to stop*	*to have stopped*
PROGRESSIVE	*to be stopping*	*to have been stopping*
PASSIVE		
SIMPLE	*to be stopped*	*to have been stopped*

The infinitival uses are often identified by other word categories. Thus, *nominal* infinitives.

NOMINAL	*To run* is healthy.
	Compare: *Exercise* is healthy.
ADJECTIVAL	Here is water *to drink*.
ADJECTIVAL	Here is water *fresh from the spring*.
ADVERBIAL	We came *to eat*.
	Compare: We came *quickly*.

Notice that all forms include the preposition *to*. It is sometimes omitted after certain verbs.

(22.1) I dare (to) *go*.

(22.2) I heard him *scream*.

(22.3) We let him *talk*.

Participles, like infinitives, are distinguished from one another by tense, aspect, and voice.

	ACTIVE	PASSIVE
PRESENT	*stopping*	*being stopped*
PAST	*stopped*	—
PERFECT	*having stopped*	*having been stopped*

Participles are formed with either the suffix *-ing* or the suffix *-ed*. They are attached either directly to the stem of the verb or to the stem of an auxiliary verb.

Participles are used like adjectives, as illustrated here.

ADJECTIVE   The *running* water must be turned off.
            Compare: The *cold* water must be turned off.

            The *faded* coat was given away.
            Compare: The *red* coat was given away.

The third nonfinite verbal form is the *gerund*. It is formed in the same way as the present, active participle, namely, with the suffix *-ing*. Gerunds are said to be verbal nouns, that is, they are forms of verbs that permit them to be used as nouns.

SUBJECT   *Swimming* is forbidden.
                      Compare: *Food* is forbidden.
OBJECT   The boy enjoys *swimming*.
                      Compare: The boy enjoys *movies*.
PREPOSITIONAL OBJECT The boy was arrested for *swimming*.
                      Compare: The boy was arrested for *theft*.

2.1.4 Adjectives

An *adjective* was defined as a word or group of words expressing a quality. An adjective may occur before a noun, in which case it is said to be its *modifier,* after a copular verb, in which case it is said to be a *predicate,* or as an objective complement.

MODIFIER   The *large* apple is on the table.
PREDICATE   The apple is *large*.
OBJECTIVE COMPLEMENT Bill considers the house *large*.

Some, but not all, adjectives can be suffixed with *-er* or *-est*. These forms are said to be the *comparative* and *superlative* forms.

COMPARATIVE   This table is *heavier* than that desk.
SUPERLATIVE   That is the *heaviest* piece of furniture here.

Other adjectives have periphrastic comparative and superlative forms.

COMPARATIVE   This table is *more attractive* than that one.
SUPERLATIVE   That is the *most attractive* piece of furniture here.

One special subcategory of adjectives is *pronominal adjectives*.

PERSONAL   mine, ours, yours, his, theirs, hers
RELATIVE   which
DEMONSTRATIVE   this, that
INTERROGATIVE   which
INDEFINITE   some, every

They can occur in the very same positions as other adjectives.

MODIFIER	*Every* book is on the table.
PREDICATE	The apple is *mine*.
OBJECTIVE COMPLEMENT	Bill considers the house *his*.

Unlike adjectives in many other languages, English adjectives do not vary for number, gender, or case.

2.1.5 Adverbs

An *adverb* was defined as a word or group of words that modifies a verb, an adjective, or another adverb. Some adverbs are formed by the suffix *-ly* added to an adjective. Some have the very same form as an adjective.

	ADJECTIVE	ADVERB
SUFFIX	quick	quickly
	poor	poorly
NO SUFFIX	fast	fast
	early	early

Here are examples of such adverbial modification.

VERBAL MODIFICATION	The visitor spoke *fluently*.
ADJECTIVAL MODIFICATION	He was *nearly* frantic.
ADVERBIAL MODIFICATION	He spoke *very* quickly.

In their modification of verbs, they can express various things about the action expressed by the verb such as its manner, its place, or its time of occurrence. In their modification of adjectives and adverbs, they can express the degree attained by the quality expressed by the adjective or adverb.

MANNER	Colleen ran *swiftly*.
PLACE	Maria is traveling *abroad*.
TIME	Bill will arrive *soon*.
DEGREE	Fred is *very* quick.
	Fred ran *very quickly*.

2.1.6 Prepositions

A *preposition* is defined as a word that shows a relation between its object and another word in the sentence. The object of a preposition is a noun.

(23.1) The capital *of* France is Paris.

(23.2) *Until* Monday, no one knew the answer.

(23.3) They worked *in* the house all day.

Prepositional phrases are units comprising a preposition and its object. Prepositional phrases can modify nouns and verbs. In that way, they are like adjectives and adverbs, respectively.

ADJECTIVAL MODIFICATION	A man *with a beard* walked into the room.
	A *bearded* man walked into the room.
ADVERBIAL MODIFICATION	Bill left *in a hurry*.
	Bill left *hurriedly*.

Prepositions can be used as adverbs, in which case they have no objects.

(24.1) The horse fell *down*.

(24.2) Come *in*.

(24.3) The suspect turned *around*.

2.1.7 Conjunctions

A *conjunction* is said to be a word that connects two sentences or two parts of a sentence. They are distinguished into *coordinating* and *subordinating* conjunctions. The coordinating conjunctions include the words *and, but,* and *or*.

(25.1) It rained *and* it snowed.

(25.2) Mary left *but* Sue stayed.

(25.3) You add wine vinegar *or* balsamic vinegar.

SUBORDINATING	until, because, if, although,
	when, while, after, before

(26.1) *Before* there is any more confusion, let us depart.

(26.2) The watchman remained at his post *until* it was light.

(26.3) *If* it does not work, read the instructions.

2.1.8 Articles

Articles were not included among the parts of speech listed previously. The category was omitted from Latin grammars, since Latin has no articles. The category was included among Greek grammars, since Classical Greek does have articles. Classical Greek grammars defined an *article* as a word inflected for case and occurring either before or after a noun. In English, the article, if it occurs with a noun, occurs only before it. English has two kinds of articles, the *definite* and the *indefinite*. The word *the* is the definite article, and the word *a* is the indefinite article.

2.1.9 Interjections

An *interjection* is a word or group of words that expresses some feeling in an exclamatory fashion. Examples are *alas, oh, well*.

2.2 Clauses

In addition to categorizing the words of a language, traditional grammar also has categorized clauses. Consider the following sentences:

(27.1) It rained.

(27.2) Bill walked his dog.

(27.3) He talked on his cellular phone.

These sentences can be compounded to form more complex sentences, which will be called *compounded clauses*.

(28.1) Bill walked his dog and he talked on his cellphone.

(28.2) Bill walked his dog, while it rained.

Sentences that are parts of sentences are called *clauses*. Sentences can be *monoclausal* or *multiclausal*. The sentences in (27) are monoclausal and those in (28) are multiclausal. Typically, a sentence comprises as many clauses as it contains verbs in finite forms.

Compounded clauses may be coordinated with one another, as are the two clauses in sentence (28.1), or one clause may be *subordinate* to the other, as is the second clause in sentence (28.2). The other clause in the sentence is usually referred to as the *main* clause or *principal* clause. One might also call it the *superordinate* clause. One and the same clause can be both superordinate and subordinate.

(29) If Bill arrives while Alice is here, Carl will leave.

Thus, in the previous sentence, the clause *Bill arrives* is subordinate to the clause *Carl will leave* and superordinate to the clause *Alice is here*.

Traditional grammar categorized subordinate clauses by an affinity these clauses bear to various word categories. Thus, there are noun, or nominal, clauses, adjectival clauses, and adverbial clauses.

Noun clauses are clauses that bear an affinity to nouns, that is, they appear where nouns may appear. Nouns appear in the subject position, and they appear in the object position.

Thus, the subject of the first sentence that follows can be replaced by clauses:

(30.0) *Bill* bothers me.

(30.1) *That Bill behaved badly* bothers me.

(30.2) *What Bill did* bothers me.

Next come clauses that can appear in the object position.

(31.0) Dan believes *Mary*.

(31.1) Dan believes *that Peter gave Mary a present*.

(31.2) Dan knows *Mary*.

(31.3) Dan knows *whether Peter gave Mary a present*.

Adjectival clauses are so-called because, like adjectives, they occur next to nouns, and like adjectives, they modify them. Adjectival clauses are more often referred to as relative clauses, since they are typically introduced by relative pronouns.

(32.0) Alan saw a man.

(32.1) Alan saw the man *who gave Mary a present*.

(32.2) Alan saw the present *which Peter gave Mary*.

(32.3) Alan saw the woman *to whom Peter gave a present*.

Adjectival clauses need not begin with a relative pronoun. They may begin with the word *that* or with no special word at all.

(33.0) Alan saw a man.

(33.1) Alan saw the present *Peter gave Mary*.

(33.2) Alan saw the woman *Peter gave a present to*.

Still another kind of clause is an adverbial clause. Such clauses are usually sub-classified by the kinds of adverbial modification they effect. Thus, there are temporal adverbial clauses, manner adverbial clauses, adverbial clauses of reason, adverbial clauses of purpose, and so on.

(34.0) The athlete showered *at five o'clock*.

(34.1) The athlete showered *after he had played the game*.

(34.2) The athlete showered *as he was directed*.

(34.3) Their friends came *because they had been invited*.

(34.4) The boy cried *so that someone would come to help him*.

(34.5) Immigrants live frugally *in order that they may save money*.

Another kind of clause identified by traditional grammar is the comparative clause. As illustrated in (35), the comparative clause is introduced by *than*.

(35) Plato wrote more dialogues than *Shakespeare wrote plays*.

Having identified clauses as sentences with finite verb forms, one might wonder about clause-like units that contain nonfinite verb forms: infinitives, participles, and gerunds. Traditionally, they have been called *phrases*. In many ways, they are like clauses, except that they are missing a subject. Here are examples of infinitival phrases.

(36.1) They went *to buy a hat*.
(36.2) *To die for one's country* is every soldier's duty.
(36.3) Dan believes Peter *to have given a present to Mary*.
(36.4) Dan knows whether *to give a present to Mary*.
(36.5) Dan wonders whether *to give a present to Mary*.
(36.6) I saw a present *to give to Mary*.
(36.7) I saw a woman *to whom to give this present*.
(36.8) The child cried out *to get help*.
(36.9) Immigrants live frugally *to save money*.

Next we have two examples of participial phrases.

(37.1) John spoke to the man *giving the present to Mary*.
(37.2) Bill saw someone *giving the present to Mary*.

And finally, we have two examples of gerundial phrases.

(38.1) *Bill's giving a present to Mary* surprised everyone.
(38.2) *Giving a present to someone* is better than receiving one.

It is a peculiarity of English that the suffix *-ing* is used both for the gerund form of a verb and for the present participle form of a verb.

This brings to an end our brief overview of traditional English grammar. We now turn to a discussion of its limitations and a presentation of the ways in which it has been improved by contemporary linguistic theory.

2.3 The Limitations of Traditional English Grammar

While traditional English grammar tells us much of importance about English, it is clearly not the whole story. First, as we learned in chapter 1, natural language expressions have a recursive structure. Nothing in traditional English grammar attempts to characterize the recursive structure of English expressions. Neither does anything in traditional grammar attempt to show how the meaning of smaller expressions contributes to the meaning of larger expressions. In addition, traditional English grammar typically does not have clear and systematic criteria whereby it characterizes its fundamental concepts. To illustrate this last point, we shall very briefly consider traditional grammar's definition of parts of speech, perhaps its most fundamental concepts.

To begin with, one would expect the characterizations of the various lexical categories to be uniform. However, some definitions characterize the part of speech by what seem to be grammatical relations the part of speech defined bears to other parts of speech, other definitions rely on what kind of thing the part of speech denotes, and still others rely on both.

Let us turn first to the characterizations that depend on grammatical relations. First, adverbs are characterized as modifying other adverbs and adjectives, but what is the relation of modification? Prepositions are said to express a relation between two things, but so do nouns such as *parent, friend, capital,* and *boss;* adjectives such as *averse, contingent, dependent, fond,* and *incumbent;* and verbs such as *abandon, catch, greet, like,* and *pursue.* In fact, even subordinating conjunctions such as *before, after,* and *because* express a relation between what is expressed by the clauses it connects. A pronoun is said to stand for a noun, but this is true only of the third-person personal pronouns such as *he, she, it,* and *they* but not of the first- and second-person personal pronouns. Conjunctions are characterized as connecting sentences as well as parts of sentences. But prepositions also connect parts of sentences. Indeed, what is it to connect parts of sentences? Finally, nouns, adjectives, adverbs, and verbs may be single words or groups of words. But not any sequence of words group together to form a noun, an adjective, or a verb.

Other characterizations depend on ontological distinctions. For example, a verb may denote an action and a noun an activity. But how is an action different from an activity? A verb may denote a state and an adjective a quality. But how is a state different from a quality? A noun and a verb may denote a condition. But how is a condition different from a state? Moreover, surely the words *peace, war, famine, drought,* and *armistice* denote states. However, they have all the grammatical properties of nouns, and none of the grammatical properties of verbs. At the same time, an adjective denotes a quality. While the words *young, sad, silly, beautiful,* and *intelligent* have the grammatical properties of other adjectives, the words *youth, sadness, silliness, beauty,* and *intelligence,* which seem to denote the same qualities, have the grammatical properties of nouns. Indeed, without some characterization of what these entities are and how they are to be distinguished from one another, they are of little or no help in deciding on parts of speech.

As we shall see in section 3, immediate constituency analysis, a purely linguistic form of analysis, fares much better, though by no means flawlessly, in helping to distinguish the parts of speech for languages.

Exercises: Limitations of traditional english grammar

Find five English words not listed in the textbook that denote either an event, a state, or a condition, and yet, from the point of view either of morphology or of syntax, pattern only with nouns. If your first language is not English, provide five such words from your first language, making clear what the grounds are for your maintaining the words you choose are nouns.

3 Syntax of English

The systematic study of language has exploited three techniques: segmentation, classification, and substitution. Long before Pāṇini's grammar of Classical Sanskrit, the *Aṣṭādhyāyī*, had appeared, his predecessors had figured out how to segment sentences into words and words into roots and endings and how to classify the resulting segments into classes of intersubstitutable expressions. The grammar comprises lists, rules, and definitions that result from this analysis of Sanskrit. European linguists and American structuralist linguists of the early twentieth century used the kinds of rules found in the *Aṣṭādhyāyī* to state the phonetic and phonological regularities of other languages and to state morphological and syntactic regularities of languages that have either little or no morphological complexity or morphological complexity very different from that found in Classical Sanskrit or other Indo-European languages.

Leonard Bloomfield's very important and influential book *Language,* published in 1933, communicated the basic ideas underlying these innovative applications to a broad audience of linguists. Within a quarter century of its publication, Bloomfield's successors—Bernard Bloch (1907–1965), Zellig Harris (1909–1992), Charles Hockett (1916–2000), Eugene Nida (1914–2011), Rulon Wells (1918–2008), among others—had developed and refined the techniques and had applied them to the analysis of a number of different languages. The form of analysis they practiced came to be known as *immediate constituency analysis*. It can be characterized as follows: To analyze a sentence, one must, as Bernard Bloch (1946, 204–205) put it, "first isolate the immediate constituents of the sentence as a whole, then the constituents of each constituent, and so on to the ultimate constituents." This means segmenting a complex expression into subexpressions, typically two, which themselves are further segmented into still less complex subexpressions, until only simple, or minimal, expressions remain. Second, one allocates each expression obtained from successive segmentation to a set of expressions with which it is intersubstitutable while preserving acceptability. And finally, one assigns distinct categories to the distinct sets of intersubstitutable expressions.

In section 3.1, we will apply the techniques of immediate constituency analysis to a sample of acceptable English expressions. What will emerge from this analysis is a number of observational regularities. In section 3.2, we shall give a formal characterization of these

Basic English Grammar

regularities, which we shall call a constituency grammar of English. This grammar will bring to light still further regularities, some of which fall within the ambit of constituency grammar and are discussed in section 3.3, and others of which that do not and are discussed in section 3.4.

3.1 Immediate Constituency Analysis

In this part, we shall undertake an immediate constituency analysis of a selected set of acceptable English expressions. Specifically, we shall segment acceptable English expressions into subexpressions, investigate their capacity for intersubstitutability, and assign them categories. In many cases herein, more than one segmentation of an expression is consistent with the data we will discuss. The presentation here is illustrative, not conclusive. Indeed, modifications to some of the specific immediate constituency analyses will be made in later chapters.

We start with the following simple, acceptable English sentence:

(39) Albert laughed.

It clearly comprises just two words: *Albert* and *laughed*. Let us ponder what other expressions can replace either of these two words and still yield an acceptable English sentence. We begin with the left-hand word *Albert*. Let us assign it the category N_p. There are other words that can replace *Albert* in sentence (39) and yield acceptable English sentences. We list some of them here, assigning them the same category.

N_p Albert, Beverly, Carl, Dan, Eric, Francine, Galileo, Ken ...

Put another way, each word listed under the category N_p may replace α in the schema in (40) to yield an acceptable English sentence.

(40) α laughed.

Next, consider another category and a different list of words.

N_c dog, enemy, friend, guest, host, man, park, visitor, woman, yard, ...

These words cannot replace α in the schema in (40) to yield an acceptable English sentence.[3]

(41) *Guest laughed.

That is to say, the words in the second list do not preserve the acceptability when they substitute for *Albert* in (39). However, should the expression *the guest* substitute for *Albert*, the result is an acceptable English sentence.

---

3. An asterisk prefixed to an expression indicates that native speakers of English judge it as unacceptable.

(42) The guest laughed.

As it happens, *the* is not the only word that can precede *guest* so that it, together with *guest*, can replace α in (40) to yield an acceptable English sentence. In fact, any word in the list

Dt   a, each, every, no, some, that, the, this, ...

may replace β in the schema and preserve acceptability.

(43) β guest laughed.

More generally, English expressions resulting from two-word sequences in which the first word is taken from words in the category Dt and the second word is taken from words in the category N_c yield expressions that, when they replace α in (40), themselves yield acceptable English sentences. Thus, we have identified two categories of expressions that can replace α in (40) to yield acceptable English sentences: the words in the category N_p and the expressions of two-word sequences in which the first word is taken from the category Dt and the second from the category N_c. Let us assign the category NP to the set of English expressions that comprise either words of the category N_p or two-word sequences where the first word comes from category Dt and the second word from the category N_c. We contract this rather cumbersome statement into the following compact notation:

NP1   NP $\Rightarrow$ N_p

NP2   NP $\Rightarrow$ Dt N_c

We shall call rules of this kind *constituency analysis rules,* or sometimes just *analysis rules,* since they are rules that relate the category of a constituent expression to the categories of its immediate constituent expressions, obtained through immediate constituency analysis.

Now consider the expression *the tall guest*. Substituted for α in (40), it yields the following acceptable English sentence.

(44) The tall guest laughed.

The word *tall,* as it happens, belongs to a category of word some of whose other members are indicated in this list:

A   friendly, hostile, old, short, surly, taciturn, tall, young, ...

This new category permits us to obtain still another schema.

(45) The γ guest.

All expressions resulting from replacing γ in (45) with words from the list in A yield expressions, all of which can replace α in (40) to yield acceptable English sentences.

Consider another category and a partial list of its words:

Dg   quite, rather, so, somewhat, too, very, ...

The expressions resulting from two-word sequences in which the first word is taken from words in the category Dg and the second word is taken from words in the category A yield expressions that, when they replace γ in (45), themselves yield English expressions that, when they replace α in (40), yield acceptable English sentences. Thus, for example, *rather* is in the category Dg and *tall* is in the category A. The two-word sequence *rather tall* may replace γ in (45) to yield *the rather tall guest,* which, in turn, may replace α in (40) to yield the acceptable English sentence:

(46)   The rather tall guest laughed.

Thus, we have identified two categories of expressions that can replace γ in (45) to yield acceptable English expressions: the words in the category A and the expressions of two-word sequences in which the first word is taken from the category Dg and the second from the category A. Let us assign these expressions the common category AP. In keeping with the notation introduced, we can sum up these results as follows:

AP1   AP $\Rightarrow$ A

AP2   AP $\Rightarrow$ Dg A

In fact, reflecting on sentence (46) and the analysis rules thus far formulated, we see that each sequence of expressions, the first an expression of category Dt, the second an expression of category AP, and the third an expression of category N_c, can also replace α to yield an acceptable English sentence. This leads to the following analysis rule:

NP3   NP $\Rightarrow$ Dt AP N_c

In a similar vein, consider sentence (47)

(47)   The guest with a dog laughed.

Note that the expression *a dog* belongs to the category NP, since the word *a* belongs to Dt and the word *dog* belongs to N_c and *a* precedes *dog*. At the same time, the expression *with* in the expression *with a dog* can be replaced by words in the following list:

P   by, near, of, on, with, ...

to form other expressions that are intersubstitutable with the expression *with a dog*. Let us call this category of expression PP. We use the categories P and PP to formulate the first analysis rule PP1. Bearing in mind sentence (47), we see that each sequence of expressions, the first an expression of category Dt, the second an expression of category N_c, and the third an expression of category PP can also replace α to yield an acceptable English sentence. This leads to the second analysis rule (NP4).

PP1   PP $\Rightarrow$ P NP

NP4   NP $\Rightarrow$ Dt N_c PP

So far we have concentrated on what can be substituted for the left-hand word in sentence (39). Let us now turn our attention to what can be substituted for the right-hand word. To do so, we consider another schema derived from sentence (39)

(48)   Albert δ.

Any word from the list V_i may replace δ to yield an acceptable English expression.

V_i barked, cried, fell, laughed, slept, sat, walked, yelled, ...

But expressions of the category V_i are not the only expressions that are acceptable immediately after an expression of the category NP. Any of the words from the category V_t,

V_t abandoned, caught, chased, greeted, insulted, saw, watched, ...

can replace ϵ in the following

(49)   Albert ϵ the host

to yield acceptable English sentences. Moreover, we note that the expression *the host* is an expression belonging to the category NP. Expressions formed by a word in the category V_t followed by an expression of the category NP are intersubstitutable with words of the category V_i. Again, two analysis rules state this intersubstitutability of expression. We shall assign that category of expression VP.

VP1   VP $\Rightarrow$ V_i

VP2   VP $\Rightarrow$ V_t NP

Still other expressions belong to the category VP. Expressions of the category V_i can be succeeded by expressions of the category PP. The expression *near the dog,* for example, belongs to the category PP and *walked* belongs to the category V_i. Together, they yield *walked near the dog,* which can certainly substitute for *laughed* in (39) to yield an acceptable expression, namely, *Albert walked near the dog.* Similarly, expressions whose first expression is a word from the category V_t, whose second expression is from the category NP, and whose third is from the category PP are also expressions of the category VP. Acceptable expressions of this form include *greeted the host in the kitchen.*

Basic English Grammar

These possible combinations of substitutions are summarized by the following analysis rules:

VP3   VP $\Rightarrow$ V$_i$ PP

VP4   VP $\Rightarrow$ VP PP

So far we have identified many categories of expression within a sentence. But we did not identify the sentence itself as a category. It is easy to see that such a category is required. First, obviously, replacing an acceptable sentence with an acceptable sentence yields an acceptable sentence. More interestingly, however, is the fact that sentences occur within sentences.

Consider, for example, the next schema, derived from the acceptable sentence *The woman thought Albert laughed*. We notice that following *thought* is an entire sentence.

(50)   The woman thought η.

Each of the sentences sanctioned by these analysis rules is an acceptable English sentence and each can replace η in (50) to yield another acceptable English sentence. The word *thought* is just one of a set of words that we place in the category V$_s$.

V$_s$ believed, hoped, knew, noticed, thought, ...

We now state two more analysis rules:

S1   S $\Rightarrow$ NP VP

VP5   VP $\Rightarrow$ V$_s$ S

The first says that each expression that is a sentence can be replaced by an expression belonging to the category NP followed by one belonging to the category VP to yield a sentence. The second says that each expression belonging to the category VP can be replaced by one belonging to the category V$_s$ followed by one belonging to the category S.

It is now time to take stock of what we have done. We began with a simple acceptable English sentence, *Albert laughed,* and we considered expressions that can be substituted for each of the two words yielding other acceptable English expressions. Some of the substituted expressions were single words, other expressions made up of words. We placed into categories expressions that were intersubstitutable with respect to a fixed expression. Categories that comprise only single words of English are known as *lexical categories*. Thus far, we identified nine lexical categories: namely, A, P, Dt, Dg, N$_c$, N$_p$, V$_i$, V$_t$, and V$_s$. In addition, we saw that words are grouped into larger expressions that are paired with other categories. These nonlexical categories include: NP, VP, PP, AP, and S.

These categories are organized by the analysis rules. Here are the analysis rules set out previously:

S1   S $\Rightarrow$ NP VP
NP1   NP $\Rightarrow$ N_p
NP2   NP $\Rightarrow$ Dt N_c
NP3   NP $\Rightarrow$ Dt AP N_c
NP4   NP $\Rightarrow$ Dt N_c PP
PP1   PP $\Rightarrow$ P NP
AP1   AP $\Rightarrow$ A
AP2   AP $\Rightarrow$ Dg A
VP1   VP $\Rightarrow$ V_i
VP2   VP $\Rightarrow$ V_t NP
VP3   VP $\Rightarrow$ V_i PP
VP4   VP $\Rightarrow$ VP PP
VP5   VP $\Rightarrow$ V_s S

What we have seen is that a complex expression, such as

(51)   The friendly visitor greeted the surly host.

belongs to a category, and the complex expression can be segmented into immediate constituent expressions each of which is assigned a category. These latter immediate constituent expression themselves can each be segmented into their respective immediate constituent expressions, to each of which is assigned a category. Each resulting immediate constituent expression can be further segmented and categorized until finally one arrives at simple words. In the case of sentence (51), one arrives at the following list of immediate constituents and their categories:

(52)   S   the friendly visitor greeted the surly host
       NP   the friendly visitor
       Dt   the
       AP   friendly
       A   friendly
       N_c visitor
       VP   greeted the surly guest
       V_t greeted
       NP   the surly guest
       Dt   the
       AP   surly
       A   surly
       N_c guest

Basic English Grammar

While this arrangement is accurate, it is not very revealing. Here is another arrangement of the same information that better exploits the fact that every analysis of an expression into its immediate constituent expressions segments the expression into its immediate constituents.

(53) $[_S [_{NP} [_{Dt} \text{ the}] [_{AP} [_A \text{ friendly}]] [_{N_c} \text{ visitor}]] [_{VP} [_{V_t} \text{ greeted}] [_{NP} [_{Dt} \text{ the}] [_{AP} [_A \text{ surly}]] [_{N_c} \text{ host}]]]]$

Thus, the left bracket labeled S in (53) and its partner, the last right bracket in (53), correspond to the first constituent and its category as listed in (52). The first left bracket labeled NP in (53) and its partner, the fifth right bracket, correspond to the second constituent and its category as listed in (52). In short, there is a bijection between each pair consisting of a constituent and its category in (52), on the one hand, and each pair of labeled matching brackets in (53), on the other. This notation is known as *labeled bracketing*.

In section 3.2, we shall show that the immediate constituency analysis brings to light a mathematical structure, which we will define and explain.

Exercises: Immediate constituency analysis

1. Use immediate constituency analysis to analyze each of the following sentences. Present the analysis using labeled brackets. In each case, list the analysis rules you use.

(a) The host slept.

(b) Each old dog barked.

(c) The visitor walked in the park.

(d) An old dog chased Dan.

(e) The very tall guest saw the host.

(f) Dan knew the guest laughed.

(g) Beverly thought the woman sat in the yard.

2. Use immediate constituency analysis to analyze each of the following sentences. Present the analysis using labeled brackets. In addition, list the analysis rules you use. If the rule is one listed in section 3.1, simply give its name. If the rule is one you have hypothesized, write it out as an analysis rule. You may have to hypothesize new categories for the words in the sentence. Also, show to which category each new word belongs by writing down the word and its category name, putting a vertical bar between.

(a) Aristotle was a philosopher.

(b) The child seemed sleepy.

(c) This boy lost his wallet.

(d) Albert grew a tomato.

(e) The soldier grew sad.

(f) Alice gave the pen to Mary.

(g) Carl sat in the first chair.

(h) Carl sat in the chair behind the first chair.

(i) The city of Paris is rather beautiful.

(j) The comedian appeared in a suit.

(k) Galileo persuaded the philosopher of his mistake.

(l) No director will approve of every proposal.

(m) Galileo persuaded the patrician he should leave.

(n) An expert will speak to this class about the languages of India.

3.2 Constituency Grammar

We shall now give a formal characterization of grammatical expressions of English based on immediate constituency analysis. As recognized by Pāṇini, the aim of a grammar is to determine all and only the grammatical expressions of a language. The immediate constituency analysis carried out in section 3.1 suggests that the grammatical English expressions are those assigned a category by the constituency analysis rules and that all other expressions are ungrammatical English expressions.

Clearly, each word of English is a grammatical expression of English. Following another insight of Pāṇini, we shall assume that one of the starting points for the grammar of a language is an inventory of the words of the language in which each word is assigned a basic category. In contemporary linguistic parlance, such an inventory is called a *lexicon*. At a bare minimum, a lexicon is a set of pairs in which each word of a language's lexicon is paired with its category. Such pairs are called *lexical entries*.

In section 3.1, we associated lists of words with various lexical categories. Here are four of those lists:

Dt	a, each, every, no, some, that, the, this ...
N_c	dog, enemy, friend, guest, host, park, visitor, woman, yard ...
A	friendly, hostile, old, short, surly, taciturn, tall, young, ...
V_t	abandoned, caught, chased, greeted, insulted, saw, watched, ...

Based on these lists, we can formulate lexical entries for all the words assigned to a category. In particular, *the* and *a* have been assigned the lexical category Dt, *host* and *guest* the lexical category N_c, *friendly* and *surly* the lexical category A, and *greeted* the lexical category V_t. Using set theoretic notation for pairs, one writes the lexical entries for these words as follows: ⟨*the*,Dt⟩, ⟨*a*,Dt⟩, ⟨*guest*,N_c⟩, ⟨*host*,N_c⟩, ⟨*friendly*,A⟩, ⟨*surly*,A⟩,

Basic English Grammar

and $\langle greeted, V_t \rangle$. Anticipating difficulties of readability, we shall deviate from standard set theoretical notation and annotate the preceding pairs slightly differently: *the*|Dt, *a*|Dt, *guest*|N_c, *host*|N_c, *friendly*|A, *surly*|A, and *greeted*|V_t.

ADAPTED NOTATION	STANDARD SET THEORETICAL NOTATION
the|Dt	$\langle the, \text{Dt} \rangle$
a|Dt	$\langle a, \text{Dt} \rangle$
guest|N_c	$\langle guest, N_c \rangle$
host|N_c	$\langle host, N_c \rangle$
friendly|A	$\langle friendly, A \rangle$
surly|A	$\langle surly, A \rangle$
greeted|V_t	$\langle greeted, V_t \rangle$

The question now arises: how does one assign a category to expressions that are not words? In immediate constituency analysis, we segment each complex expression, to which a category is assigned, into its immediate constituent expressions and assign to each of the immediate constituent expressions a category. This procedure continues until one reaches expressions that cannot be further analyzed. We now wish to do the inverse: to start with expressions that have no immediate constituent expressions, or words, and to synthesize, as it were, complex expressions from them. This can be done by applying the analysis rules in reverse.

Here is a simple example. Consider the complex expression *the guest,* which can be segmented into the two immediate constituent expressions, *the* and *guest*. If the category NP is assigned to the expression *the guest,* then, using the analysis rule NP2, or NP $\Rightarrow$ Dt N_c, and the segmentation just specified, one assigns the category Dt to the left-hand immediate constituent *the* and the category N_c to the right-hand immediate constituent *guest*. Inversely, if the expression *the* has the category Dt and the expression *guest* has the category N_c, then using the same rule, but in the inverse direction, one assigns the category NP to the expression resulting from appending the expression *guest* to the expression *the*.

In general, just as one can use an analysis rule, reading it from left to right, to analyze a complex expression with its category into its immediate constituent expressions, each of the latter with its category, so one can use the same rule, reading it from right to left, to synthesize a complex expression with its category from its immediate constituent expressions, each of the latter again having its category. Equivalently, rather than using a constituency analysis rule and reading it from right to left in order to form complex expressions from their immediate constituent expressions, one can instead invert all the rules and read the resulting rules from left to right. We shall call the inverse of a constituency analysis rule a *constituency synthesis rule*. Such rules are closely related to the formation rules used in section 2.2 of chapter 1. To reduce the risk of confusion of analysis rules with synthesis rules, we shall continue to use a double-shafted arrow ($\Rightarrow$) in the notation for analysis rules and a single-shafted arrow in the notation for synthesis rules ($\rightarrow$). The synthesis rules corresponding to the analysis rules introduced in section 3.1 follow:

	ANALYSIS RULES	SYNTHESIS RULES
S1	S $\Rightarrow$ NP VP	NP VP $\rightarrow$ S
NP1	NP $\Rightarrow$ N_p	N_p $\rightarrow$ NP
NP2	NP $\Rightarrow$ Dt N_c	Dt N_c $\rightarrow$ NP
NP3	NP $\Rightarrow$ Dt AP N_c	Dt AP N_c $\rightarrow$ NP
NP4	NP $\Rightarrow$ Dt N_c PP	Dt N_c PP $\rightarrow$ NP
PP1	PP $\Rightarrow$ P NP	P NP $\rightarrow$ PP
AP1	AP $\Rightarrow$ A	A $\rightarrow$ AP
AP2	AP $\Rightarrow$ Dg A	Dg A $\rightarrow$ AP
VP1	VP $\Rightarrow$ V_i	V_i $\rightarrow$ VP
VP2	VP $\Rightarrow$ V_t NP	V_t NP $\rightarrow$ VP
VP3	VP $\Rightarrow$ V_i PP	V_i PP $\rightarrow$ VP
VP4	VP $\Rightarrow$ VP PP	VP PP $\rightarrow$ VP
VP5	VP $\Rightarrow$ V_s S	V_s S $\rightarrow$ VP

Let us see how the synthesis rules can be applied to form sentence (52), namely, *the friendly visitor greeted the surly host,* and to assign to it the category S. The lexicon provides the following lexical entries: *friendly*|A and *surly*|A. The synthesis rule AP1 yields two pairs: the pair *friendly*|AP and the pair *surly*|AP. From the pair *friendly*|AP and the lexical entries *the*|Dt and *visitor*|N_c, the synthesis rule NP3 yields the pair *the friendly visitor*|N_c. Similarly, the synthesis rule NP3 applies to the lexical entries *the*|Dt and *host*|N_c and the pair *surly*|AP to yield the pair *the surly host*|NP. Another lexical entry is the pair *greeted*|V_t. The synthesis rule VP2 yields this pair: *greeted the surly host*|VP. Finally, the synthesis rule S1 allows one to form the pair: *the friendly visitor greeted the surly host*|S.

The foregoing can be displayed with the help of a tree diagram. A tree diagram used to display the formation of an expression in accordance with synthesis rules will be called a *labeled synthesis tree diagram,* or just a *synthesis tree* for short. The top nodes in the synthesis tree are labeled with lexical entries. A labeling pair at any other node is obtained from the labeling pairs at the nodes immediately above it as follows: the left-hand member of the lower pair is obtained by appending in left to right order the first member of the upper pairs, while the right-hand member of the lower pair is the category symbol to the right of the arrow of the synthesis rule whose category symbols to the left of the arrow corresponds to category symbols that are second members of the upper pairs, the left to right order of the category symbols in the rule being the same as the left to right order of the category symbols in the upper pairs. Though this procedure sounds complicated, reviewing it bearing in mind the following synthesis tree should make it clear.

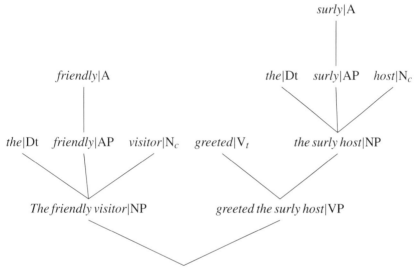

This tree diagram bears a close resemblance to the tree diagrams used in section 2.2 of chapter 1, which were used there to show the application of formation rules in the generation of expressions in SL.

This synthesis tree can be simplified. Each expression at a node that is not a top node arises from the concatenation of the expressions of the nodes immediately above it, where the concatenation starts with the expression of the left-most node and ends with the expression at the right-most node. Thus, once one is given the expressions of the top nodes, one can determine the expressions associated with all the other nodes. We therefore refrain from writing down the expressions for all the nodes that are not the top nodes. In addition, we shall write the expressions of the top nodes above their category labels. In other words, the ordered pair that composes a lexical entry is depicted by writing its first coordinate, the expression, above the second coordinate, its category symbol. The result of these simplications applied to the previous synthesis tree is the following simplified synthesis tree.

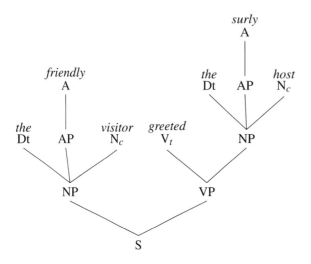

In the use of notation, one must always seek a balance between detail and clarity. On the one hand, if one tries to express in notation each aspect of the structure that is being studied, the result can be so detailed that one can no longer easily discern the relevant information. On the other, if one leaves unexpressed by notation too much of the structure, the pattern is no longer exhibited. As we see, the partially labeled synthesis tree diagram is less cluttered than the one preceding it. It is therefore easier to understand; yet no information is lost. Indeed, there is a bijection between the fully and partially labeled synthesis tree diagrams.

There is also a bijection between the labeled synthesis tree diagrams and the labeled bracketing notation introduced in section 3.1. Any grammatical expression can be equally well analyzed either by an assignment of labeled bracketing or by a synthesis tree. Here is the procedure that shows the correspondence. Recast each top node of the simplified synthesis tree, whose label has the form $\overset{e}{C}$ where e is the expression and C is its category symbol, so that the expression e is enclosed in square brackets and the category symbol C is to the immediate right of the left bracket, yielding the following: [C e]. Continue down the synthesis tree as follows: once each of the labels of the upper nodes of some node have been converted, write the labeled bracketing corresponding to the node to which the upper nodes converge by putting down a right bracket with the category symbol of the lower node, followed by each of the labeled bracketing corresponding to each of the upper nodes, abiding by the left to right order of the upper nodes, followed by a right bracket. Application of this conversion to the partially labeled synthesis tree diagram yields the labeled bracketing annotation of sentence (53), which is repeated here for convenience:

(54) [S [NP [Dt the] [AP [A friendly]] [N_c visitor]] [VP [V_t greeted] [NP [Dt the] [AP [A surly]] [N_c host]]]]

It should be clear that this conversion, applied to the labeled bracketing in (54) in reverse, yields the partially labeled synthesis tree diagram presented earlier.

As we used synthesis rules to create synthesis trees, so we could have used analysis rules to create analysis trees. Analysis rules and synthesis rules are in bijective correspondence, so it should come as no surprise that there is a bijection between the synthesis trees and analysis trees: for any given sentence, its analysis tree is just the inversion of its synthesis tree. Here is the simplified analysis tree of the previous simplified synthesis tree.

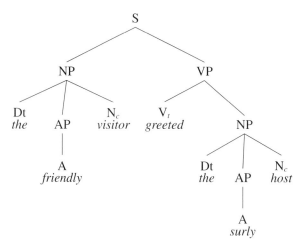

Readers with previous exposure to linguistics may have seen labeled tree diagrams in a form very similar to the form of the preceding analysis tree, and might expect to see, not that tree, but the following one.

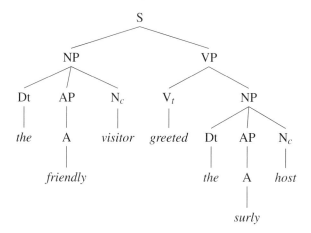

The analysis trees and the customary trees differ only with respect to the bottom nodes. Whereas the labels of the bottom nodes of analysis trees are lexical entries, displayed as a category symbol placed directly over its expression, the customary trees expand each bottom node into one, an upper node labeled with the category symbol and a lower one with the expression. This practice is a vestige of a mathematical treatment of formal grammars whose direct application to natural language has now been abandoned. In addition, this form of labeled tree diagram does not lend itself to enrichments of constituency grammar and to the additional notation that new regularities and facts will require us to make. We shall be using primarily synthesis trees, rarely analysis trees, and never the customary trees.

A constituency grammar for a language comprises a lexicon of the language's basic expressions and a set of constituency formation rules whereby all the expressions of the language can be generated. As stated, a lexicon for a language is a set of lexical entries, each of which is an ordered pair whose first member is a sequence of sounds, which we shall call an *expression,* and whose second member is a category, a lexical category. It should be stressed that the term *expression,* when used as a technical term, refers to just a sequence of phones. Thus, a lexical entry comprises the sequence of phones making up a word and its lexical category. Finally, each formation rule is a constituency synthesis rule, which is a finite sequence of the language's categories, paired with a single category. In other words, each formation rule has the form: $C_1 \ldots C_n \to C$ (where $C, C_1, \ldots, C_n$ are categories in the language). We now give a formal definition of a constituency grammar.

CONSTITUENCY GRAMMAR

Let L be a language. Then, G, or $\langle \text{BX}, \text{CT}, \text{LX}, \text{FR} \rangle$, is a (synthesis) constituency grammar of L iff,

(1) BX is a nonempty, finite set of the basic expressions of L;

(2) CT is a nonempty, finite set of the categories of L;

(3) LX is a nonempty, finite set of ordered pairs $v|C$, the lexical entries of L, where $v \in \text{BX}$ and $C \in \text{CT}$;

(4) FR is a nonempty, finite set of constituency formation rules of L, where each rule has the form $C_1 \ldots C_n \to C$ and $C, C_1, \ldots, C_n$ are in CT.

We have just defined a constituency grammar. The formation rules for a constituency grammar are constituency synthesis rules. Equivalently, we could have defined a constituency grammar to have formations rules that are constituency analysis rules instead. However, as we shall henceforth no longer use constituency analysis rules, there is no need to burden ourselves with this superfluous complication.

Equipped with a constituency grammar for L, we next define the constituents of L. A constituent of L is either a member of the lexicon in the grammar of L or it is obtainable from the lexicon of the grammar of L by a finite number of applications of the constituency formation rules in the grammar for L. Here is its formal definition.

Basic English Grammar

CONSTITUENTS of a language (CS)
Let L be a language. Let G be a constituency grammar for L.

(1)   Each lexical entry is a constituent ($LX_G \subseteq CS_G$);

(2)   if $C_1 \ldots C_n \to C$ is a rule of FR_G and $e_1|C_1, \ldots, e_n|C_n$ are in CS_G, then $e_1 \ldots e_n|C$ is in CS_G;

(3)   nothing else is in CS_G.

We wish to close this section by showing how the ideas presented in section 2.2.1 of chapter 1 can be recast using a constituency grammar. In particular, we shall consider the case of the generation of SL, the set of all strings of letters in L, or {A, B, C, D}, using the suffixation rule. We let the basic expressions be the letters of A, B, C, and D. We use the name of the set L and the name of the set SL as the names of the basic categories. We let the lexicon comprise the four basic expressions categorized as L. And we recast the first two clauses of the suffixation formation rule as constituency formation rules: namely, as L $\to$ SL and SL L $\to$ SL.

A CONSTITUENCY GRAMMAR FOR SL

BX = {A, B, C, D}
CT = {L, SL}
LX = {A|L, B|L, C|L, D|L}
FR = {L $\to$ SL, SL L $\to$ SL}

Exercises: Constituency grammar

1. Assign to each of the following sentences a synthesis tree, using the synthesis rules given in section 3.2. In each case, list the synthesis rules you use.

(a) The visitor bit the guest.

(b) A guest in the yard yelled.

(c) The taciturn host sat in a park.

(d) Beverly thought the visitor fell.

(e) Eric saw the very tall guest.

(f) This very taciturn enemy insulted the host.

(g) The friendly host yelled in the yard.

2. For each of the following, state what BX, CT, LX, and FR are.

(a) The set SL as generated by FRp (chap. 1, sec. 2.2.1),

(b) The set SL as generated by FRb (chap. 1, sec. 2.2.3 exercise 2e),

(c) The set of tallies (chap. 1, sec. 2.2.3 exercise 1),

(d) The set of Roman numerals (chap. 1, sec. 2.2.3 exercise 3b),

(e) The set IN (chap. 1, sec. 2.2.3 exercise 4a),

(f) The set WN (chap. 1, sec. 2.2.3 exercise 4b),

(g) The set ZN (chap. 1, sec. 2.2.3 exercise 4c).

3.3 Corroborating Evidence

In section 3.1, we developed an immediate constituency analysis of part of English. In section 3.2, we gave an abstract definition of a (synthesis) constituency grammar. This put us in the position of treating the work in the first part as a partial specification of a constituency grammar of English. We have thereby arrived at a (tentative) hypothesis about the grammar of English. Next, we shall show how the various constituents of English identified by our hypothetical grammar shed light on other aspects of English expressions, thereby lending some confirmation to the hypothesis.

3.3.1 Further regularities

We begin our presentation of the corroboration of our hypothesized constituency grammar of English with five empirical regularities of English that are most easily and clearly characterized in terms of constituency, in particular, in terms of the kinds of constituents identified in sections 3.1 and 3.2.[4] One regularity is the distribution of the possessive marker *s*. Two others are the regularities exhibited by cleft and pseudocleft sentences. A fourth regularity is found in the forms of answers to questions. A fifth is exhibited by transpositions of parts of sentences in which the meaning of the sentence is preserved.

We begin with the placement of the English possessive marker, *s*. One might be tempted to think that it is an inflectional suffix attached to a word. In some cases, it is; in others though, it is placed after a noun phrase (Bloomfield 1933, 178–179).

(55.1)   On her first visit to England, Susan saw [NP the Queen of England]'s hat.
         *On her first visit to England, Susan saw [NP the Queen's of England] hat.

(55.2)   [NP The student Bill talked to]'s excuse was unconvincing.
         *[NP The student's Bill talked to] excuse was unconvincing.

There is a kind of sentence in English known as a *cleft* sentence. These sentences have the form: *it is α that ...* . As the following examples show, the α is always filled in by a constituent, in particular, either a noun phrase or a prepositional phrase.

---

4. I thank Ben Shaer for sharing his notes on this subject with me.

(56.1)   Elvin owns [NP an expensive car].
         It is [NP an expensive car] that Elvin owns.
(56.2)   Francine put her paycheck [PP under the mattress].
         It is [PP under the mattress] that Francine put her paycheck.

Another pattern, known as a *pseudocleft* sentence, has the form: *what x did was α. α* again is filled in by a constituent, namely a verb phrase.

(57.1)   Elvin [VP bought a two-masted sailboat].
         What Elvin did was [VP buy a two-masted sailboat].
(57.2)   Francine [VP is most eager to travel to the Maldives].
         What Francine is [VP is most eager to travel to is the Maldives].

Next, we turn to the forms of answers to questions. When a question is asked, one may acceptably answer it with a full sentence.

(58.1)   Question   Where did the intruder hide?
(58.2)   Answer   The intruder hid [PP in the basement].

However, one may equally acceptably answer a question with an expression that is not a sentence, but merely a part of a sentence.

(59.1)   Question   Where did the intruder hide?
(59.2)   Answer   [PP in the basement].

As we shall now see, these nonsentential expressions turn out to be expressions to which the grammar nonetheless assigns a category; that is, they turn out to be constituents that are proper constituents of sentences that could be answers. Bearing in mind the following sentence, which could serve as the answer to any of the questions that follow,

(60)   [S [NP The tall guide] [VP met [NP the foreign visitor][PP at the train station] [PP at five o'clock]]]

let us consider these question and answer pairs:

(61.1)   Question   Who met the foreign visitor at the train station at five o'oclock?
         Answer   [NP The tall guide].
(61.2)   Question   What did the tall guide do?
         Answer   [VP meet the foreign visitor at the train station at five o'clock].
(61.3)   Question   Whom did the tall guide meet at the train station at five o'clock?
         Answer   [NP the foreign visitor].

(61.4) Question Where did the tall guide meet the foreign visitor at five o'clock?
       Answer   [PP at the train station].

(61.5) Question When did the tall guide meet the foreign visitor at the train station?
       Answer   [PP at five o'clock].

In each case, the expression that is the answer to the question corresponds to a constituent of the sentence that is an answer to the question: in the first and third cases, the answers are simple noun phrases; in the last two, the answers are simple prepositional phrases; and in the second case, the answer is a verb phrase (Zwicky 1978).

Finally, we turn to a regularity of English in which a subexpression of a sentence may appear in various positions within it without the sentence's basic meaning being altered. As we shall now see, such subexpressions turn out to be constituents.

(62.1) Alice likes [NP the new painting by Picasso] very much.
       [NP the new painting by Picasso] Alice likes very much.

(62.2) Ben promised to finish his paper and he will [VP finish his paper].
       Ben promised to finish his paper and [VP finish his paper] he will.

(62.3) Colleen saw a spider [PP near the door].
       [PP near the door] Colleen saw a spider.

(62.4) A scathing review [PP of the book] appeared recently.
       A scathing review appeared recently [PP of the book].

In each case, what is transposed to either the left end or right end of the clause is a constituent—a noun phrase in the case of the first sentence, a verb phrase in the case of the second, and a prepositional phrase in the third and fourth cases. We shall return to such cases of transposition later in this chapter in section 3.4.3.

The foregoing is just a sample of some of the patterns of English that come to light from the application of immediate constituency analysis and that thereby lend support to the hypothesis that acceptable English expressions are correctly analyzed by a constituency grammar. More evidence of the usefulness of constituency in the analysis of English grammar will appear in chapters 4, 8, 10, and 14 of this book.

Exercises: Corroborating evidence

1. For each of the following English expressions, find at least two lexical categories to which it belongs. In each case provide sentences to substantiate your assignment.

answer, bet, bitter, blow out, bore, bottle, brake, canoe, cash, core, cover, desire, divide, fall, hammer, humble, love, natural, saw, shelve, show off, walk, yellow (Quirk et al. 1985, appendix I, 47–50)

Basic English Grammar

2. For each sentence, analyze it twice, once using simplified synthesis tree diagrams and once using labelled bracketing. If additional lexical categories and constituency formation rules are needed, be sure to state them.

(a) Police police police.

(b) Albert grew a beard.

(c) Alice introduced Bill to Carl.

(d) Galileo argued with the philosopher about the mountains on the moon.

(e) The day after Tuesday is the day after the day after Monday.

(f) The host considered the guest a friend.

(g) The guest treated the dog badly.

(h) This rule is subject to revision.

(i) Each official averse to the proposal resigned from his position.

3.3.2 Amphiboly

The hypothesis that English sentences can be analyzed in terms of constituency grammar has other observable consequences. Consider the sentence:

(63)   Galileo saw a patrician with a telescope.

Let the circumstance of the utterance of the sentence be fixed—say a speaker uttering the sentence in 1608. Let the circumstances with respect to which the sentence is to be evaluated be these: Galileo is looking out the window of his apartment in Venice through his telescope at a patrician walking empty-handed through Saint Mark's Square. Is the sentence uttered in the fixed circumstances true or false with respect to the circumstances stipulated? The answer is both yes and no. It is yes with respect to the stipulated circumstances because a patrician has been espied by Galileo through his telescope; it is no with respect to the very same circumstances because no patrician with a telescope is espied by Galileo.

The susceptibility of sentence (63) to give rise to alternate judgments of truth and of a lack thereof obtains for other stipulated circumstances of evaluation as well. Consider circumstances just like those specified, except the patrician in question was carrying a telescope and Galileo saw him with his naked eyes. Once again, sentence (63) can be judged alternately true and false.

It is important to stress that the sentence is not being judged both to be true and to be false with respect to one and the same circumstances at one and the same moment. Rather, there are distinct judgments each evaluating the truth of the sentence differently. As noted over a half century ago by Charles Hockett (1954, sec. 3.1), the experience is similar to that of the perception of a Necker cube: one does not perceive it simultaneously from the perspective where its top forward corner is slightly tilted down and from the perspective

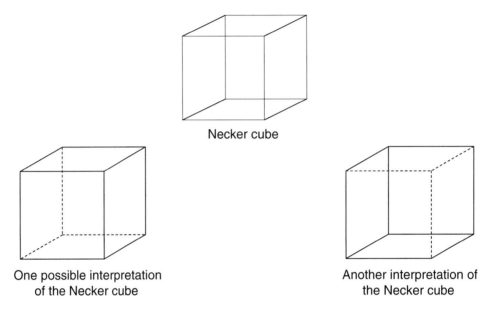

Figure 3.1

of its bottom forward corner slightly tilted upward, rather one perceives it first from one perspective and then from the other.

The same holds for sentence (63). One does not judge it simultaneously to be true and to be otherwise with respect to the circumstances stipulated, rather one judges it first to be true with respect to the circumstances and then to be otherwise with respect to the very same circumstances.

So, how is one and the same expression judged both true and false with respect to one and the same circumstances? The answer is that one and the same expression accommodates two distinct constituency analyses, and these analyses correlate with different construals of the same sentence. On the one hand, sentence (63) can be understood to say that a patrician with a telescope was seen by Galileo. This understanding correlates with the prepositional phrase *with a telescope* forming a constituent of the noun phrase *a patrician with a telescope* with the result that the prepositional phrase modifies the noun *patrician*. On the other hand, sentence (63) can be understood to say that Galileo used a telescope to see a patrician. This understanding correlates with the prepositional phrase being a constituent, not of the noun phrase, but of the verb phrase, thereby modifying the verb and its direct object.

(64.1) Galileo [VP saw [NP a patrician [PP with a telescope]]].
 A patrician with a telescope was seen by Galileo.

(64.2) Galileo [VP saw [NP a patrician] [PP with a telescope]].
 Galileo used a telescope to see a patrician.

The labeled bracketing assigned to sentence (63) correlates with the difference in the truth value judgments reported in the preceding discussion. With respect to the first circumstances, sentence (63) is true relative to the labeled bracketing in (64.1) and it is not true relative to the labeled bracketing in (64.2). With respect to the second circumstances, the sentence is not true relative to the labeled bracketing in (64.1) but it is true relative to the labeled bracketing in (64.2).

What we have here is a clear application of the method of agreement and difference, known to linguists as the method of minimal pairs. One and the same utterance is judged both true and false relative to fixed circumstances, and at the same time, the one and same utterance can be assigned two distinct labeled bracketings. Relative to one, the utterance is judged true; relative to the other, it is judged false. Clearly, constituent structure plays a role in how the value associated with a complex expression is determined by values associated with its constituent expressions. In other words, we have empirical evidence for constituent structure and for the role it plays in determining how the meaning of a complex expression depends on the meaning of its own constituents, the central question in the semantics of natural language.

When an expression accommodates more than one constituency analysis, as with sentence (63), it is said to be *amphibolous* or *structurally ambiguous*. Amphibolous expressions are a rich source of data for understanding the constituency of the expressions of a language.

Exercises: Amphiboly

1. The following sentences are ambiguous. Establish that they are ambiguous with relevant observations. Use paraphrases to elucidate the ambiguity. Provide each sentence with the relevant constituent structures.

(a) Bill said Alice arrived on Tuesday.

(b) They can fish. (Lyons 1968, 212)

(c) The old men and women left.

(d) Alice knows how good Chinese food tastes.

(e) The merchant rolled up the thick carpet.

(f) Alice greeted the mother of the girl and the boy.

(g) April bought the men's store.

(h) The security guards must stop drinking after midnight.

(i) The prosecutor requires more convincing evidence. (Lyons 1968, 212–213)

(j) Bill told Colleen Fred left on Monday.

(k) The stout major's wife is dancing. (Hockett 1958, 153)

(l) The sleigh visitor took his host's picture.

(m) Ali does not like worrying neighbors. (adapted from Leech 1970, 81)

(n) Alice met a friend of Bill's mother.

(o) Babu dropped an egg onto the concrete floor without cracking it.

2. Explain how an amphibolous sentence is a minimal pair, and more generally, how a minimal pair is an instance of Mill's method of agreement and difference, drawing a comparison between an amphibolous sentence, on the one hand, and a case of the application of Mill's method of agreement and difference outside of the area of linguistics.

3.3.3 Recursion

We observed in chapter 1 that the set of acceptable sentences of English is unbounded. In particular, we saw that a sentence, such as (65.0), can be expanded into ever longer sentences of English.

(65.0) Bill sat behind the first chair.

(65.1) Bill sat behind the chair behind the first chair.

(65.2) Bill sat behind the chair behind the chair behind the first chair.

⋮ ⋮

As it happens, any sentence in the unbounded set of sentences, a few members of which are exemplified in (65) can be generated by the following eight constituency formation rules:

S1   NP VP → S
VP3 V_i PP → VP
AP1 A → AP
NP1 N_p → NP
NP2 Dt N_c → NP
NP3 Dt AP N_c → NP
NP4 Dt N_c PP → NP
PP1 P NP → PP

Indeed, these constituency formation rules and the unbounded set of sentences alluded to in (65) provide an example of how a finite set of rules can be used to characterize an unbounded set of acceptable English sentences.

(66.0) [S [NP [N_p Bill]] [VP [V_i sat] [PP [P behind] [NP [Dt the] [AP [A first]] [N_c chair]]]]]

(66.1) [S [NP [N_p Bill]] [VP [V_i sat] [PP [P behind] [NP [Dt the] [N_c chair] [PP [P behind] [NP [Dt the] [AP [A first]] [N_c chair]]]]]]]

Basic English Grammar

(66.2) $\left[\text{S} \left[\text{NP} \left[\text{N}_p \text{ Bill}\right]\right] \left[\text{VP} \left[\text{V}_i \text{ sat}\right] \left[\text{PP} \left[\text{P behind}\right] \left[\text{NP} \left[\text{Dt the}\right] \left[\text{N}_c \text{ chair}\right] \left[\text{PP} \left[\text{P behind}\right] \left[\text{NP} \left[\text{Dt the}\right] \left[\text{N}_c \left[\text{N}_c \text{ chair}\right] \left[\text{PP}\right]\left[\text{P behind}\right] \left[\text{NP} \left[\text{Dt the}\right] \left[\text{N}_c \left[\text{AP} \left[\text{A first}\right]\right] \left[\text{N}_c \text{ chair}\right]\right]\right]\right]\right]\right]\right]\right]\right]\right]$

⋮   ⋮

In other words, these constituency formation rules, together with the lexical entries of $Bill|\text{N}_p$, $chair|\text{N}_c$, $the|\text{Dt}$, $behind|\text{P}$, $first|\text{A}$ and $sat|\text{V}_i$, generate the infinite set of sentences partially listed in (65). The constituency formation rules responsible for the proper recursion are NP3, NP4 and PP1. Notice that on the right of the first rule occurs the label PP and on the right of the second rule occurs the label NP.

Exercises: Recursion

Find another instance of recursion in the preceding constituency formation rules and illustrate the recursion with English sentences.

3.4 Problems

In preceding parts of section 3, we learned about immediate constituency analysis and saw that it brought to light many patterns in the expressions of English. We also saw that these patterns could be captured by a formally defined constituency grammar. We saw, in addition, that the constituency grammar made sense of certain forms of ambiguity that are not ascribable to the ambiguity of specific words of English. These forms of ambiguity provided evidence that constituent structure plays a role in the way in which values associated with complex expressions have their values determined by values associated with expressions making them up. And finally we saw that constituency grammar, through its constituency formation rules, accounts for the fact that the acceptable expressions of English seem to be unbounded. Taken together, the foregoing considerations suggest that constituency grammar will help us to explain two crucial features of English and other natural languages: namely, that the set of acceptable expressions is unbounded and that the values that are associated with complex expressions arise from values associated with their parts.

It would be astonishing, however, if all there is to the grammar of English is that it is a constituency grammar. As we are about to see, constituency grammar is far from all there is. In sections 3.4.1 to 3.4.4, we shall discuss some of the empirical problems that constituency grammar faces. In section 3.4.1, we shall examine three of problems that directly challenge the empirical adequacy of such grammars. These same problems will also illustrate a methodological point: namely, the importance of carrying out the formalization of one's hypotheses so as to verify that they achieve what their informal presentation suggests that they can achieve. We shall enrich constituency grammar to address these three problems, for these enrichments will prove vital to addressing the central problems of semantics. In section 3.4.2, we shall raise a general problem that faces any attempt to establish a formal or formalizable grammar of a natural language, namely, to what extent does a native speaker's judgment of the acceptability or unacceptability of an expression depend on his or her grasp of the grammar and to what extent does it depend on the oddity or the unusualness of what the expression expresses. We shall then turn to a number of other empirical problems that face even the enriched constituency grammar adopted. A number of proposals have been put forth to address these data. We shall simply note them and go on. In section 3.4.2, we shall call attention to a number of completely unresolved problems for which no solutions have been proposed.[5]

3.4.1 Enriching constituency grammar

We now turn to three empirical problems just alluded to that require that we enrich constituency grammars. The first problem is that of agreement. An example is the fact that the noun of a clause's subject noun phrase agrees with its verb in grammatical number. The second is the fact that different words in the same lexical category require different complements. And the third is the fact that each phrase of a certain category must contain a word of a certain category. Let us review them in turn.

3.4.1.1 Number agreement One fact about English is that a noun that is the subject of a clause and the clause's verb must agree in grammatical number. Many Indo-European languages have elaborate systems of inflection, one for nouns and another for verbs. This was also true of Old English. We also saw in our review of traditional English grammar that even Modern English has retained some inflection. In the case of nouns, it is limited to indicating the contrast between singular and plural; in the case of verbs, it is more comprehensive, indicating not only the contrast in singular and plural but also the contrasts in person and in tense. As we shall see, constituency grammars do not naturally accommodate the agreement in grammatical number.

Like nouns in other languages, English nouns and verbs evince a contrast between singular and plural. A regular English noun is grammatically plural, if it has the inflectional suffix *-s,* and singular otherwise. Thus, for example, *dog* is the grammatically singular form of the noun *dog,* while *dog-s* is its grammatically plural form.[6] The English simple present tense also shows a contrast between singular grammatical number and plural grammatical number. The third-person plural form of the verb *to bark* is *bark,* while the third-person singular form of the verb is *bark-s*. As noted in traditional grammars of English, not every

---

5. For some perspective on the role of exceptions to regularities, interested readers may wish to revisit section 3.2.2 of chapter 1.

6. Phonologically, the rule is more complex. The exact phonological value of the suffix is conditioned by the final phoneme of the word to which the suffix is suffixed. See chapter 1, section 2.1.1.

Basic English Grammar

combination of noun as subject and verb is acceptable. The first pair of the following sentences is acceptable, but the second pair is not.

(67.1)   The dog barks.
(67.2)   The dogs bark.
(68.1)   *The dog bark.
(68.2)   *The dogs barks.

Traditional English grammar has the rule that the noun of the subject of a clause and the verb of the same clause agree with respect to grammatical number. This rule certainly discriminates between the sentences in (67) and those in (68).

Our judicious selection of words in our exposition of immediate constituency analysis permitted us to avoid the problem posed by agreement with respect to grammatical number. We chose only common nouns in the singular for our category N_c, and we chose only verbs in the simple past tense for the categories V_i, V_t, and V_s, for the simple past tense in English shows no contrast between singular and plural. The questions arise: What category is to be assigned to plural count nouns and what categories to verbs whose forms are either singular or plural? And how, once this assignment is made, will the empirical generalization in the rule of agreement with respect to grammatical number be respected?

On the one hand, one might assign all common nouns, whether singular or plural, the lexical category N_c and assign all verbs, whether singular or plural, the lexical category their past tense counterparts have been assigned. Thus, one might hypothesize the following lexical entries: $dog|N_c$, $dogs|N_c$, $bark|V_i$, and $barks|V_i$. We require only three of the constituency formation rules used earlier to make our point.

S1   NP VP $\to$ S

VP1 $V_i \to$ VP

NP2   Dt $N_c \to$ NP

It is easy to check that these constituency formation rules, together with the hypothesized lexical entries, generate not only both sentences in (67) but also both sentences in (68). Yet, the two sentences in (68) are not acceptable.

On the other hand, one might want to have distinct lexical categories for singular common nouns (N_{cs}) and for plural common nouns (N_{cp}), as well as distinct lexical categories for singular verbs (V_{is}) and for plural verbs (V_{ip}). By way of illustration, let us hypothesize the following lexical entries: $dog|N_{cs}$, $dogs|N_{cp}$, $bark|V_{ip}$, and $barks|V_{is}$.

Once we have doubled our lexical categories, it will be necessary to double our constituency formation rules.

Ss $NP_s\ VP_s \to S$
Sp $NP_p\ VP_p \to S$
VP1s $V_{is} \to VP_s$
VP1p $V_{ip} \to VP_p$
NP2s $Dt\ N_{cs} \to NP_s$
NP2p $Dt\ N_{cp} \to NP_p$

With these additional categories and constituency formation rules, only the sentences in (67) are assigned the category S; no category is assigned to either sentence in (68). Thus, by doubling the categories, one can accommodate the fact that the noun in the subject agrees with respect to grammatical number with the verb in the predicate.

What we have done is to double the set of constituency formation rules by appending either p or s to the previously used category symbols. The p and the s are mnemonic for *plural* and *singular*. But these mnemonic symbols have no place in a formal constituency grammar. The definition of a constituency grammar makes no provision for appending symbols to symbols. To show this, let us replace the mnemonic symbols used with truly arbitrarily chosen symbols, say, the capital letters from the beginning of the Latin alphabet.

Ss $B\ C \to A$
Sp $D\ E \to A$
VP1s $F \to C$
VP1p $G \to E$
NP2s $H\ I \to B$
NP2p $H\ J \to D$

(We have retained the mnemonic labels for the constituency formation rules to facilitate comparison with the early formulation of the expanded rules.) What these completely formal constituency formation rules make clear is that the constituency grammar does not recognize that singular and plural common nouns are both common nouns. Singular common nouns are assigned the category I and plural common nouns are assigned the category J; similarly, singular intransitive verbs are assigned the category F and plural intransitive verbs are assigned the category G.

The foregoing demonstrates that simple categories, which are formally symbolized by single symbols, are too simple. The solution we shall adopt is the one suggested by the informal notation. Categories of words evincing the distinction between singular and plural will be so categorized. In particular, such words will be assigned an ordered pair as their category label: the first coordinate will indicate the lexical category, and the second will indicate the grammatical number. Thus, the word *dog* will be assigned the complex category $\langle N_c, s \rangle$, and the word *dogs* will be assigned the complex category

Basic English Grammar 149

⟨N_c,p⟩. Similarly, the words *barks* and *bark* will be assigned the complex categories ⟨V_i,s⟩ and ⟨V_i,p⟩.

Now, in many cases in English, the phonological form of a word in the plural does not result from the general rule of suffixing it with -*s*. In some cases, the word has no singular form. Thus, for example, the noun *police* permits only plural number agreement. It, then, is assigned the category ⟨N_c,p⟩. In other cases, the words have both a singular and a plural form, but the form of the latter does not result from the former by the application of the rule. Thus, nouns such as *foot, goose, louse, mouse, man,* and *woman* permit only singular number agreement, while the nouns *feet, geese, lice, mice, men,* and *women* permit only plural number agreement. So, the complex category assigned, for example, to *man* is ⟨N_c,s⟩ and the one assigned to *men* is ⟨N_c,p⟩. It has been recognized since Pāṇini that the peculiarities of words such as these must simply be stipulated. The idea resurfaced in the work of the English philologist and grammarian Henry Sweet (1913, 31) and was reiterated by Leonard Bloomfield (1933, 274), who said that "the lexicon is really an appendix of the grammar, a list of basic irregularities."

The set theoretic notation for a word, its grammatical category, and its grammatical number is given in the left-hand column, while the notation we shall adopt is given in the right-hand column.

STANDARD SET THEORETICAL NOTATION	ADAPTED NOTATION
⟨*dog*,⟨N_c,s⟩⟩	*dog*|N_c;s
⟨*dogs*,⟨N_c,p⟩⟩	*dogs*|N_c;p
⟨*barks*,⟨V_i,s⟩⟩	*barks*|V_i;s
⟨*bark*,⟨V_i,p⟩⟩	*bark*|V_i;p

Now that we have assigned some words more complex grammatical categories, we turn to the problem of ensuring the agreement, with respect to grammatical number, of the noun in the clause's subject with its verb. We do this by permitting the grammatical categories used in the constituency formation rules to be complex grammatical categories. In particular, we introduce a variable in the complex grammatical category, notated x, which ranges over the features s and p.

NP;x VP;x   →   S
Dt N_c;x   →   NP;x
V_i;x   →   VP;x

It is easy to show, using what we have introduced, that the extended grammar assigns the grammatical category S to the expressions in (67) and assigns no grammatical category to the expression in (68).

3.4.1.2 Subcategorization We now turn to another problem. As readers have undoubtedly noticed, the symbols chosen for the various categories were not chosen at random. They were deliberately chosen, because they are suggestive of the categories used in traditional English grammar. To the category of words to which is assigned the symbol A belong only adjectives, just as only prepositions belong to the category of words to which the symbol P is assigned. As we saw, traditional English grammar also distinguishes between common nouns and proper nouns. Only common nouns are assigned to the category whose symbol is subscripted with c, while only proper nouns are assigned to the category whose symbol is subscripted with p. Thus, the symbolization uses subscripts to distinguish within one lexical category two lexical categories that are subcategories of the first. In other words, the mnemonic notation has honored some important facts of English recognized by traditional grammar. Traditional grammar, as well as standard lexicographical practice, also distinguishes verbs into transitive and intransitive. Again, subscripts are used to distinguish within one lexical category two subcategories.

However, these empirical generalizations, which are recognized by traditional English grammar and by the mnemonic symbols adopted, are not reflected in the formal version of constituency grammar given thus far. To see why they are not, let us again see what happens when the mnemonic symbols are replaced with arbitrarily chosen symbols. The point can be made by considering only these four constituency formation rules: S1, VP1, VP2, and NP2. Replacing the seven mnemonic category symbols in them—namely S, NP, VP, V_i, V_t, Dt, and N_c—with the plain symbols A, B, C, D, E, F, and G, we obtain the following formulation:

S1   B C → A
VP1 D → C
VP2 E B → C
NP2 F G → B

Such a replacement in the constituency formation rules will not change the set of expressions generated by the constituency grammar. However, such a replacement makes clear that the mnemonic notation introduces something the formal notation does not, namely, that there is a category that groups all verbs together. But, in the formal grammar, we have assigned category D to what traditional grammar calls intransitive verbs and category E to what it calls transitive verbs. Thus, there is no single symbol for verbs. As a result, one has to state twice the rule that regular English verbs take the suffix -*s* in the third-person singular present indicative, once for words in category D and again for words in category E. Should we look at the full range of English verbs and continue to assign categories as required by the formal definition of constituency grammar, we would come up with dozens of categories of English verbs to which distinct symbols would have to be assigned. The result would be that the rule for the formation of third-person singular indicative would have to be stated dozens of times. This is clearly an undesirable consequence.

It might be thought that such a proliferation of lexical categories for what otherwise seems like a single lexical category is unnecessary. Why not simply assign all English verbs one lexical category, say D. Such an identification of lexical categories has two effects. First, it puts all verbs into the same lexical category. Thus, in particular, *slept* and *greeted* will belong to the same lexical category. Their lexical entries will be *slept*|D and *greeted*|D. Second, the symbol E in the constituency formation rule VP2 will be replaced by D, so that the previous four fully formalized rules will become these:

S1   B C → A

VP1   D → C

VP2   D B → C

NP   F G → B

The result is that all of the sentences that follow, including the two unacceptable sentences, will be assigned the category A and therefore will be grammatical expressions of English.

(69.1)   The visitor slept.
(69.2)   *The visitor slept the dog.

(70.1)   *The host greeted.
(70.2)   The host greeted the guest.

An analogous problem arises for English adjectives, nouns, and prepositions.

Traditional English grammar and standard lexicographical practice distinguishes between transitive and intransitive verbs. Intransitive verbs are those that refuse a direct object, and transitive verbs are those that require one. Put in terms of constituency analysis, an intransitive verb is one that refuses a noun phrase to its right and a transitive verb is one that requires a noun phrase to its immediate right.

A more careful consideration of the various kinds of English verbs shows that they are distinguished by the kinds of constituents that are required so as to *complete,* as it were, the verb; in other words, they are distinguished by the kinds of complements they require. Thus, an intransitive verb—for example, *to bloom, to die, to disappear, to elapse, to expire, to fail, to faint, to fall, to laugh, to sleep, to stroll, to vanish*—refuses any complement, while a transitive verb—for example, *to abandon, to cut, to buy, to destroy, to devour, to expect, to greet, to keep, to like, to lock, to prove, to purchase, to pursue, to vacate*—requires a noun phrase complement. Some verbs—for example, *to dash* (to), *to depend* (on), *to hint* (at), *to refer* (to), *to wallow* (in)—require a prepositional phrase complement, while some—for example, *to give* (to), *to hand* (to), *to exempt* (from), *to introduce* (to), *to place* (in/on), *to put* (in/on), *to send* (to), *to stand* (on), *to talk* (to)—require a noun phrase complement followed by a prepositional phrase complement. Still other verbs—such as *to maintain, to note, to remark*—require a clause complement.

What is required, then, is not just a two-way distinction between transitive and intransitive verbs, but a way to distinguish verbs by the kind of complements they require. In chapter 10, we shall delve more deeply into the diversity of complements not only of English verbs but of English adjectives, prepositions, and nouns, and we shall see how the problem raised by this diversity for constituency grammar can be addressed.

3.4.1.3 Phrasal and lexical categories We close this section with an explanation of the third problem, which we shall call the *projection problem*. Grammarians and linguists have long recognized that many constituents of a certain category contain a word of a related category. In particular, each adjective phrase contains an adjective, each noun phrase contains a noun, each prepositional phrase contains a preposition, and each verb phrase contains a verb. However, constituency grammar, as defined in section 3.1, cannot respect this generalization. Once again, this becomes evident the moment we replace the mnemonic informal notation with nonmnemonic formal notation. Here, then, are the constituency formation rules for phrases stated once in an informal synthesis version and once in a formal synthesis version.

	INFORMAL RULES	FORMAL RULES
NP1	$N_p \to NP$	$A \to B$
NP2	$Dt\ N_c \to NP$	$C\ D \to B$
NP3	$Dt\ AP\ N_c \to NP$	$C\ J\ D \to B$
NP4	$Dt\ N_c\ PP \to NP$	$C\ D\ H \to B$
PP1	$P\ NP \to PP$	$G\ B \to H$
AP1	$A \to AP$	$I \to J$
AP2	$Dg\ A \to AP$	$K\ I \to J$
VP1	$V_i \to VP$	$L \to M$
VP2	$V_t\ NP \to VP$	$N\ B \to M$
VP3	$V_i\ PP \to VP$	$L\ H \to M$
VP4	$VP\ PP \to VP$	$M\ H \to M$
VP5	$V_s\ S \to VP$	$P\ Q \to M$
S1	$NP\ VP \to S$	$B\ M \to Q$

In the informal constituency formation rules, a connection is displayed between the symbol on the right of the arrow and one symbol on its left. Thus, for example, in the informal versions of the rules NP1 through NP4, the symbol NP appears on the right side of the arrow and some symbol with an N, either N_p or N_c, appears on the left side. In the formal version of the constituency formation rules, no such connection is made: nothing suggests a connection between the category on the right of the arrow and any category on the left.

Basic English Grammar 153

Thus, in the case of the rules for NP1 through NP4, nothing signals a connection between category B and categories A and D. An analogous observation holds for the rule PP1 as well as for the rules AP1 through AP2 and VP1 through VP5.

In chapter 10, we shall modify our definition of constituency grammar to solve the last two problems just identified.

Exercises: Enriching constituency grammar

1. For each of the words that follow, give its plural form and identify at least two other English words with a similar plural form. Do not hesitate to consult a dictionary.

abacus, basis, criterion, deer, fish, knife, ox, hoof

2. For each of the sentences that follow, write out a lexical entry for the verb of its main clause. If further constituency formation rules are required, state them.

(a) Dan remained silent.

(b) The doctor inquired of Alice what she had done.

(c) Chunka noticed it was raining.

(d) Alice said to Bill it was cold.

(e) Angelyn decided to leave the house.

(f) The witness recalled what had happened.

(g) Alec made her friend angry.

(h) The judge accused the witness of lying.

(i) The lawyer convinced the jury of his client's innocence.

(j) Each student wondered what the answer was.

3. Restate any new constituency formation rules and lexical entries for the preceding exercise in purely formal notation, that is, using single capital letters of the Roman alphabet.

3.4.2 Grammar and belief

While the constituency formation rules generate the first of the following sentences, they also generate the second. After all, noun phrases are intersubstitutable, thereby preserving acceptability. Consider the following sentence and its constituent structure.

(71.0)   The man walks.

(71.1) $[S\ [NP\ [Dt\ The]\ [N_c\ man]]\ [VP\ walks]]$.

It contains the expression *the man,* which belongs to the category NP, and *the car* also belongs to the category NP. Hence, it should be substitutable for *the man* to yield an acceptable sentence.

(72.0) The car walks.

(72.1) $[S\ [NP\ [Dt\ The]\ [N_c\ car]]\ [VP\ walks]]$.

However, this sentence is not acceptable.

We notice, to begin with, that the word *man* denotes an animal, while the word *car* does not. Animals walk; nonanimals do not. We might, taking our cue from our solution to the problem posed to constituency grammar by grammatical number, expand the grammatical classification of English expressions and incorporate the expansion into our constituency formation rules, just as we expanded our classification of English expressions to include the distinction between singular and plural. Thus, we might postulate two kinds of common nouns: those that denote animals and those that do not. Thus, the word *man* would be assigned the complex category N_c;s;a and *car* would be assigned the complex category N_c;s;b. In addition, the word *walks* would be assigned the complex category V_i;s;a. This expansion of the grammatical classification of words would have to be extended to the constituency formation rules.

NP;n;x VP;n;x → S
Dt N_c;n;x → NP;n;x
V_i;n;x → VP;n;x

The foregoing would rule in the sentence in (72) as being a sentence of English and would rule out the sentence in (71) as not being a sentence of English.

However, consider the verb *moves*.

(73.1) The man moves.

(73.2) The car moves.

It occurs both with NP;s;a and with NP;s;b. How, then, shall we classify *moves*? To be sure, there is an answer to this question along the line we have started to pursue. But in the end, any such answer will turn out to be complicated, requiring more and more symbols in our grammar. But is this really the best analysis?

Let us consider an entirely different solution to the problem posed by the sentences in (71) and (72). Recall that our point of departure in our pursuit of the initial solution was the observation that *man* and *car* denote different kinds of things, the former is a mammal—and mammals walk; the latter is a machine—and machines do not walk. One might wonder whether or not this information falls properly within the purview of grammar.

To ponder this question, we are well advised to distinguish between acceptability and grammaticality. Speakers judge expressions to be acceptable or unacceptable. Linguists determine whether expressions are grammatical or ungrammatical. The former is an observable fact. The latter is a theoretical determination that follows from a set of hypotheses, which include, among other hypotheses, a hypothesis about the grammar of English.

In section 2.1.1 of chapter 1, we introduced a number of general hypotheses that guide modern linguistic inquiry. One general hypothesis is that the human linguistic capacity, whereby humans use and understand their language, involves the joint exercise of many capacities, which include not only a hypothesized grammatical capacity, whereby humans form and recognize grammatical expressions, but also other capacities such as reason and memory and focussing attention. Thus, the speaker's acceptance of an expression of his or her language is determined by the exercise of his or her linguistic capacity, which is the joint exercise of a number of capacities, including his or her grammatical capacity. In other words, the acceptability of an expression depends on a number of factors, including not only its grammaticality, but also its appropriateness and what it expresses.

Returning, then, to the sentences in (71) and (72), we might consider that the unacceptability of the sentence in (72) does not arise from its being ungrammatical but rather from its expressing something that speakers take to be implausible or unreasonable. Are there any reasons to believe that this is the case? Indeed, it is possible to find situations in which the sentence in (72) is perfectly acceptable, for example, in cartoons in which a car is personified.

Notice the work done, once more, by minimal pairs: one and the same sentence can be judged acceptable or unacceptable relative to one's beliefs. This suggests that it is the beliefs we have, and not the grammar we have, that is responsible for any difference in judgment by a speaker of the acceptability of the sentences in (71) and (72).

Let us now consider another example of simple lexical replacements. Recall the amphibolous sentence in (63), repeated here as (74.1). This sentence is judged to be ambiguous. With respect to a judiciously specified single circumstance, it can be judged true at one moment and not judged true at another. As we saw previously, this ambiguity correlates with two immediate constituency analyses.

(74.1)   Galileo saw a patrician with a telescope.

(74.2)   Galileo hit a patrician with a telescope.

(74.3)   Galileo hit a patrician with a beard.

Now, consider a second sentence, resulting from the first by the substitution of one transitive verb for another. It too should accommodate the same pair of immediate constituency analyses.

(75.1)   Galileo [VP hit [NP a patrician [PP with a telescope]]].

(75.2)   Galileo [VP hit [NP a patrician] [PP with a telescope]].

And therefore, one should be able to specify a single circumstance with respect to which one can judge the sentence true at one moment and not true at another. And indeed we can, as readers can do easily for themselves.

However, if one replaces the common noun *telescope* in sentence (74.2) with the common noun *beard,* yielding sentence (74.3), the resulting sentence can also accommodate the same pair of immediate constituency analyses.

(76.1) Galileo [VP hit [NP a patrician [PP with a beard]]].

(76.2) Galileo [VP hit [NP a patrician] [PP with a beard]].

And therefore, once again, one should be able to specify a single circumstance with respect to which one can judge the sentence true at one moment and not true at another. And while it is easy to imagine a circumstance in which sentence (74.2), as analyzed in (75), is judged true as well as one in which it is judged false, it is not so easy to imagine a circumstance in which sentence (74.3) as analyzed in (76.2) is judged true. After all, one does not typically use a beard as a weapon to strike someone else. Should one, then, try to revise somehow the constituency formation rules to block the consequence that sentence (74.3) is amphibolous?

Some thought reveals that a circumstance corresponding to the constituent structure given in (76.2) is not, however, impossible. Imagine that Galileo wore a fake beard and that, in a fit of anger, he hit a beardless patrician with it. Clearly, sentence (74.3) is judged true one moment and false the next with respect to that circumstance and judgments correlate with the immediate constituency analyses displayed in (76). Again, we see the usefulness of invoking the distinction between acceptability and grammaticality and of permitting ourselves to seek the explanation of unacceptability in terms of factors extrinsic to grammar.

The foregoing examples illustrate a fundamental problem that the linguist faces in assessing judgments of acceptability, namely, how to distinguish between judgments of unacceptability that arise from a speaker's grammatical capacity in his or her language and judgments of unacceptability that arise from his or her beliefs about the world. A very nice analogy is furnished by Geoffrey Leech (1974, 8). Think of the rules of grammar as rules of a game, say, soccer. Now contrast the following two reports:

(77.1) The center forward scored by heading the ball from his own goal line.

(77.2) The center forward scored by throwing the ball into the other team's goal.

Both sentences would be disbelieved by anyone knowledgable about soccer: the first because what it expresses is a physical impossibility; and the second because what it expresses does not conform to the rules of soccer.

What gives rise to a judgment of unacceptability and what the relevant factors are in any given case is not something that can be ascertained a priori. Just as there is no a priori way to determine the constituency formation rules of a grammar, except by conjecture and testing, so there is no a priori way to determine the relevant factors, except by conjecture and testing.

Basic English Grammar

Exercises: Grammar and belief

1. Consider the following English expressions that have been judged to be odd (Chomsky 1965, 75–77). Determine whether the expression is grammatical or ungrammatical. Adduce evidence to support your conclusion.

(a) Dan frightened his sincerity.

(b) Each book elapsed in quick succession.

(c) The dog looked barking.

(d) Colleen solved the pipe.

(e) The rioter dispersed.

(f) Ophthalmologists are smarter than eye doctors.

(g) Bill memorized the score of the sonata he will compose next week.

(h) The decision amazed the injustice.

(i) The repairman is wiring the entire poem this time.

(j) A sour flash disturbed the baby's sleep.

2. The American pragmatist philosopher William James, described the following puzzle:

Some years ago, being with a camping party in the mountains, I returned from a solitary ramble to find everyone engaged in a ferocious metaphysical dispute. The corpus of the dispute was a squirrel—a live squirrel supposed to be clinging to one side of a tree trunk; while over against the tree's opposite side a human being was imagined to stand. This human witness tries to get sight of the squirrel by moving rapidly round the tree, but no matter how fast he goes, the squirrel moves as fast in the opposite direction, and always keeps the tree between it and the man, so that never a glimpse of it is caught. The resultant metaphysical problem now is this: *Does the man go round the squirrel or not*? (cited in Martin 1992, 1–2)

James's point is that many so-called metaphysical problems are pseudometaphysical problems: that is, they are simply "problems of language." Do you think that this is a problem of language? If so, what is the problem?

3.4.3 Discontiguity

The use of constituency grammar to investigate the acceptable expressions of English leads to the discovery of a number of regularities that, if previously not unnoticed, was passed over unremarked. We shall set out some of these regularities that seem to go beyond the capabilities of constituency grammar, even when its categories are enriched.

In immediate constituency analysis, one segments a constituent's expression into subexpressions, each contiguous with the next and each associated with a constituency label. This analysis, as we saw, may be viewed equivalently as a synthesis in which the expression of a complex constituent arises from the concatenation of the expressions of its immediate subconstituents. This means that all expressions in a constituent are contiguous. This contiguity is less accurately described as continuity. However, as we are about to see, many complex constituents seem to contain *discontiguous* constituents. These are expressions that otherwise seem to form a constituent, yet among the expressions of the constituent, appears an expression of some nonconstituent. These patterns of discontiguity, many of which had been noticed by the structuralist linguists, were the focus of much of the research done by transformational linguists. We shall describe here nine such patterns found in English. All of them have counterparts in other languages and many of them are found in languages unrelated to English. In each case, we shall illustrate the discontiguous constituents by providing pairs of sentences that differ only insofar as the first of the pair has no discontiguous constituent, while the second does. We have annotated with suitably labeled square brackets both the smallest constituent exhibiting discontiguity and the immediate subconstituent that fails to be contiguous with it. We call the latter a *dislocated* constituent. We annotate the sentence exhibiting discontiguity with an underscore to indicate the position where, had the dislocated constituent appeared there instead, the constituent would not be discontiguous. We call such a position a *gap*.

The patterns of discontiguity in English fall into two groups, those where the discontiguous parts of the discontiguous constituent must be found within the same clause, and sometimes even within the same phrase, and those where one part is found in one clause and the other in a clause superordinate to the first. We begin our review of discontiguity with the former kind.

One form of discontiguity arises with English verbs with which are associated specific prepositions or adverbs. Thus, English has intransitive verbs such as *to break down* (cf. *to cry*), *to die down, to die off, to pass out* (cf. *to faint*), *to play around, to sound off* (cf. *to express one's opinion*), and *to turn up* (cf. *to appear*). Such *verb particle* sequences are not confined to intransitive verbs. English transitive verbs also include verb particle sequences: *to call up, to hand in, to knock out, to live down, to look up, to make out* (cf. *to understand*), *to set up,* and *to sound out*. It is natural to treat such verb particle sequences as constituents. Sometimes they form a contiguous constituent, as in sentence (78.1), but sometimes they form a discontiguous constituent, as in sentence (78.2), where the direct object of the verb intrudes between the verb and its associated preposition.

(78.1) Bill [VP woke up his friend].

(78.2) Bill $\left[\text{VP} \left[\text{V woke} \underline{} \right] \text{his friend [up]} \right]$.
(Wells 1947, sec. 60)

Such discontiguity is confined to the verb phrase.

Another form of discontiguity that is confined to a phrase is a form known as *wrapping*. It is so-called because the head of a constituent seems to be enwrapped by the head of a modifier on one side and the modifier's complement on the other. Thus, in the following example, *job* is modified by *good enough to pass inspection,* apparently a constituent, yet part of the modifying constituent, namely *good enough,* may precede the noun *job* and part of it, namely *to pass inspection,* may succeed *job.*

(79.1) a job [AP good enough [S to pass inspection]]

(79.2) a [AP good enough __] job [S to pass inspection]
(Wells 1947, sec. 56)

A form of discontiguity that is confined to a clause is called *extraposition.* Here, a modifier or a complement of a noun phrase may also appear at the end of the clause containing it. The noun phrase hosting the discontiguity may be a subject noun phrase,

(80.1) [NP An article [PP about malaria]] appeared in the newspaper yesterday.

(80.2) [NP An article __] appeared in the newspaper yesterday [PP about malaria].

or it may be a complement noun phrase.

(81.1) Beverly [VP read [NP an article [PP about malaria]] in the newspaper yesterday].

(81.1) Beverly [VP read [NP an article __] in the newspaper yesterday] [PP about malaria].

Another form of discontiguity confined to a clause is called *VP preposing,* where the entire verb phrase precedes its subject and its auxiliary verb.

(82.1) Ben promised to finish his paper and [S he will [VP finish his paper]].

(82.2) Ben promised to finish his paper and [VP finish his paper] [S he will __].

A third form of discontiguity confined to its clause is *PP preposing,* where a prepositional phrase thought to be proper to the verb phrase is found at the beginning of its clause.

(83.1) Colleen [VP saw a spider [PP near the door]].

(83.2) [PP near the door] Colleen [VP saw a spider __].

We now turn to discontiguous constituents where one part is found in one clause and the other in a clause superordinate to the first. Such forms of discontiguity are sometimes referred to as *long distance dependencies,* since there is no limit, or bound, on how many clauses intervening between the part of the discontiguous found in the superordinate clause and the part found in the subordinate clause. *Topicalization* is one form of such discontiguity.

(84.1)   Alice [VP likes [NP the new painting by Picasso] very much].

(84.2)   [NP the new painting by Picasso] Alice [VP likes __ very much].

(84.3)   [NP the new painting by Picasso] Bill thinks [S Alice [VP likes __ very much]].

Another form of discontiguity is called *easy* or *tough* movement, where the adjectives *easy* and *tough* are instances of the kinds of expressions involved in such discontiguities.

(85.1)   It was easy to prove that theorem.

(85.2)   [NP That theorem] was easy [to prove __].

(85.3)   [NP That theorem] was thought [to be easy [to prove __]].

Next, and perhaps best known, is so-called *wh movement,* the form of discontiguity most intensively studied by transformational linguists. It has three principal forms: that of a direct question that requires a phrasal response, rather than a yes or no response, that of an indirect question, and that of a relative clause.

Some questions seek a response of yes or no and others seek a phrasal response that provides information corresponding to the interrogative phrase. A yes or no answer is sufficient as a response to the first question. The second question seeks the identity of the person to whom Carol gave the key.

(86.1)   Did Carol [VP give the key [PP to Don]]?

(86.2)   Carol [VP gave the key [PP to whom]]?

More typically, and more naturally, a question requiring a phrasal response occurs as a discontiguous constituent, as shown by the last two sentences of (87).

(87.1)   Carol [VP gave the key [PP to whom]]?

(87.2)   [PP to whom] did Carol [VP give the key __]?

(87.3)   [PP to whom] does Alfred think [S Carol [VP gave the key __]]?

Indirect questions occur only as subordinate clauses. They do not ask a question and the main clause containing them need not be an interrogative clause.

(88.1)   Don knows [S [PP to whom] Carol [VP gave the key __]].

(88.2)   Don knows [S [PP to whom] Alfred thinks [S Carol [VP gave the key __]]]

The last form of wh extraction arises in relative clauses. A relative clause is a clause that typically modifies a common noun. The first constituent of a relative clause is a relative pronoun or a prepositional phrase containing a relative pronoun. Often the constituent that results from the removal of the initial constituent containing the relative pronoun lacks being a clause by a constituent corresponding to the initial constituent.

(89.1) Carol gave the key [PP to Don].

(89.2) Don is [NP the person [RC [PP to whom] Carol [VP gave the key __]]].

Thus, the incomplete clause *Carol gave the key* can be turned into a complete clause by adding in an appropriate place a phrase of the same kind as the phrase *to whom,* which is the initial phrase in the relative clause, labeled here RC. As this description makes clear, the prepositional phrase containing the relative pronoun *whom* is clearly to be construed as part of the verb phrase of the relative clause, yet it is not contiguous with the other constituents of the verb phrase.

A number of approaches to the treatment of discontinuity are found among grammatical theories that have their origins in constituency grammar. They include various versions of transformational grammar, initiated by Noam Chomsky in his *Syntactic Structures* (1957) and altered and expanded in numerous publications over the ensuing forty years (Chomsky 1962; 1965; 1970; 1976; 1981; 1991) as well as a number of rival theories, including relational grammar of David Perlmutter and Paul Postal (Perlmutter 1983), lexical functional grammar of Joan Bresnan (Bresnan 1978; 1982), generalized phrase structure grammar of Gerald Gazdar (Gazdar 1982; Gazdar, Klein, Pullum, and Sag 1985), and its subsequent avatar, head driven phrase structure grammar of Carl Pollard and Ivan Sag (Pollard and Sag 1994).

Since we shall be returning to the semantics of some kinds of relative clauses in chapter 14 and presenting briefly their syntax and semantics within a transformational grammar, we shall provide here a brief description of enough of the key ideas to convey the general idea found in its various versions.

Transformational grammar is so-called because it uses, in addition to the constituency formation rules, other rules of syntax called *transformational* rules. Transformational rules are mathematically more powerful than constituency formation rules (see Peters and Ritchie 1973). This means that the set of expressions generated by constituency formation rules is a proper subset of the expressions generated by transformational rules.

As we saw earlier, for each expression with a discontiguous constituent, there is a nearly synonymous equivalent without any discontiguous constituents and obtained from more or less the same set of words. The argument was made that were one to use constituency formation rules alone to analyze a pair of such expressions, the grammar would fail to reflect the fact that the pair of sentences are nearly synonymous. Consider, for example, the pair of sentences in (90).

(90.1) A review [PP of *Bleak House*] appeared.

(90.2) A review appeared [PP of *Bleak House*].

The first sentence can be analyzed using the constituency formation rules: S1, VP1, PP1 and NP4.

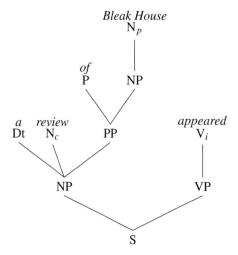

The second sentence can be analyzed with the constituency formation rules S1, VP1, and NP2, supplemented by a further constituency formation rule, say S PP → S, which is required independently of the second sentence in (90), to analyze sentences with prepositional phrases modifying them, as illustrated by the following sentences.

(91.1) Beverly jogged.

(91.2) Beverly jogged on Monday.

Here is the synthesis tree for the second sentence in (90).

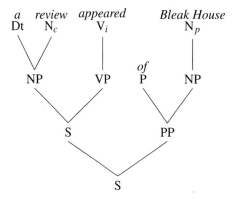

There is nothing in the constituency grammar analysis of the two sentences in (91) which suggests that they share a meaning.

One way for a grammar to show that the two sentences share their meaning is to derive the latter analysis from the former. The idea is that a transformational rule operates on the analysis of sentence (91.1) to yield the analysis of sentence (91.2). To make this more precise, one could assign to each expression a pair of analyses. The first is called the sentence's *deep structure,* the second its *surface structure*. In the case of sentence (90.1), its two analyses are the same, the one shown by the first synthesis tree; in the case of sentence (90.2), it has two analyses, the first corresponding to the first synthesis tree, the so-called deep structure, and the second corresponding to the second synthesis tree, the so-called surface structure, the latter being derived from the former by a transformational rule, which moves, so to speak, the prepositional phrase in the subject noun phrase to the end of its clause. The fact that the two sentences have a common analysis, their deep structure, reflects their common meaning.[7]

The annotation of the underscore, adopted to help describe discontiguous constituents, has a notational counterpart that is intended to be part of the syntactic analysis of discontiguous constituents: what is described as a gap is represented by a constituent with a so-called *trace* coindexed with the dislocated constituent, as shown in the second analysis here.

(92.1) [NP A review __] appeared [PP of *Bleak House*].

(92.2) [NP A review [PP,i t]] appeared [PP,i of *Bleak House*].

We bring this section to a close by briefly describing an alternative approach to the problem of discontiguous constituents. Rather than expand the set of rules with transformational rules, generalized phrase structure grammar and its later development, head driven phrase structure grammar, enrich constituency formation rules with features. We saw in section 3.4.1 that many patterns of English require that constituency grammars be enriched in some way. In the case of number agreement, it is generally thought that constituency formation rules must be enriched to include features. The idea was suggested by Gerald Gazdar that features could also be used to reflect the connection between the two parts of a discontiguous constituent.

Exercises: Discontiguity

For each sentence, determine whether or not it has discontiguous constituents. Identify them. Provide evidence to support your identification.

1. The manager sent the invitations out.
2. Clever though he may be, he is not reliable.

---

7. This idea goes back to Pāṇini's *Aṣṭādhyāyī*. Many pairs of synonymous expressions having more or less the same words but with different grammatical analyses are given a common underlying grammatical analysis.

3. Such a person we shall never meet again.
4. Alice made a similar assessment to Bill's.
5. This is a more serious matter than anyone expected.
6. We all thought Bill would be swimming laps, and swimming laps he was.
7. The proposal of which Alice approves was adopted unanimously.
8. On the way home we shall buy some milk.

3.4.4   No left embedding

One of the distinguishing features of the expressions of English is that many of them have a recursive structure. However, as noted by Chomsky (1965, 12–15), these very same expressions, which are acceptable in one syntactic environment, are not in another and yet there is no known syntactic difference to distinguish the environments. In this section, we shall see three pairs of contrasting syntactic environments.

We observed that a prepositional phrase can be a constituent of a prepositional phrase that, in turn, can be a constituent of a prepositional phrase, with no obvious limit on the number of prepositional phrases that can occur one within another. A similar pattern occurs with adjectives such as *clear, obvious, true,* to mention only a few. In each case, the adjective may host a clausal complement that is a complete sentence. Thus, for example, the expression *it is true that* may have a clause appended to it—say, the sentence *two is a prime number*—to form a complete sentence, which, in turn, may be appended to *it is true that* to form still another complete sentence, and so on. There is no limit to this iteration.

(93.1)   It is true that [S two is a prime number].

(93.2)   It is true that [S it is true that [S two is a prime number]].

(93.3)   It is true that [S it is true that [S it is true that [S two is a prime number]]].

⋮   ⋮

Just as these adjectives have clausal complements introduced by *that,* so too they can serve as adjectives predicated of a subject noun phrase comprising the same clause introduced by *that.*

(94.1)   That [S two is a prime number] is true.

(94.2)   That [S two is a prime number] is obvious.

(94.3)   That [S two is a prime number] is clear.

⋮   ⋮

Yet, while any of these adjectives admits a complement clause, which itself may comprise one of these adjectives also admitting a complement clause, none of these adjectives admits

Basic English Grammar

a subject clause, which itself comprises one of these adjectives with a subject clause. In other words, there is a limit to this iteration.

(95.1)   That [S two is a prime number] is true.

(95.2)   *That $\big[$S that [S two is a prime number] is true$\big]$ is true.

(95.3)   *That $\big[$S that $\big[$S that [S two is a prime number] is true$\big]$ is true$\big]$ is true.

Similar failure of iterative embedding arises with pseudocleft sentences

(96.1)   [S What it would buy in Germany] is amazing.

(96.2)   *$\big[$S What [S what it cost in New York] would buy in Germany$\big]$ is amazing.

(96.3)   *$\big[$S What $\big[$S what [S what he wanted] cost in New York$\big]$ would buy in Germany$\big]$ is amazing.

and with relative clauses within subject noun phrases.

(97.1)   The rat [S which the cat chased] died.

(97.2)   *The rat $\big[$S which the cat [S which the people own] chased$\big]$ died.

(97.3)   *The rat $\big[$S which the cat $\big[$S which the people [S which I met] own$\big]$ chased$\big]$ died.

4 Conclusion

In this chapter, we first briefly surveyed traditional English grammar, then we introduced immediate constituency analysis and its formalization as constituency grammar, and finally we discussed the extent to which constituency grammar correctly reflected known regularities of English and revealed new ones.

Immediate constituency analysis was the first major development in syntactic analysis since Pāṇini's pioneering work. When applied to English, it was able to do justice to many of the regularities of English already known to traditional English grammar as well as to reveal many, many more that had hitherto passed unnoticed. Moreover, its formalization in terms of constituency grammar by Chomsky brought to light the recursive structure of natural language expressions, providing for the first time a clear mathematical characterization of the observation that human languages are unbounded. In addition, the notation of labeled tree diagrams and labeled brackets resulting from Chomsky's work on formalizing grammars made the regularities revealed by immediate constituency analysis stand out with a clarity and precision not previously known in linguistics. Consequently, the correctness of the regularities expressed in terms of constituency grammar were more precise and their confirmation or disconfirmation more objectively ascertainable. Finally, constituency grammar was a first step to addressing the question first raised by Indian grammarians, namely, how do the meanings of minimal expressions of a language contribute to the meaning of the complex expression they constitute.

The mathematical wherewithal to address this question did not become available until the early part of the twentieth century, when Alfred Tarski, the discoverer of model theory, showed how the values assigned to complex expressions in a notation are determined by values assigned to the simpler expressions making them up. Indeed, it was Alfred Tarski (1932, 164) who first entertained the idea that this part of model theory might be applied to the study of the meaning of complex linguistic expressions. But Tarski was skeptical that such a project could be realized, for he thought that no satisfactory specification of the expressions of a natural language could be given, a step necesssary to carrying out his idea. The formalization of constituency grammar is that step, because constituent structure is the structure that furnishes the relationship relevant to relating the values assigned to parts to the value assign to the whole.

Constituency grammar, as we have seen, is too simple to provide a satisfactory specification of all of the expressions of English. And even enriched with features to accommodate grammatical agreement, many patterns of expressions seem beyond its reach. Still, we have excellent empirical evidence that constituency plays a crucial role in the specification of the expressions of English and other languages, namely the evidence from the fact that amphibolous expressions are be judged both true and not true.

SOLUTIONS TO SOME OF THE EXERCISES

3.1 Immediate constituency analysis

1. (a) The host slept.
    $[S\ [NP\ [Dt\ the]\ [N_c\ host]]\ [VP\ [V_i\ slept]]]$
    S1; NP2; VP1

   (c) The visitor walked in the park.
    $[S\ [NP\ [Dt\ the]\ [N_c\ visitor]]\ [VP\ [V_i\ walked]\ [PP\ [P\ in\ ]\ [NP\ [Dt\ the]\ [N_c\ park]]]]]$
    S1; NP2; VP3; PP1

   (e) The very tall guest saw the host.
    $[S\ [NP\ [Dt\ the]\ [AP\ [Dg\ very]\ [A\ tall]]\ [N_c\ guest]]\ [VP\ [V_t\ saw]\ [NP\ [Dt\ the]\ [N_c\ host]]]]$
    S1; NP2; NP3; VP2; AP2

   (g) Beverly thought the woman sat in the yard.
    $[S[NP\ [N_p\ Beverly\ ]][VP\ [V_s\ thought][S\ [NP\ [Dt\ the]\ [N_c\ woman]]\ [VP\ [V_i\ sat]\ [PP\ [P\ in]\ [NP\ [Dt\ the]\ [N_c\ yard]]]]]]]$
    S1; NP1; NP2; VP5; VP3; PP1

2. (a) Aristotle was a philosopher.
 [S [NP [N$_p$ Aristotle]] [VP [V$_t$ was] [NP [Dt a] [N$_c$ philosopher]]]]
 S1; VP2; NP1; NP2
 Aristotle|N$_p$, *was*|V$_t$, *philosopher*|N$_c$

 (c) This boy lost his wallet.
 [S [NP [Dt this] [N$_c$ boy]] [VP [V$_t$ lost] [NP [Dt his] [N$_c$ wallet]]]]
 S1, NP2, VP2
 boy|N$_c$, *lost*|V$_t$, *his*|Dt, *wallet*|N$_c$

 (e) The soldier grew sad.
 [S [NP [Dt The] [N$_c$ soldier]] [VP [V$_a$ grew] [AP [A sad]]]]
 S1; NP2; AP1; VP $\Rightarrow$ V$_a$ AP
 grew|V$_a$, *sad*|A, *soldier*|N$_c$

 (g) Carl sat in the first chair.
 [S [NP [N$_p$ Carl]] [VP [V$_i$ sat] [PP [P in] [NP [Dt the] [AP [A first]] [N$_c$ chair]]]]]
 S1; VP3; PP1; NP1; NP3; AP1
 first|A, *chair*|N$_c$

 (i) The city of Paris is rather beautiful.
 [S [NP [Dt the] [N$_c$ city] [PP [P of] [NP [N$_p$ Paris]]]] [VP [V$_a$ is] [AP [Dg rather] [A beautiful]]]]
 S1; PP1; AP2; NP1; NP4; VP $\Rightarrow$ V$_a$ AP
 is|V$_a$, *city*|N$_c$, *Paris*|N$_p$, *beautiful*|A

 (k) Galileo persuaded the philosopher of his mistake.
 [S [NP [N$_p$ Galileo]] [VP [VP [V$_i$ persuaded] [NP [Dt the] [N$_c$ philosopher]]] [PP [P of] [NP [Dt his] [N$_c$ mistake]]]]]
 S1; VP4; NP1; NP2; PP1
 mistake|N$_c$, *persuaded*|V$_i$

 (m) Galileo persuaded the patrician he should leave.
 [S [NP [N$_p$ Galileo]] [VP [V$_{ps}$ persuaded] [NP [Dt the] [N$_c$ patrician]] [S [NP [N$_{pr}$ he]] [Av should] [VP [V$_i$ leave]]]]]
 S1; NP1; VP1; VP $\Rightarrow$ V$_{ns}$ NP S; NP2; S $\Rightarrow$ NP Av VP; NP $\Rightarrow$ N$_{pr}$
 should|Av, *he*|N$_{pr}$, *leave*|V$_i$, *persuaded*|V$_{ns}$, *patrician*|N$_c$

 (n) An expert will speak to this class about the languages of India.
 [S [NP [Dt an] [N$_c$ expert]] [Av will] [VP [V$_{pp}$ speak] [PP [P to] [NP [Dt this] [N$_c$ class]]] [PP [P about] [NP [Dt the] [N$_c$ languages] [PP [P of] [NP [N$_p$ India]]]]]]]
 PP1; NP2; NP1; NP4; S $\Rightarrow$ NP Av VP; VP $\Rightarrow$ V$_{pp}$ PP PP
 will|Av, *expert*|N$_c$, *India*|N$_p$, *speak*|V$_{pp}$

3.2 Constituency grammar

1. (a) The visitor bit the guest.
   [S [NP [Dt the] [N_c visitor]] [VP [V_t bit] [NP [Dt the] [N_c guest]]]]
   NP VP → S; Dt N_c → NP; V_t NP → VP

   (c) The taciturn host sat in a park.
   [S [NP [Dt the] [AP [A taciturn]] [N_c host]] [VP [V_i sat] [PP [P in] [NP [Dt a] [N_c park]]]]]
   NP VP → S; Dt N_c → NP; Dt AP N_c → NP; V_i PP → VP; P NP → PP

   (e) Eric saw the very tall guest.
   [S [NP [N_p Eric]] [VP [V_t saw] [NP [Dt the] [AP [Dg very] [A tall]] [N_c guest]]]]
   NP VP → S; N_p → NP; Dt AP N_c → NP; Dg A → AP; V_t NP → VP

   (g) The friendly host yelled in the yard.
   [S [NP [Dt the] [AP [A friendly]] [N_c host]] [VP [V_i yelled] [PP [P in] [NP [Dt the] [N_c yard]]]]]
   NP VP → S; Dt N_c → NP; Dt AP N_c → NP; V_i PP → VP; P NP → PP

2. (a) The set SL as generated by FRp (chap. 1, sec. 2.2.1):
   BX = {A, B, C, D}
   CT = {L, SL}
   LX = {A|L, B|L, C|L, D|L}
   FR = {L → SL, SL L → SL}
   A|L

   (b) The set SL as generated by FRb (chap. 1, sec. 2.2.2):
   FR = { L → SL, SL SL → SL}

   (d) The set of roman numerals (chap. 1, sec. 2.2.3):
   The answers to the preceding questions are adapted directly from formation rules given in chapter 1, section 2.2. A constituency grammar cannot be so directly adapted. The following, while a constituency grammar, does not properly characterize the roman numerals as specified in chapter 1.

   BX = {I, V, X, L, C, D, M}

   CT = {N, RN}

   LX = {I|N, V|N, X|N, L|N, C|N, D|N, M|N}

   FR = {N → RN, RNRN → RN}

The problem is that the second formation rule generates expressions that are not roman numerals, such as, for example, IIIII, IIIV, VV, LL. The reason is that the following condition, used in the formation rule for roman numerals in the first chapter included the following condition:

Basic English Grammar   169

no symbol in z outranks any symbol in y and, in y and z taken together, V, L, and D occur at most once and I, X, C, and M occur at most four times

To incorporate this condition into a constituency grammar requires that more categories be added to CT and that many more rules be added to FR.

3.3.1 Corroborating evidence

1. *answer:* noun and verb
   Ben will *answer* the question versus The *answer* to the question was not convincing

   bitter: adjective, noun
   The drink is *bitter* versus The customer drank two pints of *bitter*

   bore: verb, noun
   A dull lecturer will *bore* his or her audience versus
   The lecturer is a *bore*

   brake: verb, noun
   The driver could not *brake* in time versus The driver did not apply the *brake* quickly enough

   cash: verb, noun
   The teller did *cash* the check versus The customer did have have enough *cash*

   divide: verb, noun
   Will they *divide* the reward versus On which side of the *divide* lie the contenders

   hammer: verb, noun
   Max *hammered* a nail versus Max used a *hammer*

   love: verb, noun
   Did Juliette *love* Romeo versus Juliette *loved* Romeo

   yellow: verb, adjective
   White cotton shirts *yellow* gradually versus The white cotton shirt turned *yellow* gradually

2. (a) Police police police.
       $[S\ [NP\ [N_b\ police]]\ [VP\ [V_t\ police]\ [NP\ [N_b\ Police]]]]$
       $N_b \to NP$

   (c) Alice introduced Bill to Carl.
       $[S\ [NP\ [N_p\ Alice]]\ [VP\ [V_d\ introduced]\ [NP\ [N_p\ Bill]]\ [PP\ [P\ to]\ [NP\ [N_p\ Carl]]]]]]$
       $V_d\ NP\ PP \to VP$

   (e) The day after Tuesday is the day after the day after Monday.
       $[S\ [NP\ [Dt\ the]\ [N_c\ day]\ [PP\ [P\ after]\ [NP\ [N_p\ Tuesday]]]]\ [VP\ [V_c\ is]\ [NP\ [Dt\ the]\ [N_c\ day]\ [PP\ [P\ after]\ [NP\ [Dt\ the]\ [N_c\ day]\ [PP\ [P\ after]\ [NP\ [N_p\ Monday]]]]]]]]]$

(g) The guest treated the dog badly.
[S [NP [Dt the] [N_c guest]] [VP [V_{na} treated] [NP [Dt the] [N_c dog]] [Ad badly]]]
V_{na} NP Ad → VP

(i) Each official averse to the proposal resigned from his position.
[S [NP [Dt each] [N_c official] [AP [A averse] [PP [P to] [NP [Dt the] [N_c proposal]]]]] [VP [V_i resigned] [PP [P from] [NP [Dt his] [N_c position]]]]]
Dt N_c AP → NP; A PP → AP

3.3.2 Amphiboly

1. Abbreviated labeled brackets are used.

   (a) Bill said Alice arrived on Tuesday.
   CONSTRUAL 1: Bill said that on Tuesday Alice arrived.
   [S [NP Bill] [VP said [S [NP Alice] [VP arrived [PP on [NP Tuesday]]]]]]
   CONSTRUAL 2: On Tuesday Bill said Alice arrived.
   [S [NP Bill] [VP said [S [NP Alice] [VP arrived]]] [PP on [NP Tuesday]]]

   (b) They can fish.
   CONSTRUAL 1: They put fish into cans.
   [S [NP they] [VP [V can] [NP fish]]]
   CONSTRUAL 2: They are able to fish.
   [S [NP they] [Ax can] [VP fish]]

   (c) The old men and women left.
   CONSTRUAL 1: The old men and the old women left.
   [S [NP [Dt the] [AP old] [N_c [N_c men] [C_c and] [N_c women]]] [VP left]]
   CONSTRUAL 2: The women and the old men left.
   [S [NP [NP [Dt the] [AP old] [N_c men]] [C_C and] [NP women]] [VP left]]

   (d) Alice knows how good Chinese food tastes.
   CONSTRUAL 1: Alice knows what good Chinese food tastes like.
   [S [NP Alice] [VP knows [S [Av how] [NP good Chinese food] [VP tastes]]]]
   CONSTRUAL 2: Alice knows how good Chinese is.
   [S [NP Alice] [VP knows [S [AP how good] [NP Chinese food] [VP tastes]]]]

   (e) The merchant rolled up the thick carpet.
   CONSTRUAL 1: The thick carpet was rolled up by the merchant.
   [S [NP the merchant] [VP [V rolled up] [NP the thick carpet]]]
   CONSTRUAL 2: Up the carpet rolled the merchant.
   [S [NP the merchant] [VP rolled [PP up [NP the thick carpet]]]]

(f) Alice greeted the mother of the girl and the boy.
CONSTRUAL 1: Alice greeted one person, the mother of the girl and the boy.
[S [NP Alice] [VP greeted [NP the mother [PP [P of] [NP the girl and the boy]]]]]
CONSTRUAL 2: Alice greeted two people, the boy and the mother of the girl.
[S [NP Alice] [VP greeted [NP [NP the mother [PP of [NP [NP the girl]]]] [C_c and] [NP the boy]]]]

(g) April bought the men's store.
CONSTRUAL 1: April bought these men's store.
Here the possessive suffix 's follows a noun phrase to form a noun phrase: that is, NP Ps N_c → NP.
[S [NP April] [VP bought [NP [NP the men] [Ps s] store]]]
CONSTRUAL 2: April bought this men's store.
Here the possessive suffix 's follows a noun to form a compound noun: that is, N_c Ps N_c → N_c.
[S [NP April] [VP bought [NP the [Nc [men] [Ps s] store]]]].

(h) The security guards must stop drinking after midnight.
CONSTRUAL 1: The security guards' drinking after midnight must stop.
[S [NP the security guards] [VP must stop [GC drinking [PP after [NP midnight]]]]]
CONSTRUAL 2: After midnight the security guards must stop drinking.
[S [NP the security guards] [VP must stop [GC drinking]] [PP after [NP midnight]]]

(i) The prosecutor requires more convincing evidence.
CONSTRUAL 1: The prosecutor requires more evidence that is convincing.
[S [NP The prosecutor] [VP requires [NP [Dt more] [AP convincing] evidence]]]
CONSTRUAL 2: The prosecutor requires evidence that is more convincing.
[S [NP The prosecutor] [VP requires [NP [AP [Av more] convincing] evidence]]]

(j) Bill told Colleen Fred left on Monday.
CONSTRUAL 1: Bill told Colleen that on Monday Fred left.
[S [NP Bill] [VP told [NP Colleen] [S [NP Fred] [VP left] [PP on Monday]]]]
CONSTRUAL 2: On Monday Bill told Colleen that Fred left.
[S [NP Bill] [VP [V told] [NP Colleen] [S [NP Fred] [VP left]]] [PP on Monday]]

(k) The stout major's wife is dancing.
CONSTRUAL 1: The wife of the stout major is dancing.
[S [NP [NP [NP The stout major] [Ps s]] wife] [VP is dancing]]
CONSTRUAL 2: The stout wife of the major is dancing.
[S [NP The stout [Nc [Nc major] [Ps s]] wife] [VP is dancing]]

(l) The tall visitor took his host's picture.
No amphiboly; *take* is ambiguous.
CONSTRUAL 1: The tall visitor removed his host's picture.
CONSTRUAL 2: The tall visitor snapped a picture of his host.
[S [NP The tall visitor] [VP took [NP his host's picture]]]

(m) Ali does not like worrying neighbors.
CONSTRUAL 1: Ali does not like neighbors who worry.
[S [NP Ali] [Ax does] [Av not] [VP like [NP [AP worrying] neighbors]]]
CONSTRUAL 2: Ali does not like to worry neighbors.
[S [NP Ali] [Ax does] [Av not] [VP like [GC worrying [NP neighbors]]]]

(n) Alice met a friend of Bill's mother.
CONSTRUAL 1: Alice met a friend of the mother of Bill.
[S [NP Alice] [VP met [NP a friend [PP of [NP [NP Bill] [Ps s] mother]]]]]
CONSTRUAL 2: Alice met the mother of Bill's friend.
[S [NP Alice] [VP met [NP [NP a friend [PP of [NP [NP Bill]]]] [Ps s] mother]]]

(o) Babu dropped an egg onto the concrete floor without cracking it.
No amphiboly; *it* may take either of two antecedents.
[S [NP Babu] [VP dropped [NP an egg] [PP onto the concrete floor] [PP without cracking it]]]

2. Consider, for example, the case where Louis Pasteur showed that body temperature is a factor in whether or not an animal succumbs to anthrax.

3.3.3 Recursion

Relative clause modification:
This is the cat that killed the rat that ate the malt that lay in the house that Jack built (Quirk et al. 1985, chap. 2.9)

3.4.1 Enriching constituency grammar

2. (a) lexical entry: *remained*|V_a;
new rule: V_a AP → VP

(c) lexical entry: *noticed*|V_s;
No new rule required.

(e) lexical entry: *decided*|V_{ic};
new rule: V_{ic} IC → VP

(g) new entry: *made*|V_{na};
new rule: V_{na} NP AP → VP

(i) lexical entry: *convinced*|V_{np};
new rule: V_{np} NP PP → VP

3.4.2 Grammar and belief

1. Dan frightened his friend/sincerity.
2. Each week/book elapsed in quick succession.
3. The dog looked terrifying/barking.
4. Colleen solved the problem/pipe.
5. The rioters/rioter dispersed.
6. Surgeons/Ophthalmologists are smarter than eye doctors.
7. Bill memorized the score of the sonata he composed/will compose last/next week.
8. The decision amazed the experts/injustice.
9. The repairman is wiring the entire house/poem this time.
10. A sudden/sour flash disturbed the baby's sleep.

4 Language and Context

1 Context

In chapter 3, we elaborated on three ideas fundamental to semantics set out in chapter 1. The first idea is that the myriad of expressions comprising a language results from the application of a finite set of finitely specifiable rules to a finite set of basic expressions. We illustrated this idea in great detail in chapter 3. The second idea is that humans understand a complex expression through an understanding of its constituent subexpressions. This idea received empirical support from the simple observation made in chapter 1 that two complex expressions, in particular two sentences, that differ from each other only by one word can thereby differ from each other in one being judged true and the other false. The third idea is that the way in which simpler expressions are put together to form more complex expressions plays a role in how one's understanding of a complex expression arises from one's understanding of constituent subexpressions making it up. Thus, as we saw illustrated in chapter 3, one and the same sentence can be judged both true and false and these judgments of truth and falsity are correlated with the different immediate constituent analyses to which it is subject.

A fourth idea, not mentioned previously but equally fundamental, is that some aspect of the understanding of an expression remains invariant over its different uses. For should any expression have no invariant understanding associated with it from occasion of use to occasion of use, no one could understand it at all. However, to say that some aspect of the understanding of an expression remains invariant from occasion of use to occasion of use does not mean that other aspects of its understanding cannot be modulated by the context in which it is used. Indeed, every natural language has expressions whose understanding is modulated from context of their use to context of their use. In this chapter, for a limited but commonly used class of expressions, we shall explore how their understanding can be modulated by the context in which they are used.

It is not news that declarative sentences, when uttered, convey knowledge about the world. Thus, if it is said

(1)   Water is composed of oxygen and hydrogen.

one learns something about the chemical composition of water. What is equally true, but perhaps not as obvious, is that knowledge of the world is often necessary to understand what knowledge declarative sentences, when uttered, are expressing.

Despite appearances, there is no circularity here; though there would be circularity should it be that to understand a sentence expressing some fact, one had to know already that very fact. But this is not the case. What is the case is that to understand a sentence expressing some fact, one has to know other facts, namely facts pertaining to the circumstances in which the sentence is uttered.

To see how knowledge of the circumstances in which a sentence is uttered contributes to the understanding of what the sentence expresses, consider the following pair of sentences:

(2.1)   Marco Polo died here.
(2.2)   Marco Polo died in Venice.

The second sentence is true, as anyone with an adequate knowledge of Marco Polo's life can judge, whereas even someone with a thorough knowledge of Marco Polo's life cannot judge whether or not the first sentence is true without knowing the circumstances of its utterance. Clearly, then, a complete understanding of the first sentence requires knowledge of the situation in which it is uttered. This is confirmed by the following. Holding sentence (2.1) fixed and changing the circumstances in which it is uttered, one sees that its truth value can change. Said in Genoa, where Marco Polo was imprisoned, the sentence is false; said in Venice, the city from which he left on his famous voyage, the sentence is true.

Let us consider another pair of sentences.

(3.1)   It is even.
(3.2)   Two is even.

These sentences, though grammatically indistinguishable, differ in one crucial respect. Anyone who knows elementary arithmetic knows that the second sentence is true, but not even the greatest mathematician can say whether or not the first sentence is true.

To appreciate better the point, consider two other sentences, each of which is true:

(4.1)   Two is a prime number.
(4.2)   Three is a prime number.

Conjoin each of the sentences in (4) with sentence (3.1), thereby obtaining the following pair of biclausal sentences.

Language and Context

(5.1) Two is a prime number and it is even.

(5.2) Three is a prime number and it is even.

Even in the absence of all knowledge of the situation in which they are uttered, one judges the second clause of the first sentence true and one judges the second clause of the second sentence false. But the second clause is the very same as sentence (3.1). Clearly, a complete understanding of sentence (3.1) depends on its preceding text.

What we just saw regarding the pronoun *it* and the word *here* is that to understand fully what sentences containing them mean, one must know either something about the circumstances in which the sentence is uttered or something about the text that either comes before or comes after them. The words we shall be studying in this chapter are English words of this kind. We shall call the text that either comes before or comes after an utterance its *cotext* and we shall call the physical circumstances in which an expression is uttered its *setting* or its *circumstances of utterance*. Our first example shows how knowledge of the setting, or circumstances of utterance, is necessary for sentence (2.1) to be understood fully; our second example shows how knowledge of the cotext of sentence (3.1) is necessary for it to be understood fully.

This chapter will provide an overview of the diversity of expressions that, like *it* and *here,* depend for their understanding on knowledge of either their setting or their cotext. The section 2 is concerned exclusively with expressions whose full understanding requires knowledge of the setting in which they are uttered; section 3 is devoted exclusively to those whose full understanding requires knowledge of the cotext surrounding them. Section 4 will turn to an important distinction between a sentence's *circumstances of utterance,* or its *setting,* and a sentence's *circumstances of evaluation,* the circumstances with respect to which it is evaluated as either true or false. Appreciating this distinction will permit us to avoid confusing an expression's ambiguity with its dependence on context, where an expression's context is either its cotext or its setting.

2 Setting and Exophora

The following amusing story illustrates extremely well both the extent to which speakers of a language take for granted knowledge of the setting of utterances and its being essential to the very possibility of communication (Rosten 1968, 443–444, cited in Levinson 1983, 68). A melamed (Hebrew teacher) discovering that he had left his comfortable slippers back in the house, sent a student after them with a note for his wife. The note read: "Send me your slippers with this boy." When the student asked why he had written "your" slippers, the melamed answered: "Yold! If I wrote 'my' slippers, she would read 'my' slippers and would send her slippers. What could I do with her slippers? So I wrote 'your' slippers, she'll read 'your' slippers and send me mine."

We shall call expressions such as *your* and *my* whose full understanding requires knowledge of the setting in which they are uttered *exophors*.[1] Before surveying English exophoric expressions, let us reflect on two quintessential examples of English exophoric expressions, the first- and second-person personal pronous *I* and *you*.

Consider the following two pairs of sentences:

(6.1) I am sad.
(6.2) You are sad.
(7.1) Reed is sad.
(7.2) Dan is sad.

Suppose Reed utters sentence (6.1) and Dan does too. Have Reed and Dan said the same thing? Well, yes and no. On the one hand, they have both uttered the same sentence and they have thereby said the same thing. Yet, on the other hand, they have said different things. For when Reed utters sentence (6.1), he or anyone else could say the same thing by uttering sentence (7.1). Similarly, when Dan uttered sentence (6.1), the very same thing would be expressed by him or anyone else uttering sentence (7.2). But clearly the sentences in (7) do not say the same thing. Indeed, sentence (7.1) could be true, while sentence (7.2) could be false.

Let us now turn to sentence (6.2). Suppose that Reed utters sentence (6.1) and Dan utters sentence (6.2), addressing it to Reed. Have Reed and Dan said the same thing? On the one hand, they have, for instead of Reed uttering sentence (6.1) and Dan uttering sentence (6.2), they could both utter sentence (7.1). Yet, on the other hand, they have not. Suppose that Reed is sad and Dan is not. If Dan utters sentence (6.2) addressing it to Reed, he has said something true. But should anyone else utter sentence (6.2) addressing it to Dan, he or she would have said something false.

What we just saw is that exphors such as *I,* the first-person singular personal pronoun, and *you,* the second-person personal pronoun, refer to different people, depending on who is uttering the pronoun and who is being addressed in the utterance. In this way, their understanding varies from circumstance of utterance to circumstance of utterance. Yet, at the same time, something remains invariant in their understanding on each occasion of use: the first-person personal pronoun always refers to the speaker of the utterance in which it occurs and the second-person one always refers to the person addressed by the utterance in which it occurs.

The philosopher David Kaplan dubbed what is invariant in the understanding of exophors *character*. Thus, associated with sentence (6.1) is a character that remains invariant, no matter who utters it when. The meaning is that whereby sadness is attributed by its utterer to whoever utters it at the time he or she utters it. Thus, each sentence in (6) has

---

1. This term is used by Ewan Klein (1980). Linguists often call such expressions *deictic,* while philosophers often call them *indexical*.

a character different from the other. In this way, the sentences say different things. What the character of each sentence is can be ascertained from the following thought experiment: imagine what you would understand from either sentence, should you come across it written on a piece of paper found in a bottle that washes up on the shore. Under those circumstances, you can see that the sentences manage to convey very little, but what they do manage to convey are different from one another. The first attributes sadness to its writer at the time of his or her writing; the second attributes sadness to some unknown addressee at the time of writing.

At the same time, the sentences in (6), used in the right setting, can come to say the same thing. In the setting just described, they both come to say the same thing as sentence (7.1). Borrowing a term from Kaplan, we shall call the meaning an utterance acquires when it is, as it were, filled out by a suitable knowledge of the setting in which it is used, its *content*.

In general, all expressions have a character, the meaning of which remains invariant from setting to setting. At the same time, many expressions, in particular exophors and expressions containing exophors, acquire a fuller meaning, which we shall call content, when they are used. So, while the character of an expression remains invariant from setting to setting, its content might very well change. Moreover, as we just saw, expressions with different characters may, on particular occasions of use, come to have the same content.

Linguists and philosophers have long been aware of the fact that certain expressions in a natural language depend for their complete understanding on knowledge of their setting. Relevant discussion by philosophers goes back at least to Charles Sanders Peirce (1839–1914), Gottlob Frege (1848–1925), Edmund Husserl (1859–1938), and Bertrand Russell (1872–1970). However, no real formal appreciation of how values assigned to exophoric expressions differ from how values are assigned to nonexophoric expressions was reached until the middle of the twentieth century when philosopher logicians such as Hans Reichenbach (1947, sec. 50), Arthur Burks (1949), Edward Lemmon (1966), and Yehoshua Bar-Hillel (1954) devoted serious attention to them. The breakthrough came in the late 1960s, when the logician Richard Montague (1968; 1970c) realized that the model theory of modal logic could be adapted to treat these problems. As a result, a number of philosopher logicians tackled the problem of exophora in detail, the most notable among them being Max Cresswell (1973), David Kaplan (1978–1979), David Lewis (1975), and Robert Stalnaker (1974; 1978). We do not, unfortunately, have space to present their treatments here.

Philosophers are not the only thinkers with an interest in exophors. They have also received attention from linguists as well. Thus, while noted by linguists such as Karl Bühler (1934) and Henri Frei (1944) in the first half of the last century, they started to get more systematic descriptive attention in the last third of the twentieth century by Charles Fillmore (1971) and John Lyons (1977), just to mention two. We now turn to their observations.

Each natural language frames the setting in which utterances are made. Though languages differ in how they do so, much of how they frame the setting is the same. Typically, the setting includes "at least two participants, one of whom is signalling in the phonic medium along the vocal-auditory channel, with all the participants present in the same actual situation able to see one another and to perceive the associated nonvocal paralinguistic features of their utterances, and each assuming the role of sender and receiver in turn" (Lyons 1977, 637).[2]

All frames agree in taking the time and the location of an utterance as the *origin*. Thus, the utterance naturally divides its cotext into those utterances that precede it and those that come after it. Moreover, the time of the utterance naturally divides events into those that precede the event of the utterance, those that succeed it, and those that are simultaneous with it. In addition, the person who produces the utterance has a privileged role in the scene to which other participants have a subordinate role. An utterer's physical location serves as a natural origin with respect to which things can be classified either as near or remote. Moreover, these roles change as the speaker changes. We shall see how different expressions and devices require knowledge of different coordinates for their content to be understood. In particular, we shall see how the various roles are expressed and how things are related temporally and spatially to the origin. In sections 2.1 to 2.4, we shall survey English exophoric expressions.

2.1 Person

We begin with the notion of grammatical person. The term *person* comes from the Latin word *persona* (mask), used by the Latin Grammarians to translate the Greek term for dramatic character or role (Lyons 1977, 638).

When a sentence is uttered, one person, the speaker of the sentence, is in the role of speaker. The pronoun used by the speaker to refer to himself or herself is the first-person (personal) pronoun. The person to whom the utterance is addressed is in the role of the addressee and the pronoun used by the speaker to refer to that person is the second-person (personal) pronoun. If there is only one addressee, then the pronoun is used in its singular form; if there are more than one addressee, then the pronoun is used in its plural form.

Other persons in the setting who are not addressed by the speaker are in the role of bystanders. The third-person (personal) pronoun may be used to refer to them. This may be done by the speaker demonstrating the relevant person. Thus, for example, a person can point to a man in a crowd and say

(8) He is a friend of mine.

The deictic usage, however, need not involve a demonstration. Sometimes, the referent may be evident from context. A medic, arriving on the scene of an accident, can very easily inquire about the state of a single injured woman, asking simply

---

2. The organization may also include the social position of participants with respect to one another (see Allan 1986, chap. 1.3; Lyons 1977, chap. 14.1–14.2).

(9) How is she?

Indeed, as Geoffrey Nunberg (1993, 23) has observed, the referent may be remote. Walking through the Taj Mahal, the speaker who says

(10) He certainly spared no expense.

can use the third-person personal pronoun, without any demonstration, to refer to Shah Jahān (1592–1666), the ruler who commissioned the building of the Taj Mahal.

To illustrate the exophoric nature of the personal pronouns, consider sentence (11.1).

(11.0) Napoleon Bonaparte is the emperor of France.
(11.1) I am the emperor of France.
(11.2) You are the emperor of France.
(11.3) He is the emperor of France.

Whether or not they express something true depends on who is in what role. If the speaker is Napoleon Bonaparte (1769–1821), then sentence (11.1) is true. If the addressee is Napoleon Bonaparte, then sentence (11.2) is true. And if Napoleon Bonaparte is the referent of the pronoun *he,* then sentence (11.3) is true. It should be obvious then that one must have correct beliefs about who is in which role to understand fully any sentence containing a personal pronoun, as the story about the Hebrew teacher makes clear.

Personal pronouns are not the only way that grammatical person is expressed. Grammatical person can also be expressed by verbal inflection. The fact that verbal inflection expresses grammatical person is particularly evident in languages such as Latin. The following verb paradigm is for the present tense active voice of the verb *amāre* (to love).

	SINGULAR	PLURAL
FIRST PERSON	am-**o** **I** love	amā-**mus** **we** love
SECOND PERSON	ama-**s** **you** love	amā-**tis** **you** love
THIRD PERSON	ama-**t** **he** loves	ama-**nt** **they** love

As shown by the English translation, the Latin verbal endings express, among other things,[3] the grammatical person. The English verb, having no verbal ending, except in the third-person singular, expresses almost nothing about grammatical person. Indeed, so expressive of grammatical person are the Latin verbal endings that Latin does not require

---

3. Inflectional morphemes encode diverse information; see Lyons (1968, chap. 5.3; 1977, 639) for more discussion.

any clause to have a subject, whereas English requires all finite clauses to have a subject. Thus, the Latin poet Marcus Valerius Martialis (ca. 40–103 CE) could say the following declarative sentence, which lacks any noun phrase corresponding to the subject of the clause,

(12)   odi   profanum   vulgus.
         I hate   common   crowd

whereas its English translation cannot lack a subject noun phrase.

(13.1)   I hate the common crowd.

(13.2)   *hate the common crowd.

In other languages, the inflectional morpheme of the verb may have the same person as the verb's object or indirect object.

2.2 Temporal Order

Sentences expressing an event usually indicate the relation between the time of the occurrence of the event expressed and the time of the utterance of the sentence expressing it. There are only three possible temporal relations: the time of an event expressed is simultaneous with the time of the utterance expressing the event; the time of the event expressed is before the time of the utterance; and the time of the event is after the time of the utterance. We shall refer to these temporal relations as *grammatical tense* and distinguish them as *present, past,* and *future*. Moreover, we distinguish what we call grammatical tense from *morphological tense,* which refers to the form of the verb. As we are about to see, grammatical tense and morphological tense are only roughly correlated.

These three temporal relationships, or grammatical tenses, may be expressed by inflectional endings on verbs, thereby giving rise to various morphological tenses (Levinson 1983, 76–78).

(14.1)   Napoleon Bonaparte *is* the emperor of France.
         True, if uttered in 1805.
         False, if uttered in 1822.

(14.2)   Napoleon Bonaparte *was* the emperor of France.
         True, if uttered in 1822.
         False, if uttered in 1804.

(14.3)   Napoleon Bonaparte *will be* the emperor of France.
         True, if uttered in 1799.
         False, if uttered in 1815.

Verbs have different morphological tenses whereby the time of what is expressed in the sentence in which the verb occurs is ordered with respect to the time of the utterance of the sentence.

It is important to emphasize that the relation between morphological tense and grammatical tense is not the simple-minded one where the present tense of the verb in a clause requires what is expressed by the clause to obtain at the time of the clause's being uttered; the past tense of the verb in a clause requires what is expressed by the clause to obtain at a time prior to the clause's being uttered; and the future tense of the verb in the clause requires what is expressed by the clause to obtain at a time subsequent to the clause's being uttered.

Thus, for example, clauses whose verbs are in the simple present tense are used to express eternal, or atemporal, truths.

(15.1) A quadratic equation *has* at most two real-valued solutions.

(15.2) Water *is composed* of two hydrogen atoms and one oxygen atom.

And verbs in the present progressive can be used to express truths about the future.

(16) Colleen *is leaving* for Boston tomorrow.

The three basic temporal relationships, as well as other more specific temporal relationships, can be expressed by adverbs.[4] For example, the adverb *now* relates the event expressed by a sentence to the time of its utterance. The adverb *then* relates the event expressed to any time other than the time of the utterance of the sentence in which it occurs. Other temporal adverbs include *later, earlier, recently, today, tomorrow, yesterday*, and so on. Moreover, even names for the days of the week can express temporal relations between the time of the setting and the time of what is expressed.[5]

(17.1) Bill arrives *on Monday*.
Monday is the first Monday after the day of the utterance.

(17.2) Bill arrived *on Monday*.
Monday is the first Monday preceding the day of the utterance.

(17.3) Bill will arrive *on Monday*.
Monday is the first Monday after the day of the utterance.

Grammatical tense conveys not only grammatical tense but also aspect. Aspect pertains to whether or not the action denoted by the verb is to be regarded as complete, with the relevant morphological tense qualified with the word *perfect,* or as incomplete, with the relevant morphological tense qualified as *imperfect, continuous,* or *progressive*. (For examples from English, readers should consult section 2.1.3 of chapter 3. These properties of grammatical tense were recognized and studied both by Classical Greek and Latin grammarians and by the Indian grammatical tradition. However, the key ideas used to

---

4. For more, consult Anderson and Keenan (1985).
5. See Leech (1970, chap. 7.1.4); also see Kamp and Reyle (1993, chap. 5.2, esp. 546, 613–635).

treat grammatical tense and aspect appeared in the middle of the last century in the work of Hans Reichenbach (1947) and in the work of Arthur Prior (1914–1969), who drew on ideas found in Medieval European logicians (Prior 1957), and in the work of Donald Davidson (1967). More properly linguistic treatments began with the work of David Dowty (1979), taking up the approach found in Prior (1957), and with that of Hans Kamp, taking up the approach found in Reichenbach (1947). Work based on Davidson's seminal paper is extensive and goes under the name of event semantics, one systematic version of which appears in Parsons (1990).

2.3 Spatial Location

An utterer's physical location serves as a natural origin with respect to which things can be classified either as near or remote. This classification is reflected in the difference between the demonstrative pronouns *this* and *that* as well as the adverbs *here* and *there*.

Suppose that Steve and Jeff are friends and that they are at a party in a conversation with one another. Suppose also that Steve has a colleague at the party whom he wishes to introduce to Jeff. On the one hand, if the colleague is facing them and within normal range to be engaged in a conversation, then the first sentence in (18) is appropriate, not the second. On the other hand, if the colleague is across the room with his or her back turned to Steve and Jeff, the second sentence alone of the two is appropriate.

(18.1) *This* is my colleague.

(18.2) *That* is my colleague.

A similar contrast is true of the adverbs *here* and *there*.

(19.1) *Here* is my colleague.

(19.2) *There* is my colleague.

Or consider a mother warning her child not to eat a lollipop. If the lollipop is in the mother's hand, the first sentence in (20) is appropriate, not the second; if, however, it is in the hand of the child she is addressing, the second sentence, not the first, is appropriate.

(20.1) Don't eat *this* lollipop.

(20.2) Don't eat *that* lollipop.

Yet, the choice of one demonstrative is not always exclusive of the choice of the other. Suppose that the lollipop is on the table in front of the child and the mother and at the same distance from both. Then either demonstrative can be used and without any evident contrast in meaning.

It should be noted that demonstrative pronouns can also be used to pick out events. If the time of the event referred to either precedes or is simultaneous with the the time of the

Language and Context

utterance, then the demonstrative *that* is appropriate, while if the time of the event succeeds the time of the utterance, then the demonstrative *this* is appropriate:

(21.1) *That* is an amazing stunt.

(21.2) *This* is an amazing stunt.

Some adverbs also permit one to refer to events that are about to happen. For example, one can refer to an act one will do to illustrate an action with the adverbs *so* and *thus*.

(22.1) One shuffles cards *so*.

(22.2) One holds chopsticks *thus*!

Besides the demonstrative adjectives *this* and *that,* their pronominal counterparts and the locational adverbs *here* and *there,* certain verbs also require knowledge of spatial relations within the setting. Consider the verbs *to go* and *to come*. The first implies that what is expressed by the subject left the place where the speaker is located at the time of utterance, while the second implies that what is expressed by the subject arrived at the place where the speaker is located at the time of utterance.

(23.1) Bill went.

(23.2) Bill came.

When these verbs are followed by locative complements, as in these sentences,

(24.1) Bill went to the airport.

(24.2) Bill came to the airport.

the sentences containing them imply that the speaker, at the time of utterance, is not at the location expressed by the locative complement and that the speaker is there, respectively.

2.4 Further Subtleties

Some exophoric expressions have characters that narrowly circumscribe what they can denote in a setting; others are not so narrowly circumscribed. Consider the sentence

(25) He is my best friend.

uttered in two settings. In the first setting, Alice and Bill are talking to one another in a room. There are only two other people in the room, Dan and Carol, who are on the other side of the room standing next to one another. Alice utters sentence (25) to Bill. In this setting, it is unequivocal that Alice has said to Bill that Dan is her best friend. In the second setting, Alice and Carol are talking to one another in the same room. The only other two people in the room are Bill and Dan, who are standing on the other side of the room. Alice utters sentence (25). But we cannot say that what she said to Carol is that Dan is her best friend. We cannot tell from the setting whether what she intended to communicate to Carol is that Dan is her best friend or that Bill is. However, should, in the second setting, Alice point to Dan, nod in Dan's direction, or orient her gaze toward Dan, in such a way as her gesture were clear to Carol, then she would have communicated that Dan is her best friend. Thus, we see that it is possible for a speaker in a setting to modify it in such a way so as to make exophoric expression successfully pick out an entity that without the modification, it would not pick out.

Notice that no amount of a speaker's modification of a setting can make the setting such that the first-person singular personal pronoun picks out anyone but the speaker or such that the second-person personal pronoun picks out any but the person addressed by the speaker. Thus, for example, in a setting in which Dan is speaking to Paul, it is not possible for Dan to use the first-person singular personal pronoun *I* to refer to Paul, nor can Paul use the second-person personal pronoun (*you*) to refer to himself.

Writing was invented over four thousand years ago. It has had a tremendous impact on civilization. It has also had an impact on how exophoric expressions are used. In speech, the setting in which something is said is the same as the setting in which it is understood. Once writing appeared, this physical limitation was overcome. Suddenly, some message could be expressed at one time and understood at another. In other words, the setting in which something is written down need not be the same as the time it is understood.

Let us consider a banal example. It is a commonplace practice for people to put notes on the door of their offices. Consider the note on which is written this sentence:

(26)   I shall be back in one hour.

What does this message convey? It is that the person who wrote it will be back at the place he wrote it one hour after his writing it. If the person had written the message at noon on April 1, 2000, then the content of the message would be that the person would be back in his or her office before 1:00 pm on April 1, 2000.

Let us contrast this with another message.

(27)   Come back in one hour.

What it conveys is that the person reading the message should return to the place where the message is placed within one hour. If someone were reading the message at 1:00 pm on April 1, 2000, then the content of the message would be that he or she should return to the office at 2:00 pm on April 1, 2000.

Thus, in one case, the content of the message is relativized to the time it is written and in the other, it is relativized to the time it is read. However, there is nothing in the grammar of English that requires one to assign the character we did to each of the sentences. Consider again sentence (26). The sentence could have been written on the sheet of paper at any time before it was put on the door, anytime between minutes beforehand to years beforehand.

Language and Context

After all, if the person is someone who is often leaving his or her office for an hour during a normal work day, it would not be unreasonable to have one note that he or she merely pins to the office door on leaving. So, what is pertinent is not the time at which the note had been written, but the time at which it is put to use.

In the case of sentence (27), the moment it is put on the door is the moment it is put into use. However, the visitor will not know when that time is. Nonetheless, whereas the visitor cannot grasp its content, the visitor can grasp enough for the message to be useful. Knowing that the person left before he or she had come to the door, the visitor concludes that if the person keeps his or her word he or she will be back within an hour from the reading of the note and will conclude that an upper bound for the person's absence is a little less than one hour from the time of the reading of the note.

Exercises: Setting and exophora

1. The following sentences have been said to be paradoxical. Explain why they are paradoxical. Provide evidence to support your explanation.

(a) I am not speaking.

(b) I am dead.

(c) I do not exist.

2. Consider the following sentence. Explain why this sentence, on the door of an office, is not as informative as one might like.

I'll be back in 5 minutes.

3. Formulate three sentences that are devoid of any dependence on setting for their full understanding.

4. Explain the humor of the following joke. Provide evidence for your explanation.

Two mountain climbers are on the top of a mountain wondering where they are. One of them looks at his map. Suddenly, he exclaims, while pointing to a distant peak: *I know where we are! We are on that mountain over there*.

5. Explain why the sentence *I am not here* seems false whenever someone utters it, yet can be either true or false when it is on an answering machine.

6. Using the concepts of setting and exophora, explain the role of pointing in the following sentences:

(a) You [pointing to A], you [pointing to B] and you [pointing to C] are fired. (Levinson 1983, 65–66)

(b) He [pointing to A] is not the duke, he [pointing to B] is. He [pointing to A again] is the butler. (Levinson 1983, 65–66)

3   Cotext: Endophora and Ellipsis

There are at least two well-known ways in which one part of an expression depends on another for its proper understanding. One way is for an expression to contain a special kind of word whose complete understanding depends on one's understanding of another expression in the former's cotext. An expression that typically has this kind of dependence is a third-person personal pronoun. As an illustration, consider sentence (28.1) as it occurs in (28.2).

(28.1)   He was driving it today.

(28.2)   Bill bought a BMW. He was driving it today.

(28.3)   Ed bought a Honda. He was driving it today.

Imagine a situation in which Reed and Dan are talking about their friend, Bill, and Dan utters to Reed sentences (28.2). Imagine further that a neighbor, say Paul, overhears only the second sentence in (28.2). Not having heard the first sentence, Paul has a less full understanding of the second sentence than Reed does. When Reed hears sentences (28.2), upon hearing the second sentence, he learns that Bill was driving that day a BMW that Bill himself had bought; whereas when Paul hears the second sentence, not having heard anything else, he learns only that someone was driving something that day. Moreover, notice that what Reed understands in hearing the second sentence varies, depending on whether what he heard immediately before is the sentence *Bill bought a BMW,* as in (28.2), or the sentence *Ed bought a Honda,* as in (28.3).

We shall call this kind of dependence *endophora,* and we shall call those words exhibiting this kind of dependence *endophors* or *proforms*. As we shall see, some endophors are not pronouns and not all pronouns are endophors.[6]

Another form of dependence of the understanding of one expression on the understanding of another in its cotext is ellipsis. The expression in (29.1) is an example.

(29.1)   Ed a Porsche.

(29.2)   Bill bought a BMW and Ed a Porsche.

(29.3)   Bill sold a BMW and Ed a Porsche.

Suppose again that Dan is talking to Reed and he utters the expression in (29.2). Suppose also that the neighbor, Paul, overhears only the part of the expression in (29.2) that is identical with the utterance in (29.1).

Now Reed, when he hears the utterance in (29.2), understands, upon hearing *Ed a Porsche,* that Ed bought a Porsche. But Paul, when he hears *Ed a Porsche,* does not

---

6. We avoid the term *anaphor,* originally used for the grammar of Classical Greek, since it has acquired, in much of linguistic theory, a sense narrower than that of the term *endophor,* as it will be used here.

understand that Ed bought a Porsche. At most, he understands that Ed and a Porsche are in some kind of relation. What accounts for the difference in understanding is that Reed heard *Bill bought a BMW* and Paul did not. Indeed, what Reed understands corresponds to the clause that can be reconstituted from the nonclause *Ed a Porsche* with the help of an expression that occurs in the preceding clause, namely, the verb *bought*. Moreover, had Reed heard sentence (29.3), he would have understood something different in hearing the utterance in (29.1). He would have understood, not that Ed had bought a Porsche, but that he had sold one.

Notice that the expression in (29.1) is not a constituent of English. Moreover, by itself, it fails to convey a proposition, a question, or a command. Yet, when preceded by a clause that expresses a proposition, as in (29.2) and (29.3), it too conveys a proposition. Moreover, what the nonconstituent in (29.1) conveys varies with what the preceding clause conveys, as illustrated by sentences (29.2) and (29.3), respectively. Finally, the proposition that the nonconstituent in (29.1) conveys when preceded by a suitable clause is expressed by a clause obtained from the nonconstituent by inserting the verb from the preceding clause.

Generalizing this example, we can characterize *ellipsis* as follows. An expression is a case of ellipsis just in case: (1) it is not a constituent, though it has constituents; (2) on its own, it fails to convey a proposition, command, or question; (3) yet, in the presence of an appropriate cotextual constituent, it does convey either a proposition, command, or question; (4) moreover, what it conveys in the presence of an appropriate cotextual constituent can be expressed by an expression constructed from the constituents of the ellipsis and other constituents from the cotext; and (5) what the nonconstituent expression conveys varies as the relevant cotextual constituent varies.

Endophora and ellipsis are different insofar as constituents containing endophors are not, ipso facto, judged to be defective, but those containing ellipsis are so judged. Thus, for example, sentence (28.1) is judged not to be defective, whereas the utterance in (29.1) is judged to be defective. At the same time, proforms and ellipsis are alike, insofar as a full understanding of the expression containing a proform or ellipsis requires an understanding of relevant portions of its cotext. We shall call the relevant portion of the cotext necessary for the full understanding the expression's *antecedent*. Whereas in many cases the antecedent does indeed precede the endophoric device, as we shall see, this is not always the case. Nonetheless, we shall adhere to this well used traditional term. We shall dub the relation of an endophoric device to its antecedent the relation of *antecedence*. We shall show it by underlining the relata of the relation. What follows in sections 3.1 and 3.2 is a brief survey of the various forms of the antecedence relation.

3.1 Endophora

Endophors, or proforms, encompass a variety of expressions. The third-person personal pronouns are the best known. In addition, there are the demonstrative pronouns and the indefinite pronoun *one*. But proforms are not confined to pronouns: noun phrases whose determiners are demonstrative adjectives or the definite article can be endophors too. Indeed, at least one adverb and at least one adjective are also proforms.

Not only do proforms fail to correspond to a single kind of constituent, the antecedents of proforms also fail to do so. Antecedents of endophoric third-person personal pronouns include noun phrases, modified nouns, verb phrases, and clauses. Noun phrases and clauses also serve as antecedents for endophoric demonstrative and definite noun phrases. Although the antecedents of proforms are drawn from such diverse syntactic classes as clauses, noun phrases, and verb phrases, they are, nonetheless, all constituents.

Let us look in detail at the various proforms and their antecedents. Third-person personal pronouns are perhaps the proforms that most readily come to mind. They often have noun phrases as antecedents. The noun phrases may be simple, comprising a lone proper noun, or they may be complex, comprising a noun, a determiner, and various modifiers.

(30.1) [NP Max] entered the room. After five minutes [NP he] left.

(30.2) [NP A man wearing a strange hat] entered the room. After five minutes [NP he] left.

While the antecedent of a third-person personal pronoun often comes before the pronoun, this is not always the case.

(31.1) When [NP Max] returned, [NP he] went straight to bed.
(31.2) When [NP he] returned, [NP Max] went straight to bed.
(32.1) You may not see [NP him], but [NP your father] is watching you.
(32.2) [NP Its] name does not feature prominently in tourist guides, but [NP Lausanne] is one of the prettiest cities in Europe.

Often, a third-person plural personal pronoun has but one noun phrase as an antecedent. But this too need not be the case. When they have more than one noun phrase as antecedents, they are said to have *split antecedents*.

(33.1) [NP Madge] wants to go to Majorca for [NP their] holidays, but [NP her husband] won't go there.

(33.2) [NP Bill] told [NP Mary] that [NP they] should leave.

Indeed, the relation of antecedence can be quite complex.

(34) [NP Each girl in the class] told [NP a boyfriend of [NP hers]] that [NP they] should meet [NP each other] for dinner.

Here, the antecedent of the pronoun, *they,* is split between the main clause's subject noun phrase, *each girl in the class,* and its indirect object, *a boyfriend of hers.* The indirect object itself contains a pronoun, *hers,* whose antecedent is also the main clause's subject

Language and Context

noun phrase. In addition, the pronoun, *they,* also serves as the antecedent to the reciprocal pronoun, *each other*.

Noun phrases are not the only constituents that can serve as antecedents to third-person personal pronouns. In particular, clauses may serve as antecedents of the singular personal pronoun *it*.

(35.1) [S My boss gave me a raise]. I can't believe [NP it].

(35.2) Even the people who believe [S John is a spy] publicly deny [NP it].

The clausal antecedents of the pronoun, *it,* may come after the pronoun.

(36.1) I can't believe [NP it]. [S My boss gave me a raise].

(36.2) Even the people who believe [NP it] publicly deny [S John is a spy].

(36.3) You may not realize [NP it], but [S Hesperus and Phosphorus are the same planet].

Verb phrases may also serve as antecedents to the pronoun *it*.

(37.1) First, John [VP fell into the lake]. Then [NP it] happened to Bill.

(37.2) People who [VP take drugs at an early age] later regret [NP it].

Notice that the antecedent of the pronoun *it* in the first example cannot be the preceding clause, for it is not John's falling into the lake that happened to Bill, it is falling into the lake that happened to him.

There are nonpersonal pronouns that are endophors. One is the indefinite pronoun *one*.[7]

(38.1) [NP Clean towels] are in the drawer, if you need [NP one].

(38.2) This [N coat] is more expensive than the [NP ones] I saw in the market.

(38.3) The [N man] who saves his money is wiser than the [N one] who spends his.

In general, there is a contrast between the third-person personal pronoun and the indefinite pronoun. Consider, for example, the following pair of sentences:

(39.1) Alice bought [NP a car] and Bill also bought [NP one].

(39.2) Alice bought [NP a car] and Bill also bought [NP it].

The truth of the first sentence is compatible with Alice and Bill buying different cars. Indeed, such is the salient construal. However, the truth of the second sentence is

---

7. The pronoun *one* raises questions about its lexical category and that of its antecedent, which we must ignore here.

incompatible with Alice and Bill buying different cars: the only construal is that they bought the very same car.

Interestingly, there are cases where *it,* like *one,* permits the pronoun and the antecedent to refer to different things.

(40.1) A lizard that loses [NP its tail] often grows [NP it] back.

(40.2) A lizard that loses [NP its tail] often grows [NP one] back.

Other kinds of nonpersonal pronouns that can serve as endophors are the demonstrative pronouns. Their antecedents may be entire clauses.

(41.1) [S Columbus reached the New World in 1492], but [NP this/that] did not convince anyone that the earth is round.

(41.2) [S George Bush became president] and [NP this/that] in spite of the fact that he received less than a majority of the vote.

The preceding examples suggest that *this* and *that* are interchangeable. But they are not. First, the pronoun *this* may precede its antecedent; *that* may not.

(42.1)   [NP This] is what we shall do: [S cancel classes …]

(42.2)   *[NP That] is what we shall do: [S cancel classes …]

Moreover, even when both *this* and *that* succeed their antecedents and either pronoun can be used, as shown by sentences in (43), still a subtle difference in meaning distinguishes the use of the two pronouns.

(43.1) [S Bill is sick]. [NP This] makes him irritable.

(43.2) [S Bill is sick]. [NP That] makes him irritable.

To see the difference, consider two circumstances. Both circumstances involve Bill and two of his employees and one of them is talking to the other about Bill. In the first circumstance, Bill is in his office and the two employees are about to enter. In the second, Bill is at home. He has called in sick and the two employees are at work. The first sentence is more appropriate to the first circumstance, while the second sentence is more appropriate to the second circumstance.

So far, we have been considering uses of the singular demonstrative pronouns where the antecedent is a clause. However, the antecedent can also be a noun phrase.

(44) His name isn't really [NP Rex]. He just calls himself [NP that].

Indeed, it is common for the antecedent to be a noun phrase when the demonstrative is a noun phrase determiner.

(45.1) [NP Ed] is coming to dinner. I detest [NP that man].

(45.2) I saw Fred in [NP his new sombrero]. [NP That/This hat] is really something.

In such cases, the noun in the demonstrative noun phrase has a broader denotation than its antecedent noun phrase.

Not only can clauses and noun phrases serve as antecedents to demonstrative pronouns, but verb phrases can as well.

(46.1) We shouldn't ask Mary [VP to bring a cake], since John has already volunteered to do [NP that].

(46.2) Do you [VP smoke or drink], or do you abstain from [NP these things]?

Definite noun phrases are also liable to having antecedents. As with demonstrative noun phrases with antecedents, the noun of the definite noun phrase has a broader denotation than its antecedent noun phrase.

(47.1) I never drink [NP milk]. I can't stand [NP the stuff].

(47.2) Fred applied for [NP the job with DOW], and George applied for [NP the same one].

A particularly pervasive use of definite noun phrases with antecedents are epithetic definite noun phrases.

(48.1) [NP Tom] brought me those flowers yesterday, [NP the sweetie].

(48.2) After [NP John] pushed me into the pool, [NP the bastard] ate my sandwich.

The adverb *so* is also an endophor. Its antecedents may be a clause, a verb phrase, an adjective phrase, or even an adverbial phrase.

(49.1) Though Bill won't say [AdvP so], [S he is quite good at poker].

(49.2) Fred [VP wants to go to Spain], and [AdvP so] does Tom.

(49.3) Paul was [AP generous] as a child and he has always remained [AdvP so].

(49.4) Carol behaved [AdvP very defensively] yesterday, but she does not always [AdvP so] behave.

Notice, however, that the adverb *so* overlaps in its endophoric use with pronouns such as *it* and *that*.

(50.1) Though Bill won't say [AdvP so], [S he is quite good at poker].

(50.2) Though Bill won't say [NP it], [S he is quite good at poker].

(51.1) We shouldn't ask Mary [VP to bring a cake], since John has already volunteered to do [AdvP so].

(51.2) We shouldn't ask Mary [VP to bring a cake], since John has already volunteered to do [NP that].

However, the adverb *so* is not interchangeable with either *it* or *that*.

(52.1) Though Bill won't admit [NP it], [S he is quite good at poker].

(52.2) *Though Bill won't admit [AdvP so], [S he is quite good at poker].

(53.1) Since John has volunteered to do [AdvP so], we shouldn't ask Mary [VP to bring a cake].

(53.2) *Since John has volunteered to do [NP that], we shouldn't ask Mary [VP to bring a cake].

Another adverb that is an endophor is *there*.

(54) Alan went to [NP Bujumbura] and stayed [AdvP there] a week.

Finally, the adjective *such* is an endophor.

(55) John [VP got drunk at his own birthday party]. His wife detests [NP such behavior].

We have just reviewed a diversity of proforms, including pronouns, adjectives, and adverbs, each of which may, and sometimes must, have an antecedent so that the expression in which the proform occurs may be fully understood. What we did not review are the syntactic patterns that help to determine whether a constituent can serve as an antecedent to a proform. We did note that, while a constituent's preceding a proform is, in some cases, a necessary condition for it to serve as the proform's antecedent, it is typically not a sufficient condition. Moreover, as we also noted, contrary to the suggestion of the technical term *antecedent*, a constituent need not precede a proform to be its antecedent. This question of the syntactic conditions under which a constituent may serve as an antecedent to a proform has been at the center of much syntactic research and analysis from the last third of the twentieth century on.

We also did not address a concomitant semantic question: how does the value of a constituent that is an antecedent for a proform determine the value of the proform? This question was first raised by philosophers, such as Willard Quine (1908–2000) (Quine 1960, sec. 28) and Peter Geach (1916–2013) (Geach 1962, sec. 68 and 84), who noticed the inadequacy of the assumption in traditional grammar, dating back to Classical Greece and Rome, that a pronoun stands in for its antecedent. Thus, while the first sentence of (56) on the construal where the noun phrase *Alice* serves at the antecedent to the proform *she*, is correctly paraphrased by the second sentence in the pair,

Language and Context   **195**

(56.1)   *Alice* believes that *she* is smart.
(56.2)   *Alice* believes that *Alice* is smart.
(57.1)   *Each girl* believes that *she* is smart.
(57.2)   *Each girl* believes that *each girl* is smart.

the second sentence in (57) does not correctly paraphrase the first sentence on the construal in which the noun phrase *each girl* serves as the antecedent of the proform *she*. The suggestion was to treat pronouns like variables in logic. (Variables are introduced and explained in chapter 11.) Subsequent research has shown the situation even with third-person personal pronouns to be much more complex.

Exercises: Endophora

1. Identify, in each of the following sentences, the endophoric expressions and their antecedents. Support your identification with evidence.

(a) If I realize later that I have made a mistake, I shall tell you so.

(b) I can't believe it. They have given me a raise.

(c) Fred dropped by yesterday to borrow a tool. The man just does not seem to have any tools of his own.

(d) I don't like it when it rains.

(e) A viscid atom and a dry one form a pair.

(f) People who see the movie first usually regret doing so.

(g) Fred wants a new toy for Christmas, and George wants the same thing.

(h) One knows just by seeing it that a cat is on the table.

(i) John is a Republican and proud of it.

(j) Can you tell a good inference from a bad one?

(k) If someone touches my computer, I'll know it.

(l) The gems that are bright are so rendered by polishing.

(m) Bill got a job. He is very happy about it.

(n) He will go far, if circumstances favor him.
    Il ira loin si les circonstances le favorisent.[8]

(o) For those of you who have children and don't know it, there is a a nursery downstairs.

(p) Bill often behaves imprudently, but he does not always behave thus.

---

8. Written of Napoleon in his *carnet de notes* by his professor of history at the École militaire de Paris.

2. Consider the proform *one*. Using what you learned about constituency in chapter 3, critically discuss the hypothesis that the antecedent of this pronoun is a constituent.

3.2 Ellipsis

The cases of proforms discussed in section 3.1 comprise pairs of expressions—a proform and one or more antecedents—where a proper understanding of the denotation of the proform depends on one's having a proper understanding of the denotation of its antecedent or antecedents. Ellipsis also involves an antecedent. But, unlike proforms, rather than having an overt expression, ellipsis consists in the absence of an expression.

Consider the following sentence:

(58) Colleen is eager to eat and Evan is too.

Notice that the expression *Colleen is eager to eat* forms a clause and that, by itself, relative to some fixed circumstances of evaluation, it can be judged either true or false. The expression *Evan is too* does not appear to form a clause and, by itself, relative to some fixed circumstances of evaluation, it cannot be judged either true or false. However, should the expression *eager to eat* be inserted into the expression *Evan is too* to yield *Evan is eager to eat too,* the result is a clause and that clause, by itself, can be judged true or false relative to some fixed circumstances of evaluation. Moreover, the result *Evan is eager to eat too* expresses precisely what *Evan is too* relative to the cotext.

Next, we shall examine six forms of ellipsis: *gapping, interrogative ellipsis, verb phrase ellipsis, copular complement ellipsis, appended coordination,* and *nominal ellipsis*.

We begin with *gapping*. Typically, but not always, gapping occurs under the following circumstances: an independent clause is followed by an expression that, though not itself a constituent, comprises two constituents, neither of which is a constituent of the other; the first of these latter two constituents corresponds to the initial constituent in the preceding clause and the second to the clause's final constituent; the point of ellipsis, or gap, is the point between the two constituents that follow the clause; and the expression between the initial and final constituent of the clause is the antecedent for the gap. For example, in sentence (59.1), for example, the noun phrases, *Susan* and *the play,* do not form a constituent; however, the noun phrase *Susan* corresponds to the first phrase in the preceding clause, *Peter,* and the noun phrase *the play* to the last phrase in the preceding clause, *the movie*.

(59.1) Peter <u>saw</u> the movie and Susan __ the play.

(59.2) The man <u>had immigrated to Canada</u> many years ago, his wife __ only last year. (based on Payne and Huddleston 2002, [8] iii)

(59.3) Alice <u>expected to receive</u> an A, Bill __ merely a C. (based on Payne and Huddleston 2002, [13] i)

Language and Context

(59.4)   The towns <u>they attacked</u> with airplanes and the villages __ with tanks. (Quirk et al. 1985, <u>975</u>)

As the last two examples make clear, the expression that is the antecedent of the gap need not be a constituent, though the two constituents flanking it are constituents and of the same category as an initial and a final constituent of the preceding independent clause.

It is also possible to have two independent constituents after the point of ellipsis, and it is also possible to have two points of ellipsis, one final and one not.

(60.1)   Ed <u>had given</u> me [NP earrings] [PP for Christmas] and Bob __ [NP a necklace] [PP for my birthday]. (Payne and Huddleston 2002, [12])

(60.2)   Carl <u>wanted</u> the Canadian <u>to win</u>, Fred __ the American __. (based on Payne and Huddleston 2002, [13] ii)

The second form of ellipsis is *interrogative ellipsis,* often called *sluicing.* Here, the antecedent is an entire clause. The ellipsis consists either in an interrogative constituent, standing in for an independent clause

(61)   A: <u>Ed bought a Mercedes.</u>
       B: Where __ ?

or in an interrogative constituent standing in for an indirect question.

(62)   The people who understand how <u>Fred killed himself</u> don't know why __.

The third form of ellipsis is *verb phrase ellipsis*. It is so-called since a verb phrase, either finite or infinite, serves as the antecedent for the point of ellipsis. The point of the ellipsis is immediately preceded either by a simple auxiliary verb or by the preposition *to*. All four cases are illustrated in (63).

(63.1)   The man who promised that he would <u>bring wine</u> will __.
(63.2)   The man who promised that he would <u>bring wine</u> will be glad to __.
(63.3)   The man who promised to <u>bring wine</u> will __.
(63.4)   The man who promised to <u>bring wine</u> was glad to __.

When the tense for the ellipted material is the simple present, the auxiliary verb *to do* is used.

(64)   Fred <u>likes fast cars</u>, Tom does __ too.

We now turn to a form of ellipsis known as *copular complement ellipsis*. It is so-called, since the antecedent of the point of ellipsis is the complement of the copular verb *to be* in the cotext.

(65.1) Bill is <u>a criminal lawyer</u> and Mary is __ too.

(65.2) Bill is <u>fond of cheese</u> and Mary is __ too.

(65.3) Bill is <u>in Bamako</u> and Mary is __ too.

This form of ellipsis should not be confused with verb phrase ellipsis. Verb phrase ellipsis requires an auxiliary verb or the preposition *to* to precede immediately the point of ellipsis. One auxiliary verb is the verb *to be*. But the copular verb *to be* is also a nonauxiliary verb and it is required to precede immediately the point of ellipsis in copular complement ellipsis. Nonetheless, the two forms of ellipsis are clearly distinguishable, for in verb phrase ellipsis, the antecedent of the point of ellipsis is a verb phrase, whereas in copular complement ellipsis, it is never a verb phrase.

Another form of ellipsis is *appended coordination,* also called *stripping*. This results when one, or possibly two, phrases are appended to a clause, often introduced by a coordinating conjunction such as *and* or *but*. In some cases, the appended phrase corresponds to a phrase in the preceding clause and what the appended phrase conveys is what would be conveyed by a clause just like the preceding clause, except that the appended phrase replaces its counterpart in the preceding clause. An example will make this much clearer.

(66.0) Albert gave <u>a book</u> to his colleague, and __ a pen __.

(66.1) Albert gave a book <u>to his colleague</u>, and *Albert gave* a pen *to his colleague*.

In other cases, the preceding clause may contain no counterpart.

(67.1) <u>The speaker lectured about the periodic table</u>, but __ only briefly.

(67.2) <u>Fred goes to the cinema</u>, but __ seldom __ with his friends.

A sixth form of ellipsis is *nominal ellipsis*. Here, the antecedent expression is a noun phrase and the defective expression is one comprising a constituent that is sister to a noun.

(68.1) Colleen ate two <u>large cookies</u> and Evan ate three (large cookies).

(68.2) Bill picked up Carol's <u>coat</u> and Ed picked up Joan's (coat).

(68.3) An article <u>on this topic</u> is more likely to be accepted than a book (on it). (Payne and Huddleston 2002, 424)

(68.4) I didn't see any <u>of the movies</u>, but Lucille saw some (of them). (Payne and Huddleston 2002, 424)

It is not always clear what is ellipsis and what is the use of a proform. Thus, for example, while there is independent evidence to show that that *so* is a proform, there is no such evidence to suggest that either *as* or *too* is.

(69.1) Fred <u>likes fast cars</u>, Tom does __ too.

(69.2) Fred <u>likes fast cars</u>, as does Tom __.

Language and Context

(69.3) Fred <u>likes fast cars</u>, and Tom does <u>so</u> too.

(69.4) Fred <u>wants to go to Spain</u>, and <u>so</u> does Tom.

Moreover, what is obligatorily expressed in one language by ellipsis requires a proform in another. For example, French requires the pronoun *le* in the construction in (70.1) and does not tolerate ellipsis in its place, whereas the English counterpart requires ellipsis and does not tolerate a pronoun instead.

(70.1) Le racisme, si vous n'<u>y mettez</u> pas <u>fin</u>, qui <u>le</u> fera?

  *Le racisme, si vous n'<u>y mettez</u> pas <u>fin</u>, qui fera __?

(70.2) *Racism, if you don't <u>put an end to it</u>, who will do <u>it</u>?

  Racism, if you don't <u>put an end to it</u>, who will __?

(71.1) Les Romains <u>aidèrent les habitants de Messine</u>, comme ils <u>l'</u>avaient promis.

  *Les Romains <u>aidèrent les habitants de Messine</u>, comme ils avaient promis__.

(71.2) *The Romans <u>aided the inhabitants of Messina</u>, as they had promised to do <u>it</u>.

  The Romans <u>aided the inhabitants of Messina</u>, as they had promised__.

  The Romans <u>aided the inhabitants of Messina</u>, as they had promised to do__.

		ENGLISH	FRENCH
(72)	Speaker A:	Are you <u>tired</u>?	Es-tu fatigué?
	Speaker B:	Yes, I am __.	*Oui, je suis__.
		Yes, I am tired.	Oui, je <u>le</u> suis.

Indeed, whether ellipsis occurs or a proform occurs may depend only on grammatical number. In English, the pronoun *one* alternates with ellipsis, depending on whether the antecedent is singular or plural.

(73.1) My <u>computer</u> works well, but not Dan's (computer).

(73.2) My <u>computer</u> works well, but not <u>those</u> of my assistants.

(74.1) Mon <u>ordinateur</u> marche bien, mais pas <u>celui</u> de Daniel.

(74.2) Mon <u>ordinateur</u> marche bien, mais pas <u>ceux</u> de mes adjoints.

The alternation may also depend on word order.

(75.1) Can you distinguish a good <u>argument</u> from a bad (one).

(75.2) Can you distinguish a good (argument) from a bad <u>argument</u>.

We have just reviewed the diversity of forms of ellipsis found in English. This review did not attempt to characterize the syntactic patterns that help to determine what may serve as an antecedent for an elliptical gap. This question, like its counterpart for proforms, has received much attention in the syntactic literature. Nor did the review attempt to characterize how an expression containing an elliptical gap comes to acquire a value, a topic that has also been at the center of much research in semantics.

Exercises: Ellipsis

1. For each of the following sentences, identify the form of ellipsis and the antecedent for the ellipted material. Provide evidence for your identification.

(a) There was a transition from a prelogical to a logical method of argument.

(b) Carl bought a car but no one knows from whom.

(c) The lieutenant had not issued soap to the new recruits, or towels.

(d) Both a Chinese and a Tibetan translation have appeared.

(e) We have two layers of prisms, one facing one way and one the other. (Thompson 1917, 107)

(f) Ask Dan to sign the petition to end the war; he will be glad to.

(g) Every woman who wants to run for office should.

(h) Mary is fond of ice cream and Zach is as well.

(i) David got a bike for his birthday and a book and a fountain pen.

(j) A generous bumpkin will always be better than a polite egoist, and an honest boor than a refined scoundrel. (Comte-Sponville 1995, 26)
Un rustre généreux vaudra toujours mieux qu'un egoïste poli. Un honnête homme incivil, qu'une fripouille raffinée.

(k) We may produce cells with thick walls or with thin and cells with plane or with curved partitions. (Thompson 1917, 105)

(l) Bill cannot remember where but he recalls that he has met Carol.

(m) Instruments are of various sorts, some living, some lifeless. In the rudder of a ship a pilot has a lifeless, in the lookout man a living instrument. (Aristotle *Politics*, 1253b27–1253b29)

2. Do any of the following sentences contain ellipsis? Justify your answer. Also, for each sentence that you think contains ellipsis, identify the ellipted material and indicate either what form of ellipsis it contains or that it contains none of the forms identified. Again, provide evidence for your answer.

(a) You may buy a car with airbags or one without.

(b) A knife and fork are on the table.

(c) Ptolemy thought that Venus is farther from the Earth and Mercury closer.

(d) He takes his work seriously, himself lightly. (Shaw 1963, 74)

Language and Context 201

3. Identify all forms of endophora, proforms and ellipsis, in the sentences that follow and indicate their antecedents.

(a) I do not think that generosity teaches us much about love nor the latter about the former.
Je ne suis pas sûr non plus que la générosité nous apprenne beaucoup sur lui, ni lui sur elle.

(b) The police wish to know whether Bill left and if so, with whom.
La police veut savoir si Guillaume est parti, et si oui, avec qui.

(c) A magnitude, if divisible one way, is a line, if two ways, a surface, and if three a body. (Aristotle, *De Caelo*, 268a9)

(d) Gaul is, as a whole, divided into three parts, one part of which inhabit the Belgians, another the Aquitainians, a third those who are called Celts in their language and Gauls in ours. (Julius Caesar, *De Bello Gallico*, I.1)

4 Context and Ambiguity

One should not confuse a sentence's variation in truth value that results from ambiguity with a sentence's variation in truth value that results from variation in context, whether it arises from a change in cotext or from a change in setting. Let us consider each way the truth value of a sentence can change.

The following sentence is ambiguous:

(76)   Galileo espied a patrician with a telescope.

Let the context for the utterance of the sentence be fixed—say a speaker uttering the sentence in 1609. Let the circumstances with respect to which the sentence is to be evaluated be these: Galileo is looking out the window of his apartment in Venice through his telescope at a patrician walking empty-handed through Saint Mark's Square. Is the sentence uttered in the fixed context true with respect to the stipulated circumstances? Yes and no. With respect to the circumstances envisaged, no patrician with a telescope is espied by Galileo, and so the sentence can be judged false. However, in the very same circumstances, a patrician has been espied by Galileo through his telescope, and so the sentence can be judged true. In this case, the variation in truth value judgment is due to the amphiboly of the sentence.

Now, let us turn to a pair of unambiguous sentences.

(77.1)   Hans Lippershey built a telescope, and so did Galileo.

(77.2)   Hans Lippershey made spectacles, and so did Galileo.

Let us take the circumstances of evaluation to be what is believed to be the historical case. Hans Lippershey (1570–1610), a lens maker of German origin who settled in Middelberg, the Netherlands, built a telescope in 1608. Galileo also built a telescope in

1609. In addition, let us suppose that Galileo never made a pair of spectacles in his life. Now, the expression *so did Galileo* expresses something true when it occurs in the cotext given in (77.1), but it expresses something false when it occurs in the cotext given in (77.2). The change in truth value arises from a change in the cotext.

Next, consider another sentence.

(78)   I made spectacles.

Let us retain the same circumstances of evaluation as we used evaluating the truth of the last pair of sentences, but let us consider two different settings. Should the setting be one in which Hans Lippershey utters sentence (78), then it is true. But should the setting be one in which Galileo utters it, it is false.[9]

Evidence for the contextual dependence of an expression comes from finding a suitable sentence containing the expression, liable to being judged either true or false, and, while holding the circumstances of evaluation fixed, one sees that it can be judged true or false, depending on the context. Evidence for the ambiguity of an expression comes from finding a suitable sentence containing the expression, liable to being judged either true and to be judged false, and, while holding the circumstances of evaluation fixed and the context, one can judge the sentence as true and as false.

Exophoric expressions may give rise to ambiguity: the very same expression may have both an exophoric construal and a nonexophoric one. The directional nouns *left* and *right* are notorious in many languages for just such an ambiguity.

Consider the following circumstances of evaluation. Three people—say, Peter, Mary, and John—are standing in a line, shoulder to shoulder, facing in the same direction, for example, as depicted here, each arrow indicating the direction in which the person is facing.

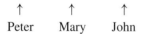

Next, consider the sentence

(79)   Peter is to the left of Mary.

with respect to two settings. In the first setting, the speaker is facing Peter, Mary and John.

---

9. The admonishment not to confuse ambiguity with deixis goes back at least to Quine (1960, sec. 27, 131–132), though, alas, it is not always heeded.

In the second, the people have remained exactly as they are—in other words, the circumstances of evaluation have remained fixed—and the speaker has relocated so that he or she is looking at the backs of the three people, as follows:

In the first setting, sentence (79) can be judged alternately true and false; in the second setting, it can only be judged true. The contrast in truth value judgments with respect to the first setting shows that the sentence is ambiguous, while the contrast in truth value judgments arising from a change in setting shows that the sentence contains an exophor.

Let us see how this is so. The distinction between right and left is a distinction made to distinguish two sides of the human body. The word *left,* like the word *right,* designates a side of the body of the person denoted by the noun that is the object of the preposition *of.* This sense is nonexophoric: that is to say, it does not depend on the setting for its interpretation. Indeed, it is invariant across settings. It is the sense whereby one judges sentence (76) true, whether the speaker is in front of the three people facing them or in back of them looking at their backs. This meaning of sentence (76) can be paraphrased as

(80)   Peter is to Mary's left.

At the same time, the noun *left* also has a exophoric sense, that is, a sense whose correct interpretation requires knowledge of the setting. On this sense, the word *left* is interpreted with respect to the orientation of the speaker with respect to the people being talked about. Under this sense, one must be able to identify the speaker in the setting in order to know which side of Mary is meant. A change in the orientation brings about a change in the truth value of the sentence. It is this meaning that accounts for the judgment of false with respect to the first setting and true with respect to the second.

It is easy to garner further evidence for this explanation. Imagine that, in the picture there appears, instead of Mary, a tree.

(81)   Peter is to the left of the tree.

This sentence can only be judged false with respect to the new state of affairs and the first setting, and true with respect to new state of affairs and the second setting. With respect to this new state of affairs, only the exophoric meaning of the noun *left* is available, since trees fail to have a left and a right.

In sum, then, use of alternate truth value judgments to determine ambiguity of a sentence relies on a fixed state of affairs and a fixed setting, whereas use of alternate truth value judgments to determine exophora and endophora relies on a change in setting and cotext, respectively.

Exercises: Context and ambiguity

1. Consider the following jokes. Explain the basis of their humor. Where it is based on ambiguity, and if it is, identify the ambiguity involved.

(a) Question: Where was the Declaration of Independence signed?
Answer: At the bottom of the page.

(b) Question: In which state does the River Ravi flow?
Answer: Liquid.

(c) Question: If you throw a red stone into the blue sea, what will it become?
Answer: It will simply become wet.

(d) Question: How can a man go eight days without sleeping?
Answer: No problem, he sleeps at night.

(e) Question: How can you lift an elephant with one hand?
Answer: You will never find an elephant that has only one hand.

(f) Question: How can you drop a raw egg onto a concrete floor without cracking it?
Answer: Any way you want, concrete floors are very hard to crack.

(g) A wife asks her husband: "Could you please go shopping for me and buy one carton of milk, and if they have avocados, get six." A short time later her husband comes back with six cartons of milk. His wife asks him: "Why did you buy six cartons of milk?" He replied: "They had avocados."

2. The following two sentences appeared at the foot of a staircase in a public area. Use what you know about exophors to explain why some people find it confusing and others do not.

Please, when using the stairs, stay to the right when going up, and stay to the left when going down. This will keep people from running into each other.

3. This argument is, as surely you will agree, specious. Use what you know about endophors and exophors to diagnose the source of its speciousness. Provide evidence for your diagnosis.

A: I shall prove that you are not here.
B: Okay. Try.
A: You are not in New York.
B: True.
A: Therefore, you are somewhere else.

B: True.
A: Therefore, you are not here.
      (Larson and Segal 1995, 225)

5 Conclusion

At the beginning of this chapter, we pointed out that an expression of a natural language must have associated with it an understanding that is invariant through its various uses. But the fact that an expression has an understanding associated with it that is invariant through its various uses does not mean that its invariant understanding cannot be modulated from use to use. Indeed, endophoric and exophoric expressions are all examples of expressions that, while having an understanding invariant through its different uses, also have their understanding modulated. The invariant understanding of exophors is modulated by an understanding of the setting in which they are used, and the invariant understanding of endophors is modulated by the understanding of their cotext, in particular, by an understanding of their antecedents.

Exercises: Conclusion

1. For each of the following sentences, identify which expressions are endophors and which are exophors. In each case, justify your answer:

(a) Bill thinks that we should leave.

(b) Dan lives nearby. But he is out of town this week.

(c) Last year I saw people celebrate New Year's downtown. This year I saw them do it again.

2. Consider the two occurrences of the expression *this e-mail* in the following passage. Which, if either, of these occurrences can be both exophoric and endophoric? Justify your answer.

This e-mail contains your password to access the extranet site for assessors. You should have already received an e-mail from us that contains your username. If you have not received this e-mail, please contact the On-line Services Support Helpdesk.

3. Suppose that you are looking at a photograph of three people who appear in it facing you and arranged as follows:

Bill   Peter | CHURCH >   John

(where the arrow indicates the sole entrance to the church). Suppose that someone asks Bill any of the following questions.

(1.1)   Who is in front of the church?

(1.2)   Who is at the front of the church?

The first question can receive two answers. The second sentence can receive only one.

(2.1)   John.
        Peter.

(2.2)   John.

Explain why this is so. Provide additional evidence to support your explanation.

SOLUTIONS TO SOME OF THE EXERCISES

2   Setting and exophora

1. (a) The sentence "I am not speaking" contains the first person personal pronoun and is in the present tense. Thus, it attributes to the speaker of the sentence at the time of speaking that he or she is not speaking. But no one can be speaking and yet not speak. Thus, the sentence can never be truly uttered.

3.1   Endophora

1. (a) If I realize later that <u>that I have made a mistake</u>, I shall tell you <u>so</u>.

    If one changes *that I have made a mistake* to, say, *that I have understood the issue,* then what is being told will change from the speaker stating that he or she has made a mistake to the speaker stating that he or she has understood the issue.

   (c) <u>Fred</u> dropped by yesterday to borrow a tool. <u>The man</u> just does not seem to have any tools of <u>his</u> own.
    If one changes *Fred* to *Bill,* then *the man* will refer to Bill, not Fred, as will *his*.

   (e) A viscid <u>atom</u> and a dry <u>one</u> form a pair.
    If one changes *atom* to *leaf,* then *one* will refer to a leaf and not to an atom.

   (g) Fred wants <u>a new toy</u> for Christmas, and George wants <u>the same thing</u>.
    If one changes *a new toy* to *a new car,* then *the same thing* will refer to a new car, not to a new toy.

   (i) John <u>is a Republican</u> and proud of <u>it</u>.
    If one changes *is a Republican* to *is a Democrat,* then *it* will refer to being a Democrat, not to being a Republican.

   (k) If <u>someone touches my computer</u>, I'll know <u>it</u>.
    If one changes *someone touches my computer* to *it is raining,* then *it* will refer to its raining, not to someone's touching the speaker's computer.

Language and Context 207

(m) Bill got a job. He is very happy about it.
If one changes *Bill* to *Carl,* then *he* will refer to Carl, not to Bill.
If one changes *got a job* to *was fired,* then *it* will refer to Bill's being fired, not to Bill's getting a job.

(o) For those of you who have children and don't know it, there is a nursery downstairs. This sentence is ambiguous. The pronoun *it* more plausibly has for its antecedent the main clause, *there is a nursery downstairs;* however, it could refer to the possession of children by those addressed.

3.2 Ellipsis

1. (a) There was a transition from a prelogical ⎯ to a logical method of argument.
   Nominal ellipsis
   Change *method of argument* to *point of view* and what is expressed as the the transition will be one from a prelogical point of view to a logical point of view.

(c) The lieutenant had not issued soap to the new recruits, ⎯ or towels ⎯.
   Appended coordination
   If one changes the underlined portions of the main clause to *The manager had not received* and *from the tenants*, respectively, then *or towels* will convey that the manager had not received towels from the tenants, not that the lieutenant had not issues towels to the new recruits.

(e) We have two layers of prisms, one facing one way and one ⎯ the other.
   Gapping

(g) Every woman who wants to run for office should ⎯.
   VP ellipsis

(i) David got a bike for his birthday and ⎯ a book ⎯ and ⎯ a fountain pen ⎯.
   Appended coordination

(j) A generous bumpkin will always be better than a polite egoist, and an honest boor ⎯ than a refined scoundrel.
   Gapping

(l) Bill cannot remember where ⎯ but he recalls that he has met Carol.
   Interrogative ellipsis

2. (a) You may buy a car with airbags or one without ⎯.
   Endophora: *one*
Ellipsis of the complement of a preposition

(c) Ptolemy thought that Venus is farther from the Earth and Mercury ⎯ closer ⎯.
   Gapping: *is*
Ellipsis of the complement of an adjective

REMARK: The antecedent is *from the Earth* but the ellipted material, if expressed, would be *to the Earth*.

3. (a) I do not think that generosity <u>teaches us much</u> about love nor <u>the latter</u> about <u>the former</u>.
   Gapping
   Endophora

(c) <u>A magnitude</u>, if (<u>it</u>) (is) divisible one way, is a line, if (<u>it</u>) (is divisible) two ways, (is) a surface, and if (<u>it</u>) (is divisible) three (ways) (is) a body.

5 Language and Belief: Implicatures and Presuppositions

1 Language, Communication, and Belief

Successful communication requires that something be shared between the communicators. In the case where the medium is language, it depends on a congruence of the linguistic capacities of the interlocutors. As we saw in section 2.1.1 of chapter 1, the linguistic capacity includes, beyond the grammatical competence, not only memory and attention but also beliefs and the capacity to reason.

It is clear, to begin with, that two people, one of whom speaks only Chinese and the other only Marathi, will never communicate with each other if each relies solely on his or her linguistic capacity; their respective grammatical competences are completely different. However, successful linguistic communication between two people depends on more than sharing a grammatical competence. Failure to have congruent beliefs and to share certain things one takes for granted can also render communication impossible.

As we saw in chapter 4, the content of many expressions vary from occasion of use to occasion of use, though something, namely their character, which determines how an expression is adapted to context, remains invariant. Thus, to understand an endophoric or exophoric expression requires not only that one understand its character, but also that one know how to use one's knowledge of the setting or the cotext in which the expression is used to arrive at its content. Should the interlocutors not have congruent beliefs about the setting and cotext, they will fail to assign the same content to the endophoric or exophoric expression.

Moreover, the setting in which expressions are used changes. After all, as the person speaking changes, so do the conversational roles. If the intelocutors are moving, the physical environment may change and the spatial origin, as it were, of the setting changes. And as the conversation proceeds, the temporal origin also changes. Moreover, as the conversation continues, the cotext changes. The interlocutors must continue to have congruent beliefs that take into account such changes. This, in turn, suggests that they must have congruent ways of incorporating beliefs about these changes into the beliefs they already have.

In this chapter, we shall investigate two other ways in which the beliefs of interlocutors, or things they take for granted, modulate how the expressions they use are understood. We shall begin by distinguishing between what is said (literally) and what is intended to be conveyed. Commentators of classical texts in a variety of civilizations have noted that what a sentence says and what it conveys may be quite distinct. The philosopher Herbert Paul Grice (1913–1988), in the latter half of the last century, gave the first systematic and sustained treatment of this discrepancy. This is the subject of section 2. The second phenomenon to be treated is that of presupposition. The observation that certain expressions carry with them presuppositions has been known since the Classical Greeks, yet once again it was only in the latter half of the last century that linguists and philosophers devoted systematic and sustained attention to this pervasive natural language phenomenon. The last phenomenon to be examined—again, one well known to all who have reflected on language use—is the fact that speakers often do not say what they intend to say.

2   Implicatures

Suppose someone, Bill, for example, says to a prospective landlord: "I own one cat." The landlord would certainly be surprised to observe the day Bill moves into his apartment that he owns twenty cats. Did Bill lie? Bill surely does own one cat. So, how can one say that he lied? Yet, clearly the landlord came to form an erroneous belief on the basis of what Bill said, a belief that anyone would have formed on the same basis. The landlord, like us, formed the belief that Bill has *exactly* one cat. But Bill did not say that he had exactly one cat. He said that he had one cat. Still, Bill seems to have conveyed to his prospective landlord and to us that he has exactly one cat. Thus, we can see the difference between what Bill said, namely that he has one cat, and what Bill conveyed, namely that he has exactly one cat.

What was conveyed in this example, namely, that Bill has exactly one cat, was called by H. P. Grice an *implicature*. In his William James Lecture at Harvard University in 1967, Grice coined the word *implicature,* rather than avail himself of the common word *implication*, because the latter word was, and indeed still is, used in logic in a specific sense that is distinct from, though somewhat similar to, the sense he wished to give to the word *implicature*. The technical notion assigned to the word *implication* in logic will be called *entailment* in this book. Though it will be discussed in detail in chapter 6, we would do well to introduce it here so as to be clear about how it differs from implicature.

Entailment is a relation between a set of statements, on the one hand, and a single statement, on the other, where the single statement must be true, whenever the statements in the set are all true. Insofar as sentences make statements, we can say that the pair of sentences (1.1) entails sentence (1.2).

(1.1) All men are mortal.
       Socrates is a man.
(1.2) Socrates is mortal.

A special case of entailment is where the set of sentences contain but one sentence.

(2.1) Wellington was taller than Napoleon.
(2.2) Napoleon was shorter than Wellington.

Then it is customary to say the one sentence, here sentence (2.1), entails another sentence, here sentence (2.2). As we shall see, entailment and implicature are fundamentally different.

An utterance's conversational implicature is the enrichment or alteration that is obtained on the basis of the utterance's (literal) meaning and the beliefs shared by the interlocutors through the use of principles of reasoning, guided by conversational maxims.

As an example, consider the question in (3.0) and its possible answers in (3.1) through (3.4). Let the situation be that when Alan addresses Bill it is half past noon.

(3.0) Alan: Can you tell me what time it is?
(3.1) Bill: Yes, I can.
(3.2) Bill: It is twelve thirty.
(3.3) Bill: Yes, I can. It is twelve thirty.
(3.4) Bill: The postman just delivered the mail.

The first answer, while clearly an answer to the question, would usually be viewed as humorous in intent but annoying in fact, being uncooperative, indeed, perverse. The second answer, in contrast, is precisely the kind of answer one would expect. The third is also a natural answer, though somewhat elaborate. The fourth answer, without further specification of the situation, might seem odd, but the oddity quickly disappears one includes in the situation that Alan and Bill are roommates, they each know roughly when the mail is delivered, and they each know the other knows this.

Let us take a closer look at the gap between what is said and what is conveyed. The first exchange, comprising (3.0) and (3.2), is noteworthy only because it is completely unsurprising. The second exchange, comprising (3.0) and (3.3), differs from the first, insofar as the reply in (3.3) to the single question in (3.0) consists of two sentences, while the one in (3.2) consists in only one. In fact, these two sentences constitute answers to two questions. These two questions are what Alan conveyed to Bill in asking the question in (3.0). Turning what Alan conveyed by his question in (3.0) into something literally said, while retaining Bill's answer as what is literally said, we arrive at the following exchange in which what is said in the question is perfectly matched by what is literally said in the reply.

(4.0) Alan: Can you tell me what time it is? If so,
              please tell me the time it is right now.
(4.1) Bill: Yes, I can. It is twelve thirty.

The explicitation, as it were, of sentence (3.0) by sentence (4.0) provides a natural bridge to an explicitation of Bill's reply in (3.4) in the fourth exchange.

(5.0) Alan: Can you tell me what time it is?
              If so, please tell me what time it is right now.
(5.1) Bill: No, I cannot. However, this information is relevant:
              the postman just delivered the mail.

Bill does not know the precise time. Nonetheless, he does have pertinent information. Rather than guess what time it is on the basis of the pertinent information, Bill passes the pertinent information along to Alan, thereby enabling him to answer the question himself with as much accuracy as Bill could have.

We stated that an utterance's conversational implicature is obtained from the utterance's literal meaning and the beliefs shared by the interlocutors. This is accomplished using the principles of reason guided by conversational maxims.

To illustrate the importance of shared beliefs, let us reconsider the first exchange. It seems marred by Bill's willful uncooperativeness. But this uncooperativeness disappears if we change the situation. Suppose that Alan wants to be able to see a clock in his kitchen from his living room. Alan asks his friend Bill to go into the living room, while he remains in the kitchen to place the clock. Alan could ask the question in (3.0) and Bill, in replying with sentence (3.1), would be fully cooperative. Their exchange could be paraphrased as

(6.0) Alan: Are you capable of telling me what time it is?
(6.1) Bill: Yes, I am.

As the paraphrase shows, in this situation, nothing is conveyed by the question in (3.0) over and above its literal meaning. That is to say, no conversational implicature arises. What this shows is that shared beliefs play an essential role in determining whether conversational implicatures arise.

Having considered conversational implicatures in a general way, let us turn to the maxims that Grice thought of as guiding us in deriving conversational implicatures from the literal meaning of utterances and shared beliefs.

2.1 Grice's Maxims

The maxims that Grice (1975) thought of as guiding interlocutors in their negotiation of the gap between what one says and what one conveys are these:

COOPERATION	One should make one's conversational contribution such as is required, at the stage at which it occurs, by the accepted purpose or direction of the conversation in which one is engaged.
QUALITY	One should have adequate evidence for what one says. One should not say what one believes to be false.
QUANTITY	One should contribute as much information as is required for the purposes of the conversation. One should not contribute more than is required.
RELEVANCE	One's remarks should be relevant.
MANNER	One should be perspicuous. One should be brief, orderly, clear and unambiguous.

It is important to understand the force of these maxims. First of all, the last four maxims are subordinated to the first maxim. In other words, when one applies the last four maxims in a conversation, one presumes that the first maxim is in force. That is to say, abiding by the last four maxims makes sense only if one is aiming to abide by the first maxim.

Before turning to the applications of the last four maxims, let us expatiate on the first maxim. There are many ways in which it can be violated. While lying is a violation of the maxim of quality, it is also a violation of the maxim of cooperation. One is certainly not being cooperative if one lies. Another way to violate the maxim of cooperation is to violate one or more of the other maxims, but in such a way that the violation remains undetected by one's interlocutor. This was illustrated in the first section with the example of prospective tenant saying that he has one cat, when in fact he has twenty. The prospective tenant has not lied—after all, he does have one cat—but he has misled his landlord.

Within the bounds of abiding by the first maxim, Grice distinguished between *observing* the other four maxims and *flouting* them. For one to observe a maxim means for one to abide by them in the spirit of cooperation. For one to flout a maxim means for one to appear to transgress it and to do so in such a way that it is clear to the interlocutors that the maxim is being transgressed.[1] As we go through the applications of these various maxims, we shall have examples of observing and of flouting.

QUALITY

The maxim of quality demands that one speaks what one believes and that one have evidence for it. Let us see how this works.

Consider the following exchange:

(7) Dan: What news do you have of John?
    Eileen: John has bought a farm.

---

1. This meaning of the verb *to flout*, which is Grice's, is different from its dictionary meaning, which is to treat with contempt.

Let the situation be that John has in fact not bought a farm and Eileen knows that he has not. In other words, Eileen's statement is false and she knows that it is false. Clearly, Eileen is violating Grice's first maxim of cooperation. Moreover, she is also violating the second maxim, for she is portraying as true something she knows to be false, that is, she is lying.

Chances are, unless Dan has some reason to suspect that Eileen is lying, he will believe her. He presumes, among other things, that Eileen believes what she has said and has adequate evidence for it, that is to say, that she is observing both the maxim of cooperation and the maxim of quality. More generally, unless we have reason to suspect the contrary, we presume that our interlocutors are cooperative and that they both believe what they say and have evidence for it.

Further evidence that the maxim of quality is at work in conversation comes from an observation made by the philosopher George Edward Moore (1873–1958). He noted that the sentences of the following kind are paradoxical.

(8) John has a farm, but I don't believe it.

Notice that the paradox does not arise when the second clause does not contain the first-person personal pronoun.

(9) John has a farm, but Sheila doesn't believe it.

People have false beliefs. Clearly, it is possible for someone, say John, to have a farm and for someone else, say Sheila, not to believe that he does. Moreover, it is perfectly possible for some person to assert that John has a farm and for the very same person not to believe it. After all, people do assert things they do not believe. The paradox arises when, in the very same breath, a person asserts something and asserts that he or she does not believe what he or she just asserted.

Grice's maxim of quality shows us how the paradox arises. The statement by some person that John has a farm and the statement by the same person that he or she does not believe that John has a farm are not ipso facto contradictory. However, when one makes a statement one is implicating that one believes it to be true. Thus, the statement that John has a farm carries the implicature that one believes that John has a farm, and it is this implicature that is contradicted by the same person's statement that he or she does not believe that John has a farm. In other words, the contradiction is not between two statements, but between an implicature and a statement.

While Grice's maxims of cooperation and quality forbid one to lie, they do not forbid one to say something one knows to be false. For lying requires not only that one say something false, but that one does so with the intention to deceive. It is possible to say things that we know to be false without having any intention to deceive. Indeed, one may have every hope and expectation that one's interlocutors will recognize what is being uttered is false.

Language and Belief

Here is an example:

(10) Fred: Teheran is in Turkey.
Gina: And London is in France.

Let the situation be the following: Gina has the average European's or North American's knowledge of world geography. Fred does not. In particular, Fred does not know much about the geography of the Middle East. He knows that Teheran is in the Middle East, but he mistakenly thinks that it is in Turkey. Furthermore, both know that London is in England and not in France.

Gina has not only said something that she knows to be false, but she crucially presumes that Fred knows that what she has said is false and that he knows that she knows. As a result of the known obviousness of her falsehood, she conveys to Fred that what he has said is false.

Notice that it is crucial that what Gina says be obviously false to her and to Fred. For suppose she says something else.

(11) Fred: Teheran is in Turkey.
Gina: And Roseau is in Saint Vincent.

To someone like Fred, with his average European or North American knowledge of geography, Gina will not have conveyed to him that what he had said is false.

So, what is the difference between the exchange in (7) and the one in (10)? In the first, when an assertion is made, the implicature is that Eileen believes the statement she has made. In the second, given in (10), it is obviously known to both Gina and Fred that what Gina has said is false and each believes the other to believe that it is false. Thus, Eileen violates the rule, intending that Dan not know that she has done so, while Gina flouts the rule, intending that her nonobservation of the rule be recognized by Fred.

QUANTITY

The maxim of quantity bears neither on the truth of what is said nor on one's belief in it, but on the informativeness of what is said. Imagine this situation. Henrietta and Ian are preparing a recipe that requires four eggs. There are four eggs in the refrigerator. The following exchange takes place:

(12) Henrietta: How many eggs are there in the refrigerator?
Ian: There are two eggs in the refrigerator.

What Ian has said is absolutely true. There are, in fact, two eggs in the refrigerator, for if there are four, then there are two. However, Ian has violated the maxim of quantity and the maxim of cooperation. The cooperative answer, which would provide the needed information, is that there are four eggs in the refrigerator.

Indeed, the following exchange would abide perfectly by Grice's maxims in general and the maxim of quantity in particular.

(13) Henrietta: How many eggs are there in the refrigerator?
      Ian: There are four eggs in the refrigerator.

Here, under the belief that that Ian is observing the maxim of quality, he conveys to Henrietta that there are *exactly* four eggs in the refrigerator, for, had he known that more than four eggs were in the refrigerator, he would not have communicated all the required information.

We now come to flouting the maxim of quantity. Before adducing an example, let us consider an exchange that is like an exchange involving flouting, but is not. Suppose Jack has been assigned a mathematical problem: he has to determine whether 9,571 is a prime number. Jack's friend Karla is pretty good at mathematics, especially arithmetic. Karla's reply in the following exchange

(14) Jack: Is 9,571 a prime number?
      Karla: Either it is prime or it isn't.

is both uninformative and uncooperative.

In contrast, consider an exchange where the reply is literally uninformative, but cooperative. Suppose Lora and Mike are expecting Neil and Neil has not yet arrived.

(15) Lora: Neil is late.
      Mike: Either Neil will come or he won't.

What this seemingly uninformative response conveys is that there is nothing Lora can do to affect whether Neil comes. Mike, then, is flouting the maxim of quantity.

RELEVANCE

The maxim of relevance shows how what out of context appears to be a non sequitur, turns out in context to be a sequitur.

Suppose that Orville and Paul are Canadians, and, like many Canadians, they are hockey enthusiasts. What Paul conveys to Orville in the following exchange

(16) Orville: Does Bill like hockey?
      Paul: Bill's from Edmonton.

is that Bill likes hockey. He does this by stating a fact, namely, that Bill is from Edmonton. Knowledge of this fact, together with shared knowledge of a premise, such as people from Edmonton like hockey, puts Paul in a position to draw a conclusion that is the answer to his question.

While one usually seeks to utter things in a conversation that are relevant, sometimes one deliberately says things that, even in context, appear to be non sequiturs. Consider the exchange:

(17) Richard: I think Prof. Smart is an idiot.
      Sandy: I am going to lift weights.

Language and Belief

Sandy's response is a non sequitur and thus apparently irrelevant. However, suppose that she sees Prof. Smart approaching and knows that Richard would not want himself to be overheard by Prof. Smart to be insulting him. In this case, Sandy would be flouting the maxim of relevance.

MANNER

We now come to the last maxim, the maxim of manner. The maxim of manner requires that one be brief, orderly, perspicuous, and clear. Consider the following example (taken from Levinson 2000)

(18.1)   Put your key in the ignition. Rotate the key until it points to the start position. Hold it until the engine starts. Then immediately release your grip.

(18.2)   Start the car.

The first example is orderly, perspicuous, and clear. It is not, however, brief. The second example is brief, perspicuous, and clear. If the first example should be addressed to a competent driver, it would violate the maxim of cooperation and manner and would, in all likelihood, be taken as an insult. However, should it be addressed to someone learning to drive, the utterer would be abiding by the Gricean maxims, and in particular, the maxim of manner.

Though implicatures often convey states of affairs, best expressed by declarative sentences, sometimes they convey requests, best expressed by sentences in the imperative mood. For example, a customer shopping in a grocery store and saying to one of its employees the declarative sentence in (19.1) conveys, not some state of affairs, but a request, expressed by the sentence in the imperative mood in (19.2).

(19.1)   I cannot find the minced garlic.

(19.2)   Please help me find the minced garlic.

2.2 Properties of Implicatures

Grice maintained that conversational implicatures have a number of properties: *nonconventionality, derivability, nondetachability,* and *cancelability*.

NONCONVENTIONALITY

Conversational implicatures are not conventional. The literal meaning of an utterance is conventional, whether it arises from the literal meanings of the individual words making up the utterance itself or it has a conventional meaning of its own, as is the case with idioms. The literal meaning of an utterance remains the utterance's meaning in all contexts. In contrast, implicatures vary from context to context and they do so as a result of changes in what beliefs the interlocutors take to be pertinent. As we saw, changes in the beliefs of the interlocutors can change which implicatures are associated with an utterance and, indeed, whether an utterance has any implicature whatsoever. Thus, the question in (3.0) has an implicature in one situation, which is made explicit by the sentences in (4.0), and has no implicature in another, which is made explicit by the question in (6.0).

Conversational implicatures should not be confused, then, with idiomatic expressions, for though one cannot grasp the meaning of such an expression by grasping the literal meaning of its constituent expressions and its constituent structure, the meaning such an expression has does not vary with changes in the shared beliefs of the interlocutors.

(20)   How do you do?

Thus, the fact that the utterance is used as a greeting is not a conversational implicature of its utterance.

NONDETACHABILITY

Conversational implicatures remain invariant under paraphrase. Grice refers to this as *nondetachability*. Thus, if one is being sarcastic, it makes no difference whether one utters the first sentence in (21) or the second. The sarcasm will be conveyed either way.

(21.1)   Bill is a genius.
(21.2)   Bill is a mental prodigy.

DERIVABILITY

Conversational implicatures, then, are not meanings attached to utterances by convention; however, they also do not arise out of thin air. Rather, they can be characterized by a form of reasoning. Thus, in the exchange in (16), the understanding that Bill likes hockey is derived from knowledge of a shared premise, such as people from Edmonton like hockey, and Paul's explicit statement that Bill is from Edmonton.

CANCELABILITY

Finally, we come to *cancelability,* or *defeasibility*. Here is where conversational implicatures differ crucially from entailments. Entailments of an utterance are not undone by the addition of further information, but implicatures can be undone or even canceled. Consider the entailment of the second sentence by the first.

(22.1)   Sheila owns a cat.
(22.2)   Sheila owns an animal.

Once one has uttered the first sentence, anything said that contradicts what it entails is taken to contradict the first sentence. Thus, what is entailed by the first clause in the next sentence is directly contradicted by what the second clause states.

(23)   Sheila owns a cat, but she does not own an animal.

Anyone uttering the previous sentence is judged to have contradicted himself or herself.

Language and Belief

In constrast, consider the pair of sentences

(24.1) Sheila owns one cat.

(24.2) Sheila owns exactly one cat.

What the second sentence expresses is an implicature of the first. It is certainly no entailment, for the first sentence can be true and the second false. After all, Sheila might very well own two cats. In that case, it is false that she owns exactly one cat, but it is true that she owns one cat. Now, notice that the following pair of sentences are contradictory:

(25.1) Sheila owns exactly one cat.

(25.2) Sheila owns two cats.

Yet, although what is implicated by the first clause of the next sentence is contradicted by the second,

(26)   Sheila owns a cat; in fact, she owns two.

no one uttering sentence (26) is judged to have contradicted himself or herself. Rather, the first clause's implicature seems to evanesce in the presence of the second clause. In this way, conversational implicatures are undoable, or defeasible, but entailments are not.

2.3 Implicatures and Ambiguity

We have been introduced to the distinction between what is said and what is conveyed. What an utterance says is its literal meaning; what it conveys is not. One might be tempted to conclude that expressions with implicatures are ambiguous between what they say and what they convey, that is, between their literal meaning and their implicatures. But this would be a mistake.

To see why, consider two typical cases of ambiguity—lexical ambiguity and amphiboly, or structural ambiguity.

(27.1) This is a pen.

(27.2) Galileo spotted a patrician with a telescope.

Each of these sentences can be paired with another sentence whose literal meaning contradicts one of the literal meanings of the ambiguous sentence with which it is paired.

(28.1) This is a pen, but it is not a writing instrument.

(28.2) Galileo spotted a patrician with a telescope, but Galileo never used a telescope.

Notice that one can judge the sentences in (28) both as contradictory and as noncontradictory. The fact that the second clause states something that contradicts one of the literal meanings of the first clause does not cancel the contradicted meaning. However, a clause or sentence does cancel the conversational implicature of a preceding utterance.

In other words, the literal meanings associated with a sentence are not undone, or canceled, when they are contradicted by the literal meaning of another sentence in its cotext, but conversational implicatures are.

Exercises: Implicatures

1. Consider the following sentences. In each case, identify the implicature. Explain how the implicature might arise using the Gricean maxims.

(a) The bus stop is some distance from my house.

(b) This flag is red.

(c) Richard got a job. I can't believe it.

(d) The gasoline station is less than fifty kilometers down the road.

(e) No one other than Peter came.

(f) Helen caused the lights to go off.

(g) Sheila got out her key and opened the door.

(h) I think we have met before.

(i) There is somebody behind you.

(j) I tried to reach Bill yesterday.

2. Consider the following exchanges. In each case, identify the implicature. Explain how the implicature might arise using the Gricean maxims.

(a) A: Are we going to the ballet?
    B: The tickets are sold out.

(b) A: What shall we do today?
    B: I'm really tired.

(c) A: How do you like this compact disc?
    B: I don't like jazz.
    (Adapted from Blakemore 1992, 126)

(d) A: Let's get the kids something.
    B: Okay, but I veto I-C-E-C-R-E-A-M.
    (From Levinson 1986, 104)

3. Years ago, so the story goes, a company that sold canned tuna increased its sales tremendously when it began using the advertising slogan: *It never turns black in the can*. Explain this using Grice's maxims. (From Martin 1992, 50.)

4. Consider the following passages. In each case, explain the basis of the humor.

(a) Captain L. had a first mate who was at times addicted to the use of strong drink, and occasionally, as the slang has it, "got full." The ship was lying in a foreign port and the

Language and Belief

mate had been on shore and had there indulged rather freely in some of the vile compounds common in foreign ports. He came on board, "drunk as a lord" and thought he had a mortgage on the whole world. The captain, who rarely ever touched liquors himself was greatly disturbed by the disgraceful conduct of his officer, particularly as the crew had all observed his condition. One of the duties of the first officer (i.e., the mate) is to write up the "log" each day, but as that worthy was not able to do it, the captain made the proper entry but added: "The mate was drunk all day." The ship left port the next day and the mate got "sobered off." He attended to his writing at the proper time, but was appalled when he saw what the captain had done. He went on deck, and soon after the following colloquy took place:

The mate: Cap'n, why did you write in the log yesterday that I was drunk all day?
The captain: It was true, wasn't it?
The mate: Yes, but what will the owners say if they see it? It will hurt me with them.

But the mate could get nothing more from the captain than "it was true, wasn't it?" The next day, when the captain was examining the book, he found at the bottom of the mate's entry of observation, course, winds and tides: "The captain was sober all day." (Trow 1905, 14–15)

(b) The following quotation is attributed to Groucho Marx:
 I've had a perfectly wonderful evening. But this wasn't it.

(c) A: Where was the Declaration of Independence signed?
 B: At the bottom of the page.

(d) A: How many months of the year have 28 days?
 B: Just one, February.
 A: Wrong. Each month of the year has 28 days.

(e) A: In which battle did Lord Nelson die?
 B: His last battle.

(f) A: What is the main reason for divorce?
 B: Marriage.

(g) A: What can you never eat for breakfast?
 B: Lunch and dinner.

(h) A: What looks like half an apple?
 B: The other half.

(i) A: If you had three apples and four oranges in one hand and four apples and three oranges in other hand, what would you have?
 B: Very large hands.

(j) A: If it took eight men ten hours to build a wall, how long would it take four men to build it?

B: No time at all, the wall is already built.

(k) I think that hyperbole is the single greatest factor contributing to the decline of society.

3 Presupposition

Let us turn to another kind of belief that plays a role in linguistic communication. Philosophers and logicians, stretching all the way back to Aristotle, have noted that some kinds of questions seem to be double questions, that is, in addition to the question actually asked, it seems to take for granted an answer to another, unasked question.

Suppose Aaron and Beth are complete strangers to one another. They meet at a party and strike up a conversation. At some point very early in their conversation, they notice a person leave the room to go outside to smoke a cigarette. Now compare the following two questions put by Beth, say, to Aaron:

(29.1) Do you smoke?

(29.2) Have you quit smoking?

In light of their joint observation, question (29.1) seems natural enough, but question (29.2) seems quite odd. Given that Aaron and Beth are complete strangers to one another, why would Beth presume that Aaron used to smoke? What is the difference between these two questions? The second question suggests that Beth takes it for granted that Aaron used to smoke. The first question takes no such thing for granted.

The oddity of question (29.2) can be dissipated, should it be broken up into two questions:

(30) Beth: Did you ever smoke?
      Aaron: Yes, I did.
      Beth: Have you quit smoking?

As the exchange in (30) shows, it seems that question (29.2) takes for granted a positive response to the question *Did you use to smoke*.

Questions of this kind—ones whose answers take for granted an answer to another question—have been said by logicians and philosophers to be instances of a fallacy called *the fallacy of many questions*. Such questions are said to *presuppose,* or to take for granted, something else, which are called their *presuppositions*. Thus, the question *Did Aaron quit smoking* has the presupposition that *Aaron used to smoke*.

(31) Question   Did Aaron quit smoking?
      Presupposition Aaron used to smoke.

Language and Belief

Questions are not the only kinds of sentences that may have presuppositions. Commands and declarative sentences may as well.

(32.1) Command   Quit smoking.
        Presupposition The addressee of the command smokes.
(32.2) Declarative   Aaron quit smoking.
        Presupposition Aaron used to smoke.

Presuppositions are not peculiar to main clauses. A sentence often retains a clause's presupposition, even when the clause is subordinated. To appreciate this point, we must first appreciate what happens to a clause when it is turned into a conditional one. Compare, for example, the force of sentence (33.1) with its force when it is converted into a subordinate, conditional clause.

(33.1) It is raining.
(33.2) If it is raining, the picnic will be canceled.

It is perfectly possible for sentence (33.1) to be false, while sentence (33.2) is true. That is, it is possible for it to be true that, if it is raining, the picnic will be canceled, though it is false that it is raining. Thus, the second sentence in (33) does not entail the first.

What is interesting is that declarative sentences that have presuppositions often retain them, even when they are turned into subordinate, conditional clauses.

(34) Sentence   If Aaron quits smoking, his health will improve.
      Presupposition Aaron smokes.

Similar observations arise with clauses containing modal expressions such as the auxiliary verb *might* or the adjective *possible* and with clauses containing the negative adverb *not*. Thus, none of the last three sentences entails the first.

(35.0) It is raining.
(35.1) It might be raining.
(35.2) It is possible that it is raining.
(35.3) It is not raining.

Yet, a clause having a presupposition retains its presupposition, even when such words are added to them.

(36.1) Sentence   Aaron might not quit smoking.
(36.2) Sentence   It is possible that Aaron will quit smoking.
(36.3) Sentence   Aaron has not quit smoking.
        Presupposition Aaron smokes.

In short, clauses have presuppositions and their presuppositions remain, whether they are asserted, denied, questioned, or turned either into a conditional clause or into a command. Thus, sentence (37) presupposes that Charles cheated on the exam.

(37)   Sentence   Charles admitted that he had cheated on the exam.
        Presupposition   Charles cheated on the exam.

This presupposition remains, even if the sentence is negated, turned into a question, turned into a command, or turned into a subordinate, conditional clause.

(38)   Negation   Charles did not admit that he had cheated on the exam.
        Modal   Charles might admit that he had cheated on the exam.
        Question   Did Charles admit that he had cheated on the exam?
        Command   Admit, Charles, that you cheated on the exam.
        Conditional   If Charles admits that he had cheated on the exam, his punishment will be mitigated.
        Presupposition   Charles cheated on the exam.

In fact, such invariance across clausal types is used to obtain prima facie evidence that the clause carries a presupposition. If something that appears to be taken for granted by a clause persists, even when the clause has been denied, turned into a question or a command, or put into a conditional clause, then one has prima facie evidence that what is taken for granted by the clause is its presupposition.

Finally, it should be noted that presupposing something is no guarantee that it is true. A person asking the question in (37) presupposes that Charles cheated on the exam. Such a question can be asked, even if Charles is innocent of any misconduct. However, anyone who is asked such a question and who believes that Charles is innocent will find the question odd, just as in our earlier example, one finds odd Beth's question regarding whether Aaron had quit smoking.

3.1   Triggers of Presupposition

Having seen that clauses have presuppositions, one might very well wonder where presuppositions come from. As it happens, a variety of expressions give rise to presuppositions. They include adverbs such as *again, even, still,* and *too*

(39)   Sentence   Pat is leaving too.
        Presupposition   Someone other than Pat is leaving or has left.

and verbs of various kinds, including aspectual verbs (e.g., *to continue, to quit, to stop*), desiderative verbs (e.g., *to desire, to wish*), factive verbs (e.g., *to admit, to know, to recognize*), implicative verbs (e.g., *to manage, to struggle*), and iterative verbs (e.g., *to return, to restate, to reconsider*).

Language and Belief

(40) Sentence   I wished Joan lived here.
      Presupposition   Joan does not live here.

(41) Sentence   Nick admitted that the Canadiens had lost.
      Presupposition   The Canadiens had lost.

(42) Sentence   John managed to open the door.
      Presupposition   John had tried to open the door.

(43) Sentence   Napoleon returned to power.
      Presupposition   Napoleon had once held power.

In addition, presuppositions accrue to certain kinds of clauses: cleft,

(44) Sentence   It was Lee who signed up for the course.
      Presupposition   Someone signed up for the course.

pseudocleft,

(45) Sentence   What John broke was his typewriter.
      Presupposition   John broke something.

subordinate clauses whose subordinator is *if,* when the verb is in the perfect subjunctive,

(46) Sentence   If Caesar had not crossed the Rubicon, Pompei would not have fled to Egypt.
      Presupposition   Caesar crossed the Rubicon.

as well as subordinate clauses whose subordinators are *when* and *since*.

(47) Sentence   There was a riot when the Canadiens beat the Kings.
      Presupposition   The Canadiens beat the Kings.

We now turn to definite noun phrases. The existence of presuppositions, which was hinted at by Gottlob Frege (1892) and forcefully argued for by Peter Strawson (1950), has been the locus of great controversy, going back to Bertrand Russell (1905). To see the basis of the controversy, let us consider this pair of sentences:

(48.1) Julius Caesar was a Roman consul.

(48.2) Julius Caesar was not a Roman consul.

Sentence (48.1) is true, while sentence (48.2) is false. One way to think of the truth of sentence (48.1) is to consider a list of all the Roman consuls. Since sentence (48.1) is true, one knows that a name for Julius Caesar must be on it. One way to think of the falsity of sentence (48.2) is to consider a list of all people who were not Roman consuls. Though admittedly a long list, you can be sure that no name for Julius Caesar will appear on it.

In general, if a (monoclausal) sentence is false, its negation is true, and if a (monoclausal) sentence is true, its negation is false.

Now consider the sentences in (49), as uttered today.

(49.1) The present King of France is bald.

(49.2) The present King of France is not bald.

As Russell (1905) pointed out, the list of everyone who is bald today would not include a name for someone who is the King of France. Hence, the first sentence is false. Moreover, the list of everyone who is today is not bald would also not include a name for someone who is the King of France. Hence, the second sentence is false. Russell's puzzle was: how is it possible for a sentence and its negation to both be false?

Russell suggested that, while both of the sentences are grammatically monoclausal, they are logically biclausal. In other words, the sentences have a *grammatical form,* determined by linguists, and they have a *logical form,* which may differ from their grammatical form. The sentences in (48) are sentences whose grammatical form indeed differ from their logical form. A sentence's logical form is given, according to Russell, by its proper translation into a notation of logic. To avoid the complications of rendering the sentences in (48) into logical notation, we can simply paraphrase them into other English sentences that more closely correspond to what Russell would call their logical form.

(50.1) There is someone who is the king of France and he is bald.

(50.2) There is someone who is the king of France and he is not bald.

Russell further observed that sentence (49.2) could also be true. In other words, Russell thought that the sentence is ambiguous and the two construals to which it is liable are two logical forms corresponding to the two sentences in (51).

(51.1) There is no bald king of France.

(51.2) There is no person who is the king of France and is bald.

Another approach to this puzzle was developped by Peter Strawson (1950) (see also Geach 1950). Anticipated by Gottlob Frege (1892) and by Edmund Husserl (1900), Strawson maintained that sentences like the ones in (49) are neither true nor false. According to Strawson, should one utter either sentence in (49), one makes no statement, by which he meant that neither sentence expresses a proposition.

3.2 The Common Ground

Let us return to the example used to introduce the notion of presupposition. We contrasted the following sentences

Language and Belief

(52.1) Do you smoke?

(52.2) Have you quit smoking?

with respect to circumstances in which two people, Aaron and Beth, complete strangers to one another, meet at a party and strike up a conversation. We further imagined that, at some point very early on in their conversation, they notice a person leave the room to go outside to smoke a cigarette. With respect to that circumstance, sentence (52.1) seems natural, while sentence (52.2) seems odd.

We noticed that sentence (52.1) carries no presupposition, but sentence (52.2) does. In every case we have looked at, a sentence's presupposition is triggered by some part of the sentence, either by a particular word or by a particular construction. The presupposition is then retrieved from the sentence in a way specifiable in terms of the words and structure of the sentence. Thus, any sentence of the form in (53.1) gives rise to the presupposition in (53.2).

(53.1) Sentence   NP quit V-ing.

(53.2) Presupposition   NP used to V.

But how does the fact that sentence (51.2) carries a presupposition and sentence (51.1) does not account for the contrast in judgments?

The answer is that it depends on whether or not the presupposition of the sentence is taken for granted, or follows from what it taken for granted, by the conversational participants. Robert Stalnaker has suggested that we refer to what is taken for granted by the conversational participants or follows from what is taken for granted as the *common ground*. The reason, then, that sentence (52.2) is odd in the circumstances specified is that Aaron's formerly being a smoker is not part of the common ground, nor does it follow from the common ground.

Let us test this explanation. Let us change the circumstances. Suppose again that Aaron and Beth are meeting for the first time, but that they are meeting at a gathering for people who are trying to quit smoking or are intending to quit smoking. In these circumstances, the question in (52.2) no longer seems odd. The reason is that it is part of the common ground that people at the gathering are those who have either quit smoking or intend to quit smoking.

The common ground is not fixed once and for all. It changes as the circumstances in which people find themselves change. The changing of the common ground, however, is not only something that happens to the participants, it is something the participants themselves can bring about. They can do so when they speak. Recall that, according to Grice, when a conversation takes place, the conversational maxims hold. Abiding, in particular, by the maxim of co operation and the maxim of quality, participants in a conversation take it for granted that what the other participants say is true. As participants speak, then, the common ground is enriched. That part of the common ground that accrues to the verbal contribution of the participants is called the *conversational record*. As the conversation record grows, the common ground is enriched.

Again, we can test this suggestion. Let us return to our example of Aaron and Beth, complete strangers to one another, standing next to one another and noticing at the same moment a person leave the room to outside to smoke a cigarette.

(54) Aaron: At one time I smoked two packs of cigarettes a day.
      Beth: Have you quit smoking?
      Aaron: No, not yet.

The first sentence uttered by Aaron puts on the conversational record the fact that he used to smoke. The presupposition to Beth's question is now part of the common ground, so her question, when posed, does not appear odd.

3.3 Presupposition, Entailment, and Implicature

How do entailments, implicatures, and presuppositions differ from one another? To answer this question, let us begin by noting that entailments, implicatures, and presuppositions are entailments, implicatures, and presuppositions of sentence utterances; while what is entailed by a sentence, what is implicated by a sentence, and what is presupposed by a sentence are not themselves sentences, it is useful to avail ourselves of the sentences that express what is entailed, what is implicated, and what is presupposed to distinguish entailments, implicatures, and presuppositions from one another.

Entailment is a relation between one or more declarative sentences susceptible of being judged either true or false, in other words, between declarative sentences and a single declarative sentence, all of which are either true or false.[2] Thus, only sentences making statements can enter into the entailment relation; sentences expressing commands, requests, and questions do not, since commands, requests, and questions are not, in principle, either true or false.

Implicature is a relation between a sentence, which may be declarative, imperative, or interrogative, and another sentence, which also may be declarative, imperative, or interrogative. Implicature, then, is a relation between a pair of sentences that may express commands, requests, or questions or make statements.

Presupposition is a relation between a sentence, which may be declarative, imperative, or interrogative, and a declarative sentence susceptible of being judged either true or false. Presupposition is also a relation between a pair of sentences; however, while the sentence having the presupposition may make a statement or express a command, request, or question, the sentence that expresses the presupposition only makes a statement, it does not

---

2. The purpose of the restriction of declarative sentences to those that are either true or false is to exclude from consideration declarative sentences such as *I promise to return the book I borrowed from you tomorrow,* which do not seem to be either true or false.

express a command, request, or question. In other words, a sentence's presupposition is always expressed by a declarative sentence that is either true or false, but the sentence itself need not be a sentence that is either true or false, for the sentence itself may express a command, request, or question.

Let us elaborate further on the difference between entailment and presupposition. While entailment and presupposition are similar insofar as what a sentence entails and what a sentence presupposes are both expressed by declarative sentences expressing a statement, they differ in three ways. First, only a single sentence carries a presupposition, whereas one or more sentences together can give rise to an entailment. Second, the single sentence that carries a presupposition may be interrogative, imperative, or declarative, whereas the sentences that together give rise to an entailment must all be declarative ones. Third, as we noted earlier, if a simple clause presupposes some statement as true, it does so even should the clause be negated, modified to become a clause expressing a possibility, or used as the protasis of a conditional sentence.[3] A simple clause's entailments disappear under such modifications, as illustrated in the following sets of sentences.

(55) Sentence   The chief constable arrested three men.
      Presupposition   There is a chief constable.
      Entailment   The chief constable arrested two men.

(56) Negation   The chief constable did not arrest three men.
      Presupposition   There is a chief constable.
      No Entailment   The chief constable arrested two men.

(57) Modal   The chief constable might have arrested three men.
      Presupposition   There is a chief constable.
      No Entailment   The chief constable arrested two men

(58) Conditional   If the chief constable arrested three men,
                       then he will be up for a promotion.
      Presupposition   There is a chief constable.
      No Entailment   The chief constable arrested two men.

Let us now elaborate on the difference between presupposition and implicature. While presupposition and implicature are similar insofar as sentences with presuppositions and sentences with implicatures may be imperative, interrogative, or declarative, they differ in many ways. To begin with, what a sentence presupposes may only be expressed by a declarative sentence, whereas what a sentence implicates may be expressed by an imperative, interrogative, or declarative sentence. Other differences arise with respect

---

3. The *protasis* of a conditional is the subordinate clause introduced by the subordinator *if,* and the *apodosis* is the subordinating clause. Another pair of terms used to refer to such clauses are *antecedent* and *consequent*, respectively. We avoid these latter terms, since we have used the term *antecedent* to denote expressions that serve to fix the denotation of endophoric expressions.

to the four properties that Grice attributes to conversational implicatures: they are nonconventionality, derivability, nondetachability, and defeasibility.

We stated that presuppositions are triggered by specific expressions, either individual words or clause types. Thus, presuppositions are conventional. Conversational implicatures are not conventional. No maxims have to be applied to the literal meaning of the sentence to derive the presupposition, whereas maxims are applied to sentences to derive their conversational implicatures. Thus, conversational implicatures are derivable, but presuppositions are not. Moreover, conversational implicatures are nondetachable, that is, they are preserved under paraphrase, but presuppositions do not seem to be preserved under paraphrase. Consider, for example, a cleftsentence and its nonclefted counterpart.

(59.1)   Lee signed the contract.

(59.2)   It was Lee who signed the contract.

These two sentences are true under exactly the same circumstances. Yet the second carries the presupposition that someone signed the contract, whereas the first does not. Aside from the difference in presupposition, they are perfectly synonymous.

Presuppositions are usually not defeasible. To follow up sentence (60.1) with sentence (60.2) is decidedly odd.

(60.1)   Brian has quit smoking.

(60.2)   In fact, he has never smoked.

However, some factive verbs that in some cotexts carry a presupposition, give them up in others.

(61.1)	Sentence	If Jim finds out that Bill is in New York, then there will be trouble.
	Presupposition	Bill is in New York.
(61.2)	Sentence	If I find out that Bill is in New York, then there will be trouble.
	No Presupposition	Bill is in New York.

Example (61) is taken from Chierchia and McConnell-Ginet (1990, 285) and the next from Soames (1989, 574).

(62.1)	Sentence	I regret that I have not told the truth.
	Presupposition	I have not told the truth.
(62.2)	Sentence	If I regret later that I have not told the truth, then I shall expose my lie to everyone.
	Presupposition	I have not told the truth.

Finally, entailments and implicatures are different. An entailment is a relation between one or more declarative sentences making statements and a single declarative sentence making a statement, whereas implicature is a relation of a pair of sentences of any kind. Second, as we have also noted, implicatures are defeasible, but entailments are not.

Exercises: Presuppositions

1. Identify the presuppositions associated with the following sentences. Provide evidence to support your claim. (Almost all of the examples are taken from Leech 1974, chap. 13.)

(a) The governor of Idaho is currently in London.

(b) I wonder whom Bill met.

(c) Tom has a bigger stamp collection than I do.

(d) Lee's surrender to Grant spelt the end of the Confederate cause.

(e) The inventor of the flying bicycle is a genius.

(f) Bill forced Fred to leave.

(g) Marion pretended that her sister was a witch.

(h) It is nice to see that Yorick has many friends.

4 What Is Intended and What Is Understood

We distinguished earlier between what is said and what is conveyed. We would now like to distinguish between what is said, on the one hand, and what one intends to say and what someone else thinks one has said, on the other. Gaps between what one says and what one intends to say or gaps between what one says and what someone else thinks one has said are common. Yet, as we shall see, these gaps pass unnoticed, for one often automatically passes over what one's interlocutor says to what one thinks the interlocutor intends to say, sometimes successfully and sometimes not.

Obvious examples of gaps between what one says and what one intends to say or between what one says and what someone else thinks one has said are those in which one or the other of the interlocutors is not a native speaker of the language of communication. Thus, native speakers are often able to understand the expressions of nonnative speakers, in spite of containing errors. Rarely, for example are native French speakers unable to figure out what a nonnative speaker intends to say by an expression in which there are errors of gender or in which there are failures of proper agreement. Similarly, adults easily accommodate the deficiencies of young children as they learn the language of the adult. It is unlikely for example that, should a nonnative speaker or child say *that was as easy as cake,* he or she would be misunderstood. We usually guess correctly what it is the speaker intends. Sometimes, though, one's guess is wrong, as illustrated by a cartoon.

A mother is in the kitchen preparing a meal and a young child is in the adjacent dining room with the door open, but neither can see the other. The child has evidently just finished sawing the leg off of one of the dining room chairs. The child says to the mother: Mommy, I sawed the chair. The mother replies: no honey, you saw the chair.

Indeed, adult native speakers often misspeak, yet their errors are overlooked in favor of the intended message. Notorious here are the sentences ascribed to Reverend William A. Spooner (1844–1930), one time head of New College, Oxford University, who is said to have made a great many, often humorous, speech errors, thereby saying something different from what he intended to say. Reverend Spooner is supposed to have said the first sentence in (63), though he apparently intended to say the second.

(63.1)   Work is the curse of the drinking class.

(63.2)   Drink is the curse of the working class.

As we remarked earlier, the interlocutors of a conversation must have congruent beliefs about the setting of a conversation and its cotext. In general, adult speakers do not get confused about which role interlocutors in a verbal exchange are in. One does not use the first-person pronoun intending to use the second and one does not misunderstand the third-person pronoun as the second-person pronoun. It is not, however, difficult to imagine people just beginning to learn a language making such errors. Indeed, precisely such errors occur with children.

Misunderstanding involving adult native speakers is more likely to arise from erroneous beliefs about the setting. In chapter 4, we assumed an ideal fit between the exophors used and the settings in which they are used. However, such a fit does not always occur. An exophor in an utterance may fail to find a suitable value. Consider, for example, a setting in which two roommates, Charles and Mark, are at a party. Charles starts up a conversation with someone next to him, say Alice. After a few minutes, it occurs to Charles that he should introduce his roommate to Alice. In the meantime, Mark has stepped away without Charles noticing it. Charles, without looking to where Mark had been, points to that place and says:

(64)   By the way, this guy is my roommate.

In this setting, the exophor *this guy* fails to find a suitable value. In all likelihood, the speaker's misbelief about the setting will not deter another native speaker from grasping what the speaker intended.

Let us turn to still another case, this one discussed by Keith Donnellan (1966). Imagine that there is a party and exactly one person at the party is holding a champagne glass containing a bubbly liquid. Suppose further that one of the people at the party utters the following sentence to another, as they both gaze at the man with the champagne glass (see Stalnaker 1970, 283–285; Soames 1989, sec. 2.2–2.3).

(65)   The man drinking champagne is a philosopher.

If it turns out that the champagne glass that the man is holding contains champagne, then the utterance of (65) has resulted in a statement. However, if it turns out that glass contains sparkling water, then no statement has been made, but the interlocutor will ascribe to the speaker a statement that the speaker in all probability intended to make, namely, that the person in question is a philosopher (Kripke 1979).

Still another source of misunderstanding is ambiguity. A person utters sentence (66.0), intending to say what is expressed by sentence (66.1), but he or she is understood to be saying what is expressed by sentence (66.2).

(66.0)   Galileo spotted a patrician with a telescope.

(66.1)   A patrician with a telescope was spotted by Galileo.

(66.2)   Using a telescope, Galileo spotted a patrician.

Again, such ambiguities often do not deter the native speaker from ascertaining what is intended, often without any conscious awareness of what the unintended meanings might be.

Underspecification is still another source of misunderstanding. It is also possible for the setting to contain too many potential values for the exophor. As the philosopher Ludwig Wittgenstein (1889–1951) noted, contrary to what one might think, pointing, or demonstration, need not uniquely identify what the person intends to be pointed out or demonstrated. Consider a situation in which someone has decanted a bottle of wine into a decanter. Should the person touch the decanter with his or her index finger and say sentence (67.0), it would not be clear whether the person intended to say what is expressed by sentence (67.1) or what is expressed by sentence (67.2).

(67.0)   This is imported.

(67.1)   This wine is imported.

(67.2)   This decanter is imported.

Notice that, should the wine be imported and the decanter domestic, sentence (67.0) can be understood to say either something true or something false. How the sentence is taken depends on what the interlocutor takes to be pointed to by the speaker.

Still other ways in which a gap may arise comes from implicatures. One may fail to convey what one intends to convey. This is especially evident when a speaker utters a sentence ironically, saying something that he or she assumes the interlocutor takes to be patently false, but the interlocutor takes the sentence to be true.

(68) A: Derek is really clever.
     (intended ironically)
   B: Oh really?

Exercises: What is intended and what is understood

1. Identify any ambiguities in the following sentences. Provide evidence to support your identification.

(a) Tom and Julie married each other.

(b) The boy put his clothes on himself.

(c) Bill knows what I know.

(d) The volunteers replaced the cushions. (Taken from Blakemore 1992, 11)

(e) There are too many marks in this book. (Taken from Blakemore 1992, 6)

(f) There is a plant right in front of the window.

2. Consider the following pairs of sentences. Identify for each pair the contrasting implicatures. Identify the likely background beliefs, which, when coupled with Grice's maxims, lead to the difference in implicatures.

(a) Fred appeared in a tie.
    Fred appeared in his underwear.

(b) Bill has had breakfast.
    Bill has had caviar.

5 Conclusion

Not only does successful communication between two interlocuters using natural language require some shared grammatical competence, it also requires congruent beliefs pertaining to the setting and the context. In this chapter, we saw how what a sentence says and what it conveys need not be the same and how Grice's conversational maxims permit what a sentence says to convey something, namely a conversational implicature, different from what it says. We also saw how a sentence, through a certain kind of expression in it, signals that its utterer is taking something for granted, that is, making a presupposition. We distinguished between a sentence's entailments, its implicatures, and its presuppositions and showed that both a sentence's implicature and a sentence's presupposition are different from the various meanings a sentence may have by dint of an ambiguity in it. We finished by pointing out what one says, what one intends to say, and how one is understood need not be the same thing, and we noted that interlocutors are often able to understand what someone intends to say, even if what he or she intended to say is not what he or she actually said.

Language and Belief

SOLUTIONS TO SOME OF THE EXERCISES

2 Implicatures

1. (a) The bus stop is some distance from my house.

The sentence is bound to be true, for unless the speaker's house is the bus stop, there is a distance separating the house from the bus stop. By observing the maxim of relevance and flouting the maxim of quantity, the speaker conveys that the distance from the bus stop to his or her house is greater than the speaker thinks the person addressed expects it to be.

(c) Richard got a job. I can't believe it.

The first sentence conveys, by the maxim of quality, that the speaker believes that Richard got a job. In uttering the second sentence, the speaker makes it clear that he or she is flouting the maxim, thereby conveying that he or she did not expect that Richard would get a job.

(d) The gasoline station is less than fifty kilometers down the road.

The implicature is that the gasoline station is not much less than fifty kilometers down the road. This is conveyed by observing the maxim of quantity; after all, if the gasoline station were, say, a kilometer down the road, the speaker would not have provided sufficient information to the person addressed by telling him or her that it is less than fifty kilometers down the road.

(f) Helen caused the lights to go off.

The implicature is that Helen did not turn off the lights in the usual way, that is, by simply flipping the switch. This is conveyed by the speaker's statement being more elaborate than what would be warranted had Helen turned off the lights in the usual fashion. This follows from the maxim of manner.

(h) I think we have met before.

The speaker conveys that he is not certain that he and the addressee have met before. This is conveyed by speaker's statement being less strong than the statement *we have met before*. Since the speaker has not made the stronger statement, the implicature follows by the maxim of quantity.

2. (a) A: Are we going to the ballet?
    B: The tickets are sold out.

The implicature is that A and B are not going to the ballet. What B says is not a direct answer to the question put to B by A. By the maxim of relevance, B has said something enabling A to answer the question. If the tickets are sold out, then it is unlikely that B got tickets to the ballet. Without tickets, one cannot attend the ballet. Therefore, B has conveyed that A and B are not going to the ballet.

(c) A: How do you like this compact disc?
    B: I don't like jazz.

The implicature is that B does not like the compact disc. What B says is not a direct answer to the question put to B by A. By the maxim of relevance, B has said something that enables A to answer the question. Now, presumably, the compact disc is one of jazz. By saying that he or she does not like jazz, it is improbable that B likes the music recorded on the compact disc. Therefore, B has conveyed that he or she does not like the compact disc.

3 Presupposition

1. (a) The presupposition is that Idaho has a governor.

   (c) The presupposition is that the speaker has a stamp collection.

   (e) The presupposition is that someone invented the flying bicycle.

   (g) The presuppositions are that Marion has a sister and that her sister is not a witch.

4 What is intended and what is understood

1. (a) Tom and Julie married each other.

   CONSTRUAL 1   Tom and Julie became spouses of each other.
   CONSTRUAL 2   Tom officiated Julie's marriage to someone and Julie officiated Tom's marriage to someone.

   (c) Bill knows what I know.

   CONSTRUAL 1   Bill knows everything I know.
   CONSTRUAL 2   Bill knows something I know.

   (e) There are too many marks in this book.

   CONSTRUAL 1   This books has too many grade entries.
   CONSTRUAL 2   This books has too many markings.

2. (b) Implicature Bill has had breakfast *that day*.

   Implicature Bill has had caviar *at least once in his life*.

6 Classical Propositional Logic: Notation and Semantics

1 Arguments

Arguments consist of declarative sentences, spoken or written, one of which is the conclusion and the remainder of which are premises. While the sentences composing an argument are either true or false, the argument itself is not either true or false, rather, it is either valid or invalid. An argument is valid if, whenever its premises are true, its conclusion is true. Another way to put the very same point is to say that an argument is valid if it is impossible for its premises to all be true yet its conclusion false. In other words, a valid argument is such that the truth of its premises guarantees the truth of its conclusion.

An argument is valid if, and only if, its conclusion must be true, if its premises are true.

An alternative definition is

An argument is valid if, and only if, it could never be the case that, at the same time, its premises are true and its conclusion false.

Consider the following argument. It has the property that, if its premises are true, its conclusion must be true: that is to say, it has the property of being a valid argument.

(1.1) If team A won its game today,
then team B is not in the play-offs.

(1.2) If team C lost its game today,
then team D is in the play-offs.

(1.3) Either team A won its game today or
team C lost its game today.
_____

(1.4) Either team B is not in the play-offs or
team D is in the play-offs.

One does not know whether or not its premises are true. They may all be true, some may be true and others false, or they may be all false. If at least one of them is false, then the conclusion may, though need not, be false. If the conclusion is false, then one, or possibly all, of the premises is false.

The fact that an argument is valid does not mean that any of its premises are true. They might all be false, though the argument is nonetheless valid. If its premises were true, its conclusion would be true.

The following is not a valid argument, even though its premises are all true:

(2.1)   Every man is human.

(2.2)   Some humans have black hair.

(2.3)   Some men have black hair.

For it could have been that the premises are true and the conclusion false. For example, it could very well have been the case that, while every man is human, only female humans have black hair.

It has been known since the time of Aristotle that an argument such as the one in (2) is fallacious: that is, it is not valid, for its premises can be true while its conclusion is false.

(3.1)   If it is raining, then it is cold.

(3.2)   It is not the case that it is raining.

(3.3)   It is not the case that it is cold.

How does one determine whether an argument is valid? One attempts to find an argument of the same form, but whose premises are clearly true and whose conclusion is equally clearly false.

(4.1)   If it is snowing, then it is cold.

(4.2)   It is not the case that it is snowing.

(4.3)   It is not the case that it is cold.

It is clear that the premises of this argument can be true and yet the conclusion is false. After all, consider any cold winter's day when it does not snow.

It is easy to see that the arguments in (3) and (4) share a form:

(5.1)   If p, then q.

(5.2)   It is not the case that p.

Classical Propositional Logic: Notation and Semantics 239

(5.3)   It is not the case that q.

What we have done is to replace clauses in (3) and (4) with propositional variables, thereby abstracting a schema, the skeleton of which involves such special expressions as *if* and *it is not the case*. It is the distribution of such words in sentences that helps to determine whether an argument is valid. Consider this argument.

(6.1)   If it is snowing, then it is cold.

(6.2)   It is not the case that it is cold.
        _____

(6.3)   It is not the case that it is snowing.

This argument is valid and differs in a subtle, though important, way from the one in (4), as shown by abstracting to its form.

(7.1)   If p, then q.

(7.2)   It is not the case that q.
        _____

(7.3)   It is not the case that p.

Arguments of the form in (5) are fallacies. The fallacy is known as *denial of the antecedent*. Arguments of the form in (7) are valid, and the form of argument is sometimes called by its Latin name, *modus tollens*.

How does one determine whether an argument is valid or invalid? One way to demonstrate that an argument is valid is to set out a chain of arguments where, first, the conclusion of the last argument in the chain is the conclusion of the argument whose validity is being demonstrated, where, second, each argument in the chain is self-evidently valid, and where, finally, any premise of any argument in the chain is either a premise of the argument whose validity is being demonstrated or the conclusion of a preceding argument in the chain. Paradigmatic examples of such demonstrations are the proofs found in Euclid's *Elements*. An alternative way to demonstrate that an argument is valid is to show that the form of its sentences are such that the truth of the premises compels the truth of the conclusion.

The sentences expressing the preceding arguments are in English. So, one should conclude, naturally enough, that we should be looking at the syntax of English. And we shall be, in chapter 7, where we shall study the syntax of coordinated and subordinated independent clauses and how the value of a complex clause is determined by the value of the clauses making it up. Here, however, we shall be studying classical propositional logic, for it furnishes us with some of the tools required for the treatment the syntax and semantics of coordinated and subordinated independent clauses.

Exercises: Arguments

1. In each of the following passages, indicate whether an argument is expressed. Then, if so, identify its premises and its conclusion.

(a) With regard to good and evil, these terms indicate nothing positive in things considered in themselves, nor are they anything else than modes of thought, or notions which we form from the comparison of one thing with another. For one and the same thing may at the same time be both good and evil or indifferent. Music, for example, is good to a melancholy person, bad to one mourning, while to a deaf man it is neither good nor bad. (*Ethics,* Baruch Spinoza; cited in Copi 1953, 40)

(b) ...we are told that this God, who prescribes forbearance and forgiveness of every fault, exercises none himself, but does the exact opposite; for a punishment which comes at the end of all things, when the world is over and done with, cannot have for its object either to improve or deter, and is therefore pure vengeance. (*The Christian System,* Arthur Schopenhauer; cited in Copi 1953, 40)

(c) Particles and their antiparticles quickly annihilate upon meeting, their combined mass converting to energy as described in Einstein's famous equation. Thus no serious thought has been given to manufacturing large amounts of antimatter, since it should be impossible to store. ("Antimatter—the Ultimate Explosive?," Paul Preuss, *Science* v. 80; cited in Copi 1953, 40)

(d) Now, as soon as landowners are deprived of their strong sentimental attachment to the land, based on memories and pride, it is certain that sooner or later they will sell it, for they have a powerful pecuniary interest in so doing, since other forms of investment earn a higher rate of interest and liquid assets are easily used to satisfy the passions of the moment. (*Democracy in America,* Alexis de Tocqueville; cited in Copi 1953, 41)

(e) ...if materialism is true, all our thoughts are produced by purely material antecedents. These are quite blind, and are just as likely to produce falsehood as truth. We have thus no reason for believing any of our conclusions—including the truth of materialism, which is therefore a self-contradictory hypothesis. (*Philosophical Studies,* John M. E. McTaggart; cited in Copi 1953, 41)

2. In each of the following passages, identify the argument and state whether it is valid.

(a) The patient will die unless we operate. We will operate. Therefore, the patient will not die.

(b) John and Bill left. Hence, Bill left.

(c) Fred did not meet anyone. Therefore, Fred did not meet Ted.

(d) Most Canadians like hockey. Everyone who likes hockey likes curling. Thus, most Canadians like curling.

Classical Propositional Logic: Notation and Semantics

(e) Either the government will call an election or it will raise taxes. Therefore, if the government raises taxes, it will not call an election.

2 Classical Propositional Logic

Classical propositional logic (CPL) is concerned with the validity of arguments that depend on so-called *propositional connectives*. Since these connectives are the linchpin of CPL, we precede its exposition with a brief discussion of the connectives, relating them to the expressions in English that, to some extent, parallel them. The connectives of CPL are: $\neg$, $\rightarrow$, $\leftrightarrow$, $\vee$, and $\wedge$. The English counterparts to them are *it is not the case that, if, if and only if, or,* and *and,* respectively.

EXPRESSIONS OF ENGLISH	PROPOSITIONAL CONNECTIVES
It is not the case that ...	$\neg$ ...
... and ...	... $\wedge$ ...
... or ...	... $\vee$ ...
if ... , then ...	... $\rightarrow$ ...
... if and only if ...	... $\leftrightarrow$ ...

We shall explore both the connectives and their correspondence with English expressions in great detail in chapter 8. For now, however, let us get a better grasp of the connectives by taking for granted the correspondence.

NEGATION

Consider the following pair of sentences.

(8.1) It is raining.

(8.2) It is not the case that it is raining.

It should be obvious that, if one is true, the other is false and vice versa. This observation is summed up in the following table:

It is raining.	It is not the case that it is raining.
T	F
F	T

Now, this observation is not limited to the proposition that it is raining. Indeed, it is true of any proposition. To express this generalization, let p denote the proposition denoted by *it is raining*. Then $\neg p$ denotes the proposition *it is not the case that it is raining*. We can then generalize the preceding table as follows:

p	$\neg p$
T	F
F	T

CONJUNCTION

Consider next the pair of sentences *it is raining* and *it is cold*. It is clear that the conjunction of this pair of sentences, *it is raining and it is cold,* is true only in the case where each sentence in the pair is true. Again, we can use a table to recapitulate this observation.

It is raining.	It is cold.	It is raining and it is cold.
T	T	T
T	F	F
F	T	F
F	F	F

Moreover, as before, the observation is general, and the generalization can be expressed with propositional variables.

p	q	$p \wedge q$
T	T	T
T	F	F
F	T	F
F	F	F

DISJUNCTION

It is raining.	It is cold.	It is raining or it is cold.
T	T	T
T	F	T
F	T	T
F	F	F

p	q	$p \vee q$
T	T	T
T	F	T
F	T	T
F	F	F

MATERIAL IMPLICATION

To understand what is meant by *material implication,* let us avail ourselves of an obvious analogy between truth and falsity, on the one hand, and to abide by one's promise and not to abide by one's promise, on the other. Clearly either one abides by one's promise or one does not. Analogously, either a formula of CPL is true or it is not. Consider, now, a promise made by an instructor to a class of students.

Classical Propositional Logic: Notation and Semantics

(9) If you pass the final exam, then you will pass the course.

Under what circumstances does the instructor not abide by the promise? The answer is clear: only under the circumstances in which the student passes the final but the instructor fails the student in the course. In all other circumstances, the instructor has abided by the promise. In particular, should the student fail the final but otherwise have done well in the course, the instructor, in passing the student in the course, will not have reneged on the promise; nor will the instructor have reneged, if the instructor fails the student, the student having done poorly in class work and other exams. In other words, once the student has failed the final exam, no matter what the instructor does, the instructor will not have reneged on the promise.

You pass the final.	You pass the course.	If you pass the final, then you pass the course.
T	T	T
T	F	F
F	T	T
F	F	T

By analogy, only in the case where the propositional variable to the left of the arrow is assigned true and the one to the right of the arrow is assigned false is the entire formula assigned false; in all other cases, it is assigned true.

p	q	$p \to q$
T	T	T
T	F	F
F	T	T
F	F	T

MATERIAL EQUIVALENCE

Let us continue the analogy between abiding by promises on the one hand and truth values on the other, for it also sheds light on material equivalence, and in doing so, it will shed still further light on material implication. In the case of the promise in (9), we observed that the only circumstances in which the instructor fails to abide by the promise are those in which the student passes the final but the instructor fails the student in the course; in all other circumstances, the instructor has abided by the promise. This contrasts with the following promise.

(10) If, and only if, you pass the final, will you pass the course.

In the case of the promise in (9), once the student has failed the final exam, no matter what the instructor does, the instructor will not have reneged on the promise. However, in the case of the promise in (10), should the student fail the final exam, the instructor is obliged to fail the student in the course. Should the student pass the final, the promise in (10), like the promise in (9), obliges the instructor to pass the student in the course.

You pass final.	You pass course.	You pass final, iff you pass course.
T	T	T
T	F	F
F	T	F
F	F	T

p	q	$p \leftrightarrow q$
T	T	T
T	F	F
F	T	F
F	F	T

2.1 Notation

In this section, we wish to explain the notation of CPL. Its notation is built from two disjoint sets of symbols: the set of propositional connectives, referred to as PC and containing precisely five distinct symbols, and a nonempty set of propositional variables, referred to as PV. From these sets of symbols are built all the formulae of CPL.

The symbols used here as the five propositional connectives are the one unary connective ($\neg$) and the four binary connectives ($\land, \lor, \rightarrow, \leftrightarrow$).

The symbols for the propositional connectives vary from book to book. So the reader might find it useful to take notice of some of the other symbols used for the propositional connectives.

1. $\neg$ is the negation symbol. It is a unary propositional connective. Other symbols used to serve in the role which the negation symbol serves in here include $\sim$.
2. $\land$ is the conjunction symbol. Other symbols used are & and $\cdot$.
3. $\rightarrow$ is the material implication symbol. An alternate for it is $\supset$.
4. $\leftrightarrow$ is the material equivalence symbol, for which the symbol $\equiv$ is often an alternate.
5. There are other propositional connectives. They include: $\top$, $\bot$, and $|$, among others. These are not alternates to the other propositional connectives but have roles of their own to serve.

Classical Propositional Logic: Notation and Semantics

The symbols we shall use for the propositional variables are the lowercase letters of the Roman alphabet, starting with p. A small set of propositional variables is the set $\{p,q,r\}$. Should a large number of propositional variables be required, we shall use the letter p, subscripted with positive integers. For example, if we required fifteen propositional variables, we would use the symbols $p_1, \ldots, p_{15}$.

The formulae of CPL are obtained from the set of propositional connectives and a set of propositional variables. Since the set of propositional connectives is fixed, while the set of propositional variables is not, the set of formulae of CPL one obtains will depend on the set of propositional variables chosen. Thus, any set of formulae of CPL is defined relative to a set of propositional variables.

As we shall see presently, the set of formulae of CPL, or FM, can be defined in two different ways. One way, typical of logicians, is to use a *syncategorematic* form of definition; the other, more congenial to comparisons with constituency grammar, is to use a *categorematic* definition. We present each of these definitions in sections 2.1.1 and 2.1.2.

2.1.1 Definition of formulae: Syncategorematic

The syncategorematic definition of FM states that all propositional variables are formulae. It also states that any formula prefixed with $\neg$ is a formula and that a pair of formulae, infixed with $\land$, $\lor$, $\rightarrow$, or $\leftrightarrow$ and enclosed with parentheses is a formula. Finally, it says that nothing else is a formula. Here is the definition stated formally.

Definition 1 Formulae of CPL (syncategorematic version)
Let PV be a set of propositional variables. Then, FM, the formulae of *CPL* based on PV, is the set defined as follows:

(1) $\text{PV} \subseteq \text{FM}$;

(2.1)   if $\alpha \in \text{FM}$, then $\neg \alpha \in \text{FM}$;

(2.2.1) if $\alpha, \beta \in \text{FM}$, then $(\alpha \land \beta) \in \text{FM}$;

(2.2.2) if $\alpha, \beta \in \text{FM}$, then $(\alpha \lor \beta) \in \text{FM}$;

(2.2.3) if $\alpha, \beta \in \text{FM}$, then $(\alpha \rightarrow \beta) \in \text{FM}$;

(2.2.4) if $\alpha, \beta \in \text{FM}$, then $(\alpha \leftrightarrow \beta) \in \text{FM}$;

(3)   nothing else is.

This definition is said to be syncategorematic, because it does not invoke explicitly the category of propositional connective. The only categories to be explicitly invoked are PV and FM; PC is simply not referred to.

Suppose that p, q, and r are propositional variables. By clause (1) of the definition, each of these propositional variables is a formula. By clause (2.1) $\neg p$ is a formula. Moreover, by clause (2.2.1), $(q \wedge r)$ is a formula. Since $\neg p$ and $(q \wedge r)$ are formulae, it follows by clause (2.2.3) that $\bigl(\neg p \to (q \wedge r)\bigr)$ is a formula. We can display this reasoning in a diagram of the sort we saw in chapters 1 and 3.

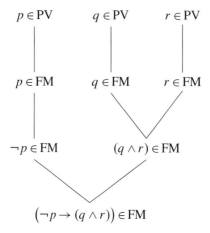

However, it is customary to simplify the diagram, first by restricting our attention to expressions insofar as they are formulae, thereby omitting the nodes labeled with PV, and second, by omitting the indication that the formula is a member of FM, as all the remaining expressions are in FM. We shall call the resulting simplified diagram a *syncategorematic synthesis tree diagram*:

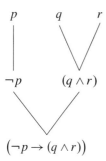

It can be turned upside down, becoming thereby what we shall call a *syncategorematic analysis tree diagram*.

Classical Propositional Logic: Notation and Semantics

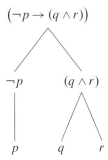

Different sets of propositional variables can be used to build a set of formulae. It should be obvious that, if we use the set $\{p,q,r\}$ as a set of propositional variables, we shall obtain one set of formulae, and if we use the set $\{s,t,u\}$ as propositional variables, we shall obtain another set of formulae. In fact, the two sets will be disjoint. Which symbols are being used as propositional variables, on any given occasion, will be either stated explicitly or clear from context.

2.1.2 Definition of formulae: Categorematic

The categorematic definition of FM is very similar to the syncategorematic one. The difference is that, in addition to the sets PV and FM, we explicitly identify two other sets: UC, the set of unary propositional connectives, which contains only one member, namely $\neg$, and BC, the set of binary propositional connectives, which contains four members, namely, $\wedge$, $\vee$, $\rightarrow$, and $\leftrightarrow$. Like the syncategorematic definition, it states that all propositional variables are formulae. It also states that any formula prefixed with a unary propositional connective ($\neg$) is a formula and that a pair of formulae, infixed with a binary propositional connective and enclosed with parentheses is a formula. Finally, like the syncategorematic definition, it says that nothing else is a formula. Here is its formal definition.

Definition 2 Formulae of CPL (categorematic version)

Let PV be a set of propositional variables. Then, FM, the formulae of *CPL* based on PV, is the set defined as follows:

(1) $PV \subseteq FM$;
(2.1)   if $\alpha \in FM$ and $* \in UC$, then $*\alpha \in FM$;
(2.2)   if $\alpha, \beta \in FM$ and $\circ \in BC$, then $(\alpha \circ \beta) \in FM$;
(3)   nothing else is.

Comparing this definition to the syncategorematic one, we note that the clauses (1) and (3) are the same. The difference lies with the clauses in (2). Let us begin with the clause in (2.1). The categorematic version of the clause has replaced $\neg$ with $*$ and it contains a further condition in its protasis, namely, that $* \in \mathrm{UC}$. The other change is that the four clauses in (2.2) of the syncategorematic definition are reduced to one clause in the categorematic one. This is done by the addition of the condition that $\circ \in \mathrm{BC}$ in the protasis and the replacement of the various propositional connectives in the apodosis by the single symbol $\circ$, thereby allowing the four clauses to be collapsed into one.[1]

Let us return to the formula whose formation we established earlier using the syncategorematic definition of formulae and see that it can be equally obtained using the categorematic definition. Suppose, as we did previously, that p, q, and r are propositional variables. By clause (1) of the definition, each of these propositional variables is a formula. By clause (2.1), $\neg p$ is a formula. Moreover, by clause (2.2), $(q \wedge r)$ is a formula. Since $\neg p$ and $(q \wedge r)$ are formulae, it follows again by clause (2.2) that $(\neg p \to (q \wedge r))$ is a formula. Here is the *categorematic synthesis tree diagram* for this formula.

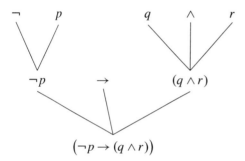

Here is the corresponding *categorematic analysis tree diagram*.

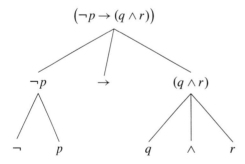

---

1. As explained in chapter 5, the subordinate *if* clause is the protasis and the clause to which it is subordinate is the apodosis.

Classical Propositional Logic: Notation and Semantics

Exercises: Formation rules

1. For each of the following formulae, provide both a syncategorematic tree diagram and a categorematic tree diagram, one analytic and one synthetic.
   (a) $((p \land q) \to p)$
   (b) $(p \land (q \to p))$
   (c) $\neg(p \land (\neg p \land q))$
   (d) $(\neg p \land (\neg p \land q))$
   (e) $((p \to q) \to (\neg q \to \neg p))$
   (f) $(\neg(p \land (q \lor r)) \to (\neg r \lor q))$

2. Identify, among the following strings of symbols, which are formulae and which are not. Justify your answer.
   (a) (p)
   (b) $(p \land (p \lor p))$
   (c) $p \to (p \land p)$
   (d) q
   (e) $(\neg q \land (p \leftarrow q))$
   (f) $\neg\neg(\neg\neg q \lor \neg\neg\neg r)$
   (g) $((p \land q) \to (r \lor p))$
   (h) $p \lor q$
   (i) $((p \land (q \to r)) \land p)$
   (j) $((r \land q \land r) \to p)$
   (k) $\neg((p \leftrightarrow q) \lor r)$
   (l) $\neg(\neg(p \land r))$

2.1.3 Formulae and subformulae

In this section, we shall classify formulae into various kinds. We shall also explore the relation a formula bears to formulae that are its parts.

As we saw in sections 2.1.1 and 2.1.2, all propositional variables (PV) are formulae. Because they, together with the propositional connectives, are the building blocks for other formulae, they are also known as *atomic formulae* (AF), when considered as formulae. Formulae that are not atomic are *composite formulae* (CF). $\neg\neg p$, $(q \land \neg p)$, and $\neg(p \to q)$ are examples of CF. Clearly, no propositional variable is a composite formula.

Finally, there is a class of propositional formulae that overlap the CF and the AF, namely, *basic formulae* (BF). BF are AF and their negations. p and $\neg q$ are BF. Neither $(r \lor q)$ nor $\neg\neg p$ is. Such formulae are also called *literals*.

It is often useful to be able to speak about the formulae that are parts of a larger formula. To do this, we must introduce two relations: the relation of being an immediate subformula and the relation of being a subformula.

One formula is an immediate subformula of another just in the case where the first can be either prefixed by the negation symbol to yield the second or paired with another formula and a binary propositional connective and enclosed with parentheses to yield the second.

Definition 3 Immediate subformulae of CPL
Let α and γ be members of FM. α is an immediate subformula of γ iff either $\neg \alpha = \gamma$ or there is a formula β such that one of the following obtains: $(\alpha \wedge \beta) = \gamma$, $(\beta \wedge \alpha) = \gamma$, $(\alpha \vee \beta) = \gamma$, $(\beta \vee \alpha) = \gamma$, $(\alpha \rightarrow \beta) = \gamma$, $(\beta \rightarrow \alpha) = \gamma$, $(\alpha \leftrightarrow \beta) = \gamma$, or $(\beta \leftrightarrow \alpha) = \gamma$.

The relation of one formula's being an immediate subformula of another is well depicted by its tree diagram, either categorematic or syncategorematic. The immediate subformulae of a formula α are the formulae immediately above α in its tree diagram. Thus, for example, $(p \wedge q)$ is an immediate subformula of $(\neg r \rightarrow (p \wedge q))$.

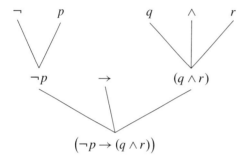

It is not, however, an immediate subformula of $(p \vee ((p \wedge q) \rightarrow r))$.

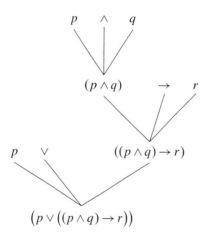

It is useful to have a broader notion than that of being an immediate subformula, one in which, for example, $(p \wedge q)$ is a subformula of $(p \vee ((p \wedge q) \rightarrow r))$, though, as we just saw, the former is not an immediate subformula of the latter. Let us dub this broader notion that of being a *proper subformula*. Roughly, a formula α is a proper subformula of β if α is

Classical Propositional Logic: Notation and Semantics   251

an immediate subformula of a formula β, it is an immediate subformula of an immediate subformula of β, an immediate subformula of an immediate subformula of an immediate subformula of β, or ... Here is the formal definition.

Definition 4 Proper subformulae of CPL
Let α and γ be members of FM. α is a proper subformula of γ iff one of the following holds:

(1) α is an immediate subformula of γ;

(2)   there is a formula β such that β is an immediate
      subformula of γ and α is a proper subformula of β.

The relation of one formula's being a proper subformula of another is also well depicted by the formula's tree diagram, whether it is a syncategorematic one or a categorematic one. α is a proper subformula of β only in the case where α is above β in β's synthesis tree. So, though $(p \wedge q)$ is not an immediate subformula of $\left(p \vee \left((p \wedge q) \to r\right)\right)$, it is a proper subformula.

In fact, logicians and mathematicians prefer a still broader notion of subformulae, namely, one that allows a formula to be a subformula of itself. The advantage of this broader notion is that it facilitates the definition of other concepts, that will be introduced presently.

Definition 5 Subformulae of CPL
Let α and β be members of FM. α is a subformula of β iff either α is β or α is a proper subformula of β.

Notice that the relations of being an immediate subformula, of being a proper subformula, and of being a subformula are just like the relations of being an immediate constituent, of being a proper constituent, and of being a constituent, as discussed in chapter 3, section 3.1 and following.

In talking about formulae and their parts, one must attend to the fact that the very same propositional variable or propositional connective may occur more than once in the same formula. For example, p occurs twice in the formula $\left(p \wedge (q \vee p)\right)$ and $\wedge$ occurs twice in the formula $\left(p \wedge (q \wedge r)\right)$. Sometimes it is necessary to be able to distinguish between the occurrences. In that event, one speaks, not of a symbol, but of its occurrence. We shall abide by this distinction, only when we feel that there is a risk of confusion; otherwise, we shall continue with the usual practice of using such terms to refer to the symbol itself as well as to its occurrences.

One case where the distinction just made must be observed is in connection with the relation of scope.

Definition 6 Scope of a propositional connective of CPL

The scope of an occurrence of a propositional connective in a formula α is the subformula of α that contains the propositional connective's occurrence but whose immediate subformulae do not.

One must speak of the scope of a propositional connective relative to its occurrence. Consider the formula $((p \wedge q) \wedge r)$. It does not make sense to inquire into the scope of $\wedge$; it makes sense only to inquire into the scope either of its first occurrence, in which case its scope is $(p \wedge q)$, or of its second occurrence, in which case its scope is $((p \wedge q) \wedge r)$.

Definition 7 Formula being within the scope of CPL

A formula α occurs within the scope of a propositional connective's occurrence in a formula β iff α is a subformula of the scope of the propositional connective's occurrence in formula β.

Thus, for example, both p and $(q \vee \neg r)$ occur within the scope of $\rightarrow$ in the propositional formula $(p \rightarrow (q \vee \neg r))$. Notice p is not within the scope of $\vee$ in the same formula. Moreover, in the formula $(p \vee (q \vee \neg r))$, p is within the scope of the first occurrence of $\vee$ but not within the scope of the second.

Definition 8 Propositional connective's occurrence being within the scope of CPL

One propositional connective's occurrence in a formula occurs within the scope of another's iff the first occurs within a formula that is within the scope of the second.

In the formula $(p \rightarrow \neg(r \rightarrow q))$, the only occurrence of $\neg$ occurs within the scope of the first occurrence of $\rightarrow$, but the second occurrence of $\rightarrow$ occurs within the scope of the only occurrence of $\neg$.

Definition 9 Formula's main propositional connective of CPL

A formula's main propositional connective is that propositional connective whose occurrence in the formula has the formula for its scope.[2]

The main propositional connective of the formula $\neg(p \vee \neg q)$ is the first occurrence of $\neg$, while the main propositional connection of $((\neg p \rightarrow q) \vee (r \wedge p))$ is the only occurrence of $\vee$.

Notice that parentheses are not listed among the symbols of CPL; rather, they are insinuated into formulae through the definition for the set of formulae. For this reason, they are sometimes said to be syncategorematic. Their purpose, like that of commas and periods in English, is to eliminate ambiguity; for this reason, they are sometimes referred to as punctuation. The rule governing good punctuation is to omit it, when otiose. The same

---

2. The definition has been framed in terms of a propositional connective, instead of its occurrence. Phrasing the definition in terms of the latter leads to awkward English.

Classical Propositional Logic: Notation and Semantics

applies in logic and in mathematics. So parentheses are often omitted. To forestall confusion, conventions are adopted governing their omission. These are the conventions adopted herein:

(i) Omit the parentheses enclosing an entire formula. Thus, $p \wedge q$ is an abbreviation of $(p \wedge q)$.

(ii) Omit the parentheses enclosing a conjunct, if it itself is a conjunction. Thus, $p \vee (p \wedge q \wedge r)$ abbreviates both $p \vee (p \wedge (q \wedge r))$ and $p \vee ((p \wedge q) \wedge r)$, which are themselves, by the previous convention, abbreviations.

(iii) Omit the parentheses enclosing a disjunct, if it itself is a disjunction. Thus, $p \wedge (p \vee q \vee r)$ abbreviates both $p \wedge (p \vee (q \vee r))$ and $p \wedge ((p \vee q) \vee r)$, which are themselves, by the previous convention, abbreviations.

Exercises: Formulae and subformulae

1. Identify, among the following strings of symbols, which are formulae and which are formulae by convention.

(a) p

(b) $(p \wedge (p \vee p))$

(c) $p \wedge q$

(d) $(p \wedge q \wedge r)$

(e) $\neg\neg(\neg\neg q \vee \neg\neg\neg r)$

(f) $((p \wedge q) \rightarrow (r \vee p))$

(g) $(p \wedge (q \rightarrow r)) \wedge p$

(h) $\neg((p \leftrightarrow q) \vee r)$

(i) $p \wedge \neg q \wedge (p \rightarrow q)$

2. Identify, among the following formulae, which are AF, which are BF, and which are CF.

(a) $(p \wedge (p \vee p))$

(b) p

(c) $((q \wedge p) \leftarrow q)$

(d) $\neg q$

(e) $\neg\neg\neg r$

(f) r

(g) $p \vee q$

(h) $(r \wedge q \wedge r) \rightarrow p$

(i) $\neg p$

(j) $\neg((p \leftrightarrow q) \vee r)$

3. State for each of the following whether it is true or false. If it is false, provide a counterexample.

(a) $AF = PV$

(b) $AF \subseteq BF$

(c) $BF \subseteq FM$

(d) $BF \subseteq AF$

(e) $CF \subseteq BF$

(f) $BF \subseteq CF$

(g) $BC \subseteq FM$

4. For each of the following formulae, identify its main propositional connective and all of its subformulae.

(a) $((q \leftrightarrow p) \lor p)$ (d) r

(b) $\neg((r \lor p) \land q)$ (e) $(\neg(r \lor (q \land r)) \to p)$

(c) $\neg\neg\neg r$

5. Consider this formula: $(((p \lor q) \to q) \to ((q \land r) \lor p))$

(a) What is the scope of the first occurrence of $\to$?
(b) What is the scope of the second occurrence of $\to$?
(c) What is the scope of the first occurrence of $\lor$?
(d) What is the scope of the second occurrence of $\lor$?
(e) What is the scope of $\land$?

6. Consider another formula: $((p \land (q \leftrightarrow r)) \to (\neg p \lor (q \land r)))$

(a) Is the first occurrence of r within the scope of the only occurrence of $\leftrightarrow$?
(b) Is the second occurrence of r within the scope of the only occurrence of $\lor$?
(c) Is the first occurrence of p within the scope of the second occurrence of $\land$?
(d) Is the only occurrence of $\leftrightarrow$ within the scope of the only occurrence of $\to$?
(e) Is the second occurrence of q within the scope of the first occurrence of $\land$?
(f) Is the only occurrence of $\lor$ within the scope of the only occurrence of $\neg$?
(g) Is the first occurrence of q within the scope of the first occurrence of $\land$?
(h) Is the first occurrence of p within the scope of the first occurrence of $\land$?
(i) Is the only occurrence of $\leftrightarrow$ within the scope of the first occurrence of $\land$?
(j) Is either occurrence of $\land$ within the scope of the other's?

2.2 Semantics

The fundamental ideas underlying propositional semantics are those of a truth value assignment and a valuation. Both a truth value assignment and a valuation are functions.

A *truth value assignment* is any function from a domain, which is the set of propositional variables (PV), to a codomain, which is a set containing T and F, where T and F are intuitively the value of being true and the value of being false. A *bivalent truth value assignment*—or, a bivalent assignment, for short—is a truth value assignment whose codomain contains only T and F.[3]

---

3. *Bivalent* is another word for two-valued.

Classical Propositional Logic: Notation and Semantics

Suppose for the moment that PV contains only three propositional variables: p, q, and r. Then the following is a bivalent assignment:

$p \mapsto T$, $q \mapsto T$, $r \mapsto F$.

However, neither of the following are bivalent assignments:

$p \mapsto T$, $r \mapsto F$;

$p \mapsto T$, $q \mapsto N$, $r \mapsto F$.

The former is not a function whose domain is PV, and the latter, though a function whose domain is PV, is not bivalent.

The following are all the bivalent assignments of the truth values $\{T,F\}$ to the propositional variables in the set $\{p, q, r\}$.

		a_1	a_2	a_3	a_4	a_5	a_6	a_7	a_8
p	$\mapsto$	T	T	T	T	F	F	F	F
q	$\mapsto$	T	T	F	F	T	T	F	F
r	$\mapsto$	T	F	T	F	T	F	T	F

The bivalent assignment a_1 assigns T to each of the three propositional variables, while the bivalent assignment a_2 assigns T to p and q and F to r. Since truth value assignments are functions, one can also write that $a_1(p) = a_1(q) = a_1(r) = T$ and that $a_2(p) = a_2(q) = T$ and $a_2(r) = F$.

It is noteworthy that, whenever the number of propositional variables is finite, the number of assignments from the set $\{T,F\}$ is finite. After all, for each propositional variable, one has but two choices—T or F. So, if one has three propositional variables, then there are but $2 \cdot 2 \cdot 2$, or 2^3, possible assignments, and if one has n propositional variables, then the number of assignments from $\{T,F\}$ is 2^n.

Let us turn to *valuations*. A valuation is any function whose domain is the set of formulae (FM) and whose codomain is a set containing T and F. If the only values in the codomain are T and F, then the valuation is bivalent. FM is infinite. (FM would be infinite, by the way, even if PV contained only one propositional variable.) This fact has two consequences. First, the number of valuations—even, the number of bivalent valuations—is uncountable. Second, no valuation's graph can be written down in list notation.

Nonetheless, to have an idea of how diverse bivalent valuations are, suppose again that PV contains only three propositional variables: p, q, and r. The following are four partial specifications of bivalent valuations whose domain is the set of formulae obtained from $\{p,q,r\}$.

		f_1	f_2	f_3	f_4
p	$\mapsto$	T	T	T	T
q	$\mapsto$	T	T	F	F
r	$\mapsto$	T	F	T	F
$\neg p$	$\mapsto$	T	F	F	F
$\neg q$	$\mapsto$	F	F	T	T
$\neg r$	$\mapsto$	F	T	F	T
$p \wedge q$	$\mapsto$	T	T	T	F
$p \vee r$	$\mapsto$	T	T	T	T
$\neg p \vee q$	$\mapsto$	F	F	F	F
$\neg q \wedge r$	$\mapsto$	T	F	T	F
$\vdots$	$\vdots$	$\vdots$	$\vdots$	$\vdots$	$\vdots$

If one were to complete the list of formulae and ensure that each position under one or another of the f's is filled in with precisely one of T or F, each f will provide a bivalent valuation. However, three of them certainly do not comply with the interpretation given earlier to the symbols $\neg$, $\wedge$, and $\vee$. (See section 2.) f_1, for example, assigns T to both p and $\neg p$. Moreover, f_2 assigns F to $\neg p \vee q$, though it also assigns T to q. Finally, f_3 assigns T to $p \wedge q$, even though it assigns F to q. Only f_4, at least to the extent it is specified, actually complies with our earlier interpretation of the propositional connectives.

Bivalent valuations that agree with the interpretation of the propositional connectives given earlier are known as *classical* valuations. They can be defined as follows:

Definition 10 Classical valuation for CPL

A valuation v is classical iff v is bivalent and complies with these conditions: for each $\alpha, \beta \in \mathrm{FM}$,

(1) $v(\neg \alpha) = T$ iff $v(\alpha) = F$;

(2.1) $v(\alpha \wedge \beta) = T$ iff $v(\alpha) = v(\beta) = T$;

(2.2) $v(\alpha \vee \beta) = T$ iff $v(\alpha) = T$ or $v(\beta) = T$;

(2.3) $v(\alpha \rightarrow \beta) = T$ iff $v(\alpha) = F$ or $v(\beta) = T$;

(2.4) $v(\alpha \leftrightarrow \beta) = T$ iff $v(\alpha) = v(\beta)$.

As we have noted, the set of all bivalent valuations is infinite. However, should a restriction be put on the set of bivalent valuations, one might wonder whether any bivalent valuations satisfy the definition, that is, whether the set of classical valuations is empty or not.

In fact, the number of classical valuations is exactly the same as the number of bivalent assignments. Moreover, as we shall see, though we shall not prove, each classical valuation

Classical Propositional Logic: Notation and Semantics 257

is determined by a bivalent assignment and distinct bivalent assignments determine distinct classical valuations. As one can see from inspecting the definition, the formulae specified on the right-hand side of the definition are the immediate subformulae of those on the left-hand side, and the values assigned to the formulae on the left-hand side are defined in terms of the values assigned to their immediate subformulae. Thus, each composite formula has its truth value determined by the truth values of its immediate subformulae. This means that any assignment of truth values to the atomic subformulae of a formula immediately propagate, through the definition, truth values to larger and larger subformulae until, at last, a truth value is assigned to the formula itself.

How, precisely, this propagation of truth values to a formula from its atomic subformulae, or propositional variables, is done depends on whether one takes as a point of departure the syncategorematic definition of formulae or the categorematic one. The first point of departure leads to a syncategorematic definition of a canonical extension of a bivalent assignment to a classical valuation and the second to a categorematic definition of a canonical extension. In sections 2.2.1 and 2.2.2, we shall pursue the first and second alternatives, respectively.

Exercises: Assignments and valuations

1. Describe three bivalent valuations.

2. Suppose the number of truth values were three. How many truth value assignments are there to a set of n propositional variables?

3. Which of the following rules of association determines a bivalent valuation? In each case, justify your answer.

(a) $\alpha \mapsto T$, if α has an occurrence of a propositional connective;
$\alpha \mapsto F$, otherwise.

(b) $\alpha \mapsto T$, if α has an even number of occurrences of binary propositional connectives;
$\alpha \mapsto F$, if α has an odd number of occurrences of the unary propositional connective.

(c) $\alpha \mapsto T$, if α has any occurrences of a binary propositional connective;
$\alpha \mapsto F$, if α has any occurrences of the unary propositional connective.

(d) $\alpha \mapsto T$, if α has an even number of propositional variables;
$\alpha \mapsto F$, if α has an odd number of propositional variables.

4. Which of the following sets is empty? In each case, justify your answer.

(a) The set of functions from FM to $\{T, F\}$ where each $\alpha \in$ FM that has an occurrence of the unary propositional connective is assigned T.

(b) The set of functions from FM to $\{T, F\}$ where each $\alpha \in$ FM with an even number of occurrences of binary propositional connectives is assigned F.

(c) The set of functions from FM to $\{T,F\}$ where each $\alpha \in$ FM with no occurrence of the unary propositional connective is assigned T and each $\alpha \in$ FM with any occurrences of binary propositional connectives is assigned F.

(d) The set of functions from FM to $\{T,F\}$ where each $\alpha \in$ FM is assigned T if α contains three distinct propositional variables and is assigned F otherwise.

2.2.1 Syncategorematic approach

We begin our presentation of the syncategorematic definition of the extension of a bivalent assignment to a classical valuation with an illustration of how a_3 can assign a truth value to $\neg p \to (q \wedge r)$ that respects the interpretation of the propositional connectives stated in section 2.

Let us begin by bearing in mind the formula's syncategorematic synthesis tree diagram.

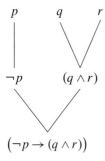

Recall the interpretation given earlier to $\neg$.

α	$\neg \alpha$
T	F
F	T

This interpretation is equivalent to clause (1) of the definition of a classical valuation (definition 10). Applying it to our tree, one assigns the value F to the subformula $\neg p$.

Next, recall the interpretation we gave for $\wedge$.

α	β	$\alpha \wedge \beta$
T	T	T
T	F	F
F	T	F
F	F	F

This is equivalent to clause (2.1) of the definition of a classical valuation. It requires us to assign F to the subformula $q \wedge r$.

Classical Propositional Logic: Notation and Semantics

Finally, recall the interpretation of $\to$.

α	β	$\alpha \to \beta$
T	T	T
T	F	F
F	T	T
F	F	T

This is equivalent to clause (2.4) of the definition of a classical valuation. Taking α here as $\neg p$ in the tree and β here as $q \wedge r$ in the tree, one assigns T to the formula $\neg p \to (q \wedge r)$.

We display on the syncategorematic synthesis tree diagram for $\bigl(\neg p \to (q \wedge r)\bigr)$ the foregoing assignments of truth values imposed by a_3 on its various subformulae by the conditions stated previously.

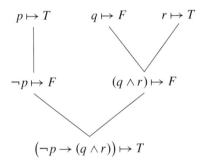

Note that once the values are assigned to the top nodes on the tree, the values of the lower nodes are determined, for each lower node is calculated from the values of the nodes immediately above it. What guarantees that the nodes other than the top nodes are assigned unique values are the three clauses in definition 10 corresponding to the propositional connectives $\neg$, $\wedge$, and $\to$.

What the preceding example illustrates, but does not prove, is that each bivalent assignment, by respecting the clauses in definition 10, assigns a truth value on each formula, thereby inducing a classical valuation.

We can now state, in its full generality, the procedure whereby one extends a bivalent assignment to a classical valuation.

Definition 11 Extension to a classical valuation for CPL (syncategorematic version)
Let a be a bivalent assignment for PV. Then v_a classically extends a, if and only if v_a is a function of FM conforming to the following conditions:

(1) for each $\alpha \in \mathrm{AF}$, $v_a(\alpha) = a(\alpha)$;
(2) for each $\alpha, \beta \in \mathrm{FM}$,

(2.1) $v_a(\neg \alpha) = T$ iff $v_a(\alpha) = F$;
(2.2.1) $v_a(\alpha \wedge \beta) = T$ iff $v_a(\alpha) = v_a(\beta) = T$;
(2.2.2) $v_a(\alpha \vee \beta) = T$ iff $v_a(\alpha) = T$ or $v_a(\beta) = T$;
(2.2.3) $v_a(\alpha \to \beta) = T$ iff $v_a(\alpha) = F$ or $v_a(\beta) = T$;
(2.2.4) $v_a(\alpha \leftrightarrow \beta) = T$ iff $v_a(\alpha) = v_a(\beta)$.

Diligent readers may wonder whether the definition of a classical extension of a bivalent assignment results in a valuation and indeed a classical valuation. The answer is that it does. The clause in (1) guarantees that the classical extension assigns a truth value to each atomic formula. The clauses in (2), which merely reformulate those in the definition of a classical valuation (definition 10), require that the extension respects the interpretation of the propositional connectives, given in section 2. Moreover, as we have noted, the value of a formula given on the left-hand side of any biconditional clause is determined by the value or values of its immediate subformula or its immediate subformulae given on the right-hand side.

What the foregoing makes clear, but does not prove, is that each bivalent assignment classically extends to a unique classical valuation. Indeed, we can prove two other facts: distinct bivalent assignments result in distinct classical valuations and each classical valuation is the classical extension of some bivalent assignment. These facts taken together mean that there is a bijection between bivalent assignments and classical valuations. And this is why, in the definition of a classical extension of a bivalent assignment, the function symbol for a classical valuation v is subscripted by a function symbol for the bivalent assignment a.

Using the notation of a classical extension, we can recast the syncategorematic synthesis tree diagram for $\neg p \to (q \wedge r)$ as follows.

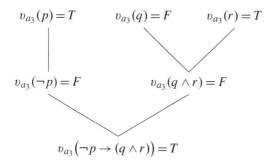

Recall that it is not possible to write down a classical valuation, since its domain is infinite. However, what the definition of a classical extension enables one to do, given

Classical Propositional Logic: Notation and Semantics

a bivalent assignment, is to calculate, for any formula, the value the classical valuation assigns to it, and to do so in a finite number of steps.

While this calculation, carried out as previously shown, with a syncategorematic synthesis tree diagram is perspicuous, it takes up lots of space. A more compact way of doing the same thing is to do the following. First, write out the formula (with good spacing between the symbols). Then, write under each atomic formula its truth value. Write under any propositional connective that has only an atomic subformula within its scope the truth value of the subformula. Continue in this way until a truth value is written under the main propositional connective of the formula. The truth value under the formula's main propositional connective is the truth value of the formula (with respect to the truth values assigned to the formula's atomic subformulae).

	$\neg$	p	$\rightarrow$	$(q$	$\wedge$	$r)$
a_3	F	T	T	F	F	T

Assembling the displays for each of the eight bivalent assignments to the propositional variables of the formula $(\neg p \rightarrow (q \wedge r))$ yields the formula's *truth table*.

TVA	$\neg$	p	$\rightarrow$	$(q$	$\wedge$	$r)$
a_1	F	T	T	T	T	T
a_2	F	T	T	T	F	F
a_3	F	T	T	F	F	T
a_4	F	T	T	F	F	F
a_5	T	F	T	T	T	T
a_6	T	F	F	T	F	F
a_7	T	F	F	F	F	T
a_8	T	F	F	F	F	F

Exercises: Interpreted syncategorematic synthesis trees

1. Give the syncategorematic synthesis tree diagram for each of the following formulae and assign to each node a value using the indicated bivalent assignment.

(a) $\neg p \vee q$ at a_1.

(b) $p \vee \neg p$ at a_2.

(c) $p \wedge \neg p$ at a_3.

(d) $p \leftrightarrow \neg\neg p$ at a_4.

(e) $(p \wedge q) \rightarrow \neg p$ at a_5.

(f) $(p \vee \neg q) \rightarrow r$ at a_6.

(g) $(p \rightarrow q) \rightarrow (\neg q \rightarrow \neg p)$ at a_7.

(h) $(p \vee q) \leftrightarrow (p \vee (\neg p \wedge q))$ at a_8.

2. Provide a truth table for each of the formulae in the previous exercise.

3. How many lines are there in a formula's truth table?

2.2.2 Categorematic approach

The classical extension of a bivalent assignment in accordance with definition 11 permits one to calculate the value of the application of the extension to any formula. In the case of the syncategorematic version of the definition, such a calculation is done on the basis of a formula's syncategorematic synthesis tree diagram. In the case of the categorematic version, it is done on the basis of the formula's categorematic synthesis tree diagram. To see how this is done, let us now consider the categorematic synthesis tree diagram for the formula $\neg p \to (q \wedge r)$, which was previously treated syncategorematically.

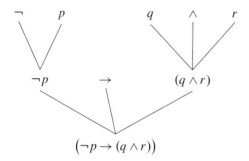

Again using the bivalent assignment a_3, one assigns F to q and T to p and r. In the syncategorematic synthesis tree diagrams, no nodes were labeled with propositional connectives, hence one had no need to assign them values. However, in the categorematic synthesis tree diagram, there are nodes for each of the propositional connectives in the formula. These nodes too must be assigned values.

Now, a glance at the interpretation given earlier of the propositional connectives reveals that they are functions from truth values, or pairs of truth values, to truth values: that is, they are truth functions. This becomes clear when the interpretations given earlier are restated in the following format. Here is the truth function that interprets $\neg$ as $o_\neg$, defined as follows.

$o_\neg$	DOMAIN		CODOMAIN
	T	$\mapsto$	F
	F	$\mapsto$	T

Here are the four two-place functions to interpret the four remaining propositional connectives.

Classical Propositional Logic: Notation and Semantics

$o_\wedge$	DOMAIN		CODOMAIN
	$\langle T,T \rangle$	$\mapsto$	T
	$\langle T,F \rangle$	$\mapsto$	F
	$\langle F,T \rangle$	$\mapsto$	F
	$\langle F,F \rangle$	$\mapsto$	F

$o_\vee$	DOMAIN		CODOMAIN
	$\langle T,T \rangle$	$\mapsto$	T
	$\langle T,F \rangle$	$\mapsto$	T
	$\langle F,T \rangle$	$\mapsto$	T
	$\langle F,F \rangle$	$\mapsto$	F

$o_\rightarrow$	DOMAIN		CODOMAIN
	$\langle T,T \rangle$	$\mapsto$	T
	$\langle T,F \rangle$	$\mapsto$	F
	$\langle F,T \rangle$	$\mapsto$	T
	$\langle F,F \rangle$	$\mapsto$	T

$o_\leftrightarrow$	DOMAIN		CODOMAIN
	$\langle T,T \rangle$	$\mapsto$	T
	$\langle T,F \rangle$	$\mapsto$	F
	$\langle F,T \rangle$	$\mapsto$	F
	$\langle F,F \rangle$	$\mapsto$	T

Let us apply this to the formula $\neg p \rightarrow (q \wedge r)$. $\neg$ and p are the immediate parts of $\neg p$. $\neg$ is assigned the truth function $o_\neg$ and p is assigned T. The value assigned to $\neg p$ is the value obtained from applying the truth function $o_\neg$, the value assigned to $\neg$, to T, the value assigned to p: that is to say, the value assigned to $\neg p$ is $o_\neg(T)$, or F. Similarly, q, $\wedge$, and r are the immediate parts of $(q \wedge r)$. $\wedge$ is assigned $o_\wedge$ and by the bivalent assignment a_3, q is assigned F, and r is assigned T. The value of $(q \wedge r)$ is obtained by applying the truth function $o_\wedge$, the value $\wedge$, to the pair of values $\langle F,T \rangle$, which are the values of q and r, respectively: that is to say, the value assigned to $(q \wedge r)$ is $o_\wedge(F,T)$, or F. Finally, the immediate parts of $\bigl(\neg p \rightarrow (q \wedge r)\bigr)$ are $\neg p$, $\rightarrow$, and $(q \wedge r)$. The value of $\rightarrow$ is $o_\rightarrow$ and, as just calculated, F is the value of $\neg p$ and $(q \wedge r)$. Applying the truth function $o_\rightarrow$ to the pair of truth values $\langle F,F \rangle$ yields the truth value T (that is, $o_\rightarrow(F,F) = T$), which is assigned to $\bigl(\neg p \rightarrow (q \wedge r)\bigr)$.

This calculation is shown with the aid of $\neg p \rightarrow (q \wedge r)$'s categorematic synthesis tree diagram.

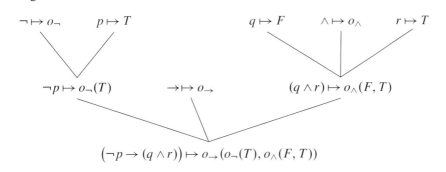

Notice that, once the values are assigned to the top nodes on the synthesis tree, the values of the nodes beneath can be calculated, for each lower node is calculated from the truth function, which is the value of one of the nodes immediately above it, and a truth value or a pair of truth values, which are the values of the other nodes immediately above it.

The generalization of this example yields the categorematic version of the definition of a classical extension of a bivalent assignment.

Definition 12 Extension to a classical valuation for CPL (categorematic version)
Let a be a bivalent assignment for PV. Then v_a classically extends a, if and only if v_a is a function of FM conforming to the following conditions:

(1)   for each $\alpha \in \mathrm{AF}$, $v_a(\alpha) = a(\alpha)$;

(2)   for each $\alpha, \beta \in \mathrm{FM}$,

(2.1) $v_a(\neg \alpha) \;=\; o_\neg(v_a(\alpha))$;

(2.2.1) $v_a(\alpha \wedge \beta) \;=\; o_\wedge(v_a(\alpha), v_a(\beta))$;

(2.2.2) $v_a(\alpha \vee \beta) \;=\; o_\vee(v_a(\alpha), v_a(\beta))$;

(2.2.3) $v_a(\alpha \to \beta) \;=\; o_\to(v_a(\alpha), v_a(\beta))$;

(2.2.4) $v_a(\alpha \leftrightarrow \beta) \;=\; o_\leftrightarrow(v_a(\alpha), v_a(\beta))$.

Exercises: Interpreted categorematic synthesis trees

1. Give the categorematic synthesis tree diagram for each of the following formulae and assign to each node a value using the indicated bivalent assignment.

(a) $\neg p \vee q$ at a_2.

(b) $p \vee \neg p$ at a_3.

(c) $p \wedge \neg p$ at a_4.

(d) $p \leftrightarrow \neg\neg p$ at a_5.

(e) $(p \wedge q) \to \neg p$ at a_6.

(f) $(p \vee \neg q) \to r$ at a_7.

(g) $(p \to q) \to (\neg q \to \neg p)$ at a_8.

(h) $(p \vee q) \leftrightarrow (p \vee (\neg p \wedge q))$ at a_1.

2. In this section, one truth function from $\{T,F\}$ to $\{T,F\}$ was given. Find all the other truth functions with the same domain and codomain and write them out in the same format.

3. In this section, four truth functions from $\{T,F\} \times \{T,F\}$ to $\{T,F\}$ were given. How many others are there? Write them out in the same format.

4. Reexpress the truth functions for $o_\wedge$, $o_\vee$, $o_\to$, and $o_\leftrightarrow$ as matrices.

2.2.3 Semantic properties and relations

Formulae have semantic properties, as do sets of formulae. Similarly, pairs of formulae bear semantic relations to each other, as do pairs of FMs. Finally, sets of formulae bear

Classical Propositional Logic: Notation and Semantics

a semantic relation to single formulae. We shall learn about some of these properties and relations in this section.

We shall define these various properties and relations in terms of a fundamental relation, the *satisfaction* relation, which relates bivalent assignments either to single formulae or to single sets of formulae. A bivalent assignment is said to *satisfy* a formula just in case its classical extension assigns T to it, and it is said to *satisfy* a set of formulae just in case its classical extension assigns T to each formula in the set. Here is the definition formulated with some notation.

Definition 13 Satisfaction for CPL

(1) For bivalent assignment a, and for each formula α,
    a satisfies α iff $v_a(\alpha) = T$;

(2) For bivalent assignment a, and for each set of formulae Γ,
    a satisfies Γ iff, for each $\alpha \in \Gamma$, $v_a(\alpha) = T$.

Readers should verify that a_3 satisfies the formula $\neg p \to (q \wedge r)$ but that a_6 does not and that a_4 satisfies the set of formulae $\{p \to \neg q, r \to \neg p\}$ but that a_3 does not.

Here are two important facts about satisfaction. First, a bivalent assignment satisfies a formula just in case it satisfies the singleton set whose sole member is the formula in question. Let us see why. Suppose that a bivalent assignment satisfies a formula. Then, by clause (1) of the definition of satisfaction, its classical extension assigns the formula T. It therefore assigns T to each member of the singleton set whose sole member is the formula in question. Conversely, suppose that a bivalent assignment satisfies a singleton set whose sole member is a formula. Then, by clause (2) of the definition of satisfaction, the assignment assigns T to the formula that is the sole member of the singleton set in question.

Second, each bivalent assignment satisfies the empty set of formulae. Why is this? Suppose that an arbitrarily chosen bivalent assignment does not satisfy the empty set. Then, by clause (2) of the definition of satisfaction, the bivalent assignment fails to satisfy some formula in the set. But this is impossible, since the empty set has no members. Therefore, no bivalent assignment fails to satisfy the empty set. In other words, each bivalent assignment satisfies the empty set.

We record these two facts for later reference.

Fact 1 Facts about satisfaction

(1) For each bivalent assignment a, and for each formula α,
    a satisfies α iff a satisfies $\{\alpha\}$;

(2) Each bivalent assignment satisfies the empty set.

We now turn to properties of formulae.

PROPERTIES OF FORMULAE

Propositions are customarily divided into three kinds: *tautologies, contingencies,* and *contradictions*. A tautology is a proposition that is true, no matter what: for example, *if Socrates was Greek, then Socrates was Greek*. This proposition simply cannot fail to be true. However, consider the proposition *Aristotle was a philosopher*. That is a contingent proposition, for, though it is true, it could have been false. Aristotle might have been persuaded by his parents to pursue some other vocation. A contingent proposition is one which, if true, still might have been false, and if false, might have been true. A contradiction is a proposition that can never be true, no matter what. *Every circle is a square* is a contradiction: it simply can never be true.

These same terms are applied to formulae in CPL, and their definitions within the context of CPL follow. A formula to which each classical valuation assigns T is a *tautology*. The same thing can be stated in terms of satisfaction. Here is our definition.

Definition 14 Tautology of CPL
A formula is a tautology iff each bivalent assignment satisfies it.

Some examples of tautologies are $p \vee \neg p$, $p \to p$, $(p \wedge \neg p) \to q$, $p \to (q \to p)$, $p \to (p \vee q)$, and $(p \wedge q) \to q$.

Next, a formula to which each classical valuation assigns F is a *contradiction*; or equivalently,

Definition 15 Contradiction of CPL
A formula is a contradiction iff each bivalent assignment does not satisfy it.

Contradictions include such formulae as these: $p \wedge \neg p$, $(q \vee \neg q) \to (p \wedge \neg p)$, $p \wedge \neg(\neg p \to q)$, and $p \leftrightarrow \neg p$.

Finally, a formula to which some classical valuation assigns T and to which some classical valuation assigns F is a *contingency*.

Definition 16 Contingency of CPL
A formula is a contingency iff some bivalent assignment satisfies it and some other does not.

As readers should verify for themselves, these formulae are contingencies: $(p \wedge q) \to \neg q$, $p \vee \neg q$, and $p \to (p \wedge r)$.

PROPERTIES OF FORMULAE AND OF SETS OF FORMULAE

There are two properties that can be ascribed to either a set of propositions or single propositions: *satisfiability* and *unsatisfiability*. A proposition is satisfiable if it is possible for it to be true, and it is unsatisfiable if it is impossible for it to be true. It is possible for the proposition that it is raining to be true. So, it is a satisfiable proposition. However, it is impossible for the proposition that every square is round to be true. Hence, it is unsatisfiable. Next, a

Classical Propositional Logic: Notation and Semantics

set of propositions is satisfiable if it is possible for all of the propositions in the set to be true at once. The set of propositions, { it is raining, it is cold }, is also satisfible, since both propositions in the set can be true at the same time. In contrast, the set { it is raining, it is not raining } is unsatisfiable, for it is impossible for both propositions in it to be true at the same time. One says that a single proposition is satisfiable if it can be true and that it is unsatisfiable if it cannot be true.

We now define these terms as they will be used in CPL and we do so in terms of the relation of satisfaction.

Definition 17 Satisfiability for CPL

(1) A formula is satisfiable iff some bivalent assignment satisfies it;

(2) a set of formulae is satisfiable iff some bivalent assignment satisfies it.

These sets of formulae are satisfiable: $\{p, q, p \to q\}$, $\{p\}$, $\{p \vee \neg q, p \wedge q, r \vee (p \to \neg r)\}$, and $\{p \vee q, \neg p, p \to q\}$. These formulae are also satisfiable: $p \wedge q$, r, and $p \to (q \wedge \neg p)$. Another term for satisfiability is *semantic consistency*.

Here are three important facts about satisfiability:

Fact 2 Facts about satisfiability

(1) For each formula α, α is satisfiable iff $\{\alpha\}$ is satisfiable.

(2) Each subset of a satisfiable set of formulae is satisfiable.

(3) The empty set is satisfiable.

Clearly the first fact holds. According to fact (1.1), a formula and the singleton set containing it are satisfied by precisely the same bivalent assignments. So, each bivalent assignment satisfying a formula satisfies the singleton set containing it and vice versa. To see that the second fact holds, consider a satisfiable set of formulae. Since it is satisfiable, there is a bivalent assignment that satisfies each member of the set. So, the very same bivalent assignment satisfies each formula in any subset of the initial set. The third fact follows from the second, since the empty set is a subset of every set.

The second property of single formulae and of single sets of formulae, unsatisfiability, is the opposite of satisfiability.

Definition 18 Unsatisfiability for CPL

(1) A formula is unsatisfiable iff no bivalent assignment satisfies it;

(2) a set of formulae is unsatisfiable iff no bivalent assignment satisfies all the set's formulae.

Two unsatisfiable sets of formulae are $\{p, \neg q, p \to q\}$ and $\{p \wedge q, r \vee (p \to \neg r), p \wedge \neg q\}$. Two unsatisfiable formulae are $\neg q \wedge q$ and $(p \wedge \neg q) \wedge (p \to q)$. In fact, there are infinitely many others. Not surprisingly, another term for unsatisfiability is *semantic inconsistency*.

There are three facts pertaining to unsatisfiability corresponding to the three facts pertaining to satisfiability that we just saw.

Fact 3   Facts about unsatisfiability

(1)   For each formula α, α is unsatisfiable iff $\{\alpha\}$ is unsatisfiable.

(2)   Each superset of an unsatisfiable set of formulae is unsatisfiable.

(3)   FM, or the set of all formulae, is unsatisfiable.

The first fact obtains in virtue of fact (1.1) which guarantees that a formula and the singleton set containing it are satisfied by precisely the same bivalent assignments. So, each bivalent assignment that fails to satisfy a formula fails to satisfy the singleton set containing it and vice versa.

Why does the second fact hold? Consider an unsatisfiable set of formulae. Since it is unsatisfiable, no bivalent assignment satisfies it, that is, each bivalent assignment fails to satisfy some formula of the set. Consider any bivalent assignment and a formula that it fails to satisfy. The bivalent assignment fails to satisfy the formula, even if it is a member of a larger set. Thus, each superset of an unsatisfiable set of formulae is unsatisfiable. The third fact follows from the second, since the set of formulae is a superset of every unsatisfiable set.

RELATIONS BETWEEN FORMULAE AND SETS OF FORMULAE

At last, we come to the relations of *semantic equivalence* and of *entailment*. We turn our attention first to semantic equivalence. This relation holds between a pair of sets of formulae, a pair of formulae or a pair comprising a set of formulae, and a single formula.

Definition 19   Semantic equivalence for CPL

A set of formulae or a formula, on the one hand, and a set of formulae or a formula, on the other, are semantically equivalent iff (1) each bivalent assignment satisfying the former satisfies the latter and (2) each bivalent assignment satisfying the latter satisfies the former.

The following two sets of formulae are semantically equivalent: $\{p \wedge q, p \rightarrow q\}$ and $\{p, q, \neg p \vee q\}$; as are the following two formulae: $(\neg p \vee q) \vee q$ and $p \rightarrow q$. In addition, the following formula and the following set of formulae are semantically equivalent: $p \wedge q$ and $\{p, q\}$.

The following facts interconnect semantic equivalence with tautologies and contradictions.

Fact 4   Facts about semantic equivalence

(1)   All tautologies and sets of tautologies are semantically equivalent.

(2)   All tautologies and all sets of tautologies are semantically equivalent to the empty set.

Classical Propositional Logic: Notation and Semantics

(3) All contradictions and all sets of contradictions are semantically equivalent.

(4) All contradictions and all sets of contradictions are semantically equivalent to FM, or the set of all formulae.

By the definition of tautology, each bivalent assignment satisfies a tautology. Clearly, then, each bivalent assignment satisfies every set of tautologies. Thus, the first fact holds. Moreover, according to fact (1.2), each bivalent assignment satisfies the empty set. Thus, the second fact holds. The last two facts are left to readers to establish.

The second relation to be discussed in this section is the semantic relation of *entailment*, a relation that a set of formulae bears to a single formula. A set of propositions entails a single proposition just in case whenever every proposition in the set is true the single proposition is true too. Within CPL, entailment is defined as a relation borne by an set of formulae to a single formula.

Definition 20 Entailment for CPL
The set of formulae Γ *entails* a formula α, or $\Gamma \models \alpha$, iff each bivalent assignment that satisfies the set Γ, satisfies the formula α.

The set of formulae $\{r \wedge \neg p, q \to p\}$ entails the formula $\neg q$, that is, $\{r \wedge \neg p, qc \to p\} \models \neg q$; the set of formulae $\{p \wedge \neg q, q \to r\}$, however, does not entail the formula $r \wedge p$, that is, $\{p \wedge \neg q, q \to r\} \not\models r \wedge p$.

The following are important facts about entailment.

Fact 5 Facts about entailment
Let α and β be formulae. Let Γ and Δ be sets of formulae.

(1) If $\alpha \in \Gamma$, then $\Gamma \models \alpha$.

(2) If $\Gamma \subseteq \Delta$ and $\Gamma \models \alpha$, then $\Delta \models \alpha$.

(3) $\Gamma \cup \{\alpha\} \models \beta$ iff $\Gamma \models \alpha \to \beta$.

(4) $\Gamma \models \alpha \wedge \beta$ iff $\Gamma \models \alpha$ and $\Gamma \models \beta$.

The first statement is true. After all, each bivalent assignment satisfying a set of formulae satisfies each formula in the set. To see that the second fact holds, suppose that $\Gamma \models \alpha$ and that $\Gamma \subseteq \Delta$. Suppose further that a is an arbitrarily chosen bivalent assignment satisfying Δ. Thus, a satisfies each formula in Δ and thereby satisfies Γ, since each formula in Γ is in Δ. Moreover, since we have supposed that $\Gamma \models \alpha$, it follows that a satisfies α. Since a is arbitrarily chosen, we conclude that $\Delta \models \alpha$.

To establish the third fact, we have to establish two claims: first, if $\Gamma \cup \{\alpha\} \models \beta$, then $\Gamma \models \alpha \to \beta$; and second, if $\Gamma \models \alpha \to \beta$, then $\Gamma \cup \{\alpha\} \models \beta$. To show that the first holds, suppose that $\Gamma \cup \{\alpha\} \models \beta$. Suppose further that a is an arbitrarily chosen bivalent assignment that satisfies Γ. Now, two cases arise: either a satisfies α or a does not satisfy α. CASE 1: Suppose that a satisfies α. Then, a satisfies each formula in $\Gamma \cup \{\alpha\}$. It follows from the principal supposition that a satisfies β. Therefore, a satisfies $\alpha \to \beta$. CASE 2: Suppose that a does not satisfy α. Then, $v(\alpha) = F$. So, $v(\alpha \to \beta) = T$. Thus, a satisfies $\alpha \to \beta$. Either way, a satisfies $\alpha \to \beta$. Since a is an arbitrarily chosen bivalent assignment, it follows that each bivalent assignment that satisfies Γ satisfies $\alpha \to \beta$. Therefore, by the definition of entailment, $\Gamma \models \alpha \to \beta$.

The second claim of the third statement and the two claims that constitute the fourth statement are left to readers to explain.

Next, we draw readers' attention to two facts connecting entailment with tautologies.

Fact 6 Facts about entailment and tautologies
Let α be a formula.

(1) α is a tautology iff $\emptyset \models \alpha$.
(2) α is a tautology iff, for each Γ, a subset of FM, $\Gamma \models \alpha$.

The first fact holds because every bivalent assignment satisfies a tautology and every bivalent assignment satisfies the empty set. The second fact follows from the first. Suppose that α is a tautology. By the first fact, it follows that $\emptyset \models \alpha$. Now every set is a superset of the empty set. So, it follows by fact (5.2) that, for each set of formulae Γ, $\Gamma \models \alpha$. Conversely, suppose that, for each set of formulae Γ, $\Gamma \models \alpha$. So, $\emptyset \models \alpha$. And, by the first fact, one concludes that α is a tautology.

Finally, we come to three facts connecting unsatisfiability and entailment.

Fact 7 Facts about entailment and unsatisfiability
Let α be a formula. Let Γ be a set of formulae.

(1) $\Gamma \models \alpha$ iff $\Gamma \cup \{\neg \alpha\}$ is unsatisfiable.
(2) Γ is unsatisfiable iff, for any α, $\Gamma \models \alpha$.
(3) Γ is unsatisfiable iff, for at least one α, a contradiction, $\Gamma \models \alpha$.

The first is true. Here is why. Suppose that $\Gamma \models \alpha$. It follows from the definition of $\models$ that each bivalent assignment that satisfies Γ satisfies α. Thus, no bivalent assignment that satisfies Γ satisfies $\neg \alpha$. Thus, $\Gamma \cup \{\neg \alpha\}$ is unsatisfiable. Conversely, suppose that $\Gamma \cup \{\neg \alpha\}$ is unsatisfiable. So, there is no bivalent assignment that satisfies it. Thus, each classical assignment that satisfies Γ does not satisfy $\neg \alpha$, and therefore satisfies α. It follows, then, by the definition of entailment, that $\Gamma \models \alpha$.

Readers are left to grapple with establishing the truth of the last two statements.

Exercises: Semantic properties and relations

1. Determine which of the following formulae are tautologies, which are contingencies, and which are contradictions.

(a) $\neg(p \wedge (\neg p \wedge q))$
(b) $(p \wedge q) \to \neg p$
(c) $(p \to q) \to (\neg q \to \neg p)$
(d) $(p \vee q) \leftrightarrow (p \vee (\neg p \wedge q))$
(e) $(p \vee \neg q) \to q$
(f) $p \vee \neg p$
(g) $p \leftrightarrow \neg\neg p$
(h) $(p \vee q) \leftrightarrow (p \vee (\neg p \to q))$
(i) $((p \to p) \to q) \to p$
(j) $p \wedge \neg p$

2. Determine whether the following sets of formulae are satisfiable or unsatifiable.

(a) $\{(p \vee q) \wedge (p \vee r),\ p \wedge (q \vee r)\}$
(b) $\{(p \to q) \to (r \to s),\ \neg(p \wedge q) \to s,\ \neg s \to \neg p\}$
(c) $\{p \leftrightarrow q,\ p \leftrightarrow \neg q\}$
(d) $\{(p \vee q) \wedge r,\ r \to s,\ \neg p \to (r \to s)\}$

3. Establish the truth or falsity of the following claims.

(a) $p \wedge (q \vee r) \models (p \vee q) \wedge (p \vee r)$
(b) $p \leftrightarrow (p \wedge q) \models q \to p$
(c) $((p \wedge q) \to r) \models p \to (q \to r)$
(d) $q \to p \models \neg q \to \neg p$
(e) $\{p \vee q,\ p\} \models q$
(f) $(p \wedge q) \vee (\neg p \wedge \neg q) \models p \to q$

4. For each of the following claims, indicate whether it is true or false. If it is true, explain why it is true; if it is false, provide a counterexample.

(a) Every function from FM into $\{T, F\}$ is a classical valuation.
(b) Each unsatisfiable set contains a contradiction.
(c) No set of contingencies can form a satisfiable set.
(d) Each set of tautologies is satisfiable.
(e) No satisfiable set contains a tautology.
(f) Some satisfiable sets contain a tautology.
(g) α is a contingency if and only if $\{\alpha\}$ is satisfiable.
(h) A contingent formula's negation may be a tautology.
(i) $\Gamma \models \alpha \vee \beta$ if and only if $\Gamma \models \alpha$ or $\Gamma \models \beta$.
(j) $\{\alpha\}$ and $\{\beta\}$ are semantically equivalent if and only if $\alpha \leftrightarrow \beta$ is a tautology.
(k) If $\Delta \subseteq \Gamma$ and Γ is unsatisfiable, then Δ is unsatisfiable.
(l) If $\Delta \subseteq \Gamma$ and Δ is satisfiable, then Γ is satisfiable.

(m) If $\Gamma \models \alpha$, then $\alpha \in \Gamma$.

(n) If $\Delta \subseteq \Gamma$ and $\Gamma \models \alpha$, then $\Delta \models \alpha$.

SOLUTIONS TO SOME OF THE EXERCISES

2.1.2 Formation rules

(a) No   (c) No   (e) No

(g) Yes   (i) Yes   (k) Yes

2.1.3 Formulae and subformulae

1. (a) Yes   (c) By convention   (e) Yes

    (g) By convention   (i) By convention

2. (a) Composite formula   (c) Not a formula

    (e) Composite formula   (g) Composite formula

    (i) Both a basic and composite formula

3. (a) True   (c) True   (e) False   (g) False

4. (a) $\vee$ (c) The first $\neg$ (e) $\rightarrow$

5. (a) $((p \vee q) \rightarrow q)$ (c) $(p \vee q)$ (e) $(q \wedge r)$

6. (a) Yes   (c) No   (e) No   (g) Yes   (i) Yes

2.2 Assignments and valuations

2. 3^n

3. (a) Bivalent valuation   (c) Not a bivalent valuation, not left total

4. (a) Nonempty   (c) Empty

2.2.1 Interpreted syncategorematic synthesis trees

1. (a) T (c) F (e) T (g) T

2.2.2 Interpreted categorematic synthesis trees

1. (a) T (c) F (e) T (g) T

Classical Propositional Logic: Notation and Semantics

2.2.3 Semantic properties and relations

1. (a) Tautology   (c) Tautology   (e) Contingency
   (g) Tautology   (i) Contingency

2. (a) Satisfiable   (c) Unsatisfiable

3. (a) True   (c) True   (e) False

4. (a) False   (c) False   (e) False   (g) False
   (i) False   (k) False   (m) False

REMARKS

(g) The left condition implies the right condition, but not conversely.

(i) The right condition implies the left condition, but not conversely.

7 Classical Propositional Logic: Deduction

1 Deduction

We now turn to the aspect of CPL that concerns *deduction*. Before defining deduction, let us return to the notions of an argument and its validity, raised at the beginning of chapter 6.
Consider once again the argument.

(1.1)   If team A won its game today,
        then team B is not in the play-offs.

(1.2)   If team C lost its game today,
        then team D is in the play-offs.

(1.3)   Either team A won its game today or
        team C lost its game today.

(1.4)   Either team B is not in the play-offs or
        team D is in the play-offs.

We asserted that the argument is valid, that is, that whenever its premises are true its conclusion is true. To the inexperienced eye, it is not obvious that this is so. One way to demonstrate that an argument is valid is to construct a series of arguments, each of which is clearly valid and which links one or more of the premises to a new conclusion or links earlier intermediate conclusions and the premises to a new conclusion, with the net result that the desired final conclusion is eventually reached. Such is, in essence, a deduction.
Let us undertake to deduce the conclusion of the preceding argument from its premises.

1. Suppose, on the one hand, that team A won its game today.
- Then, by premise (1.1), team B is not in the play-offs.
- But if it is true that team B is not in the play-offs, then it is true either that team B is not in the play-offs or that team D is in the play-offs.
- So, if team A won its game today, then either team B is not in the play-offs or team D is in the play-offs.

2. Suppose, on the other hand, that team C lost its game today.

- Then, by premise (1.2), team D is in the play-offs.

- But if it is true that team D is in the play-offs, then it is true either that team B is not in the play-offs or that team D is in the play-offs.

- So, if team C lost its game today, then either team B is not in the play-offs or team D is in the play-offs.

3. Now, we established that if team A won its game today, then either team B is not in the play-offs or team D is in the play-offs. And we established that if team C lost its game today, then either team B is not in the play-offs or team D is in the play-offs. But, we have a premise, namely (1.3), that either team A won its game today or team C lost its game today.

4. Either way, it turns out that either team B is not in the play-offs or team D is in the play-offs.

There are many different ways of presenting deduction for CPL. We shall study five, which may be called *natural deduction* presentations. Natural deduction presentations are distinguished by the fact that there is a pair of rules for each propositional connective. These presentations divide into two kinds: those that show how a formula may be deduced from a set of formulae and those in which a pair comprising a set of formulae and a formula may be derived from a set of such pairs. We shall call the former kind of presentation *formula natural deduction* and the latter *sequent natural deduction,* where a *sequent* is a pair comprising a set of formulae and a single formula. (Sequents will be discussed more in section 3.) Both formula natural deductions and sequent natural deductions may be displayed as either a column of expressions or a tree of expressions. The first four presentations are natural deduction with the formulae in a column, with the formulae in a tree, with the sequents in a column, and with the sequents in a tree. The fifth presentation is also a form of sequent natural deduction; however, rather than having an elimination rule and an introduction rule for each propositional connective, it has two introduction rules. This is known as the *Gentzen sequent calculus*. Its derivations are set out as a tree of expressions, though it could, in principle, be set out as a column of expressions.

We first set out two presentations of natural deduction with formulae, then two presentations of natural deduction with sequents and finally the Gentzen sequent calculus. (See Pelletier 1999 for a brief history of the development of various presentations of natural deduction for CPL.) We conclude this chapter with a brief discussion of substructural logics, a generalization of the notion of logic that has grown out of the sequent natural deduction presentations.

The presentation of natural deduction of formulae in a column was first suggested by the Polish logician Stanisław Jaśkowski (1906–1965) (Jaśkowski 1934). It is perhaps the most commonly taught presentation of natural deduction. For this reason, it is given first. However, it is the presentation using trees that lend themselves most readily to the study of the

Classical Propositional Logic: Deduction

syntax and semantics of natural language. We nonetheless also set out the presentation of natural deduction of sequents in a column since it will help readers make the transition from the more intuitive presentations of formula natural deductions to the less intuitive presentations of sequent natural deductions.

There are two important further benefits to be drawn from studying various presentations of the same mathematical or logical concept. Different presentations of the same mathematical concept, while mathematically equivalent, have different advantages and disadvantages and thereby vary in the suitability of their applications. Thus, one presentation may be well suited to one application and another presentation to another. The second lesson is that different presentations lend themselves to different ways of generalizing the mathematical concept expressed. Both of these lessons are especially useful to students of formal accounts of natural language. We shall elaborate on this point in section 5, at the end of this chapter.

2 Formula Natural Deduction

As we stated, natural deduction with formula can be displayed either in a column format or a tree format. We begin with the column format.

2.1 Formulae in a Column

One can think of an argument, such as the one set out in section 1, as comprising a series of steps, each step being a simple argument whose validity is unimpeachable. Jaśkowski (1934) sought to formalize this. Here is the relevant definition.

Definition 1 Deduction for CPL (formula column format)
A deduction of a formula from a set of formulae, the premises of the deduction, is a sequence of formulae each one of which is either a premise, in which case it is taken from among the set of premises, or obtained from earlier formulae of the sequence by one of the rules specified below, or it can be deduced in an ancillary deduction.

While a natural deduction is just a sequence of formulae, it is customary to annotate the sequence with an enumeration of the formulae and an indication of how each formula in the sequence is obtained. This is done by writing the formulae comprising a deduction in a column. The basic column of formulae is customarily supplemented with book keeping notation to ease the task of verifying that the sequence of formulae in question is indeed a deduction. To this end, one writes on the left-hand side of a formula a numeral indicating the formula's position in the sequence, and, on the right-hand side, one writes the justification for the formula being in its position in the sequence. If the formula is a premise of the deduction, one writes that it is a premise; if it is deduced from earlier formulae by a rule, one indicates the numbers of the formulae and the rule used. The display of a deduction, then, will comprise three columns: a column of numerals where "n" enumerates the nth formula in the column, a column of formulae, and a column of justifications. Each line of the display, then, comprises a numeral, a formula, and its justification. (See EXAMPLE 1.)

Every deduction that occurs within a longer deduction is an *ancillary deduction*. It always has exactly one *supposition,* or *ancillary premise*. The deduction that is not part of any longer deduction is the *principal deduction*. We shall permit a principal deduction to have more than one premise, and we shall require that any premise that is used be listed right at the beginning.

Having described the general features of how a deduction is displayed, let us turn to the rules used in a deduction. There are ten fundamental rules for the presentation of CPL adopted in this book. These rules are of two kinds: some of them are rules that permit a formula to appear in the deductive sequence by dint of an immediately preceding ancillary deduction; the others are specifications of the basic self-evidently valid arguments, which have premises and a conclusion.

We turn to the latter rules first. Like a deduction, each of these rules has premises and a conclusion. We shall make sure to distinguish between premises of a rule and premises of the deduction in which the rule is used, as well as between the conclusion of a rule and the conclusion of the deduction. The rules of deduction that have premises fall into two types: those in which the formula of a rule's conclusion has as an immediate subformula a formula of one of its premises and those in which the formula of a rule's conclusion is an immediate subformula of one of its premises. The former rules are known as *introduction rules*. An introduction rule introduces one occurrence of a connective, where the connective is the main connective of the formula of the rule's conclusion and at least one of its immediate subformulae is a formula found among the rule's premises. The latter rules are known as *elimination rules*. An elimination rule eliminates one occurrence of the main connective of one of the formulae in the rule's premises to yield an immediate subformula that is the formula of the rule's conclusion.

We now come to the second kind of rule, the rule that permits a formula to appear in the deductive sequence by dint of an immediately preceding ancillary deduction. These rules either introduce a formula whose immediate subformulae appear in the immediately preceding ancillary deduction or eliminate a main propositional connective from a formula in the ancillary deduction.

With this general description of the deductive rules complete, we shall now learn about each of them in detail. We begin with the rules that have premises; the rules of introduction and elimination for $\land$, $\lor$, and $\leftrightarrow$; as well as the rule for the elimination of $\rightarrow$. Then we shall turn to those that require an ancillary deduction.

2.1.1 $\land$ Elimination

This rule reflects the fact that, if one is justified in asserting a conjunctive proposition, that is, a proposition formed from two propositions connected by *and,* then one is justified in asserting either constituent proposition on its own. Thus, one is justified in asserting the proposition that *it is raining,* if one is justified in asserting the conjunctive proposition that

Classical Propositional Logic: Deduction

it is raining **and** *it is cold*. And on the basis of the same conjunctive proposition, one is equally justified in asserting *it is cold*.

The following display expresses the rule of $\wedge$ Elimination by showing in a schematic way the result of applying the rule as a last step in a deduction. Thus, the last line of the deduction, numbered n, is obtained from an earlier line, numbered m, and the formula of the line numbered n is an immediate subformula of the formula in the line numbered m. Since the formula in the line numbered m is the conjunction of two formulae $\alpha \wedge \beta$, and either of its immediate subformulae can be deduced from it, the rule has two versions.

$\wedge$ Elimination			
Version 1		Version 2	
$\vdots$ $\vdots$		$\vdots$ $\vdots$	
m $\alpha \wedge \beta$		m $\alpha \wedge \beta$	
$\vdots$ $\vdots$		$\vdots$ $\vdots$	
n α	m $\wedge$ E	n β	m $\wedge$ E

2.1.2 $\wedge$ Introduction

This rule is the inverse of the preceding rule. If one is justified in asserting each of two propositions, then one is justified in asserting the conjunctive proposition formed from them. Thus, if one is justified in asserting the proposition that *it is raining* and one is justified in asserting the proposition that *it is cold,* then one is justified in asserting the proposition that results from connecting the one proposition with the other, regardless of the order in which they appear in either the deduction or the resulting conjunctive proposition.

In the schematic display of the rule of $\wedge$ Introduction, the last line of the deduction, numbered r, is obtained from earlier lines, numbered m and n, and the formulae on the lines numbered m and n are immediate subformulae of the formula in the line numbered r. Since the order of the occurrence of the formulae α and β, the immediate subformula of $\alpha \wedge \beta$, in the column of formulae is immaterial, this rule too has two versions.

$\wedge$ Introduction			
Version 1		Version 2	
$\vdots$ $\vdots$		$\vdots$ $\vdots$	
m α		m β	
$\vdots$ $\vdots$		$\vdots$ $\vdots$	
n β		n α	
$\vdots$ $\vdots$		$\vdots$ $\vdots$	
r $\alpha \wedge \beta$	m, n $\wedge$ I	r $\alpha \wedge \beta$	m, n $\wedge$ I

An illustration of the application of these rules is provided by the formal deduction that shows the validity of an argument whose single premise is $p \wedge (q \wedge r)$ and whose conclusion is $(p \wedge q) \wedge r$.

Example 1 (formula column format)

1	$p \wedge (q \wedge r)$	premise
2	p	1 $\wedge$ E
3	$q \wedge r$	1 $\wedge$ E
4	q	3 $\wedge$ E
5	r	3 $\wedge$ E
6	$p \wedge q$	2, 4 $\wedge$ I
7	$(p \wedge q) \wedge r$	5, 6 $\wedge$ I

It is important to stress that deduction rules apply to entire formulae, not to their subformulae. In other words, the main propositional connective of the formula that results from the use of $\wedge$ Introduction must be $\wedge$ and the immediate subformulae must be the formulae of two earlier lines. Thus, the following is not a deduction of $(p \wedge q) \wedge r$ from $p \wedge (q \wedge r)$. It contains two illegitimate applications of the rule of $\wedge$ Elimination. The offending lines are marked with an asterisk.

1	$p \wedge (q \wedge r)$	premise
2	p	1 $\wedge$ E
*3	q	1, $\wedge$ E
4	$p \wedge q$	2, 3 $\wedge$ I
*5	r	1 $\wedge$ E
6	$(p \wedge q) \wedge r$	4, 5 $\wedge$ I

Line 3 cannot be obtained from line 1, since q is not an immediate subformula of $p \wedge (q \wedge r)$. For the same reason, line 5 cannot be obtained from line 1.

2.1.3 $\rightarrow$ Elimination

$\rightarrow$ Elimination was first explicitly identified as valid by the Stoic philosopher Chrysippus (ca. 279–ca. 360 BCE) and is now also often called by its Latin name, *modus ponens*. The idea is this: If one is justified in asserting a conditional and one is also justified in asserting

Classical Propositional Logic: Deduction

the conditional's *protasis,* then one is justified in asserting the conditional's *apodosis.* This is a very common form of deduction. It occurs twice in the informal argument in (1). Here too, because the order of the rule's premises is immaterial, two versions are given.

$\rightarrow$ Elimination			
Version 1		Version 2	
⋮ ⋮		⋮ ⋮	
m	$\alpha \rightarrow \beta$	m	α
⋮ ⋮		⋮ ⋮	
n	α	n	$\alpha \rightarrow \beta$
⋮ ⋮		⋮ ⋮	
r	β m, n $\rightarrow$ E	r	β m, n $\rightarrow$ E

The next deduction uses the rule for $\rightarrow$ Elimination to establish that an argument whose premises are $p \rightarrow (q \rightarrow r)$, p, and q and whose conclusion is r is valid.

Example 2 (formula column format)

1	$p \rightarrow (q \rightarrow r)$	premise
2	p	premise
3	q	premise
4	$q \rightarrow r$	1, 2 $\rightarrow$ E
5	r	3, 4 $\rightarrow$ E

It is important to note that the previous deduction could not be carried out as follows:

1	$p \rightarrow (q \rightarrow r)$	premise
2	p	premise
3	q	premise
*4	$p \rightarrow r$	1, 3 $\rightarrow$ E
5	r	2, 4 $\rightarrow$ E

Line 4 is a misapplication of the rule of → Elimination. Its misapplication consists in its application to a subformula of the formula in line 1.

The rule of → Introduction does not draw on earlier lines in the deduction, rather it draws on an ancillary deduction. Therefore we postpone its exposition and discussion until later in this section, after all the rules that draw only on earlier lines have been presented and illustrated.

2.1.4 ∨ Elimination

This rule, which is the last rule used in the informal deduction given earlier, is also unquestionably valid, but it may require a moment's reflection to see that this is so. Suppose that one is justified in asserting three propositions: if α, then γ; if β, then γ; and either α or β. Then, the rule says, one is justified in asserting γ all by itself. Think of it this way. Suppose that one is told, to begin with, two things: first, that if switch A is up, then the light in the next room is on; and second, that if switch B is up, then the very same light in the next room is on. At this point, the person addressed does not know whether the light in the next room is on. But now suppose the person is told that either switch A is up or switch B is up. Surely, he or she can now conclude that the light in the next room is on. And this is all the ∨ Elimination rule is.

Since the order in which the three formulae—α, β, and $\alpha \vee \beta$—appear is immaterial to the validity of the rule, we have to state the rule in such a way that any of the six orders in which they might appear are covered by the statement of the rule.

∨ Elimination
Let Σ be $\{\alpha \vee \beta, \alpha \to \gamma, \beta \to \gamma\}$
$\vdots \quad \vdots$
m $\quad \pi_1 \quad$ (where $\pi_1 \in \Sigma$)
$\vdots \quad \vdots$
n $\quad \pi_2 \quad$ (where $\pi_2 \in \Sigma - \{\pi_1\}$)
$\vdots \quad \vdots$
r $\quad \pi_3 \quad$ (where $\pi_3 \in \Sigma - \{\pi_1, \pi_2\}$)
$\vdots \quad \vdots$
s $\quad \gamma \quad\quad\quad\quad\quad\quad\quad\quad\quad\quad$ m, n, r ∨ E

As an illustration of the application of the rule, consider this deduction of the conclusion s from the premises $p \wedge (q \vee r)$, $p \to (q \to s)$, and $p \to (r \to s)$.

Classical Propositional Logic: Deduction

Example 3 (formula column format)

1	$p \land (q \lor r)$	premise
2	$p \to (q \to s)$	premise
3	$p \to (r \to s)$	premise
4	p	1 $\land$ E
5	$q \lor r$	1 $\land$ E
6	$q \to s$	2, 4 $\to$ E
7	$r \to s$	3, 4 $\to$ E
8	s	5, 6, 7 $\lor$ E

Notice that the following is not a deduction of the conclusion from the premises.

1	$p \land (q \lor r)$	premise
2	$p \to (q \to s)$	premise
3	$p \to (r \to s)$	premise
4	p	1 $\land$ E
5	$q \lor r$	1 $\land$ E
*6	$p \to s$	2, 3, 5 $\lor$ E
7	s	4, 6 $\to$ E

since the $\alpha \to \gamma$ and the $\beta \to \gamma$ of the rule of $\lor$ Elimination do not correspond to the formulae of line 2 and line 3 but to proper subformulae of each.

2.1.5 $\lor$ Introduction

The $\lor$ Introduction rule is surely valid. It says that if one is justified in asserting a proposition, then one is justified in asserting the disjunctive proposition that results from that proposition being disjoined with any proposition whatsoever. Surely, if it is true that *it is raining*, then it is true **either** *it is raining* **or** *it is cold*. Granted it has the counterintuitive result that it permits one to conclude something less informative, namely, **either** *it is raining* **or** *it is cold*, from something more informative, namely, *it is raining*. But remember, the task at hand is merely to give arguments that are unquestionably valid. And surely this rule is. Moreover, this rule is rarely applied without heed to the deductive goal, and that goal usually determines the choice of proposition to be connected by *or* with the proposition one has already deduced.

Since the assertion of either of two propositions warrants the assertion of the two propositions connected by *or*, two versions of this rule also are stated.

∨ Introduction			
Version 1		Version 2	
⋮	⋮	⋮	⋮
m	α	m	β
⋮	⋮	⋮	⋮
n	α ∨ β m ∨ I	n	α ∨ β m ∨ I

Here is an illustration of a formal argument using ∨ Introduction to establish that an argument with the premise $p \wedge q$ and with the conclusion $p \wedge (q \vee r)$ is valid.

Example 4 (formula column format)

1. $p \wedge q$ premise
2. p 1 ∧ E
3. q 1 ∧ E
4. $q \vee r$ 3 ∨ I
5. $p \wedge (q \vee r)$ 2, 4 ∧ I

Notice that this conclusion cannot be obtained directly from the premise by an application of ∨ Introduction.

1. $p \wedge q$ premise
*2. $p \wedge (q \vee r)$ 1 ∨ I

since ∨ is not the main propositional connective of the formula in line 2.

2.1.6 ↔ Elimination and Introduction

This pair of rules interconnects conditional propositions and biconditional propositions. In essence, what these rules depend on is the fact that a biconditional is just that: two conditionals. The elimination rule says, therefore, that one can deduce either of the two conditionals from a biconditional. There are, therefore, two versions of the rule.

Classical Propositional Logic: Deduction

↔ Elimination	
Version 1	Version 2
⋮ ⋮	⋮ ⋮
m $\alpha \leftrightarrow \beta$	m $\alpha \leftrightarrow \beta$
⋮ ⋮	⋮ ⋮
n $\alpha \rightarrow \beta$ m ↔ E	n $\beta \rightarrow \alpha$ m ↔ E

The introduction rule says the inverse: one can deduce a biconditional from a suitable pair of conditionals. Again, since the order of the occurrence that the conditionals appear in a deduction is immaterial, two versions are given.

↔ Introduction	
Version 1	Version 2
⋮ ⋮	⋮ ⋮
m $\alpha \rightarrow \beta$	m $\beta \rightarrow \alpha$
⋮ ⋮	⋮ ⋮
n $\beta \rightarrow \alpha$	n $\alpha \rightarrow \beta$
⋮ ⋮	⋮ ⋮
r $\alpha \leftrightarrow \beta$ m, n ↔ I	r $\alpha \leftrightarrow \beta$ m, n ↔ I

Next, we present two deductions. The first illustrates the use of ↔ Elimination, and the second the use of ↔ Introduction.

The first deduction establishes that the validity of an argument whose premises are p, r, and $p \rightarrow (q \leftrightarrow r)$ and whose conclusion is q.

Example 5 (formula column format)

1	p	premise
2	r	premise
3	$p \rightarrow (q \leftrightarrow r)$	premise
4	$q \leftrightarrow r$	1, 3 → E
5	$r \rightarrow q$	4 ↔ E
6	q	2, 5 → E

This following is not an admissible deduction of the same conclusion:

1	p	premise
2	r	premise
3	$p \to (q \leftrightarrow r)$	premise
*4	$p \to (r \to q)$	3 $\leftrightarrow$ E
5	$r \to q$	1, 4 $\to$ E
6	q	2, 5 $\to$ E

The second deduction establishes the validity of an argument whose premises are $p \to (q \to r)$, $p \to (r \to q)$, and p and whose conclusion is $q \leftrightarrow r$.

Example 6 (formula column format)

1	$p \to (q \to r)$	premise
2	$p \to (r \to q)$	premise
3	p	premise
4	$q \to r$	1, 3 $\to$ E
5	$r \to q$	2, 3 $\to$ E
6	$q \leftrightarrow r$	4, 5 $\leftrightarrow$ I

This deduction could not have been shortened to the following:

1	$p \to (q \to r)$	premise
2	$p \to (r \to q)$	premise
3	p	premise
*4	$p \to (q \leftrightarrow r)$	1, 2 $\leftrightarrow$ I
5	$q \leftrightarrow r$	3, 4 $\to$ E

for the $\alpha \to \beta$ and the $\beta \to \alpha$ of the rule of $\leftrightarrow$ Introduction do not correspond to the formulae of line 1 and line 2 but to their proper subformulae.

The rules described so far all work in essentially the same way. The rules of $\wedge$ Elimination, $\leftrightarrow$ Elimination, and $\vee$ Introduction yield a new line in a deduction on the basis of one earlier line. The rules of ∧ Introduction, ↔ Introduction, and → Elimination yield a new line in a deduction on the basis of two earlier lines. And the rule of ∨ Elimination yields a new line in a deduction on the basis of three earlier lines. The next three rules do not avail themselves of any specific number of earlier lines, they avail themselves of an immediately preceding ancillary deduction, undertaken precisely with a view to applying the rule in question. There are three such rules: the rule for the introduction of → and the rules for the introduction and elimination of ¬.

2.1.7   → Introduction

Let us begin with the rule of → Introduction, which was used twice in the informal deduction previously presented. The idea behind this rule is very simple: to prove a conditional statement, one must furnish a deduction of the apodosis of the conditional from its protasis. In other words, to prove a formula of the form $\alpha \to \beta$, one must furnish a deduction whose sole premise is α and whose conclusion is β. The rule can be expressed as follows.

	→ Introduction	
m	α	
⋮	⋮	
n	β	
n+1	$\alpha \to \beta$	m–n → I

A deduction that furnishes the basis for an application of the rule of → Introduction is an *ancillary deduction* and its premise is an *ancillary premise,* or *supposition*. To identify the lines that form the ancillary deduction, one draws a vertical line to the immediate left of the entire length of the ancillary deduction from its initiating supposition α to its terminating conclusion β.

Once an ancillary deduction is terminated, its ancillary premise, or supposition, is thereby *discharged*. Neither the ancillary deduction nor any of its lines can be used in any subsequent application of a rule. This is the significance of the left line. In other words, each step in an ancillary deduction is *inaccessible* once the deduction is terminated. No deduction is complete unless each ancillary deduction is terminated, that is, each supposition has been discharged.

Here is an example of an application of the rule to show that an argument with the premise $p \to (q \to r)$ and the conclusion $(p \land q) \to r$ is valid.

Example 7 (formula column format)

1	$p \to (q \to r)$	premise
2	$p \land q$	supposition
3	p	$2 \land E$
4	$q \to r$	$1, 3 \to E$
5	q	$2 \land E$
6	r	$4, 5 \to E$
7	$(p \land q) \to r$	$2\text{–}6 \to I$

Notice that the formula in line 2, the supposition, is identical with the immediate subformula to the left of the $\to$ in the formula in line 7 and that the formula in line 6 is identical with the immediate subformula to the right of the $\to$ in the formula in line 7. If either the immediate subformula to the left of the formula inferred by $\to$ Introduction is not identical with the formula that is the supposition, or ancillary premise, of the ancillary deduction occurring immediately above the application of $\to$ Introduction, or the immediate subformula to the right of the formula inferred by $\to$ Introduction is not identical with the formula that which is the conclusion of the same ancillary deduction, then the rule of $\to$ Introduction has been misapplied. Such a misapplication is shown here:

1	$p \leftrightarrow q$	premise
2	$p \to (r \land s)$	premise
3	p	supposition
4	$r \land s$	$2, 3 \to E$
5	r	$4 \land E$
6	$p \to q$	$1 \leftrightarrow E$
*7	$p \to r$	$3\text{–}6 \to I$

The error in this purported deduction occurs in line 7. The formula in line 7 is supposed to be deduced by the rule of $\to$ Introduction from the ancillary deduction, which begins at line 3 and ends at line 6. While the formula in line 7 has $\to$ as its main propositional connective, and while its protasis and the supposition at line 3 are the same formula, the apodosis of the formula in line 7 is not the same as the formula in line 6, the last line of the ancillary deduction. Though the formula in line 6 is correctly deduced from the one

Classical Propositional Logic: Deduction

in line 1, the application of the rule of $\to$ Introduction at this point should result in the formula $p \to (p \to q)$, not in the formula $p \to r$.

We stated earlier that once an ancillary deduction had been terminated, none of its lines is available for further use. The following deduction illustrates a violation of this requirement.

1	$p \to (r \wedge s)$	premise
2	p	supposition
3	$r \wedge s$	1, 2 $\to$ E
4	r	3 $\wedge$ E
5	$p \to r$	2–4 $\to$ I
*6	s	3 $\wedge$ E
7	$s \wedge (p \to r)$	5, 6 $\wedge$ I

2.1.8 $\neg$ Elimination and Introduction

The last two rules are more usually known by their Latin name reductio ad absurdum (literally, reduction to absurdity). Indeed, the earliest recorded argument is just such an argument. It is an argument attributed to Pythagoras by Aristotle in which it is established that the square root of 2 cannot be expressed as the ratio of two whole numbers. Because of its historical significance and of its simplicity, we shall state it.

To appreciate more fully the deduction, two arithmetical facts should be borne in mind: First, an even (natural) number that is a number that is a multiple of 2: n is even if and only if there is a natural number k such that $n = 2k$. Second, each fraction can be reduced to a fraction that is equal to it and whose numerator and denominator have no factor in common other than 1. For example, the fraction $\frac{15}{27}$ can be reduced to the fraction $\frac{5}{9}$.

1. Suppose that $\sqrt{2} = \frac{m}{n}$, where m and n are positive integers with no common factors other than 1.
2. $\frac{m^2}{n^2} = 2$.
3. $2n^2 = m^2$.
4. m^2 is even (since it is equal to a multiple of 2).
5. m is even (since the square of an odd number is never even).
6. $m = 2k$, for some k (since m is even).
7. $2n^2 = 4k^2$.
8. $n^2 = 2k^2$.
9. n^2 is even (since it is equal to a multiple of 2).

10. *n* is even (since the square of an odd number is never even).

11. *m* and *n* are both even numbers.

12. *m* and *n* have a common factor other than 1, namely, 2.

13. But it was supposed that *m* and *n* have only 1 as a common factor.

14. Therefore, it is not the case that there are positive integers *m* and *n* with no common factors other than 1 such that $\sqrt{2} = \frac{m}{n}$.

Having seen this example of an application of the rule of reductio ad absurdum in actual mathematical reasoning, let us now formulate the rules. First, we give the elimination rule. Notice that there are two versions. This arises from the fact that the order of the appearance of the contradictory immediate subformulae occuring in the last line of the ancillary deduction is immaterial.

¬ Elimination	
Version 1	Version 2
m ¬α ⋮ ⋮ n β ∧ ¬β n+1 α m–n ¬ I	m ¬α ⋮ ⋮ n ¬β ∧ β n+1 α m–n ¬ I

Second, we give the introduction rule. Again, there are two versions.

¬ Introduction	
Version 1	Version 2
m α ⋮ ⋮ n β ∧ ¬β n+1 ¬α m–n ¬ I	m α ⋮ ⋮ n ¬β ∧ β n+1 ¬α m–n ¬ I

Notice that, like the rule of → Introduction, these rules require for their application ancillary deductions. They differ from the rule of → Introduction only with respect to the forms of the ancillary deductions.

Classical Propositional Logic: Deduction

An example of the application of the rule of $\neg$ Introduction is given in the deduction of the conclusion $\neg(q \wedge \neg(p \wedge q))$ from the premise p.

Example 8 (formula column format)

1	p	premise
2	$q \wedge \neg(p \wedge q)$	supposition
3	q	$2 \wedge E$
4	$\neg(p \wedge q)$	$2 \wedge E$
5	$p \wedge q$	$1, 3 \wedge I$
6	$(p \wedge q) \wedge \neg(p \wedge q)$	$4, 5 \wedge I$
7	$\neg(q \wedge \neg(p \wedge q))$	$2\text{–}6 \neg I$

We now come to the last rule, which is also the simplest. It is the rule of reiteration. It is a rule that allows one to reiterate a formula that has occurred earlier in the deduction, provided that it is not found within a preceding ancillary deduction that has been terminated.

2.1.9 The rule of reiteration

A formula in a deduction may be reiterated later in the deduction, provided the formula that is reiterated occurs at least once outside of any preceding terminated ancillary deduction.

Here is one example of its use.

Example 9 (formula column format)

1	$\neg p$	premise
2	$r \wedge (p \wedge q)$	supposition
3	$p \wedge q$	$2 \wedge E$
4	p	$3 \wedge E$
5	$\neg p$	1 reit
6	$p \wedge \neg p$	$4, 5 \wedge I$
7	$\neg(r \wedge (p \wedge q))$	$2\text{–}6 \neg I$

Unlike the other rules, this rule can be dispensed with, for any deduction that uses the rule, there is an equivalent deduction that dispenses with it. While the rule's dispensibility is obvious in the example just given, there is at least one case where the dispensibility is not obvious. A deduction for which its dispensibility is not obvious has been included in the exercises.

Here is an example of a deduction that misapplies this rule.

1	$p \to (r \wedge s)$	premise
2	p	supposition
3	$r \wedge s$	1, 2 $\to$ E
4	r	2 $\wedge$ E
5	$p \to r$	2–4 $\to$ I
*6	r	4 reit

2.1.10 Deducibility: Formula column format

We now introduce the formal notation that expresses that a formula can be deduced from a set of premises. We denote a set of premises by uppercase letters from the Greek alphabet.

Definition 2 Deducibility for CPL (formula column format)
α is deducible from Γ, or $\Gamma \vdash \alpha$, iff there is a deduction (in the formula column format) in which each ancillary premise, or supposition, has been discharged, whose only premises are formulae in Γ and which terminates with α.

One feature of this definition, which is important to note, is the second condition requires that Γ have as its formulae only formulae that are the premises used in the deduction; it does not exclude as its formulae formulae not used. Thus, for example, not only is $\neg p$ deducible from $\{p \to q, \neg q\}$, it is also deducible from $\{p \to q, \neg q, r\}$, or indeed any other set of formulae that has the set $\{p \to q, \neg q\}$ as a subset.

While the relation of deducibility is a binary relation between a set of formulae, on the one hand, and a single formula, on the other, if the set has a finite number of formulae, it is customary to omit the braces used to denote a set if the set has a finite number of formulae and simply to write down the formulae alone. For example, rather than to write $\{\neg(p \wedge r), \neg(\neg(p \wedge q) \wedge \neg p), r\} \vdash \neg s$, we shall write $\neg(p \wedge r), \neg(\neg(p \wedge q) \wedge \neg p), r \vdash \neg s$.

The definition of deducibility obtains only in the case where there is a deduction of the single formula in which the only premises are the formulae in the set. The question arises: can the set of premises be empty? That is, does it ever occur that $\emptyset \vdash \alpha$? The answer is yes. And here is an example.

Classical Propositional Logic: Deduction

1	$p \land (r \land \neg p)$	supposition
2	p	$1 \land E$
3	$(r \land \neg p)$	$1 \land E$
4	$\neg p$	$3 \land E$
5	$p \land \neg p$	$2, 4 \land I$
6	$\neg(p \land (r \land \neg p))$	$1\text{-}5 \neg I$

This deduction has no premises. Notice, however, it arises from an ancillary deduction that does have a supposition.

Any formula that can be deduced from no premises—that is, any formula that can be deduced from the empty set of formulae—is known as a *theorem* of CPL.

Definition 3 Theorem for CPL

Let α be a formula. α is a theorem iff $\emptyset \vdash \alpha$.

It is customary to omit the symbol for the empty set, writing $\vdash \alpha$, instead of $\emptyset \vdash \alpha$.

Besides using the notion of deduction to identify a property of formulae, one can use it to identify a property of sets of formulae.

Definition 4 Consistency for CPL

Let Γ be a set of formulae. Γ is consistent iff there is no formula α such that $\Gamma \vdash \alpha$ and $\Gamma \vdash \neg \alpha$.

Here are some examples of sets of formulae that are consistent: $\{p, q, p \to q\}$, $\{p\}$, $\{p \lor \neg q, p \land q, r \lor (p \to \neg r)\}$.

Definition 5 Inconsistency for CPL

Let Γ be a set of formulae. Γ is inconsistent iff there is a formula α such that $\Gamma \vdash \alpha$ and $\Gamma \vdash \neg \alpha$.

The following sets of formulae are inconsistent: $\{p, \neg q, p \to q\}$, $\{p \land q, r \lor (p \to \neg r), p \land \neg q\}$.

Definition 6 Interdeducibility for CPL

Let α and β be formulae. α and β are interdeducible iff both $\alpha \vdash \beta$ and $\beta \vdash \alpha$.

The following pairs of formulae are interdeducible: $p \lor p$ and p; $p \lor \neg q$ and $q \to p$; and $\neg(p \land q)$ and $\neg p \lor \neg q$.

Exercises: Natural deduction: formulae in a column

1. For each of the following displays, determine which are deductions and which are not. If the display is a deduction, say so; if it is not, identify the point at which it fails to be a deduction.

(a)
	1	$(p \wedge q) \wedge r$	premise
	2	r	$1 \wedge E$
	3	q	$1 \wedge E$
	4	$q \wedge r$	$2, 3 \wedge I$
	5	p	$1 \wedge E$
	6	$p \wedge (q \wedge r)$	$4, 5 \wedge I$

(b)
	1	$p \wedge s$	premise
	2	$q \wedge s$	premise
	3	p	$1 \wedge E$
	4	q	$2 \wedge E$
	5	$r \vee (p \wedge q)$	$3, 4 \wedge I$

(c)
	1	$p \wedge (q \rightarrow (r \wedge s))$	premise
	2	p	$1 \wedge E$
	3	$q \rightarrow (r \wedge s)$	$1 \wedge E$
	4	$q \rightarrow r$	$3 \wedge E$

(d)
	1	$(p \vee q) \rightarrow r$	premise
	2	p	premise
	3	$p \vee q$	$2 \vee I$
	4	r	$1, 3 \rightarrow E$
	5	$p \vee r$	$4 \vee I$

(e)
1	p	premise
2	r	premise
3	$p \to (q \leftrightarrow r)$	premise
4	$q \leftrightarrow r$	1, 3 $\to$ E
5	$r \to q$	4 $\leftrightarrow$ E
6	q	2, 5 $\to$ E

(f)
1	$p \wedge (q \to r)$	premise
2	p	1 $\wedge$ E
3	$q \to r$	1 $\wedge$ E
4	$q \to (s \vee r)$	3 $\vee$ I

(g)
1	$p \wedge (q \to (r \wedge s))$	premise
2	p	1 $\wedge$ E
3	$q \to (r \wedge s)$	1 $\wedge$ E
4	$q \to r$	3 $\wedge$ E
5	$p \wedge (q \to r)$	2, 4 $\wedge$ I

2. Establish the following, by providing a deduction.

(a) $(p \wedge q) \wedge r \vdash p \wedge (q \wedge r)$
(b) $p \wedge (q \to (r \wedge s)) \vdash p \wedge (q \to r)$
(c) $p \wedge (q \to r) \vdash p \wedge (q \to (p \vee r))$
(d) $p, s, s \to (p \to r), r \to (s \to t) \vdash v \vee t$
(e) $p \to q, q \to r, r \to s, p \vdash s$
(f) $p \to q, q \to r, r \to s \vdash p \to s$
(g) $p \wedge \neg q, \neg r \wedge s \vdash p \wedge s$
(h) $\neg q \vdash \neg(\neg p \wedge \neg q)$
(i) $p \to q, \neg q \vdash \neg p$
(j) $\neg p \to \neg q \vdash q \to p$

(k) $\neg(p \wedge r), \neg(\neg(p \wedge q)), r \vdash s$
(l) $p \to q, q \to r, \neg r \vdash \neg p$
(m) $p \to q, q \to r \vdash \neg r \to \neg p$
(n) $p \leftrightarrow (p \wedge q) \vdash p \to q$
(o) $(p \wedge q) \to r \vdash p \to (q \to r)$
(p) $\neg(\neg s \wedge q), \neg(p \wedge (\neg q \wedge \neg r)), \neg(r \wedge \neg s) \vdash \neg(\neg s \wedge p)$
(q) $p \leftrightarrow q \vdash \neg p \leftrightarrow \neg q$
(r) $p \wedge (q \vee r) \vdash (p \wedge q) \vee (p \wedge r)$
(s) $\neg p \wedge \neg q \vdash \neg(p \wedge q)$
(t) $p \vee q, \neg p \vdash q$
(u) $(p \vee q) \wedge r, q \to s \vdash \neg p \to (r \to s)$
(v) $(p \wedge q) \vee (\neg p \wedge \neg q) \vdash p \leftrightarrow q$
(w) $(p \leftrightarrow q) \leftrightarrow (q \leftrightarrow r) \vdash p \leftrightarrow r$
(x) $p \vdash p$

3. Establish the following, by providing a deduction.

(a) $\vdash \neg(p \wedge (\neg p \wedge q))$
(b) $\vdash (p \wedge q) \to p$
(c) $\vdash (p \to q) \to (\neg q \to \neg p)$
(d) $\vdash \neg((p \wedge q) \wedge (r \wedge \neg q))$
(e) $\vdash (p \vee q) \leftrightarrow (p \vee (\neg p \wedge q))$
(f) $\vdash (p \wedge \neg p) \to q$
(g) $\vdash p \vee \neg p$
(h) $\vdash p \leftrightarrow \neg\neg p$
(i) $\vdash (p \vee q) \leftrightarrow (p \vee (\neg p \to q))$
(j) $\vdash ((p \to q) \to p) \to p$
(k) $\vdash (\neg p \vee \neg q) \leftrightarrow \neg(p \wedge q)$
(l) $\vdash p \to p$

4. Establish that the following pairs of formulae are interdeducible.

(a) $p \to q$ and $\neg p \vee q$.
(b) $p \to q$ and $\neg(p \wedge \neg q)$.
(c) $p \vee q$ and $\neg(\neg p \wedge \neg q)$.
(d) $p \wedge q$ and $\neg(\neg p \vee \neg q)$.
(e) $p \leftrightarrow q$ and $(p \wedge q) \vee (\neg p \wedge \neg q)$.
(f) $p \leftrightarrow q$ and $(p \to q) \wedge (q \to p)$.

Classical Propositional Logic: Deduction

(g) $p \to q$ and $p \leftrightarrow (p \wedge q)$.
(h) $p \to q$ and $\neg q \to \neg p$.

2.2 Formulae in a Tree

One way to mitigate the complexity that arises from the display of ancillary deductions as columns of formulae is to display them as trees of formulae. On this view, each rule extends one or more trees of formulae to a new tree of formulae. A rule's conclusion defines a tree in terms of trees specified in its premise. In most cases, to identify the form of the trees of the premise, one must identify merely the form of the last formula of the tree in question. In a few cases, more must be identified. This kind of display was first used out by the German logician Gerhard Gentzen (1909–1945) (Gentzen 1934) and later developed by Swedish logician Dag Prawitz (born 1936) (Prawitz 1965).

Definition 7 Deduction for CPL (formula tree format)
A deduction of a formula from a set of formulae, the premises of the deduction, is a tree of formulae each one of which either is a formula at the top of the tree, in which case it is taken from among the set of premises, or else is a formula obtained from formulae immediately above it in the formula tree by one of the rules specified below.

The simplest formula tree comprises a single formula: it is a deduction of the formula from itself, a perfectly logical deduction. To obtain more useful and interesting deductions, we require rules similar to those expounded in section 2.1 for deductions in the formula column format, but adapted to formula tree format.

2.2.1 ∧ Elimination and Introduction

We start with the rules governing ∧. The elimination rule says that any deduction tree that terminates in a formula either of the form $\alpha \wedge \beta$ or of the form $\beta \wedge \alpha$ may be extended to a tree that terminates in a formula either of the form α or of the form β, respectively. The introduction rules say that any pair of deduction trees that one of which terminates in a formula of the form α and the other in a formula of the form β may be extended to a tree that terminates in a formula of the form $\alpha \wedge \beta$.

∧	Elimination		Introduction
	Version 1	Version 2	
	$\alpha \wedge \beta$	$\alpha \wedge \beta$	α, β
	———	———	———
	α	β	$\alpha \wedge \beta$

Notice that the elimination rule has two versions, as did its counterpart for deductions in column format, whereas the introduction rule has only one version, in contrast to its counterpart for deductions in column format, which has two versions. This compression in the formulation of the rule arises from notation of the comma, placed between the occurrences of the two formulae above the horizontal line. The comma indicates that the order of the two formulae is irrelevant. In other words, in a tree display, α may appear either to the left or to the right of β.

Example 1 (formula tree format)

$$\cfrac{\cfrac{p \wedge (q \wedge r)}{p} \wedge E \qquad \cfrac{\cfrac{p \wedge (q \wedge r)}{q \wedge r} \wedge E}{q} \wedge E}{p \wedge q} \wedge I \qquad \cfrac{\cfrac{p \wedge (q \wedge r)}{q \wedge r} \wedge E}{r} \wedge E}{(p \wedge q) \wedge r} \wedge I$$

It is worth taking a moment to compare this deduction in tree format with its counterpart in column format. In the two deductions, precisely the same seven formulae appear. However, in the deduction displayed in a column, each formula appears precisely once, while in the deduction displayed in a tree, one formula appears three times and another appears twice. The reason for this is that, in a deduction displayed in a column, the very same appearance of a formula may serve in different applications of the same rule, whereas in a deduction displayed in a tree, distinct applications of a rule require distinct appearances of the formulae used in them. Thus, since $\wedge$ Elimination applies twice to the formula $q \wedge r$, the formula appears twice in the tree display. In addition, $p \wedge (q \wedge r)$ is required to deduce p and to deduce $q \wedge r$, so it appears three times, once for the deduction of p and twice for the deduction of the two appearances of $q \wedge r$.

2.2.2 ∨ Elimination and Introduction

We now turn to ∨ Introduction and ∨ Elimination. While there are two versions of ∨ Introduction for deductions in tree format, just as there are two for deductions in column format, the comma notation permits us to simplify the formulation of ∨ Elimination for deductions in tree format, eschewing the extra notation required to make clear its six versions.

∨	Elimination	Introduction	
		Version 1	Version 2
	$\alpha \vee \beta, \alpha \rightarrow \gamma, \beta \rightarrow \gamma$	α	β
	γ	$\alpha \vee \beta$	$\alpha \vee \beta$

The elimination rule says that any three deduction trees terminating in formulae of the form $\alpha \vee \beta$, $\alpha \rightarrow \gamma$, and $\beta \rightarrow \gamma$ may be combined and extended to a tree that terminates in

Classical Propositional Logic: Deduction

a formula of the form γ. The introduction rule says that any deduction that terminates in a formula may be combined and extended to a new one terminating in a formula whose main propositional connective is $\vee$ and one of whose immediate subformulae is the formula above it in the tree.

Readers are encouraged to compare the following two deductions in tree form with their counterparts in column format.

Example 3 (formula tree format)

$$\cfrac{\cfrac{p \wedge (q \vee r)}{q \vee r} \wedge E \quad \cfrac{\cfrac{p \wedge (q \vee r)}{p} \wedge E \quad p \to (q \to s)}{q \to s} \to E \quad \cfrac{\cfrac{p \wedge (q \vee r)}{p} \wedge E \quad p \to (r \to s)}{r \to s} \to E}{s} \vee E$$

Example 4 (formula tree format)

$$\cfrac{\cfrac{p \wedge q}{p} \wedge E \quad \cfrac{\cfrac{\cfrac{p \wedge q}{q} \wedge E}{q \vee r} \vee I}{p \wedge (q \vee r)}}{} \wedge I$$

2.2.3 $\leftrightarrow$ Elimination and Introduction

The rule for $\leftrightarrow$ Elimination states that any tree terminating in a formula of the form $\alpha \leftrightarrow \beta$ may be extended to one terminating in a formula either of the form $\alpha \to \beta$ or of the form $\beta \to \alpha$. The rule for $\leftrightarrow$ Introduction states that any pair of trees terminating in formulae of the forms $\alpha \to \beta$ and $\beta \to \alpha$ may be combined and extended to a tree terminating in a formula of the form $\alpha \leftrightarrow \beta$.

$\leftrightarrow$	Elimination		Introduction
	Version 1	Version 2	
	$\alpha \leftrightarrow \beta$	$\alpha \leftrightarrow \beta$	$\alpha \to \beta, \beta \to \alpha$
	$\alpha \to \beta$	$\beta \to \alpha$	$\alpha \leftrightarrow \beta$

Example 5 (formula tree format)

$$\cfrac{r \quad \cfrac{\cfrac{p \quad p \to (q \leftrightarrow r)}{q \leftrightarrow r} \to E}{r \to q} \leftrightarrow E}{q} \to E$$

Example 6 (formula tree format)

$$\frac{\dfrac{p \quad p \to (q \to r)}{q \to r} \to E \quad \dfrac{p \quad p \to (r \to q)}{r \to q} \to E}{q \leftrightarrow r} \leftrightarrow I$$

2.2.4 → Elimination and Introduction

We now come to the rules pertaining to →. The rule for → Elimination states that any pair of trees terminating in formulae of the forms α and $\alpha \to \beta$ and may be combined and extended to a tree terminating in a formula of the form β. The rule for → Introduction states that any tree terminating in a formula of the form β and having one or more instances of a formula of the form α at the top of the tree may be extended to a tree that terminates in a formula of the form $\alpha \to \beta$ and which has each instance of the formula of the form α at the top of the tree enclosed within square brackets.

→	Elimination	Introduction
		$[\alpha]$ $\vdots$
	$\alpha, \alpha \to \beta$	β
	β	$\alpha \to \beta$

Example 2 (formula tree format)

$$\frac{p \quad \dfrac{p \to (q \to r) \quad q}{q \to r} \to E}{r} \to E$$

The rule of → elimination is sufficiently similar to the other elimination rules in the formula tree format that we do not need to comment on it or its illustration; however, we shall comment on the rule for → introduction since this rule requires the discharge of an assumption, something we have not yet seen in the formula tree format.

Consider the following deduction in formula column format whose last step uses the rule of → introduction.

Example 7 (formula tree format)

$$\frac{\dfrac{p \to (q \to r) \quad \dfrac{[p \wedge q]}{p} \wedge E}{q \to r} \to E \quad \dfrac{[p \wedge q]}{q} \wedge E}{\dfrac{r}{(p \wedge q) \to r} \to I}$$

Precisely the same seven formula appear in both formats. However, in the deduction in column format, each formula appears exactly once, whereas in the deduction in tree format, one formula, $p \wedge q$, appears twice, and as explained, it must appear twice in the deduction in tree format since the rule of $\wedge$ Elimination is used twice.

The $\to$ Introduction rule for deductions in column format requires an ancillary deduction, the first formula of which is an ancillary premise, supposition, to the left of which appears a vertical line indicating the start of the ancillary deduction. The ancillary deduction terminates upon the application of the $\to$ Introduction, at which point the premise to this ancillary deduction is discharged. The termination of the ancillary deduction is indicated by the termination of the vertical line that began to the immediate left of the premise of the ancillary deduction. The $\to$ Introduction rule for deductions in tree format also requires an ancillary deduction. The required ancillary deduction is in the form of a tree and the discharge of the ancillary premise is indicated by enclosing all appearances of the premise in square brackets. In this case, both occurrences of the formula $p \wedge q$.

2.2.5 ¬ Elimination and Introduction

Both rules for ¬ are like the rule for $\to$ Introduction in that they both involve the discharge of an assumption. The ¬ Introduction rule states that a tree terminating in a contradiction is extended to tree terminating in the negation of some formula occurring at the top of the tree identified by the premise. When the rule is applied, each instance of the formula that is negated is enclosed in square brackets. The ¬ Elimination rule specifies the same thing as the ¬ Introduction rules, except that the formula discharged has the form $\neg \alpha$ and the formula of the extended tree has the form α.

¬	Elimination		Introduction	
	Version 1	Version 2	Version 1	Version 2
	$[\neg \alpha]$ $\vdots$ $\beta \wedge \neg \beta$ $\overline{\alpha}$	$[\neg \alpha]$ $\vdots$ $\neg \beta \wedge \beta$ $\overline{\alpha}$	$[\alpha]$ $\vdots$ $\beta \wedge \neg \beta$ $\overline{\neg \alpha}$	$[\alpha]$ $\vdots$ $\neg \beta \wedge \beta$ $\overline{\neg \alpha}$

Example 8 (formula tree format)

$$\cfrac{p \quad \cfrac{\cfrac{[q \wedge \neg(p \wedge q)]}{q} \wedge E}{p \wedge q} \wedge I \quad \cfrac{[q \wedge \neg(p \wedge q)]}{\neg(p \wedge q)} \wedge E}{\cfrac{(p \wedge q) \wedge \neg(p \wedge q)}{\neg(q \wedge \neg(p \wedge q))} \neg I} \wedge I$$

2.2.6 Deducibility: Formula tree format

In light of the fact that a deduction defined for the formula column format is different from a deduction defined for a formula tree format, the definition for deducibility in terms of deduction has a different definition.

Definition 8 Deducibility for CPL (formula tree format)
α is deducible from Γ, or $\Gamma \vdash \alpha$, iff there is a deduction (in formula tree format) that terminates with α and in which each formula at the top of the tree is either a formula in Γ or is enclosed in square brackets.

Though we shall not prove it, the ways of defining a deduction for the formula column format and for the formula tree format are equivalent, as are the corresponding ways of defining deducibility.

We conclude this section with an important observation. It could have been made in section 2.1.10, but we make it here, since its truth is more clearly evident when thinking of deductions in a tree format, rather than in a column format.

Fact 1 The cut theorem

$$\frac{\Gamma \vdash \alpha, \ \Delta \cup \{\alpha\} \vdash \beta}{\Gamma \cup \Delta \vdash \beta}$$

This states that should one have a deduction of a formula, say β, from a set of formulae that includes a formula, say α, that is to say, a deduction of β from $\Delta \cup \{\alpha\}$, and should one also have a deduction of α from a possibly different set of premises Γ, then one has a deduction of β from the combined set of formulae, $\Gamma \cup \Delta$, where α has been, as it were, cut out. Thinking of deductions displayed in the format of trees, one can easily see that this is true. Any deduction of α from Γ appears as a tree whose root node is labeled with α. Any deduction of β from $\Delta \cup \{\alpha\}$, will have α, if it is actually used in the deduction, appear at least once as a top node in its tree. The deduction of β from $\Gamma \cup \Delta$ will appear as a tree in which the tree displaying the deduction of α from Γ replaces each occurrence of an undischarged instance of α in a top node of the tree displaying the deduction of β from $\Delta \cup \{\alpha\}$.

Exercises: Natural deduction: formulae in a tree

Using the deduction rules for natural deductions of formulae in a tree to establish the results in exercises 2 through 4 of section 2.1.

3 Sequent Natural Deduction

We now turn to natural deduction done with sequents. As stated, natural deduction with sequents may be displayed as a column of sequents or as a tree of sequents. We begin with the presentation of sequents in a column.

3.1 Sequents in a Column

In natural deduction, one distinguishes between the deduction of a formula from a set of formulae and the deducibility of a formula from a set of formulae. Deducibility is defined in terms of deduction. Deduction, in turn, is defined in terms of a set of rules. Some of the rules used to define deduction relate one or more formulae to a single formula, while others relate an ancillary deduction to a single formula. It is the rules relating an ancillary deduction to a single formula that, though intuitive, nonetheless usher in complications, especially in the display of deductions.

There are several ways to eschew the identification of, and the appeal to, ancillary deductions. The idea was put forth by the German logician Gerhard Gentzen (1909–1945) in his doctoral dissertation (1934). His insight was to take the deducibility relation as fundamental. Recall that deducibility is a binary relation between a set of formulae and a single formula. Gentzen called the expression for instances of this relation a *sequent*. His idea is to define all instances of the deducibility relation by taking some instances of the deducibility relation as basic and to obtain the other instances using rules, not for formulae, but for sequents. We shall consider Gentzen's original proposal later. Here we shall consider an alternative way of exploiting his insight, suggested by the American logician Patrick Suppes (born 1922; Suppes 1957, chap. 2).

The fundamental idea is that of a *sequent deduction*. A natural deduction in the column format comprises a sequence of formulae, each with a numeral to its left and a justification to its right. A sequence deduction in column format comprises a sequence of sequents,[1] each with a numeral to its left and a justification to its right. Here is the definition of a sequent deduction.

Definition 9 Sequent deduction for CPL (sequent column format)
A deduction of a sequent is a sequence of sequents each one of which is either of the form $\Gamma \vdash \alpha$, where $\alpha \in \Gamma$, or obtained from earlier sequents in the sequence by one of the rules specified below.

Sequents of the form $\Gamma \vdash \alpha$, where $\alpha \in \Gamma$, are called *axioms*. Here are two examples of axioms: $p \wedge q, r \rightarrow p \vdash p \wedge q$ and $r \vdash r$. Axioms may appear at any point in a sequent deduction, and such sequents are justified as axioms. Next come the rules where one sequent is obtained from earlier sequents in the sequence. The main idea here is to forego special conventions tracking suppositions by listing at each step all the suppositions and premises that the formula at that step in the deduction depends on. Not surprisingly, the rules for sequent deduction in column format are very similar to those for natural deduction in column format. Indeed, any rule for natural deduction that does not involve an ancillary deduction and its counterpart for sequent deduction in the column format are essentially alike. If one suppresses from the middle column of the latter rules the expression $\vdash$ and the expressions for sets of formulae, one obtains the former rules. The rule of sequent

---

1. It is unfortunate that the two English words *sequence* and *sequents* are homophonic.

deduction corresponding to a rule of natural deduction appealing to an ancillary deduction has the elimination of a formula from the set of formulae on the left side of the turnstile and a suitable change in the formula on the right side.

Bearing these general explanatory remarks in mind, readers are invited to read through the rules and the illustrations of their application. Readers are also encouraged to compare the illustrative deductions with their counterparts in the preceding sections. We start with the rules for $\wedge$.

3.1.1 $\wedge$ Elimination

$\wedge$ Elimination	
Version 1	Version 2
$\vdots \quad \vdots \quad \vdots \quad \vdots$ m $\quad \Gamma \vdash \alpha \wedge \beta$ $\vdots \quad \vdots \quad \vdots \quad \vdots$ n $\quad \Gamma \vdash \alpha \qquad$ m $\wedge$ E	$\vdots \quad \vdots \quad \vdots \quad \vdots$ m $\quad \Gamma \vdash \alpha \wedge \beta$ $\vdots \quad \vdots \quad \vdots \quad \vdots$ n $\quad \Gamma \vdash \beta \qquad$ m $\wedge$ E

3.1.2 $\wedge$ Introduction

$\wedge$ Introduction	
Version 1	Version 2
$\vdots \quad \vdots \quad \vdots \quad \vdots$ m $\quad \Gamma \vdash \alpha$ $\vdots \quad \vdots \quad \vdots \quad \vdots$ n $\quad \Delta \vdash \beta$ $\vdots \quad \vdots \quad \vdots \quad \vdots$ r $\quad \Gamma \cup \Delta \vdash \alpha \wedge \beta \quad$ m, n $\wedge$ I	$\vdots \quad \vdots \quad \vdots \quad \vdots$ m $\quad \Delta \vdash \beta$ $\vdots \quad \vdots \quad \vdots \quad \vdots$ n $\quad \Gamma \vdash \alpha$ $\vdots \quad \vdots \quad \vdots \quad \vdots$ r $\quad \Gamma \cup \Delta \vdash \alpha \wedge \beta \quad$ m, n $\wedge$ I

We now establish that $p \wedge (q \wedge r) \vdash (p \wedge q) \wedge r$, using sequents in a column. In fact, the following presentation is obtained from the presentation of the deduction establishing the same sequent done in the presentation of formula in a column by replacing the formula in each line in the first presentation with a sequent. This yields the following proof.

Classical Propositional Logic: Deduction

Example 1 (sequent column format)

1	$p \wedge (q \wedge r)$	$\vdash$	$p \wedge (q \wedge r)$	axiom
2	$p \wedge (q \wedge r)$	$\vdash$	p	$1 \wedge E$
3	$p \wedge (q \wedge r)$	$\vdash$	$q \wedge r$	$1 \wedge E$
4	$p \wedge (q \wedge r)$	$\vdash$	q	$3 \wedge E$
5	$p \wedge (q \wedge r)$	$\vdash$	r	$3 \wedge E$
6	$p \wedge (q \wedge r)$	$\vdash$	$p \wedge q$	$2, 4 \wedge I$
7	$p \wedge (q \wedge r)$	$\vdash$	$(p \wedge q) \wedge r$	$5, 6 \wedge I$

Next we turn to the rules for $\vee$.

3.1.3 $\vee$ Elimination

$\vee$ Elimination
Let Σ be $\{\alpha \vee \beta, \alpha \to \gamma, \beta \to \gamma\}$
$\vdots$
m $\Gamma \vdash \pi_1$ (where $\pi_1 \in \Sigma$)
$\vdots$
n $\Delta \vdash \pi_2$ (where $\pi_2 \in \Sigma - \{\pi_1\}$)
$\vdots$
r $\Theta \vdash \pi_3$ (where $\pi_3 \in \Sigma - \{\pi_1, \pi_2\}$)
$\vdots$
s $\Gamma \cup \Delta \cup \Theta \vdash \gamma$ m, n, r $\vee$ E

Example 3 (sequent column format)

1	$p \wedge (q \vee r)$	$\vdash$	$p \wedge (q \vee r)$	axiom
2	$p \to (q \to s)$	$\vdash$	$p \to (q \to s)$	axiom
3	$p \to (r \to s)$	$\vdash$	$p \to (r \to s)$	axiom
4	$p \wedge (q \vee r)$	$\vdash$	p	$1 \wedge E$
5	$p \wedge (q \vee r)$	$\vdash$	$q \vee r$	$1 \wedge E$
6	$p \wedge (q \vee r), p \to (q \to s)$	$\vdash$	$q \to s$	$2, 4 \to E$
7	$p \wedge (q \vee r), p \to (r \to s)$	$\vdash$	$r \to s$	$3, 4 \to E$
8	$p \wedge (q \vee r), p \to (q \to s), p \to (r \to s)$	$\vdash$	s	$5, 6, 7 \vee E$

3.1.4 ∨ Introduction

∨ Introduction	
Version 1	Version 2
$\vdots \quad \vdots \quad \vdots \quad \vdots$ m $\quad \Gamma \ \vdash\ \alpha$ $\vdots \quad \vdots \quad \vdots \quad \vdots$ n $\quad \Gamma \ \vdash\ \alpha \vee \beta \quad$ m ∨ I	$\vdots \quad \vdots \quad \vdots \quad \vdots$ m $\quad \Gamma \ \vdash\ \beta$ $\vdots \quad \vdots \quad \vdots \quad \vdots$ n $\quad \Gamma \ \vdash\ \alpha \vee \beta \quad$ m ∨ I

Example 4 (sequent column format)

1	$p \wedge q$	$\vdash$	$p \wedge q$	axiom
2	$p \wedge q$	$\vdash$	p	1 ∧ E
3	$p \wedge q$	$\vdash$	q	1 ∧ E
4	$p \wedge q$	$\vdash$	$q \vee r$	3 ∨ I
5	$p \wedge q$	$\vdash$	$p \wedge (q \vee r)$	2, 4 ∧ I

3.1.5 ↔ Elimination

↔ Elimination	
Version 1	Version 2
$\vdots \quad \vdots \quad \vdots \quad \vdots$ m $\quad \Gamma \ \vdash\ \alpha \leftrightarrow \beta$ $\vdots \quad \vdots \quad \vdots \quad \vdots$ n $\quad \Gamma \ \vdash\ \alpha \to \beta \quad$ m ↔ E	$\vdots \quad \vdots \quad \vdots \quad \vdots$ m $\quad \Gamma \ \vdash\ \alpha \leftrightarrow \beta$ $\vdots \quad \vdots \quad \vdots \quad \vdots$ n $\quad \Gamma \ \vdash\ \beta \to \alpha \quad$ m ↔ E

Example 5 (sequent column format)

1	p	$\vdash$	p	axiom
2	r	$\vdash$	r	axiom
3	$p \to (q \leftrightarrow r)$	$\vdash$	$p \to (q \leftrightarrow r)$	axiom
4	$p, p \to (q \leftrightarrow r)$	$\vdash$	$q \leftrightarrow r$	1, 3 → E
5	$p, p \to (q \leftrightarrow r)$	$\vdash$	$r \to q$	4 ↔ E
6	$p, r, p \to (q \leftrightarrow r)$	$\vdash$	q	2, 5 → E

Classical Propositional Logic: Deduction

3.1.6 ↔ Introduction

↔ Introduction	
Version 1	Version 2
⋮ ⋮ ⋮ ⋮ m Γ ⊢ α → β ⋮ ⋮ ⋮ ⋮ n Δ ⊢ β → α ⋮ ⋮ ⋮ ⋮ r Γ ∪ Δ ⊢ α ↔ β m, n ∧ I	⋮ ⋮ ⋮ ⋮ m Δ ⊢ β → α ⋮ ⋮ ⋮ ⋮ n Γ ⊢ α → β ⋮ ⋮ ⋮ ⋮ r Γ ∪ Δ ⊢ α ↔ β m, n ∧ I

Example 6 (sequent column format)

```
1                           p → (q → r) ⊢ p → (q → r)     axiom
2                           p → (r → q) ⊢ p → (r → q)     axiom
3                                     p ⊢ p               axiom
4               p → (q → r), p ⊢ q → r                    1, 3 → E
5   p → (q → r), p → (r → q) ⊢ r → q                      2, 3 → E
6   p → (q → r), p → (r → q), p ⊢ q ↔ r                   4, 5 ↔ I
```

3.1.7 → Elimination

→ Elimination	
Version 1	Version 2
⋮ ⋮ ⋮ ⋮ m Γ ⊢ α ⋮ ⋮ ⋮ ⋮ n Δ ⊢ α → β ⋮ ⋮ ⋮ ⋮ r Γ ∪ Δ ⊢ β m, n → E	⋮ ⋮ ⋮ ⋮ m Δ ⊢ α → β ⋮ ⋮ ⋮ ⋮ n Γ ⊢ α ⋮ ⋮ ⋮ ⋮ r Γ ∪ Δ ⊢ β m, n → E

Example 2 (sequent column format)

1	$p \to (q \to r)$	$\vdash$	$p \to (q \to r)$	axiom
2	p	$\vdash$	p	axiom
3	q	$\vdash$	q	axiom
4	$p, p \to (q \to r)$	$\vdash$	$q \to r$	1, 2 $\to$ E
5	$p, q, p \to (q \to r)$	$\vdash$	r	3, 4 $\to$ E

3.1.8 $\to$ Introduction

$\to$ Introduction				
$\vdots$	$\vdots$ $\vdots$ $\vdots$			
m	$\Gamma \cup \{\alpha\}$	$\vdash$	β	
$\vdots$	$\vdots$ $\vdots$ $\vdots$			
n	Γ	$\vdash$	$\alpha \to \beta$	m $\to$ I

Example 7 (sequent column format)

1	$p \to (q \to r)$	$\vdash$	$p \to (q \to r)$	axiom
2	$p \wedge q$	$\vdash$	$p \wedge q$	axiom
3	$p \wedge q$	$\vdash$	p	2 $\wedge$ E
4	$p \to (q \to r), p \wedge q$	$\vdash$	$q \to r$	1, 3 $\to$ E
5	$p \to (q \to r), p \wedge q$	$\vdash$	q	2 $\wedge$ E
6	$p \to (q \to r), p \wedge q$	$\vdash$	r	4, 5 $\to$ E
7	$p \to (q \to r)$	$\vdash$	$(p \wedge q) \to r$	6 $\to$ I

3.1.9 $\neg$ Elimination

$\neg$ Elimination		
Version 1		Version 2
$\vdots$ $\quad$ $\vdots$ $\vdots$ $\vdots$		$\vdots$ $\quad$ $\vdots$ $\vdots$ $\vdots$
m $\quad$ $\Gamma \cup \{\neg\alpha\}$ $\vdash$ $\beta \wedge \neg\beta$		m $\quad$ $\Gamma \cup \{\neg\alpha\}$ $\vdash$ $\neg\beta \wedge \beta$
$\vdots$ $\quad$ $\vdots$ $\vdots$ $\vdots$		$\vdots$ $\quad$ $\vdots$ $\vdots$ $\vdots$
n $\quad$ Γ $\vdash$ α $\quad$ m $\neg$ E		n $\quad$ Γ $\vdash$ α $\quad$ m $\neg$ E

Classical Propositional Logic: Deduction

3.1.10 ¬ Introduction

¬ Introduction	
Version 1	Version 2
$\begin{array}{llll} \vdots & \vdots & \vdots & \\ m & \Gamma \cup \{\alpha\} & \vdash \beta \wedge \neg\beta & \\ \vdots & \vdots & \vdots & \\ n & \Gamma & \vdash \neg\alpha & m \neg I \end{array}$	$\begin{array}{llll} \vdots & \vdots & \vdots & \\ m & \Gamma \cup \{\alpha\} & \vdash \neg\beta \wedge \beta & \\ \vdots & \vdots & \vdots & \\ n & \Gamma & \vdash \neg\alpha & m \neg I \end{array}$

Example 8 (sequent column format)

$$
\begin{array}{rrcll}
1 & p & \vdash & p & \text{axiom} \\
2 & q \wedge \neg(p \wedge q) & \vdash & q \wedge \neg(p \wedge q) & \text{axiom} \\
3 & q \wedge \neg(p \wedge q) & \vdash & q & 2 \wedge E \\
4 & q \wedge \neg(p \wedge q) & \vdash & \neg(p \wedge q) & 2 \wedge E \\
5 & p, q \wedge \neg(p \wedge q) & \vdash & p \wedge q & 1, 3 \wedge I \\
6 & p, q \wedge \neg(p \wedge q) & \vdash & (p \wedge q) \wedge \neg(p \wedge q) & 4, 5 \wedge I \\
7 & p & \vdash & \neg(q \wedge \neg(p \wedge q)) & 6 \neg I
\end{array}
$$

Exercises: Natural deduction: sequents in a column

Using the deduction rules for natural deductions of sequents in a column to establish the results in exercises 2 through 4 of section 2.1.

3.2 Sequents in a Tree

Not surprisingly, one can recast the deduction of sequents in a sequent column format into one in a sequent tree format. In the deduction format of formula tree, the top nodes of a tree displaying a deduction are labeled with formulae, the plain formulae are the formulae used in the deduction and serving as premises and the formulae enclosed in square brackets are suppositions, or ancillary premises, which are eventually discharged in the course of the deduction. Here, the top nodes of the tree displaying a deduction are the sequents of the form: $\Gamma \vdash \alpha$ where $\alpha \in \Gamma$, the axioms of section 3.1. As with deductions in the formula tree format, each rule extends one or more trees into a new tree and specifies the sequent terminating the new tree in terms of the sequents terminating the tree or trees being extended.

Definition 10 Deduction for CPL (sequent tree format)
A deduction of a sequent is a tree of sequents each one of which either is a sequent at the top of the tree, in which case it is an axiom, or else is a sequent obtained from sequents immediately above it in the sequent tree by one of the rules specified.

This format is a simple recasting of the formula tree format. Indeed, the turnstile used here is just another way of characterizing the turnstile defined in definition 1 in section 2.1.10 and the one defined in definition 8 in section 2.2.6. This format is here to ease coming to grips with the next and last presentation.

It is customary to generalize the notation for rules so as to encompass the axioms. The axioms serve to label the top nodes in a tree displaying a deduction. To treat the axiom as a rule, one uses the form of a rule, leaving what is above the rule blank, after all, no node dominates it, and to write the form of the axiom below the line.

Axiom

$$\overline{\Gamma \vdash \alpha} \quad \text{(where } \alpha \in \Gamma\text{)}$$

Recall that any formula by itself is a deduction in the formula tree format of it from itself. Hence, by the definition of deducibility (formula tree format), a formula is deducible from any set of formulae of which it is a member.

The rules that follow are just like the elimination rules and introduction rules used for the formula tree format. The difference is that, whereas in the formulation of the rules for the formula tree format no symbols were used to denote the premises on which the deduction depends, here such symbols are used.

3.2.1 ∧ Elimination and Introduction

∧	Elimination		Introduction
	$\dfrac{\Gamma \vdash \alpha \wedge \beta}{\Gamma \vdash \alpha}$	$\dfrac{\Gamma \vdash \alpha \wedge \beta}{\Gamma \vdash \beta}$	$\dfrac{\Gamma \vdash \alpha, \ \Delta \vdash \beta}{\Gamma \cup \Delta \vdash \alpha \wedge \beta}$

The ∧ Elimination rule here says that if there is a deduction of a formula $\alpha \wedge \beta$ whose premises are among the formulae of Γ, there is a deduction of α from the same set of premises. Indeed, suppose that there is a deduction in the formula tree format of the formula $\alpha \wedge \beta$ from formulae in the set Γ. Then, there is a deduction in the formula tree format

Classical Propositional Logic: Deduction

whose undischarged top nodes are labeled with formulae from Γ and whose bottom node is labeled with $\alpha \wedge \beta$. Now, by the rule of $\wedge$ Elimination of section 2.2.1, one extends the tree by one node and labels it α thereby obtaining a deduction of the formula α all of whose undischarged top nodes are labeled with formulae from Γ. The same reasoning applies to the second version of the $\wedge$ Elimination rule.

The $\wedge$ Introduction rule says that if there is a deduction of a formula α in the formula tree format whose premises are among the formulae of Γ and a deduction of a formula β in the formula tree format whose premises are among the formulae of Δ, there is a deduction of $\alpha \wedge \beta$ in the formula tree format whose premises are among those of Γ and Δ taken together. One simply combines the formula tree for each deduction, extending them to a formula tree terminating in $\alpha \wedge \beta$ invoking the rule of $\wedge$ Introduction in section 2.2.1.

Example 1 (sequent tree format)

$$\dfrac{\dfrac{p\wedge(q\wedge r)\vdash p\wedge(q\wedge r)}{p\wedge(q\wedge r)\vdash p}\wedge E \quad \dfrac{\dfrac{p\wedge(q\wedge r)\vdash p\wedge(q\wedge r)}{p\wedge(q\wedge r)\vdash q\wedge r}\wedge E}{p\wedge(q\wedge r)\vdash q}\wedge E}{p\wedge(q\wedge r)\vdash p\wedge q}\wedge I \quad \dfrac{\dfrac{p\wedge(q\wedge r)\vdash p\wedge(q\wedge r)}{p\wedge(q\wedge r)\vdash q\wedge r}\wedge E}{p\wedge(q\wedge r)\vdash r}\wedge E$$
$$p\wedge(q\wedge r)\vdash (p\wedge q)\wedge r \quad \wedge I$$

3.2.2 ∨ Elimination and Introduction

∨	Elimination	Introduction	
		Version 1	Version 2
	$\Gamma \vdash \alpha \to \gamma,\ \Delta \vdash \beta \to \gamma,\ \Theta \vdash \alpha \vee \beta$	$\Gamma \vdash \beta$	$\Gamma \vdash \alpha$
	$\Gamma \cup \Delta \cup \Theta \vdash \gamma$	$\Gamma \vdash \alpha \vee \beta$	$\Gamma \vdash \alpha \vee \beta$

Readers are encouraged to use the preceding exposition of the equivalence of the $\wedge$ Elimination and Introduction rules with their counterparts in section 2.2.1 to see the equivalence of the $\vee$ Elimination and Introduction rules here with their counterparts in section 2.2.2.

Example 3 (sequent tree format)

$$\cfrac{\cfrac{p\wedge(q\vee r)\vdash p\wedge(q\vee r)}{p\wedge(q\vee r)\vdash q\vee r}\wedge E \qquad \cfrac{\cfrac{p\wedge(q\vee r)\vdash p\wedge(q\vee r)}{p\wedge(q\vee r)\vdash p}\wedge E \qquad p\to(q\to s)\vdash p\to(q\to s)}{p\wedge(q\vee r),\, p\to(q\to s)\vdash q\to s}\to E \qquad \cfrac{\cfrac{p\wedge(q\vee r)\vdash p\wedge(q\vee r)}{p\wedge(q\vee r)\vdash p}\wedge E \qquad p\to(r\to s)\vdash p\to(r\to s)}{p\wedge(q\vee r),\, p\to(r\to s)\vdash r\to s}\to E}{p\wedge(q\vee r),\, p\to(q\to s),\, p\to(r\to s)\vdash s}\vee E$$

(Incomplete derivation — middle $\to E$ step: $p\wedge(q\vee r),\, p\to(q\to s)\vdash p\to s$; right branch similarly yields $p\to s$.)

Classical Propositional Logic: Deduction

Example 4 (sequent tree format)

$$\dfrac{\dfrac{p\wedge q\vdash p\wedge q}{p\wedge q\vdash p}\wedge E \quad \dfrac{\dfrac{p\wedge q\vdash p\wedge q}{p\wedge q\vdash q}\wedge E}{p\wedge q\vdash q\vee r}\vee I}{p\wedge q\vdash p\wedge(q\vee r)}\wedge I$$

3.2.3 ↔ Elimination and Introduction

↔	Elimination		Introduction
	$\dfrac{\Gamma\vdash \alpha\leftrightarrow\beta}{\Gamma\vdash \alpha\to\beta}$	$\dfrac{\Gamma\vdash \alpha\leftrightarrow\beta}{\Gamma\vdash \beta\to\alpha}$	$\dfrac{\Gamma\vdash \alpha\to\beta,\ \Delta\vdash\beta\to\alpha}{\Gamma\cup\Delta\vdash\alpha\leftrightarrow\beta}$

We hope the equivalence between these rules and their counterparts in section 2.2.3 is clear.

Example 5 (sequent tree format)

$$\dfrac{\dfrac{\dfrac{\dfrac{p\to(q\leftrightarrow r)\vdash p\to(q\leftrightarrow r) \quad p\vdash p}{p\to(q\to r),\ p\vdash q\leftrightarrow r}\to E}{p\to(q\leftrightarrow r),\ p\vdash r\to q}\leftrightarrow E \quad r\vdash r}{p,\ r,\ p\to(q\leftrightarrow r)\vdash q}\to E$$

Example 6 (sequent tree format)

$$\dfrac{\dfrac{p\to(q\to r)\vdash p\to(q\to r) \quad p\vdash p}{p\to(q\to r),\ p\vdash q\to r}\to E \quad \dfrac{p\to(r\to q)\vdash p\to(r\to q) \quad p\vdash p}{p\to(r\to q),\ p\vdash r\to q}\to E}{p\to(r\to q),\ p\vdash q\leftrightarrow r}\leftrightarrow I$$

3.2.4 → Elimination and Introduction

→	Elimination	Introduction
	$\dfrac{\Gamma\vdash\alpha,\ \Delta\vdash\alpha\to\beta}{\Gamma\cup\Delta\vdash\beta}$	$\dfrac{\Gamma\cup\{\alpha\}\vdash\beta}{\Gamma\vdash\alpha\to\beta}$

The reasoning showing the equivalence between the → Elimination rules for natural deduction in the formula tree format and the sequent tree format continues to be similar to what was said previously. However, the equivalence for the → Introduction rules warrants a word or two of elucidation.

The → Introduction rule says is that, if there is a deduction of a formula β in the formula tree format whose premises are among the formulae of $\Gamma \cup \{\alpha\}$, there is a deduction of a formula $\alpha \to \beta$ in the formula tree format whose premises are among those of Γ alone. One extends the deduction of β in the formula tree format to a formula tree terminating in $\alpha \to \beta$ and one encloses all the top nodes labeled by α in square brackets. The result is a deduction in the formula tree format of the formula $\alpha \to \beta$ from formulae among those in Γ.

Example 2 (sequent tree format)

$$\dfrac{\dfrac{p \to (q \to r) \vdash p \to (q \to r) \qquad p \vdash p}{p \to (q \to r), p \vdash q \to r} \to E \qquad q \vdash q}{p \to (q \to r), p, q \vdash r} \to E$$

Example 7 (sequent tree format)

$$\dfrac{\dfrac{p \to (q \to r) \vdash p \to (q \to r) \qquad \dfrac{p \wedge q \vdash p \wedge q}{p \wedge q \vdash p} \wedge E}{p \to (q \to r), p \wedge q \vdash q \to r} \to E \qquad \dfrac{p \wedge q \vdash p \wedge q}{p \wedge q \vdash q} \wedge E}{\dfrac{p \to (q \to r), p \wedge q \vdash r}{p \to (q \to r) \vdash (p \wedge q) \to r} \to I} \to E$$

¬ Elimination and Introduction

¬	Elimination		Introduction	
	Version 1	Version 2	Version 1	Version 2
	$\dfrac{\Gamma \cup \{\neg \alpha\} \vdash \beta \wedge \neg \beta}{\Gamma \vdash \alpha}$	$\dfrac{\Gamma \cup \{\neg \alpha\} \vdash \neg \beta \wedge \beta}{\Gamma \vdash \alpha}$	$\dfrac{\Gamma \cup \{\alpha\} \vdash \beta \wedge \neg \beta}{\Gamma \vdash \neg \alpha}$	$\dfrac{\Gamma \cup \{\alpha\} \vdash \neg \beta \wedge \beta}{\Gamma \vdash \neg \alpha}$

It is hoped that the exposition of the equivalence of the → Introduction rule of section 2.2.4 and the → Introduction rule just stated is sufficient to make clear the equivalence of the ¬ Elimination and Introduction rules here with the rules of ¬ E and ¬ I of section 2.2.5.

Example 8 (sequent tree format)

$$\dfrac{p \vdash p \quad \dfrac{\dfrac{q \wedge \neg(p \wedge q) \vdash q \wedge \neg(p \wedge q)}{q \wedge \neg(p \wedge q) \vdash q} \wedge E}{p, q \wedge \neg(p \wedge q) \vdash p \wedge q} \wedge I \quad \dfrac{\dfrac{q \wedge \neg(p \wedge q) \vdash q \wedge \neg(p \wedge q)}{q \wedge \neg(p \wedge q) \vdash \neg(p \wedge q)} \wedge E}{\dfrac{p, q \wedge \neg(p \wedge q) \vdash (p \wedge q) \wedge \neg(p \wedge q)}{p \vdash \neg(q \wedge \neg(p \wedge q))} \neg I} \wedge I$$

Exercises: Natural deduction: sequents in a tree

Using the deduction rules for natural deductions of sequents in a column to establish the results in exercises 2 through 4 of section 2.1.

4 Gentzen Sequent Calculus

We come at last to the Gentzen sequent calculus, which we shall describe only very briefly. In the Gentzen sequent calculus, there are no elimination rules, there are only introduction rules. The elimination rules used in the two preceding sequent forms of natural deduction are replaced with introduction rules, where the introduction is done, not in the consequent of the sequent, but in the antecedent. Thus, there are only introduction rules, and they divide into those in which the introduction is performed in the antecedent and those in which the introduction is done in the consequent. Rules that introduce a propositional connective in a formula in the antecedent, that is to say, rules that alter a formula to the left of the turnstile by adding a propositional connective to it, are called *left introduction rules* and rules that introduce a propositional connective in the formula of the consequent, that is, rules that alter a formula to the right of the turnstile by adding a propositional connective to it, are called *right introduction rules*. The right introduction rules are precisely the same as the introduction rules we encountered in section 3.2.

To carry out fully Gentzen's idea for his sequent calculus, one requires a more general notion of sequent, one in which, not a single formula appears to the right of the turnstile, but a set of formulae, possibly the empty set. However, this more general notion of a sequent is required for the deduction rules governing negation. Since the rules of negation have no counterparts that play a role in the subject matter of this book, we shall dispense with the more general notion of sequent and the rules of negation.

Definition 11 Gentzen sequent calculus deduction for CPL (tree format)
A Gentzen sequent calculus deduction of a sequent is a tree of sequents each one of which either is a sequent at the top of the tree, in which case it is an axiom, or else is a sequent obtained from sequents immediately above it in the sequent tree by one of the rules specified below.

The axiom for the Gentzen sequent calculus is just like the axiom for natural sequent sequent deduction.

Axiom

$$\overline{\Gamma \vdash \alpha} \quad \text{(where } \alpha \in \Gamma\text{)}$$

We turn to the other rules. Since the rules for introduction on the right are the same as those for natural deduction with sequents in the tree format, we shall confine our comments to the rules for introduction on the left.

4.1 ∧ Left and Right Introduction

∧	Introduction L	Introduction R
	$\dfrac{\Gamma \cup \{\alpha, \beta\} \vdash \gamma}{\Gamma \cup \{\alpha \wedge \beta\} \vdash \gamma}$	$\dfrac{\Gamma \vdash \alpha, \ \Delta \vdash \beta}{\Gamma \cup \Delta \vdash \alpha \wedge \beta}$

The left ∧ introduction states that if there is a deduction of γ from a set of formulae that includes α and β, then there is a deduction from the same set of formulae, without α and β, but with $\alpha \wedge \beta$ in their stead. To see that this rule is valid, let us think of deductions in the format of formulae in a tree. Now suppose that there is a deduction of γ from a set of formulae that includes α and β. We further suppose, without loss of generality, that α and β are actually used in a deduction. It follows that α and β, without square brackets, each label at least one top node in the deduction tree. To obtain a deduction tree from the same set of formulae, but with $\alpha \wedge \beta$ instead of α and β, one places above each occurrence of α or β without square brackets in a top node another node labeled $\alpha \wedge \beta$ with the intervening deduction line labeled ∧ E. The resulting deduction tree is a deduction of γ from the same set of formulae used in the original deduction, without α and β, but with $\alpha \wedge \beta$ in their stead.

Now that we understand the validity of the left ∧ Introduction rule, let us see it and its partner, the right ∧ Introduction rule, applied.

Example 1 (Gentzen sequent calculus tree)

$$\cfrac{\cfrac{\cfrac{\cfrac{\cfrac{p \vdash p \quad q \vdash q}{p, q \vdash p \wedge q} \wedge R \quad r \vdash r}{p, q, r \vdash (p \wedge q) \wedge r} \wedge R}{p, q \wedge r \vdash (p \wedge q) \wedge r} \wedge L}{p \wedge (q \wedge r) \vdash (p \wedge q) \wedge r} \wedge L}$$

Classical Propositional Logic: Deduction

4.2 ↔ Left and Right Introduction

↔	Introduction L	Introduction R
	$\Gamma \cup \{\alpha \to \beta, \beta \to \alpha\} \vdash \gamma$	$\Gamma \vdash \alpha \to \beta,\ \Delta \vdash \beta \to \alpha$
	$\Gamma \cup \{\alpha \leftrightarrow \beta\} \vdash \gamma$	$\Gamma \cup \Delta \vdash \alpha \leftrightarrow \beta$

We ask readers to take the time to explain to themselves why the left ↔ Introduction rule is valid.

Example 5 (Gentzen sequent calculus tree)

$$\dfrac{p \vdash p \quad \dfrac{\dfrac{r \vdash r \quad q \to r, q \vdash q}{r, q \to r, r \to q \vdash q} \to L}{\dfrac{r, q \leftrightarrow r \vdash q}{} \leftrightarrow L}}{p, r, p \to (q \to r) \vdash q} \to L$$

Example 6 (Gentzen sequent calculus tree)

$$\dfrac{\dfrac{p \vdash p \quad q \to r \vdash q \to r}{p, p \to (q \to r) \vdash q \to r} \to L \quad \dfrac{p \vdash p \quad r \to q \vdash r \to q}{p, p \to (r \to q) \vdash r \to q} \to L}{p, p \to (q \to r), p \to (r \to q) \vdash q \leftrightarrow r} \leftrightarrow R$$

4.3 → Left and Right Introduction

→	Introduction L	Introduction R
	$\Gamma \vdash \alpha,\ \Delta \cup \{\beta\} \vdash \gamma$	$\Gamma \cup \{\alpha\} \vdash \beta$
	$\Gamma \cup \{\alpha \to \beta\} \cup \Delta \vdash \gamma$	$\Gamma \vdash \alpha \to \beta$

The left → introduction rule states that, if there are deductions of α from a set of formulae and of γ from a possibly different set of formulae which includes β, then there is a deduction of γ from the two sets of formulae, without β, but with $\alpha \to \beta$ instead.

Let us see how the left → introduction rule is equivalent to the → E rule of section 2.2.4. Suppose that there are two deductions in the formula tree format, one of α from a set of formulae, that is, from Γ, and another of γ from a set of formulae which includes β, that is, from $\Delta \cup \{\beta\}$. Suppose too, without loss of generality, that β is actually used in the latter deduction. To obtain a deduction in the formula tree format of the formula γ from

$\Gamma \cup \Delta \cup \{\alpha \to \beta\}$, one constructs a deduction tree whose terminal node is labelled by γ as follows. First, one uses the formula $\alpha \to \beta$ as an assumption and the deduction tree whose terminal node is labelled by α and which depends on assumptions in Γ to obtain a deduction tree whose terminal node is labelled by β and which depends on assumptions in $\Gamma \cup \{\alpha \to \beta\}$. The rule which justifies the transition to the terminal node of the new deduction tree is the rule of $\to$ E. Second, one replaces each occurrence of $[\beta]$ in the deduction tree whose terminal node is labelled by γ and which depends on assumptions in $\Delta \cup \{\beta\}$ with the deduction tree obtained in the first step. (Recall that its terminal node is labelled by β.) The result is a deduction tree whose terminal node is labelled by γ and which depends on assumptions in $\Gamma \cup \Delta \cup \{\alpha \to \beta\}$.

Example 2 (Gentzen sequent calculus tree)

$$\cfrac{p \vdash p \quad \cfrac{q \vdash q \quad r \vdash r}{q, q \to r \vdash r} \to L}{p, q, p \to (q \to r) \vdash r} \to L$$

Example 7 (Gentzen sequent calculus tree)

$$\cfrac{\cfrac{\cfrac{p \vdash p \quad \cfrac{q \vdash q \quad r \vdash r}{q, q \to r \vdash r} \to L}{p, q, p \to (q \to r) \vdash r} \to L}{p \wedge q, p \to (q \to r) \vdash r} \wedge L}{p \to (q \to r) \vdash (p \wedge q) \to r} \to R$$

4.4 ∨ Left and Right Introduction

∨	Introduction L	Introduction R	
	$\Gamma \cup \{\alpha\} \vdash \gamma \,,\, \Delta \cup \{\beta\} \vdash \gamma$	$\Gamma \vdash \alpha$	$\Gamma \vdash \beta$
	$\Gamma \cup \Delta \cup \{\alpha \vee \beta\} \vdash \gamma$	$\Gamma \vdash \alpha \vee \beta$	$\Gamma \vdash \alpha \vee \beta$

The left ∨ introduction rule states that, if there are deductions of γ in the formula tree format both from a set of formulae which includes α and from a set which includes β, then there is a deduction in the formula tree format from the same set of formulae, without α and β, but with $\alpha \vee \beta$.

Let us see how the left ∨ introduction rule is equivalent to the ∨ E rule of section 2.2.2. Suppose that there are two deductions of γ, one from a set of formulae which includes α that is, from $\Gamma \cup \{\alpha\}$, and another from a set which includes β, that is, from $\Delta \cup \{\beta\}$. Again,

Classical Propositional Logic: Deduction

suppose, without loss of generality, that α and β appear in their respective deductions. By $\to$ I of section 2.2.4, we know that there are two deductions in the formula tree format, one of $\alpha \to \gamma$ from Γ and another of $\beta \to \gamma$ from Δ. We now construct a deduction tree for γ using as undischarged premises formulae from $\Gamma \cup \Delta \cup \{\alpha \vee \beta\}$: place above γ the deduction tree for $\Gamma \vdash \alpha \to \gamma$, the deduction tree for $\Delta \vdash \beta \to \gamma$ and the formula $\alpha \vee \beta$ and label the intervening deduction line with $\vee$ I.

Example 3 (Gentzen sequent calculus tree)

$$\cfrac{\cfrac{p \vdash p \quad \cfrac{q \vdash q \quad s \vdash s}{q, q \to s \vdash s} \to L}{p, q, p \to (q \to s) \vdash s} \to L \quad \cfrac{p \vdash p \quad \cfrac{r \vdash r \quad s \vdash s}{r, r \to s \vdash s} \to L}{p, r, p \to (r \to s) \vdash s} \to L}{\cfrac{p, q \vee r, p \to (q \to s), p \to (r \to s) \vdash s}{p \wedge (q \vee r), p \to (q \to s), p \to (r \to s) \vdash s} \wedge L} \vee L$$

Example 4 (Gentzen sequent calculus tree)

$$\cfrac{\cfrac{p, q \vdash p}{p \wedge q \vdash p} \wedge L \quad \cfrac{\cfrac{p, q \vdash q}{p, q \vdash q \vee r} \vee R}{\cfrac{p \wedge q \vdash q \vee r}{} \wedge L}}{p \wedge q \vdash p \wedge (q \vee r)} \wedge R$$

4.5 ¬ Left and Right Introduction

As stated earlier, the two rules governing negation require a more general notion of sequent, one in which a set of formulae appear to the right of the turnstile. Since the counterparts to these rules play no role in this book, we shall state the two rules here for readers who are curious to see what these rules look like and say nothing more.

¬	Introduction L	Introduction R
	$\Gamma \vdash \{\alpha\} \cup \Delta$	$\Gamma \cup \{\alpha\} \vdash \Delta$
	$\Gamma \cup \{\neg \alpha\} \vdash \Delta$	$\Gamma \vdash \{\neg \alpha\} \cup \Delta$

Exercises: Gentzen sequent calculus

1. Explain why the left $\leftrightarrow$ Introduction rule is valid.

2. Using the deduction rules of the Gentzen sequent calculus to establish the results in exercises 2 through 4 of section 2.1.

5 Substructural Logics

In section 1, we defined a sequent to be a relation between a set of formulae and a single formula. However, Gentzen thought of what is on the left side of the symbol $\vdash$, not as a name for a set of formulae, but as a finite list, which could be an empty list. Moreover, his axioms have the form $\gamma \vdash \gamma$. To attain a deduction relation equivalent to the one we have given for sequent deduction, Gentzen added to his deduction rules three so-called *structural rules*. In the statement of these rules that follows, the uppercase Greek letters denote a finite list of formulae that are possibly empty. Moreover, to stress the fact that we have lists, and not sets, we use no punctuation on the left side of the turnstile. Thus, the expression $\Delta\ \alpha\ \Theta$ denotes a list of formulae, which begins with the finite list of formulae denoted by Δ, possibly empty, followed by the formula α, followed by another finite, possibly empty, list of formulae, denoted by Θ.

Permutation	Contraction	Weakening
$\Delta\ \alpha\ \beta\ \Theta \vdash \gamma$	$\Delta\ \alpha\ \alpha\ \Theta \vdash \beta$	$\Delta \vdash \beta$
$\Delta\ \beta\ \alpha\ \Theta \vdash \gamma$	$\Delta\ \alpha\ \Theta \vdash \beta$	$\Delta\ \alpha \vdash \beta$

A list is an ordered set. As we explained in chapter 2, in list notation, a set is identified merely by the members named, without regard to order or repetition. The structural rules employed by Gentzen mean that a deduction holds without regard to the order of the formulae in the list, as a result of rule of permutation, and without regard to repetitions, as a result of contraction. The discrepancy between Gentzen's axioms, which have the form $\gamma \vdash \gamma$, and the axioms adopted here, which have the form the form $\Gamma \vdash \alpha$, where $\alpha \in \Gamma$, is bridged by the use of weakening, for it permits the addition of any formula to the premises, since in the weakening rule, α need not be in Δ.

All three structural rules are required for classical logic as well as for intuitionistic logic. Another logic, known as relevance logic, gives up weakening, but retains permutation and contraction. Still another logic, known as linear logic, gives up weakening and contraction and retains only permutation. The Lambek calculus (see chapter 13) drops all of these structural rules. These three logics, and others of similar inspiration, are known as *substructural logics*.

8 English Connectors

1 Introduction

In section 3.1 of chapter 1, it was stated that the fundamental question to be explored is this: how is the value of a complex expression of English determined by the values of the expressions making it up? It was also stated that some guidance in answering this question comes from logic, for that part of logic known as model theory asks the question: how is the value of complex formula determined by the values of the parts making it up? In sections 2.2.1 and 2.2.2 of chapter 6, we saw how a truth value assigned to a composite formula is determined by the truth values assigned to its immediate subformulae, and ultimately, by the truth values assigned to its atomic subformulae, or propositional variables.

The analogs to the propositional variables of CPL are the simple declarative sentences of English. Just as formulae can be assigned truth values, so declarative sentences are liable to being judged either true or false. Moreover, just as a composite formula has its truth value determined by the truth values of its immediate subformulae, so a compounded declarative sentence has its truth value determined by the truth values of the declarative clauses that are its immediate constituents. Consider, by way of an example, the compound declarative sentence *it is raining and it is cold*. It seems to be that we judge the entire sentence to be true when and only when we judge each of its immediate constituent declarative clauses to be true.

It is raining.	It is cold.	It is raining *and* it is cold.
T	T	T
T	F	F
F	T	F
F	F	F

But this is exactly the truth function $o_\wedge$ that interprets the propositional connective $\wedge$. It seems natural then to hypothesize that the truth function $o_\wedge$ characterizes at least part of the meaning of the English connector *and*.

Just as one cannot obtain a value for a complex expression of notation on the basis of an assignment of values to its basic symbols without knowing how the notation is structured, so one cannot obtain an interpretation of a compound declarative sentence of a natural language such as English on the basis of interpretations of its constituent clauses without knowing how the constituent clauses combine. For this reason, we must ascertain how connectors combine with other constituent clauses to form compound clauses or sentences. We start, then, with a review of the principal syntactic facts pertaining to English connectors and their clausal syntax.

2 English Connectors and Clauses

In English, it is possible to use two independent clauses to form a compound clause. For example, the two independent clauses

(1.1)   John plays the guitar.

(1.2)   Mary plays the piano.

can form a compounded independent clause of the kind illustrated next.

(2.1)   [S John plays the guitar] *moreover* [S Mary plays the piano].

(2.2)   [S John plays the guitar] *and* [S Mary plays the piano].

(2.3)   [S John plays the guitar] *when* [S Mary plays the piano].

Words that permit such compounding are called *connectors*. They include the *coordinators*—*and, or,* and *but*—and the *subordinators,* among which are counted such words as *when, because,* and *if.* Adopting terminology used in Quirk et al. (1985, chap. 8.134), we shall call connectors that are neither coordinators nor subordinators *conjuncts*. They include such words as *moreover* and *however*.

In general, English coordinators and subordinators do not intrude into the clauses they are connecting. In particular, they do not intrude into the second constituent clause. In this respect, they differ from other English connectors such as *moreover,* for, while *moreover* may be situated between the independent clauses it connects, as illustrated in sentence (3.1), it may also intrude into the second connected clause, as illustrated in (3.2). Indeed, connectors such as *moreover* can occur at the end of the second clause.

(3.1)   [S John plays the guitar] *moreover* [S his sister plays the piano].

(3.2)   [S John plays the guitar] [S his sister *moreover* plays the piano].

(3.3)   [S John plays the guitar] [S his sister plays the piano] *moreover*.

English coordinators and English subordinators, however, neither intrude into the second clause nor occur at its right edge.

English Connectors 323

(4.1) [S John plays the guitar] *and* [S his sister plays the piano].

(4.2) *[S John plays the guitar] [S his sister *and* plays the piano].

(4.3) *[S John plays the guitar] [S his sister plays the piano] *and*.

(5.1) [S John plays the guitar] *when* [S his sister plays the piano].

(5.2) *[S John plays the guitar] [S his sister *when* plays the piano].

(5.3) *[S John plays the guitar] [S his sister plays the piano] *when*.

No logical necessity requires that either coordinators or subordinators intrude into the second clause. For, in some languages, coordinators are prohibited from appearing between the two clauses, rather they must appear with some constituent of the second clause to their left. This is true, for example, of the coordinators *ca* (*and*), *vā* (*or*), and *tu* (*but*) of Classical Sanskrit.

(6.1) *[S Devadattaḥ vīṇām vādayati] *ca* [S Yajñadattaḥ odanam pacati].

(6.2) [S Devadattaḥ vīṇām vādayati] [S Yajñadattaḥ *ca* odanam pacati].

(6.3) [S Devadattaḥ vīṇām vādayati] [S Yajñadattaḥ odanam pacati] *ca*.

(6.4) Devadatta is playing the lute and Yajñadatta is cooking rice.

In still other languages, the connector may occur either initially in the clause or right after the clause's subject. This is the case, for example, in Chinese.

(7.1) [S *suīrán* wǒ xiǎng fā cái] kě shì bù gǎn mào xiǎn.

(7.2) [S wǒ **suīrán** xiǎng fā cái] kě shì bù gǎn mào xiǎn.

(7.3) Although I plan to get rich, I am not willing to take risks.
(Chao 1968, chap. 2.12.6)

Though English coordinators and subordinators are alike in their contrast with other connectors such as *moreover,* English coordinators and subordinators do not pattern uniformly. To begin with, coordinators may connect verb phrases, whereas subordinators may not.

(8.1) Dan [VP drank his coffee] *and* [VP left quickly]

(8.2) *Dan [VP did not drink his coffee] *when* [VP left quickly]

Another difference manifests itself in gapping. (See section 2.2 of chapter 4.) When two English clauses of parallel syntactic structure and sharing the same verb occur connected by a coordinator, the verb of the second clause may be omitted. This is not the case when the connector is a subordinator.

(9.1) Alice encouraged Beth, and Carl encouraged Dan.

(9.2) Alice encouraged Beth, and Carl __ Dan.

(10.1) Alice encouraged Beth, when Carl encouraged Dan.

(10.2) *Alice encouraged Beth, when Carl __ Dan.

Third, coordinators never occur one immediately after another. However, subordinators may.

(11.1) *John is unhappy *and but* he does what he is told.

(11.2) Bill left, *because if* he hadn't, he would have been in trouble.

(11.3) We don't need to worry about Carol, *because if when* she arrives we are not home, she can let herself in with the key I lent her.

Nonetheless, a coordinator may immediately precede a subordinator.

(12) Dan asked to be transferred, *because* he was unhappy *and because* he saw no chance of promotion.

Finally, while the English subordinator may occur at the beginning of the pair of connected clauses, the English coordinator may not.

(13.1) [S It is cold] *and* [S it is raining]

(13.2) *And* [S it is raining] [S it is cold]

(14.1) [S It is cold] *because* [S it is raining]

(14.2) *Because* [S it is raining] [S it is cold]

We summarize the foregoing in the following table.

	COORDINATORS	SUBORDINATORS
Intrusion	No	No
May connect verb phrases	Yes	No
Admits gapping	Yes	No
May iterate	No	Yes
Initial in compound clause	No	Yes

The foregoing observations show that subordinators and coordinators occur in different patterns. They also show that subordination and coordination give rise to different patterns of constituency. A coordinated clause consists of two clauses, one on either side of a coordinator. This structure is nicely depicted by the following synthesis tree:

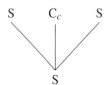

English Connectors 325

(where S is the category of a clause and C_c is that of a coordinator). It is convenient to specify such a structure with the constituency formation rule: S C_c S → S. This rule says that two clauses flanking a coordinator form a clause.

Confirming evidence for this rule comes from the ambiguity of compound declarative sentences that contain clauses coordinated by *and* and *or* as shown in the following discussion.

(15.0)   It is cold and it is overcast or it is windy.

Consider the circumstances in which it is false that it is cold but it is true that it is overcast and that it is windy. With respect to such circumstances, sentence (15.0) can be judged both true and false. It is judged true when it is understood as in (15.1), and false when it is understood as in (15.2).

(15.1)   Either it is cold and overcast or it is windy.

(15.2)   It is cold and it is either overcast or windy.

Whereas the structure of coordinated clauses can be characterized by one constituency formation rule, that of subordinate clauses requires three rules. This is because a subordinate clause may appear either to the right or to the left of the clause to which it is subordinate. These alternate patterns are shown by the pair of synthesis trees given next.

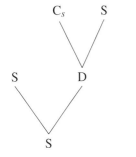 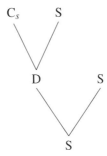

(where S is a clause, D is a subordinate clause, and C_s is a subordinator). These two synthesis trees can be specified with three constituency formation rules: S D → S, D S → S, C_s S → D. We regroup all the rules as follows.

CONSTITUENCY FORMATION RULES

S2   S C_c S → S
S3   S D → S
S4   D S → S
D1 C_s S → D

(where C_c is the lexical category of coordinators and C_s is the lexical category of subordinators.)

Exercises: English connectors and clauses

1. Consider the following list of English connectors:

after, although, as, before, but, nevertheless, once, since, so, therefore, though, unless, until, whereas, while, and *yet*

For each word in the list, use the criteria set out in section 2 to determine whether it is a coordinator, a subordinator, or neither.

2. Use the constituency formation rules given in section 2 to analyze the clausal structure of the following sentences (taken or adapted from Gamut 1991 v. 1, 40–41).

(a) John is going to swim, and unless it rains, Bill will too.

(b) If it rains while the sun shines, a rainbow will appear.

(c) If it rains or too many people are sick, the picnic will be canceled.

(d) If you do not help me if I need you, I shall not help you if you need me.

(e) Although once it rains, people will leave, while it is not raining, people will stay.

(f) If you stay with me if I won't drink anymore, then I won't drink anymore.

(g) We don't need to worry about Carol, because if when she arrives we are not home, she can let herself in with the key.

(h) Though while we are picnicing, everyone will be happy, before it starts or after it ends, everyone will be preoccupied.

3. For each of the following sentences, determine whether or not they are acceptable. State what bearing, if any, the acceptability or unacceptability has for the criteria for distinguishing between English coordinators and subordinators. State also what bearing, if any, the acceptability or unacceptability has on the constituency formation rules given in section 2.

(a) You may have coffee and or you may have tea.

(b) If it rains or if too many people are sick, the picnic will be canceled.

(c) He asked to be transferred, for he was unhappy and for he saw no chance of promotion.

(d) Because if when Carol arrives, we are not at home, she can let herself in with the key, we don't need to worry about Carol.

(e) For his sister plays the piano John plays the guitar.

(f) So that we arrived home late, the rush hour traffic delayed us.

English Connectors 327

4. Make up constituency formation rules to analyze these sentences.

(a) Yesterday the sun was very warm and the ice melted.

(b) Yesterday the sun was very warm and today the ice melted.

(c) Yesterday, the sun was very warm and the ice melted.

5. To which category does the word *nor* belong?

2.1 Open Problems for English Connectors

Not all of the observations made pertaining to coordinated and subordinated clauses can be accounted for only by the constituency formation rules given earlier. To begin with, some of the subordinators evince peculiarities not implied by the constituency formation rules. For example, a subordinate clause introduced by the subordinator *for* never precedes the clause to which it is subordinate.

(16.1) Rick purchased a suitcase, *for* he wishes to travel abroad.

(16.2) **For* he wishes to travel abroad, Rick purchased a suitcase.

This peculiarity is not unique to the subordinator *for;* sometimes, subordinate clauses introduced by the subordinator *so that* may precede the clause to which they are immediately subordinate. Thus, while the first sentence in (17) is acceptable, the second is not.

(17.1) The rush hour traffic delayed us *so that* we arrived home late.

(17.2) **So that* we arrived home late, the rush hour traffic delayed us.
        (Quirk et al. 1985, chap. 13.8)

As it happens, the subordinator *so that* is ambiguous. It may introduce a so-called purpose clause or it may introduce a so-called result clause. The first sentence in (18) contains a purpose clause introduced by *so that*. Sometimes purpose clauses introduced by *so that* can be paraphrased by infinitival clauses, which themselves can be introduced by *in order*. The first sentence in (19) contains a result clause. Result clauses introduced by *so that* can be paraphrased by the same sentence, except that the subordinator *so that* is replaced by *as a result of which*.

(18.1) Alice sold her stamp collection *so that* she could buy a car.

(18.2) Alice sold her stamp collection *in order* to buy a car.

(19.1) The rush hour traffic delayed us *so that* we arrived home late.

(19.2) The rush hour traffic delayed us, *as a result of which* we arrived home late.
        (ibid.)

Subordinate clauses introduced by *for* and by *so that* (resultative) are not the only subordinate clauses that may not precede their subordinating clauses, neither may a subordinated clause which itself has a subordinate clause to its left. (See section 3.4.4 of chapter 3.)

(20.1)   A rainbow appeared, *because* it rained *while* the sun was shining.

(20.2)   *Because* it rained *while* the sun was shining, a rainbow appeared.

(20.3)   **Because while* the sun was shining it rained, a rainbow appeared.

The next problem pertains to coordination. Coordinated clauses can be put together both with the coordinator *and* and without any coordinator whatsoever. Coordination brought about with a coordinator is known as *syndetic* coordination; coordination brought about with no coordinator is known as *asyndetic* coordination. In sentence (21.1), there are three clauses and two occurrences of the coordinator *and*. In sentence (21.2), the first occurrence of the coordinator has been omitted, resulting in asyndetic coordination.

(21.1)   The wind roared *and* the trees shook *and* the sky grew dark.

(21.2)   The wind blew, the trees shook *and* the sky grew dark.
(cf. Quirk et al. 1985, chap. 13.17)

Not all cases of simple juxtaposition of clauses are cases of asyndetic coordination. Consider this sentence:

(22)   The wind roared *and* the trees shook, the sky grew dark.

Sentence (22) is equivalent to the ones in (21). However, whereas the first two clauses in (21.2) are asyndetically coordinated, the syndetically coordinated clauses in (22) are not asyndetically coordinated with the last clause. Here is why: asyndetic coordination can occur in subordinate clauses; merely juxtaposed clauses cannot.

(23.1)   That the wind roared *and* the trees shook *and* the sky grew dark frightened everyone.

(23.2)   That the wind blew, the trees shook, *and* the sky grew dark frightened everyone.

(23.3)   *That the wind roared *and* the trees shook, the sky grew dark frightened everyone.

Readers are invited to explore how constituency formation rules fail to properly characterize asyndetic coordination.

Finally, while the connector *but* has many properties of a coordinator, unlike the coordinators *and* and *or*, which may occur repeatedly in a sequence of coordinated clauses, *but* may occur at most once and it must immediately precede the last clause in the sequence of coordinated clauses.

(24.1)   The wind roared *and* the trees shook *but* the sky did not grow dark.

(24.2)   *The wind roared *but* the house stood *but* the sky did not grow dark.

This fact cannot be handled by constituency formation rules.

Exercises: Open problems for the English adverb *not*

1. It was said that a subordinate clause whose subordinator is *for* may not precede the clause to which it is subordinate. Explain why the following sentence is not a counterexample.

   For travel abroad, Rick purchased a suitcase.

2. Formulate a constituency formation rule for asyndetic coordination and then provide counterexamples.

3. Show in detail how the constituency formation rules for coordination and the assignment of the category C_c to *but* lead to unacceptable expressions of English.

4. Describe the distribution of the coordinator *nor*.

3   Truth and Independent, Declarative Clauses

In chapter 6, we observed that a composite formula is obtained from its immediate subformulae through their combination with an appropriate propositional connective in accordance with a rule. For example, given the formulae $\neg p$ and $(q \vee r)$, one obtains the formula $(\neg p \wedge (q \vee r))$ by the rule that if $\alpha, \beta \in FM$, then $(\alpha \wedge \beta) \in FM$. In section 2.1, we observed that a coordinated clause is obtained from its constituent clauses through their combination with a suitable grammatical connector in accordance with a constituency formation rule. So, the independent clauses *it is raining* and *it is cold* combine with the coordinator *and*, according to the constituency formation rule S2 (viz. S C_c S $\rightarrow$ S), to yield the coordinated clause *it is raining and it is cold*.

We also observed in sections 2.2.1 and 2.2.2 of chapter 6 that a composite formula has its truth value determined by the truth values of its immediate subformulae. Thus, if the formulae $\neg p$ and $(q \vee r)$ are assigned T and F, respectively, and $\wedge$ is assigned $o_\wedge$, then by the definition of a classical valuation, $(\neg p \wedge (q \vee r))$ is assigned F. These formulae, in turn, have their truth values determined by their immediate subformulae. This continues until one arrives at the propositional variables out of which the entire formula is constructed. But where do these propositional variables get their truth values from? From truth value assignments.

As we have seen, if independent clauses such as *it is raining* and *it is cold* are judged true, then the coordinated clause *it is raining and it is cold* is judged true. This suggests that a coordinated clause, like a formula, receives its truth value through its structure from the truth values of its immediate constituent clauses. Setting out exactly how this is done is the subject matter of the remainder of this chapter.

One question that naturally arises at the outset is: how do simple clauses get their truth values? The answer is that they are determined by the circumstances of evaluation, those circumstances about which purport to express a fact. If what a clause expresses about a circumstance is the case, then the clause is true; otherwise, it is false. The circumstances with respect to which a clause is to be assessed for its truth or falsity are called its *circumstances of evaluation*. Those circumstances of evaluation that render a clause true are its *truth conditions*.

Truth conditions are that by dint of which one can form the counterparts of the logical concepts of satisfaction, satisfiability, semantic equivalence, and entailment for declarative sentences. Circumstances of evaluation satisfy a declarative sentence just in case they make it true, and a declarative sentence is satisfiable just in case there are circumstances of evaluation that satisfy it. A declarative sentences is, respectively, tautologous, contradictory, and contingent, just in case each circumstance of evaluation satisfies it, no circumstance of evaluation satisfies it, and some do and some do not. A pair of declarative sentences is semantically equivalent just in case the same circumstances of evaluation satisfy them. And a set of declarative sentences entails a single declarative sentence just in case whatever circumstances of evaluation satisfy the set of declarative sentences satisfy the single declarative sentence true. We stress that the concepts of satisfaction, satisfiability, tautology, contradiction, contingency, semantic equivalence, and entailment for natural language are distinct from, but analogous to, those for CPL. After all, the latter are explicitly defined for the notation of CPL, whereas the former will have to be defined for the declarative sentences of English.

4 The English Coordinator *and*

The English coordinator *and* bears striking similarities to the propositional connective $\wedge$. First, the truth of a clause that is formed from two independent clauses coordinated by *and* depends on the truth of its constituent clauses in the same way as the truth of a composite formula formed from two formulae connected by $\wedge$ does. Consider, for example, the two simple independent clauses: *it is raining* and *it is cold*. It is indisputable that the clause formed from these two independent clauses with the coordinator *and* is true under the circumstances in which both coordinated clauses are true and not otherwise.

It is raining.	It is cold.	It is raining *and* it is cold.
T	T	T
T	F	F
F	T	F
F	F	F

English Connectors

But this is exactly the truth function that interprets $\wedge$ of CPL, as can be seen from its truth table:

α	β	$\alpha \wedge \beta$
T	T	T
T	F	F
F	T	F
F	F	F

This truth function enjoys several logical properties. To begin with, it is commutative. Commutativity is also a property of the coordinator *and*.

(25.1) It is raining *and* it is cold.

(25.2) It is cold *and* it is raining.

(26.1) Mary studies at McGill *and* John studies at Concordia.

(26.2) John studies at Concordia *and* Mary studies at McGill.

In addition, just as one can infer either of the immediate subformula of a formula whose main connective is $\wedge$, so one can infer either independent clause of a pair of independent clauses forming a clause connected by the coordinator *and*. Thus, in CPL, the following hold:

(27.1) $\alpha \wedge \beta \models \alpha$

(27.2) $\alpha \wedge \beta \models \beta$

(27.3) $\{\alpha, \beta\} \models \alpha \wedge \beta$

Similarly, in English, the following also holds:

(28.1) it is raining *and* it is cold   ENTAILS   it is raining.

(28.2) it is raining *and* it is cold   ENTAILS   it is cold.

(28.3) {it is raining, it is cold}   ENTAILS   it is raining *and* it is cold

Finally, the structure of the categorematic tree diagram in which the propositional connective $\wedge$ occurs is the same as the structure of the constituency in which the coordinator *and* coordinates independent, declarative clauses. To see this, consider the following categorematic tree diagram (left) for the formula schema $\alpha \wedge \beta$. Replace each formula in it with the name of the set to which it belongs—namely, FM—and the propositional connective by the name of the set to which it belongs—namely, BC. This yields the middle tree. Next, replace the label FM with the label S and the label BC with the label C_c. One thereby obtains the synthesis tree below (right) corresponding to S2 (that is, S C_c S $\rightarrow$ S).

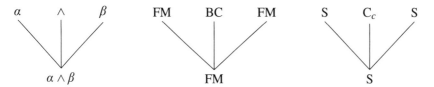

More precisely, the synthesis tree has the following form:

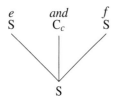

Let us now turn to how values are assigned in these structures. It should be evident that the same valuation rule, which applies to the categorematic tree diagram to assign a value to the mother node on the basis of values assigned to the daughter nodes, applies to the corresponding part of the preceding synthesis tree so that the top mother node acquires a value determined by the values assigned to its daughter nodes, as shown here:

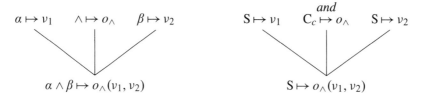

The constituency formation rule S C_c S → S characterizes the structure of the constituent. The value to be assigned to $and|C_c$ is $o_\wedge$.

S2 Constituency formation rule

If $e|S$, $f|C_c$, and $g|S$, then $efg|S$.

S2 Constituency valuation rule

Let a be an assignment of truth values to the basic clauses.

$v_a(efg|S) = v_a(f|C_c)(v_a(e|S), v_a(g|S))$.

In the case where the coordinator is *and*, the value assigned is $o_\wedge$: in other words, $v_a(\,and\,|C_c) = o_\wedge$.

4.1 Apparent Problems for the English Coordinator *and*

In this section, we shall explore subtleties of the coordinator *and*. In particular, we shall further investigate the hypothesis that the truth function $o_\wedge$ characterizes at least part of the meaning of the English coordinator *and* by investigating what appear to be counterexamples to the hypothesis. The point of this section is not just to alert readers to subtleties pertaining to the use of the coordinator but also to illustrate how one investigates hypotheses about the meaning of natural language expressions. We shall be applying, then, reasoning set out in section 3.2 of chapter 1.

A consequence that one might attribute to the previous hypothesis is that arbitrarily chosen independent clauses should be liable to coordination by *and*. However, such clausal coordination by *and* does not appear to have such freedom. Consider the following sentences.

(29.1)   *Two is an even number and the concept of intentionality is inexplicable.

(29.2)   *Six men are throwing stones at a frog and seven men can fit in a Ford.

   (Lakoff 1971, 117)

No doubt these sentences are odd. However, their oddity is not attributable to the hypothesis stated previously, for these sentences are judged equally odd when the same clauses are merely juxtaposed.

(30.1)   *Two is an even number; the concept of intentionality is inexplicable.

(30.2)   *Six men are throwing stones at a frog; seven men can fit in a Ford.

Thus, whatever renders sentences (29) odd is independent of the syntactic and semantic properties of the coordinator *and*.

In fact, a single, independent principle explains the oddity of both sentences in (29), which are coordinated by *and*, and of those in (30), which are merely juxtaposed. The explanation is that these sentences violate Grice's maxim of relevance, which requires that contributions to a conversation be relevant. (See section 2 of chapter 5.) In the absence of any special context, it is difficult to imagine to what topic the subject matter of both constituent clauses could be relevant.

Confirming evidence that Grice's maxim of relevance is apposite is the fact that once a suitable context is established for sentences such as those in (29) and (30), they lose their air of oddity. Imagine, for example, that the person uttering sentences (29) or (30) is a respondent in a quiz game and that he or she has been requested to state the first two facts that come to mind.

Violations of the maxim of relevance are not the only case where coordination by *and* may seem odd. Consider this sentence.

(31)   *Buddhists are vegetarians, and Buddhist men are vegetarians.

Notice again that the oddity is not peculiar to coordination by *and*. It appears, even when the syndetic form of coordination is replaced by mere juxtaposition.

(32)   *Buddhists are vegetarians; Buddhist men are vegetarians.

Here, the oddity is explained by another of Grice's maxims, namely, his maxim of quantity, which states that one's contribution to a conversation should be informative. As elaborated by Stalnaker (1978, 321–322), when one makes a series of statements, one is increasing information. If an earlier statement entails a later one, then the later one fails to increase the information already expressed, and hence it fails to be informative. In sentences (31) and (32), what is said in the first clause obviously entails what is said in the second. Hence, once the first clause has been uttered and understood, the second fails to make an informative contribution.

Thus far, we have seen how violations of Grice's maxims of relevance and quantity give rise to infelicitous coordinated independent clauses. We now turn to a third apparent consequence of the hypothesis and data that are incompatible with it.

As we have remarked, $o_\wedge$ is a commutative truth function. This means that two formulae whose main propositional connective is $\wedge$ and whose immediate sub formulae are the same, but differ from each other by the order of the immediate sub formulae, are equivalent, as shown here:

α	β	$\alpha \wedge \beta$	$\beta \wedge \alpha$
T	T	T	T
T	F	F	F
F	T	F	F
F	F	F	F

This appears to imply that, if the coordinator *and* has as part of its meaning the truth function for $\wedge$, then the transposition of two clauses coordinated by *and* in a coordinated clause will result in an equivalent coordinated clause. Yet, as a matter of fact, this does not always appear to be the case. Thus, sentences (33) can be obtained one from the other by a simple transposition of constituent clauses.

(33.1)   Peter owns a car and Mary does too.

(33.2)   *Mary does too and Peter owns a car.

Yet the first sentence is fine, while the second is decidedly unacceptable.

Again, it is easily shown that the unacceptability of the second sentence is independent of the occurrence of the coordinator *and*, for the transposition of the merely juxtaposed clauses in (33.1) gives rise to a sentence as odd as the one in (33.2).

(34.1)   Peter owns a car; Mary does too.

(34.2)   *Mary does too; Peter owns a car.

The explanation for the oddity of the sentences in (33.2) and (34.2) is obvious: there is no antecedent for the elliptical gap. As we saw in chapter 4, section 3.2, antecedents of elliptical gaps must precede the gap. This condition is satisfied by the sentences in (33.1) and (34.1). However, once the clauses in these sentences are transposed, the condition is no longer satisfied and the resulting sentences become ungrammatical.

The truth of this explanation is easily confirmed. With the elimination of the endophoric dependence, commutativity is restored.

(35.1)   Peter owns a car and Mary owns a car.

(35.2)   Mary owns a car and Peter owns a car.

(36.1)   Peter owns a car; Mary owns a car.

(36.2)   Mary owns a car; Peter owns a car.

Disruption of the conditions necessary to a phrase's serving as an antecedent for an endophoric element is not the only way in which commutativity can be inhibited. Consider this pair of sentences:

(37.1)   Carl has children, and all of Carl's children are asleep.

(37.2)   *All of Carl's children are asleep, and Carl has children.

Again, the problem does not lie with the coordinator *and*, for the same contrast in acceptability appears when the clauses in either sentence in (37) are merely juxtaposed.

(38.1)   Carl has children; all of Carl's children are asleep.

(38.2)   *All of Carl's children are asleep; Carl has children.

As we learned in section 3 of chapter 5, certain kinds of expressions carry with them presuppositions. In particular, a noun phrase having as an immediate constituent a possessive noun phrase carries with it the presupposition that the relationship by dint of which the noun phrase denotes is a fact. Thus, the noun phrase *Carl's children* presupposes that Carl has children. In sentences (37.1) and (38.1), the first clause states what the partitive noun phrase found in the second clause presupposes. Thus, the first clause ensures that the presupposition required by the second clause is in place. In sentences (37.2) and (38.2), the clauses are transposed. The result is that what is presupposed by the first clause is explicitly stated by the second. Now, either the presupposition carried by the first clause is in place for the hearer or, if it is not, the hearer has accommodated it. Either way, the second clause is redundant, violating Grice's maxim of quantity.

A third way in which commutativity can fail is illustrated by the next pair of sentences.

(39.1)   It rained, and it rained hard.

(39.2)   *It rained hard, and it rained.

While the first sentence is completely idiomatic, the second, which results from the first by a simple transposition of its coordinated clauses, is odd.

Again, one might be tempted to conclude that the oddity is a consequence of our assumptions regarding the coordinator *and;* however, it is easy to show that it is not. Indeed, sentence (39.2) is odd for the very same reason that sentences (31) and (32) are odd. In transposing the first clause in (39.1) with the second, one obtains a sentence in which the first clause implies the second, which violates Grice's maxim of quantity.

We now turn to a fourth way in which the coordinator *and* appears to fail to be commutative. Consider such pairs of sentences as:

(40.1) Robert died and he was buried in a cemetery.

(40.2) Robert was buried in a cemetery and he died.

(41.1) Evan heard an explosion and he telephoned the police.

(41.2) Evan telephoned the police and he heard an explosion.

The sentences in each pair are judged to be inequivalent. But, the truth function $o_\wedge$, which is hypothesized to be part of the meaning of the coordinator *and*, is commutative.

However, the failure of clauses to preserve equivalence under transpositon is not peculiar to clauses coordinated by the coordinator *and;* it also fails for the same clauses merely juxtaposed.

(42.1) Robert died; he was buried in the cemetery.

(42.2) Evan heard an explosion; he telephoned the police.

The explanation of this failure of transpositional equivalence lies not with our assumptions pertaining to the coordinator *and,* but with certain implicatures that arise from coordination. As is well known,[1] frequently associated with clauses coordinated by *and* are the construals of temporal and causal sequence. These construals are highlighted by appending *then* and *as a result,* respectively, to the coordinator *and*.

(43.1) Robert died and then he was buried in a cemetery.

(43.2) Robert died; then he was buried in a cemetery.

(44.1) Evan heard an explosion and as a result he telephoned the police.

(44.2) Evan telephoned the police; as a result, he heard an explosion.

Clearly, the order of the clauses conveys the temporal or causal order between what the clauses express; this temporal or causal order inhibits the commutativity of the clauses. After all, temporal and causal ordering is asymmetric. Transpose the clauses and one thereby conveys an inverted order, temporal or causal.

---

1. See Quirk et al. 1985, chap. 13.23–13.25

There is independent evidence for this explanation. In chapter 5, section 2.2, it was shown that implicatures are cancelable. Cancellation can even be done explicitly, without pain of contradiction. Thus, if a sentence is ambiguous and has as part of its immediate cotext a sentence whose literal meaning contradicts one of the literal meanings of the ambiguous sentence, then the contradiction is detectable; however, if a sentence has an implicature and has as part of its immediate cotext a sentence whose literal meaning contradicts the implicature, then the implicature is canceled without detectable contradiction.

We see that those aspects of sequentiality and resultativity associable with *and* can be explicitly canceled, without detectable contradiction.

(45.1) Robert died and he was buried in a cemetery—but not necessarily in that order.

(45.2) Evan heard an explosion and he telephoned the police—but not necessarily in that order.

Notice that, when this is done, commutativity is restored.

Another way to cancel implicatures, which is peculiar to clauses coordinated by *and*, is to pair them with the correlative *both*. We observed that *both* cannot be used when *and* coordinates unsubordinated, independent clauses; however, it can be used to coordinate subordinated, independent clauses:

(46.1) It is true both that Robert died and that he was buried in a cemetery.

(46.2) It is true both that Robert was buried in a cemetery and that he died.

Notice that, still again, commutativity is restored.

(47.1) It is true both that Evan heard an explosion and that he telephoned the police.

(47.2) It is true both that Evan telephoned the police and that he heard an explosion.

Finally, it should be noted that the mere juxtaposition of clauses is not the result of some rule deleting the coordinator *and,* for if it were, syndetic coordination and the mere juxtaposition should share all the same implicatures. But they do not. While sentences (48) carry the same implicature—namely, that the event expressed by the second clause is the result of the state expressed by the first—

(48.1) The road was icy: the car spun out of control.

(48.2) The road was icy and the car spun out of control.

sentences (49) do not.[2]

---

2. See Gazdar (1979, 4) as well as Bar-Lev and Palacas (1980) for further discussion.

(49.1) The car spun out of control: the road was icy.

(49.2) The car spun out of control and the road was icy.

The transposition of the clauses in (48.1), where the clauses are merely juxtaposed, leaves the implicature that the event resulted from the state, whereas their transposition in the sentence with its clauses coordinated syndetically, as in (48.2), cancels the implicature.

In this section, we have looked at a number of unacceptable sentences that might be thought to comprise data at odds with the hypothesis that truth function $o_\wedge$ characterizes at least part of the meaning of the English coordinator *and*. However, closer consideration of the data has shown that such sentences do not constitute counterexamples to the hypothesis. Rather, we have seen that the same sentences, where the coordinated clauses become merely juxtaposed sentences are equally unacceptable and that the unacceptability of both kinds of sentences can be explained by independently established principles either of grammar or of pragmatics.

Exercises: The English Coordinator *and*

The point of these exercises is to have readers bring to bear hypothesis testing, as discussed in section 3.2 of chapter 1, on the material covered in this section. The material covered in this section illustrates how this is done.

1. For each of the following sentences, determine whether its unacceptability militates against the hypothesis that the English coordinator *and* denotes the truth function for $o_\wedge$.

(a) *John is a strict vegetarian and he eats lots of meat.
(Lakoff 1971, 116, ex. 7)

(b) *Bill has a PhD in linguistics and he can read and write.
(Lakoff 1971, 124, ex. 30)

(c) *Alice is a vegetarian and Fred washed the clothes.

(d) *Cassius Clay eats apples and Muhammed Ali eats apples.

2. The following sentences are not judged to be perfectly equivalent. Why might such judgments be thought to be relevant to the hypothesis that the English coordinator *and* denotes the truth function for $o_\wedge$? Do you think, after careful reflection, that the judged inequivalence warrants rejecting the hypothesis?

(a) Bill owns a BMW and he owns a yacht.
     He owns a yacht and Bill owns a BMW.

(b) Bill washed the dishes and Carol dried them.
     Carol dried the dishes and Bill washed them.

(c) A wolf might get in and it would eat you.
     A wolf would eat you and it might get in.

English Connectors   339

(d) These three men share a condominium and each owns a BMW.
     Each owns a BMW and these three men share a condominium.

(e) I had suspected the solution was elusive, and I was right.
     I was right, and I had suspected the solution was elusive.
(adapted from Schmerling 1975, ex. 6a)

(f) That is what Bill says, and we all know how reliable he is.
     We all know how reliable Bill is, and that is what he says.
(Schmerling 1975, ex. 20)

(g) Joan sings ballads and she accompanies herself on guitar.
     Joan accompanies herself on guitar and she sings ballads.
(Schmerling 1975, ex. 7)

4.2 Open Problems for the English Coordinator *and*

We end this treatment of the coordinator *and* with brief discussions of four open problems, in each case the constituents coordinated by the coordinator *and* are not ones to which truth conditions can be attributed.

4.2.1 Coordination of nondeclarative clauses by *and*

First and foremost is the problem presented by coordinated, independent clauses in either the imperative or the interrogative mood. Such clauses do not have truth values. After all, the command *clean your room* is neither true nor false; nor is the question *is your room clean* either true or false. Yet both clauses in the imperative mood and clauses in the interrogative mood can be coordinated by *and*.

(50.1)   Clean your room and wash the car.

(50.2)   Did John eat his breakfast and did Mary take out the garbage?

However, the function $o_\wedge$ operates on a pair of truth values to return a truth value. If no truth value can be associated with clauses in either the imperative mood or the interrogative mood, then the truth function for $o_\wedge$ cannot be the precise meaning of the coordinator *and*, at least when it coordinates such clauses.

  What is its meaning when it occurs with such clauses? This question cannot be answered here, but a description of a promising approach can be sketched. We remarked that the truth of a declarative clause is determined by the circumstances with respect to which it is purported to express a fact. Such circumstances were dubbed the declarative clause's truth conditions. In a similar vein, one can speak of an imperative clause's *compliance conditions* and of an interrogative clause's *answerhood conditions*. An imperative clause's compliance conditions are the circumstances with respect to which the command expressed by the clause is said to have been complied with. Thus, the imperative clause *clean your room!* is said to be complied with in any circumstance in which the room of the person addressed is clean. An interrogative clause's answerhood conditions are the answers that can be said to have answered the question. An independent, interrogative clause such as *Did Mary take out the garbage?* is said to be answered by a statement entailing that Mary has taken out the garbage.

Compliance conditions and answerhood conditions permit us to see that the meaning the coordinator *and* contributes to the coordination of independent, imperative, and interrogative clauses is very similar to the meaning it contributes to the coordination of independent, declarative clauses. Recall that the circumstances that make a declarative clause coordinated by *and* true makes both constituent clauses true, and that the circumstances that make both constituent clauses true make the coordinated clause true. Similarly, the circumstances in which an imperative clause coordinated by *and* is complied with are circumstances in which both constituent clauses are complied with. And, the circumstances in which both constituent clauses are complied with are the circumstances in which the coordinated clauses are complied with. Finally, the responses that answer an interrogative clause coordinated by *and* are responses that answer both coordinated clauses, and responses that answer both constituent clauses are also answers to the coordinated clause.

Moreover, these notions of compliance conditions and answerhood conditions permit one to define a kind of entailment relation for clauses in the imperative mood and those in the interrogative mood. To distinguish these relations from the relation of entailment defined for clauses in the declarative, we shall call relation for commands *compliance entailment,* or *c-entailment* for short. A set of clauses in the imperative mood entails by compliance a single clause in the imperative mood, precisely if, when all the clauses in the imperative mood in the set are complied with, so is the single clause in the imperative mood. Similarly, we shall speak of the relation for clauses in the interrogative mood *answerhood entailment,* or *a-entailment* for short. A set of clauses in the interrogative mood entails by answerhood a single clause in the interrogative mood, precisely if, when all the clauses in the interrogative mood in the set are answered, so is the single clause in the interrogative mood.

One notices further that coordinated imperative and interrogative clauses exhibit entailments analogous to those noted for coordinated declarative clauses.

(51.1)   Clean your room *and* wash the car! C-ENTAILS Clean your room!

(51.2)   Clean your room *and* wash the car! C-ENTAILS Wash the car!

(52.1)   Did John eat his breakfast and did Mary take out the garbage?
         A-ENTAILS Did John eat his breakfast?

(52.2)   Did John eat his breakfast and did Mary take out the garbage?
         A-ENTAILS Did Mary take out the garbage?

Finally, the notions of compliance conditions and answerhood conditions permit one to define an equivalence between a pair of clauses in the imperative mood and a pair of clauses in the interrogative mood, respectively. A pair of imperative clauses are equivalent just in case they have the same compliance conditions. Two clauses in the interrogative mood are equivalent just in case they have the same answerhood conditions.

The characterization of the coordinator *and* in terms of compliance conditions and answerhood conditions implies that it is commutative. This is confirmed by examples such as those given next.

(53.1) Clean your room and wash the car.

(53.2) Wash the car and clean your room.

(54.1) Did John eat his breakfast and did Mary take out the garbage?

(54.2) Did Mary take out the garbage and did John eat his breakfast?

In short, we have seen that, just as independent declarative clauses have truth conditions, so independent imperative and interrogative clauses have compliance conditions and answerhood conditions, respectively. Appealing to these conditions, we see that independent imperative clauses or independent interrogative clauses, when coordinated by *and*, have properties, like independent declarative clauses, analogous to entailment and semantic equivalence.

Exercises: Open problems for the English coordinator *and*

1. Find apparent counterexamples similar to those found for the assumption pertaining to coordinated declarative clauses for coordinate imperative clauses and coordinated interrogative clauses. Verify that the same principles used to explain away the apparent counterexamples for coordinated, declarative clauses serve to explain away the apparent counterexamples for imperative and interrogative clauses.

4.2.2 Coordination of an imperative and a declarative clause by *and*

In section 4.2.1, it was suggested that independent imperative clauses have compliance conditions and independent declarative clauses have truth conditions. What kind of conditions do clauses compounded from an imperative clause and a declarative clause, as illustrated by the next pair of sentences, have?

(55.1) Go by air and you will save time.

(55.2) Join the Navy and you will see the world.

In other words, do the compounded clauses in (55) have truth conditions or compliance conditions or something else?

It might be the thought that, in spite of appearances, the second clausal constituent of each sentence is not a declarative clause, but an imperative one. Evidence for this comes from the fact that sentences (55) can be paraphrased by sentences (56).

(56.1) Go by air and save time.

(56.2) Join the Navy and see the world.

If so, then the compounded clauses have compliance conditions and our question is answered.

However, this attempt to reduce the problem of the coordination of an imperative clause with a declarative one to the problem of the coordination of two imperative clauses does not succeed, for it faces several insurmountable obstacles. First, the paraphrase of sentences (55) by sentences (56) is not sufficiently general. The form of the paraphrase of the second compounded declarative clause does not hold when the subject of the declarative clause is not the second person. For example, sentence (57.2) is not a paraphrase of (57.1).

(57.1) Give me some money and I shall help you escape.

(57.2) Give me some money and help you escape.

Second, two imperative clauses coordinated by *and* C-ENTAILS each of the clauses on their own, as shown in (51). The same cannot be said for compounded clauses of the kind in (56).

(58.1) Go by air and save time.

(58.2) Go by air.

(58.3) Save time.

Neither imperative clause in (58.1) C-ENTAILS either the clause in (58.2) or the clause in (58.3). In fact, it seems that sentences compounded from an imperative clause and a declarative clause do not have compliance conditions at all, but truth conditions. Consider sentence (59).

(59)   Move and I'll shoot.

The utterer is certainly not ordering the addressee to move! Rather, the utterer is ordering the addressee not to move. As a matter of fact, the sentence is rather well paraphrased by the following sentence:

(60)   If you move, then I shall shoot.

This paraphrase implies that sentences like (59) should be liable to *modus tollens*–like inferences, which it is.

(61)   A: Move and I'll shoot.
       B: Don't shoot!
       A: So, don't move!

English Connectors

Moreover, this paraphrase extends naturally to all the sentences comprising an imperative clause conjoined by *and* with a declarative one.

(62.1) If you go by air, you will save time.

(62.2) If you join the Navy, then you will see the world.

(62.3) If you give me money, I shall help you escape.

These paraphrases suggest, then, that *and* used to coordinate a clause in the imperative mood with one in the declarative mood does not have as part of its meaning the truth function $o_\wedge$, but rather the truth function $o_\rightarrow$. Whether *and* does indeed have the meaning of the truth function $o_\rightarrow$ or some other analysis can be provided in which *and* retains as its meaning the truth function $o_\wedge$, we leave as an open question.

4.2.3 Coordination of a noun phrase and a declarative clause by *and*

Another problem is the one that arises from the coordination of simple noun phrases with a declarative clause.

(63.1) Any noise and I shall shoot.

(63.2) Another beer and I'll leave.

These sentences, depending on the context of utterance, could be paraphrased as follows.

(64.1) If any noise is made, I shall shoot.

(64.2) If another beer is had, I shall leave.

Again, these coordinated sentences permit neither *and* elimination nor commutativity. Moreover, they seem paraphrasable by sentences in which the noun phrase forms part of the protasis and the second clause forms the apodosis of a conditional sentence.[3]

4.2.4 Coordination of phrases

Just now we saw that a noun phrase can be coordinated with a declarative clause. The problem is that while a declarative clause has truth conditions, noun phrases do not. Indeed, neither noun phrases nor adjective phrases nor prepositional phrases nor verb phrases have truth conditions. Yet each kind of phrase can be coordinated with another phrase of the same kind.

(65.1) Chunka [VP hit the ball] [C_c and] [VP ran to first base].

(65.2) My friend seemed [AP rather tired] [C_c and] [AP somewhat cross].

---

3. The *protasis* of a conditional is the subordinate clause introduced by the subordinator *if,* and the *apodosis* is the subordinating clause. Another pair of terms used to refer to such clauses are *antecedent* and *consequent,* respectively. we avoid these latter terms, since we have used the term *antecedent* to denote expressions that serve to fix the denotation of endophoric expressions.

(65.3)   [NP The man in the yellow hat] [C_c and] [NP the monkey] left in a car.

(65.4)   Bill remained [PP in the house] [C_c and] [PP on the telephone].

We shall return to this problem in chapter 14.

5   The English Coordinator *or*

Just as the English coordinator *and* bears striking similarities to the propositional connective $\wedge$, the coordinator *or* bears striking similarities to the propositional connective $\vee$. First, the truth of a clause, formed from two independent clauses coordinated by *or*, depends on the truth of its constituent clauses in the same way that the truth of a composite formula, formed from two immediate subformula connected by $\vee$, depends on the truth of its two immediate subformula. Consider again the two simple independent clauses: *it is raining* and *it is cold*. The clause formed from these two clauses with the coordinator *or* is false only under the circumstances in which the constituent clauses are both false and not otherwise.

It is raining.	It is cold.	It is raining *or* it is cold.
T	T	T
T	F	T
F	T	T
F	F	F

But this is just the function $o_\vee$, which is used to interpret the propositional connective $\vee$.

α	β	$\alpha \vee \beta$
T	T	T
T	F	T
F	T	T
F	F	F

Like $o_\wedge$, this truth function is commutative. And commutativity also seems to be a property of the coordinator *or*:

(66.1)   It is raining *or* it is cold.

(66.2)   It is cold *or* it is raining.

(67.1)   Mary studies at McGill *or* John studies at Concordia.

(67.2)   John studies at Concordia *or* Mary studies at McGill.

In addition, just as one can infer a composite formula whose main connective is $\vee$ from either of its immediate subformula,

English Connectors

(68.1) $\alpha \models \alpha \vee \beta$,

(68.2) $\beta \models \alpha \vee \beta$, and

(68.3) $\{\alpha \vee \beta, \alpha \rightarrow \gamma, \beta \rightarrow \gamma\} \models \gamma$.

so one can infer a clause coordinated by *or* from either of its constituent clauses.

(69.1) it is raining ENTAILS it is raining *or* it is cold.

(69.2) it is cold ENTAILS it is raining *or* it is cold.

(69.3) {either switch A is on or switch B is on; if switch A is on, the light is on; if switch B is on, the light is on} ENTAILS the light is on.

We also note that neither of the immediate subformulae of a formula whose main connective is $\vee$ entails the formula.

(70.1) $\alpha \vee \beta \not\models \alpha$,

(70.2) $\alpha \vee \beta \not\models \beta$

The same holds for declarative clauses coordinated by *or*:

(71.1) it is raining *or* it is cold NOT ENTAIL it is raining.

(71.2) it is raining *or* it is cold NOT ENTAIL it is cold.

Finally, since the propositional connective $\vee$ is a binary one and since *or* is a coordinator, the clause formation and valuation rule S2 applies here, where $o_\vee$ is assigned to *or*, that is, $v_a(or \mid C_c) = o_\vee$.

5.1 Apparent Problems for the English Coordinator *or*

In this section, we shall explore subtleties of the coordinator *or*. Once again, the aim is not only to bring to readers, attention subtleties pertaining to the use of the coordinator but also to illustrate how one goes about investigating hypotheses about the meaning of natural language expressions, in this case, the hypothesis that the truth function $o_\vee$ characterizes at least part of the meaning of the English coordinator *or*. We begin with a problem initially raised by the Stoic philosophers, over two thousand years ago, that of the apparent exclusivity of the coordinator *or*.

5.1.1 Apparent exclusivity of *or*

Though all participants in the controversy have agreed that a sentence comprising a pair of clauses connected by *or* is false when both of the connected clauses are false,

(72.1) Either it is raining or it is cold.

(72.2) Either George is a doctor or George is a lawyer.

(72.3) Either your car has been stolen or it has been moved.

many have thought that *or* must be given a distinct meaning in sentences comprising a pair of clauses connected by *or* in which both clauses cannot be true together. Examples of *or* having this latter meaning are given here:

(73.1) Either it is Monday or it is Tuesday.

(73.2) Either Dan is in San Francisco or Dan is in Boston.

This meaning is referred to as the *exclusive* meaning of the word *or,* and it is opposed to its so-called *inclusive* meaning, which permits the compound clause to be true when both coordinated clauses are true.

As frequently cited as sentences such as those in (73) are, they are not evidence of any ambiguity in the word *or* between an exclusive and an inclusive meaning. The reason is that the construal of exclusivity is clearly attributable to the states of affairs expressed by the coordinated clauses. Thus, no day can be both Monday and Tuesday, and no one can be in San Francisco and in Boston simultaneously.

A more convincing candidate for exclusivity is one that comes originally from Alfred Tarski (1941, 21). He invites his reader to consider a father making the following promise to his son:

(74) Either we are going on a hike or we are going to the theater.

Tarski maintains that the father can be deemed to have thereby excluded the possibility of taking the son for a hike in the morning and to the theater in the afternoon.

The question is: is the construal of exclusivity what is actually said or is it merely what is conveyed? That is, is the construal of exclusivity part of the meaning of the coordinator *or* or is it an implicature of the coordinator?

The evidence supports the view that the construal is an implicature, for it is defeasible.

(75) Either we are going on a hike *or* we are going to the theater.
      In fact, we are going to do both.

The exclusivity is canceled by the follow-up sentence.

Another kind of example adduced to support the conclusion that the coordinator *or* is ambiguous between an inclusive meaning and an exclusive meaning are sentences involving permission.

(76) Either you may have coffee *or* you may have tea.

It is claimed that, in most settings—say, when uttered by a waiter or waitress in a restaurant—one understands the alternatives to be exclusive. However, we observe that, once again, the construal of exclusivity is defeasible.

(77) Either you may have coffee *or* you may have tea.
      Indeed, you may have both.

There is another consideration that militates against the view that the coordinator *or* is ambiguous between an exclusive and an inclusive meaning. Recall that the best indication that an expression is ambiguous is that one can, with respect to a fixed circumstance of evaluation, both truly affirm and truly deny a declarative sentence containing it. Thus, with respect to the circumstances in which Albert has one desk and in that desk he keeps a folder containing papers of various kinds but he does not keep in his desk any instrument with a series of small ridges used for smoothing surfaces, one can both truly affirm and truly deny the sentence

(78)   Albert has a file in his desk.

Indeed, one can both truly affirm and truly deny the sentence

(79)   A file is not a file.

It is judged unfailingly false, when one takes each occurrence of the word *file* to have the same meaning, but it can be judged true, when one takes each occurrence to have a different meaning.

Now, let us suppose that the coordinator *or* has two meanings, an inclusive one, given by the truth function $o_\vee$, and an exclusive one, given the truth function o_+.[4]

α	β	$\alpha + \beta$
T	T	F
T	F	T
F	T	T
F	F	F

Now consider one of the sentences in (72), say the second sentence, and evaluate it with respect to the circumstances in which George is both a doctor and a lawyer. Then, if the coordinator *or* is ambiguous between a meaning in which it denotes the truth function $o_\vee$ and another in which it denotes the truth function o_+, one should be able to judge the sentence true, insofar as *or* denotes $o_\vee$, and false, insofar as the coordinator *or* denotes o_+. Or consider the third sentence and evaluate it with respect to the circumstances in which a thief has driven off with the car. In that case, the car has been both stolen and moved. If the coordinator *or* is ambiguous between an inclusive and an exclusive meaning, then on the exclusive meaning one should be able to judge sentence (72.3) as false, since both clauses are true. But no such judgment is made.

In fact, if the coordinator *or* is ambiguous between an exclusive and an inclusive meaning, one should be able to understand the following sentence as expressing a

---

4. I have used the symbol +, since this is the symbol often used to denote symmetric difference, the set theoretic counterpart to this function.

contradiction, for it is impossible for the constituent clauses to have different truth values. Hence, on the exclusive meaning of the coordinator *or*, it is a contradiction.

(80)   Either it is raining *or* it is raining.

But this sentence is not judged to express a contradiction.

A further argument against the coordinator *or* having an additional exclusive meaning is one due to Ray Jennings (1994). As he points out, a composite formula all of whose connectives are $+$ is false if and only if an even number of the atomic subformula is true. Thus, should *or* have an exclusive meaning, then sentence (81) would be judged false, if Dan happens not to be a lawyer and he is both a doctor and an accountant.

(81)   Either Dan is a doctor or he is a lawyer or he is an accountant.

But, in fact, a sentence such as the one in (81) is judged true, even if two of its clauses are judged true.

5.1.2   Apparent noncommutativity of *or*

Like the connective $\wedge$, the connective $\vee$ is commutative.

α	β	$\alpha \vee \beta$	$\beta \vee \alpha$
T	T	T	T
T	F	T	T
F	T	T	T
F	F	F	F

And, like the coordinator *and*, the coordinator *or* gives rises to sentences that are judged inequivalent to those that arise from the mere transposition of their coordinated clauses. We begin with the puzzle presented by the following pair.

(82.1)   Dan is laughing or Dan appears to be laughing.

(82.2)   *Dan appears to be laughing or Dan is laughing.

We noted in the previous section that the coordinator *and* can be used to add a clause that strengthens its preceding clause. The coordinator *or* can be used to add a clause to weaken, or even to correct, its preceding clause. This usage can be highlighted by the addition of the focus word *rather*.

(83.1)   Dan ran to the store or he walked.

(83.2)   Dan ran to the store, or rather he walked.

The failure of commutativity here is clearly ascribable to this special usage. While it makes perfect discourse sense to salvage the truth of a false or inaccurate clause by appending to it with the coordinator *or* a true one, it makes no discourse sense to append to a true clause a false one with *or*.

Another case where commutativity of the coordinator *or* is inhibited is illustrated by the two pairs of sentences given next.

(84.1)   *Either* little Seymour eats his dinner *or* his mother complains to her neighbors.

(84.2)   *Either* little Seymour's mother complains to her neighbors *or* little Seymour eats his dinner.

(85.1)   They must have liked the apartment *or* they would not have stayed so long.

(85.2)   They would not have stayed so long *or* they must have liked the apartment.

The failure of commutativity here is a more subtle matter. We are not in a position here to adduce all of the evidence; however, the following will make the proposed treatment plausible.

To begin with, observe that *or* can be paraphrased by *or else* (see *Webster's Third New International Dictionary,* s.v. ^1or 3).

(86.1)   *Either* little Seymour eats his dinner, *or else* his mother complains to her neighbors.

(86.2)   They must have liked the apartment, *or else* they would not have stayed so long.

Next, observe that the adverb *else* is an endophor whose antecedent can be a preceding clause coordinated by the coordinator *or*. This usage can be made more explicit by the gloss *if not* (see *Webster's Third New International Dictionary,* s.v. ^1else 2a).

(87.1)   *Either* little Seymour eats his dinner, *or if not*, his mother complains to her neighbors.

(87.2)   They must have liked the apartment, *or if not*, they would not have stayed so long.

Finally, as the sentences in (88) show, the protasis in a conditional is sometimes understood as expressing a reason for what is expressed by its apodosis.

(88.1)   If you put the baby down, Bill will scream.

(88.2)   If Alice does not leave, Bill will scream.

This construal is an implicature, though we shall not try to establish that here.

Combining the foregoing, we conclude that the coordinator *or* implicates a tacit protasis for the second clause whose content is the negation of the preceding clause, whereby arises a further implicature that a failure of what is expressed in the first clause is a reason for the occurrence of what is expressed in the second.

(89.1) Either little Seymour eats his dinner *or,* if he does not (eat his dinner), his mother complains to the neighbors.

(89.2) Either little Seymour's mother complains to the neighbors *or,* if she does not (complain to the neighbors), little Seymour eats his dinner.

(90.1) They must have liked the apartment, *or* if they had not liked the apartment, they would not have stayed so long.

(90.2) They would not have stayed so long, *or* if they had stayed so long, they must have liked the apartment.

Exercises: The English coordinator *or*

1. Is the formula $\alpha + \beta$ semantically equivalent to $\alpha \leftrightarrow \neg\beta$? Prove your answer.
2. Compute the truth tables for $\alpha \vee (\beta \vee \gamma)$ and $\alpha + (\beta + \gamma)$.
3. Is the formula $\alpha \vee \beta$ semantically equivalent to $\alpha \vee (\neg\alpha \to \beta)$? Prove your answer.

5.2 Open Problems for the English Coordinator *or*

The open problems for the coordinator *or* include not only problems similar to the open problems for the coordinator *and* but also an additional problem. We discuss the similar problems first and conclude with an exposition of the additional problem.

5.2.1 Coordination of nondeclarative clauses by *or*

To begin with, imperative and interrogative clauses can be coordinated by *or*.

(91.1) Clean your room or wash the car.

(91.2) Did John eat his breakfast or did Mary take out the garbage?

However, as we noted earlier, imperative and interrogative clauses do not have truth conditions. Thus, $o_\vee$ cannot be the meaning of *or* in such clauses.

We also noted that imperative clauses have compliance conditions and interrogative clauses have answerhood conditions. Just as in the case of the coordinator *and,* compliance conditions and answerhood conditions permit us to see *or*'s meaning in imperative and interrogative clauses as analogous to its meaning in coordinating declarative clauses. Thus, we noted that if either of two clauses coordinated by *or* is true, then the coordinated clause is true, and that if both are false, then the coordinated clause is false. In the same way, if either of two imperative clauses are complied with, then the coordinated clause is complied with; if neither is complied with, then the coordinated clause is not complied with either. And, if either of two interrogative clauses are answered, then the coordinated clause is answered, and if neither is, then the coordinated clause is not either.

English Connectors

Using the kinds of entailment relations defined in section 4.2.1 for compliance conditions and answerhood conditions, we see that a clause comprising two imperative clauses coordinated by *or* is entailed by either of its components.

(92.1) Clean your room! C-ENTAILS Clean your room *or* wash the car!

(92.2) Wash the car! C-ENTAILS Clean your room *or* wash the car!

In the same way, a clause comprising two interrogative clauses coordinated by *or* is entailed by either of its components.

(93.1) Did John eat his breakfast? A-ENTAILS
Did John eat his breakfast or did Mary take out the garbage?

(93.2) Did Mary take out the garbage? A-ENTAILS
Did John eat his breakfast or did Mary take out the garbage?

It is important to note that the converse entailments do not hold.

(94.1) Clean your room *or* wash the car! NOT C-ENTAIL
Clean your room!

(94.2) Clean your room *or* wash the car! NOT C-ENTAIL
Wash the car!

(95.1) Did John eat his breakfast or did Mary take out the garbage?
NOT A-ENTAIL Did John eat his breakfast?

(95.2) Did John eat his breakfast or did Mary take out the garbage?
NOT A-ENTAIL Did Mary take out the garbage?

Finally, the characterization of the coordinator *or* in terms of compliance conditions and answerhood conditions implies that it is commutative. This is confirmed by examples such as the following:

(96.1) Clean your room or wash the car.

(96.2) Wash the car or clean your room.

(97.1) Did John eat his breakfast or did Mary take out the garbage?

(97.2) Did Mary take out the garbage or did John eat his breakfast?

5.2.2 Coordination of an imperative and a declarative clause by *or*

Just as an imperative clause can be coordinated with a declarative one by *and,* so it can be coordinated with one by *or*.

(98.1) Move or you will die.

(98.2) Move or I'll shoot.

(98.3) Move or he will shoot.

As the second and third sentences of (98) show, the declarative clause need not be in the second person. Hence, such sentences cannot be reduced to the coordination by *or* of two independent, imperative clauses.

However, it is possible to convert the first clause in each sentence into a declarative clause and thereby obtain a good paraphrase of each.

(99.1) You move or you will die.

(99.2) You move or I'll shoot.

(99.3) You move or he will shoot.

These paraphrases fit well with the fact that the sentences in (98) give rise to inferences analogous to an inference form of CPL known as *disjunctive syllogism*: $\{\alpha \vee \beta, \neg \beta\} \models \alpha$.

(100) A: Move or I'll shoot.
      B: Don't shoot!
      A: So, move!

5.2.3 Coordination of phrases by *or*

Just as with the coordinator *and,* the coordinator *or* may coordinate a pair of phrases of the same kind—a noun phrase with a noun phrase, a verb phrase with a verb phrase, a prepositional phrase with a prepositional phrase, or an adjective phrase with an adjective phrase—and, it may coordinate a noun phrase with a declarative clause.

(101.1) Bill [VP sold his car] [C_c or] [VP bought another one].

(101.2) Carl appeared [AP somewhat annoyed] [C_c or] [AP somewhat surprised].

(101.3) [NP The host] [C_c or] [NP his guest] went upstairs.

(101.4) Doris hid [PP in the basement] [C_c or] [PP in the attic].

And, not surprisingly, as with *and,* it is possible to connect a simple noun phrase with a declarative clause.

(102) Your money or I'll shoot.

5.2.4 Apparent synonymity of *or* and *and*

Beyond the problems raised by the coordinator *or* that are analogous to those raised by the coordinator *and, or* raises a problem of its own. We observed that a clause comprising two declarative clauses coordinated by *or* does not entail either of the constituent clauses.

English Connectors 353

(103.1)   it is raining *or* it is cold NOT ENTAIL it is raining.

(103.2)   it is raining *or* it is cold NOT ENTAIL it is cold.

However, this does not always seem to be the case. Sentence (104.0), in which *or* coordinates two declarative clauses, does seem to entail each of its constituent clauses.

(104.0)   Either you may have coffee *or* you may have tea.

(104.1)   ENTAILS You may have coffee.

(104.2)   ENTAILS You may have tea.

Though no solution to this puzzle is known, the fact that the implications are defeasible suggests that what appear to be entailments are in fact implicatures.

(105.0)   Either you may have coffee *or* you may have tea.
          But I don't remember which.

(105.1)   NOT ENTAILS You may have coffee.

(105.2)   NOT ENTAILS You may have tea.

Exercises: Open problems for the English coordinator *or*

1. Discuss the noncommutativity of the following sentences:

(a) Give me liberty *or* give me death.

(b) Don't be too long *or* you will miss the bus.

2. Which binary truth function is best assigned to the English connector *nor*?

6   The English Subordinator *if*

We have discussed the two English coordinators *and* and *or*. We now turn to the English subordinator *if*. We have already shown that coordinated clauses have a different syntactic structure from subordinated clauses. But syntactic structure is not the only way in which coordinated and subordinated clauses differ from one another. Subordinate clauses and their subordinating clauses are sometimes limited in the verb forms they admit.

We begin with one of the peculiarities of the verb of a subordinate clause in the indicative mood, when the verb of its subordinating clause is in the indicative mood as well. For the sake of comparison, consider subordinating and subordinate clauses whose verbs are in the past tense of the indicative mood.

(106.1)   Aaron *returned* the keys [S before he *left*].

(106.2)   [S While Fred *was washing*], we *prepared* dinner.

(106.3)   [S If it *rained*] the ground *was* wet.

Now, should the verb of the subordinating clause be in the future tense, the verb of the subordinate clause is usually in the present tense, even if it is understood, like the main clause, to be picking out a state of affairs future to the time of utterance.

(107.1) Aaron *will return* the keys [S before he *leaves*].

(107.2) [S While Fred *is washing*], we *will prepare* dinner.

(107.3) [S If it *rains*] the ground *will be* wet.

Let us confine our attention, now, to subordinate clauses whose subordinator is *if*. We shall call sentences comprising a main clause and a subordinate clause whose subordinator is *if conditional* (*clauses*). The subordinate clause will be called the *protasis* and the main clause the *apodosis*.

Sentences (108) are two conditional sentences that differ only in the form of the verb in the apodosis: one has the present tense form of the modal *will* and the other its past tense form *would*.

(108.1) [S If it *rained*] the ground *will be* wet.

(108.2) [S If it *rained*] the ground *would be* wet.

The form *rained* in (108.1) is in the past tense of the indicative mood; the form *rained* in (108.2) is in the past tense of the subjunctive mood. In fact, the past tenses in the two moods pretty much coincide, except in the case of the verb *to be,* where *were* replaces *was*.

(109.1) [S If I *were* you], I *would leave*.

(109.2) [S If Richard *were* here], Beth *would be* upset.

Having distinguished indicative from subjunctive conditionals, let us turn our attention to the meaning of *if* in indicative conditionals. We shall return to the meaning of *if* in subjunctive conditionals later.

The meaning of *if* in indicative clauses corresponds fairly well to the truth function $o_\rightarrow$, though, as we shall see, the correspondence is not perfect. This is illustrated by sentence (110), which readers should imagine to be uttered by an instructor to a student in his or her class.

(110) If you pass the final, you pass the course.

Under what circumstances would the instructor be deemed not to have kept his or her word? Precisely under the circumstances where the student passes the final, but the instructor fails him. Under the assumption that the instructor has kept his or her promise just in case he or she has not failed to keep it, then the following truth value assignments properly characterize the use of *if* in sentence (110).

You pass the final.	You pass the course.	*If* you pass the final, you pass the course.
T	T	T
T	F	F
F	T	T
F	F	T

But this is exactly the truth function that interprets → of CPL, as can be seen from its truth table:

α	β	α → β
T	T	T
T	F	F
F	T	T
F	F	T

Moreover, two of the most well-known rules of entailment for → are *modus ponens* and *modus tollens*:

MODUS PONENS $\{\alpha \to \beta, \alpha\} \models \beta$.

MODUS TOLLENS $\{\alpha \to \beta, \neg\beta\} \models \neg\alpha$.

Their counterparts hold for indicative conditionals of English. Thus,

(111.1) {if it is raining, it is cold, it is raining} ENTAILS it is cold.

(111.2) {if it is raining, it is cold, it is not cold} ENTAILS it is not raining.

These observations hold no less for sentence (110). Thus, suppose an instructor promises a student what is said in (110) and the student passes the final. Then, on pain of breaking his or her promise, the instructor must pass the student in the course. Moreover, should the student not pass the course, then the student must not have passed the final.

Still further evidence that the truth function $o_\to$ provides the meaning of the subordinator *if* arises from its fit with the hypotheses adopted with regard to the coordinators *and* and *or*. To show this, we must adopt an auxiliary assumption that *not* has its meaning given by the truth function $o_\neg$.

To begin with, we note the equivalence of $\alpha \to \beta$ and $\neg(\alpha \wedge \neg\beta)$.

α	β	α → β	¬(α ∧ ¬β)
T	T	T	T
T	F	F	F
F	T	T	T
F	F	T	T

This equivalence, together with the hypothesis that the meaning of *and* is given by the truth function $o_\wedge$ and auxiliary assumption that an English clause containing *not* can be adequately treated by the application of the truth function $o_\neg$ to the truth value of the clause containing it, implies that the following pair of sentences should be judged equivalent.

(112.1) If London is in China, I am a monkey's uncle.

(112.2) It is not the case both that London is in China and that I am not a monkey's uncle.

Another equivalence is that of $\alpha \to \beta$ and $\neg \alpha \vee \beta$.

α	β	$\alpha \to \beta$	$\neg \alpha \vee \beta$
T	T	T	T
T	F	F	F
F	T	T	T
F	F	T	T

Again, this equivalence, together with the auxiliary assumption regarding *not* and the hypothesis that the meaning of *or* is given by the truth function $o_\vee$, implies that the next pair of sentences be judged as equivalent.

(113.1) If London is in China, I am a monkey's uncle.

(113.2) Either London is not in China or I am a monkey's uncle.

And, indeed, they are so judged.

A final equivalence to be noted is that of a formula whose main connective is $\to$ and its contrapositive:

α	β	$\alpha \to \beta$	$\neg \beta \to \neg \alpha$
T	T	T	T
T	F	F	F
F	T	T	T
F	F	T	T

Once again we note that the following corresponding English sentences are judged equivalent.

(114.1) If London is in China, I am a monkey's uncle.

(114.2) If I am not a monkey's uncle, then London is not in China.

6.1 Apparent Problems for the English Subordinator *if*

Many objections have been raised to the hypothesis the truth function $o_\rightarrow$ corresponds well to the meaning of the subordinator *if* in indicative conditionals. Some of the objections are more serious than others. Here we shall consider only a few objections that seem susceptible of a plausible explanation. Later, we shall turn to deeper objections.

6.1.1 Some apparent nonequivalences

Let us return to sentence (110) and consider it in light of the equivalences discussed thus far. The first equivalence we observed applies perfectly.

(115.0)   If you pass the final, you pass the course.

(115.1)   It is not the case both that you pass the final and that you do not pass the course.

The second equivalence seems less clear.

(115.0)   If you pass the final, you pass the course.

(115.2)   Either you do not pass the final or you pass the course.

Notice that the difficulty is not that one judges the sentences to be inequivalent, rather one has difficulty in seeing the equivalence. Yet, the equivalence is clear when the clauses in sentence (115.2) are commuted.

(116)   Either you pass the course or you do not pass the final.

This fact suggests that the problem of the lack of clarity in one's judgment of the equivalence between sentences (115.0) and (115.2) is not a problem with the hypothesis regarding the meaning of *if*, but a problem arising from some other aspect of the sentence.

McCawley (1981, 221–222) noted that not all the equivalences corresponding to the logical equivalence of $\alpha \rightarrow \beta$ and $\neg \alpha \vee \beta$ are felicitous. In particular, he noted that while Grice's example of an equivalence

(117.1)   If Labour doesn't win the next election, there'll be a depression.

(117.2)   Either Labour will win the next election or there'll be a depression.

is perfectly acceptable, his own example, which removes the negation from the protasis, is not clearly equivalent.

(118.1)   If Labour wins the next election, there'll be a depression.

(118.2)   Either Labour won't win the next election or there'll be a depression.

Notice that in Grice's example, the first clause of the pair coordinated by *or* does not contain *not*, whereas McCawley's example, the first clause contains *not*. We note further that in sentence (115.2), *not* occurs in the first clause. When the clauses are commuted, as in sentence (116), the initial clause no longer contains *not* and the equivalence to sentence (115.0) is restored. Further confirmation that the culprit is the occurrence of *not* in the first clause comes from the following paraphrases of sentences (115.0) and (115.2) where *not* is again eliminated from the first clause of the clauses connected by *or*.

(119.1) If you do not fail the final, you pass the course.

(119.2) Either you fail the final or you pass the course.

Why should *not* make a difference? This is an open question that we leave to readers to ponder.

Let us turn to the equivalence of a conditional and its contrapositive.

(115.0) If you pass the final, you pass the course.

(115.3) If you do not pass the course, then you do not pass the final.

Here again, the equivalence is unclear. Now, contraposition is not always blocked. Recall that it worked fine for sentences (9). Moreover, many other sentences permit contraposition:

(120.1) If Bill is in the car, then he is safe.

(120.2) If Bill is not safe, then he is not in the car.

Notice that the clauses in the sentences in (120) are not temporally ordered, whereas those in (115.0) and (115.4) are. The event expressed in the protasis of sentence (115.0) precedes the one expressed in the apodosis. Through altering the tenses in the clause, this effect can be countered.

(121.1) If you pass the final, you pass the course.

(121.2) If you do not pass the course, you will not have passed the final.

6.1.2 Subordination

We saw that the constituency of coordinated clauses is different from that of subordinated clauses. A coordinator has two sister clauses, while a subordinator has only one sister clause. Thus, when a binary truth function is assigned to the node C_c, its sister nodes supply the pair of truth values the function requires so that a truth value can be assigned to the mother node.

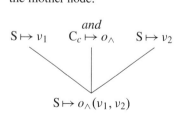

English Connectors

However, should a binary truth function be assigned to the node labeled C_s in the structure of a subordinate clause, its sister node supplies only one of the pair of truth values the function requires so that a truth value can be assigned to the mother node.

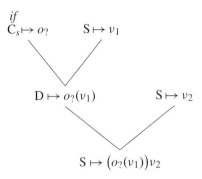

The solution is to find a truth function that is equivalent to $o_\to$ but can apply in the structure of a subordinate clause. The following shows how this might be done. We are going to consider two cases: the case where the protasis, or *if* clause, is assigned T and the case where it is assigned F. Let us begin with the case where the protasis is assigned T. Now the apodosis, or main clause, may be assigned either T or F. If the apodosis is assigned T, then the bottom S node must also be assigned T; if the apodosis is assigned F, then the bottom S node must be assigned F as well. In short, should the protasis be assigned T, then the value of the bottom S node is precisely the same as the value of the apodosis. This means that, should the protasis be assigned T, then the D node must be assigned the unary function that assigns T to T and F to F; that is, the D node must be assigned the unary identity truth function, which we shall designate o_i.

Next, let us turn to the case where the protasis is assigned F. We know, then, that the bottom node must be assigned T, no matter what is assigned to the apodosis. This means that the D node must be assigned the constant unary truth function that maps both T and F to T. We shall call this function o_t.

In sum, the subordinator *if* is interpreted by a function that maps T to the o_i and F to o_t. Readers should check that this interpretation of *if* works, regardless of whether the subordinate clause precedes or succeeds the main clause.

We now state the relevant formation and valuation rules. The first rule forms a subordinate clause and assigns it a value, based on the value assigned to the subordinator and its sister clause.

D1 Constituency formation rule
If $e|C_s$ and $F|S$, then $ef|D$.
D1 Constituency valuation rule
Let a be an assignment of truth values to the basic clauses.
$v_a(ef|D) = v_a(e|C_s)v_a(f|S)$.

Since a subordinate clause may either come before or go after a clause to form a more complex clause, we require two rules.

S3 Constituency formation rule
If $e|S$ and $f|D$, then $ef|S$.

S3 Constituency valuation rule
Let a be an assignment of truth values to the basic clauses.
$v_a(ef|S) = v_a(e|S)v_a(f|D)$.

S4 Constituency formation rule
If $e|D$ and $f|S$, then $ef|S$.

S4 Constituency valuation rule
Let a be an assignment of truth values to the basic clauses.
$v_a(ef|S) = v_a(e|D)v_a(f|S)$.

Exercises: The English subordinator *if*

The point of the first two exercises is to have readers bring to bear hypothesis testing, as discussed in section 3.2 of chapter 1, on the material covered in this section. The material covered in this section illustrates how this is done.

1. Consider the following sentence: *If it rained, it did not rain hard*. What problems, if any, does it raise for the hypothesis that $o_\rightarrow$ corresponds to the meaning of the subordinator *if*?

2. Explain why the following pair of sentences is thought to be problematic for the hypothesis that $o_\rightarrow$ corresponds to the meaning of the subordinator *if*:

- If its main switch is on and its safety switch is on, the machine runs.
- The machine runs, if its main switch is on, or the machine runs, if its safety switch is on.

3. Find functions that can be used to give the meaning of a subordinator corresponding to the following truth functions:

(a) $o_\wedge$

(b) $o_\vee$

4. Establish whether *unless* is a subordinator or a coordinator. Find a suitable paraphrase of sentences of the form α *unless* β to determine a binary truth function for *unless*. State whether such a binary truth function is suited to *unless*'s syntactic status as a subordinator or a coordinator. If the function is not suited, find one that is.

6.2 Open Problems for the English Subordinator *if*

We now turn to four open problems. The first pertains to the denial of sentences with the subordinator *if*. The second is the problem of how to treat sentences whose main clause is either in the imperative mood or in the interrogative and whose subordinate clause is one introduced by *if*. The third is the well-known problem of subjunctive conditionals, and the last is the problem posed by so-called Austinian conditionals.

6.2.1 Negation of conditionals

The problem of the denial of indicative conditionals was first raised by L. Jonathan Cohen (1971), reacting critically to Grice's theory of implicatures. Cohen observed that, while the following entailment,

(122) $\neg(\alpha \to \beta) \models \alpha$,

holds in logic, the corresponding entailment in English,

(123)   It is not the case that, if God is dead, everything is permitted ENTAIL God is dead.

does not hold. To date, there has been no satisfactory treatment of this problem under the hypothesis that the meaning of *if* corresponds to $o_\to$ and that the meaning of *it is not the case that* corresponds to $o_\neg$.

6.2.2 Subordination within nondeclarative clauses

It should come as no surprise that nondeclarative clauses may have subordinated to them conditional clauses. Conditional sentences may be in the imperative mood

(124.1)   If it rains, take an umbrella and wear a hat.

(124.2)   If it rains, either take an umbrella or wear a hat.

or the interrogative mood.

(125.1)   If it snows, will Paul snowshoe or will he ski?

(125.2)   If it rains, will Anne swim or will she jog?

Again, the ideas sketched earlier for coordinated imperative and interrogative clauses seem to be applicable here as well.

6.2.3 Subjunctive conditionals

The best known residual problem is that posed by subjunctive conditionals. These conditionals are ones where the truth of the protasis either is unknown or is presupposed to be false. Subjunctive conditionals of the latter kind are known as counterfactual conditionals.

As we shall see shortly, subjunctive conditionals expressing counterfactuals are not truth functional, that is, unlike indicative conditionals where the truth values of the constituent clauses determine the truth value of the conditional itself, the truth values of the constituent clauses of counterfactual conditionals do not determine the truth value of the counterfactual conditional.

To appreciate this point, let us consider an example of an English connector that is clearly not truth functional. For a connector to be truth functional is for the following to obtain: the truth values of the clauses connected determine at least one truth value for the complex clause and at most one truth value. Thus far, we have seen good evidence that *or* and *and* are truth functional. We have also seen some instances where *if* is also truth functional. The subordinator *because,* however, is not truth functional, for it is possible for both of its constituent clauses to be true and for the clause itself to be either true or false, depending on what precisely the constituent clauses state.

Consider the following circumstances of evaluation. Richard was married to Linda and he is now married to Beth. Richard and Beth attend a party together. Richard avoids the company of his former wife. Linda, his former wife, also attends the party. When Richard discovers that Linda is at the party, he leaves the party earlier than he had planned. Now, the following sentences are true:

(126.1) Richard left the party early.

(126.2) Beth was at the party.

(126.3) Linda was at the party.

Now consider two sentences comprising a main clause and a subordinate clause containing the subordinator *because:*

(127.1) Richard left the party early, because Beth was at the party.

(127.2) Richard left the party early, because Linda was at the party.

Although the constituent clauses are all true, only sentence in (127.2) is true, not sentence (127.1). This shows that the subordinator *because* is not truth functional. The pair of truth values corresponding to the pair of clauses in sentence (127.1) is the same as the pair of truth values assigned to pair of corresponding clauses in sentence (127.2), yet the truth value of sentence (127.1) is false, while the truth value of sentence (127.2) is true. No function can map the same pair of values—here T and T—to two distinct values.

Bearing this in mind, let us turn to counterfactual conditionals. Let us begin by recalling what counterfactual conditionals are. Counterfactual conditionals are conditional sentences, the verb of whose protasis has the form of a past tense of the subjunctive mood and the verb of whose apodosis has a past tense form of the modal *will*. In (128) are two sentences. The first sentence is in the indicative mood: its protasis in the past tense and its apodosis in the present tense. The second sentence is not in the indicative mood: the protasis is in the past tense of the subjunctive mood, whose forms are precisely those of the pluperfect tense of the indicative mood, and the apodosis is in the conditional mood, signaled by the auxiliary verb *would*.

(128.1) If it rained, then the ground is wet.

(128.2) If it had rained, then the ground would be wet.

The first sentence would be false, if its protasis is true and its apodosis false, and true otherwise. The truth of the second sentence is more tricky to assess, as we are about to see.

Let us return to the circumstances specified concerning Richard, Beth, and Linda, except slightly modified: imagine that Richard is at the party, but neither Beth nor Linda is. In these circumstances, all the sentences in (126) are all false. Yet, consider the following pair of counterfactual sentences. The first counterfactual has sentence (126.1) as its apodosis and sentence (126.2) as its protasis, while the second counterfactual, which also has sentence (126.1) as its apodosis, has sentence (126.3) as its protasis.

(129.1)   Richard would have left the party early, if Beth had been at the party.

(129.2)   Richard would have left the party early, if Linda had been at the party.

Sentence (129.1) is judged false, while sentence (129.2) is judged true. Yet, each of the sentences in (126), the nonsubjunctive versions of the clauses of the sentences in (129), are false. This means that counterfactual conditionals are not truth functional. No function can map the same pair of values—here F and F—to two distinct values.

6.2.4 Austinian conditionals

Another use of the subordinator *if* that is not truth functional is the so-called *Austinian conditional,* named after the English philosopher John Langshaw Austin (1911–1960), who first brought them to the attention of those interested in the semantics of natural language.

(130)   If it snows, there is a shovel in the trunk.

John Austin observed that the subordinator *if* in the preceding sentence is not to be understood as having for its meaning $o_\rightarrow$.

To begin with, such conditionals sound very odd as premises in arguments of the forms either of *modus ponens* or of *modus tollens*.

(131.1)   It is snowing. If it snows, there is a shovel in the trunk.
          Therefore, there is a shovel in the trunk.

(131.2)   There is no shovel in the trunk. If it snows, there is a shovel
          in the trunk. Therefore, it is not snowing.

Moreover, when Austinian conditionals are reformulated into the form of the equivalences we have noted, the results are very odd sounding sentences.

(132.0)   If it snows, there is a shovel in the trunk.

(132.1)   *It is not the case both that it will snow and that there is no
          shovel in the trunk.

(132.2)   *Either it will not snow or there is a shovel in the trunk.

(132.3)   *If there is no shovel in the trunk, it will not snow.

The question arises: is there any independent evidence to suggest that the subordinator *if* in Austinian conditionals is distinct from the truth functional subordinator *if* discussed previously? Indeed, there is. As discussed by Quirk et al. (1985, chap. 15.20–15.21), English subordinators are distinguishable according to their tolerance of focusing devices.

Consider, to begin with, this pair of sentences, which are exactly alike, except that where one has *because,* the other has *since.*

(133.1)   Bill likes Carol *because* she is always helpful.

(133.2)   Bill likes Carol *since* she is always helpful.

Now, only the first sentence can give rise either to a cleft construction

(134.1)   It is *because* she is always helpful that Bill likes Carol.

(134.2)   *It is *since* she is always helpful that Bill likes Carol.

or to a pseudocleft construction.

(135.1)   The reason that Bill likes Carol is *because* she is always helpful.

(135.2)   *The reason that Bill likes Carol is *since* she is always helpful.

Similarly, *because,* but not *since,* may form a single clause reply to a question.

(136.1)   A: Why does Bill like Carol?
          B: *Because* she is always helpful.

(136.2)   A: Why does Bill like Carol?
          B: **Since* she is always helpful.

The difference between the two subordinators also surfaces in their tolerance of focusing adverbs such as *only, just, simply,* and *mainly.*

(137.1)   Bill likes Carol only *because* she is always helpful.

(137.2)   *Bill likes Carol only *since* she is always helpful.

The difference between *because* and *since* is further reflected in terms of question formation:

(138.1)   Does Bill like Carol *because* she is always helpful or *because* she never complains?

(138.2)   *Does Bill like Carol *since* she is always helpful or *since* she never complains?

Finally, we come to the sensitivity of these subordinators to negation.

English Connectors 365

(139.1) Bill dislikes Carol not *because* she is unhelpful, but *because* she always complains.

(13.2) *Bill dislikes Carol not *since* she is unhelpful, but *since* she always complains.

In the foregoing, we concentrated on a pair of subordinators that have the same meaning. We now observe that all temporal subordinators pattern with *because,* while all concessive subordinators pattern with *since*.

Note, however, that some subordinators are clearly ambiguous. For example, the subordinator *since* is ambiguous between a meaning that is shared with *because* and one that is temporal.

(140.1) Dan put up his umbrella, *since* it was raining.

(140.2) Dan has been relaxing, *since* he has gone on holiday.

As readers can check for themselves, the subordinator *since* in the temporal meaning patterns with *because*.

Adopting the terminology used in Quirk et al. (1985, chap. 15.20), let us distinguish between subordinators that pattern with the temporal ones and subordinators that pattern with the concessive ones as *adjunctive* and *disjunctive,* respectively.

Next, we remark on an important usage of these subordinators, be they adjunctive or disjunctive, which makes them pattern with the subordinators in the disjunctive column. To see the difference, consider the following pair of sentences.

(141.1) McGill University has no electricity, *because* its trunk line failed.

(141.2) McGill University has no electricity, *because* I just called there.

As can be easily checked, only the first usage is an adjunctial one.

(142.1) It is *because* its trunk line failed that McGill University has no electricity.

(142.2) *It is *because* I just called there that McGill University has no electricity.

What distinguishes the usage found in (141.1) from the usage found in (141.2) is that subordinate clause in the former case bears on the content of the clause to which it is subordinate, whereas the subordinate clause in the latter bears on some relationship the denotation of the subject of the subordinate clause bears to the content of the main clause.

(143.1) McGill University has no electricity, *because* I just called there.

(143.2) I know that McGill University has no electricity, *because* I just called there.

(144.1) I know that McGill University has no electricity, *because* I called there.

(144.2) It is *because* I called there that I know that McGill University has no electricity.

Here are some other examples.

(145.1)   *Since* you are so smart, what does this word mean?

(145.2)   *While* we are on the subject, where does Fred live?

(145.3)   *Before* you say anything, you have the right to remain silent.

We sum up our discussion of the distinction between adjunctive and disjunctive subordinators with the following table.

	ADJUNCTIVE SUBORDINATORS	DISJUNCTIVE SUBORDINATORS
Cleft	Yes	No
Pseudocleft	Yes	No
Forms a single clause answer	Yes	No
Immediately preceded by focus adverb	Yes	No
May form disjunctive question	Yes	No
Immediately preceded by *not*	Yes	No

Armed with this distinction, we can easily see that Austinian conditionals constitute a disjunctive usage of the subordinator *if*, as borne out by the syntactic tests.

(146.1)   It is if London is in China that I am a monkey's uncle.

(146.2)   *It is if it snows that there is a shovel in the trunk.

(147.1)   The condition under which I am a monkey's uncle is if London is in China.

(147.2)   *The condition under which there is a shovel in the trunk is if it snows.

Readers are invited to carry out the other such tests on their own.

Finally, it has been noted that the adjunctive *if* and the disjunctive *if* differ in their tolerance of the adverb *then* in the apodosis.

(148.1)   If London is in China, then I am a monkey's uncle.

(148.2)   *If it snows, then there is a shovel in the trunk.

Exercises: Open problems for the English subordinator *if*

1. Consider the following list of English subordinators:

after, although, as, before, once, since, so that, though, until, when, whereas, while, unless

For each word in the list, use the criteria in section 6.2.4 to determine whether it is a disjunctive or an adjunctive subordinator or neither.

2. For each subordinator in exercise 1 with a disjunctive use, give a sentence illustrating its disjunctive use.

7 The English Adverb *not*

At first sight, it appears that there is a close correspondence between *not* of English and the $\neg$ of propositional logic. After all, just as one can prefix *it is not the case that* in front of a clause to obtain its negation, so one can prefix $\neg$ in front of a formula to obtain a formula.

(149.1) It is raining α
(149.2) It is not the case that it is raining $\neg \alpha$

Moreover, it is indisputable that when the first is true, the second is false, and when the second is true, the first is false.

It is raining.   It is *not* the case that it is raining.
    T   F
    F   T

But this is precisely the truth function $o_\neg$ of CPL.

α $\neg \alpha$
T   F
F   T

Moreover, the subordinating clause *it is not the case* satisfies the law of double negation that is enjoyed by $\neg$.

(150.1) $\alpha \models \neg \neg \alpha$

(150.2) $\neg \neg \alpha \models \alpha$

(151.1) it is raining ENTAILS
          it is *not* the case that it is *not* the case that it is raining.

(151.2) it is *not* the case that it is *not* the case that it is raining
          ENTAILS it is raining.

Furthermore, on the assumption that the English coordinator *or* is accurately characterized by the truth function $o_\vee$, the subordinating clause *it is not the case* also satisfies the law of excluded middle

(152) $\models \alpha \vee \neg \alpha$

for it is impossible for it to be false that

(153) Either it is raining or it is not the case that it is raining.

Similarly, on the assumption that the English coordinator *and* is accurately characterized by the truth function $o_\wedge$, the subordinating clause *it is not the case* also satisfies the law of noncontradiction

(154) $\not\models \alpha \wedge \neg \alpha$

for it is impossible for it to be true that

(155)   It is raining and it is not the case that it is raining.

However, while the truth function $o_\neg$ does seem to characterize the subordinating clause *it is not the case,* it is far from clear that it characterizes the adverb *not*. For, unlike the subordinating clause *it is not the case,* which is prefixed to a clause to yield a clause, the adverb *not* is inserted in the middle of a clause.

(156.1)   *It is not the case* that it is raining.

(156.2)   It is *not* raining.

And while sentences (156) are equivalent, we have not answered the question of what the semantic contribution of the English adverb *not* is. To do this, we must first determine its syntactic properties.

7.1   The Syntax of *not*

Close examination of the distribution of the adverb *not* shows the syntactic structures in which it is found are quite disanalogous from the structure in which the propositional connective $\neg$ is found. We undertake this close examination now.

In traditional grammar, *not* is classified as an adverb. Its pattern of distribution is virtually identical to that of the adverb *never*. In nonfinite clauses, they occur to the immediate left of the verb, if the verb is in a nonperiphrastic form, and they occur anywhere to the left of the verb up to the immediate left of the leftmost auxiliary, if the verb is in a periphrastic form.[5] In short, at least one verbal element must occur to their immediate right.

(157.1)   Joan regrets [*never/not* having attended parties]

(157.2)   Joan regrets [having *never/not* attended parties]

(157.3)   *Joan regrets [having attended *never/not* parties]

---

5. Periphrases are forms of expression in which auxiliary words are used, instead of a suffix. For example, some English comparative adjectives, such as *tall,* are formed by the addition of the suffix *-er,* that is, nonperiphrastically; others, such as *admirable,* are formed periphrastically, that is, with the use of the word *more*. Periphrastic forms of verbs are those forms that contain auxiliary verbs; nonperiphrastic ones are those which do not.

In finite clauses, the distribution of *not* and *never* obey an additional constraint: they require at least one verbal element (that is, an auxiliary verb) to their immediate left.[6]

(158.1)   Joan *?never/*not* has attended parties.

(158.2)   Joan has *never/not* attended parties.

(158.3)   *Joan has attended *never/not* parties.

The one exception arises when the main verb is a nonperiphrastic form of the verb *to be*. Both adverbs give up the requirement, otherwise found, that some verbal element occur to their right.[7]

(159.1)   Joan *?never/*not* is late.

(159.2)   Joan is *never/not* late.

Notice that the simple description given accounts for the fact that the adverb *not* never occurs in main clauses without an auxiliary verb, whereas *never* does.[8]

We summarize these observations[9] in the following table.

VERB	Preceded by a verbal element	Succeeded by a verbal element
Nonfinite	Not required	Required
Finite	Required	Required
Copular	Required	Excluded

These observations give rise to the following constituency formation rules for the adverb *not* in finite clauses:

S5   NP Av Adv VP $\rightarrow$ S

VP6 V_c Adv AP $\rightarrow$ VP

VP7 V_c Adv NP $\rightarrow$ VP

At the beginning of this section on the English adverb *not*, we noted certain similarities between the unary connective $\neg$ and the adverb *not*. These similarities suggest that the adverb *not* has as part of its meaning the truth function $o_\neg$. However, once we turn to

---

6. This additional restriction is relaxed in the case of *never*, if it is focused. Thus, the following sentence is perfectly acceptable, when *never* is emphasized: *Bill never left the house.*

7. Again, this additional restriction is relaxed in the case of *never*, if it is focused.

8. The description just given for the distribution of the adverbs *not* and *never* requires modification for interrogative clauses.

9. See Baker (1989, chap. 11.3.5) for more details about the distribution of *not*.

structures involved in formulae, on the one hand, and clauses containing the adverb *not,* on the other, we encounter a problem.

On the one hand, any formula containing ¬ has an immediate subformula that is sister to ¬. Since formulae have truth values, once the propositional variables are assigned a truth value, the function interpreting the unary connective has a suitable value to apply to. On the other hand, the same truth function used to interpret the adverb *not* does not have a suitable value to apply to. The reason is this: truth values are assigned only to declarative clauses. It makes no sense to assign truth values to noun phrases, verb phrases, adjective phrases, or the copular verb. After all, what would it mean to say that any of the following are either true or false: the noun phrase *the visitor,* the verb phrase *walked in the park,* the adjective phrase *courageous,* or the copular verb *is*? However, as the constituency formation rules for simple clauses with *not* show, its only sisters are noun phrases, verb phrases, adjective phrases, or the copular verb. There are two options: one is to take the constituent structure at face value and to assign *not* a function that can apply to the kind of value assigned to the verb phrase and that, though distinct from the truth function $o_\neg$, nonetheless yields an equivalent result. This option can be fully presented only after we have completed chapter 10, where we shall see how to assign values to phrases. The other option is to assign to the sentence a constituent structure amenable to assigning the adverb *not* the value $o_\neg$. This can done with what are called transformation rules, which will be presented in chapter 14. In other words, at this point, we cannot provide an adequate treatment of the adverb *not*.

7.2 Open Problems for the English Adverb *not*

Three further problems remain regarding the adverb *not*. First, it may occur in positions different from the ones characterized in the previous section. For example, *not* may occur at the left-hand margin of any constituent. This is known as constituent negation and it works best when the negated constituent is balanced by an unnegated counterpart, as shown in (160).

(160.1)   Bill bought, *not* hot dogs, *but* hamburgers.

(160.2)   Bill bought hamburgers, *not* hot dogs.

(160.3)   Bill bought hamburgers; he did *not* buy hot dogs.

Second, *not* may occur at the beginning of a finite clause, provided it is followed noun phrase whose determiner is a quantificational one. However, not any quantificational determiner will do.

(161.1)   Not every bird flies.

(161.2)   *Not some bird flies.

(161.3)   *Bill spotted not every bird.

This will be discussed further in chapter 14, section 3.5.3.

Finally, we come to an aspect of the use of *not* uncovered by Laurence Horn (1989, chap. 6) and dubbed *metalinguistic negation*. Such uses are not truth functional. Indeed, its use pertains to nonsemantic aspects of a sentence. Thus, the metalinguistic use of the adverb *not* can be used to deny an implicature associated with a sentence, as illustrated next.

(162) Ann does not have three children; she has four.

It can also be used to deny a presupposition.

(163) Bill has not quit smoking, for he never smoked in the first place.

Indeed, it can even be used to correct errors in grammar and errors in pronunciation.

(164.1) It is not concertos, but concerti.
It is not concertos; it is concerti.

(164.2) This is not a tomayto; it is a tomahto.

Exercises: The English adverb *not*

1. Assign a synthesis tree to the following sentences:

(a) Bill did not sleep for two days.

(b) Carl will not be in town before midnight.

8 Conclusion

In this chapter, we have investigated the extent to which the coordinators *and* and *or* and the subordinator *if* as well as the adverb *not* have as their meaning the truth functions $o_\wedge$, $o_\vee$, $o_\rightarrow$, and $o_\neg$, respectively. We discovered that $o_\wedge$ and $o_\vee$ characterize the coordinators *and* and *or*, respectively, fitting well with speakers' judgments regarding the truth or falsity of declarative clauses formed from coordinated declarative clauses. Moreover, we noticed that the formation rule for binary propositional connectives, of which $\wedge$ and $\vee$ are two, and the constituency formation rules for coordinators, of which *and* and *or* are two, are exactly alike. As a result, we found that we can adopt directly the values of $\wedge$ and $\vee$, namely $o_\wedge$ and $o_\vee$, as values for *and* and *or*, at least when they are coordinating declarative clauses.

We also saw that the situation turned out to be more complicated when it comes to the subordinator *if*. Setting aside its uses in disjunctive clauses, so-called Austinian conditionals, and in subjunctive clauses, we found that speakers' judgments regarding the truth or falsity of declarative clauses subordinated by *if,* when one controls for confounding contextual factors, fit reasonably well with the truth function $o_\rightarrow$. However, we also noticed that the formation rule for binary propositional connectives, of which $\rightarrow$ is one, and the constituency formation rules for subordinators, of which *if* is one, are not alike. As a result, we could not adopt the function $o_\rightarrow$ off the logical shelf of CPL as the value to be assigned to the subordinator *if*. But we could assign *if* the function o_{if}, though mathematically distinct from $o_\rightarrow$, nonetheless equivalent to it, and adapted to the constituent formation rules required for the formation of a main clause with a main clause and a subordinate one.

Finally, we saw that, aside from metalinguistic uses of the adverb *not* in declarative clauses and its uses in constituent negation, speakers' judgments regarding the truth or falsity of declarative clauses containing the adverb *not* fits very well with the truth function $o_\neg$. Yet, we also saw that the position of the adverb within the clause precludes it from being assigned the truth function $o_\neg$. We did not attempt to present either of the two options proposed to address this problem.

SOLUTIONS TO SOME OF THE EXERCISES

2 English connectors and clauses

1. Subordinator: *as, once, whereas*
Coordinator: *but*
Conjunct: *nevertheless, so, therefore*

2. Here, partial labeled bracketing is given.

   (b) [S [D [C$_s$ if] [S [S it rains] [D [C$_s$ while] [S the sun shines]]]] [S a rainbow will appear]]

   (c) [S [D [C$_s$ if] [S [S it rains] [C$_c$ or] [S too many people are sick]]] [S the picnic will be canceled]]

   (e) [S [D [C$_s$ although] [S [D [C$_s$ once] [S it rains]] [S people will leave]]] [S [D [C$_s$ while] [S it is not raining]] [S people will stay]]]

   (g) [S [S we don't need to worry about Carol] [D [C$_s$ because] [S [D [C$_s$ if] [S [D [C$_s$ when] [S she arrives]] [S we are not home]] [S she can let herself in with the key]]]]

3. (a) *And or* is an attested expression of English. If you judge it acceptable in the exercise sentence, then it constitutes a prima facie counterexample to our assertion that English coordinators do not iterate. Yet, if it is a single word, as are such words as *into, upon,* and such, then it is not a counterexample. Notice that *or and* is not an acceptable expression.

   (c) If you judge the exercise sentence as unacceptable, then it shows that subordinate clauses whose subordinators are *for* are not coordinated.

English Connectors 373

(e) If you judge the exercise sentence as unacceptable, then it shows that a subordinate clause introduced by *for* cannot precede the clause to which it is subordinate.

4. The following single rule, together with the other S rules, is sufficient to generate all three sentences. Adv S → S.

4 The English coordinator *and*

1. None of the sentences militates against the hypothesis that *and* denotes the truth function for $o_\wedge$.

2. Under the hypothesis that *and* denotes $o_\wedge$, a commutative truth function, one would expect any two independent clauses conjoined by *and* to transpose, without disturbing the acceptability or equivalence. Yet, this is not the case. In each case, the failure can be plausibly attributed to other factors. Moreover, in each case, it is possible to adduce evidence to support the attribution.

5 The English coordinator *or*

(1.) The formula $p + q$ is semantically equivalent to the formulae $p \leftrightarrow \neg q$. You can prove this using truth tables.

(3.) The formula $p + q$ is semantically equivalent to the formula $p \leftrightarrow \neg q$. You can prove this using truth tables.

9 Classical Predicate Logic

1 Introduction

In this chapter, we turn to *classical predicate logic* (CPDL). CPL is called a propositional logic, because its basic expressions, aside from the propositional connectives and parentheses, are propositional variables. CPDL is called a predicate logic, because its basic expressions, besides the logical constants and parentheses, are predicates and individual symbols, rather than propositional variables. CPDL, as described here, is only a proper part of what is usually called *first-order predicate logic,* or *predicate logic* for short. What is usually called first-order predicate logic is called here *classical quantificational logic*. In other words, in this book, we have broken up what is customarily called first-order predicate logic into two parts, the part that involves predicates and the part that involves, in addition, quantification. The reason for this division is to facilitate exposition.

The term *predicate* is an unfortunate term. The term, an invention of Aristotle's, now refers in traditional grammar to that part of a monoclausal sentence that remains when its subject is removed. It corresponds, then, to what today one calls a verb phrase.

	SENTENCE	PREDICATE
(1.1)	Bill *is tall*.	_ *is tall*.
(1.2)	The visitor *is a Canadian*.	_ *is a Canadian*.
(1.3)	Each tourist *bought a ticket*.	_ *bought a ticket*.
(1.4)	Which guest *gave Mary a present*?	_ *gave Mary a present*?

In this way, a predicate is a portion of a monoclausal sentence that lacks being a sentence by one constituent, namely, the subject noun phrase. But in logic, a predicate is a part of an atomic formula that lacks being an atomic formula by lacking one or more individual symbols.

A better analogy for the predicates of logic, then, are the verbs of natural language. Imagine that English comprises only verbs and proper nouns. Some verbs require the addition of only one proper noun to become a sentence. These verbs are intransitive verbs. Some verbs require the addition of two proper nouns to become a sentence. These verbs are transitive verbs. And finally some require the addition of three proper nouns to become a sentence. And these are known as ditransitive verbs. Verbs, then, can be thought of as having slots that, to become sentences, must be filled by proper nouns. Intransitive verbs come with one slot; transitive verbs come with two; and ditransitive verbs come with three. (The notion of "slot" is merely a metaphor.)

	VERB	SENTENCE
Intransitive	_ walk	John walks
Transitive	_ sees _	Mary sees Bill
Ditransitive	_ gave _ _	Mary gave Bill Fido

We now turn to describing the notation of CPDL.

2 Notation

Recall that the basic expressions of CPL comprise the propositional connectives and a set of propositional variables, disjoint from the set of propositional connectives. Like CPL, predicate logic includes among its basic expressions the propositional connectives; however, in place of propositional variables, predicate logic has two sets of basic expressions, the set of *individual symbols,* often called *constants,* and the set of *relational symbols,* often called *predicates*. These latter sets of basic expressions are not only disjoint from each other but they are also disjoint from the set of propositional connectives. Just as the set of propositional variables may be either finite or infinite, so the set of individual symbols, or constants, may be either finite or infinite and the set of relational symbols, or predicates, also may be either finite or infinite.

We use lowercase letters from the beginning of the Roman alphabet for individual symbols, such as $a, b,$ and c. Should a large number of individual symbols be required, we shall use the letter c, subscripted with positive integers.

The notation for relational symbols is more complicated. While we shall use uppercase letters of the Roman alphabet starting from P as relational symbols, more needs to be said. As we noted, individual symbols are the analogues in CPDL of the proper nouns in English, while relational symbols are the analogues of verbs. Just as verbs can be divided, or subcategorized, into intransitive, transitive, and ditransitive verbs, so the relational symbols of CPDL can be divided, or subcategorized, into those with one place, those with two places, those with three places, and so on. The verbs of English differ from the relational symbols of CPDL, insofar as English verbs rarely have more than three places and never more than four, whereas the relational symbols of CPDL may have any number of places. How many places a relational symbol has can be conveyed in three different, but equivalent, ways.

The first, and most intuitive, way to convey how many places a relational symbol has is to annotate each relational symbol with a number of underscores corresponding to the relational symbols places. Thus, a one-place relational symbol, say P, may given as $P\_$, a two-place relational symbol, say R, as $R\_\_$, and a three-place relational symbol, say T, as $T\_\_\_$. While this annotation is fine when dealing with a small number of relational symbols, each with no more than a few places, it does not lend itself to expressing the general case.

A way to speak about how many places a relational symbol has in a general way uses the neologism *adicity*.[1] A one-place relational symbol is said to have an adicity of one, a two-place relational symbol an adicity of two, and in general, an n-place relational symbol an adicity of n. What is evident is that each relational symbol has an adicity, which is a positive integer. In other words, adicity is just a function from relational symbols into the positive integers. We shall symbolize this function as ad. Thus, in lieu of the underscore notation added to the symbols $P\_$, $R\_\_$, and $T\_\_\_$, we express what their adicity is by writing $ad(P) = 1$, $ad(R) = 2$, and $ad(T) = 3$.

A specification of the notation for CPDL as comprising a set of individual symbols (IS), a set of relational symbols (RS), and an adicity function (ad) from the relational symbols into the positive integers, is called a *signature*.[2] Here is its formal definition.

Definition 1 Signature
Let ⟨IS, RS, ad⟩ be a signature iff IS and RS are disjoint sets and ad is a function from RS into $\mathbb{Z}^+$.

Notice that the fact that ad is a function into the positive integers guarantees that no relational symbol is assigned distinct adicities.

There is still another way to convey the adicity of relational symbols.[3] We can consider the set of relational symbols as segregated into disjoint sets according to their adicity and then label each set by RS annotated by a numerical subscript denoting the adicity of the relational symbols in the set. Not only might the set of relational symbols be infinite, but each segregated set might be infinite as well. We display what such a set might look like. We use subscripts to indicate a relational symbol's adicity, that is, the number of empty places associated with it; we use superscripts to distinguish one relational symbol of one adicity from another relational symbol with the same adicity.

---

1. This neologism is formed by suffixing the suffix *-ity* to the adjectival suffix *-adic*, found in such adjectives as *monadic, dyadic, triadic*, and treating the result, *adicity*, as a noun.
2. Many authors use the word *language* to refer to the set of nonlogical symbols of CPDL.
3. Notation of this sort will be used in chapter 13.

$$\text{RS}_1 = \{P_1^1, P_1^2, \ldots, P_1^m, \ldots\}$$
$$\text{RS}_2 = \{P_2^1, P_2^2, \ldots, P_2^m, \ldots\}$$
$$\vdots \quad \vdots$$
$$\text{RS}_n = \{P_n^1, P_n^2, \ldots, P_n^m, \ldots\}$$
$$\vdots \quad \vdots$$

In this case, $\text{RS} = \bigcup_n \text{RS}_n$.

Just as a proper noun and an intransitive verb combine to yield a sentence, so an individual symbol and a one-place relational symbol combine to yield an atomic formula of CPDL, and, just as a transitive verb combines with two proper nouns to yield a sentence, so two individual symbols combine with a two-place relational symbol to form an atomic formula of CPDL.

RELATIONAL SYMBOL	INDIVIDUAL SYMBOL	FORMULA
W _	b	W <u>b</u>
S _ _	a, b	S <u>a</u> <u>b</u>
T _ _ _	a, b, c	T <u>c</u> <u>b</u> <u>a</u>

Generalizing on this, we define an atomic formula of CPDL as any n-place relational symbol followed by n occurrences of individual symbols. Here is the formal definition.

Definition 2 Atomic formulae of CPDL

Let $\langle \text{IS}, \text{RS}, \text{ad} \rangle$ be a signature. α is an atomic formula (of CPDL), that is, $\alpha \in \text{AF}$, iff $\Pi \in \text{RS}$, $\text{ad}(\Pi) = n$ and there are n occurrences of individual symbols from IS — $c_1, \ldots, c_n$ — such that $\alpha = \Pi c_1 \ldots c_n$.

Examples of how this rule definition between atomic formulae and nonformulae follow.

RELATIONAL SYMBOL	ATOMIC FORMULA	NONFORMULA
ad(W) = 1	Wb, Wa	Wba, W
ad(S) = 2	Sab, Saa	Sa, Sabc
ad(T) = 3	Tcba, Taba	Tc, Tab, Tabca

This same information can be recast in terms of the metaphor of slots and the use of dashes as follows.

RELATIONAL SYMBOL	ATOMIC FORMULA	NONFORMULA
W _	W <u>b</u>, W <u>a</u>	W <u>b</u> a, W _
S _ _	S <u>a</u> <u>b</u>, S <u>a</u> <u>a</u>	S <u>a</u> _, S <u>b</u> <u>a</u> c
T _ _ _	T <u>c</u> <u>b</u> <u>a</u>, T <u>a</u> <u>b</u> <u>a</u>	T <u>c</u> _ _, T <u>d</u> <u>b</u> _, T <u>c</u> <u>b</u> <u>a</u> c

2.1 Formulae of CPDL

In addition to atomic formulae, CPDL has composite formulae. They are just the formulae that can be obtained from the atomic formulae by combining atomic formulae with the propositional connectives. Here is a categorematic definition of a formula of CPDL.

Definition 3 Formulae of CPDL (categorematic version)
FM, the formulae of CPDL, is the set defined as follows:

(1)   AF $\subseteq$ FM;
(2.1) if $\alpha \in$ FM and $* \in$ UC, then $*\alpha \in$ FM;
(2.2) if $\alpha, \beta \in$ FM and $\circ \in$ BC, then $(\alpha \circ \beta) \in$ FM;
(3)   nothing else is.

Just as the definition of a formula of CPL was also formulated syncategorematically, so the definition of a formula of CPDL can also be formulated syncategorematically. Readers are asked to do so as an exercise.

It should be noted that we have recycled, as it were, the name AF. In chapter 6, AF denotes the set of atomic formulae of CPL, which is a set of single symbols, the propositional variables; whereas in this chapter, AF denotes the set of atomic formulae of CPDL, which are not single symbols, but composite symbols, comprising an initial relational symbol followed by a string of individual symbols. The atomic formulae of CPL are disjoint from the atomic formulae of CPDL. Thus, the name AF is ambiguous. We could have disambiguated the name, using AF_{CPL} for the atomic formulae of CPL and AF_{CPDL} for the atomic formulae of CPDL. Instead, as is customary in mathematics and logic, we rely on the context to make clear which set of expressions AF denotes.

Similarly, we have recycled the name FM, which, in chapter 6, denotes the set of formulae of CPL and, in this chapter, denotes the formulae of CPDL. As the atomic formulae of CPL are disjoint from the atomic formulae of CPDL, so the formulae of CPL are disjoint from those of CPDL. And this is so, despite the fact that the categorematic definition of a formula of CPDL, just given, is nearly identical with the categorematic definition of a formula of CPL (definition 2 in chapter 6). However, they differ crucially with respect to the first clause in their respective definitions: where categorematic definition of a formula of CPL contains PV, the first clause in the categorematic definition of a formula of CPDL contains AF. And as we just were reminded, the atomic formula of CPDL is disjoint from the atomic formulae of CPL, which are propositional variables. Thus, even though the formulae are constructed using the same logical constants and parentheses, the fact that the basic building blocks for the formulae are disjoint sets means that the resulting set of formulae are disjoint. So, the name FM is also ambiguous. Again, we could have disambiguated FM, using FM_{CPL} for the formulae of CPL and FM_{CPDL} for the formulae of CPDL. Instead, we have again opted for the custom in mathematics and logic to rely on the context to make clear which set of expressions FM denotes.

Given the definition of the formulae of CPDL, one naturally wonders how one decides whether or not a string of symbols constitute a legitimate formulae of CPDL. According to the previous definitions, any formula must be one of the following: either an atomic formula, comprising a relational symbol Π, of some adicity n, followed by n occurrences of individual symbols from IS; or a formula of the form $\neg \alpha$; or a formula of one of the forms $(\alpha \wedge \beta)$, $(\alpha \vee \beta)$, $(\alpha \rightarrow \beta)$, $(\alpha \leftrightarrow \beta)$.

As an example, let IS be $\{a, b, c\}$ and let RS be $\{P, R\}$ where $\mathrm{ad}(P) = 1$ and $\mathrm{ad}(R) = 2$. Now, consider the following list of strings of symbols: Pa, $\neg Rbc$, $(\neg Pa)$, $(\neg Rbc \rightarrow Pa)$, $\big((\neg Pa) \vee (Pa \wedge Rbc)\big)$. The first string of symbols comprises a one-place relational symbol followed by exactly one occurrence of an individual symbol. Thus, by definition 2, it is an atomic formula of CPDL, and by clause (1) of definition 3, it is a formula of CPDL. None of the remaining strings of symbols constitute an atomic formula, since each contains at least one occurrence of at least one propositional connective. Now, the string Rbc comprises a two-place relational symbol followed by two occurrences of two individual symbols. Thus, by definition 2, it is an atomic formula. Thus, by clause (1) of definition 3, it is a formula, and so, by clause (2.1) of definition 3, $\neg Rbc$ is a formula of CPDL. The third string is not a formula, since it corresponds to none of the possibilities just enumerated. The fourth string, however, is a formula of CPDL. We have already established that $\neg Rbc$ and Pa are formulae. Invoking clause (2.2) of definition 3, we conclude that $(\neg Rbc \rightarrow Pa)$ is a formula. Finally, we come to the last string: $\big((\neg Pa) \vee (Pa \wedge Rbc)\big)$. Were it a formula, then by clause (2.2) of definition 3, both $(\neg Pa)$ and $(Pa \wedge Rbc)$ would be formulae. But we showed previously that $(\neg Pa)$ is not a formula. Hence, we conclude that the last string is not a formula.

2.2 Formulae and Subformulae

We have already singled out one special subset of the set of formulae of CPDL, namely, the atomic formulae. As with CPL, we shall distinguish atomic formulae from composite formulae. Examples of composite formulae are $\neg\neg Rba$, $(Pc \wedge \neg Saa)$, and $\neg(Rba \rightarrow Pa)$. Clearly, no atomic formula is a composite formula.

Finally, there is a class of formulae that overlap the composite formulae and the atomic ones, namely, *basic formulae*. Basic formulae are atomic formulae and their negations. Rbc and $\neg Rbc$ are basic formulae. Neither $(Pa \vee Sba)$ nor $\neg\neg Pa$ are.

The composite formulae of CPDL, like those of CPL, all have immediate subformulae. The same definition as the one for CPL (definition 3 in chapter 6) can be used. As in CPL, so in CPDL, which formulae are a composite formula's immediate subformulae is well depicted by the formula's tree diagram. In addition, the same definition used in CPL for proper subformulae (definition 4 in chapter 6) can also be used in CPDL. Again, which formula is a subformula of which is well depicted by a tree diagram. Finally, the definition for subformula (definition 5 in chapter 6) can also be used in CPDL.

Classical Predicate Logic

Three other definitions also carry over directly from CPL: the scope of a propositional connective, a symbol's being within the scope of a propositional connective and a formula's main propositional connective.

As in the case of CPL, one must speak of the scope of a propositional connective, relative to its occurrence. Consider the formula $((Pb \to Rab) \to Qc)$. It does not make sense to inquire into the scope of $\to$; it makes sense only to inquire into the scope either of its first occurrence, in which case its scope is $(Pb \to Rab)$, or of its second occurrence, in which case its scope is $((Pb \to Rab) \to Qc)$.

The definition of being within the scope of a propositional connective permits one to say that the individual symbol a, for example, is within the scope of $\vee$ in the formula $(Pa \vee \neg Rcb)$, but a is not within the scope of $\neg$. Moreover, $\neg$ is within the scope of $\vee$, but not vice versa.

The composite formulae of CPDL, like those of CPL, have main propositional connectives. The main propositional connective of $(Pa \vee \neg Rcb)$ is $\vee$.

Finally, notice that, as with CPL, parentheses are not listed among the symbols of CPDL; rather, they are insinuated into formulae through the definition for the set of formulae. The same abbreviatory conventions apply to the formulae of CPDL as apply to the formulae of CPL. Readers should refresh their memories by consulting the final paragraphs of section 2.1.3 of chapter 6.

Exercises: Formulae and subformulae

1. Provide a syncategorematic definition of the set of formulae of CPDL.

2. Provide a definition for each of the following relations between parts of formulae in CPDL:

(a) scope of a propositional connective,

(b) formula being within the scope of a propositional connective,

(c) propositional connective's occurrence being within the scope of a propositional connective,

(d) a formula's main propositional connective.

3. Let IS be $\{a, b, c, d\}$. Let RS be $\{P, Q, R, S, G, H\}$. Let $\text{ad}(P) = 1$, $\text{ad}(Q) = 1$, $\text{ad}(R) = 2$, $\text{ad}(S) = 2$, $\text{ad}(G) = 3$, and $\text{ad}(H) = 3$. Now, identify, among the following strings of symbols, which are formulae of CPDL, which are conventional abbreviations for formulae of CPDL, and which are neither. In the case of conventional abbreviations, identify the formula or formulae for which a conventional abbreviation is a conventional abbreviation.

(a) $Haab$

(b) $((Sab \wedge)Rba \vee Qb)$

(c) (Qb)

(d) $((Gbaa \to \neg Paa) \leftrightarrow Qa)$

(e) $\neg\neg(\neg\neg Sab \leftrightarrow \neg\neg\neg Rdb)$

(f) $((Pc \land Qa) \leftarrow Ra) \lor Pb)$

(g) $Pd \lor Qb$

(h) $((Pc \land (Qa \to Sbaa)) \land Pc)$

(i) $(Pc \lor Qa \lor Rdb)$

(j) Sd

(k) $(\neg((Pc \leftrightarrow Qa) \lor Rab))$

(l) $((Hab \land Rba) \to Qbc)$

4. Identify, among the following formulae, which are atomic formulae, which are basic formulae, and which are composite formulae.

(a) $(Pc \lor Rab)$

(b) $((Sdb \lor (Qa \land Hadb)) \to Pc)$

(c) $\neg Qb$

(d) Sba

(e) $((Gbba \to Pa) \lor Rba)$

(f) $\neg Pc$

(g) $\neg((Pc \leftrightarrow Qa) \lor Rdb)$

(h) $(Pc \to (Qc \lor Rba))$

(i) $\neg\neg\neg Rdb$

(j) Saa

5. For each of the formulae given, identify all of its subformulae and its main connective.

(a) $(Qa \to (Pc \land Pc))$

(b) $\neg\neg\neg Hbcd$

(c) $\neg((\neg Rdb \lor Pc) \land \neg Qa)$

(d) Rdb

(e) $(\neg(Rdb \lor (Qa \land Sdd)) \to Pc)$

6. Consider this formula:

$(((Pb \to Qc) \lor (Gabc \land Qa)) \land (Scc \to Pc))$

(a) What is the scope of the first occurrence of $\to$?

(b) What is the scope of the second occurrence of $\to$?

(c) What is the scope of the first occurrence of $\land$?

(d) What is the scope of the second occurrence of $\land$?

(e) What is the scope of $\lor$?

Classical Predicate Logic

7. Consider another formula:
$$((Pc \land ((Qc \to Rdb) \leftrightarrow \neg Habc)) \lor (Qa \land Rbc))$$

(a) Is the only occurrence of Rdb within the scope of the only occurrence of $\leftrightarrow$?

(b) Is the only occurrence of Rbc within the scope of the only occurrence of $\lor$?

(c) Is the first occurrence of c within the scope of the first occurrence of $\land$?

(d) Is the only occurrence of $\leftrightarrow$ within the scope of the only occurrence of $\to$?

(e) Is the second occurrence of a within the scope of the first occurrence of $\land$

(f) Is the only occurrence of $\neg$ within the scope of the only occurrence of $\lor$?

(g) Is the first occurrence of b within the scope of the first occurrence of $\land$

(h) Is the first occurrence of Pc within the scope of the first occurrence of $\land$?

(i) Is the only occurrence of $\leftrightarrow$ within the scope of the first occurrence of $\land$?

(j) Is either occurrence of $\land$ within the scope of the other?

3 Semantics

Recall that a bivalent truth value assignment is any function that assigns to the propositional variables of CPL either T or F. Since the atomic formulae of CPL are precisely the propositional variables, a bivalent truth value assignment assigns values to the atomic formula. But it does not assign any value to the composite formulae. The definition of a classical valuation extends any bivalent truth value assignment to an assignment of truth values to all the formulae of CPL and it does so in a unique way.

Since the composite formulae of CPDL are built in the same way as the composite formulae of CPL, one expects that, once truth values are assigned to the atomic formulae of CPDL, the truth values of the composite formulae of which they are constituents can be computed in the very same way as those of the composite formulae of CPL are computed on the basis of a bivalent truth value assignment to the propositional variables. So, the question arises: how are truth values to be assigned to the atomic formulae of CPDL?

Recall that an atomic formulae of CPDL is not a basic expression, rather it is a complex one, made up of a single occurrence of a relational symbol and one or more occurrences of individual symbols. The idea, now, is to assign values to relational and individual symbols in such as way so as to be able to assign a truth value to atomic formulae built from them. The assignment of values to the relational and individual symbols is called an *interpretation* function. Its domain comprises the set of individual symbols and the set of relational symbols (IS ∪ RS). Its codomain comprises a set of individuals, called a *universe,* and set theoretic entities built from them. However, if an assignment to the symbols of values from the codomain is to be an interpretation function, it must respect certain conditions. In particular, an interpretation function is an assignment that assigns to each member of IS a member of U, and it assigns to each relational symbol of adicity 1 a subset of U and to each relational symbol of adicity n, where $n \geq 2$, a set of n-tuples drawn from U.

Let us make this more precise. Let U, the *universe*, be a nonempty set. The set of all ordered pairs formed from elements of U is denoted by U^2; the set of all ordered triples formed from the elements of U is denoted by U^3. In general, the set of all n-tuples is referred to as U^n. However, in the case where n equals 1, we shall stipulate that U^1 is just U, rather than the set of one-tuples. Then, if a symbol is an individual symbol, an interpretation function must assign it a value from U and if a symbol is an n-place relational symbol, the interpretation function must assign it a value from $Pow(U^n)$.

We now define a *structure*, or a *model*, for a signature.

Definition 4 Structure for a signature
M, or $\langle U, i \rangle$, is a structure for a signature, $\langle \text{IS}, \text{RS}, \text{ad} \rangle$ iff
(1) U is a nonempty set;
(2) i is a function where
(2.1)   its domain is $\text{IS} \cup \text{RS}$,
(2.2)   for each $c \in \text{IS}$, $i(c) \in U$, and
(2.3)   for each $\Pi \in \text{RS}$ such that $\text{ad}(\Pi) = n$, $i(\Pi) \in Pow(U^n)$.

Let us consider a case where $\text{IS} = \{a, b, c\}$ and $\text{RS} = \{P, R\}$, where $\text{ad}(P) = 1$ and $\text{ad}(R) = 2$. Let U be the set $\{1, 2, 3\}$. Both of the following are functions from $\text{IS} \cup \text{RS}$ to $U \cup Pow(U) \cup Pow(U^2)$, but only the first, i, is an interpretation function.

i a $\mapsto$ 1
     b $\mapsto$ 2
     c $\mapsto$ 3
     P $\mapsto$ $\{2, 3\}$
     R $\mapsto$ $\{\langle 1, 2 \rangle, \langle 2, 3 \rangle, \langle 3, 1 \rangle\}$
j a $\mapsto$ 1
     b $\mapsto$ 2
     c $\mapsto$ 3
     P $\mapsto$ $\{\langle 1, 2 \rangle, \langle 2, 3 \rangle, \langle 3, 1 \rangle\}$
     R $\mapsto$ $\{2, 3\}$

j fails to be an interpretation function on two counts: first, the one-place relational symbol P is not interpreted as a subset of U; second, R is not interpreted as a subset of the set of ordered pairs formed from U.

We now want to define a classical valuation for the formulae of CPDL. Just as a classical valuation for the formulae of CPL extends a bivalent truth value assigment, which is a function from propositional variables, or atomic formulae, into truth values, to a function from all formulae into truth values, so a classical valuation for the expressions of CPDL

Classical Predicate Logic

extends an interpretation function, which is a function from the nonlogical symbols into a universe, into subsets of the universe or into subsets of n-tuples of the universe, to a function that also assigns values to all the expressions, both nonlogical symbols as well as the formulae built from them. Thus, just as a classical valuation for formulae of CPL contains within it the bivalent truth value assignment that it extends, so a classical valuation for the formulae of CPDL contains within it the interpretation function it extends. The definition of a classical valuation for the formulae of CPDL has three parts. The first part stipulates that the classical valuation, which extends an interpretation function, agrees with the interpretation function when the classical valuation is applied to nonlogical symbols, just as a classical valuation for the formulae of CPL agrees with the bivalent truth value assignment it extends.

Definition 5 Classical valuation for CPDL

Let M, or $\langle U, i \rangle$, be a structure for a signature, $\langle \mathrm{IS}, \mathrm{RS}, \mathrm{ad} \rangle$. v_M is a classical valuation (for CPDL) if and only if it is a function whose domain is $\mathrm{IS} \cup \mathrm{RS} \cup \mathrm{FM}$ and which meets the following conditions:

(0) NONLOGICAL SYMBOLS

If c is a member of IS, then $v_M(c) = i(c)$.

If Π be a member of RS, then $v_M(\Pi) = i(\Pi)$.

(1) ATOMIC FORMULAE

Let Π be a member of RS, let $\mathrm{ad}(\Pi) = 1$ and let c be a member of IS. Then,

(1.1) $\quad v_M(\Pi c) = T$ iff $v_M(c) \in v_M(\Pi)$.

Let Π be a member of RS, let $\mathrm{ad}(\Pi) = n$ (where $n > 1$) and let $c_1, \ldots, c_n$ be occurrences of members of IS. Then,

(1.2) $\quad v_M(\Pi c_1 \ldots c_n) = T$ iff $\langle v_M(c_1), \ldots, v_M(c_n) \rangle \in v_M(\Pi)$.

(2) COMPOSITE FORMULAE

Then, for each α and for each β in FM,

(2.1) $\quad v_M(\neg \alpha) \quad = o_\neg(v_M(\alpha))$;

(2.2.1) $\quad v_M(\alpha \wedge \beta) \quad = o_\wedge(v_M(\alpha), v_M(\beta))$;

(2.2.2) $\quad v_M(\alpha \vee \beta) \quad = o_\vee(v_M(\alpha), v_M(\beta))$;

(2.2.3) $\quad v_M(\alpha \rightarrow \beta) \quad = o_\rightarrow(v_M(\alpha), v_M(\beta))$;

(2.2.4) $\quad v_M(\alpha \leftrightarrow \beta) \quad = o_\leftrightarrow(v_M(\alpha), v_M(\beta))$.

Let us see how an atomic formula of CPDL receives a truth value on the basis of a structure and the definition of an atomic formula (definition 2). An atomic formula of CPDL has a relatively simple structure. Consider, for example, the formula *Rab*. Its synthesis tree diagram has exactly two generations, as it were: the upper nodes *R, a,* and *b* are labeled and the lower node is labeled *Rab*. The structure $\langle U, i \rangle$ assigns values to each of these elements, and clause (1.2) of definition 5 determines the value of the lower node on the basis of the values of the three nodes immediately above it.

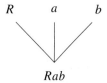

As careful readers will have noted, the definition of a classical valuation for the formulae of CPDL is a categorematic one. It is a straightforward matter to restate this definition as a syncategorematic one. This restatement is left as an exercise for the readers.

Though we shall not prove it here, interpretation functions give rise to classical valuations for the set of formulae. For this reason, the classical valuation is indexed by the interpretation function that gives rise to it, just as the classical valuations of CPL are indexed by truth value assignments.

Exercises: Interpretation functions and structures

1. Let IS be $\{a, b, c\}$; let RS be $\{P, R, T\}$; and let $\text{ad}(P) = 1$, $\text{ad}(R) = 2$, and $\text{ad}(T) = 3$. Also, let U be $\{1, 2, 3\}$. Determine, for each of the following functions, whether it, together with U, forms a structure for the signature given.

i_1 a $\mapsto$ $\{1\}$
    b $\mapsto$ 2
    c $\mapsto$ 3
    P $\mapsto$ $\{1, 2, 3\}$
    R $\mapsto$ $\{\langle 1, 2\rangle, \langle 2, 3\rangle, \langle 3, 1\rangle\}$
    T $\mapsto$ $\{\langle 2, 1, 2\rangle, \langle 2, 3, 1\rangle, \langle 1, 3, 1\rangle\}$

i_2 a $\mapsto$ 3
    b $\mapsto$ 2
    c $\mapsto$ 1
    P $\mapsto$ $\{1, 3\}$
    R $\mapsto$ $\{\langle 1, 1\rangle, \langle 2, 2\rangle, \langle 3, 3\rangle\}$
    T $\mapsto$ $\{\langle 1, 2, 3\rangle, \langle 2, 3, 1\rangle, \langle 3, 1, 2\rangle\}$

i_3 a $\mapsto$ 1
    b $\mapsto$ 2
    P $\mapsto$ $\{2, 3\}$

Classical Predicate Logic

$$
\begin{aligned}
&\quad R \mapsto \{\langle 2,1\rangle, \langle 2,2\rangle, \langle 3,1\rangle\} \\
&\quad T \mapsto \{\langle 1,1,3\rangle, \langle 3,3,1\rangle, \langle 2,3,2\rangle\}
\end{aligned}
$$

i_4
- $a \mapsto 1$
- $b \mapsto 1$
- $c \mapsto 1$
- $P \mapsto \{1, 2\}$
- $R \mapsto \{\langle 1,3\rangle\}$
- $T \mapsto \{\langle 1,1,1\rangle, \langle 2,2,2\rangle, \langle 3,3,3\rangle\}$

i_5
- $a \mapsto 1$
- $b \mapsto 2$
- $c \mapsto 2$
- $P \mapsto \{1, 1\}$
- $R \mapsto \{\langle 2,1\rangle, \langle 2,2\rangle, \langle 3,1\rangle\}$
- $T \mapsto \{\langle 1,3,1\rangle, \langle 3,3,2\rangle\}$

i_6
- $a \mapsto \emptyset$
- $b \mapsto 2$
- $c \mapsto 2$
- $P \mapsto \{1\}$
- $R \mapsto \{\langle 2,1\rangle, \langle 2,2\rangle, \langle 3,1\rangle\}$
- $T \mapsto \{\langle 1,1,3\rangle, \langle 3,3,1\rangle, \langle 2,3,2\rangle\}$

i_7
- $a \mapsto 1$
- $b \mapsto 1$
- $c \mapsto 1$
- $P \mapsto \emptyset$
- $R \mapsto \emptyset$
- $T \mapsto \emptyset$

i_8
- $a \mapsto 4$
- $b \mapsto 2$
- $c \mapsto 2$
- $P \mapsto \{3, 1\}$
- $R \mapsto \{\langle 1,1\rangle, \langle 2,2\rangle, \langle 3,3\rangle\}$
- $T \mapsto \{\langle 3,2,1\rangle, \langle 1,3,2\rangle, \langle 2,1,3\rangle\}$

i_9
- $a \mapsto 3$
- $b \mapsto 2$
- $c \mapsto 2$
- $P \mapsto \{1\}$
- $R \mapsto \{\langle 3,2\rangle, \langle 2,1\rangle, \langle 1,4\rangle\}$
- $T \mapsto \{\langle 2,2,1\rangle, \langle 1,3,3\rangle, \langle 1,1,3\rangle, \langle 2,1,2\rangle\}$

i_{10} a $\mapsto$ 3
    b $\mapsto$ 2
    c $\mapsto$ 2
    P $\mapsto$ $\{1, 1\}$
    R $\mapsto$ $\{\langle 3, 2\rangle, \langle 2, 1\rangle, \langle 1, 3\rangle\}$

2. Let $\langle U, j\rangle$ be a structure for the signature where IS is $\{a, b, c\}$, RS is $\{P, R, T\}$, $\mathrm{ad}(P) = 1$, $\mathrm{ad}(R) = 2$, and $\mathrm{ad}(T) = 3$ and where U is $\{1, 2, 3\}$ and j can be defined as follows.

j a $\mapsto$ 1
    b $\mapsto$ 2
    c $\mapsto$ 3
    P $\mapsto$ $\{3, 1\}$
    R $\mapsto$ $\{\langle 1, 2\rangle, \langle 3, 2\rangle, \langle 3, 3\rangle\}$
    T $\mapsto$ $\{\langle 3, 2, 1\rangle, \langle 1, 3, 2\rangle, \langle 2, 1, 3\rangle\}$

For each of the following formulae, state its truth value, as determined by the structure $\langle U, j\rangle$.

(a) $Pc \lor Rab$

(b) $\neg Taab$

(c) $\big((Rab \lor (Pa \land Tccc)) \to Pb\big)$

(d) $Pb \lor \neg Pb$

(e) Rba

(f) $\big((Tbba \to Pa) \lor Rbb\big)$

(g) $\neg Pc$

(h) $\neg\big((Pc \leftrightarrow Gcba) \lor Raa\big)$

(i) $\big(Pa \to (Tbab \lor Rcc)\big)$

(j) $\neg\neg\neg Rcc$

(k) Raa

3. Consider the definition of a classical valuation for CPDL (definition 5).

(a) Explain why clause (2.1) cannot be eliminated in favor of the single clause (2.2), modified by the simple elimination of the restriction that $n > 1$?

(b) What has to be done to permit clause (2.1) to be reduced to clause (2.2) by permitting n to be any positive integer? (HINT: reconsider the definition of U^n.)

(c) Notice that the domain of the classical valuation that extends an interpretation function is a proper superset the domain of the interpretation function. Define a classical valuation

Classical Predicate Logic

so that its domain is just the set of formulae, thereby making its domain and that of the interpretation function disjoint.

(d) Is the redefined classical valuation an extension of an interpretation function?

3.1 Semantic Properties and Relations

In chapter 6, several properties and relations were defined for the formulae and for sets of formulae of CPL. Properties and relations analogous to them are defined here for the formulae of CPDL and for sets of such formulae. The fundamental relation is the satisfaction relation, defined here in analogy with the way it was defined for CPL. The reason we say that the satisfaction relation defined here for CPDL is analogous to the one defined for CPL is because the relata, or the things related, of the satisfaction relation are different. In CPL, the satisfaction relation is between bivalent truth value assignments, on the one hand, and a formula or a set of formulae of CPL, on the other. In CPDL, the satisfaction relation is between structures for a signature, on the one hand, and a formula or a set of formulae of CPDL, on the other. Recall that the formulae of CPL and the formulae of CPDL are disjoint sets of formulae.

Definition 6 Satisfaction for CPDL

(1) For each structure M, and for each formula α,
  M satisfies α iff $v_M(\alpha) = T$.

(2) For each structure M, and for each set of formulae, Γ,
  M satisfies Γ iff, for each $\alpha \in \Gamma$, $v_M(\alpha) = T$.

Here are a pair of facts about the satisfaction relation.

Fact 1 Facts about satisfaction

(1) For each structure M, and for each formula α,
  M satisfies α iff M satisfies $\{\alpha\}$.

(2) Each structure satisfies the empty set.

In light of what was said earlier to the effect that the satisfaction relation of CPL and the satisfaction relation of CPDL are analogous, it should come as no surprise that these two facts pertaining to the satisfaction relation for CPDL are analogous to two facts pertaining to the satisfaction relation for CPL.

Let us see why these two analogous facts hold for CPDL. First, let us see why a structure for a signature satisfies a formula just in case it satisfies the singleton set whose sole member is the formula in question. Suppose that a structure for a signature, M, satisfies a formula, α. Then, by clause (1) of the definition of satisfaction, v_M assigns T to α. But since α is the only formula in the set, $\{\alpha\}$, v_M assigns T to each member of $\{\alpha\}$. Conversely, suppose that a structure for a signature, M, assigns T to each formula in $\{\alpha\}$, a set comprising a single formula. Then, by clause (2) of the definition of satisfaction, the v_M assigns T to α. Second, we show that each signature satisfies the empty set of formulae. To begin with, suppose that an arbitrarily chosen signature does not satisfy the empty set. Then, by clause (2) of the definition of satisfaction, the bivalent assignment fails to satisfy some formula in the set. But this is impossible, since the empty set has no members. Therefore, no bivalent assignment fails to satisfy the empty set. In other words, each bivalent assignment satisfies the empty set.

Readers should take the time to see that the way in which the two facts are established here is virtually identical to the way in which their counterparts in section 2.2.3 of chapter 6 are established. Indeed, the analogy between the satisfaction relation in CPL and the satisfaction relation in CPDL underpins not only all the properties and relations defined in this section, but also the logical facts pertaining to them. Readers are urged to take the time to establish these facts on their own. If, after a serious attempt or two, readers does not see how to establish some fact, they should consult the establishment of its analogue in section 2.2.3 of chapter 6.

3.1.1 Properties of formulae

As with CPL, so with CPDL, the formulae are divided into tautologies, formulae that each structure satisfies, contradictions, formulae that no structure satisfies, and contingencies, formulae that some structures satisfy and others do not.

Definition 7 Tautology of CPDL
A formula is a tautology iff each structure satisfies it.

Here are some tautologies of CPDL: $Pa \vee \neg Pa$, $Rab \to Rab$, $(Pa \wedge \neg Pa) \to Rab$, $Pb \to (Pa \to Pb)$. Notice that each of the formulae can be obtained from a tautology of CPL by substituting for the latter's atomic propositional variables a suitable formula of CPDL. For example, substituting Rab for p in $p \vee \neg p$, a tautology of CPL, yields the first of the preceding tautologies, while substituting Pb and Pa for p and q, respectively, in $p \to (q \to p)$ yields the last of the preceding tautologies.

Definition 8 Contradiction of CPDL
A formula is a contradiction iff each structure does not satisfy it.

$Pa \wedge \neg Pa$, $(Rab \vee \neg Rab) \to (Gaab \wedge \neg Gaab)$, $Rba \leftrightarrow \neg Rba$ are contradictions of CPDL. Again, each of these contradictions can be obtained through a suitable substitution of formulae of CPDL for propositional variables in formulae that are contradictions in CPL.

Definition 9 Contingency of CPDL
A formula is a contingency iff some structure satisfies it and some other does not.

Contingencies of CPDL include such formulae as $Pa \vee \neg Rab$, $Sa \to (Sa \wedge Rbb)$. They, like the other contingencies of CPDL, can be obtained by the substitution of formulae

Classical Predicate Logic

of CPDL for suitably chosen propositional variables in suitably chosen contingencies of CPL.

3.1.2 Properties of formulae and of sets of formulae

Next we turn to the properties of satisfiability and unsatisfiability. We shall define these properties so as to apply both to single formula and to sets of formula, as we did with their counterparts in CPL.

Definition 10 Satisfiability for CPDL

(1) A formula is satisfiable iff some structure satisfies it;

(2) a set of formulae is satisfiable iff some structure satisfies each formula in the set.

These sets of formulae are satisfiable: $\{Rab, Pb, Rab \to Pb\}$, $\{Rca\}$, $\{Pa \vee \neg Rab, Pa \wedge Rab, Pc \vee (Pa \to \neg Pc)\}$. Satisfiability is also known as *semantic consistency*.

Not surprisingly, facts analogous to those that hold for satisfiability in CPL hold in CPDL.

Fact 2 Facts about satisfiability

(1) For each formula α, α is satisfiable iff $\{\alpha\}$ is satisfiable.

(2) Each subset of a satisfiable set of formulae is satisfiable.

(3) The empty set is satisfiable.

We now turn to *unsatisfiability*, also known as *semantic inconsistency*.

Definition 11 Unsatisfiability for CPDL

(1) A formula is unsatisfiable iff no structure satisfies it;

(2) a set of formulae is unsatisfiable iff no structure satisfies all of the set's formulae.

Here are two unsatisfiable sets: $\{Pa, \neg Rab, Pa \to Rab\}$ and $\{Pa \wedge Rab, Pc \vee (Pa \to \neg Pc), Pa \wedge \neg Rab\}$.

Once again, facts analogous to those that hold for satisfiability in CPL hold in CPDL.

Fact 3 Facts about unsatisfiability

(1) For each formula α, α is unsatisfiable iff $\{\alpha\}$ is unsatisfiable.

(2) Each superset of an unsatisfiable set of formulae is unsatisfiable.

(3) FM, or the set of all formulae, is unsatisfiable.

Readers are urged to establish these facts for themselves.

We note, without establishing it, that each satisfiable set of formulae of CPDL can be obtained through a suitable substitution of formulae of CPDL for propositional variables in formulae in a satisfiable set of formulae of CPL. The same thing can be said for each unsatisfiable set of formulae of CPDL.

3.1.3 Relations between formulae and sets of formulae

Finally, we come to the two analogues in CPDL of the relations of *semantic equivalence* and of *entailment* defined for CPL.

Definition 12 Semantic equivalence for CPDL

A set of formulae or a formula, on the one hand, and a set of formula or a formula, on the other, are semantically equivalent iff (1) each structure satisfying the former satisfies the latter and (2) each structure satisfying the latter satisfies the former.

The following pairs of formulae are semantically equivalent: $Pa \vee Pa$ and Pa, $Pa \vee \neg Rab$ and $Rab \rightarrow Pa$, and $\neg(Pa \wedge Rab)$ and $\neg Pa \vee \neg Rab$.

Still again, analogous facts obtain for the semantic equivalence of CPDL as obtain for the semantic equivalence of CPL.

Fact 4 Facts about semantic equivalence

(1) All tautologies and sets of tautologies are semantically equivalent.

(2) All tautologies and all sets of tautologies are semantically equivalent to the empty set.

(3) All contradictions and all sets of contradictions are semantically equivalent.

(4) All contradictions and all sets of contradictions is semantically equivalent to FM, or the set of all formulae.

The second relation, entailment, is a relation that a set of formulae bears to a single formula.

Definition 13 Entailment for CPDL

A set of formulae *entails* a formula, or $\Gamma \models \alpha$, iff each structure that satisfies the set Γ satisfies the formula α.

Whereas the set of formulae, $\{Pa, Pa \rightarrow Rab\}$ entails the formula Rab, that is, $\{Pa, Pa \rightarrow Rab\} \models Rab$, the set of formulae, $\{Rab, Pa \rightarrow Rab\}$ does not entail the formula Pa, that is, $\{Rab, Pa \rightarrow Rab\} \not\models Pa$.

Fact 5 Facts about entailment

Let α and β be formulae. Let Γ and Δ be sets of formulae.

(1) If $\alpha \in \Gamma$, then $\Gamma \models \alpha$.
(2) If $\Gamma \subseteq \Delta$ and $\Gamma \models \alpha$, then $\Delta \models \alpha$.
(3) $\Gamma \cup \{\alpha\} \models \beta$ iff $\Gamma \models \alpha \rightarrow \beta$.
(4) $\Gamma \models \alpha \wedge \beta$ iff $\Gamma \models \alpha$ and $\Gamma \models \beta$.

Fact 6 Facts about entailment and tautologies

Let α be a formulae. Let Γ be a set of formulae.

Classical Predicate Logic

(1) α is a tautology iff $\emptyset \models \alpha$.

(2) α is a tautology iff, for each Γ, $\Gamma \models \alpha$.

Fact 7 Facts about entailment and unsatisfiability
Let α be a formulae. Let Γ be a set of formulae.

(1) $\Gamma \models \alpha$ iff $\Gamma \cup \{\neg \alpha\}$ is unsatisfiable.

(2) Γ is unsatisfiable iff, for any α, $\Gamma \models \alpha$.

(3) Γ is unsatisfiable iff, for at least one α, there is a contradiction, $\Gamma \models \alpha$.

We close this section by pointing out that the relation of semantic equivalence in CPDL can be obtained through a suitable substitution of formulae of CPDL for the propositional variables in formulae that are themselves, in CPL, semantically equivalent or form sets that are semantically equivalent. The same holds for entailment in CPDL.

Exercises: Semantic properties and relations

1. Determine which of the following formulae are tautologies, which are contingencies, and which are contradictions.

(a) $\neg(Pa \wedge (\neg Pa \wedge Qc))$

(b) $(Rab \wedge Qc) \rightarrow \neg Rab$

(c) $(Pb \rightarrow Qa) \rightarrow (\neg Qa \rightarrow \neg Pb)$

(d) $(Pc \vee Qb) \leftrightarrow (Pc \vee (\neg Pc \wedge Qb))$

(e) $(Pa \vee \neg Rab) \rightarrow Rab$

(f) $Gabc \vee \neg Gabb$

(g) $Pa \leftrightarrow \neg\neg Pc$

(h) $(Pa \vee Qc) \leftrightarrow (Pa \vee (\neg Pa \rightarrow Qc))$

(i) $((Pa \rightarrow Pa) \rightarrow Gabc) \rightarrow Pa$

(j) $Pa \wedge \neg Pb$

2. Determine whether the following sets of formulae are satisfiable (semantically consistent) or unsatisfiable (semantically inconsistent).

(a) $\{(Pb \vee Qb) \wedge (Pb \vee Rba),\ Pb \wedge (Qa \vee Rba)\}$

(b) $\{(Pb \rightarrow Qb) \rightarrow (Rbc \rightarrow Sbb),\ \neg(Pb \wedge Qb) \rightarrow Scc,\ \neg Scc \rightarrow \neg Pb\}$

(c) $\{Pb \leftrightarrow Qa,\ Pb \leftrightarrow \neg Qa\}$

(d) $\{(Pb \vee Qb) \wedge Rcb,\ Rcb \rightarrow Sca,\ \neg Pb \rightarrow (Rcb \rightarrow Sca)\}$

3. Establish the truth or falsity of the following claims.

(a) $Pa \land (Qc \lor Rab) \models (Pa \lor Qa) \land (Pa \lor Rab)$

(b) $Pa \leftrightarrow (Pa \land Qb) \models Qc \to Pb$

(c) $(Pa \land Qc) \to Rab \models Pa \to (Qc \to Rab)$

(d) $Qc \to Pa \models \neg Qc \to \neg Pa$

(e) $\{Pa \lor Qc, \neg Pa\} \models \neg Qc$

(f) $(Pa \land Qb) \lor (\neg Pa \land \neg Qb) \models Pa \to Qa$

4. For each of the following claims, indicate whether it is true or false. If it is true, explain why it is true; if it is false, provide a counterexample.

(a) Each unsatisfiable set contains a contradiction.

(b) No set of contingencies can form a satisfiable set.

(c) Each set of tautologies is satisfiable.

(d) No satisfiable set contains a tautology.

(e) Some satisfiable sets contain a tautology.

(f) α is a contingency iff $\{\alpha\}$ is satisfiable.

(g) A contingent formula's negation may be a tutology.

(h) $\Gamma \models \alpha \lor \beta$ iff $\Gamma \models \alpha$ or $\Gamma \models \beta$.

(i) $\{\alpha\}$ and $\{\beta\}$ are semantically equivalent iff $\alpha \leftrightarrow \beta$ is a tautology.

(j) If $\Delta \subseteq \Gamma$ and Γ is unsatisfiable, then Δ is unsatisfiable.

(k) If $\Delta \subseteq \Gamma$ and Δ is satisfiable, then Γ is satisfiable.

(l) If $\Gamma \models \alpha$, then $\alpha \in \Gamma$.

(m) If $\Delta \subseteq \Gamma$ and $\Gamma \models \alpha$, then $\Delta \models \alpha$.

4 Deduction

Deduction in CPDL is precisely the deduction of CPL.

Exercises: Deductions

Assuming that all the following formulae are formulae of CPDL, justify each of the claims with a deduction.

1. $\{Pa \to Qb,\ Qb \to Rab,\ Rab \to Scc\} \vdash Pa \to Scc$

2. $Pb \to Gabc \vdash \neg Gabc \to \neg Pb$

3. $\neg Gabc \to \neg Pb \vdash Pb \to Gabc$

Classical Predicate Logic 395

4. $\{Pc \to Qa, Qa \to Rbb\} \vdash \neg Rbb \to \neg Pc$
5. $Pa \to Qb \vdash Pa \leftrightarrow (Pa \land Qb)$
6. $Pc \to (Qa \to Rac) \vdash (Pc \land Qa) \to Rac$
7. $\neg Gaac \leftrightarrow \neg Rca \vdash Gaac \leftrightarrow Rca$
8. $\{\neg Pa \lor \neg Qa, Qa\} \vdash Pa$
9. $(Pa \land Qb) \lor (Pa \land Rbc) \vdash (Pa \land (Qb \lor Rbc))$

SOLUTIONS TO SOME OF THE EXERCISES

2.2 Formulae and subformulae

1. Formulae of CPDL (syncategorematic version) FM, the formulae of CPDL, is the set defined as follows:

(1)   AF ⊆ FM;

(2.1)   if $\alpha \in$ FM, then $\neg\alpha \in$ FM;

(2.2.1) if $\alpha, \beta \in$ FM, then $(\alpha \land \beta) \in$ FM;

(2.2.2) if $\alpha, \beta \in$ FM, then $(\alpha \lor \beta) \in$ FM;

(2.2.3) if $\alpha, \beta \in$ FM, then $(\alpha \to \beta) \in$ FM;

(2.2.4) if $\alpha, \beta \in$ FM, then $(\alpha \lor \beta) \in$ FM;

(3)   nothing else is.

2. (a) The scope of an occurrence of a propositional connective in a formula α is the subformula of α that contains the propositional connective's occurrence but whose immediate subformulae do not.

(c) A formula α occurs within the scope of a propositional connective's occurrence in a formula β iff α is a subformula of the scope of the propositional connective's occurrence in formula β.

3. (a) Yes   (e) Yes   (i) Conventional
   (c) No   (g) Conventional (k) No

4. (a) Composite (e) Composite   (i) Composite
   (c) Basic   (g) Composite

5. (a) $(Qa \to (Pc \land Pc))$, $(Pc \land Pc)$, Qa, Pc
   (c) $\neg((\neg Rdb \lor Pc) \land \neg Qa)$, $((\neg Rdb \lor Pc) \land \neg Qa)$, $(\neg Rdb \lor Pc)$, $\neg Qa$, Qa, $\neg Rdb$, Pc, Rdb
   (e) $(\neg(Rdb \lor (Qa \land Sdd)) \to Pc)$, $\neg(Rdb \lor (Qa \land Sdd))$, Pc, $(Rdb \lor (Qa \land Sdd))$, Rdb, $(Qa \land Sdd)$, Qa, Sdd.

6. (a) $(Pb \to Qc)$ (d) $\bigl(\bigl((Pb \to Qc) \lor (Gabc \land Qa)\bigr) \land (Scc \to Pc)\bigr)$
   (c) $(Gabc \land Qa)$

7. (a) Yes   (e) No   (i) Yes
   (c) Yes   (g) Yes

3 Interpretation functions and structures

1. For the signature $\langle \text{IS, RS, ad} \rangle$, $\langle U, i_1 \rangle$, $\langle U, i_3 \rangle$ and $\langle U, i_9 \rangle$ are not structures; whereas $\langle U, i_5 \rangle$ and $\langle U, i_7 \rangle$ are.

2. (a) True   (e) False   (i) True
   (c) False   (g) False   (k) False

3.1.3 Semantic properties and relations

1. (a) Tautology   (e) Contingency   (i) Contingency
   (c) Tautology   (g) Contingency

2. (a) Satisfiable   (c) Unsatisfiable

3. (a) Yes   (c) Yes   (e) No

4. (a) False   (e) True   (i) True   (m) False
   (c) True   (g) False   (k) False

(f): The left condition implies the right condition, but not conversely.
(h): The right condition implies the left condition, but not conversely.

10 Grammatical Predicates and Minimal Clauses in English

1 Introduction

In chapter 9, to help readers get an intuitive grasp of the notions of relational symbols and individual symbols, fundamental to CPDL, we pointed out an analogy between the relational symbols and individual symbols of CPDL, on the one hand, and the verbs and proper nouns of English, on the other. In particular, we noticed that minimal, independent English clauses—consisting, say, of a proper noun and an intransitive verb, or of two proper nouns and a transitive verb, or of three proper nouns and a ditranstive verb—are analogous to an atomic formula consisting of a one-place relational symbol and an individual symbol, a two-place relational symbol and two occurrences of individual symbols, and a three-place relational symbol and three occurrences of individual symbols, respectively. It is our familiarity with minimal, independent English clauses that helps us to grasp how the notation of CPDL is intended to be understood.

What we shall see in this chapter is that the matters pertaining to minimal, independent English clauses are more complicated than matters pertaining to the atomic formulae of CPDL. In particular, the categorization of the basic expressions of CPDL is much simpler than the categorization of the basic expressions of minimal English clauses, and the formation of the atomic formulae of predicate logic is also much simpler than the formation of the minimal, independent English clauses.

Recall that in CPDL we start with a stock of basic expressions that are categorized by what is called a signature. We then use the stock of basic expressions to define a set of complex expressions, which comprise the atomic formulae and the composite formulae built from them. In addition, we permit a range of values to be assigned to the basic expressions. However, not just any assignment of values is an interpretation function. What is crucial is that the simple symbols are assigned values in light of how they are categorized. Finally, interpretation functions are extended to valuations that assign truth values first to the atomic formula and thereafter to all composite formulae.

In chapter 3, we introduced the notion of a constituency grammar. A constituency grammar comprises a nonempty, finite set of basic expressions, a nonempty, finite set of categories, a lexicon, and a set of constituency formation rules. Now, just as a signature provides a set of categories and categorizes symbols into individual symbols and relational symbols of various adicities, so a constituency grammar's lexicon pairs each basic expression with a category. In addition, just as the formation rules of CPDL state how to form atomic formulae using the categories of the signature and how to obtain composite formulae from the atomic ones, so the formation rules of a constituency grammar, constituency formation rules, provide for the formation of more complex constituents from simpler constituents furnished by its lexicon. The problem we shall be addressing here is that of assigning values not only to the members of the lexicon but also to the constituents formed from them. In other words, we shall determine what constitutes an interpretation function for the basic expressions of the language, analogous to the interpretation function for the set of the individual and relational symbols of CDPL, as well as define a valuation based on the interpretation function that assigns a value to each complex constituent, analogous to the valuation that assigns a truth value to each formulae.

Constituency grammars are inadequate in a number of ways, as we learned in section 3.4 of chapter 3. We promised to address two forms of inadequacy set out in section 3.4.1, namely their failure to recognize that words of the same lexical category are nonetheless distinguished by the kinds of complements they admit and their failure to recognize that any expression assigned the syntactic category informally annotated as XP contains a word assigned the syntactic category informally annotated as X. As we shall see, the solution to the first problem is also a solution to the second. Another inadequacy of constituency grammars, not discussed earlier but to be discussed herein, is that constituency grammars do not permit the definition of a structure. As we shall see, the solution to the first two problems is also a solution to this last problem. In other words, there is a single solution to all three of these problems. Moreover, this single solution paves the way for addressing other inadequacies to be set out later in this chapter.

2 Minimal English Clauses

Let us define a *minimal English clause* to be a clause in the declarative mood each of whose constituents is minimal. Recall that each independent English clause in the declarative mood comprises a subject noun phrase and a verb phrase. A minimal noun phrase is one that comprises just a proper noun. But what is a minimal verb phrase? To answer this question, we must distinguish between a head, a modifier, and a complement. In general, a constituent has a *head*. A noun phrase has as its head a noun, a verb phrase a verb, an adjective phrase an adjective, a prepositional phrase a preposition, and so forth.

Grammatical Predicates and Minimal Clauses in English

In addition to the head, a constituent may contain immediate constituents known as *complements*. From the point of view of traditional grammar, complements were thought of as constituents that complete, as it were, the head. Complements, then, are coconstituents of the head that the head seems to require, whereas modifiers are coconstitutents that are not required. As we shall see, noncomplements are, indeed, optional, but complements may or may not be optional. A minimal English clause then comprises a minimal subject noun phrase and a verb phrase headed by an inflected verb and accompanied by its complements alone.

In this section, we shall explore the syntax of minimal constituents as well as the syntax of the minimal clauses that can be obtained from such constituents. This exploration naturally divides into an investigation of verbs and their complements, adjectives and their complements, prepositions and their complements, and nouns and their complements.

2.1 Verb Phrases: Verbs and Their Complements

In previous chapters, we saw how the same expression shows up as a constituent in many different larger expressions. In chapter 3, we saw that expressions which we labeled with VP appear as answers to questions, as the constituent following the copular verb in so-called pseudocleft sentences as well as the dislocated constituent in so-called VP fronting.

(1.1) Question What did the tall guide do?
      Answer   [VP meet the foreign visitor at the train station at five o'clock].
(1.2)   What the tall guide did was [VP meet the foreign visitor at the train station at five o'clock].
(1.3)   Ben promised to finish his paper and [VP finish his paper] he will.

In chapter 4, we saw that expressions that serve as the antecedents to certain forms of ellipsis as well as those that serve as the antecedents of the proforms such as *so* and *the same thing,* when taken with the verb *to do,* are VP constituents.

(2.1)   Fred [VP drove a fast car], and Tom did __ too.
(2.2)   Fred [VP drove a fast car], as did Tom __ .
(2.3.1) Fred [VP drove a fast car], and Tom did so too.
(2.3.2) Fred [VP drove a fast car], and so did Tom.
(2.4)   Fred [VP drove a fast car], and Tom did the same thing.

Having identified some patterns that warrant the hypothesis that a verb phrase is a constituent of English expressions, let us turn to the problem of distinguishing constituents that are complements of a verb and those that are not.

2.1.1 Distinguishing verb complements from noncomplements

It has long been thought that some constituents are essential to verbs while others are not. In sentence (3.1),

(3.1)   Alice [VP greeted [NP the visitor] [PP at the door]].

(3.2)   Alice [VP greeted [NP the visitor]].

(3.3)   *Alice [VP greeted [PP at the door]].

the prepositional phrase *at the door* can be omitted, as shown in (3.2), without the resulting expression becoming unacceptable, whereas omission of the noun phrase *the visitor*, as shown in (3.3), results in unacceptability.

However, as we shall see, while the inomissibility of a verb's coconstituent is good prima facie evidence that it is a complement of the verb, the omissibility of a coconstituent is not evidence that it is not a complement of the verb. In other words, the inomissibility of a verb's coconstituent is a prima facie sufficient condition on its being a complement of the verb; it is not, however, a necessary condition.

With the advent of transformational grammar, various patterns were discovered that seem to help discriminate complements from the noncomplements of verbs. For example, Klima (1965) noticed that the clause that is subject of a pseudocleft sentence admits noncomplements of the sentence's verb phrase and excludes its complements. Thus, for example, the prepositional phrase, *in the bathtub,* may appear in the subject clause of the pseudocleft in sentence (4.2), but not the noun phrase, *dirty dishes*.

(4.1)   Alice [VP washes [NP dirty dishes] [PP in the bathtub]].

(4.2)   [NP What Alice does in the bathtub] [VP is wash [NP dirty dishes]].

(4.3)   *[NP What Alice does dirty dishes] [VP is wash [PP in the bathtub]].

In other words, the coconstituents of a verb that may appear in the subject clause of a pseudocleft sentence to the immediate right of the verb *to do* are the noncomplements of the verb.

Another pattern is found with endophoric expressions with verb phrase antecedents. A coconstituent of a verb in the verb phrase serving as the antecedent of an endophor is the verb's complement if a constituent of the same type cannot be added to the endophoric clause without disturbing the endophoric clause's acceptability; whereas it is not a complement if a constituent of the same type can be added without disturbing the endophoric clause's acceptability.

Consider the following sentences evincing verb phrase ellipsis and the verb phrase pro-forms *so* (Huddleston 2002, sec. 1) and *the same thing* (Baker 1989, chap. 3.1), preceded by the auxiliary verb *to do*.

(5.1) Alice [VP washes clothes in the bathtub] and Bill does too.
(5.2) Alice [VP washes clothes in the bathtub] and Bill does too in the kitchen sink.
(5.3) Alice [VP washes clothes] and *Bill does too dishes.
(6.1) Alice [VP washes clothes in the bathtub] and Bill does *so* too.
(6.2) Alice [VP washes clothes in the bathtub] and Bill does *so* too in the kitchen sink.
(6.3) Alice [VP washes clothes] and *Bill does *so* too dishes.
(7.1) Alice [VP washes clothes in the bathtub] and Bill does *the same thing*.
(7.2) Alice [VP washes clothes in the bathtub] and Bill does *the same thing* in the kitchen sink.
(7.3) Alice [VP washes clothes] and *Bill does *the same thing* dishes.

In the second sentence of each of the triples, the prepositional phrase *in the kitchen sink*, parallel to the prepositional phrase *in the bathtub* in the antecedent verb phrase, is appended to the endophoric expression without disturbing the acceptability of the resulting sentence. In the third sentence, the noun phrase *dishes*, parallel to the noun phrase *clothes* in the antecedent verb phrase, is appended to the endophoric expression with the result that the clause containing the endophor becomes unacceptable.

2.1.2 Basic patterns of English verb complements

Here we survey the range of verb complements identified in the principal descriptive grammars of English (see Quirk et al. 1985, chap. 16; Huddleston 2002.) As we saw in chapter 3, traditional English grammar distinguished between intransitive and transitive verbs. English intransitive verbs are those verbs that resist any complement.

(8.1) The smoke vanished.
(8.2) *Bill vanished the salad.

NO COMPLEMENT (intransitive verbs)

to bloom, to crawl, to die, to disappear, to elapse, to expire, to fail, to faint, to fall, to laugh, to sleep, to stroll, to vanish, ...

English transitive verbs are those that require a noun phrase complement.

(9.1) Bill abandoned [NP his teammate].
(9.2) *Bill abandoned

NP COMPLEMENT (transitive verbs)

to abandon, to buy, to cut, to catch, to consider (the proposal), *to destroy, to devour, to expect, to greet, to guarantee* (the product), *to keep* (the gift), *to lack, to like, to lock, to maintain* (one's health), *to note* (the error), *to prove, to purchase, to pursue, to vacate,* ...

But noun phrases are not the only complements for English verbs, adjective phrases and prepositional phrases are as well. These include what traditional grammar calls *copular,* or *linking,* verbs.

(10.1) Carl remained [AP proud of his accomplishments].

(10.2) *Carl remained

AP COMPLEMENT (copular verbs)
to be, to become, to appear, to feel, to get, to look, to seem, to smell, to sound, to taste, to remain, to keep, to stay, ...

Some English verbs have single prepositional phrase complements.

PP COMPLEMENT
to approve of, *to dash* to, *to depend* on, *to dispose* of, *to hint* at, *to refer* to, *to rely* on, *to wallow* in, ...

(11.1) Dan relied [PP on Carl's advice].

(11.2) *Dan relied

Clauses as well as infinitival and gerundial phrases may serve as complements. Clauses and infinitival and gerundial phrases all contain verbs. The verb of a clause has a finite form, while the verb of an infinitival phrase has a nonfinite form, known as an infinitive, and the verb of a gerundial phrase also has a nonfinite form, known as a gerund. Clauses have subjects and are optionally preceded by the word *that,* which, in that role, is called a *complementizer*. Clauses and infinitival phrases may take either a declarative form or an interrogative form. Such a distinction does not apply to gerundial phrases. We shall assign the category S to clauses, infinitival phrases, and gerundial phrases. However, when useful, we shall distinguish among these five types of constituents. We shall distinguish gerundial phrases from the others with the label Sg. We shall distinguish among the remaining four as follows: Sfd for declarative clause, which has a finite verb, Sfi for an interrogative clause, which has a finite verb, Snd for a declarative infinitival phrase, which has a nonfinite verb, and Sni for an interrogative infinitival phrase, which also has a nonfinite verb.

(12.1) Alice noted [Sfd it was raining].

(12.2) *Alice noted

(13.1) Alice wondered [Sfi who was leaving].

(13.2) *Alice wondered

(14.1) Bill decided [Snd to spend the night in the garage].

(14.2) *Bill decided

(15.1) Alice recalled [Sni where to drop off her keys].

(15.2) *Alice recalled

Grammatical Predicates and Minimal Clauses in English   403

(16.1)   Carl enjoyed [Sg spending the night in the garage].
(16.2)   *Carl enjoyed

Some English verbs take two complements: the first a noun phrase, the second either a noun phrase, an adjective phrase, a prepositional phrase, or an adverbial phrase.

(17.1)   Carl made [NP his friend] [AP angry at him].
(17.2)   *Carl made [NP his friend].
(17.3)   *Carl made [AP angry at him].
(18.1)   The members proclaimed [NP their leader] [NP the president].
(18.2)   *The members proclaimed [NP their leader].
(18.3)   *The members proclaimed [NP the president].
(19.1)   The judge accused [NP the defendant] [PP of malfeasance].
(19.2)   *The judge accused [NP the defendant].
(19.3)   *The judge accused [PP of malfeasance].
(20.1)   The client treated [NP the waiter] [AdvP shabbily].
(20.2)   *The client treated [NP the waiter].
(20.3)   *The client treated [AdvP shabbily].

We now turn to those cases where the second complement is a clause. In many cases, one of the complements is optional and sometimes both are.

(21.1)   Alice convinced [NP Bill] [Sfd it was raining].
(21.2)   Alice asked [NP Bill] [Sfi where it was raining].
(21.3)   Alice persuaded [NP Bill] [Snd to sell his car].
(21.4)   Alice asked [NP Bill] [Sni what to say].
(21.5)   Alice hates [NP Bill] [Sg jumping up and down on the bed].
(22.1)   Bill said [PP to Alice] [Sfd it is raining].
(22.2)   Bill inquired [PP of Alice] [Sfi where she had gone].
(22.3)   Alice waited [PP for Bill] [Snd to wash the dishes].
(22.4)   Bill inquired [PP of Alice] [Sni where to go].
(22.5)   Bill thought [PP of Alice] [Sg winning the prize].

Next we turn to three complement verbs. These are relatively rare. And for the most part, two of the three complements are optional.

(23.1)   Bill fined [NP Alice] [NP one hundred dollars] [PP for the speeding violation].
(23.2)   Bill transferred [NP money] [PP from one bank account] [PP to another].
(23.3)   Alice bet [NP Bill] [NP ten dollars] [S that she would win the race].

We now tabulate the range of complements to English verbs without regard for the refined distinctions among clauses, infinitival phrases, and gerundial phrases.

VERB COMPLEMENT PATTERN

NONE	ONE	TWO	THREE
	AP	NP AP	
	NP	NP NP	NP NP PP
	PP	NP PP	NP PP PP
	AdvP	NP AdvP	
	S	NP S	NP NP S
		PP S	
		PP AP	
		PP PP	

2.1.3 Complications

We now turn to some of the complications of verbs and their complements. We shall consider three: one is unique to verbs and their complements and two occur also with adjectives, prepositions, and nouns and their complements.

2.1.3.1 Verbs and prepositions We begin with the one confined to verbs. Prepositions that may immediately follow a verb often give rise to a well-known complication of the English verb phrase. To illustrate, consider the verb *to call* and the two prepositions *up* and *on*.

It might seem that the two sentences in (24) have the very same constituent structure, comprising the verb *to call* when immediately followed by the prepositions *up* and *on*. It is easy to show that these prepositions are required by the verb *to call*.

(24.1)   Dan called up his boss.

(24.2)   Dan called on his boss.

The pseudocleft version of these sentences make it clear that what follows the expression *call* is its complement.

(25.1)   *What Dan did up his boss is call.

(25.2)   *What Dan did on his boss is call.

One might therefore be tempted to conclude that the verb takes a prepositional phrase complement and assign the same constituent structure to both sentences.

(26.1)   Dan [VP called [PP up his boss]].

(26.2)   Dan [VP called [PP on his boss]].

Grammatical Predicates and Minimal Clauses in English

However, such a view is belied by the contrast in acceptability of the patterns into which they can enter (Quirk et al. 1985, chap. 16.4, 16.6; Huddleston 2002, sec. 6.2), as shown here:

(27.1) Dan called his boss up/*on.

(27.2) On/*up which person did Dan call?

(27.3) Dan called on/*up Bill and on/*up Carl.

(27.5) Dan called angrily on/*up his boss.

(27.6) Dan called on/*up him.

To address this difference in patterns, it is useful to distinguish transitive verbs associated with prepositions in their complements into two types: those where the preposition is compounded with the verb and the complement of the compounded verb is a noun phrase, as is the case in (24.1), and those where the verb is not compounded and the complement is indeed a prepositional phrase, as is the case in (24.2).

(28.1) Dan [VP [V called up] [NP his boss]].

(28.2) Dan [VP [V called] [PP on his boss]].

We shall refer to verbs compounded with a preposition as *compounded verbs*. Other compounded verbs include *to hand in, to knock out, to live down, to look up, to make out, to set up,* and *to sound out*.

This distinction and its analysis pay two dividends. First, many English expressions comprising a lone verbal expression followed by a preposition and a noun phrase are ambiguous. Consider the next two sentences.

(29.1) Bill got over the message.

(29.2) Bill got the message over.

The first sentence conveys either that Bill succeeded in communicating a message or that Bill overcame an emotional upset resulting from a message communicated to him. The second sentence only conveys that Bill succeeded in communicating the message; it does not convey that Bill overcame an emotional upset resulting from a message communicated to him. There are many such expressions, including *to shout down* (an opponent/a hall), *to turn in* (a fugitive/a wrong direction), *to run off* (a copy/a road), *to turn on* (the light/his supporters), *to take in* (the box/his friends).

The second dividend, which comes as no surprise, is that, among the intransitive verbs of English, are verbs that must be followed by a preposition but cannot be further followed by a noun phrase. In other words, English has intransitive compounded verbs. Here are some examples: *to break down* (cf. *to cry*), *to die down, to die off, to pass out* (cf. *to faint*), *to play around, to sound off* (cf. *to express one's opinion*), *to turn up* (cf. *to appear*).

The fact that there are compounded intransitive verbs invites the question of whether there are compounded verbs that take complements. Indeed there are, some taking noun phrase complements, as we saw in some of the preceding examples, and some taking prepositional phrase complements, as exemplified by such verbs as *to touch down* (on something), *to take off* (from something), *to catch on* (to something), *to get on* (with someone), *to give in* (to something).

These facts, well known since the inception of transformational grammar, are without any accepted treatment.

2.1.3.2 Complement polyvalence   Readers may have noticed that many English verbs with a complement admit different syntactic categories as the complement. We shall say that such verbs have *polyvalent complements*. To stress just how pervasive this is, we shall briefly survey the cases. We start with verbs taking just one complement.

The English copular verb *to be* requires a complement yet its complement may be either an adjective phrase, a noun phrase, a prepositional phrase, or even an adverbial phrase.

(30.1)   Dan is [AP silent].

(30.2)   Carl is [NP a scholar].

(30.3)   Beth is [PP in Paris].

(30.4)   Alice is [AdvP downstairs].

There are at least two other copular verbs in English with the same variability. Still other copular verbs require a complement that may be either an adjective phrase, a noun phrase, a prepositional phrase headed by *like,* an infinitival phrase headed by the verb *to be,* or a declarative clause introduced by *as though.*

(31.1)   Bill appeared [AP foolish].

(31.2)   Bill appeared [NP a fool].

(31.3)   Bill appeared [PP like a fool].

(31.4)   Bill appeared [Snd to be a fool].

(31.5)   Bill appeared [Sfd as though he were a fool].

Besides copular verbs, many verbs that take a clausal complement of any kind alternate with noun phrase complements. Examples of each kind of clausal complement follow.

(32.1)   Bill believes [NP the claim that π is transcendental].

(32.2)   Bill believes [Sfd π is transcendental].

(33.1)   Bill asked [NP the time].

(33.2)   Bill asked [Sfi what time it is].

(34.1)   Bill started [NP the book].

(34.2)   Bill started [Snd to read the book].

(35.1)   Bill tried [NP the door].
(35.2)   Bill tried [Sg opening the door].
(36.1)   Bill inquired [PP about a place to sleep].
(36.2)   Bill inquired [Sni where to sleep].

Double complement verbs also show alternations with respect to the second complement. We begin with one alternation recognized by grammarians and extensively studied by linguists.

(37.1)   Bill [VP considers [NP Alice] [NP a friend]].
(37.2)   Bill [VP considers [NP Alice] [AP quite competent]].

Several of these verbs permit the second complement to be an infinitival phrase headed by the verb *to be*.

(38.1)   Bill considers [VP [NP Alice] [Snd to be [NP a friend]]].
(38.2)   Bill considers [VP [NP Alice] [Snd to be [AP quite competent]]].

In fact, these verbs usually permit their two complements to alternate with a clausal complement.

(39.1)   Bill [VP considers (that) [Sfd [NP Alice] [VP is a friend]]].
(39.2)   Bill [VP considers (that) [Sfd [NP Alice] [VP is quite competent]]].

By dint of the fact that the sentences in (39) are paraphrases of their counterparts in (37) and (38), transformational linguists have described the relationship of the second complement to the first in the sentence in (37) as one of *secondary predication* and call the pair of complements taken together as a *small clause*.

It is interesting to note that not all verbs that admit both a noun phrase and an adjective phrase as a second complement also admit an infinitival phrase as a second complement or admit a lone clause.

(40.1)   Alice [VP keeps [NP Bill] [AP poor]].
(40.2)   Alice [VP keeps [NP Bill] [NP a pauper]].
(40.3)   *Alice [VP keeps [NP Bill] [Snd to be poor/a pauper]].
(40.3)   *Alice [VP keeps [Snd Bill is poor/a pauper]].

Another alternation for a second complement is between a prepositional phrase and a noun phrase. There are two patterns. In one pattern, suppression of the preposition in the second complement turns the second complement into a noun phrase with the resulting sentence being a paraphrase of the first.

(41.1)   Alice [VP appointed [NP Bill] [PP as her assistant]].
(41.2)   Alice [VP appointed [NP Bill] [NP her assistant]].

We note, however, that the verb *to choose,* which is nearly synonymous in one usage with *to appoint,* does not permit the same alternation.

(42.1)   Alice [VP chose [NP Bill] [PP as her assistant]].

(42.2)   *Alice [VP chose [NP Bill] [NP her assistant]].
         (Allerton 1982, 138)

Another, better known alternation between a prepositional phrase and a noun phrase occurs with respect to the second complement of certain verbs, sometimes called *ditransitive* verbs. Perhaps the most commonly cited example of such a verb is the verb *to give.*

(43.1)   Bill [VP gave [NP a dog] [PP to Alice]].

(43.2)   Bill [VP gave [NP Alice] [NP a dog]].

We observe that sentences (43), like those in (42), are synonymous. However, unlike sentences (42), where the second is obtained from the first by merely suppressing the preposition, the second sentence in (43) is obtained from the first by both suppressing the preposition and permuting the complements.

Similar alternations occur with verbs whose second complement is headed by the preposition *for.*

(44.1)   Bill [VP bought [NP a dog] [PP for Alice]].

(44.2)   Bill [VP bought [NP Alice] [NP a dog]].

Again, the permutation is crucial, if equivalence is to be preserved.[1]

This pattern is confined to polyvalent verbs whose complement list specifies a noun phrase complement in the first coordinate of its complement list and specifies either both a noun phrase and a prepositional phrase complement or two prepositional phrase complements, each with a different preposition from the other. Moreover, they give rise to pairs of synonymous sentences, provided that the noun phrases in the complements are permuted and that either the preposition is suppressed or the prepositions are exchanged.[2] We shall refer to such verbs as ones that admit *complement permutation.*[3]

Not all verbs that specify a noun phrase complement in their complement list and either both a noun phrase and a prepositional phrase complement or two prepositional phrase complements give rise, through permutation, to such pairs of synonymous sentences.

---

1. Green (1974, chap. 4B) discusses some verbs, such as *to teach* and *to show,* that admit such a permutation but, in her judgment, are not fully equivalent.

2. The pioneering work here is Fillmore (1965) and Green (1974).

3. Early transformational linguists treated such permutation by a transformation called *dative shift.* The word *dative* was used because the prepositions *for* or *to* often mark what is called in traditional English grammar the indirect object and the indirect object in Indo-European languages with case is often marked by the dative case.

Grammatical Predicates and Minimal Clauses in English

Indeed, there are many pairs of verbs that are near synonyms, one of which admits the permutation the other of which does not. Thus, for example, *to give* permits the permutation, but its near synonyms *to donate* and *to contribute* do not.

(45.1.1) Alice gave ten dollars to the United Way.

(45.1.2) Alice gave the United Way ten dollars.

(45.2.1) Alice donated ten dollars to the United Way.

(45.2.2) *Alice donated the United Way ten dollars.

Other such verbs are *to tell, to send, to show, to teach,* and *to throw*.

Most verbs with these alternations have their prepositional phrase complements headed either by the preposition *to*, as in sentence (43.1), or by the preposition *for*, as in sentence (44.1). However, the same alternation arises with other prepositions, such as *of, toward,* and *with*.

(46.1) Alice asked [NP a favor] [PP of Bill].

(46.2) Alice asked [NP Bill] [NP a favor].
(Allerton 1982, 102–104)

(47.1) Alice bears [NP ill will] [PP toward Bill].

(47.2) Alice bears [NP Bill] [NP ill will].
(Huddleston 2002, 311)

(48.1) Alice played [NP a game of chess] [PP with/against Bill].

(48.2) Alice played [NP Bill] [NP a game of chess].
(Huddleston 2002, 311)

Finally, there are verbs where the equivalence of the permuted complements involves pairs of prepositional phrases. The pairs of prepositions include *to* and *with, for* and *on, from* and *of, into* and *from* or *out of, with* and a locative preposition such as *in, into, on, onto, over, against,* and so on (see Huddleston 2002, chap. 8.3.1 for further details.)

(49.1) Alice credited the discovery to Bill.

(49.2) Alice credited Bill *(with) the discovery.

(50.1) Alice blamed Bill for the accident.

(50.2) Alice blamed the accident *(on) Bill.

(51.1) Bill cleared the dishes from the table.

(51.2) Bill cleared the table *(of) the dishes.

(52.1) Bill hunted the deer in the woods.

(52.2) Bill hunted the woods *(for) the deer.

(53.1) Alice built a shelter out of stones.

(53.2) Alice built stones *(into) a shelter.

(54.1) Bill sprayed the wall with the paint.
(54.2) Bill sprayed paint *(on) the wall.
(55.1) Bill engraved the ring with his initials.
(55.2) Bill engraved his initials *(on) the ring.
(56.1) Alice banged the fence with a stick.
(56.2) Alice banged a stick *(against) the fence.

2.1.3.3 Complement polyadicity Not only do many English verbs admit complements whose syntactic category may be one of several, but many English verbs also have optional complements. We shall call words with optional complements *polyadic words*. In this section, we shall investigate polyadic verbs. As we shall see, the omission of an optional complement of a polyadic English verb is correlated with a shift in the verbs construal and the range of this shift is quite limited. We turn to a survey of this variation now.

We begin with a relatively restricted class of verbs. Consider the verb *to dress*.

(57.1) Alice dressed the doll.
(57.2) Alice dressed.
(57.3) Alice dressed herself.

It is easy to show the noun phrase *the doll* to be a complement of the verb *to dress,* yet the complement can be omitted without the resulting expression becoming unacceptable. Moreover, the meaning of the resulting sentence is synonymous with the one that results from the replacement of the complement noun phrase in the first sentence by a suitable reflexive pronoun, as shown in sentence (57.3).

While many of these verbs are verbs denoting grooming, grooming is neither necessary nor sufficient, as shown by these examples.

(58.1) Alice clothed *(herself).
(58.2) Bill behaved (himself).

We shall call these verbs *reflexive polyadic* verbs. Here are some others.

REFLEXIVE POLYADIC VERBS
to bathe (oneself), *to shave* (oneself), *to shower* (oneself), *to wash* (oneself), *to disrobe* (oneself), *to dress* (oneself), *to undress* (oneself), *to strip* (oneself) naked, *to behave* (oneself), ...

We come next to a larger class of verbs, which we shall call *reciprocal polyadic* verbs.

RECIPROCAL POLYADIC VERBS
to court (each other), *to divorce* (each other), *to embrace* (each other), *to equal* (each other), *to fight* (each other), *to hug* (each other), *to kiss* (each other), *to marry* (each other), *to match* (each other), *to meet* (each other), *to touch* (each other), ...

(59.1)   Carol met Bill.

(59.2)   Carol and Bill met.

(59.3)   Carol and Bill met (each other).

As the next example shows, it is not the meaning of the verb that determines whether or not its complement can be omitted.

(60.1)   The socks match (each other).

(60.2)   The socks resemble *(each other).

Besides reflexive and reciprocal polyadic verbs, there are also *indefinite polyadic verbs*. These are, in fact, quite numerous.

INDEFINITE POLYADIC VERBS
to bake (pastry), *to carve* (wood), *to clean, to cook* (food), *to crochet, to dig, to draw, to drink* (alcohol), *to drive* (a car), *to eat* (food), *to embroider, to file, to hoe, to hunt, to iron, to knit, to paint, to plow, to read* (a book), *to sew, to smoke* (a cigarette), *to sow, to study, to sweep, to telephone, to type, to wash, to weave, to weed, to whittle, to worship* (a deity), *to write*, ...

(61.1)   Bill read (the sign).

(61.2)   Bill read (something).

In many cases, one can find synonymous verbs, one of which is an indefinite polyadic verb, the other is not polyadic.

(62)   Bill perused *(the book).

A fourth kind of polyadic verb is *contextual*. These are verbs that, when the relevant complement is omitted, the verb behaves either endophorically or exophorically.

CONTEXTUAL POLYADIC VERBS
to approach, to choose, to call (in the sense of *to telephone*), *to close, to enter, to find out, to fit, to follow, to interrupt, to lead, to leave, to lose, to match, to obey, to oppose, to pass, to pull, to push, to visit, to watch, to win*, ...

The first example in (63) is a case where, when the complement is omitted, the sentence gives rise to an endophoric construal, while the first sentence in (64) is a case where, when the complement is omitted, the sentence gives rise to an exophoric construal.

(63.1)   Bill drove to Toronto. He arrived (there) an hour ago.

(63.2)   Bill drove to Toronto. He reached *(there) an hour ago.

(64.1)   A: When did you arrive (here)?

(64.2)   A: When did you reach *(here)?

(where the sentences are uttered by A upon meeting B, the addressee of the sentences.)

Indeed, the omissibility of a complement for a contextual polyadic verb may depend on subtle differences in construal. The verb *to leave* is one example.

(65.1) Bill left (the house).

(65.2) Bill left *(the package).

(Fillmore 1986, 101)

A fifth class of polyadic verbs is *causative* polyadic verbs. They are so-called, since, when the relevant complement is omitted, the correct paraphrase with the same verb with the complement present is one with the verb *to cause,* as illustrated in (66).

CAUSATIVE POLYADIC VERBS
to bake, to balance, to bend, to bleed, to blow up, to boil, to break, to burn, to close, to cook, to dissolve, to drop, to dry, to explode, to fill, to float, to gallop, to grow, to hang, to ignite, to improve, to march, to melt, to move, to open, to rock, to roll, to shake, to shine, to sink, to spill, to spread, to stretch, to tear, to walk, to withdraw, ...

(66.1) The butter melted.

(66.2) Bill melted the butter.

(66.3) Bill caused the butter to melt.

The last kind of polyadic verb is unsatisfactorily known as *middle*. English has only two voices, *active* and *passive,* and they are morphologically different from one another. Some languages have a third voice, called the *middle* voice, that can be morphologically distinguished from both the active and the passive voice. However, in the case of English, the so-called middle voice has the same form as the active voice, though it is construed as though it were in the passive voice. Thus, in sentence (67.1), the verb *sold* is in the active voice, the first noun phrase denotes the agent of the verb, the second noun phrase the thing being sold; in sentence (67.2), however, the verb has the very same form and the first and only noun phrase denotes the thing being sold.

(67.1) Dan sold the book easily.

(67.2) The book sold easily.

MIDDLE POLYADIC VERBS
to alarm, to amuse, to demoralize, to embarrass, to flatter, to frighten, to intimidate, to offend, to pacify, to please, to shock, to unnerve, to clean, to cut, to hammer, to iron, to read, to wash, ... (Huddleston 2002, 308).

Exercises: Complications

1. For each of these polyvalent verbs, find a near synonym that is not polyvalent:

    to send, to show, to teach, to tell, to throw

2. For each of these polyadic verbs, identify what kind of polyadic verb it is and find a near synonym that is not polyadic:

    to accept, to approach, to drink, to follow, to hide, to eat, to meet, to write, to hunt, to leave, to shake

2.2 Adjective Phrases: Adjectives and Their Complements

English adjectives may occur either attributively, that is, as a modifier within a noun phrase, or predicatively, that is, as the complement of a copular verb, as illustrated here.

(68.1)   The [AP bald] man fainted.

(68.2)   The man who fainted is [AP bald].

While the overwhelming majority of adjectives occur both attributively and predicatively, some occur only predicatively and others occur only attributively.

ATTRIBUTIVE ONLY

damn, drunken, ersatz, erstwhile, eventual, former, frigging, future, latter, lone, maiden, main, marine, mere, mock, only own, premier, principal, putative, self-confessed, self-same, self-styled, sole, utter, veritable, ... (Pullum and Huddleston 2002, 553)

PREDICATIVE ONLY

ablaze, afloat, afoot, afraid, aghast, agleam, aglimmer, aglitter, aglow, agog, ajar, akin, alight, alike, alive, alone, amiss, askew, asleep, averse, awake, aware, awash, awry, ... (Pullum and Huddleston 2002, 559)

An adjective that occurs attributively may either precede or succeed the head noun of the noun phrase in which it occurs. But the distribution is not free.

(69.1)   The [AP bald] man fainted.
         *The man [AP bald] fainted.

(69.2)   *A [AP happy to have been elected] candidate mounted the podium.
         A candidate [AP happy to have been elected] mounted the podium.

English adjectives, whether occurring attributively or predicatively, like English verbs, may have complements. We limit our survey to adjectives used predicatively, that is, adjective phrases that occur as complements to copular verbs such as *to be, to become, to seem, to appear,* and *to remain*. The question arises: how does one distinguish within an adjective phrase, the constituents that are complements of the adjective from those that are not? Unfortunately, this question has not been addressed by linguists, either theoretical or descriptive. However, we shall abide by the generally held consensus that adjectives have at most one complement (Baker 1989, chap. 3.9.1).

2.2.1 Basic patterns of English adjective complements

Complements to adjectives are of two major kinds: prepositional phrases and clauses.[4] However, not all predicative adjectives admit complements. Here are some that resist any complement:

NO COMPLEMENT

ambulatory, bald, concise, dead, despondent, enormous, farcical, friendly, gigantic, hasty, immediate, intelligent, light, lovely, main, nefarious, ostentatious, purple, quiet, regular, salty, surly, tentative, unreliable, urban, vivid, wild, young, ... (Pullum and Huddleston 2002, 543).

When prepositional phrases serve as complements to predicative adjectives, the prepositions heading such phrases are more or less confined to these: *about, at, by, for, from, in, of, on, upon, to, toward,* and *with*.

A few English adjectives have been identified as requiring prepositional phrase complements. They are *averse to, contingent on, dependent on, due to, fond of, incumbent on, intent on, liable to, loath to, mindful of, reliant on,* and *subject to*.

(70.1)   *Max is averse.

(70.2)   Max is averse to games.
         (Quirk et al. 1985, chap. 16.69)

Complements may also be clauses, though ones such as *likely*, which require a clausal complement, are rare.

(71.1)   *Carol is unlikely.

(71.2)   Carol is unlikely to resign.
         (Quirk et al. 1985, chap. 2.32)

Still, adjectives do admit clausal complements of the five varieties we found with verbs. The clauses may be either finite or infinite or declarative or interrogative, and they may also be gerundial.

(72.1)   Carl is wrong [Sfd that it is raining].
(72.2)   Carl is unsure [Sfi where it is raining].
(73.1)   Carl was unwilling [Snd to attend the ceremony].
(73.2)   Carl was unsure [Sni where to find the ceremony].
(74)   Carl was busy [Sg washing the dishes].

---

4. I know of only one case where an adjective has a noun phrase complement, namely *worth*, which was noted by Allerton (1975, 225).

ADJECTIVE COMPLEMENT PATTERN

NONE   ONE
            PP
            S

2.2.2 Complications

We now turn to some of the complications of adjectives and their complements. Like verbs, many adjectives are polyvalent, that is, admitting different complements, and many are polyadic, that is, having optional complements.

Not only may the same adjective admit both a clausal complement and a prepositional phrase complement, but it may also admit prepositional phrases headed by different prepositions. Consider, for example, the adjective *angry*.

(75.1)   Bill is angry [Sfd (that) he did not get a raise].

(75.2)   Bill is angry [PP at his boss].

(75.3)   Bill is angry [PP with his boss].

(75.4)   Bill is angry [PP about his raise].

Indeed, many, many adjectives have single complements that alternate between clauses and prepositional phrases as well as between prepositions: *busy, clear, cross, glad, good, happy, mad, ready, sure,* and *unsure,* to mention but a few.

In addition, many adjectives that take a complement may also have their complements omitted. As we are about to see, while there are reciprocal, contextual and indefinite polyadic adjectives, there seem to be no reflexive polyadic adjectives.

Reciprocal polyadic adjectives are fairly common. They include such adjectives as the following.

RECIPROCAL POLYADIC ADJECTIVES

compatible (with), distinct (from), divergent (from), equivalent (to), identical (to or *with), incompatible (with), parallel (to* or *with), perpendicular (to), similar (to), simultaneous (with),* and *separate (from), . . .*

The adjective *similar* is especially interesting, as it forms one of a minimal triple. The adjectives *similar* and *alike* and the preposition *like* are synonymous words. Thus, *similar* conveys a reciprocal sense, even when used with no complement, as shown by the synonymy of the pair of sentences in (76.1). Yet, the preposition *like,* which shares a sense with the adjective *similar,* requires its complement, as shown by the pair of sentences in (76.2). And finally, the adjective *alike,* which also shares a sense with the adjective *similar,* excludes any complement, as shown by the pair of sentences in (76.3).

(76.1)   Bill and Carol are similar.
            Bill and Carol are similar to each other.

(76.2) *Bill and Carol are like.
Bill and Carol are like each other.

(76.3) Bill and Carol are alike.
*Bill and Carol are alike (to) each other.

Also common are contextual polyadic adjectives. Here we have such adjectives as *close, faraway, foreign, local,* and *near.* Consider the adjective *faraway.* Thus, in certain situations, the sentences in (77.1) are synonymous, as are those in (77.2).

(77.1) Bill lives faraway.
Bill lives faraway from here.

(77.2) Although Bill lives faraway, he visits his parents regularly.
Although Bill lives faraway from them, he visits his parents regularly.

Cotextual material needed for the full understanding of a contextual polyadic adjective is not restricted to noun phrases, it may be an entire clause.

(78) Bill left early. Alice was glad.
Bill left early. Alice was glad that he did.

A further complication with polyadic adjectives is that many adjectives are ambiguous between one that is polyadic and one that is not. Let us begin with a case where the meanings are easily distinguished, the adjective *sick*. In the first sentence in (79), *sick* has no complement and it is synonymous with the adjective *ill;* in the second, it has a complement yet it is not synonymous with the adjective *ill*.

(79.1) Bill is sick.
Cf. Bill is ill.

(79.2) Bill is sick of school.
Cf. Bill has a strong distaste of school.

Notice that Bill can be sick, or ill, without being sick of anyone or anything, and Bill can be sick of someone or something without being sick, or ill. Thus, neither sentence in (79) entails the other.

Another adjective with distinguishable meanings is the adjective *proud*.

(80.1) Bill is proud.
Cf. Bill is arrogant.

(80.2) Bill is proud of his success.
Cf. Bill is highly satisfied with his success.

Once again we note the lack of entailment: Bill can be proud, or arrogant, without being proud of, or highly satisfied with, anyone or anything, and Bill can be proud of, or highly satisfied with, his success without being proud, or arrogant.

Grammatical Predicates and Minimal Clauses in English

We conclude this discussion of adjectival complements with this example.[5] The adjective *familiar* has two different senses, indeed, one being the converse of the other. These senses are distinguished by whether the prepositional phrase complement is headed by the preposition *to* or *with*.

(81.1) These facts are familiar to the expert.
        Cf. These facts are known to the expert.
(81.2) The expert is familiar with these facts.
        Cf. The expert knows these facts.

But observe that when the preposition *to* heads the complement prepositional phrase, the complement may be omitted, but not when the preposition *with* heads it.

(82.1) These facts are familiar.
        Cf. These facts are known.
(82.2) *The expert is familiar.
        Cf. The expert knows.

The second sentence is acceptable, provided the adjective *familiar* is construed as *is known*.

Summarizing, we see that the range of complements for adjectives is much smaller than that for verbs. Either adjectives admit no complement or they admit just one, in which case, the complement is either a prepositional phrase or a clause. In addition, adjectival complements, like verbal complements, may vary, some adjectives taking either prepositional phrase complements or clausal complements. Adjectival complements may also be optional. When they are optional, we see that polyadic adjectives are either contextual or reciprocal; none are reflexive.

2.3 Prepositions

Next, we come to prepositions. It has long been thought that prepositions take only noun phrase complements.

(83.1) *Dan stood on.
(83.2) Dan stood on the porch.

Indeed, many do.

NP COMPLEMENT
about, at, by, during, for, from, in, into, near, of, on, onto, out, upon, to, toward, under, up and *with*, ...

However, we also know that some subordinators double as prepositions.

---
5. I thank Andrew Reisner for bringing this example to my attention.

S COMPLEMENT
after, before, since, until

(84.1) Dan came after lunch.

(84.2) Dan came after lunch had been served.

Yet, there is nothing about the meaning of a word that determines whether its complement is a clause or a noun phrase. Thus, as noted by Sag et al. (1999, 99), *during* and *while* form a minimal pair.

(85.1) The storm arrived during [NP the picnic].
*The storm arrived during [S we were eating the picnic].

(85.2) *The storm arrived while [NP the picnic].
The storm arrived while [S we were eating the picnic].

In addition, traditional grammarians and descriptive linguists have long known that many prepositions may have no complement whatsoever (see Quirk et al. 1985, chap. 9.65–66). They spoke of prepositions so used as prepositions used adverbially.

POLYADIC PREPOSITIONS
aboard, about, above, across, after, along, alongside, around, before, behind, below, beneath, besides, between, beyond, by, down, in, inside, near, off, on, opposite, out, outside, over, past, round, since, through, throughout, under, underneath, up, within, without,
. . . (Quirk et al. 1985, chap. 9.65)

(86) In only a few years an English clergyman, Joseph Priestley, isolated and studied more new gases than any person before or since.
(Ihde 1964, 40)

Nothing about the meaning of a word that determines whether its complement is optional, as shown by the minimal pair *in* and *into*.

(87) Dan stood in front of the house. When the phone rang,
*he suddenly ran into.
he suddenly ran in.

As with verbs and adjectives, so with prepositions: complements may vary and their adicity may vary.

Polyadic prepositions are only contextual. The contextual value may be determined either with respect to the setting

(88) A: Is your sister in?
      B: No, she is out.
         (Quirk et al. 1985, 715)

Grammatical Predicates and Minimal Clauses in English   419

or with respect to an antecedent. This latter point is confirmed by the perfect paraphrase of the first sentence in (89), by the second.

(89.1) In the third century AD, if not before, this Greek intellectual conception ... served to crystallize ...
(Robinson 1948, 196)

(89.2) In the third century AD, if not before then, this Greek intellectual conception ... served to crystallize ...

We close this discussion with a final observation about adverbs. The adverb *afterward* can frequently be paraphrased by the prepositional phrase *after that,* as shown in (90).

(90) Alice lived in Montreal until 2010.
Afterward, she moved to Vancouver.
After that, she moved to Vancouver.

Unlike *after,* however, the adverb *afterward* admits no complement. The adverb *afterward* expresses a relation, just as the preposition *after,* but, unlike the preposition that can admit an overt expression of both of its relata, *afterward* cannot. This adverb is not the only such adverb:

COMPLEMENTLESS

afterward, ago, beforehand, downstairs, downtown, earlier, later, overhead, presently, previously, shortly, soon, subsequently, upstairs, uptown, ...

PREPOSITION COMPLEMENT PATTERN

NONE   ONE
       NP
       S

2.4   Nouns

As we shall discuss English nouns in some detail in chapter 14, we shall confine our attention here to the complements that nouns admit. It is well known that nouns derived from verbs admit complements; it is less well known that underived nouns also admit complements. Common nouns such as *table, man, virtue, water, air, picnic,* ... , admit no complements. But common nouns such as *friend, brother, husband, enemy, neighbor, top,* and *bottom* do. Their complements are all prepositional phrases with the preposition *of.*

(91.1) Bill is [NP the father [PP of the bride]].

Some nouns have a clause as a complement—*report, rumor, statement.*

(91.2) Alice heard [NP the rumor [S Bill resigned]].

Others have two complements: a prepositional phrase and an adjectival phrase, for example, *rendering,* or two prepositional phrases, for example, *dismissal,* or a prepositional phrase and a clause, for example, *report, statement.*

(91.3) [NP the rendering [PP of seawater] [AP potable]].

(91.4) The voters were amazed [PP at [NP the dismissal [PP of the minister] [PP from the position]].

(91.5) [NP The statement [PP by Alice] [S that all is well]] is not believable.

Finally, some admit three complements: either three prepositional phrases, for example, *gift,* or two prepositional phrases and a clause, for example, *persuasion.*

(91.6) [NP The gift [PP to Alice] [PP of the book] [PP by Bill]] was touching.

(91.7) [NP The persuasion [PP of the patrician] [PP by Galileo] [S that the moon has craters]] was crucial.

In short, English common nouns have the following distribution of complements.

COMMON NOUN COMPLEMENT PATTERN

NONE	ONE	TWO	THREE
	PP	PP PP	PP PP PP
	S	PP AP	PP PP S
		PP S	

3 A Structure for an English Lexicon

Recall that one of the principal aims of this chapter is to determine what constitutes an interpretation function for the set of lexical entries, analogous to the interpretation function for the set of the simple expressions of CPDL. In light of the fact that we are using a constituency grammar to define the set of expressions for English, it is natural to ask what aspect of a constituency grammar will do for the basic expressions of a natural language what a signature does for nonlogical symbols of CPDL. One might be tempted to answer that it is the basic lexical categories that do for the basic expressions what a signature does for the nonlogical symbols of CPDL. After all, the basic lexical categories serve to determine which complex expressions can be formed from the basic expressions. But do they determine what kind of set theoretic object can be assigned to which basic expressions?

One might think that basic expressions categorized as adjectives are assigned one kind of value, those categorized as nouns another, those categorized as verbs still another, and those categorized as prepositions a fourth. But a little thought shows that this is not right. Some verbs, such as *to sleep,* are naturally thought to be assigned a subset of a universe, in the case of the verb *to sleep,* the set of sleepers, while other verbs, such as *to greet,* are naturally thought to be assigned a set of pairs that is a subset of the Cartesian product of the universe with itself, in case of the verb *to greet,* a set of pairs the first member of which is a greeter and the second element of which is the person the greeter greets. Similarly, some adjectives, such as *happy,* are naturally thought to be assigned a subset of the universe, in the case of the adjective *happy,* the set of people who are happy, while other adjectives, such as *fond,* are most naturally thought to be assigned a set of pairs, which is a subset of the Cartesian product of the universe with itself, in case of the adjective *fond,* a set of pairs, the first member of which is fond of someone or something and the second member of which is the person or thing of which the person is fond. Thus, simply knowing that a basic expression is a verb or an adjective is not sufficient to determine what kind of value is assigned to it.

It turns out that a solution to this problem follows in the wake of a solution to two problems already discussed in chapter 3. One problem we noted there is that constituency grammars fail to formalize the well-known assumption that any expression assigned the syntactic category informally annotated as XP contains a word assigned the syntactic category informally annotated as X. In other words, such grammars fail to express the generalization that every noun phrase contains a noun, every verb phrase contains a verb, every adjective phrase contains an adjective, and every prepositional phrase contains a preposition. (See section 3.4.1.3.) A second inadequacy is the failure of such grammars to formalize the division of syntactic categories assigned to words into subcategories.[6] (See section 3.4.1.2.)

We illustrated the latter inadequacy in section 3.4.1.2 where we pointed out a dilemma. On the one hand, English verbs form a single syntactic category, since, as recognized by traditional grammar, English verbs are the only English words that are conjugated. On the other hand, traditional grammar and traditional lexicography has long recognized that English verbs are divided into transitive and intransitive verbs. Yet formalized constituency grammars cannot express such facts. Any constituency grammar that assigns all verbs the same syntactic category will generate unacceptable English expressions, and any constituency grammar that assigns the different English verbs different syntactic categories will fail to assign all verbs a common category.

Indeed, the problem is far worse. What we have seen, following our investigation of English verb complements, is that English verbs are distinguishable into at least thirty categories of verbs. In other words, the traditional distinction between transitive and intransitive verbs is just a special case of a broader distinction arising from differences of complements admitted by the verb in question. Moreover, this distinction among verbs is itself a special case of a still more general distinction, in which words of each major

---

6. The problem was first noted in unpublished work by George H. Matthews and by Robert P. Stockwell done in the 1960s as well as in published work by Paul Schachter (1962) and by Emmon Bach (1964). (See Chomsky 1965, 79n13, 213.) In fact, *Aspects of the Theory of Syntax,* Noam Chomsky's well-known book from 1965, offers two answers to the syntactic question. Neither of Chomsky's answers has been retained, though his discussion did serve to popularize Emmon Bach's term for the relevant phenomenon, namely *subcategorization.*

lexical category are distinguishable by the different complements they admit, the English verb having the widest diversity of complements.

What is required is an enrichment of the labels for the syntactic categories assigned to words that takes into account how different words of the same syntactic category admit different complements. The notation for this should not be *ad hoc,* with a subscript here and a subscript there, as was the wont of early transformational linguists, rather some systematic notation has to be introduced that is capable of capturing the full range of diversity of complements. We now turn to that problem.

As we shall see, there is a single solution to these two problems, which, at the same time, provides a solution to the problem of how to assign correct values to expressions on the basis of their syntactic categories. We shall formulate the solution, first in terms of the simplified description of complements for words, and then modify the solution to account for the complications discussed.

3.1 Constituency Grammar with Enriched Categories

To address the problem of how to formulate a systematic notation for the syntactic categories assigned to the words of English, let us first reflect on the patterns we garnered for the complements of English verbs, adjectives, and prepositions. A little reflection shows that each word can be categorized by a pair comprising a syntactic category of the kind encountered informally in chapter 3, namely one of the lexical categories of adjective (A), noun (N), preposition (P), or verb (V), on the one hand, and a list of the syntactic categories of its complements given in the order in which they appear, on the other. To define the set of syntactic categories required for the discussion in this chapter, we shall assume that there is a distinguished category S and five *basic categories:* A, Adv, N, P, and V. In addition, we have a *phrasal category,* which comprises a lexical category and the empty sequence, annotated by $\langle \rangle$. In the following table, we present the four phrasal categories. The first column contains the customary, informal notation found in chapter 3, the second column contains the official set theoretical notation we shall use, and the third contains the formal notation slightly modified to enhance readability.

PHRASAL CATEGORIES

CUSTOMARY NOTATION	SET THEORETICAL NOTATION	MODIFIED NOTATION
AP	$\langle A, \langle \rangle \rangle$	$A:\langle \rangle$
NP	$\langle N, \langle \rangle \rangle$	$N:\langle \rangle$
PP	$\langle P, \langle \rangle \rangle$	$P:\langle \rangle$
VP	$\langle V, \langle \rangle \rangle$	$V:\langle \rangle$

Next we define a *complement list*. A complement list is either an empty sequence or a sequence of categories that are either a phrasal category or the distinguished category S.

Grammatical Predicates and Minimal Clauses in English

Here are some examples: ⟨ ⟩, ⟨AP⟩, ⟨NP⟩, ⟨PP⟩, ⟨S⟩, ⟨NP, AP⟩, ⟨NP, PP, S⟩, and so on. Since a complement list is a sequence, we can speak of the specification of a coordinate. For example, NP is the specification of the first coordinate of the complement list of ⟨NP, PP, S⟩, PP is the specification of its second coordinate and S is the specification of its third coordinate.

We also define a *lexical category* to be an ordered pair comprising a lexical category (A, Adv, N, P, V) and a complement list. The lexical categories of the words *slept, greeted, fond,* and *introduced* are ⟨V:⟨ ⟩⟩, ⟨V:⟨NP⟩⟩, ⟨A:⟨PP⟩⟩, and ⟨V:⟨NP, PP⟩⟩, respectively. And finally, we define a *lexicon,* as before, as a set of *lexical entries,* each one of which is an ordered pair comprising a basic expression and its lexical category. Here is a table presenting a sample of lexical entries, including entries for the four words just mentioned.

STANDARD SET THEORETICAL NOTATION	MODIFIED NOTATION
⟨Alice, ⟨N, ⟨ ⟩⟩⟩	Alice|N:⟨ ⟩
⟨Bill, ⟨N, ⟨ ⟩⟩⟩	Bill|N:⟨ ⟩
⟨Carl, ⟨N, ⟨ ⟩⟩⟩	Carl|N:⟨ ⟩
⟨slept, ⟨V, ⟨ ⟩⟩⟩	slept|V:⟨ ⟩
⟨greeted, ⟨V, ⟨N, ⟨ ⟩⟩⟩⟩	greeted|V:⟨NP⟩
⟨introduced, ⟨V, ⟨⟨N, ⟨ ⟩⟩, ⟨P, ⟨ ⟩⟩⟩⟩⟩	introduced|V:⟨NP, PP⟩
⟨of, ⟨P, ⟨⟨N, ⟨ ⟩⟩⟩⟩⟩	of|P:⟨NP⟩
⟨to, ⟨P, ⟨⟨N, ⟨ ⟩⟩⟩⟩⟩	to|P:⟨NP⟩
⟨asleep, ⟨A, ⟨ ⟩⟩⟩	asleep|A:⟨ ⟩
⟨fond, ⟨A, ⟨⟨N, ⟨ ⟩⟩⟩⟩⟩	fond|A:⟨NP⟩
⟨was, ⟨V, ⟨⟨A, ⟨ ⟩⟩⟩⟩⟩	was|V:⟨AP⟩

We use this new notation to restate the patterns of complements found with verbs, adjectives, and prepositions. (Note that in the following tables, S distinguishes neither between finite and nonfinite clauses nor between interrogative and declarative clauses.)

VERB COMPLEMENT PATTERN

NONE	ONE	TWO	THREE
V:⟨ ⟩	V:⟨AP⟩	V:⟨NP, AP⟩	
	V:⟨NP⟩	V:⟨NP, NP⟩	V:⟨NP, NP, PP⟩
	V:⟨PP⟩	V:⟨NP, PP⟩	V:⟨NP, PP, PP⟩
	V:⟨AdvP⟩	V:⟨NP, AdvP⟩	
	V:⟨S⟩	V:⟨NP, S⟩	V:⟨NP, NP, S⟩
		V:⟨PP, S⟩	
		V:⟨AP, PP⟩	
		V:⟨PP, PP⟩	

ADJECTIVE COMPLEMENT PATTERN

NONE   ONE

A:⟨ ⟩   A:⟨PP⟩
         A:⟨S⟩

PREPOSITION COMPLEMENT PATTERN

NONE   ONE

P:⟨ ⟩   P:⟨NP⟩
         P:⟨S⟩

COMMON NOUN COMPLEMENT PATTERN

NONE	ONE	TWO	THREE
N:⟨ ⟩	N:⟨PP⟩	N:⟨PP, PP⟩	N:⟨PP, PP, PP⟩
	N:⟨S⟩	N:⟨PP, AP⟩	N:⟨PP, S⟩
		N:⟨PP, S⟩	

We are now in a position to draw a generalization from the foregoing. At a minimum, a phrase of a particular kind comprises a word of the same kind together with all of its complements. This leads to the following constituency formation rules, which we shall state twice, once using the notation of chapter 3 and once using our new notation.

VERB PHRASE FORMATION RULES

OLD NOTATION	NEW NOTATION
$V_1 \to VP$	V:⟨ ⟩ → V:⟨ ⟩
V_2 AP → VP	V:⟨AP⟩ AP → V:⟨ ⟩
V_3 NP → VP	V:⟨NP⟩ NP → V:⟨ ⟩
V_4 PP → VP	V:⟨PP⟩ PP → V:⟨ ⟩
V_5 S → VP	V:⟨S⟩ S → V:⟨ ⟩
V_6 NP AP → VP	V:⟨NP, AP⟩ NP AP → V:⟨ ⟩
V_7 NP NP → VP	V:⟨NP, NP⟩ NP NP → V:⟨ ⟩
V_8 NP PP → VP	V:⟨NP, PP⟩ NP PP → V:⟨ ⟩
V_9 NP S → VP	V:⟨NP, S⟩ NP S → V:⟨ ⟩
V_{10} NP NP PP → VP	V:⟨NP, NP, PP⟩ NP NP PP → V:⟨ ⟩
V_{11} NP NP S → VP	V:⟨NP, NP, S⟩ NP NP S → V:⟨ ⟩

Let us recall how the informal notation is misleading. The symbols V_1 through V_{11} are, in fact, distinct symbols, and, as far as the formal definition of a constituency grammar is concerned, the words with these labels bear no relationship to one another, in spite of the use of indices on the letter V.

Grammatical Predicates and Minimal Clauses in English

The new notation is designed to reflect the fact that words that in the old informal notation are all verbs, they differ from one another by the kinds of complements they take.

The same point holds for adjectives and their complements, prepositions and their complements, and nouns and their complements.

ADJECTIVE PHRASE FORMATION RULES

OLD NOTATION	NEW NOTATION
$A_1 \to AP$	$A:\langle\rangle \to A:\langle\rangle$
$A_2\ PP \to AP$	$A:\langle PP\rangle\ PP \to A:\langle\rangle$
$A_3\ S \to AP$	$A:\langle S\rangle\ S \to A:\langle\rangle$

PREPOSITIONAL PHRASE FORMATION RULES

OLD NOTATION	NEW NOTATION
$P_1 \to PP$	$P:\langle\rangle \to P:\langle\rangle$
$P_2\ NP \to PP$	$P:\langle NP\rangle\ NP \to P:\langle\rangle$
$P_3\ S \to PP$	$P:\langle S\rangle\ S \to P:\langle\rangle$

NOUN PHRASE FORMATION RULES

OLD NOTATION	NEW NOTATION
$N_1 \to NP$	$N:\langle\rangle \to N:\langle\rangle$
$N_2\ PP \to NP$	$N:\langle PP\rangle\ PP \to N:\langle\rangle$
$N_3\ S \to NP$	$N:\langle S\rangle\ S \to N:\langle\rangle$
$N_4\ PP\ PP \to NP$	$N:\langle PP, PP\rangle\ PP\ PP \to N:\langle\rangle$
$N_5\ PP\ S \to NP$	$N:\langle PP, S\rangle\ PP\ S \to N:\langle\rangle$
$N_6\ PP\ PP\ PP \to NP$	$N:\langle PP, PP, PP\rangle\ PP\ PP\ PP \to N:\langle\rangle$

The constituency grammar enriched with the new notation obviates the problem of subcategorization. All words of the same lexical category share the first coordinate of their constituency categories. Yet, the constituency formation rules ensure that a word combines with its proper complementary constituents. The enriched grammar also reflects the generalization that each phrase categorized as XP contains a word categorized as X: more precisely, that is, each phrase categorized as XP, or more properly as $X:\langle\rangle$, is a constituent none of whose subexpressions requires a complement. In addition, the four rules of the form $X \to XP$, which now have the form $X:\langle\rangle \to X:\langle\rangle$, are superfluous and can be dropped.

Indeed, all the rules given thus far are all instances of the following schema.

PHRASE FORMATION RULE SCHEMA: (abbreviated version)

$X:\langle C_1, \ldots, C_n\rangle\ C_1 \ldots C_n \to X:\langle\rangle.$

What the schema requires is that the number of a word's complements in the phrase of which it is the head be the same as the number of coordinates in its complement list and the category of its ith complement be a member of the specification of its nth complement list coordinate ($i \in \mathbb{Z}_n^+$). The index of a coordinate in a complement list is the coordinate's rank and the index of a complement is the complement's rank. English is the special case where the number of coordinates in the complement list is less than or equal to three. Rules of this form, by the way, are well known from categorial grammar, a topic we shall return to in chapter 13.

However, the rule for the constituency formation of minimal English clauses do not fall under the schema and so must be retained in its own right.

CLAUSE FORMATION RULE

OLD NOTATION   NEW NOTATION

NP VP $\to$ S   N:$\langle\rangle$ V:$\langle\rangle$ $\to$ S

We now have twenty constituency formation rules, though, as we pointed out, the first nineteen all fall under one rule schema.

Let us now define an enriched constituency grammar. Readers might recall that a constituency grammar had four ingredients: the basic expressions (BX), the categories (CT), the lexicon (LX), which pairs a basic expression with a category, and the constituency formation rules (FR). An enriched constituency grammar has eight ingredients: the basic expressions (BX), a distinguished category S, the basic categories (BC), the phrasal categories (PC),[7] the complement lists (CL), the lexical categories (LC), the lexicon (LX), and the constituency formation rules (FR).

ENRICHED CONSTITUENCY GRAMMAR

Let L be a language. Then, G, or $\langle$BX, S, BC, PC, CL, LC, LX, FR$\rangle$, is an enriched constituency grammar of L iff,

(1) BX is a nonempty, finite set that are the basic expressions of L;
(2) S is a distinguished category of L;
(3) BC is a nonempty set of basic categories of L;
(4) PC is the set of pairs X : $\langle\rangle$, where X is in BC and $\langle\rangle$ is the empty sequence;
(5) CL is a finite set of sequences, where a sequence is either the empty sequence or a finite sequence $\langle C_1, \ldots, C_n \rangle$, C_i being either S or in PC;
(6) LC is the set of ordered pairs $\langle x, y \rangle$, where x is in BX and y in CL;
(7) LX is a set of ordered pairs $v|C$, the lexical entries of L, where $v \in$ BX and $C \in$ LC;

---

7. In chapter 6, we used BC and PC to denote the set of binary connectives and the set of propositional connectives, respectively. Here they are used to denote the set of basic categories and the set of phrasal categories, respectively. Readers will have no problem using the context to discern which is meant.

(8) FR is the set of constituency formation rules of L, including the rule NP VP → S as well as instances of the phrase formation rule schema, where X is in BC and C_i is either in PC or is S.

Next, we define the constituents of an enriched constituency grammar.

CONSTITUENTS OF A LANGUAGE (CS)
Let L be a language. Let G be an enriched constituency grammar for L.

(1) Each lexical entry is a constituent ($LX_G \subseteq CS_G$);
(2) if $C_1 \ldots C_n \to C$ is a rule of FR_G and $e_1|C_1, \ldots, e_n|C_n$ are in CS_G, then $e_1 \ldots e_n|C$ is in CS_G;
(3) nothing else is in CS_G.

Having defined an enriched constituency grammar for a language, we turn to the problem of assigning values to the language's various constituents. As before, we must show what kinds of values are assigned to the lexical entries of a language and then show how, on the basis of the values assigned to the language's lexical entires, values are assigned to the composite constituents composed of them. This, then, is the topic of the next section.

3.2 Semantics

A lexicon, as we stated, is the counterpart of a signature in CPDL. However, unlike the signature in CPDL, where the categorization of the basic symbol determines the type of value an interpretation function may assign to it, the categorization of the basic expressions in a lexicon fails to determine the type of value an interpretation function may assign to it. However, once these basic lexical categories are enriched with complement lists, an obvious correspondence emerges: any word with n complements is assigned a set of $n + 1$ sequences of members of the universe.

STRUCTURE FOR A LEXICON
Let U be a nonempty set. Then, $\langle U, i \rangle$ is a structure for a lexicon L iff i is a function on the lexical entries satisfying the following conditions:

$X:\langle \rangle \to \text{Pow}(U)$;

$X:\langle C_1, \ldots, C_n \rangle \to \text{Pow}(U^{n+1})$.

Recall that in this chapter we set aside any detailed treatment of nouns. In what follows, we shall treat only noun phrases comprised solely of proper nouns. Now proper nouns are nouns that take no complements. This suggests that they belong to the category N: $\langle \rangle$. According to the definition of a structure for a lexicon just given, proper nouns should be assigned subsets of the structure's universe. However, in CPDL, an interpretation function assigns to each individual symbol, the counterpart in CPDL of a proper noun in natural language, an individual from the universe, not a subset of the universe. In conformity with the definition just given, we shall assign complementless nouns subsets of the universe; however, in the case of proper nouns, we shall require that they be assigned singleton sets. We note that individuals in the universe and singleton subsets of the universe are in bijective correspondence, as established by the following theorem of set theory, stated in chapter 2.

$x \in X$ iff $\{x\} \subseteq X$.

Next, we shall avail ourselves of the structure for a sample English lexicon $\langle U, i \rangle$, where $U = \{1, 2, 3, 4, 5\}$ and i is defined as follows:

PROPER NOUNS

$Alice|\text{N}:\langle\rangle$ $\mapsto$ $\{1\}$
$Bill|\text{N}:\langle\rangle$ $\mapsto$ $\{2\}$
$Carl|\text{N}:\langle\rangle$ $\mapsto$ $\{3\}$
$Dalian|\text{N}:\langle\rangle$ $\mapsto$ $\{4\}$

VERBS

$slept|\text{V}:\langle\rangle$ $\mapsto$ $\{1, 3\}$
$greeted|\text{V}:\langle\text{NP}\rangle$ $\mapsto$ $\{\langle 1, 2\rangle, \langle 2, 3\rangle, \langle 3, 2\rangle\}$
$relied|\text{V}:\langle\text{PP}\rangle$ $\mapsto$ $\{\langle 2, 1\rangle, \langle 3, 1\rangle, \langle 3, 2\rangle\}$
$introduced|\text{V}:\langle\text{NP, PP}\rangle$ $\mapsto$ $\{\langle 1, 2, 3\rangle, \langle 1, 3, 2\rangle, \langle 3, 2, 1\rangle\}$

ADJECTIVES

$asleep|\text{A}:\langle\rangle$ $\mapsto$ $\{1, 3\}$
$fond|\text{A}:\langle\text{PP}\rangle$ $\mapsto$ $\{\langle 2, 1\rangle, \langle 3, 1\rangle, \langle 1, 2\rangle, \langle 3, 5\rangle\}$

PREPOSITIONS

$in|\text{P}:\langle\text{NP}\rangle$ $\mapsto$ $\{\langle 1, 4\rangle, \langle 2, 4\rangle, \langle 3, 5\rangle\}$

Exercises: Interpretation functions for a lexicon

1. For each of the following words,
   admired, Alice, coughed, demonstrated, disliked, fainted, laughed, Osaka, transferred
   write out its lexical entry and indicate which of the values listed could be assigned to it by an interpretation function whose structure has the set $\{a, b, c, d, e\}$ as its universe.

   (a) e
   (b) $\{c\}$
   (c) $\{\langle a, a\rangle, \langle b, a\rangle, \langle e, d\rangle\}$

Grammatical Predicates and Minimal Clauses in English

(d) $\{\langle a, b\rangle, e, \langle b, c\rangle\}$

(e) $\emptyset$

(f) $\{\langle e, b\rangle, \langle a, c\rangle, \emptyset\}$

(g) $\{\langle e, a, b\rangle, \langle b, a, c\rangle, \langle e, d\rangle\}$

(h) $\{\langle a, b, c\rangle, \langle b, a, e\rangle, \langle c, e, d\rangle\}$

(i) $\{a, c\}$

3.3 Defining Constituency Valuation Rules

We now turn to the question of how to extend an interpretation function defined for the lexical entries of a language's lexicon to a valuation function defined for all the constituents generated by the language's grammar. We shall proceed step by step, beginning with minimal clauses where the verbs are intransitive and transitive, respectively. We shall then turn to minimal clauses where the verb is the copular verb *to be*. And finally, we shall turn to minimal clauses whose verbs have prepositional phrase complements.

3.3.1 Intransitive and transitive verbs

We begin with stating how a constituency valuation rule assigns a truth value to a minimal clause. It comprises a subject noun phrase and a verb phrase. Since in our enriched constituency grammar, an intransitive verb is itself a verb phrase, and since an intransitive verb requires no complement, we consider a minimal clause whose verb is intransitive, as illustrated in (92).

(92.1) Alice slept.

(92.2) Bill slept.

The customary view, dating back to Aristotle, is that a sentence is true just in case the thing named by the proper noun, the sentence's subject, is in the set each of whose members the verb is true. However, we have adopted the view that a proper noun denotes a singleton set. In light of this change in denotation for a proper noun, we must modify the statement of the conditions under which sentences such as those in (92) are true. Instead of saying the thing named by the proper noun, which is the subject, is a member of the set of things each member of which the verb is true, we say that the singleton set that is the denotation of the proper noun, which is the subject, is a subset of the set of things each member of which the verb is true. In effect, the sentences in (92) are taken to say that everyone who is Alice slept and everyone who is Bill slept. In fact, this is precisely how some European medieval logicians treated such sentences. In modern times, the idea is taken up again by Willard Quine (1960, sec. 38).

MINIMAL CLAUSE:
Constituency formation rule
If $e|$NP and $f|$VP, then $ef|$S.
Constituency valuation rule

Let $\langle U, i \rangle$ be a structure for an English lexicon. If $e|\text{NP}$ and $f|\text{VP}$, then $v_i(ef|\text{S}) = T$ iff $v_i(e|\text{NP}) \subseteq v_i(f|\text{VP})$.

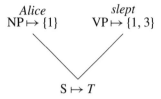

We now turn our attention to a minimal clause whose verb is transitive (V:⟨NP⟩), as illustrated in (93). Matters become more complicated.

(93.1) Alice greeted Bill.

(93.2) Bill greeted Carl.

Just as one assigns to the word *slept* the set of things each of which the word *slept* is true, in other words, the set of sleepers, so one assigns to the verb phrase *greeted Bill* the set of things each of which the verb phrase *greeted Bill* is true: in other words, one assigns to *greeted Bill* the set of Bill greeters. Given a list of who greeted whom and having identified Bill, how does one find the set of things that greeted Bill? One finds all those things that are paired with Bill in the list of who greeted whom. Thus, in the structure with which we are working, $Bill|\text{N:}\langle\rangle$ is assigned the singleton set $\{2\}$ and $greeted|\text{V: }\langle\text{NP}\rangle$ is assigned the set of ordered pairs $\{\langle 1, 2\rangle, \langle 2, 3\rangle, \langle 3, 2\rangle\}$, or G.

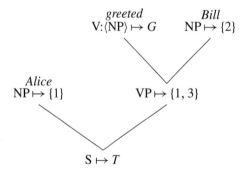

So, the set of things that greeted Bill are those things in the set assigned to *greeted* that are paired with the single thing in the set assigned to *Bill*. Only 1 and 3 are paired with 2 in this structure. Therefore, $\{1, 3\}$ is assigned to *greeted Bill*. In other words,

(94) $v_i(greeted\ Bill|V: \langle\rangle) = \{x: \langle x, y\rangle \in v_i(greeted|\text{V:}\langle\text{NP}\rangle) \text{ and } y \in v_i(Bill|\text{NP})\}$.

Grammatical Predicates and Minimal Clauses in English

We can generalize from the expression on the right-hand side of the equation in (94) to arrive at a formulation of the semantic rule for any verb phrase formed from a transitive verb and a noun phrase, replacing the expression *greeted* with e and the expression *Bill* with f.

VERB PHRASE WITH A TRANSITIVE VERB
Constituency formation rule
If $e|\text{V}:\langle\text{NP}\rangle$ and $f|\text{NP}$, then $ef|\text{V}:\langle\ \rangle$.
Constituency valuation rule
Let $\langle U, i \rangle$ be a structure for an English lexicon.
If $e|\text{V}:\langle\text{NP}\rangle$ and $f|\text{NP}$, then $v_i(ef|\text{V}:\langle\ \rangle) = \{x: \langle x, y\rangle \in v_i(e|\text{V}:\langle\text{NP}\rangle) \text{ and } y \in v_i(f|\text{NP})\}$.

Readers should note that this treatment of the valuation rule, paired with the constituency formation rule whereby a verb phrase is formed from a transitive verb and a noun phrase, is similar to the one we encountered in chapter 8. Recall that, while a coordinated independent clause of English has a structure identical to that of a composite formulae of CPL whose main connective is a binary connective, an independent clause containing a subordinate clause has a different structure, for the subordinator combines with an independent clause to form a dependent clause, which in turn combines with another independent clause to form the independent clause. A similar disparity obtains between a minimal clause with a transitive verb and its corresponding atomic formula comprising a two-place relational symbol and two individual symbols, for a two-place relational symbol combines with two individual symbols to form an atomic formula, while the complement of the verb combines with a single complement-taking verb to form a verb phrase, which in turn combines with the subject noun phrase to form the minimal clause.

3.3.2 Copular verb
As we saw, the English verb *to be* may take any of various complements: adjective phrases, prepositional phrases, noun phrases, and adverbial phrases. We shall consider only three of these four cases here: the cases of adjective phrase complements, prepositional phrase complements, and of noun phrase complements.

Let us begin by considering the adjectives. Surely sentence (95.1) is true in precisely the same circumstances as sentence (92.1) is true. Equally, sentence (95.2) is true in precisely the same circumstances as sentence (92.2) is true.

(95.1) Alice was asleep.
(95.2) Bill was asleep.

Indeed, the same holds for any pair of sentences such as the first in (95) and the first in (92) that differ from each other only in the choice of the verb phrases *slept* and *was asleep*. It follows, then, that these two verb phrases are true of the same set of things. In addition, the adjective *asleep* is true of precisely the same things of that the verb *slept* is true.

One, therefore, concludes that the denotation of the adjective *asleep* and the denotation of the verb phrase *was asleep* are identical. This identity of denotation results, if we treat the verb *was* as a transitive verb and assign it the identity relation on a structure's universe, that is, the set of ordered pairs where each member of the structure's universe is pair with itself, and itself alone.[8] In fact, there is a long tradition in European thought stretching back to Aristotle that takes the copular verb to express the identity relation. In logic, as we shall see in chapter 12, the identity relation is taken to be a logical relation.

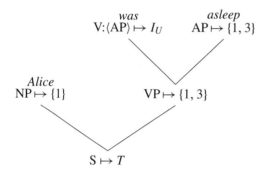

VERB PHRASE WITH A COPULAR VERB
Constituency formation rule
If $was|\text{V:}\langle\text{AP}\rangle$ and $f|\text{AP}$, then $was\ f|\text{V:}\langle\ \rangle$.
Constituency valuation rule
Let $\langle U, i \rangle$ be a structure for an English lexicon.
Let I_U be the identity relation on U.
If $was|\text{V:}\langle\text{AP}\rangle$ and $f|\text{AP}$, then $v_i(was\ f|\text{V:}\langle\ \rangle) = \{x\colon \langle x, y\rangle \in v_i(was|\text{V:}\langle\text{AP}\rangle)$ and $y \in v_i(f|\text{AP})\}$, where $v_i(was|\text{V:}\langle\text{AP}\rangle) = I_U$.

As should be obvious from the following examples, English prepositions express binary relations.

(96.1) Carl is in Dalian.

(96.2) Alice is from Europe.

(96.3) Dan is on the porch.

Clearly, the prepositions in the preceding sentences denote a binary relation between a person and a place, the preposition *in* denoting the binary relation of some thing being located in some place, the preposition *from* denoting the binary relation of some thing having some place as its origin, and the preposition *on* denoting the binary relation of some thing being on the upper surface of some thing. In the structure we adopted for our

---

8. Readers may wish review section 6.1 of chapter 2 before going further.

Grammatical Predicates and Minimal Clauses in English

previous examples, $in|\text{P:}\langle\text{NP}\rangle$ is assigned the set of ordered pairs $\{\langle 1, 4\rangle, \langle 2, 4\rangle, \langle 3, 5\rangle\}$, abbreviated as L in the following diagram.

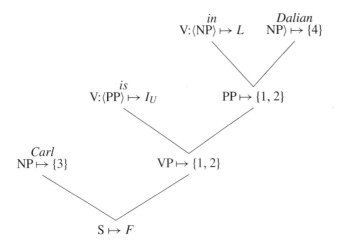

PREPOSITIONAL PHRASE

Constituency formation rule

If $e|\text{P:}\langle\text{NP}\rangle$ and $f|\text{NP}$, then $ef|\text{P:}\langle\,\rangle$.

Constituency valuation rule

Let $\langle U, i\rangle$ be a structure for an English lexicon.

If $e|\text{P:}\langle\text{NP}\rangle$ and $f|\text{NP}$, then $v_i(ef|\text{P:}\langle\,\rangle) = \{x\colon \langle x, y\rangle \in v_i(e|\text{P:}\langle\text{NP}\rangle) \text{ and } y \in v_i(f|\text{NP})\}$.

Readers should observe that the constituency valuation rule here is an almost exact copy of the constituency valuation rule for transitive verbs.

Finally comes the case where the complement of the copular verb is a noun phrase. It has long been common to distinguish between the minimal copular clauses that express predication and those that express identity. The first sentence in (97) expresses an identity, while the second expresses a predication.

(97.1)   George Orwell was Eric Blair.

(97.2)   George Orwell was an author.

Minimal copular clauses that express identity contain two noun phrases each of which are definite noun phrases. A singular definite noun phrase in English is a singular noun phrase that comprises either just a proper noun or a common noun and one of the demonstrative adjectives (*this* or *that*) or the definite article (*the*). In each such case, the subject noun phrase and noun phrase following the copular verb can be transposed, preserving not only acceptability but also truth.

(98.1)   George Orwell was Eric Blair.

(98.2)   Eric Blair was George Orwell.

(99.1) Eric Blair was the author of *Nineteen Eighty-Four*.
(99.2) The author of *Nineteen Eighty-Four* was Eric Blair.

In contrast, when the two noun phrases in sentence (97.2) are transposed, the resulting sentence is unacceptable.

(100) *An author was George Orwell.

Moreover, when a noun phrase is predicational, it must have the indefinite article *a* as its determiner; should it be replaced by its near synonym, *some,* the resulting sentence, the second,

(101.1) Eric Blair was an author.
(101.2) Eric Blair was some author.

acquires a special construal, which conveys its utterer's view that the denotation of the subject noun phrase is somehow exceptional with respect to attribute expressed by the common noun modified by *some*. This construal is especially salient when the word *some* is stressed.

We conclude from the foregoing that, when the indefinite article follows a copular verb, it has a special denotation, that of the identity relation.

3.3.3 Obligatory prepositional phrase complements

We now turn our attention to minimal clauses containing prepositional phrases that are required complements of either verbs or adjectives. The question we must address is what value is assigned to such prepositions. The answer we shall adopt here is that they denote the identity relation. This is exemplified by the first occurrence of the preposition *of* in the first sentence of (102), as borne out by the paraphrase of the first sentence by the second sentence in which the preposition *of* is omitted and the proper noun *Paris* stands in apposition to the noun phrase *the city*.

(102.1) The city of Paris is the capital of France.
(102.2) The city, Paris, is the capital of France.

Although the preposition *on* is part of the verb *to rely on*, like the preposition *of* used with common nouns such as *city,* it makes no substantive semantic contribution to the meaning of the verb. Such prepositions can also be treated as denoting the identity relation on the universe of the structure interpreting it.

(103.1) Alice relied on Bill.
(103.2) Bill depended on Carl.

In the structure we adopted earlier, $relied|V:\langle PP \rangle$ was assigned the denotation $\{\langle 2, 1 \rangle, \langle 3, 1 \rangle, \langle 3, 2 \rangle\}$, which we shall refer to in the following diagram as R.

Grammatical Predicates and Minimal Clauses in English 435

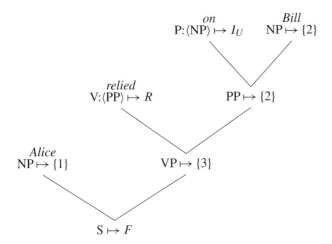

VERB PHRASE WITH A PREPOSITIONAL PHRASE COMPLEMENT
Constituency formation rule
If $e|V:\langle PP\rangle$ and $f|PP$, then $ef|V:\langle\,\rangle$.
Constituency valuation rule
Let $\langle U, i\rangle$ be a structure for an English lexicon.
If $e|V:\langle PP\rangle$ and $f|PP$, then $v_i(ef|V:\langle\,\rangle) = \{x : \langle x, y\rangle \in v_i(e|V: \langle PP\rangle)$ and $y \in v_i(f|PP)\}$.

Finally, we consider verbs with two complements, as exemplified by the verb *to introduce,* where, in our example structure, $introduced|V:\langle NP, PP\rangle$ is assigned the denotation $\{\langle 1, 2, 3\rangle, \langle 1, 3, 2\rangle, \langle 3, 2, 1\rangle\}$, abbreviated as K in the following diagram.

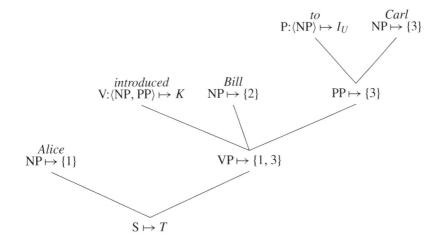

VERB PHRASE WITH TWO COMPLEMENTS
Constituency formation rule
If $e|V:\langle NP, PP\rangle$, $f|NP$ and $g|PP$, then $efg|V:\langle\ \rangle$.
Constituency valuation rule
Let $\langle U, i\rangle$ be a structure for an English lexicon.
If $e|V:\langle NP, PP\rangle$, $f|NP$ and $g|PP$, then $v_i(efg|V:\langle\ \rangle) =$
$\{x: \langle x, y, z\rangle \in v_i(e|V: \langle NP, PP\rangle), y \in v_i(f|NP) \text{ and } z \in v_i(g|PP)\}$.

3.3.4 Summary

In the preceding section, readers were presented with several constituency formation rules and their corresponding constituency valuation rules, these include a rule for clause formation, several rules for verb phrase formation and a rule for prepositional phrase formation. The clause formation rule, formalized in (104), states that any expression categorized as a noun phrase ($e|NP$) followed by any expression categorized as a verb phrase ($f|VP$) form a expression categorized as a clause ($ef|S$).

(104) CLAUSE FORMATION RULE
    If $e|NP$ and $f|VP$, then $ef|S$.

The clause formation rule has a corresponding clause valuation rule. The rule assigns a value to a complex expression that is categorized as a clause, and it does so on the basis of the values assigned to the clause's immediate constituents, a noun phrase and a verb phrase. Which values are assigned to a clause's immediate constituents depend on the values assigned the immediate constituents of the clause's immediate constituents and ultimately on the values assigned to the lexical items making up the clause. Thus, the first line of the rule refers to a structure for an English lexicon, which, of course, fixes the values assigned to each lexical item in English. The second line identifies the immediate constituents of the complex constituent to which a value is to be assigned. And the third line states precisely how the value of the complex constituent, the clause, is obtained from the values assigned to its immediate constituents, the noun phrase and the verb phrase.

(105) CLAUSE VALUATION RULE
    Let $\langle U, i\rangle$ be a structure for an English lexicon.
    If $e|NP$ and $f|VP$, then
    $v_i(ef|S) = T$ iff $v_i(e|NP) \subseteq v_i(f|VP)$.

Next we come to the phrase formation rules and their corresponding phrase valuation rules. We saw only a few such rules in section 3.3.3, but there are, in fact, dozens of such rules. Recall that the same basic lexical category—either an A, an Adv, an N, a P, or a V—is associated with different complement lists, thereby giving rise to fifty or more distinct

Grammatical Predicates and Minimal Clauses in English

lexical categories. Thus, for example, some expressions have the lexical category P:⟨NP⟩, some have the lexical category N:⟨PP, PP⟩ and some V:⟨NP, NP, PP⟩. For each of these categories, we have a rule to the effect that an expression whose lexical category requires complements, as specified by its complement list, must be followed by expressions of the lexical categories specified in its list, if the entire sequence of expressions is to form an expression of the corresponding phrase. The phrase formation rule formalized here is a rule schema.

(106) PHRASE FORMATION RULE SCHEMA

If $e|\text{X}:\langle C_1, \ldots, C_n \rangle$ and $f_j|C_j$, for each $j \in \mathbb{Z}_n^+$, then $ef_1 \ldots f_n|\text{X}:\langle \rangle$.

While every instance of phrase formation in English is an instance of this schema, not all instances of this schema are instances of phrase formation in English. In particular, it is generally thought that no English word has more than three complements. It follows therefore that no instance of the schema where n is greater than or equal to 4 is an instance of phrase formation in English.

Corresponding to the phrase formation rule schema is the phrase valuation rule schema, which provides, for any instance of the phrase formation rule schema, the corresponding phrase valuation rule. This rule schema assigns a value to any complex expression that is categorized as a phrase, and it does so on the basis of the values assigned to the clause's immediate constituents, a word and its complements, if it has any. Again, which values are assigned to a phrase's immediate constituents depend on the values assigned to the immediate constituents of the phrase's immediate constituents, and ultimately on the values assigned to the lexical items making up the phrase. As before, and as always, the first line of the rule refers to a structure for an English lexicon, for reasons already mentioned. The second line identifies the immediate constituents of the complex constituent to which a value is to be assigned. And the third line states precisely how the value of the complex constituent, here a phrase, is obtained from the values assigned to its immediate constituents, the noun phrase and the verb phrase.

(107) PHRASE VALUATION RULE SCHEMA

Let $\langle U, i \rangle$ be a structure for an English lexicon.

If $e|\text{X}:\langle C_1, \ldots, C_n \rangle$ and $f_j|C_j$, for each $j \in \mathbb{Z}_n^+$, then
$v_i(ef_1 \ldots f_n|\text{X}:\langle \rangle) = \{x: \langle x, y_1, \ldots, y_n \rangle \in v_i(e|\text{X}:\langle C_1, \ldots, C_n \rangle)$ and $y_j \in v_i(f_j|C_j)$, for each $j \in \mathbb{Z}_n^+\}$.

Readers are encouraged to work through the two phrase rule schemata to see how the various specific instances of the rule schemata introduced in section 3.3.3 are indeed instances.

Exercises: Defining constituency valuation rules

1. Let the basic lexical categories be A, Adv, N, P, and V. Let the complement categories be AP, AdvP, NP, PP, VP, and S. How many lexical categories with at most three complements are there? Explain how you have arrived at your answer.

2. To which of the following does the phrase formation rule schema apply so as to yield a phrase? In each case, justify your answer.

   (a) A:⟨PP⟩ PP   (h) V:⟨PP⟩ S
   (b) N:⟨PP⟩ PP   (i) A:⟨NP, PP⟩ NP PP
   (c) A:⟨NP⟩ PP   (j) P:⟨NP, S⟩ NP S
   (d) N:⟨PP⟩ NP   (k) A:⟨PP, NP⟩ NP PP
   (e) P:⟨S⟩ S   (l) V:⟨S, PP⟩ PP S
   (f) V:⟨S⟩ S   (m) S:⟨S, S, S⟩ S S S
   (g) P:⟨NP⟩ S   (n) A:⟨PP⟩ PP

3. Consider the following four sentences.

   (1) Beth runs.

   (2) Alan is courageous.

   (3) Beth approves of Carl.

   (4) Carl dispatched Alan to Beth.

   (a) Using the constituency grammar with enriched syntactic categories (see section 3 of this chapter), provide a synthesis tree for each of the preceding sentences.

   (b) Using the structure $\langle \{1, 2, 3\}, j \rangle$ (where j is defined as follows), provide values for each of the nodes in the synthesis trees and, for each instance in which a synthesis rule applies, write out its corresponding semantic rule.

   j $Alan|N:\langle\rangle$ $\mapsto$ $\{1\}$
        $Beth|N:\langle\rangle$ $\mapsto$ $\{2\}$
        $Carl|N:\langle\rangle$ $\mapsto$ $\{3\}$
        $of|P:\langle NP\rangle$ $\mapsto$ I_U
        $courageous|A:\langle\rangle$ $\mapsto$ $\{2, 3\}$
        $runs|V:\langle\rangle$ $\mapsto$ $\{1, 2\}$
        $approves|V:\langle PP\rangle$ $\mapsto$ $\{\langle 1, 2\rangle, \langle 2, 3\rangle, \langle 1, 3\rangle, \langle 3, 2\rangle\}$
        $dispatched|V:\langle NP; PP\rangle$ $\mapsto$ $\{\langle 1, 2, 3\rangle, \langle 1, 3, 2\rangle, \langle 2, 3, 1\rangle, \langle 3, 1, 2\rangle\}$

3.4 Further Extensions

We have now seen how a rather obvious enrichment of the syntactic categories of a constituency grammar solves simultaneously three problems, two mentioned in chapter 3, namely the projection problem and the subcategorization problem, and a third explained in this chapter, namely the problem of defining a structure of an English lexicon. We shall now show how three conservative extensions of this enrichment can account for three complications arising for words with complements described earlier: namely, those of complement polyvalence, of complement permutation, and of complement polyadicity. We shall state these extensions schematically.

3.4.1 The problem of complement polyvalence

Many English words have polyvalent complements, that is, they admit complements of different categories in the very same complement position. The most prominent example in English, as we saw previously, is the verb *to be*. It admits as its single complement an adjectival phrase, a noun phrase, a prepositional phrase, and even an adverbial phrase. If one assumes that each word that takes a complement may specify only one category for the complement, then one must conclude that the verb *to be* must be four ways ambiguous, one taking an adjectival phrase complement (V:⟨AP⟩), another a noun phrase complement (V:⟨NP⟩), a third a prepositional phrase complement (V:⟨PP⟩), and a fourth an adverbial phrase complement (V:⟨AdvP⟩). If this complement polyvalence were unique to the verb *to be,* one might be inclined to stipulate simply that these verbs have the same meaning, appealing to the special logical status of the verb *to be*. However, English, like other languages, abounds in polyvalent words.

How shall we address the problem of complement polyvalence? What we need to do is to devise notation that permits the specification of alterate categories for the same complement position. The simplest way to do this is to replace each complement category in a word's complement list with a set of complement categories, each member of which is a category permitted by the word in that complement position. Thus, a word's complement list is no longer a list of complement categories, but a list of nonempty sets of complement categories, each set having as its members just the alternate categories of expressions that may fill that position. Thus, for example, the verb *to be* takes but one complement, but its complement may be either an adjective phrase, a noun phrase, a prepositional phrase, or an adverbial phrase. Its complement list is, therefore, a list whose sole member is the set {AP, NP, PP, AdvP}. Its lexical entry would be: *be*|V:⟨{AP, NP, PP, AdvP}⟩.

Here is a sample of the lexical entries for a few of the polyvalent verbs discussed earlier.

(108.1)   *be*|V:⟨{AP, NP, PP, AdvP}⟩

(108.2)   *appear*|V:⟨{AP, NP, PP}⟩

(108.3)   *keep*|V:⟨{NP}, {AP, NP}⟩

(108.4)   *consider*|V:⟨{NP}, {AP, NP}⟩

Once we adopt this change in the notation for the lexical category of a word, we must make a suitable adjustment in our phrase formation rules.[9] The phrase formation rules given previously require that the categories in a word's complement list be the categories of the expressions following the word and that the corresponding categories appear in the same order. The extended rules require that the categories of the expressions following a word be among the categories in each set of the word's category list and that the corresponding categories appear in the same order. Rather than state a revision of each of the phrase formation rules already given, we state a revision simply of the phrase formation rule schema.

(109) PHRASE FORMATION RULE SCHEMA
(for words with polyvalent complements)

For each $j \in \mathbb{Z}_n^+$, let C_j be a complement category, let $\mathcal{C}_j$ be a nonempty subset of the complement categories and let C_j be a member of $\mathcal{C}_j$.

If $e|X:\langle \mathcal{C}_1, \ldots, \mathcal{C}_n \rangle$ and, for each $j \in \mathbb{Z}_n^+$, $f_j|C_j$, and $C_j \in \mathcal{C}_j$, then $ef_1 \ldots f_n|X:\langle \rangle$.

The preceding formation rule schema, together with the following lexical categories for the verbs *to choose* and *to appoint*,

(110.1) $appoint|V:\langle\{NP\}, \{NP, PP\}\rangle$

(110.2) $choose|V:\langle\{NP\}, \{PP\}\rangle$

allows the derivation of the first three of the following sentences and does not allow the derivation of the fourth.

(111.1) Dan appointed Alice as chief minister.
(111.2) Dan appointed Alice chief minister.
(112.1) Dan chose Alice as chief minister.
(112.2) *Dan chose Alice chief minister.

We must now adjust the phrase valuation rule schema paired with the revised version of the phrase formation rule schema so that the phrase valuation rule schema fits the revision in the specification of the complement list. Readers should notice that the value assigned is nevertheless the same as before.

(113) PHRASE VALUATION RULE SCHEMA
(for words with polyvalent complements)

---

9. These changes and similar changes in sections 3.4.2 and 3.4.3 require a revision in the definition of an enriched constituency grammar. We shall not bother to do so.

Grammatical Predicates and Minimal Clauses in English

Let $\langle U, i \rangle$ be a structure for an English lexicon.

For each $j \in \mathbb{Z}_n^+$, let C_j be a complement category, let $\mathcal{C}_j$ be a nonempty subset of the complement categories and let C_j be a member of $\mathcal{C}_j$.

If $e|X{:}\langle \mathcal{C}_1, \ldots, \mathcal{C}_n \rangle$ and, for each $j \in \mathbb{Z}_n^+$, $f_j|C_j$, and $C_j \in \mathcal{C}_j$, then
$v_i(ef_1 \ldots f_n|X{:}\langle\,\rangle) = \{x \colon \langle x, y_1, \ldots, y_n \rangle \in v_i(e|X{:}\langle \mathcal{C}_1, \ldots, \mathcal{C}_n \rangle)$ and $y_j \in v_i(f_j|C_j)$, for each $j \in \mathbb{Z}_n^+\}$.

Exercises: The problem of complement polyvalence

1. Again, let the basic lexical categories be A, Adv, N, P, and V and let the complement categories be AP, AdvP, NP, PP, VP, and S. Under the new definition of a lexical category, how many lexical categories are there with at most three complements? Explain how you have arrived at your answer.

2. To which of the following does the extended phrase formation rule schema apply so as to yield a phrase? In each case, justify your answer.

(a) A:⟨{PP}⟩ PP
(b) N:⟨{PP, S}⟩ AP
(c) A:⟨{NP, PP}⟩ PP
(d) N:⟨{PP, NP}⟩ NP
(e) P:⟨{AP, NP}⟩ S
(f) V:⟨{S, S}⟩ S
(g) P:⟨{NP, PP}⟩ S
(h) V:⟨{PP, S}⟩ S
(i) A:⟨{NP, PP}⟩ NP PP
(j) P:⟨{NP, S}, {PP, AP}⟩ NP S
(k) A:⟨{PP, NP}, {PP}⟩ NP PP
(l) V:⟨{S, PP}⟩ PP S
(m) S:⟨{S, NP, PP}⟩ PP
(n) A:⟨{PP}⟩ PP

3. Use the new notation to write lexical entries for the verbs *to try* and *to strive* that respect the acceptability judgments for the following sentences.

Dan tried to run.
Dan strove to run.
Dan tried running.
∗Dan strove running.

3.4.2 The problem of complement permutation

In our discussion of polyvalence, we observed that some polyvalent verbs admit complement permutation. We noted that all verbs that admit complement permutation have complement lists specifying in the first coordinate a noun phrase and in the second coordinate either both a noun phrase and a prepositional phrase complement or two prepositional phrase complements, each with a different preposition from the other. (For reasons that we shall make clear later, in section 3.5, we shall only treat those that specify both a noun phrase and a prepositional phrase in the second coordinate.)

We observed earlier that not all verbs whose first coordinate specifies the category NP and whose second coordinate specifies either the category NP or the category PP admit complement permutation. For example, the verb *to give* and its near synonym *to contribute* both have first coordinates specifying the category NP and second coordinates specifying either the category NP or the category PP, yet only *to give* admits a permutation of its complements.

(114.1)   Dan gave five dollars to a charity.
(114.2)   Dan gave a charity five dollars.
(115.1)   Dan contributed five dollars to a charity.
(115.2)   *Dan contributed a charity five dollars.

In light of the fact that *to give* and *to contribute* are near synonyms, it follows that the difference in the complementation of the two verbs is a lexical fact, not derivable from anything more general.

Notice that from a distributional point of view, the minimal pair of *to appoint* and *to chose* are parallel to the minimal pair of *to give* and *to contribute:* the first in each pair may take either a noun phrase or a prepositional phrase as a second complement, whereas the second in each pair may only take a prepositional phrase as a second complement.

(116.1)   *give*|V:⟨{NP}, {NP, PP}⟩

(116.2)   *contribute*|V:⟨{NP}, {PP}⟩

However, there is crucial difference between *to appoint* and *to give*. In the case of the two sentences in (111), all the words in the second sentence appear in the first sentence and in the same order and they are synonymous; whereas, in the case of the two sentences in (117), all the words in the second sentence appear in the first sentence, and in the same order, but they are not synonymous.

(117.1)   Alice gave Fido to Bill.
(117.2)   Alice gave Fido Bill.

To make a sentence synonymous with sentence (117.1), the two complement noun phrases are permuted and the preposition must be suppressed, as shown in (118).

(118.1)   Alice gave Fido to Bill.
(118.2)   Alice gave Bill Fido.

In other words, when the categories of the first and second complements of the verb *to give* are NP and PP, respectively, the first complement denotes the gift, or thing given, and the second complement the recipient of the gift; however, when the categories of the two

Grammatical Predicates and Minimal Clauses in English

complements are both NP, the first complement denotes the recipient and the second the gift.

To handle pairs of sentences such as these, we introduce a diacritic to distinguish those two complemented verbs that admit permutation from those that do not. In particular, we enrich the notation for the complement lists to permit the addition of a superscript p to the NP complement category.

(119.1) $appoint|V:\langle\{NP\}, \{NP, PP\}\rangle$

(119.2) $give|V:\langle\{NP\}, \{NP^p, PP\}\rangle$

(119.3) $donate|V:\langle\{NP\}, \{PP\}\rangle$

Once this diacritic is included in the notation for lexical entries, we must revise the phrase formation rule schema to provide for the special case where the complement category with the diacritic p in a lexical entry's second coordinate is canceled by NP. Since the relevant verbs are ones with precisely two complements, we state the rule schematically for verbs with two complements.

(120) PHRASE FORMATION RULE SCHEMA
(for two-place permuting complement verbs)

Let the complement categories include, for each complement category XP, a complement category XP^p.
Let $\mathcal{C}_1$ and $\mathcal{C}_2$ be nonempty sets of complement categories where no set includes both XP and XP^p.
Let C_1 and C_2 be complement categories.
Let C_1 be in $\mathcal{C}_1$ and let either C_2 or C_2^p be in $\mathcal{C}_2$.
$V:\langle \mathcal{C}_1, \mathcal{C}_2 \rangle\ C_1\ C_2 \rightarrow V:\langle\ \rangle$.

Naturally, we must state the phrase valuation rule schema that accompanies the phrase formation rule schema just formulated so as to assign one value to the verb phrase when the category of the second complement is in the lexical entry's second coordinate and another when the category of the second complement is NP and NP^p in the second coordinate.

(121) PHRASE VALUATION RULE SCHEMA
(for two-place permuting complement verbs)

Let M, or $\langle U, i \rangle$, be a structure for a lexicon for English.
Let the complement categories include, for each complement category XP, a complement category XP^p.
Let $\mathcal{C}_1$ and $\mathcal{C}_2$ be nonempty sets of complement categories where no set includes both XP and XP^p.
Let C_1 and C_2 be complement categories.

Let C_1 be in $\mathcal{C}_1$ and let either C_2 or C_2^p be in $\mathcal{C}_2$. Then,

CASE 1
If $C_1 \in \mathcal{C}_1$ and $C_2 \in \mathcal{C}_2$, then $v_i(V: \langle C_1, C_2 \rangle\, C_1 C_2) = \{x\colon \langle x, y_1, y_2 \rangle \in v_i(V: \langle C_1, C_2 \rangle)$ and $y_1 \in v_i(C_1)$ and $y_2 \in v_i(C_2)\}$.

CASE 2
If $C_1 \in \mathcal{C}_1$ and $C_2^p \in \mathcal{C}_2$, then $v_i(V: \langle C_1, C_2 \rangle\, C_1 C_2) = \{x\colon \langle x, y_1, y_2 \rangle \in v_i(V: \langle C_1, C_2 \rangle)$ and $y_2 \in v_i(C_1)$ and $y_1 \in v_i(C_2)\}$.

Readers should look carefully at the two cases to see that the first case is illustrated by the verb phrase *give Fido to Bill*, as found in sentence (118.1), and the second case is illustrated by the verb phrase *give Bill Fido*, as found in sentence (118.2).

The pattern of complement permutation commonly goes by the name *dative shift*, a name given to it by transformational linguists who treated the equivalence arising from the complement permutation with transformations.

Exercises: The problem of complement permutation

1. We have seen only four instances of the phrase formation and valuation rule schemata for complement permutation. Write out the two pairs of rules corresponding to the two instances in which the second coordinate of the words permitting permutation specify only one PP.

2. For each of the following English verbs, find one to pair with it so that one has a minimal pair similar to the pair *to give* and *to donate*: *to tell, to show, to teach,* and *to throw.* Provide evidence to show that the pair provided is a minimal one.

3. Suppose that *blicked* were an English word. Let $\langle U, j \rangle$ be the structure for an English lexicon where $U = \{1, 2, 3, 4, 5\}$ and j is partially specified as follows.

$$
\begin{array}{lll}
j & Alice|\text{N:}\langle\,\rangle & \mapsto \{1\} \\
& Beth|\text{N:}\langle\,\rangle & \mapsto \{2\} \\
& Carl|\text{N:}\langle\,\rangle & \mapsto \{3\} \\
& to|\text{P:}\langle\text{NP}\rangle & \mapsto I_U \\
& blick|\text{V:}\langle\{\text{NP}\}, \{\text{PP}, \text{NP}^p\}\rangle & \mapsto \{\langle 1,2,3\rangle, \langle 4,2,3\rangle, \langle 5,2,3\rangle, \langle 2,4,5\rangle, \langle 2,2,3\rangle, \\
& & \qquad \langle 5,1,2\rangle\}
\end{array}
$$

Provide a synthesis tree for each of the two following sentences and assign values to each of the nodes in the tree using the structure.

(1) Alice blicked Beth to Carl.
(2) Alice blicked Carl Beth.

3.4.3 The problem of complement polyadicity

The third complication pertaining to complements of words is that many words have optional complements. As we saw, when optional complements are omitted from a clause, the resulting clause can be paraphrased with a canonical paraphrase, since the omission of the complement gives rise to a construal characteristic of the omission. For example, when the complement of the verb *to eat* is omitted, it is construed as though it had an indefinite direct object. When the complement of the verb *to arrive* is omitted, it is construed as if it had either the pronoun *there* or the pronoun *here* as its complement. The adjective *similar* yields a reciprocal construal when its complement is omitted. Finally, on the omission of its complement, the verb *to dress* yields a reflexive construal.

Each of the verbs discussed that have optional complements have near synonyms whose corresponding complements are obligatory. What this shows is that the optionality of the complement for these words is a feature peculiar to the choice of word, independent of the word's meaning.

(122.1) Alice ate (something).
(122.2) Alice devoured *(something)
(123.1) Dan arrived (here) this morning.
(123.2) Dan reached *(here) this morning.
(123.3) Carol took a plane to Mumbai. She arrived (there) this morning.
(123.4) Carol took a plane to Mumbai. She reached *(there) this morning.
(124.1) Peter and Bill are similar (to each other).
(124.2) Peter and Bill resemble *(each other).
(125.1) Carl dressed (himself).
(125.2) Carl clothed *(himself).

A survey of English words with optional complements shows the following pattern. Each of the four construals noted earlier—contextual, indefinite, reciprocal, and reflexive—arises in connection with one or another English verb. Only three of the construals—contextual, indefinite, and reciprocal—arise for one or another English adjective and for one or another English noun. And only the contextual construal arises for English prepositions. This is summarized in the following table.

ENGLISH	CONTEXTUAL	INDEFINITE	RECIPROCAL	REFLEXIVE
Verbs	Yes	Yes	Yes	Yes
Adjectives	Yes	Yes	Yes	No
Nouns	Yes	Yes	Yes	No
Prepositions	Yes	No	No	No

To handle these cases, we permit the nonempty set of complement categories to include, besides the usual complement categories of AdvP, AP, NP, PP, and S, four features: ind (indefinite), ref (reflexive), rec (reciprocal), and cnt (contextual). Their presence in a set of categories for a complement position permits the corresponding complement to be omitted. Their difference from one another ensures that omissions lead to different construals, depending on the feature permitting the omission.

Here is a sample of the lexical entries for the polyadic words appearing in the sentences (122) through (125) and for the near synonyms paired with them in the examples.

(126.1) $eat|V:\langle\{NP, ind\}\rangle$
(126.2) $devour|V:\langle\{NP\}\rangle$
(127.1) $arrive|V:\langle\{PP, cnt\}\rangle$
(127.2) $reach|V:\langle\{NP\}\rangle$
(128.1) $dress|A:\langle\{NP, ref\}\rangle$
(128.2) $clothe|V:\langle\{NP\}\rangle$
(129.1) $meet|V:\langle\{NP, rec\}\rangle$
(129.2) $encounter|V:\langle\{NP\}\rangle$

Having introduced these features, we again must revise the phrase formation rule schema to provide for words with such lexical entries. Each version of the phrase formation rule schema requires that the number of a word's complements be the same as the number of coordinates of its complement list and that the category of the ith complement be a member of the specification of the complement list's ith coordinate. None of these schemata works in the case of phrases in which an optional complement has been omitted, since the number of complements is strictly less than the number of coordinates in the head word's complement list. Moreover, when the number of complements in a phrase is strictly less than the number of coordinates in the head word's complement list, it is no longer desirable that the category of the ith complement be a member of the specification of the complement list's ith coordinate, as we now illustrate.

Consider the verb *to trade,* which takes three complements, the second of which is optional, and which, when omitted, gives rise to an indefinite construal.

(130.1)   Alice traded [NP a car] [PP to Bill] [PP for a sailboat].

(130.2)   Alice traded [NP a car] [PP for a sailboat].

Its lexical entry is $trade|V:\langle\{NP\}, \{PP, ind\}, \{PP\}\rangle$. When all of its complements appear, as with sentence (130.1), the phrase formation rule schema for words with polyvalent complements applies, thereby assigning VP, or $V:\langle\ \rangle$, to *traded a car to Bill for a sailboat.* However, the schema does not assign VP, or $V:\langle\ \rangle$, to *traded a car for a sailboat,* for the third coordinate remains uncanceled and so the expression does not form a verb phrase.

Grammatical Predicates and Minimal Clauses in English 447

What we desire is that the first complement cancel the first coordinate and that the second complement skip, so to speak, the second coordinate and cancel the third.

More generally, we wish to prohibit a higher ranked complement from canceling a lower ranked coordinate, but to permit a lower ranked complement canceling a higher ranked coordinate, but doing so only when any uncanceled coordinate has one of the four features, ind, cnt, ref, and rec. This is accomplished if and only if the following conditions in a phrase are met: first, each complement's category is a member of some complement list specification of the head word; second, the categories of different complements are members of different complement list coordinates' specifications; third, if one complement precedes a second, then the specification of which the former is a member precedes the specification of which the latter is a member; and fourth, if one of the specifications does not correspond to one of the complements then the specification contains at least one of the four features, cnt, ind, ref, and rec. The first two conditions require, in effect, that there be an injection from the complements into the coordinates of the complement list. The third condition requires that the injection from the complements to the coordinates of the complement list be monotone increasing with respect to the ranking of the complements and the ranking of the coordinates. What this means is that, as the rank of the complement increases, the rank of the coordinate increases, permitting the rank of the complement to go up by one but the rank of the corresponding coordinate to go up by more than one. The fourth condition ensures that any coordinate that is not the image under the injection of some complement contain a feature.

(131) PHRASE FORMATION RULE SCHEMA
(for words with polyadic complements)

Let $\mathcal{C}_1, \ldots, \mathcal{C}_n$ be nonempty subsets of complement categories to which have been added the features ind, con, ref, and rec.

Let $C_1, \ldots, C_m$ be complement categories where $m \leq n$.

Let r be a monotonically increasing injection from $\mathbb{Z}_m^+$ to $\mathbb{Z}_n^+$ satisfying two conditions:
(1) for each $i \in \mathbb{Z}_m^+$, $C_i \in \mathcal{C}_{r(i)}$, and
(2) for each $j \in \mathbb{Z}_n^+$ for which $\mathcal{C}_j$ is not in the range of r,
  $\mathcal{C}_j$ contains either ind, con, ref, or rec.

If $e|X:\langle \mathcal{C}_1, \ldots, \mathcal{C}_n \rangle$ and, for each $j \in \mathbb{Z}_m^+$, $f_j|C_j$, and $C_j \in \mathcal{C}_{r(j)}$, then $ef_1 \ldots f_n|X:\langle\,\rangle$.

(The phrase formation rule schema for words with polyadic complements includes as a special case the phrase formation rule schema for words with polyvalent complements, but it does not include the phrase formation rule schema for verbs whose first and second complements permute.)

As readers have come to expect, we must formulate a phrase valuation rule schema. The schema is somewhat complex. This is because different values are assigned depending on which features occur in the uncanceled coordinates of the word heading the phrase.

As with all preceding constituency valuation rules, the first clause specifies that there is a structure for the English lexicon. The four following clauses are just a repetition on the first three clauses of the phrase formation rule and the apodosis of its fifth clause. Finally, comes the statement of the value assigned to the phrase. The value depends not only on the values assigned to its immediate constituents but also on which feature is chosen for those complements of the word that are omitted. Since there are four features, ind, cnt, ref, and rec, the four cases must be stated.

(132) PHRASE VALUATION RULE SCHEMA
(for words with polyadic complements)

Let $\langle U, i \rangle$ be a structure for an English lexicon.

Let $\mathcal{C}_1, \ldots, \mathcal{C}_n$ be nonempty subsets of complement categories to which have been added the features ind, con, ref, and rec.

Let $C_1, \ldots, C_m$ be complement categories where $m \leq n$.

Let r be a monotonically increasing injection from $\mathbb{Z}_m^+$ to $\mathbb{Z}_n^+$ satisfying two conditions:
(1) for each $i \in \mathbb{Z}_m^+$, $C_i \in \mathcal{C}_{r(i)}$, and
(2) for each $j \in \mathbb{Z}_n^+$ for which $\mathcal{C}_j$ is not in the range of r,
$\mathcal{C}_j$ contains either ind, con, ref, or rec.

Let $e|X:\langle\mathcal{C}_1, \ldots, \mathcal{C}_n\rangle$ and, for each $j \in \mathbb{Z}_m^+$, $f_j|C_j$, and $C_j \in \mathcal{C}_{r(j)}$.

Then, for some choice of feature, one for each $j \in \mathbb{Z}_n^+$ for which $\mathcal{C}_j$ is not in the range of r,
$v_i(X:\langle\mathcal{C}_1, \ldots, \mathcal{C}_n\rangle\, C_1 \ldots C_m) = \{x: \langle x, y_1, \ldots, y_n \rangle \in v_i(X:\langle\mathcal{C}_1, \ldots, \mathcal{C}_n\rangle)\}$ where
for each $i \in \mathbb{Z}_m^+$, $y_i \in v_i(C_i)$,
for each $j \in \mathbb{Z}_n^+$ for which $\mathcal{C}_j$ is not in the range of r,
(1) if the chosen feature is ind, then $y_i \in U$;
(2) if the chosen feature is cnt, then $y_i = d$
(where d is stipulated when the structure is),
(3) if the chosen feature is ref, then $y_i = x$,
(4) if the chosen feature is rec, then
$y_i : \langle x, y_1, \ldots, y_{i-1}, x, y_{i+1}, y_n \rangle \in v_i(X:\langle\mathcal{C}_1, \ldots, \mathcal{C}_n\rangle)$
(only a very rough approximation).

Let us consider examples of each of these four clauses for the features. Consider the verb *to eat*. Since the verb has just one coordinate in its complement list, it is assigned a set of ordered pairs. Since it has the feature ind, its complement may be omitted. If the complement is omitted, then the phrase valuation rule assigns to the verb phrase that comprises only the word *eat* all the members of the structure's universe that appear in the first coordinate of the set of ordered pairs assigned to the verb *eat* in the structure. In other words, the verb phrase *eat* is assigned the set of things in the structure's universe that eat. This means, then, that the verb phrases *eat* and *eat something* are assigned exactly the same subset of a structure's universe.

Next, consider the verb *to arrive*. It too has just one coordinate in its complement list. So, it too is assigned a set of ordered pairs. Since it has the feature cnt, its complement may be omitted. If it is omitted, then the phrase valuation rule assigns to the verb phrase that comprises just *arrive* all the members of the structure's universe that appear in the first coordinate of the set of ordered pairs assigned to the verb *arrive* in the structure and are paired with some distinguished member d, of the structure's universe. If, for example, the distinguished member of the universe is Paris, then the verb phrases *arrive* and *arrive in Paris* are assigned exactly the same subset of the structure's universe.

What happens when the verb is the verb *to dress*? As with the other two verbs, it has just one coordinate in its complement list. So, it too is assigned a set of order pairs. It has the feature ref. So its complement may be omitted. If it is, then the phrase valuation rule assigns to the verb phrase that comprises only the word *dress* all the members of the structure's universe that are paired with themselves in the set of ordered pairs assigned to the verb *dress* in the structure. As a result, the clauses *Bill dressed* and *Bill dressed himself* have the same truth value; indeed, the verb phrases *dressed* and *dressed himself* are assigned exactly the same subset of the structure's universe.

Finally, we come to words with the feature rec. To a first-order approximation, these verbs do not admit as subjects whose head noun is a singular count noun, rather they require plural subjects.[10] Thus, we find the following contrast:

(133.1)   *Bill met.

(133.2)   Bill and Carol met.

However, the semantics of plural count nouns falls outside the scope of this book. For this reason, we have given a rough and inaccurate statement of the semantics of this fourth feature. The basic idea is that a verb such as *to meet* is assigned a set of ordered pairs from the members of the structure's universe and that the feature rec assigns to a verb phrase comprising just the word *meet* those members of the structure's universe that are distinct from one another and occur with each other both as the first coordinate and as the second.[11]

---

10. They do admit of collective singular count nouns and some mass nouns.

11. This characterization of reciprocity as a symmetric, irreflexive relation is, as is well known to semanticists, too strong.

Exercises: The problem of complement polyadicity

1. Explain how the conditions in (131) apply to phrases that meet the following conditions:

(a) the verb takes one optional complement and the phrase has no complements;

(b) the verb takes two complements, one of which is optional, the phrase has just one complement;

(c) the verb takes three complements, two of which are optional, the phrase has either one or two complements.

3.4.4 Passive voice

We close section 3.4 by showing how the enriched constituency grammar, when conservatively extended to handle words with polyvalent, permuting, and polyadic complements, seems to be to handle a pattern that has retained the attention of transformational linguists since its inception, namely, the pattern evinced by the passive voice. It is well known that minimal clauses with transitive verbs in the active voice have equivalent minimal clauses with the transitive verbs in the passive voice, where the subject of the active voice clause becomes an optional prepositional phrase complement and the object of the active voice clause becomes a subject noun phrase, as illustrated here.

(134.1)   Alice greeted Bill.

(134.2)   Bill was greeted by Alice.

It is also well known that, when the optional prepositional phrase complement of the passivized verb is omitted, the construal is an indefinite one. Indeed, the resulting passive version of the clause without the prepositional phrase complement is judged equivalent to the same clause with an indefinite prepositional phrase complement, as exemplified by the next pair of sentences.

(135.1)   Bill was greeted.

(135.2)   Bill was greeted by someone.

This is the same paraphrasal equivalence we have seen previously with respect to polyadic words that give rise to an indefinite construal when its complement is omitted.

In short, passivization of an English transitive verb involves a change in the form of the verb and the replacement of its noun phrase complement by an optional prepositional phrase complement, headed by the preposition *by*, which, when omitted, is construed as indefinite, just like other indefinite polyadic English verbs. Moreover, transitive verbs that passivize give rise to equivalent construals, under the permutation of the subject and direct object noun phrases, reminiscent of the equivalence that arises with verbs with permuting complements.

Grammatical Predicates and Minimal Clauses in English

In ending our brief excursus into English passivization, it is worth bringing to readers' attention another similarity between a passivized English verb and indefinite polyadic English verbs. A minimal clause whose verb is an indefinite polyadic verb and one whose verb is in the passive voice are liable to similar construals when their optional complements are omitted and the clauses are negated. Such sentences have the first kind of construal illustrated in (136) and not the second.

(136) Bill did not read.

   CONSTRUAL 1
   It is not the case that there is something Bill read.

   CONSTRUAL 2
   There is something which Bill did not read.

(137) Bill was not greeted.

   CONSTRUAL 1
   It is not the case that there is someone who greeted Bill.

   CONSTRUAL 2
   There is someone who did not greet Bill.

More generally, and in short, like all omitted optional complements with an indefinite construal, the indefinite force of the omitted complement is always construed as though it is subordinate to negative adverb.

3.5 Open Problems

We draw this chapter on English grammatical predicates to a close with an enumeration of some of the many open problems raised by complementation. First, one problem is the precise characterization of the order of a word's complements. Sometimes the order is rigid.

(138.1) Alice asked [NP Bill] [NP a favor].

(138.2) *Alice asked [NP a favor] [NP Bill].

Sometimes it is flexible.

(139.1) Alice asked [NP a very big favor] [PP of Bill].

(139.2) Alice asked [PP of Bill] [NP a very big favor].

Second, while the preceding treatment of polyadic complements covers much of what is known, two kinds of complement polyadicity remain unaddressed: that of middle polyadic verbs, illustrated in (140), and that of causative polyadic verbs, illustrated by the first two sentences in (141).

(140.1) Dan sold the book easily.

(140.2) The book sells easily.

(141.1) The butter melted.

(141.2) Bill melted the butter.
Bill caused the butter to melt.

Third, there are many aspects of complement specification that are beyond the reach of what we have formalized up to this point. To begin with, we have not shown in sufficient detail how to handle prepositional phrase complements. Words that have prepositional phrases as complements do not admit just any prepositional phrase as their complements, they typically admit a prepositional phrase headed by a particular preposition. For example, in none of the following prepositional phrases can the preposition in them be replaced by another preposition, preserving the meaning, or indeed the acceptability, of the sentences.

(142.1) Alice disposed [PP of Carl's money].
*Alice disposed [PP to Carl's money].

(142.2) Bill dashed [PP to the car].
*Bill dashed [PP of the car].

(142.3) The water buffalo wallowed [PP in the mud].
*The water buffalo wallowed [PP at the mud].

It is for this reason that we did not try to address the facts pertaining to the permutation of the first and second complements of verbs whose second specifies two categories of PP. Nor, for this reason, did we go into the details of how passivization is to be handled.

In addition, some verbs permit a range of prepositions among its complement prepositional phrases: for example, the verb *to dash,* which takes one complement, allows prepositional phrases headed by the prepositions *into* (*to dash into the room*), *out of* (*to dash out of the house*), *up* (*to dash up the street*), *through* (*to dash through the hall*), and *from* (*to dash from the shower*); and the verb *to put,* which takes two complements, allows prepositional phrases headed by the prepositions *above* (*to put the key above the door*), *in* (*to put the shoes in the closet*), *on* (*to put the glass on the floor*), and *near* (*to put the chair near the table*) as well as adverbial phrases such as *downstairs* (*to put the box downstairs*).

Some verbs must specify that, if they take a complement, they take a reflexive pronoun as a complement. (See Stirling and Huddleston 2002, sec. 3.1.1, for more examples.)

(143.1) Alice prides herself on her open-mindedness.

(143.2) *Alice prides Bill on his open-mindedness.

Further associated with complements are restrictions of *valency,* also known as *thematic* or *semantic* roles. They include such roles as agent, patient, instrument, beneficiary, source,

Grammatical Predicates and Minimal Clauses in English 453

and location.[12] What these various roles are, how they are to be distinguished from one another, and how one knows when to attribute them are vexing questions. (This is discussed at length by David Dowty 1991, for example.)

Indeed, there is an entire miscellany of restrictions, often lumped together under the rubric of *selection restrictions,* ranging from very general restrictions, regarding whether the complement must denote something human (human vs. nonhuman), something animate (animate vs. inanimate), or something with gender (male vs. female), to very specific restrictions, which are often extremely difficult to specify. Consider, to begin with, the second example, discussed by Indian grammarians.

(144.1) Devadatta sprinkled water on the ground.

(144.2) *Devadatta sprinkled fire on the ground.

What is it about the verb *to sprinkle* that, when combined with *water,* yields a perfectly acceptable verb phrase, *sprinkle water,* but when combined with *fire,* yields a perfectly unacceptable verb phrase, *sprinkle fire*?

Similar observations were made by philosophers and linguists throughout the last century. Here is a pair of contrasts in the restrictions on the complements of *to drink* and *to eat.*

(145.1) Carl ate the cake.

(145.2) *Carl ate the water.
         (based on Leech 1974, 141)

(146.1) The chimpanzee drank water.

(146.2) *The chimpanzee drank footwear.
         (based on Lyons 1968, 152)

Indeed, as noted by Sally Rice (1988), the following transitive verbs take a very limited range of noun phrase complements.

to unplug (a plug), *to mail* (a letter or package), *to father* (a child), *to stub* (a toe), *to bark* (one's shin), *to purse* (one's lips), *to pucker* (one's lips), *to blow* (one's nose), *to crook* (one's neck), *to turn* (the wheel), *to gun* (an engine), *to rev* (an engine), *to floor* (a gas pedal), *to brush* (one's teeth), *to comb* (one's hair), *to shampoo* (one's hair)

In fact, as noted by McCawley (1968, 139), the transitive verb *to devein,* which applies only to shrimp, for it means to remove from shrimp its dark, dorsal vein (cf. *Merriam-Webster* s.v.).

---

12. These are the ones used by Pāṇini in his *Aṣṭādhyāyī.* In contemporary linguistics, the term *theme* has come to supplant the term *patient.*

Finally, the foregoing treatment does not address the fact that many of the so-called selection restrictions apply not only to complements but also to subjects.

(147) *The wall sees.
       (Baruch Spinoza, cited in Horn 1989, 41)
(148.1) The meat rotted.
(148.2) The food rotted.
(148.3) *The milk rotted.
(149.1) The horse neighed.
(149.2) The animal neighed.
(149.3) *The mouse neighed.
(150.1) The deadline elapsed.
(150.2) *Bill elapsed.

The grammar, as thus far developed, does not accommodate such facts.

Exercises: Open problems

1. Find verbs other than *to pride* that require a reflexive pronoun as one of its complements. If necessary, indicate in which sense of the verb its complement is reflexive.

2. Give more examples of the range of words that can serve as complements of the verbs *to sprinkle, to eat,* and *to drink.*

4 Conclusion

In this chapter, we set out to provide a syntactic and semantic treatment of minimal clauses, English declarative clauses with a minimal number of constituents and whose noun phrases comprise single proper nouns. This led us to distinguish between the head of a phrase from its complements and modifiers. We then concentrated on words and their complements, surveying the rich diversity of word complements in English. This survey showed just how crucial to the syntax of English are two problems identified in chapter 3, section 3.4.1, namely the subcategorization problem and the projection problem. To address these problems, we enriched the notation for lexical categories so that, in addition to categorizing a word by its part of speech, it was also categorized by its complements. We saw by enriching a constituency grammar with complement lists another problem, the problem of assigning a structure to the lexicon of English, was also solved by the incorporation of complement lists as part of a word's lexical entry. Our survey of words and their complements also turned up several further patterns: namely, those of complement polyvalence, complement permutation, and complement polyadicity. Conservative extensions of the notation for complement lists as well as the constituency formation and valuation rules permitted these patterns, at least to a good first-order approximation, to be brought within the ambit of the enriched constituency grammar. Many complexities, of course, were set aside. These complexities included the specification of constituency formation and valuation rules for polyadic causative and middle verbs as well as a variety of specifications that go beyond those of complement categories, such as the specification of particular prepositions and the specification of so-called selection restrictions.

11 Classical Quantificational Logic

1 Notation

In this chapter, we begin *classical quantificational logic,* or CQL. It is quantificational logic, because what distinguishes it from predicate logic is the addition of two sets of symbols: a set of *variables* (VR) and a set of *quantificational connectives* (QC), or *quantifiers,* which has just two symbols, ∃ and ∀. In other words, CQL comprises not only the set of propositional connectives (PC) and a signature, which itself consists of a set of individual symbols (IS), a set of relational symbols (RS), and an adicity function, indicating for each relational symbol what its adicity is, as does CPDL, but also the set of variables (VR) and the set of two quantificational connectives (QC).

Readers will recall from chapter 9 that we used as individual symbols lower-case letters from the beginning of the Roman alphabet for individual symbols, such as *a, b,* and *c*. If a large number of individual symbols were required, we used the letter *c,* subscripted with positive integers. The symbols we shall use for variables are the lower-case letters of the end of the Roman alphabet: *w, x, y,* and *z*. Should a large number of variables be required, we follow the convention analogous for the one we use for individual symbols: we use the letter *v,* subscripted with positive integers.

The sets—PC, QC, VR, IS, and RS—are of course, disjoint from one another. We continue to define a signature to be ⟨IS, RS, ad⟩, ad is a function from RS into $\mathbb{Z}^+$. It is convenient to group together the individual symbols and the variables, calling either a *term,* and to group together the PC and the quantificational connectives, calling them the logical connectives of CQL. Thus, TM, the set of terms, is just IS ∪ VR, and LC, the set of logical connectives, is just PC ∪ QC.

Now that we have identified all the basic symbols of CQL, we are in a position to define its atomic formulae.[1]

---

1. Readers should note that we are reusing the name AF, this time to denote the set of atomic formulae of CQL, as we shall reuse presently the name FM to denote the set of formulae of CQL.

Definition 1 Atomic formulae of CQL

Let ⟨IS, RS, ad⟩ be a signature. Let VR be a nonempty set of variables. Let TM be IS ∪ VR. α is an atomic formula of CQL ($\alpha \in$ AF) iff $\Pi \in$ RS, ad(Π) $= n$ and there are n occurrences of terms from TM—$t_1, \ldots, t_n$—such that $\alpha = \Pi t_1 \ldots t_n$.

Here are some examples of atomic formulae as well as sequences of symbols that may look like atomic formulae, but in fact are not. In the examples here and that follow, the individual symbols include a and b and the relational symbols P, R, and S.

RELATIONAL SYMBOL	ATOMIC FORMULA	NONFORMULA
ad(P) = 1	Pb, Px	Pxa, P
ad(R) = 2	Rzb, Rxz, Rzz	$Rz, Ra, Rxby$
ad(S) = 3	$Sxbz, Saxx$	$Sz, Sxy, Sxyzw$

Now that we understand what an atomic formula is, we define the set of formulae (FM) of CQL. We state here its categorematic definition, leaving it to readers to write out the corresponding syncategorematic definition.

Definition 2 Formulae of CQL (categorematic version)

FM, the set of formulae of CQL, is defined as follows:

(1) AF ⊆ FM;
(2.1) if $\alpha \in$ FM and $* \in$ UC, then $*\alpha \in$ FM;
(2.2) if $\alpha, \beta \in$ FM and $\circ \in$ BC, then $(\alpha \circ \beta) \in$ FM;
(3) if $Q \in$ QC, $v \in$ VR and $\alpha \in$ FM, then $Qv\alpha \in$ FM;
(4) nothing else is.

The following are some expressions using only the symbols of CQL. Those in the left-hand column are formulae of CQL, while those in the right-hand column are not.

FORMULA	NONFORMULA
$\exists x Px$	$\exists a Pa$
$\exists x Rxz$	$\exists xz Rxz$
$\forall y Rxz$	$\forall Rxz$
$\exists x(Rxy \wedge Px)$	$\exists z(Raz)$
$\neg \exists z Sazy$	$(\neg \exists z Sazy)$
$(\exists x Rxy \wedge Px)$	$\forall zy Ryz$

Here is the categorematic synthesis tree of the last formula, $(\exists x Rxy \wedge Px)$.

Classical Quantificational Logic

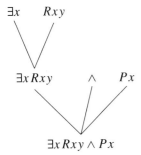

We now turn to a number of notions pertaining to the structure of formulae: the scope of a quantifier, free and bound variables, and finally open and closed formulae.

In section 2.1.3 of chapter 6, we encountered the notion of the *scope* of (an occurrence of) a logical connective: the scope of (an occurrence of) a logical connective within a formula is the smallest subformula containing (the occurrence of) the logical connective. The very same notion was found again in chapter 9 altered to accommodate the expressions of CPDL. Here too it appears still again, and still again it is altered to accommodate the symbols of CQL.

Definition 3 Scope of a logical connective for CQL
The scope of an occurrence of a logical connective in a formula α is the subformula of α that contains the logical connective's occurrence but whose immediate subformulae do not.

In particular, then, the definition allows a quantifier (occurrence) to have scope. The scope of (an occurrence of) a quantifier within a formula is the smallest subformula containing (the occurrence of) the quantifier. For example, the scope of the first occurrence of the universal quantifier in the formula $\big(\forall x (Px \rightarrow \exists y Rxy) \land \forall x (Gx \lor \exists z Rxz)\big)$ is the subformula $\big(\forall x (Px \rightarrow \exists y Rxy)\big)$, whereas the scope of the second occurrence is the subformula $\forall x(Gx \lor \exists z Rxz)\big)$. The scope of the first occurrence of the existential quantifier is the subformula $\exists y Rxy$ and of the second the subformula $\exists z Rxz$.

As careful readers will have noticed, no formula contains an occurrence of a quantifier without an occurrence of a variable to its immediate right. It is useful to divide a formula of the form $Qv\alpha$ into two parts, the first part comprising the quantifier and the variable, Qv, which we shall call the *(quantifier) prefix,* and the second part comprising the formula's immediate subformula α, which we shall call *(quantifier) matrix*. It should be clear that the scope of (an occurrence of) a quantifier and the scope of (the occurrence of) the quantifier prefix that the quantifier helps to constitute are always the same.

We now come to a crucial relation that obtains between occurrences of quantifier prefixes and occurrences of variables in a formula: binding.

Definition 4 Binding for CQL

Let α be in FM. Let Qv be a quantifier prefix that occurs in α and let w be a variable in α. (An occurrence of) the quantifier prefix Qv binds (an occurrence of) the variable w iff

(1) $v = w$ (i.e., v and w are the same variable);
(2) (the occurrence of) w is within the scope of (the occurrence of) the quantifier prefix Qv; and
(3) no occurrence of either $\exists v$ or $\forall v$, distinct from (the occurrence of) the quantifier prefix Qv, has the relevant occcurrence of w within its scope and is itself within the scope of the relevant occurrence of Qv.

The first two conditions stated in the definition are easier to grasp than the third. So, let us consider the definition, first in light of the first two conditions, and then in light of the third.

Consider again the formula $\exists x R x y \wedge P x$. The second occurrence of the variable x is bound by the only occurrence of the quantifier prefix $\exists x$. Why? Because the second occurrence of variable x is within the scope of the only occurrence of the quantifier prefix $\exists x$ and because the occurrences of the variables are occurrences of the same variable. In other words, the first two conditions in the definition are met and since there is no occurrence of any other quantifier matrix, the third condition is met, as it were, vacuously.

In contrast, the only occurrence of the variable y is not bound by the only occurrence the quantifier prefix $\exists x$, for although the only occurrence of the variable y is within the scope of the only occurrence of the quantifier prefix $\exists x$, the occurrences of the variables are occurrences of distinct variables. In other words, though the second condition is met, the first is not.

At the same time, the third occurrence of the variable x in the very same formula is not bound by the quantifier prefix $\exists x$, for although the occurrences of the variables are occurrences of the same variable, the third occurrence of the variable x is not within the scope of the only occurrence of the quantifier prefix $\exists x$.

Let us now turn to the third condition. It becomes operative only when a formula contains two or more occurrences of a quantifier prefix, each occurrence having the same variable but not necessarily having the same quantifier. Consider this formula $\forall x (R x a \rightarrow \exists x P x)$. The last occurrence of the variable x satisfies the first two conditions of the definition, both with respect to the first quantifier prefix $\forall x$ and with respect to the second quantifier prefix $\exists x$. It is undesirable to have any occurrence of a variable bound by two distinct occurrences of quantifier prefixes. To avoid such an undesirable situation, one adopts condition (3), which will, so to speak, break any tie between competing occurrences of quantifier prefixes. It requires that only the occurrence of a quantifier prefix with the smallest scope bind the occurrence of the variable. Thus, in the previous example, the last occurrence of x is bound only by the quantifier prefix $\exists x$.

Classical Quantificational Logic

One question naturally arises: what is the status of (the occurrence of) the variable within (an occurrence of) a quantifier prefix? A careful reading of the definition of binding (definition 4) reveals that (the occurrence of) the quantifier prefix binds (the occurrence of) the variable within it. Thus, for example, the occurrence of v in the quantifier prefix $\exists v$ is bound by that very occurrence of the quantifier prefix.

Definition 5 Bound variable for CQL
(An occurrence of) a variable v is bound in a formula α iff (an occurrence of) a quantifier prefix in the formula α binds it.

Definition 6 Free variable for CQL
(An occurrence of) a variable v is free in a formula α iff no (occurrence of a) quantifer prefix in the formula α binds it.

Consider the formula $Px \wedge \exists x \forall y (Rax \rightarrow Sxy)$. The first occurrence of x is free. The occurrence of x immediately following the only occurrence of a is bound by the only occurrence of the quantifier prefix $\exists x$, as is the occurrence of x immediately following the only occurrence of S. The last occurrence of y is bound by the only occurrence of the quantifier prefix $\forall y$. Notice that while a variable may be both bound and free in a formula, no single variable occurrence may be both. Moreover, each occurrence must be either free or bound.

The distinction between variables that are bound and variables that are free is used to distinguish two kinds of formulae: those that are closed and those that are open. A closed formula is one in which no free variables occur.

Definition 7 Closed formula for CQL
A formula is closed iff it has no (occurrences of a) free variable.

Thus, each of the following formulae is closed: $\forall z P z$; $\exists x \forall y R x y$; $\exists x \forall y (Rax \rightarrow Sxy)$. We shall refer to the set of closed formulae as FM^c.

An open formula is one in which at least one variable occurrence is free.

Definition 8 Open formula for CQL
A formula is open iff it has at least one (occurrence of a) free variable.

Here are some examples: Pz; $\forall y R x y$; $\exists x \forall y (Rax \rightarrow Sxz)$. We shall refer to the set of open formulae as FM^o.

It is useful to define a function that assigns to each formula the set of variables free in it. The definition is recursive.

Definition 9 Free variables in a formula of CQL
(1) ATOMIC FORMULAE

Let α be an atomic formula of the form $\Pi t_1 \ldots t_n$.
$Fvr(\alpha) = \{v \colon v \in VR$ and, for some $i \in \mathbb{Z}_n^+$, t_i is an instance of $v\}$.

(2) COMPOSITE FORMULAE

(2.1) Let α be a composite formula of the form $*\beta$, where $* \in$ UC; that is, let α be a formula of the $\neg\beta$. Then, $\text{Fvr}(\alpha) = \text{Fvr}(\beta)$.

(2.2) Let α be a composite formula of the form $(\beta \circ \gamma)$, where $\circ \in$ BC is a binary connective. Then, $\text{Fvr}(\alpha) = \text{Fvr}(\beta) \cup \text{Fvr}(\gamma)$.

(2.3) Let α be a composite formula of the form $Qv\beta$, where $Q \in$ QC and $v \in$ VR. Then, $\text{Fvr}(\alpha) = \text{Fvr}(\beta) - \{v\}$.

It follows from these definitions that a formula is closed if $\text{Fvr}(\alpha) = \emptyset$, and it is open if $\text{Fvr}(\alpha) \neq \emptyset$.

The last syntactic notion required is that of substitution. It is defined in terms of the replacement of every occurrence of a free variable by an occurrence of an individual symbol. It is defined basically for terms and, through recursion, extended to all formulae of CQL. In fact, there are as many different such operations as there are different pairs of individual symbols and variables. We shall denote such operations as follows: $[c/v]$. We shall call the individual symbol c, which substitutes for, or replaces, the free variable v, the *substituens* and the variable v, which the individual symbol c substitutes for, or replaces, the *substituendum*. These two neologisms will prove useful in chapter 13.

Definition 10 Substitution of an individual symbol for a free variable in CQL
Let c be in IS and let v be in VR. Then,

(0.1) $[c/v]t = t$,   if $t \in$ IS;

(0.2) $[c/v]t = t$,   if $t \in$ VR and $t \neq v$;

(0.3) $[c/v]t = c$,   if $t \in$ VR and $t = v$;

(1.1) $[c/v]\Pi t_1 \ldots t_n = \Pi [c/v]t_1 \ldots [c/v]t_n$

(2.1) $[c/v] * \alpha = *[c/v]\alpha$

(2.2) $[c/v](\alpha \circ \beta) = ([c/v]\alpha \circ [c/v]\beta$

(2.3.1) $[c/v]Qw\alpha = Qw[c/v]\alpha$ if $v \neq w$;

(2.3.2) $[c/v]Qw\alpha = Qw\alpha$ if $v = w$.

To see how this operation works, let us consider a specific case: $[a/x]$.

FORMULA	RESULT OF SUBSTITUTION
α	$[a/x]\alpha$
Rxy	Ray
Rxx	Raa
$\forall x Rxx$	$\forall x Rxx$
Py	Py
$Rxx \wedge \exists x Px$	$Raa \wedge \exists x Px$
$\exists x \exists y Rxy \rightarrow Px$	$\exists x \exists y Rxy \rightarrow Pa$

Classical Quantificational Logic

It is useful to be able to treat one formula as a substitution instance of another. Roughly, one formula, say α, is a substitution instance of another formula β, if β is the scope of a quantifier prefix Qv and α is just like the immediate subformula of β, except that all free occurrences of v in the subformula have been replaced by occurrences of some individual symbol. This idea will become clearer after one inspects the following.

FORMULA	SUBSTITUTION INSTANCE	NONSUBSTITUTION INSTANCE
$\forall x Px$	Pa	Px
$\forall x Rxx$	Raa	Rab
$\exists x \forall y Rxy$	$\forall y Ray$	Rab
$\exists x (\exists y Rxy \to Px)$	$\exists y Ray \to Pa$	$\exists x (Rxb \to Px)$
$\forall y \exists x Rxy \to Py$	$\exists x Rxa \to Py$	$\forall y Ray \to Py$

The notion of one formula being a substitution instance of another can be made precise using substitution functions.

Definition 11 Substitution instance for CQL
A formula α is a substitution instance of a formula β iff, for some quantifier prefix Qv, some formula γ, and some individual symbol c, $\beta = Qv\gamma$ and $[c/v]\gamma = \alpha$.

The last concept to be introduced in this section is that of one formula being an alphabetical variant of another. Let us consider a simple example of such a pair: $\forall x Px$ and $\forall y Py$. Intuitively, both formulae say the same thing: every thing has property P. The first uses the variable x and the second the variable y. A pair of formulae are alphabetical variants of one another just in case, first, they have been formed in the same way, second, their corresponding relational and individual symbol occurrences are occurrences of the same relational and individual symbols, and third, their corresponding free variable occurrences are occurrences of the same variable. Therefore, if they differ at all, they differ only with respect to the variables bound by corresponding quantifier prefixes: where one has a quantifier prefix Qv binding the occurrences of the variable v, and the other has a quantifier prefix Qw binding the occurrences of the variable w in positions corresponding to the occurrences of v bound by the occurrence of Qv. Here are formulae that are alphabetical variants and pairs of formulae that are not.

ALPHABETICAL VARIANTS		NONALPHABETICAL VARIANTS	
Px	Px	Px	Py
$\exists x Px$	$\exists y Py$	$\forall x Px$	$\exists x Px$
$\forall x \exists y Rxy$	$\forall x \exists w Rxw$	$\forall x \exists y Rxy$	$\forall x \exists w Ryw$
$\forall x \exists y Rxy$	$\forall y \exists w Ryw$	$\forall x \exists y Rxy$	$\exists y \forall x Ryx$
$\forall x \exists y Rxy$	$\forall y \exists x Ryx$	$\forall x \exists y Rxy$	$\forall y \exists x Rxy$
$\forall x Rxw$	$\forall y Ryw$	$\forall x Rxy$	$\forall x Rxx$
$\forall x (Rxw \lor \exists y Qy)$	$\forall y (Ryw \lor \exists y Qy)$		

We shall return to the notion of two expressions being alphabetical variants of one another when we present the Lambda calculus (chapter 13).

Exercises: Notation

For these exercises, assume that IS = $\{a, b, c\}$, VR = $\{x, y, z\}$, RS$_1$ = $\{F, G, H\}$, and RS$_2$ = $\{R, S, T\}$.

1. For each sequence of symbols given, determine whether it is a formula in CQL, a formula in CQL by convention only, or not a formula at all. If it is a formula by convention, state the relevant convention. If it is not a formula, explain why it is not.

    (a) $\exists x H x$
    (b) $\forall y \exists z T y z$
    (c) $\exists x F a$
    (d) $\exists a H z$
    (e) $\forall y (\exists F G y \vee R z y)$
    (f) $\forall x \exists y (R a x \vee S a x)$
    (g) $\forall y (\exists z G y \to S y z)$
    (h) $(F a \to \forall y (G b \vee R b))$
    (i) $\exists y (R a y \wedge \forall x F a x)$
    (j) $\forall x \exists y \forall z R x z \vee S y z$
    (k) $(\exists x (R a x \wedge \forall y F y) \to S z y)$
    (l) $\neg \exists x (\exists z (F z) \wedge R a x)$

2. For each variable occurrence in each formula, state which quantifier prefix occurrence binds which variable occurrence and state which variable occurrences are free.

    (a) $\exists z G z$
    (b) $\forall z R z x$
    (c) $\exists x \forall z S y z$
    (d) $\exists y G a$
    (e) $\exists x (\forall x R x a \wedge H x)$
    (f) $\forall x (F x \to \exists y R x y)$
    (g) $\forall x (\forall y \exists z R y z \leftrightarrow (G y \wedge T z x))$
    (h) $(T a y \leftrightarrow (\exists x R y x \wedge G y))$
    (i) $\exists x \exists z \exists y (R z x \vee S y x)$
    (j) $((\exists z \forall x R y x \wedge G z) \wedge R z z)$
    (k) $(F z \vee \exists y G x) \to \forall x S a y$
    (l) $\exists x \exists y (R z x \vee \exists z S y x)$
    (m) $(\exists z F z \to \exists x G x) \wedge \forall x (G x \vee S a x)$
    (n) $\forall z \exists x (R z x \to S y x) \vee (\exists y F x \leftrightarrow T a y)$
    (o) $\exists z F z \vee \forall y \exists x ((T a x \to S y x) \wedge \forall x (G x \vee S a y))$

3. In each case, the substitution function indicated applies to the entire formula. Determine which formula results from the application of the substitution function.

    (a) $[a/x] R a x$
    (b) $[b/x] S a x$
    (c) $[c/z] T a x$
    (d) $[a/y] \exists x R y x$
    (g) $[a/z] ((\exists z \forall x R y x \wedge G z) \wedge R z z)$
    (h) $[a/x] (\exists x R a x \wedge P x)$
    (i) $[c/y] (\exists y R y x \wedge (P x \to P a))$
    (j) $[a/z] [c/y] R z y$

(e) $[b/x] \exists x Syx$ (k) $[c/y] (\exists x Rxy \land \exists y Py)$

(f) $[c/z] Pb$ (l) $[a/z] (\exists x Rzx \land \exists z Sxy)$

4. State which formulae in the right-hand column are substitution instances of which formulae in the left-hand column:

(a) $\forall x Rax$ Raa

(b) $\forall x Rxx$ $\exists x Rxb$

(c) $\exists x Rxb$ Rba

(d) $\exists y Rcy$ $\exists z Rzb$

(e) $\forall x \exists y Rxy$ Rab

(f) $\forall x \forall y Rxy$ $\forall x Rax$

(g) $\forall z Rzz$ Rbb

(h) $\exists z Rzb$ Rba

(i) $\forall y Rcy$ $\exists y Rcy$

2 Classical Valuations for CQL

A classical valuation is defined for CPL on the basis of a truth value assignment to the atomic formulae, or propositional variables, out of which are formed all the formulae of CPL, and five clauses, which extend the truth value assignment from the atomic formulae to all the formulae. Similarly, a classical valuation for predicate logic is defined on the basis of an assignment of truth values to the atomic formulae and five clauses, which extend the truth value assignment from the atomic formulae to all formulae. The difference between them is that, since the atomic formulae CPL have no syntactic structure, there is nothing to limit the assignment of truth values to them, whereas the assignment of truth values to the atomic formulae of predicate logic, each of which comprises a relational symbol followed by a number of individual symbols, is determined by an interpretation function's assignment of values to the relational and individual symbols.

The notation of CQL is more elaborate than that of CPDL. In addition to individual symbols (IS), it has variables; in addition to the unary and binary connectives, it has quantifiers. Variables and quantifiers pose a problem for defining a classical valuation. To see how, consider the formula $\forall x Px$. This formula comprises a quantifier prefix $\forall x$ and a subformula Px. Should we proceed as before, we would expect that the truth value of $\forall x Px$ would be determined on the basis of a truth value assigned to Px and some values assigned to $\forall$ and x. However, as we shall see, this is not how one usually proceeds. To begin with, it is not clear how to assign a truth value to Px; for, even given a structure, one has no more of an idea whether Px is true or false in a structure than whether $x \leq 5$ is true or false. Moreover, it is also unclear what kinds of values to assign either to $\forall$ or to x.

The problem posed by open formulae such as Px and the quantifier prefix has been addressed in a variety of ways, some syncategorematic, others categorematic. We shall survey a number of different ways in which the problem has been addressed, starting with various syncategorematic ways.

2.1 Syncategorematic Definitions of Valuations for CQL

There are several syncategorematic solutions to assigning truth values to formulae of CQL, though they can be conveniently divided into two approaches. In one approach, no value is assigned to any open formula. Thus, the truth value of $\forall x Px$, for example, is determined, not on the basis of its subformula Px, but on the basis of one or more of its substitution instances, that is, on the basis of such a formula as Pa. We shall refer to classical valuations based on this approach as substitutional definitions.

An alternative approach is to use variable assignments so that open formulae can be assigned a truth value in a structure. Thus, just as $x \leq 5$ can be assessed as true or false, depending on what value is assigned to x—true, if it is assigned 4; false, if it is assigned 6— so Px can be assigned true or false in a structure, depending on what value in the structure's universe is assigned x. We shall refer to classical valuations based on this approach as variable assignment definitions.

Next, we shall learn about three substitutional definitions and one variable assignment definition. The first substitutional definition is based on an observation made by medieval European logicians. The other three definitions are of more recent vintage, dating from the end of the nineteenth century and the beginning of the twentieth.

2.1.1 Medieval valuations for CQL

Medieval European logicians noticed an equivalence between sentences formulated with such words as *each* and others formulated with *and,* as well as between those formulated with such words as *some* and others formulated with *or*. For example, it was observed that, if one were talking about four boys, say, Andy, Billy, Charlie, and Danny, the first sentence in each pair is equivalent to the second.

(1) Each boy ran.

(2) Andy ran and Billy ran and Charlie ran and Danny ran.

(3) Some boy ran.

(4) Andy ran or Billy ran or Charlie ran or Danny ran.

For this to work in general, there must be a name for each object one is talking about. For example, should there be an additional boy for whom there is no name, then neither would sentence (1) be equivalent to sentence (2), nor would sentence (3) be equivalent to sentence (4). Moreover, there must be only a finite number of things being talked about, since every sentence has only a finite number of clauses.

This insight can be adopted to formulate a classical valuation for CQL. To do so, we must make the assumption that a structure's universe U is a finite set and that its interpretation function i assigns each member of the universe to some individual symbol. In other words, the interpretation function i, restricted to IS, is a surjection onto U. In this way, a universally quantified formula is true in a structure if and only if the conjunction of each of its substitution instances is true in it; that is, $\forall v \alpha$ is true in a structure if and only if $[c_1/v]\alpha \wedge \ldots \wedge [c_n/v]\alpha$ is true in it. Similarly, an existentially quantified formula is true in a structure if and only if the disjunction of each of its substitution instances is true in it; that is, $\exists v \alpha$ is true in a structure if and only if $[c_1/v]\alpha \vee \ldots \vee [c_n/v]\alpha$ is true in it. In fact, these two conditions are the only ones that have to be added to the definition of a classical valuation for the formulae of CPDL to obtain a definition of a classical valuation of the closed formulae of CQL. In other words, the definition here simply adds the two clauses to the definition of a classical valuation for CPDL: they are the two clauses in (3) in definition 12.

Definition 12 Medieval valuation for CQL

Let M, or $\langle U, i \rangle$, be a structure for a signature, $\langle \text{IS}, \text{RS}, \text{ad} \rangle$, where U is a finite set and i, restricted to IS, is a surjection onto U. Then, v_M is a function from the set of closed formulae (FM^c) to the set of truth values $\{T, F\}$, satisfying the following clauses.

(1) ATOMIC FORMULAE

Let Π be a member of RS, let ad(Π) be 1 and let c be a member of IS. Then,

(1.1) $v_M(\Pi c) = T$ iff $i(c) \in i(\Pi)$.

Let Π be a member of RS, let ad(Π) be n (where $n > 1$) and let $c_1, \ldots, c_n$ be occurrences of members from IS. Then,

(1.2) $v_M(\Pi c_1 \ldots c_n) = T$ iff $\langle i(c_1), \ldots, i(c_n) \rangle \in i(\Pi)$.

(2) COMPOSITE FORMULAE

Then, for each α and for each β in FM,

(2.1) $v_M(\neg \alpha) = T$ iff $v_M(\alpha) = F$;
(2.2.1) $v_M(\alpha \wedge \beta) = T$ iff $v_M(\alpha) = T$ and $v_M(\beta) = T$;
(2.2.2) $v_M(\alpha \vee \beta) = T$ iff   either $v_M(\alpha) = T$ or $v_M(\beta) = T$;
(2.2.3) $v_M(\alpha \to \beta) = T$ iff   either $v_M(\alpha) = F$ or $v_M(\beta) = T$;
(2.2.4) $v_M(\alpha \leftrightarrow \beta) = T$ iff $v_M(\alpha) = v_M(\beta)$.

(3) QUANTIFIED COMPOSITE FORMULAE

Let v be a member of VR, let α be a member of FM and let $c_1, \ldots, c_n$ be occurrences of members from IS such that each member of U is assigned to some c_i ($i \in \mathbb{Z}_n^+$).

(3.1) $v_M(\forall v\, \alpha) = T$ iff $v_M([c_1/v]\alpha \wedge \ldots \wedge [c_n/v]\alpha) = T$;
(3.2) $v_M(\exists v\, \alpha) = T$ iff $v_M([c_1/v]\alpha \vee \ldots \vee [c_n/v]\alpha) = T$.

To see how the clauses in (3) work, let us work with a signature of just four individual symbols and four relational symbols and the following structure for this restricted notation M, or $\langle U, j \rangle$, where $U = \{1, 2, 3, 4\}$ and j is defined thus:

j				
	$a \mapsto 1$		$E\_$	$\mapsto \{2, 4\}$
	$b \mapsto 2$		$O\_$	$\mapsto \{1, 3\}$
	$c \mapsto 3$		$P\_$	$\mapsto \{1, 2, 3\}$
	$d \mapsto 4$		$D\_\_$	$\mapsto \{\langle 1, 1\rangle, \langle 1, 2\rangle, \langle 1, 3\rangle, \langle 1, 4\rangle,$
				$\langle 2, 2\rangle, \langle 2, 4\rangle, \langle 3, 3\rangle, \langle 4, 4\rangle\}$

Consider the formula $\exists z(Pz \wedge Ez)$. According to the structure and clause (1) of definition 12, $v_M(Pb) = T$ and $v_M(Eb) = T$. So, by clause (2.2.1), $v_M(Pb \wedge Eb) = T$. Next, by repeated application of clause (2.2.2), one obtains that the following formula is true: $(Pa \wedge Ea) \vee (Pb \wedge Eb) \vee (Pc \wedge Ec) \vee (Pd \wedge Ed)$. And finally, by clause (3.2), one obtains that $\exists z(Pz \wedge Ex)$ is true.

Next, consider $\forall x(Ox \rightarrow Px)$. By clause (1) of definition 12, each of the following is true in the structure: $Oa \rightarrow Pa$ is since 1 is a member of $j(O)$ and of $j(P)$; $Ob \rightarrow Pb$ is since 2 is not a member of $j(O)$; $Oc \rightarrow Pc$ is since 3 is a member both of $j(O)$ and of $j(P)$; and $Od \rightarrow Pd$ is since 4 is not a member of $j(O)$. By clause (2.2.1), $(Oa \rightarrow Pa) \wedge (Ob \rightarrow Pb) \wedge (Oc \rightarrow Pc) \wedge (Od \rightarrow Pd)$ is true. And, by clause (3.1), $\forall x(Ox \rightarrow Px)$ is true.

Finally, consider $\forall x(Px \rightarrow Dbx)$. By clause (1) of definition 12, $Pa \rightarrow Dba$ is false, since 1 is a member of $j(P)$ but $\langle 2, 1\rangle \notin j(D)$. Since $Pa \rightarrow Dba$ is false, it follows by clause (2.2.1) $(Pa \rightarrow Dba) \wedge (Pb \rightarrow Dbb) \wedge (Pc \rightarrow Dbc) \wedge (Pd \rightarrow Dbd)$ is false. And, by clause (3.1), $\forall x(Px \rightarrow Dbx)$ is false.

To underscore the importance of the requirement that each individual in the universe be assigned to some individual symbol, let us consider a structure in which a member of the universe is assigned to no individual symbol. The structure M', or $\langle U, j'\rangle$, which is just like $\langle U, j\rangle$ except that $j'(d) = 3$ whereas $j(d) = 4$, is such a structure.

j'				
	$a \mapsto 1$		$E\_$	$\mapsto \{2, 4\}$
	$b \mapsto 2$		$O\_$	$\mapsto \{1, 3\}$
	$c \mapsto 3$		$P\_$	$\mapsto \{1, 2, 3\}$
	$d \mapsto 3$		$D\_\_$	$\mapsto \{\langle 1, 1\rangle, \langle 1, 2\rangle, \langle 1, 3\rangle, \langle 1, 4\rangle,$
				$\langle 2, 2\rangle, \langle 2, 4\rangle, \langle 3, 3\rangle, \langle 4, 4\rangle\}$

Now consider the formula $\forall x Px$. On the one hand, it is false in the structure $\langle U, j\rangle$, for, according to clause (3.1), $v_M(\forall x Px) = T$ if and only if $v_M(Pa \wedge Pb \wedge Pc \wedge Pd) = T$. And, by clause (2.2.1), $v_M(Pa \wedge Pb \wedge Pc \wedge Pd) = T$ iff $v_M(Pa \wedge Pb \wedge Pc) = T$ and

$v_M(Pd) = T$. But, $v_M(Pd) = F$, since by clause (1.1), $j(d) \notin j(P)$. On the other hand, it is true in the structure M', or $\langle U, j' \rangle$. Again, according to clause (3.1), $v_{M'}(\forall x Px) = T$ if and only if $v_{M'}(Pa \wedge Pb \wedge Pc \wedge Pd) = T$. And again, by clause (2.2.1), $v_{M'}(Pa \wedge Pb \wedge Pc \wedge Pd) = T$ iff $v_{M'}(Pa \wedge Pb \wedge Pc) = T$ and $v_{M'}(Pd) = T$. However, $v_{M'}(Pd) = T$, since by clause (1.1), $j'(d) \in j'(P)$.

How could it be that the very same formula, $\forall x Px$, whose relational symbols are interpreted in the very same way by both structures, be true in one and false in the other? The answer is that the structure M has a name for 4, namely d, whereas the structure M' does not have a name for 4; instead, the individual symbol d, used to name 4 in M, names 3 in M'.

Before closing this section, we should take notice of how the medieval valuation defined here (definition 12) differs from valuations defined for CPL and CPDL. In CPL and CPDL, the valuation is a function from FM into $\{T, F\}$; the valuation defined here is not, for it assigns no truth value to any open formula; for example, it assigns no truth value to Px. Rather, it is a function from the closed formulae, or FM^c, into $\{T, F\}$. The reason for this is that the clauses in (3) are of a different character from the other clauses. The clauses in (1) assign a truth value to a closed atomic formula on the basis of truth values assigned by the interpretation function to the relational symbol and individual symbols making it up. The clauses in (2) assign a truth value to a composite formula on the basis of truth values assigned to its immediate subformulae. The clauses in (3), however, assign a truth value to a composite formula, not on the basis of a truth value assigned to its immediate subformula, but on the basis of a truth value assigned to a conjunction or a disjunction of substitution instances of the formula, where it is assumed that each member of the universe is assigned to some individual symbol. Thus, for example, the truth value of the formula, $\forall x Px$ is determined, not on the basis of the truth values of its immediate subformula Px, but on the basis of the truth value assigned to the formula $Pa \wedge Pb \wedge Pc \wedge Pd$, which is not a subformula of $\forall x Px$, but a conjunction of its substitution instances, whereas the truth value of the formula $\exists x Px$ is determined by the truth value of $Pa \vee Pb \vee Pc \vee Pd$, which is not a subformula of $\exists x Px$. In other words, unlike the valuations defined for CPL and CPDL, where the truth value assignments are calculated in conformity with a formula's synthesis tree, here a valuation departs from the synthesis tree of a formula once a truth value has to be assigned a subformula whose main logical connective is a quantifier.

Exercises: Medieval valuations for CQL

1. Determine the truth values of the following formulae with respect to the structure M, or $\langle U, j \rangle$

(a) $v_M(Pa)$
(b) $v_M(Od)$
(f) $v_M(Dba)$
(g) $v_M(Dca)$
(k) $v_M(\exists y Dcy)$
(l) $v_M(\forall x Dxx)$

(c) $v_M(Ec)$ (h) $v_M(\forall z Pz)$ (m) $v_M(\forall x \exists y Dxy)$
(d) $v_M(Pb)$ (i) $v_M(\exists x Ox)$ (n) $v_M(\forall x \exists y Dyx)$
(e) $v_M(Dab)$ (j) $v_M(\forall z Dzd)$ (o) $v_M(\exists y \forall x Dxy)$

Justify your answer using the definition of a medieval valuation for CQL (definition 12).

2. Determine the truth values of the formulae in exercise 1 with respect to the structure M', or $\langle U, j' \rangle$, using the clauses of the definition of a medieval valuation for CQL (definition 12).

2.1.2 Marcus valuations for CQL

The definition of a medieval valuation for CQL (definition 12) is restricted to structures with finite universes. The reason is that, even if individual symbols are infinite in number, it is not possible to form formulae with either an infinite number of conjunctions or an infinite number of disjunctions, because each formula has a finite number of atomic subformulae.

It is possible to circumvent this problem. One dispenses with the formation of conjunctions and disjunctions of substitution instances and simply determines the truth of the substitution instances themselves. The idea, first put forth by Ruth Barcan Marcus (1961), is that a universally quantified formula is true in a structure if and only if each of its substitution instances is true in it; that is to say, $\forall v \alpha$ is true in a structure if and only if for each $c \in$ IS, $[c/v]\alpha$ is true in it. Similarly, an existentially quantified formula is true in a structure if and only if at least one of its substitution instances is true in it; that is to say, $\exists v \alpha$ is true in a structure if and only if for some $c \in$ IS, $[c/v]\alpha$ is true in it. For this to work correctly, each member of the universe must be assigned to some individual symbol.

The resulting definition of a classical valuation for CQL is just like the definition for a medieval valuation for CQL, except in two respects: first, it modifies the clauses in (3) and second, it omits the restriction that a structure's universe be finite. The following abbreviated statement of the definition states only the parts of the definition that differ from those in definition 12.

Definition 13 Marcus valuation for CQL

Let M, or $\langle U, i \rangle$, be a structure for a signature $\langle$IS, RS, ad$\rangle$, where i, restricted to IS, is a surjection onto U. Then, v_M is a function from the set of closed formula (FM^c) to the set of truth values $\{T, F\}$, satisfying the following clauses:

(3) QUANTIFIED COMPOSITE FORMULAE

Let v be a member of VR and let α be a member of FM.

(3.1) $v_M(\forall v\, \alpha) = T$ iff for each $c \in$ IS, $v_M([c/v]\alpha) = T$;
(3.2) $v_M(\exists v\, \alpha) = T$ iff for some $c \in$ IS, $v_M([c/v]\alpha) = T$.

Let us consider some examples of structures with infinite domains. To do this, we retain the same relational symbols as before but replace the four individual symbols with the individual symbols from the infinite set IS. For this expanded notation, we adopt the following structure: $\langle \mathbf{Z}^+, k \rangle$ (where $\mathbf{Z}^+$ is the set of positive integers).

k	$c_1 \mapsto 1$	E_	$\mapsto$	$\{2, 4, \ldots\}$
				(i.e., the even, positive integers)
	$\ldots$	O_	$\mapsto$	$\{1, 3, \ldots\}$
				(i.e., the odd, positive integers)
	$c_n \mapsto n$	P_	$\mapsto$	$\{2, 3, 5, 7, 11, \ldots\}$
				(i.e., the prime, positive integers)
	$\ldots$	D__	$\mapsto$	$\{\langle 1, 1\rangle, \langle 1, 2\rangle, \ldots, \langle 2, 2\rangle, \langle 2, 4\rangle, \ldots,$
				$\langle 3, 3\rangle, \langle 3, 6\rangle, \ldots\}$
				(i.e., the set of pairs of positive integers
				such that the first evenly divides the second)

Let us use the same formulae for illustration as we used in section 2.1.1. The formula $\exists z(Pz \wedge Ez)$ is true in this structure because the formula has a substitution instance the structure, namely, $Pc_2 \wedge Ec_2$, that is true in the same structure. By clause (3.2), this is sufficient for the formula $\exists z(Pz \wedge Ez)$ to be true in this structure. The formula $\forall x(Px \rightarrow Ox)$ is false in this structure, since, according to clause (3.1), for the formula to be true in the structure, each of the formula's substitution instance must be true in the structure. But, one of the formula's substitution instances is false in this structure, namely, $Pc_2 \rightarrow Oc_2$. Finally, the formula $\forall x\big((Px \wedge \neg Dc_2 x) \rightarrow Ox\big)$ is true in the structure, since each of the formula's substitution instances is true in the structure, and according to clause (3.1), that is a sufficient condition for the formula to be true in this structure.

It should also be clear that, for finite universes each of whose members is assigned to some individual symbol from a finite set of such symbols, the clauses in (3) of definition 13 and those of definition 12 are equivalent. Let us see how. First, let us think about the clauses in (3.1) of each definition. Consider a universally quantified formula $\forall v\alpha$. Suppose this formula is true in a structure with a finite universe. Then, according to the definition of a medieval valuation for CQL (definition 12), the formula is true in the structure if and only if the conjunction of all of its substitution instances is true in the structure. But, if the conjunction of all of the formula's substitution instances is true in the structure, then each substitution instance is true in the structure. So, according to the definition of the Marcus valuation for CQL (definition 13), $\forall v\alpha$ is also true in the same structure. Conversely, suppose that $\forall v\alpha$ is true in some structure with a finite universe. According to the definition of the Marcus valuation for CQL (definition 13), all of the formula's substitution instances are true in the structure. Since the universe is finite and each member is assigned to an individual symbol, the substitution instances form a conjunction that is true in the structure. So, according to the definition of the medieval valuation for CQL (definition 12), $\forall v\alpha$ is true in the same structure. In brief, the formula $\forall v\alpha$ is true in a structure with a finite universe on one definition if and only if the formula is true in the same structure on the other definition. Similar reasoning establishes the equivalence of the clauses in (3.2) of each definition. Thus, for finite universes each of whose members is assigned to some individual symbol from a finite set of symbols, the clauses in (3.1) of each definition are equivalent and so are the ones in (3.2).

However, the clauses in (3) of each definition are not, in general, equivalent because the clauses in (3) of definition 12 cannot assign a truth value to quantified formulae in structures with infinite domains, since, as we noted, infinitely long conjunctions and infinitely long disjunctions are not formulae.

Notice that the Marcus valuation defined here (definition 13), like the medieval valuation defined previously (definition 12), differs from the valuations defined for CPL and CPDL. Like the medieval valuation, the Marcus valuation is not a function from FM into $\{T, F\}$, since it assigns no truth value to any open formula. Rather, like the medieval valuation, it is a function from the closed formulae, or FM^c, into $\{T, F\}$. The reason for this here, as before, lies with the clauses in (3). Here too the clauses in (3) assign a truth value to a closed formula whose main connective is a quantifier prefix, not on the basis of a truth value assigned to its immediate subformula, but on the basis of a truth value assigned to its various substitution instances, where it is assumed, as it was in the definition of the medieval valuation, that each member of the universe is assigned to some individual symbol. Yet, unlike the clauses in the definition of the medieval valuation, neither a conjunction nor a disjunction of substitution instances is formed. Rather, one considers the set of substitution instances, where each member of the universe is denoted by at least one individual symbol used in forming a substitution instance. Should the formula be prefixed by a universal quantifier, then each substitution instance in the set must be true for the formula itself to be true, and should the formula be prefixed by an existential quantifier, then one substitution instance in the set must be true for the formula itself to be true. For example, should the universe have five members, assigned to c_1 through c_5, respectively, then should at least one formula in the set $\{Pc_1, Pc_2, Pc_3, Pc_4, Pc_5\}$ be true, then $\exists x Px$ would be true; should all of them be true, then $\forall x Px$ would be true. Again, unlike the valuations defined for CPL and CPDL, and like the medieval valuation, the Marcus valuation departs from the synthesis tree of a formula once a truth value has to be assigned a subformula whose main logical connective is a quantifier.

Exercises: Marcus valuations for CQL

1. Determine the truth values of the following formulae with respect to the structure M, or $\langle U, j \rangle$, defined in section 2.1.1. Use the definition of a Marcus valuation for CQL (definition 13) to justify your answer.

(a) $v_M(Pb)$
(b) $v_M(Oa)$
(f) $v_M(Dcb)$
(g) $v_M(Dac)$
(k) $v_M(\exists y Dcy)$
(l) $v_M(\forall z \forall x Dzx)$

Classical Quantificational Logic

(c) $v_M(Ed)$ (h) $v_M(\forall z Pz)$ (m) $v_M(\forall x \exists y Dxy)$
(d) $v_M(Pc)$ (i) $v_M(\exists x Ox)$ (n) $v_M(\forall x \exists y Dyx)$
(e) $v_M(Dbc)$ (j) $v_M(\forall z Dcz)$ (o) $v_M(\exists y \forall x Dxy)$

2. Determine the truth values of the next FM with respect to the structure $\langle U, k \rangle$, defined in this section. Again, use the definition of a Marcus valuation for CQL (definition 13) to justify your answer.

(a) $v_M(Pc_3)$ (f) $v_M(Dc_{17}c_{37})$ (k) $v_M(\exists y Dc_{13}y)$
(b) $v_M(Oc_8)$ (g) $v_M(Dc_3c_9)$ (l) $v_M(\forall x Dxx)$
(c) $v_M(Ec_2)$ (h) $v_M(\exists z Pz)$ (m) $v_M(\forall x \exists y Dxy)$
(d) $v_M(Pc_6)$ (i) $v_M(\forall x Ox)$ (n) $v_M(\forall x \exists y Dyx)$
(e) $v_M(Dc_9c_8)$ (j) $v_M(\forall z Dc_1z)$ (o) $v_M(\exists y \forall x Dxy)$

2.1.3 Fregean valuations for CQL

To determine the truth of a formula over which a quantifier prefix has scope according to the last definition of a classical valuation for CQL, one determines the truth value of the formula's substitution instances. The approach here is to consider but one substitution instance. But one substitution instance says nothing about whether its universal generalization is true or its existential generalization is false. To see the problem, recall the structure $\langle U, j \rangle$, whose universe is $\{1, 2, 3, 4\}$ and whose interpretation function follows.

j $a \mapsto 1$ $E\_ \mapsto \{2, 4\}$
   $b \mapsto 2$ $O\_ \mapsto \{1, 3\}$
   $c \mapsto 3$ $P\_ \mapsto \{1, 2, 3\}$
   $d \mapsto 4$ $D\_\_ \mapsto \{\langle 1, 1 \rangle, \langle 1, 2 \rangle, \langle 1, 3 \rangle, \langle 1, 4 \rangle,$
                    $\langle 2, 2 \rangle, \langle 2, 4 \rangle, \langle 3, 3 \rangle, \langle 4, 4 \rangle\}$

The fact that Pa is true in $\langle U, j \rangle$ does not imply that $\forall x Px$ is true. Indeed, it is false in $\langle U, j \rangle$. Similarly, the fact that Pd is false in M does not imply that $\exists x Px$ is false, for indeed it is true.

How, then, is it possible to work with a single substitution instance? The answer is that one must assign different values to the individual symbol introduced by the substitution instance. Instead of determining whether Pa, Pb, Pc, and Pd is true in the structure, as one does in the last definition, one determines whether Pa is true, holding the structure's interpretation of every relational symbol and every individual symbol other than a fixed, allowing only the values assigned to a to vary. Applied here to the formula $\forall x Px$, this means to retain the assignment of $\{1, 2, 3\}$ to P and to consider a, as it is assigned 1, 2, 3, and 4, respectively.

Let us determine the truth of the formulae $\forall x P x$ with respect to the structure M, or $\langle U, j \rangle$, using first the Marcus valuation and then the Fregean valuation. Recall that, intuitively, the formula $\forall x P x$ is true in a structure if and only if P holds of each member of its universe. So, to ascertain whether $\forall x P x$ is true in this structure, one must ascertain whether P is true of 1, of 2, of 3, and of 4. Using the Marcus valuation (definition 13), one determines whether each of the substitution instances of $\forall x P x$ is true in the structure. As is clear from inspection of the structure, while P is true of 1, of 2, and of 3, P is not true of 4. This is exactly what clause (3.1) of definition 13 shows, for while $v_M(Pa) = v_M(Pb) = v_M(Pc) = T$, $v_M(Pd) = F$. Another way to do this is to use but one substitution instance of $\forall x P x$, Pa. Then, we ask whether or not Pa is true, should a be assigned 1, should a be assigned 2, should a be assigned 3, and should a be assigned 4—all the while holding what is assigned P fixed. Pa is true, should a be assigned 1, 2, or 3, but it is false, should it be assigned 4. In this way, we see that P is not true of everything in U and that $\forall x P x$ is false. The difference in the two approaches is illustrated here.

MARCUS VALUATION

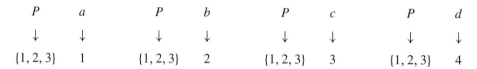

FREGEAN VALUATION

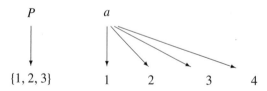

To express formally and generally this alternative use of substitution instances in the definition of a classical valuation, we must formalize the notion of holding part of an interpretation fixed and letting part vary. In each case, every assignment in the interpretation function is held fixed, except the value of the individual symbol that is used to form the substitution instance of the quantified formula. To express such interpretation functions, we avail ourselves of the notion of a near-variant function and its notation, introduced in section 6.3.2 of chapter 2.

Consider these near-variant functions of the function j: $j_{a \mapsto 1}$, $j_{a \mapsto 2}$, $j_{a \mapsto 3}$, and $j_{a \mapsto 4}$. They all assign the same value to P: that is, $j_{a \mapsto 1}(P) = j_{a \mapsto 2}(P) = j_{a \mapsto 3}(P) = j_{a \mapsto 4}(P)$. Indeed, these interpretation functions assign the same value to all the relational symbols and all the individual symbols except a. These functions hold the interpretation of P fixed, but, between them, they assign each of the four different members of U to a. Thus, we

Classical Quantificational Logic

can determine whether P is true of each or some member of U by checking the following: first, whether $j_{a \mapsto 1}(a) \in j_{a \mapsto 1}(P)$ (that is, whether Pa is true, if a is assigned 1); second, whether $j_{a \mapsto 2}(a) \in j_{a \mapsto 2}(P)$ (that is, whether Pa is true, if a is assigned 2); third, whether $j_{a \mapsto 3}(a) \in j_{a \mapsto 3}(P)$ (that is, whether Pa is true, if a is assigned 3); and fourth, whether $j_{a \mapsto 4}(a) \in j_{a \mapsto 4}(P)$ (that is, whether Pa is true, if a is assigned 4). We have, in fact, checked to see whether Pa is true in four structures: $\langle U, j_{a \mapsto 1}\rangle$, $\langle U, j_{a \mapsto 2}\rangle$, $\langle U, j_{a \mapsto 3}\rangle$, and $\langle U, j_{a \mapsto 4}\rangle$. ($\langle U, j_{a \mapsto 1}\rangle$ is just $\langle U, i \rangle$, since $j_{a \mapsto 1} = i$.) It is convenient to refer to such structures as $M_{a \mapsto 1}$, $M_{a \mapsto 2}$, $M_{a \mapsto 3}$, and $M_{a \mapsto 4}$, where $M_{a \mapsto 1}$ is just M.

The third definition of a classical valuation for CQL drops the restrictions on structures incorporated in the previous two definitions and recasts the clauses in (3) along the lines just illustrated. We state this third definition in an abbreviated form.

Definition 14 Fregean valuation for CQL

Let M, or $\langle U, i \rangle$, be a structure for a signature $\langle IS, RS, ad \rangle$. Then, v_M is a function from the set of closed formula (FM^c) to the set of truth values $\{T, F\}$, satisfying the following clauses.

(3) QUANTIFIED COMPOSITE FORMULAE

Let v be a member of VR and let α be a member of FM.

(3.1) $\quad v_M(\forall v\, \alpha) = T \quad$ iff $\quad$ for each e in U, $v_{M_{c \mapsto e}}([c/v]\alpha) = T$,

$\qquad$ where c is an individual symbol not occurring in α;

(3.2) $\quad v_M(\exists v\, \alpha) = T \quad$ iff $\quad$ for some e in U, $v_{M_{c \mapsto e}}([c/v]\alpha) = T$,

$\qquad$ where c is an individual symbol not occurring in α.

This definition of a valuation for CQL works even for structures members of whose universe are unassigned to any individual symbol. Consider the structure $\langle \mathbf{Z}^+, l \rangle$.

l	$a \mapsto 1$	E _	$\mapsto$	$\{2, 4, \ldots\}$
				(i.e., the even, positive integers)
	$b \mapsto 3$	O _	$\mapsto$	$\{1, 3, \ldots\}$
				(i.e., the odd, positive integers)
	$c \mapsto 3$	P _	$\mapsto$	$\{2, 3, 5, 7, 11, \ldots\}$
				(i.e., the prime, positive integers)
		D _ _	$\mapsto$	$\{\langle 1, 1\rangle, \langle 1, 2\rangle, \ldots, \langle 2, 2\rangle, \langle 2, 4\rangle, \ldots,$
				$\langle 3, 3\rangle, \langle 3, 6\rangle, \ldots\}$
				(i.e., the set of pairs of positive integers
				such that the first evenly divides the second)

We turn again to the same three formulae to illustrate how the clauses in our latest definition work. The formula $\exists z(Pz \wedge Ez)$ is true, because it has a true substitution instance, namely, $Pa \wedge Ea$ where it is true in a minimal variant with respect to a of $\langle \mathbf{Z}^+, l \rangle$, namely,

$\langle \mathbf{Z}^+, l_{a \mapsto 2} \rangle$. The formula $\forall x(Px \to Ox)$ is false, since, according to clause (3.1), for it to be true, $Pa \to Oa$ must be true in each minimal variant with respect to a of $\langle \mathbf{Z}^+, l \rangle$. But it is false in the minimal variant $\langle \mathbf{Z}^+, l_{a \mapsto 2} \rangle$. At last, the formula $\forall x((Px \land \neg Dbx) \to Ox)$ is true, since it has a substitution instance that is true in each of the relevant minimal variants of $\langle \mathbf{Z}^+, l \rangle$.

Readers undoubtedly noticed that each clause in (3) has a restriction to the effect that the individual symbol chosen to form the substitution instance may have no occurrence in the original formula. One example should make clear why this is so. Consider the formula $\forall x Dxc$ with respect to the structure $\langle \mathbf{Z}^+, l \rangle$. Dcc is true in it, since $l(c) = 3$ and 3 evenly divides 3. But it is also true in every structure $\langle \mathbf{Z}^+, l_{c \mapsto n} \rangle$, since every positive integer n evenly divides itself. And so, $\forall x Dxc$ is true in $\langle \mathbf{Z}^+, l \rangle$. But the formula $\forall x Dxc$ should not be true in $\langle \mathbf{Z}^+, l \rangle$, since it is not true that every positive integer evenly divides 3. This problem is avoided if the individual symbol chosen for instantiation is different from any individual symbol occurring in the formula.

The Fregean valuation (definition 14), like the medieval and the Marcus valuations (definitions 12 and 13, respectively), differs from the classical valuations defined for CPL and CPDL. The Fregean valuation (definition 14) is not a function from FM into $\{T, F\}$, but from FM^c into $\{T, F\}$. The reason, of course, lies with the clauses in (3). Once again, the clauses in (3) use a substitution instance, and not a subformula, to assign a truth value to a closed formula whose main connective is a quantifier prefix. However, the clauses in (3) here differ in an important way from their counterparts in definitions 12 and 13. In those definitions, one and the same structure is used to determine the value assigned to closed formula with its quantifier prefix and to determine the value assigned either, in the case of medieval valuations, to the conjunction or disjunction of substitution instances, and in the case of Marcus valuations, to its various substitution instances. Here, the truth value assigned by a valuation to a closed formula whose main logical connective is a quantifier prefix is obtained by consulting a variety of valuations, all but one of which is the same as the valuation being applied to the closed formula. The different valuations correspond to different structures, all of which are near variants of the structure with respect to which the truth of the formula with the quantifier prefix is being determined. Again, unlike the valuations defined for CPL and CPDL, and like the medieval and Marcus valuations, the Fregean valuation departs from the synthesis tree of a formula once a truth value has to be assigned a subformula whose main logical connective is a quantifier.

Exercises: Fregean valuations for CQL

1. Determine the truth values of the next set of formulae with respect to the structure $\langle U, l \rangle$, defined in this section. Use the definition of a Fregean valuation for CQL (definition 13) to justify your answer.

Classical Quantificational Logic

(a) $v_M(Dbc)$
(b) $v_M(Dcb)$
(c) $v_M\big(\exists z(Pz \wedge Ez)\big)$
(d) $v_M\big(\forall y(Ey \to Dby)\big)$
(e) $v_M(\forall z Daz)$
(f) $v_M(\exists y Dcy)$
(g) $v_M(\forall x \forall y Dxy)$
(h) $v_M(\forall x \exists y Dxy)$
(i) $v_M\big(\exists x(Ox \wedge \exists y(Ey \wedge Dyx))\big)$
(j) $v_M\big(\exists x(Ox \wedge \exists y(Ey \wedge Dxy))\big)$

2.1.4 Tarskian valuations for CQL

The three previous definitions of a classical valuation determine the truth value of a quantified formula, not on the basis of a truth value assigned to its matrix, but on the basis of a truth value assigned to one or more of its substitution instances. This is a departure from how classical valuations are defined for CPL and CPDL. The question arises: is it possible to determine the truth value of a quantified formula using only its proper parts, in particular, using its matrix or immediate, proper subformula?

Although the expression $x \leq 5$ cannot be determined to be true or false, it can be if we assign a value to x. Thus, if x is assigned 5, it is true, but if it is assigned 6, it is false. Similarly, the formula Px can be assigned a truth value in a structure, provided that some element of the structure's universe is assigned to x. The problem is that a truth value cannot be determined for a quantified formula such as $\forall x Px$ or $\exists x Px$, at least in general, on the basis of a single value assigned to x and a single value assigned to P. We saw that the truth value of these quantified formulae cannot be determined on the basis of the value assigned to a single substitution instance. Using the structure $\langle U, j \rangle$, we saw also, however, that, by holding fixed the value its interpretation function assigns to P and by assigning different values to the instantiating individual symbol, one can determine the truth value of such formulae in a structure.

FREGEAN VALUATION

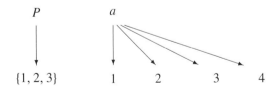

What is proposed here is to hold the structure's interpretation of P fixed and, instead of letting an instantiated individual symbol be assigned different members of the structure's universe, letting the variable that becomes free when the formula's quantifier prefix is removed from it be assigned different members of the structure's universe.

TRUTH IN A STRUCTURE UNDER A VARIABLE ASSIGNMENT

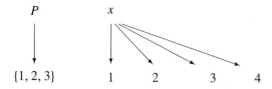

Consider again the truth of the formula $\forall x Px$ with respect to the structure $\langle U, j \rangle$. The basic idea used to ascertain the truth of such formulae with respect to a structure is this: the formula $\forall x Px$ is true in a structure if and only if P holds of each member of the structure's universe. So, to ascertain whether or not $\forall x Px$ is true in this structure, one must ascertain whether or not P is true of 1, of 2, of 3, and of 4. The way to achieve this, using definition 14, is to determine the truth of but one substitution instance of $\forall x Px$, Pa, inquiring into whether Pa is true, should a be assigned 1, should a be assigned 2, should a be assigned 3, and should a be assigned 4—all the while holding what is assigned P fixed. But the very same result can be obtained using, not a substitution instance of $\forall x Px$, but its subformula Px. Instead of letting a be assigned value after value from the universe, let x be assigned value after value. And just as we had tracked the different values of a using the notation $a \mapsto e$, where e is an element of the universe, we can track different values of x using the notation $x \mapsto e$.

Before we illustrate this, let us become acquainted with some commonly used notation. First, it is common to use the name of a structure as an index to identify its universe and its interpretation function. For example, if one refers to a structure M, then its universe is denoted by U_M and its interpretation function by i_M. In other words, $M = \langle U_M, i_M \rangle$. Second, we introduce an alternate way of writing functions, namely, by writing the name of the function to the right of the name of the value to which it is applied, rather than writing it to the left. Moreover, rather than enclosing the name of the value to which the function is applied in parentheses, we shall write it enclosed by square brackets. For example, where we wrote $v_a(\alpha)$ and $v_M(\alpha)$ for a classical valuation of CPL and of CPDL, respectively, we shall write we shall write $[\alpha]^M$.

Having introduced this alternate format of function notation, let us return now to the formulae $\forall x Px$ and $\exists x Px$. The formula $\forall x Px$ is true in a structure just in case Px is true in the structure, no matter what value from the structure's universe is assigned to x. We can state this more formally as follows.

PROVISIONAL CLAUSE (single prefixed closed formula)

(3.1') $[\forall x Px]^M = T$ iff, for each $e \in U_M$, $[Px]^M_{x \mapsto e} = T$.

Let us consider a structure M whose universe comprises $\{1, 2, 3\}$ and whose interpretation function assigns to P the set $\{1, 2, 3\}$. To ascertain whether $\forall x Px$ is true in this structure, we must ascertain whether Px is true, no matter what value from the universe is assigned to x. We have three cases to consider: the case where x is assigned 1, the case where x is assigned 2, and the case where x is assigned 3. If x is assigned 1, then Px is true, since 1, which is assigned to x, is a member of $\{1, 2, 3\}$, which is assigned to P; that is, $[Px]^M_{x \mapsto 1} = T$. If x is assigned 2, Px is true again, since 2, which is assigned to x, is a member of $\{1, 2, 3\}$, the set that the structure assigned to P: that is, $[Px]^M_{x \mapsto 2} = T$. Finally, if x is assigned 3, then Px is true still again, since 3, which is assigned to x, is a member of $\{1, 2, 3\}$, the value assigned to P: that is, $[Px]^M_{x \mapsto 3} = T$. Thus, no matter what value is assigned to x from the structure's universe, Px is true: in other words, for each $e \in U_M$, $[Px]^M_{x \mapsto e} = T$. So, by the provisional clause in (3.1'), $[\forall x Px]^M = T$, that is, $\forall x Px$ is true in the structure.

Let us see what truth value is assigned to the formula $\forall x Px$ when it is evaluated in a different structure. This structure has the same universe, but its interpretation function assigns the set $\{2, 3\}$ to P, instead of $\{1, 2, 3\}$. In this structure, which we shall also call M, the formula $\forall x Px$ is false, that is, $[\forall x Px]^M = F$. The reason is that, even though Px is true when x is assigned 2, and even though it is still true when x is assigned 3, Px is false, when x is assigned 1. Therefore, it is not the case that, for each $e \in U_M$, $[Px]^M_{x \mapsto e} = T$. So, by the provisional clause in (3.1'), $[\forall x Px]^M = F$; that is to say, $\forall x Px$ is false in this structure.

Let us turn to the formula $\exists x Px$ and ask the question, under what circumstances is it true in a structure? In other words, when is it the case that $[\exists x Px]^M = T$? It is true just in case Px is true with respect to M for an assignment of some or other value to x from M's universe. This is restated formally as follows.

PROVISIONAL CLAUSE (single prefixed closed formula)

(3.2') $[\exists x Px]^M = T$ iff, for some $e \in U_M$, $[Px]^M_{x \mapsto e} = T$.

To illustrate how the provisional clause in (3.2') works, let us turn to a structure where the universe is still the set $\{1, 2, 3\}$ but where P is assigned the set $\{2, 3\}$. Px, when x is assigned 1, is false, since 1, which is assigned to x, is not a member of the set $\{2, 3\}$, which the structure assigns to P. However, should x be assigned either 2 or 3, Px is true, since 2 and 3 are members of the set assigned by the structure to P. Thus, there is a value in the structure's universe that, when assigned to x, makes Px true; that is, for some $e \in U_M$, $[Px]^M_{x \mapsto e} = T$. So, by the provisional clause in (3.2'), $[\exists x Px]^M = T$, that is, $\exists x Px$ is true in the structure.

In contrast, let us consider a structure where the universe is still the set $\{1, 2, 3\}$ but where P is assigned the empty set. Consider Px when x is assigned 1. Px is false, since 1, which is assigned to x, is not a member of the set that the structure's interpretation function assigns to P, namely, the empty set. Nor is Px true, when x is assigned 2, since 2 is not a member of the empty set either. Nor is Px true, when x is assigned 3. In short, no matter what is assigned to x from the universe of this structure, Px is false. Thus, it is not the case that there is a member of the structure's universe, which, when assigned to x, makes Px true; that is, it is not the case that, for some $e \in U_M$, $[Px]^M_{x \mapsto e} = T$. Therefore, by the provisional clause in (3.2′), $[\exists x Px]^M = F$, that is, the formula $\exists x Px$ is false in the structure.

Happily, the procedure just applied to a closed formula with one quantifier prefix applies to a closed formula with two quantifier prefixes, and indeed, to a closed formula with any number of quantifier prefixes. To see that this is so, consider a closed formula with two quantifier prefixes, like the one in (5), and a structure M whose universe comprises only three elements, $\{1, 2, 3\}$ and in which R is assigned the set $\{\langle 1, 2\rangle, \langle 2, 3\rangle, \langle 3, 1\rangle\}$.

(5) $\forall x \exists y Rxy$.

Just as with the formula $\forall x Px$, we know that $[\forall x \exists y Rxy]^M = T$ if and only if, for each $e \in \{1, 2, 3\}$, $[\exists y Rxy]^M_{x \mapsto e} = T$: that is, we must determine the truth value of $\exists y Rxy$, the immediate subformula of $\forall x \exists y Rxy$, with respect to three functions, as shown here:

(5.1) $[\exists y Rxy]^M_{x \mapsto 1} = T$.

(5.2) $[\exists y Rxy]^M_{x \mapsto 2} = T$.

(5.3) $[\exists y Rxy]^M_{x \mapsto 3} = T$.

How do we arrive at the truth value given in (5.1)? Let us proceed just as we did with respect to the formula in (5). We eliminate the quantifier prefix $\exists y$ and assign to y the various values from U. This gives us these three equalities:

(5.1.1) $[Rxy]^M_{x \mapsto 1; y \mapsto 1} = F$.

(5.1.2) $[Rxy]^M_{x \mapsto 1; y \mapsto 2} = T$.

(5.1.3) $[Rxy]^M_{x \mapsto 1; y \mapsto 3} = F$.

How does one arrive at these values? Let us consider the equality in (5.1.1) first. The atomic formula Rxy is true when x is assigned 1 and y is assigned 1 just in case the ordered pair $\langle 1, 1\rangle$ is a member of the value the structure assigns to R. Now the structure assigns to R the set $\{\langle 1, 2\rangle, \langle 2, 3\rangle, \langle 3, 1\rangle\}$, which does not contain the ordered pair $\langle 1, 1\rangle$. So, when x is assigned 1 and y is assigned 1, Rxy is assigned F in the structure. Next, consider the equality in (5.1.2). Here, we are interested in the atomic formula Rxy when x is assigned 1 and y is assigned 2. Since the ordered pair $\langle 1, 2\rangle$ is in the set that the structure assigns to R, Rxy is assigned T, when x is assigned 1 and y is assigned 2. Finally, consider the equality in (5.1.3). The atomic formula Rxy, when x is assigned 1 and y is assigned 3, is assigned F, since the ordered pair $\langle 1, 3\rangle$ is not in the set assigned to R by the structure. Still, if x is assigned 1 and y is assigned 2, then $\exists y Rxy$ is true in the structure and we arrive, thereby, at the truth value T on the right-hand side of the equality in (5.1).

We now turn to the equality in (5.2). Proceeding in precisely the same way, we establish the following truth value assignments for the atomic formula Rxy, each calculated by a different assignment of values to the free variables x and y.

(5.2.1) $[Rxy]^M_{x \mapsto 2; y \mapsto 1} = F$.

(5.2.2) $[Rxy]^M_{x \mapsto 2; y \mapsto 2} = F$.

(5.2.3) $[Rxy]^M_{x \mapsto 2; y \mapsto 3} = T$.

Of the three ordered pairs arising from the different assignment of values to x and y, namely, the ordered pair $\langle 2, 1 \rangle$, the ordered pair $\langle 2, 2 \rangle$, and the ordered pair $\langle 2, 3 \rangle$, only the ordered pair $\langle 2, 3 \rangle$ is in the set assigned to R by the structure. Thus, F appears in the first two equalities and T appears in the last.

Finally, we come to the equality in (5.3). It holds just in case one of the following holds:

(5.3.1) $[Rxy]^M_{x \mapsto 3; y \mapsto 1} = T$.

(5.3.2) $[Rxy]^M_{x \mapsto 3; y \mapsto 2} = F$.

(5.3.3) $[Rxy]^M_{x \mapsto 3; y \mapsto 3} = F$.

T appears only in the equality in (5.3.1), since, of the three ordered pairs, only the ordered pair $\langle 3, 1 \rangle$ is in the set assigned to R.

We have now seen that T must appear in each of the equalities in (5.1), (5.2), and (5.3). Thus, we conclude by clause (3) of definition 14 that $[\forall x \exists y Rxy]^M = T$, that is, formula (5) is true in the structure M.

The foregoing procedure is clear and intuitive, though certainly tedious. Confronted with a formula with a quantifier prefix, eliminate the quantifier prefix and determine the truth or falsity of the matrix for each assignment of a value from the universe to the free variable corresponding to the variable in the eliminated quantifier prefix. If the quantifier prefix contains the universal quantifier, the formula is true just in case the matrix is true, no matter what value from the universe is assigned to the variable rendered free by the elimination of the quantifier prefix; it is false otherwise. If the quantifier prefix contains the existential quantifier, the formula is true just in case the matrix is true for at least one value from the universe assigned to the variable rendered free by the elimination of the quantifier prefix; it is false otherwise.

One might be tempted to think that the following generalization of the provisional clauses in (3.1') and (3.2') will adequately reflect the steps that we took to determine the truth of formula (1) in the specified structure.

PROVISIONAL CLAUSES GENERALIZED

(3.1'') $[\forall v \alpha]^M = T$ iff, for each $e \in U_M$, $[\alpha]^M_{v \mapsto e} = T$.

(3.2'') $[\exists v \alpha]^M = T$ iff, for some $e \in U_M$, $[\alpha]^M_{v \mapsto e} = T$.

However, these generalized provisional clauses are not sufficiently general to permit us to carry out all the steps. To see why, look carefully at the function symbols on the left and right sides of each clause. The function sign on the left-hand side of the biconditional is $[\,]^M$, while the one on the right-hand side is $[\,]^M_{v \mapsto e}$. The first function sign, $[\,]^M$, says nothing about how to assign truth values to a formula with free variables. Thus, whereas we can apply the schematic clause in (3.1″) to formula (1), obtaining thereby three functions, $[\,]^M_{x \mapsto 1}$, $[\,]^M_{x \mapsto 2}$, and $[\,]^M_{x \mapsto 3}$, each of which applies to $\exists y Rxy$, the immediate subformula of (1), as shown in (1.1), (1.2), and (1.3), respectively, we cannot apply the schematic clause in (3.2″) to the formula $\exists y Rxy$ to obtain functions that can apply to Rxy, the immediate subformula of $\exists y Rxy$. The reason is that no instance of the left-hand side of the schema in (3.2″) yields a function that specifies a value for the free variable x in $\exists y Rxy$. We could add a further pair of clauses:

(3.1‴) $[\forall w\alpha]^M_{v \mapsto e} = T$ iff, for each $d \in U_M$, $[\alpha]^M_{v \mapsto e; w \mapsto d} = T$;

(3.2‴) $[\exists w\alpha]^M_{v \mapsto e} = T$ iff, for some $d \in U_M$, $[\alpha]^M_{v \mapsto e; w \mapsto d} = T$.

However, this additional pair will not help us when it comes to closed formula with three quantifier prefixes.

To address this problem, Tarski used a *variable assignment* and defined truth in a structure at a variable assignment. A variable assignment is simply any function from the variables into the universe of a structure. By way of example, suppose that our notation had only three variables, say, x, y, and z. Using the structure $\langle U, j \rangle$ (given earlier) and a variable assignment such as g (where, say, $x \mapsto 1$, $y \mapsto 2$, and $z \mapsto 4$), one can assign truth values to any open formula, including such formulae as Rxy, Pz, Rzz, and Px (R as just defined). Thus, for example, Pz is false, since $g(z) = 4$, $j(P) = \{1, 2, 3\}$, and $g(z) \notin j(P)$. Px, in contrast, is true, since $g(x) = 1$, $j(P) = \{1, 2, 3\}$, and $g(x) \in j(P)$. Moreover, Rxy is true, since $g(x) = 1$, $g(y) = 2$, $j(R) = \{\langle 1, 2 \rangle, \langle 2, 3 \rangle, \langle 3, 1 \rangle\}$, and $\langle g(x), g(y) \rangle \in j(R)$. It is easy to check that Rzz is false. In general, then, once one has both a structure and a variable assignment, any formula, open or closed, can be assigned a truth value. Modifying slightly the notation we introduced earlier, we write $[Pz]^M_g = F$ and $[Px]^M_g = T$. We subscript the right-hand bracket with g, since both g and M are used to determine the truth value of an open formula. We use the interpretation function of the structure to assign values to individual symbols and relational symbols, and we use g to assign values to variables.

How does one adapt the procedure for determining the truth of formulae with quantifier prefixes to the use of variable assignments? Recall that, according to the procedure outlined, when confronted with a formula with a quantifier prefix, one eliminates the quantifier prefix and determines the truth or falsity of the matrix for each assignment of a value from the universe to the free variable corresponding to the variable in the eliminated quantifier prefix. To adapt the procedure to the use of variable assignments, when one eliminates the quantifier prefix from a formula, one determines the truth or falsity of the matrix with

Classical Quantificational Logic

respect to each variable assignment that differs from the one used for the original formula at most by what the new variable assignment assigns to the free variable resulting from the eliminated quantifier prefix. In other words, one consults all variable assignments that are the near variants of the variable assignment used to assign a truth value to the original formula.

We now formalize the use of variable assignments to determine the truth value of formulae with quantifier prefixes.

CLAUSES TO BE ADOPTED

(3.1) $[\forall v \alpha]_g^M = T$ iff, for each $e \in U_M$, $[\alpha]_{g_{v \mapsto e}}^M = T$.

(3.2) $[\exists v \alpha]_g^M = T$ iff, for some $e \in U_M$, $[\alpha]_{g_{v \mapsto e}}^M = T$.

Notice that both sides of the biconditional appeal to both a structure and a variable assignment, the left-hand side to $[\ ]_g^M$ and the right-hand side to $[\ ]_{g_{v \mapsto e}}^M$. The left-hand side, like the right-hand side, applies to open formulae, regardless of how many free variables they have.

The relativization of the function $[\ ]^M$ to a variable, as for example $[\ ]_x^M$, and its relativization to a variable assignment, as for example $[\ ]_g^M$, makes no difference in determining the truth of a closed formula, so long as the value assigned to a variable directly, as it were, is the same as the value the variable assignment function assigns to the variable. For illustration, consider again the formulae $\forall x P x$ and $\exists x P x$ and the structure whose universe is the set $\{1, 2, 3\}$ and that assigns the set $\{1, 2\}$ to P.

$[Px]_{x \mapsto 1}^M = T$ $\qquad$ $[Px]_{g_{x \mapsto 1}}^M = T$

$[Px]_{x \mapsto 2}^M = T$ $\qquad$ $[Px]_{g_{x \mapsto 2}}^M = T$

$[Px]_{x \mapsto 3}^M = F$ $\qquad$ $[Px]_{g_{x \mapsto 3}}^M = F$

$[\forall x P x]^M = F$ $\qquad$ $[\forall x P x]_g^M = F$

$[\exists x P x]^M = T$ $\qquad$ $[\exists x P x]_g^M = T$

The function we are about to define, is known, somewhat cumbersomely, as defining *truth in a structure under a variable assignment*. We shall refer to it more briefly as a *Tarski extension*, in honor of Alfred Tarski, who first defined such a function (Tarski 1932, definition 23). The definition comprises four sets of clauses. The clauses in (1) and (2) are just like the clauses we encountered in the previous definitions of valuations. The provisional clauses in (3) are the clauses indicated previously as the ones to be adopted. The clauses in (0) are added to ensure that the clauses in (1) remain as simple as their counterparts in the definitions of medieval valuation (definition 12), Marcus valuation (definition 13), and Fregean valuation (definition 14). These clauses define a new function that augments each variable assignment with the interpretation function, restricted to individual symbols, to yield a term assignment.

Definition 15 Tarski extension for CQL (syncategorematic version)
Let M, or $\langle U, i \rangle$, be a structure for a signature $\langle IS, RS, ad \rangle$. Let VR be a set of variables. Let TM be $IS \cup VR$. And let g be a variable assignment from VR into U. Then, $[\ ]_g^M$, a Tarski extension for CQL, is a function satisfying the following conditions.

(0) SYMBOLS

Let v be a member of VR, let c be a member of IS and let Π be a member of RS.

(0.1) $[v]_g^M = g(v)$;

(0.2) $[c]_g^M = i(c)$;

(0.3) $[\Pi]_g^M = i(\Pi)$.

(1) ATOMIC FORMULAE

Let Π be a member of RS, let $ad(\Pi)$ be 1 and let t be a member of TM. Then,

(1.1) $[\Pi t]_g^M = T$ iff $[t]_g^M \in [\Pi]_g^M$.

Let Π be a member of RS, let $ad(\Pi)$ be n (where $n > 1$) and let $t_1, \ldots, t_n$ be occurrences of members of TM. Then,

(1.2) $[\Pi t_1 \ldots t_n]_g^M = T$ iff $\langle [t_1]_g^M, \ldots, [t_n]_g^M \rangle \in [\Pi]_g^M$.

(2) COMPOSITE FORMULAE

Let α and β be members of FM.

(2.1) $[\neg \alpha]_g^M = T$ iff $[\alpha]_g^M = F$;

(2.2.1) $[\alpha \wedge \beta]_g^M = T$ iff $[\alpha]_g^M = T$ and $[\beta]_g^M = T$;

(2.2.2) $[\alpha \vee \beta]_g^M = T$ iff   either $[\alpha]_g^M = T$ or $[\beta]_g^M = T$;

(2.2.3) $[\alpha \to \beta]_g^M = T$ iff   either $[\alpha]_g^M = F$ or $[\beta]_g^M = T$;

(2.2.4) $[\alpha \leftrightarrow \beta]_g^M = T$ iff $[\alpha]_g^M = [\beta]_g^M$.

(3) QUANTIFIED COMPOSITE FORMULAE

Let v be a member of VR and let α be a member of FM.

(3.1) $[\forall v\, \alpha]_g^M = T$ iff   for each e in U, $[\alpha]_{g_{v \mapsto e}}^M = T$;

(3.2) $[\exists v\, \alpha]_g^M = T$ iff   for some e in U, $[\alpha]_{g_{v \mapsto e}}^M = T$.

Unlike the medieval, Marcus, and Fregean valuations whose domain is the set of closed formulae, or FM^c, a Tarski extension has for its domain the set of all formulae, open and closed. As a result, unlike a Fregean valuation, which assigns a truth value to a formula whose main logical connective is a quantifier on the basis of truth values assigned to a substitution instance of the formula, rather than to its immediate subformula, a Tarski extension assigns a truth value to such a formula on the basis of truth values assigned to the immediate subformula. However, a Tarski extension and a Fregean valuation are alike insofar as the truth value assigned to the composite formula with a quantifier as its main logical connective is based on the application of various functions to the immediate subformula, in the case of a Tarski extension, and to a substitution instance, in the case of a Fregean valuation. In both cases, the various functions are the near variant functions of the function being applied to the composite formula. Thus, a Tarski extension determines the value of a composite formula of the form $Qv\alpha$ by determining the truth values assigned to the α by its near variants: all of them if Q is $\forall$ and the formula is true, and all of them if Q is $\exists$ and the formula is false. Note that there are as many near variants as there are members of the universe of the structure whose interpretation function the extension extends. This contrasts with the determination of the value of a composite formula of the form $\neg\alpha$: here a single function $o_\neg$ is applied to the truth value assigned to α.

Having contrasted the Fregean valuation, which is a function from closed formulae into truth values, with the Tarski extension, which is a function from all formulae, open and closed, into truth values, one might wonder whether the counterpart to the Fregean valuation can be defined using the Tarski extension. We now turn to showing that a Tarski valuation, the counterpart of a Fregean valuation, can be defined. To appreciate how this can be done, we must bear in mind some subtleties pertaining to variable assignments.

Let M be a structure whose universe is $\{1, 2, 3\}$ and whose interpretation function i assigns $\{1, 2\}$ to P. Consider two variable assignments: g, which assigns 1 to x, 2 to y, and 3 to z, and h, which assigns 1 to x, 3 to y, and 2 to z. It is obvious that $[Px]_g^M = T$ and that $[Px]_h^M = T$. Moreover, should g assign 3 to x, 2 to y, and 1 to z, and h assign 3 to x, 1 to y, and 2 to z, then $[Px]_g^M$ would equal F, as would $[Px]_h^M$. More generally, two Tarski extensions that are based on the same structure and differ only with respect to the values their variable assignments assign to variables absent from a formula assign the same truth value to the formula.

Having seen that the value assigned by a variable assignment to a variable absent from a formula is irrelevant to determining its truth value, let us see what happens when two variable assignments differ only with respect to the values they assign to variables that are bound in a formula. Let us evaluate the closed formulae $\forall x\, Px$ and $\exists x\, Px$ with respect to the structure given in the previous paragraph, but using two variable assignments that differ in the values they assign to x, say, the two variable assignments g, which assigns 1 to x, 3 to y, and 2 to z, and h, which assigns 3 to x, 1 to y, and 2 to z.

$[Px]_{g_{x \mapsto 1}}^M = T$ $\qquad$ $[Px]_{h_{x \mapsto 1}}^M = T$

$[Px]_{g_{x \mapsto 2}}^M = T$ $\qquad$ $[Px]_{h_{x \mapsto 2}}^M = T$

$[Px]_{g_{x \mapsto 3}}^M = F$ $\qquad$ $[Px]_{h_{x \mapsto 3}}^M = F$

$[\forall x\, Px]_g^M = F$ $\qquad$ $[\forall x\, Px]_h^M = F$

$[\exists x\, Px]_g^M = T$ $\qquad$ $[\exists x\, Px]_h^M = T$

As should be clear from the example just given, the value assigned by a variable assignment to a variable that occurs in a formula, but only bound, is irrelevant to determining its truth value. In other words, one would arrive at the very same result no matter what variable assignment is chosen.

A consequence of the irrelevance of the value assigned to a variable that only occurs bound in the formula to the determination of its truth value that formulae that are alphabetical variants of one another are assigned the same value. We illustrate the truth of this claim for the two pairs of alphabetical variants, the pair $\forall x\, Px$ and $\forall z\, Pz$ and the pair $\exists x\, Px$ and $\exists z\, Pz$. To do so, we again turn to the structure given previously.

$[Px]^M_{g_{x \mapsto 1}} = T$ $\qquad$ $[Pz]^M_{g_{z \mapsto 1}} = T$

$[Px]^M_{g_{x \mapsto 2}} = T$ $\qquad$ $[Pz]^M_{g_{z \mapsto 2}} = T$

$[Px]^M_{g_{x \mapsto 3}} = F$ $\qquad$ $[Pz]^M_{g_{z \mapsto 3}} = F$

$[\forall x\, Px]^M_g = F$ $\qquad$ $[\forall z\, Pz]^M_g = F$

$[\exists x\, Px]^M_g = T$ $\qquad$ $[\exists z\, Pz]^M_g = T$

We restate the first two observations just illustrated in the first statement of fact and the third illustrated observation in the second statement of fact.

Fact 1 Facts about variable assignments

Let M be a structure, g and h variable assignments, and α a formula.

(1) If $g(v) = h(v)$ for each free variable $v \in \mathrm{Fvr}(\alpha)$, then $[\alpha]^M_g = [\alpha]^M_h$;

(2) If α and β are alphabetical variants of one another, then $[\alpha]^M_g = [\beta]^M_h$.

The irrelevance of variable assignments to the truth or falsity of a closed formula permits one to speak of the truth or falsity of a closed formula in a structure *simpliciter*. Moreover, variable assignments that differ only in what they assign to variables that either do not occur in a formula or, if they do, are bound, assign the same values to any formula. We now define a classical valuation for the closed formulae of CQL using a Tarski extension.

Definition 16 Tarski valuation for CQL

Let M, or $\langle U, i \rangle$, be a structure for a signature $\langle IS, RS, ad \rangle$. Let VR be a set of variables. Then, a Tarski valuation is the function $[\ ]^M$ from FM^c into $\{T, F\}$ such that for each $\alpha \in FM^c$,

(1) $[\alpha]^M = T$ iff, for each variable assignment g from VR into U_M, $[\alpha]^M_g = T$; and

(2) $[\alpha]^M = F$ iff, for each variable assignment g from VR into U_M, $[\alpha]^M_g = F$.

It can be shown, though we shall not do so, that this definition provides a classical valuation for the closed formulae of CQL and that the Tarski classical valuation and Fregean classical valuation (definition 14) agree on the truth values they assign to the closed formulae of CQL.

Classical Quantificational Logic

Exercises: Tarski extensions and Tarski classical valuations

1. Let $\{x, y, z\}$ be the set of variables for the notation. Let $\{1, 2, 3, 4\}$ be the universe of a structure. Which of the following are variable assignments? In each case, justify your answer.

   (a) $x \mapsto 1, y \mapsto 2, z \mapsto 3$
   (b) $x \mapsto 1, y \mapsto 5, z \mapsto 3$
   (c) $x \mapsto 1, y \mapsto 2, z \mapsto 4$
   (d) $x \mapsto 1, y \mapsto 1, z \mapsto 1$

2. Write all variable assignments to the set of variables $\{x, y, z\}$ for a structure whose universe is the set $\{1, 2\}$.

3. State the formula that gives the number of variable assignments for the VR and a structure whose universe is U.

4. Determine the truth values of the next set of formulae with respect to the structure $\langle U, j \rangle$, defined in section 2.1.1, and the variable assignment g in which $x \mapsto 2, y \mapsto 3$, and $z \mapsto 4$. Use the definition of a Tarski extension (definition 15) to justify your answer.

   (a) $[Pa]_g^M$
   (b) $[Od]_g^M$
   (c) $[Ex]_g^M$
   (d) $[Py]_g^M$
   (e) $[Oz]_g^M$
   (f) $[Dab]_g^M$
   (g) $[Day]_g^M$
   (h) $[Dzx]_g^M$
   (i) $[Dyz]_g^M$
   (j) $[Dxx]_g^M$
   (k) $[\exists z Pz]_g^M$
   (l) $[\forall x Ox]_g^M$
   (m) $[\forall x \exists y Dxy]_g^M$
   (n) $[\forall x \exists y Dyx]_g^M$
   (o) $[\exists y \forall x Dxy]_g^M$

5. Determine the truth values of the next set of formulae with respect to the same structure, but with respect to the variable assignment h, in which $x \mapsto 1, y \mapsto 1$, and $z \mapsto 1$. Use the definition of a Tarski extension (definition 15) to justify your answer.

   (a) $[Pa]_h^M$
   (b) $[Od]_h^M$
   (c) $[Ex]_h^M$
   (d) $[Py]_h^M$
   (e) $[Oz]_h^M$
   (f) $[Dab]_h^M$
   (g) $[Day]_h^M$
   (h) $[Dzx]_h^M$
   (i) $[Dyz]_h^M$
   (j) $[Dxx]_h^M$
   (k) $[\exists z Pz]_h^M$
   (l) $[\forall x Ox]_h^M$
   (m) $[\forall x \exists y Dxy]_h^M$
   (n) $[\forall x \exists y Dyx]_h^M$
   (o) $[\exists y \forall x Dxy]_h^M$

6. Determine the truth values of the next set of formulae with respect to the following structure: $M = \langle U, i \rangle$, where $U = \{1, 2\}$ and where $a \mapsto 1, b \mapsto 2, S \mapsto \emptyset$, and $R \mapsto \{\langle 1, 1 \rangle, \langle 2, 2 \rangle\}$. Again, use the definition of a Tarski classical valuation (definition 16) to justify your answer.

(a) $\exists x\, Rss$
(b) $\exists x\, Sxx$
(c) $\forall x\, Rxa$
(d) $\exists x\, Rxa$
(e) $\forall x (Rxb \rightarrow Sxb)$
(f) $\forall x \forall y (Rxy \rightarrow Sxy)$
(g) $\forall x \forall y (Sxy \rightarrow Rxy)$
(h) $\forall x \forall y ((Rxy \wedge Sxy) \rightarrow Syx)$
(i) $\forall x \forall y ((Rxy \wedge \neg Sxy) \rightarrow Ryx)$
(j) $\forall x \forall y ((Rxy \wedge Ryx) \rightarrow \exists z (Rxz \vee Syz))$

7. Using the Tarski classical valuation (definition 16), determine the truth values of the next set of formulae with respect to the following structure: $M = \langle U, i \rangle$, where $U = \mathbf{Z}_{10}^{+}$ (that is, the positive integers 1 through 10) and R and S are interpreted as $<$ and $\leq$, respectively (that is, Rab is interpreted as $a < b$ and Sab is interpreted as $a \leq b$).

(a) $\forall x\, Rxx$
(b) $\forall x\, Sxx$
(c) $\forall x \forall y (Rxy \rightarrow \neg Ryx)$
(d) $\forall x \forall y (Rxy \rightarrow Sxy)$
(e) $\forall x \forall y (Sxy \rightarrow Rxy)$
(f) $\forall x \forall y \forall z ((Rxy \wedge Ryz) \rightarrow Rzx)$
(g) $\forall x \forall y ((Rxy \wedge \neg Sxy) \rightarrow Ryx)$
(h) $\forall x \exists y\, Rxy$
(i) $\exists x \forall y\, Rxy$
(j) $\exists x \forall y\, Syx$

8. Consider the following formulae

(1) $\forall x \exists y (Rxy \wedge \neg \exists z (Rxz \wedge Rzy))$.
(2) $\exists x \forall y \exists z ((Px \rightarrow Rxy) \wedge Py \wedge \neg Ryz)$.
(3) $\exists x \exists y ((Rxy \rightarrow Ryx) \rightarrow \forall z\, Rxz)$.
(4) $\forall x (\exists y \forall z (Rzy \wedge \forall v (Rvx \rightarrow \neg Rvx)$.
(5) $\exists x \forall y ((Py \rightarrow Ryx) \wedge (\forall z (Pz \rightarrow Rzy) \rightarrow Rxy))$.
(6) $\forall x \forall y ((Px \wedge Rxy) \rightarrow ((Py \wedge \neg Ryx) \rightarrow \exists z (\neg Rzx \wedge \neg Ryz)))$.
(7) $\forall x \forall y \exists z \forall v ((Py \wedge Rzv) \rightarrow (Pv \rightarrow Rxz))$.

Determine which of these formulae is true in a structure

(a) whose universe is $\mathbb{N}$ and where R is interpreted as $\leq$ and P as the set of even natural numbers.

(b) whose universe is $\text{Pow}(\mathbb{N})$ and where R is interpreted as $\subseteq$ and P as the set of finite subsets of $\mathbb{N}$.

(c) whose universe is $\mathbb{R}$ and where R is interpreted as $\{\langle x, x^2 \rangle : x \in \mathbb{R}\}$ and P as the set of rational numbers, or $\mathbb{Q}$.

2.2 Categorematic Definition of Valuations for CQL

The definitions presented in section 2.1 were all syncategorematic ones. In particular, both the clauses in (2) and those in (3) were syncategorematic. We know that the clauses in (2) need not have been syncategorematic. These clauses are precisely like the clauses used in

Classical Quantificational Logic

the definition of a classical valuation for the formulae of CPDL, and we saw that these clauses could be given the same categorematic definition as the one used in CPL. The question that now arises is whether it is possible to provide categorematic versions of those in (3). The answer is yes. We shall show two ways of providing a categorematic definition of valuations. One way uses what we shall call a *satisfaction set;* the other uses what is called a *cylindric algebra*. We start with the first way.

2.2.1 Satisfaction sets

To state the first categorematic version of the clauses in (3), we require what we shall call a *satisfaction set*. Consider the simple open formula Px. The set of all those members of the structure's universe whose assignment to x renders Px true is the satisfaction set of Px. The set

$$\{e \in U_M : [Px]^M_{x \mapsto e} = T\},$$

which we shall also refer to as the *set of satisfiers* of Px, can be notated more compactly as

$$[Px]^M_x$$

What can we conclude if the set of satisfiers of Px, drawn from a structure's universe, is empty? Surely that $\exists x\, Px$ is false in the structure. After all, if the set of satisfiers for Px, drawn from the structure's universe, is empty, then no member of the structure's universe, assigned as a value to x, makes Px true in the structure. And what can we conclude if the set of satisfiers of Px, drawn from a structure's universe, is not empty? Well, if the set is not empty, then some member of the structure's universe, when assigned to x, renders Px true. So $\exists x\, Px$ is true in the structure. In other words,

$$[\exists x\, Px]^M = T \text{ iff } [Px]^M_x \neq \emptyset.$$

An equivalent formulation of this biconditional, which we shall use in our categorematic definition of a Tarski extension, is

$$[\exists x\, Px]^M = T \text{ iff } [Px]^M_x \in \text{Pow}(U) - \emptyset.$$

Next, let us ponder the case where the set of satisfiers of Px, drawn from a structure's universe, is equal to the structure's universe. Then, $\forall x\, Px$ is true in the structure. After all, if the set of satisfiers for Px is a structure's universe, then Px is true in the structure for any value assigned to x from the structure's universe. Yet, should the set of satisfiers fail to equal the structure's universe, then $\forall x\, Px$ is not true in the structure. For, in that case, the structure's universe contains at least one member that, if assigned to x, renders Px false. That is,

$$[\forall x\, Px]^M = T \text{ iff } [Px]^M_x = U.$$

We recast this biconditional into the an equivalent one:

$[\forall x\, Px]^M = T$ iff $[Px]^M_x \in \{U\}$.

To adapt these ideas so as to formulate a categorematic definition of a Tarski extension, we must define satisfaction sets in terms of near variants of variable assignments. Here is such a definition of a satisfaction set.

Definition 17 Satisfaction set for CQL

Let M, or $\langle U, i \rangle$, be a structure for CQL. Let g be a variable assignment from VR into U. Then, for each $\alpha \in FM$, $[\alpha]^M_{g;v} = \{e \in U : [\alpha]^M_{g_{v \mapsto e}} = T\}$.

Equipped with this definition, we can now give a categorematic definition of a Tarski extension.

Definition 18 Tarski extension for CQL (categorematic version)

Let M, or $\langle U, i \rangle$, be a structure for a signature $\langle IS, RS, ad \rangle$. Let VR be a set of variables. Let TM be $IS \cup VR$. And let g be a variable assignment from VR into U. Then, $[\ ]^M_g$, a Tarski extension, is a function satisfying the following conditions.

(0) SYMBOLS

Let v be a member of VR, let c be a member of IS, and let Π be a member of RS.

(0.1) $[v]^M_g = g(v);$

(0.2) $[c]^M_g = i(c);$

(0.3) $[\Pi]^M_g = i(\Pi).$

(1) ATOMIC FORMULAE

Let Π be a member of RS, let $ad(\Pi)$ be 1 and let t be a member TM. Then,

(1.1) $[\Pi t]^M_g = T$ iff $[t]^M_g \in [\Pi]^M_g$.

Let Π be a member of RS, let $ad(\Pi)$ be n (where n > 1), and let $t_1, \ldots, t_n$ be occurrences of members of TM. Then,

(1.2) $[\Pi t_1 \ldots t_n]^M_g = T$ iff $\langle [t_1]^M_g, \ldots, [t_n]^M_g \rangle \in [\Pi]^M_g$.

(2) COMPOSITE FORMULAE

Let α and β be members of FM.

(2.1) $[\neg \alpha]^M_g$ $= o_\neg([\alpha]^M_g);$

(2.2.1) $[\alpha \wedge \beta]^M_g$ $= o_\wedge([\alpha]^M_g, [\beta]^M_g);$

(2.2.2) $[\alpha \vee \beta]^M_g$ $= o_\vee([\alpha]^M_g, [\beta]^M_g);$

(2.2.3) $[\alpha \to \beta]^M_g$ $= o_\to([\alpha]^M_g, [\beta]^M_g);$

(2.2.4) $[\alpha \leftrightarrow \beta]^M_g$ $= o_\leftrightarrow([\alpha]^M_g, [\beta]^M_g).$

Classical Quantificational Logic

(3) QUANTIFIED COMPOSITE FORMULAE

Let v be a member of VR and let α be a member of FM.

(3.1) $\quad [\forall v\, \alpha]_g^M = T \quad$ iff $\quad [\alpha]_{g;v}^M \in \{U\}$;

(3.2) $\quad [\exists v\, \alpha]_g^M = T \quad$ iff $\quad [\alpha]_{g;v}^M \in Pow(U) - \{\emptyset\}$.

This definition is equivalent to its syncategorematic version (definition 15) and hence can be used to define a Tarski classical valuation for the closed formulae of CQL.

To illustrate how the last two clauses are used to determine the truth value of a formula, let us determine the truth of the formula

(6) $\quad \forall x \exists y\, Rxy$,

treated in section 2.1.4, with respect to the same structure as presented there. Recall that the structure M has $\{1, 2, 3\}$ as a universe and that its interpretation function assigns to R the set $\{\langle 1, 2\rangle, \langle 2, 3\rangle, \langle 3, 1\rangle\}$. Let g be a variable assignment.[2]

According to clause (3.1), $[\forall x \exists y\, Rxy]_g^M = T$ if and only if $[\exists y\, Rxy]_{g;x}^M \in \{\{1, 2, 3\}\}$, that is to say, if and only if $[\exists y\, Rxy]_{g;x}^M = \{1, 2, 3\}$. To determine whether the last equality holds, we must calculate the membership of $[\exists y\, Rxy]_{g;x}^M$. This means determining which of the following equalities hold:

(6.1) $\quad [\exists y\, Rxy]_{g_{x \mapsto 1}}^M = T?$

(6.2) $\quad [\exists y\, Rxy]_{g_{x \mapsto 2}}^M = T?$

(6.3) $\quad [\exists y\, Rxy]_{g_{x \mapsto 3}}^M = T?$

To determine which, if any, of these equalities hold, we must proceed in the way set out by clause (3.2). Thus, the equality in (2.1) holds if and only if $[Rxy]_{g_{x \mapsto 1};y}^M \in Pow(\{1, 2, 3\}) - \{\emptyset\}$; in other words, $[Rxy]_{g_{x \mapsto 1};y}^M \neq \emptyset$. We now determine which of these equalities hold:

(6.1.1) $\quad [Rxy]_{g_{x \mapsto 1; y \mapsto 1}}^M = T?$

(6.1.2) $\quad [Rxy]_{g_{x \mapsto 1; y \mapsto 2}}^M = T?$

(6.1.3) $\quad [Rxy]_{g_{x \mapsto 1; y \mapsto 3}}^M = T?$

As one can see from inspection, while the first and third equalities do not hold, the second one does. This means that the satisfaction set $[Rxy]_{x \mapsto 1; y}^M = \{2\}$. Since it is not the empty set, we conclude that the equality in (6.1) holds. This means, in turn, that 1 is a member of the satisfaction set for $\exists y\, Rxy$, that is, that $1 \in [\exists y\, Rxy]_x^M$.

---

2. Since the formula is closed, the variable assignment does no work and hence is chosen arbitrarily.

Next, we wish to determine whether the equality in (6.2) holds. It holds if and only if $[Rxy]^M_{g_{x\mapsto 2;y}} \neq \emptyset$. To determine the right-hand side of this biconditional, we must calculate the satisfaction set $[Rxy]^M_{g_{x\mapsto 2;y}}$. Thus, we must determine which of these equalities hold:

(6.2.1) $[Rxy]^M_{x\mapsto 2; y\mapsto 1} = T$?

(6.2.2) $[Rxy]^M_{x\mapsto 2; y\mapsto 2} = T$?

(6.2.3) $[Rxy]^M_{x\mapsto 2; y\mapsto 3} = T$?

By inspection, we see that only the last equality holds. This means that the satisfaction set $[Rxy]^M_{g\mapsto 2y} = \{3\}$. Thus, 2 is a member of the satisfaction set for $\exists y Rxy$, that is, that $2 \in [\exists y Rxy]^M_x$.

Finally, we come to the equality in (6.3). Again, taking our lead from the clause in (3.2), we see that the equality in (6.3) holds if and only if $[Rxy]^M_{x\mapsto 3;y} \neq \emptyset$. To calculate the membership of $[Rxy]^M_{g_{x\mapsto 3;y}}$, we must determine which of the following equalities hold:

(6.3.1) $[Rxy]^M_{x\mapsto 3; y\mapsto 1} = T$?

(6.3.2) $[Rxy]^M_{x\mapsto 3; y\mapsto 2} = T$?

(6.3.3) $[Rxy]^M_{x\mapsto 3; y\mapsto 3} = T$?

Only the first equality holds. So $[Rxy]^M_{g_{x\mapsto 3;y}} = \{1\}$, which is nonempty. Thus, 3 is a member of the satisfaction set $[\exists y Rxy]^M_{g;x}$.

We have now determined that 1, 2, and 3 are all members of the satisfaction set $[\exists y Rxy]^M_{g;x}$. But that set is the universe in this structure. Thus, $[\exists y Rxy]^M_x \in \{\{1, 2, 3\}\}$. So, by clause (3.1) we conclude that $[\forall x \exists y Rxy]^M = T$, that is, the closed formulae $\forall x \exists y Rxy$ is true in M.

Exercises: Satisfaction sets

1. Use the structure M, or $\langle U, j \rangle$ (defined in section 2.1.1), and the variable assignment g (where $x \mapsto 1$, $y \mapsto 2$, and $z \mapsto 2$) to determine the following satisfaction sets.

(a) $[Px]^M_{g;x}$

(b) $[Oy]^M_{g;y}$

(c) $[Ez]^M_{g;z}$

(d) $[Pa]^M_{g;x}$

(e) $[Ob]^M_{g;y}$

(f) $[Ec]^M_{g;z}$

(g) $[Dxy]^M_{g;y}$

(h) $[Dxy]^M_{g;x}$

(i) $[Dzx]^M_{g_{z\mapsto 3;x}}$

(j) $[Dyx]^M_{g_{y\mapsto 1;x}}$

(k) $[Dbx]^M_{g_{x\mapsto 2;y}}$

(l) $[Dxz]^M_{g_{x\mapsto 3;z}}$

2. Use satisfaction sets to determine the truth value of the following formulae with respect to the structure M, or $\langle U, j \rangle$ (defined in section 2.1.1).

Classical Quantificational Logic

(a) $\exists x Px$

(b) $\forall y Oy$

(c) $\exists z Ez$

(d) $\forall x Px$

(e) $\exists y Oy$

(f) $\forall z Ez$

(g) $\exists x \forall y Dxy$

(h) $\forall y \exists x Dxy$

(i) $\exists z \exists x Dzx$

(j) $\forall x \exists y Dyx$

(k) $\exists y \forall x Dyx$

(l) $\forall z \forall x Dxz$

2.2.2 Cylindric algebras

In CPL and CPDL, the truth value of a composite formula is determined entirely by the truth values assigned to its immediate subformulae. In CQL, as presented thus far, the truth value of a composite formula that comprises a quantifier prefix and a quantifier matrix is not determined entirely by the truth values assigned to its immediate subformulae. Indeed, in the case of medieval, Marcus, and Fregean valuations, the truth value of such a composite formula is determined not by the truth value of its immediate subformulae, but rather by the truth values of various substitution instances of the formula, none of which are subformulae of the composite formula. In the case of the Tarski extension, the truth value is determined by the function $[\,]_g^M$, where M is a structure and g is a variable assignment, applied to the composite formula, is determined, not just by the very same function applied to its immediate subformula, but by a variety of functions that differ from the function applied to the composite formula insofar as the variable assignment is a near variant of g with respect to the variable of the quantifier prefix. Thus, for example, to determine the value of the function $[\,]_g^M$ applied to the composite formula $\forall x Px$, one applies the functions $[\,]_{g_{x \mapsto e}}^M$ to the subformula Px, there being as many different functions as there are members of the structure's universe.

The question arises: is it possible to define a classical valuation that applies the same function to the composite formula as applies to its immediate subformula, as one does with the classical valuations of CPL and CPDL The answer is that there is such a function and these functions are defined on algebraic structures known as *cylindric algebras*. The idea, due to Alfred Tarski, is this: associated with each formula is a subset of the set of variable assignments. A formula is true if it is assigned the set of all variable assignments, and a formula is false if it is assigned the empty set. The unary connective is assigned set complementation, while the binary connectives for conjunction and disjunction are assigned the familiar set theoretic operations of intersection and union. Finally, each quantifier prefix is assigned a unary operation that maps a set of variable assignments into a set of variable assignments. Crucial here will be the binary relation of variable assignments wherein variable assignments are near variants of one another. We shall express with the notation $g[v]h$ the relation whereby the variable assignments g and h differ at most in the value they assign to v, that is to say, the relation whereby variable assignment g is a near variant of variable assignment h with respect to the variable v.

Here is the definition of a classical valuation for the closed formulae of CQL, which uses cylindric algebras. In this algebra, the set of all variable assignments stands in for truth and the empty set stands in for falsehood.

Definition 19 Classical valuation for CQL (using cylindric algebras)
Let M, or $\langle U, i \rangle$, be a structure for a signature $\langle IS, RS, ad \rangle$. Let VR be a set of variables. Let TM be $IS \cup VR$. And let VA be the set of variable assignments from VR into U.

(1) ATOMIC FORMULAE

Let Π be a member of RS, let $ad(\Pi)$ be 1, and let t be a member of TM. Then,

(1.1) $\quad [\Pi t]_M \;=\; \{g: g \in VA \text{ and } [t]_g^M \in [\Pi]_g^M\}$

Let Π be a member of RS, let $ad(\Pi)$ be n (where n > 1), and let $t_1, \ldots, t_n$ be occurrences of members of TM. Then,

(1.2) $\quad [\Pi t_1 \ldots t_n]_M \;=\; \{g: g \in VA \text{ and } \langle [t_1]_g^M, \ldots, [t_n]_g^M \rangle \in [\Pi]_g^M\}$.

(2) COMPOSITE FORMULAE
Let α and β be members of FM.

(2.1) $\quad [\neg \alpha]_M \;=\; VA - [\alpha]_M$
(2.2.1) $\quad [\alpha \wedge \beta]_M \;=\; [\alpha]_M \cap [\beta]_M$
(2.2.2) $\quad [\alpha \vee \beta]_M \;=\; [\alpha]_M \cup [\beta]_M$
(2.2.3) $\quad [\alpha \to \beta]_M \;=\; (V - [\alpha]_M) \cup [\beta]_M$
(2.2.4) $\quad [\alpha \leftrightarrow \beta]_M \;=\; ([\alpha]_M - [\beta]_M) \cup ([\beta]_M - [\alpha]_M)$

(3) QUANTIFIED COMPOSITE FORMULAE
Let v be a member of VR and let α be a member of FM.

(3.1) $\quad [\forall v\, \alpha]_M \;=\; [\alpha]_M$, if, for each $h \in [\alpha]_M$ such that $h[v]g$, $g \in [\alpha]_M$
$ \;=\; \emptyset$, otherwise;
(3.2) $\quad [\exists v\, \alpha]_M \;=\; [\alpha]_M \cup \{g: g[v]h \text{ and } h \in [\alpha]_M\}$

To make this more comprehensible, let us consider a structure M, whose universe U is $\{1, 2, 3\}$ and where P is assigned $\{1, 2\}$, R is assigned $\{\langle 1, 2\rangle, \langle 2, 3\rangle, \langle 3, 1\rangle\}$, and a and b are assigned 1 and 3 respectively. Let there be just two variables x and y. Since the universe has only three members, there are exactly nine variable assignments for the two variables x and y. They are set out in the following table, together with a names for each.

Classical Quantificational Logic

	g_{11}	g_{12}	g_{13}	g_{21}	g_{22}	g_{23}	g_{31}	g_{32}	g_{33}
x	1	1	1	2	2	2	3	3	3
y	1	2	3	1	2	3	1	2	3

Recall that an atomic formula is assigned the set of variable assignments that render the formula true. So, let us see what sets are assigned to the following six formulae.

(7.1) $[Pa]_M = \{g_{11}, g_{12}, g_{13}, g_{21}, g_{22}, g_{23}, g_{31}, g_{32}, g_{33}\}$

(7.2) $[Pb]_M = \emptyset$

(7.3) $[Px]_M = \{g_{11}, g_{12}, g_{13}, g_{21}, g_{22}, g_{23}\}$

(7.4) $[Py]_M = \{g_{11}, g_{21}, g_{31}, g_{12}, g_{22}, g_{32}\}$

(7.5) $[Rxy]_M = \{g_{12}, g_{23}, g_{31}\}$

(7.6) $[Ryx]_M = \{g_{21}, g_{32}, g_{13}\}$

Readers should note that Pa is assigned the entire set of variable assignments with respect to the structure, since for each variable assignment $[a]_g^M \in [P]_g^M$, and that Pb is assigned the empty set, since for each variable assignment $[b]_g^M \notin [P]_g^M$, whereas Px, Py, Rxy, and Ryx, are assigned nonempty sets that are neither the empty set nor the entire set of variable assignments.

We now turn to the formulae $\forall x Px$, $\exists x Px$, $\forall y Py$, and $\exists y Py$, which result from those in (7.3) and (7.4) by prefixing to them quantifier prefixes that render them closed formulae. We wish to see which set of variable assignments is assigned to each formula. Since the clauses in (3) of definition 17 avail themselves of variable assignments that are variants, let us see what, for each variable assignment, its x-variants and its y-variants are.

VARIABLE ASSIGNMENT	NEAR VARIANTS WITH RESPECT TO x	NEAR VARIANTS WITH RESPECT TO y
g_{11}	g_{11}, g_{21}, g_{31}	g_{11}, g_{12}, g_{13}
g_{12}	g_{12}, g_{22}, g_{32}	g_{11}, g_{12}, g_{13}
g_{13}	g_{13}, g_{23}, g_{33}	g_{11}, g_{12}, g_{13}
g_{21}	g_{11}, g_{21}, g_{31}	g_{21}, g_{22}, g_{23}
g_{22}	g_{12}, g_{22}, g_{32}	g_{21}, g_{22}, g_{23}
g_{23}	g_{13}, g_{23}, g_{33}	g_{21}, g_{22}, g_{23}
g_{31}	g_{11}, g_{21}, g_{31}	g_{31}, g_{32}, g_{33}
g_{32}	g_{12}, g_{22}, g_{32}	g_{31}, g_{32}, g_{33}
g_{33}	g_{13}, g_{23}, g_{33}	g_{31}, g_{32}, g_{33}

We now apply the clauses in (3) of the definition to the atomic formula in (7.3) and (7.4) to determine the relevant sets:

(7.3.1) $[\exists x Px]_M = \{g_{11}, g_{12}, g_{13}, g_{21}, g_{22}, g_{23}, g_{31}, g_{32}, g_{33}\}$.

(7.3.2) $[\forall x Px]_M = \emptyset$.

(7.4.1) $[\exists y Py]_M = \{g_{11}, g_{12}, g_{13}, g_{21}, g_{22}, g_{23}, g_{31}, g_{32}, g_{33}\}$.

(7.4.2) $[\forall y Py]_M = \emptyset$.

The formula $\exists x Px$ has as its immediate subformula Px. But as we saw, $[Px]_M = \{g_{11}, g_{12}, g_{13}, g_{21}, g_{22}, g_{23}\}$. The clause in (3.2) of the definition says that $[\exists x Px]_M = [Px]_M \cup \{g: g[v]h$ and $h \in [Px]_M\}$. It is not a bad idea to carry out the laborious calculation to verify that the equality in (7.3.1) is correct. One can also note that $[Px]_M$ is three variable assignments short of being the set of all variable assignments. They are g_{31}, g_{32}, g_{33}. But, g_{31} is an x-variant of g_{11}, which is in $[Px]_M$, g_{32} is an x-variant of g_{12}, which is also in $[Px]_M$, and g_{33} is an x-variant of g_{31}, which itself is in $[Px]_M$. Readers are invited to carry out a similar verification for the formula in (7.4.1).

We turn next to the formula $\forall x Px$. Its immediate subformula is also Px, and as we have seen, $[Px]_M = \{g_{11}, g_{12}, g_{13}, g_{21}, g_{22}, g_{23}\}$. The clause in (3.1) of the definition says that $[\forall x Px]_M = \emptyset$, unless, for each variable assignment in $[Px]_M$, all of its x-variants are in $[Px]_M$, but while g_{11} is in $[Px]_M$, its x-variant, g_{31}, is not. Nor, for that matter, are the x-variants g_{32} and g_{31}, of g_{12} and of g_{11}, respectively. Readers are urged to carry out the verification for formula (7.4.2).

We end our presentation of classical valuations using cylindric algebras with a display of the determination of the values of the closed formulae $\forall x \exists y Rxy$, $\exists y \forall x \exists y Rxy$ $\forall x \exists y Ryx$, and $\exists y \forall x \exists y Ryx$, corresponding to the atomic formulae in (7.5) and (7.6).

(7.5) $[Rxy]_M = \{g_{12}, g_{23}, g_{31}\}$
$[\exists y Rxy]_M = \{g_{11}, g_{12}, g_{13}, g_{21}, g_{22}, g_{23}, g_{31}, g_{32}, g_{33}\}$

(7.5.1) $[\forall x \exists y Rxy]_M = \{g_{11}, g_{12}, g_{13}, g_{21}, g_{22}, g_{23}, g_{31}, g_{32}, g_{33}\}$

(7.5) $[Rxy]_M = \{g_{12}, g_{23}, g_{31}\}$
$[\forall x Rxy]_M = \emptyset$

(7.5.2) $[\exists y \forall x Rxy]_M = \emptyset$

(7.6) $[Ryx]_M = \{g_{21}, g_{32}, g_{13}\}$
$[\exists y Ryx]_M = \{g_{21}, g_{22}, g_{23}, g_{31}, g_{32}, g_{33}, g_{11}, g_{12}, g_{13}\}$

(7.6.1) $[\forall x \exists y Ryx]_M = \{g_{21}, g_{22}, g_{23}, g_{31}, g_{32}, g_{33}, g_{11}, g_{12}, g_{13}\}$

(7.6) $[Ryx]_M = \{g_{12}, g_{23}, g_{31}\}$
$[\forall x Ryx]_M = \emptyset$

(7.6.2) $[\exists y \forall x Ryx]_M = \emptyset$

2.3 Semantic Properties and Relations

In chapter 9, several properties and relations were defined for the formulae and for sets of formulae of CPDL, analogous to, but distinct from, properties and relations defined for the formulae and sets of formulae of CPL. In the case of CPL, the satisfaction relation relates bivalent truth value assignments to the formulae or sets of formulae of CPL; in the case of CPDL, the satisfaction relation is by structures of signatures to formulae or sets of formulae of CPDL. As readers surely know, bivalent truth value assignments and structures for signatures are different kinds of functions, while the formulae of CPL and those of CPDL are different sets of expressions. In CQL, we shall define the satisfaction relation as one that obtains between a structure for a signature and a variable assignment, on the one hand, and the formulae or sets of formulae of CQL, on the other hand. The relata of the satisfaction relation for CQL are different from the relata either of the satisfaction relation for CPL or of the one for CPDL.

Definition 20 Satisfaction in CQL
Let α be a formula, Γ a set of formulae, M a structure, and g a variable assignment. Then,

(1) $\langle M, g \rangle$ satisfies α iff $[\alpha]_g^M = T$;
(2) $\langle M, g \rangle$ satisfies Γ iff, for each $\alpha \in \Gamma$, $[\alpha]_g^M = T$.

In light of the analogy between the satisfaction relation of CQL and the satisfaction relations of CPDL and CPL, readers will not be surprised to learn that the properties and relations defined for the formulae and sets of formulae of CPL and of CPDL have their analogues for formulae and sets of formulae of CQL. Moreover, the logical facts pertaining to these properties and relations for CPL and CPDL have analogues for CQL.

Where the logical facts pertaining to these properties and relations for CQL can be established in a way perfectly analogous to the way their counterparts are established for CPL and CPDL, we omit to show how the facts are established, though we strongly urge readers to do so themselves, referring back to section 2.2.3 of chapter 6 for similar demonstrations only after a serious effort has been made to demonstrate the facts on their own.

Fact 2 Facts about satisfaction
Let M be a structure and g a variable assignment. Then,

(1) For each formula, α, $\langle M, g \rangle$ satisfies α iff $\langle M, g \rangle$ satisfies $\{\alpha\}$;
(2) $\langle M, g \rangle$ satisfies $\emptyset$.

2.3.1 Properties of formulae

As with CPDL, so with CQL, the formulae are partitioned into three sets: tautologies,[3] contradictions, and contingencies. Here a *tautology* is a formula of CQL that each structure

---

3. Such formulae are often called *valid formulae*. To emphasize that these properties and relations are analogous to those in CPL and CPDL, we shall, contrary to custom, call them *tautologies*.

variable assignment pair satisfies; a *contradiction* is a formula that no structure variable assignment pairs satisfy; and a *contingency* is a formula that some structure variable assignment pairs satisfy and others do not.

Definition 21 Tautology of CQL
A formula is a tautology iff each structure variable assignment pair satisfies it.

Here are some tautologies of CQL: $Rab \vee \neg Rab$, $Px \to Px$, $\forall x(Px \vee \neg Px)$, and $\exists y \forall x Rxy \to \forall x \exists y Rxy$. Notice that the first two formulae, though formulae of CQL, can be obtained by substituting atomic formulae of CQL for propositional variables of formulae in CPL. Thus, substituting Rab for p in $p \vee \neg p$ yields the first tautology and substituting Px for p in $p \to p$ yields the second. The third and fourth tautologies cannot be so obtained.

Definition 22 Contradiction of CQL
A formula is a contradiction iff each structure variable assignment pair does not satisfy it.

The contradictions of CQL include $Rab \wedge \neg Rab$, $Px \wedge \neg Px$, $\exists x(Px \wedge \neg Px)$, and $\exists y \forall x Rxy \wedge \exists x \forall y \neg Rxy$. The first two can be obtained by substituting atomic formulae of CQL for propositional variables of formulae of CPL; the last two cannot.

Definition 23 Contingency of CQL
A formula is a contingency iff some structure variable assignment pair satisfies it and some other does not.

And here are some contingencies of CQL: $Rab \wedge \neg Rba$, $Pz \vee \neg Rab$, $\exists x Px$, and $\forall x \exists y Rxy$.

2.3.2 Properties of formulae and of sets of formulae

We now define the properties of satisfiability and unsatisfiability for CQL, which, like the counterparts for CPL and CPDL, apply both to single formulae and to sets of formulae.

Definition 24 Satisfiability for CQL

(1) A formula is satisfiable iff some structure variable assignment pair satisfies it;
(2) a set of formulae is satisfiable iff some structure variable assignment pair satisfies each formula in the set.

These sets of formulae are satisfiable: $\{\exists z Pz, \neg Py\}$, $\{\exists z Rzb, \forall y Py, Rab \to Pb\}$, and $\{\forall x(Px \vee \neg \exists y Rxy), \exists z Pz \wedge Rab, Rca, \forall y \forall z(Py \to \neg Syz)\}$. Satisfiability is also known as *semantic consistency*.

Not surprisingly, facts analogous to those that hold for satisfiability in CPL and CPDL hold also for satisfiability in CQL, as readers should verify.

Classical Quantificational Logic

Fact 3 Facts about satisfiability

(1) For each formula α, α is satisfiable iff $\{\alpha\}$ is satisfiable.
(2) Each subset of a satisfiable set of formulae is satisfiable.
(3) The empty set is satisfiable.

We now define *unsatisfiability,* also known as *semantic inconsistency.*

Definition 25 Unsatisfiability for CQL

(1) A formula is unsatisfiable iff no structure variable assignment pair satisfies it;
(2) a set of formulae is unsatisfiable iff no structure variable assignment pair satisfies all of the set's formulae.

Here are three unsatisfiable sets: $\{\neg Rxy, Rxy\}$, $\{\forall x Px, \neg \exists x Rxx, Pa \rightarrow \exists y Ryy\}$, and $\{\forall x (Px \wedge Rxb), \forall y(\neg Ryy \vee \exists x \neg Px), Pa \wedge \neg Rab\}$.

Once again, facts analogous to those that hold for satisfiability in CPL and CPDL hold for satisfiability in CQL.

Fact 4 Facts about unsatisfiability

(1) For each formula α, α is unsatisfiable iff $\{\alpha\}$ is unsatisfiable.
(2) Each superset of an unsatisfiable set of formulae is unsatisfiable.
(3) FM, or the set of all formulae, is unsatisfiable.

2.3.3 Relations between formulae and sets of formulae

Finally, we come the two analogues in CQL of the relations of *semantic equivalence* and of *entailment* defined for CPL and for CPDL.

Definition 26 Semantic equivalence for CQL

A set of formulae or a formula, on the one hand, and a set of formula or a formula, on the other, are semantically equivalent iff (1) each structure variable assignment pair satisfying the former satisfies the latter and (2) each structure variable assignment pair satisfying the latter satisfies the former.

The following pairs of formulae are semantically equivalent: $\exists z Pz \vee \exists y Py$ and $\exists x Px$; $\neg \forall x \exists y Rxy$ and $\exists x \forall y \neg Rxy$; and $\neg(\exists x Px \wedge \forall y Ryy)$ and $\neg \exists x Px \vee \neg \forall z Rzz$. Moreover, two sets of formulae are also semantically equivalent: $\{\forall x \exists y Rxy \wedge \exists z Pz, \exists z Pz \rightarrow \exists w \forall y Syx\}$ and $\{\forall x \exists y Rxy, \exists x Px, \forall z \neg Pz \vee \exists x \forall y Syx\}$.

Still again, analogous facts obtain for the semantic equivalence of CQL as obtain for the semantic equivalence of CPL and of CPDL.

Fact 5 Facts about semantic equivalence

(1) All tautologies and sets of tautologies are semantically equivalent.
(2) All tautologies and all sets of tautologies are semantically equivalent to the empty set.

(3) All contradictions and all sets of contradictions are semantically equivalent.
(4) All contradictions and all sets of contradictions are semantically equivalent to FM, or the set of all formulae.

In addition, there are important semantic equivalences for pairs of formulae with quantifier prefixes of which readers should take note.

Fact 6 Important semantically equivalent formulae of CQL
Let α and β be formulae of CQL.

(1.1) $\neg \forall v \alpha$ and $\exists v \neg \alpha$;
(1.2) $\neg \exists v \alpha$ and $\forall v \neg \alpha$;
(2.1) $\forall v \forall w \alpha$ and $\forall w \forall v \alpha$;
(2.2) $\exists v \exists w \alpha$ and $\exists w \exists v \alpha$;
(3.1) $\forall v (\alpha \wedge \beta)$ and $\forall v \alpha \wedge \forall v \beta$;
(3.2) $\exists v (\alpha \vee \beta)$ and $\exists v \alpha \vee \exists v \beta$;
(4) $\exists v (\alpha \to \beta)$ and $\forall v \alpha \to \exists v \beta$.

These facts are not analogous to any facts pertaining to either CPD or CPDL. For this reason, we shall take the time to establish that some of them hold, leaving the others to readers to establish. We begin with a demonstration of the equivalence in (1.1). To establish that formulae of CQL of the form $\neg \forall v \alpha$ and $\exists v \neg \alpha$ are semantically equivalent, we must show two things: first, that each structure variable assignment pair satisfying a formula of the form $\neg \forall v \alpha$ satisfies the corresponding formula of the $\exists v \neg \alpha$, and second, that each structure variable assignment pair satisfying the formula of the form $\exists v \neg \alpha$ satisfies the formula of the form $\neg \forall v \alpha$. Once these two claims are established, we can use the definition of semantic equivalence (definition 26) to conclude that the two formulae are indeed semantically equivalent.

Let us now establish that each structure variable assignment pair satisfying $\neg \forall v \alpha$ satisfies $\exists v \neg \alpha$. To do so, we suppose that $\langle M, g \rangle$ is an arbitrarily chosen structure variable assignment pair satisfying $\neg \forall v \alpha$. We wish to show that it satisfies $\exists v \neg \alpha$. From the supposition that $\langle M, g \rangle$ satisfies $\neg \forall v \alpha$, we infer, by the definition of satisfaction (definition 20), that $[\neg \forall v \alpha]_g^M = T$. Hence, by clause (2.1) $[\forall v \alpha]_g^M = F$. By its clause (3.1) we know that U_M has a member—call it e—such that $[\alpha]_{g_{v \mapsto e}}^M = F$. Again, by clause (2.1), $[\neg \alpha]_{g_{v \mapsto e}}^M = T$. And by clause (3.2), $[\exists v \neg \alpha]_g^M = T$. Finally, by the definition of satisfaction (definition 20), $\langle M, g \rangle$ satisfies $\exists v \neg \alpha$. In short, since $\langle M, g \rangle$ is an arbitrarily chosen structure variable assignment pair satisfying $\neg \forall v \alpha$, we conclude that each structure variable assignment pair satisfying $\neg \forall v \alpha$ satisfies $\exists v \neg \alpha$. We have now established the first clause in the definition of semantic equivalence (definition 26) for the pair of formulae $\neg \forall v \alpha$ and $\exists v \neg \alpha$.

We next establish the second clause of the definition of semantic equivalence (definition 26). To this end, we suppose that $\langle M, g \rangle$ is an arbitrarily chosen structure variable assignment pair satisfying $\exists v \neg \alpha$. We do so with a view to showing that it satisfies $\neg \forall v \alpha$. From our supposition that $\langle M, g \rangle$ satisfies $\exists v \neg \alpha$, we infer, by the definition of satisfaction (definition 20), that $[\exists v \neg \alpha]_g^M = T$. Hence, by clause (3.2) of the definition of a Tarski extension (definition 15), U_M has a member—call it e—such that $[\neg \alpha]_{g_{v \mapsto e}}^M = T$. By clause its (2.1), $[\alpha]_{g_{v \mapsto e}}^M = T$. Again, by its clause (3.1), $[\forall v \alpha]_{g_{v \mapsto e}}^M = F$. And by clause (2.1), $[\neg \forall v \neg \alpha]_g^M = T$. Finally, by the definition of satisfaction (definition 20), $\langle M, g \rangle$ satisfies $\neg \forall v \alpha$. In short, since $\langle M, g \rangle$ is an arbitrarily chosen structure variable assignment pair satisfying $\exists v \neg \alpha$, we conclude that each structure variable assignment pair satisfying $\exists v \neg \alpha$ satisfies $\neg \forall v \alpha$. We have now established the second clause in the definition of semantic equivalence (definition 26) for the pair of formulae $\neg \forall v \alpha$ and $\exists v \neg \alpha$. As a result, by the definition of semantic equivalence (definition 26), we conclude that the claim in (1.1) holds.

Readers should establish the truth of the claim in (1.2) by imitating the reasoning just used to establish the truth of the claim in (1.1).

Next, we turn to the pair of semantic equivalences in (2). We shall establish the first pair and we leave the establishment of the second to the readers. To establish the first equivalence, we establish, first, that each structure variable assignment that satisfies the formula $\forall v \forall w \alpha$ satisfies $\forall w \forall v \alpha$ and, second, that each structure variable assignment that satisfies $\forall w \forall v \alpha$ satisfies $\forall v \forall w \alpha$. Once these two claims are established, we can again use the definition of semantic equivalence (definition 26) to conclude that these two formulae are indeed semantically equivalent.

To establish that each structure variable assignment that satisfies $\forall v \forall w \alpha$ satisfies $\forall w \forall v \alpha$, we suppose that an arbitrarily chosen structure variable assignment pair $\langle M, g \rangle$ satisfies $\forall v \forall w \alpha$. We wish to show that it satisfies $\forall w \forall v \alpha$. From the supposition that $\langle M, g \rangle$ satisfies $\forall v \forall w \alpha$, we infer, by the definition of satisfaction (definition 20), $[\forall v \forall w \alpha]_g^M = T$. Let e be an arbitrarily chosen member of U_M. By clause (3.1) of the definition of the Tarski extension (definition 15), $[\forall w \alpha]_{g_{v \mapsto e}}^M = T$. Let e' be also an arbitrarily chosen member of U_M. Again, by its clause (3.1), $[\alpha]_{g_{v \mapsto e;\, w \mapsto e'}}^M = T$. Since e is arbitrarily chosen in U_M, we infer that, for each $e \in U_M$, $[\forall w \alpha]_{g_{v \mapsto e;\, w \mapsto e'}}^M = T$. But this means, by clause (3.1), that $[\forall v \alpha]_{g_{w \mapsto e'}}^M = T$. And again, since e' is also arbitrarily chosen in U_M, we infer that, for each $e' \in U_M$, $[\forall v \alpha]_{g_{w \mapsto e'}}^M = T$. So, by clause (3.1), $[\forall w \forall v \alpha]_g^M = T$. Finally, by the definition of satisfaction (definition 20), $\langle M, g \rangle$ satisfies $\forall w \forall v \alpha$. In short, since $\langle M, g \rangle$ is an arbitrarily chosen structure variable assignment pair satisfying $\forall v \forall w \alpha$, we conclude that each structure variable assignment pair satisfying $\forall v \forall w \alpha$ satisfies the formula $\forall w \forall v \alpha$.

We have now established the first clause in the definition of semantic equivalence (definition 26) for the pair of formulae $\forall v \forall w \alpha$ and $\forall w \forall v \alpha$. We leave it to readers to use the preceding demonstration as an example to prove the second clause in the definition of semantic equivalence (definition 26). Once readers have done this, they will have established the first equivalence in (2). Readers should then establish the second equivalence.

Leaving the establishment of the truth of the claim in (3.1) to the readers, we next establish the truth of the one in (3.2). To do so, we must establish, in accordance with the definition of semantic equivalence (definition 26), first, that each structure variable assignment that satisfies $\exists v(\alpha \vee \beta)$ satisfies $\exists v\alpha \vee \exists v\beta$ and, second, that each structure variable assignment that satisfies $\exists v\alpha \vee \exists v\beta$ satisfies $\exists v(\alpha \vee \beta)$.

Suppose, to begin with, that an arbitrarily chosen pair comprising a structure and a variable assignment, say $\langle M, g \rangle$, satisfies $\exists v(\alpha \vee \beta)$. We wish to show that $\langle M, g \rangle$ satisfies $\exists v\alpha \vee \exists v\beta$. From the supposition that $\langle M, g \rangle$ satisfies $\exists v(\alpha \vee \beta)$, we infer, by the definition of satisfaction (definition 20), $[\exists v(\alpha \vee \beta)]_g^M = T$. It follows from clause (3.2) of the definition of a Tarski extension (definition 15) that U_M has a member—call it e—such that $[\alpha \vee \beta]_{g_{v \mapsto e}}^M = T$. By clause (2.2.2), then, either $[\alpha]_{g_{v \mapsto e}}^M = T$ or $[\beta]_{g_{v \mapsto e}}^M = T$. We now consider each of these two cases. Case 1: Suppose, on the one hand, that $[\alpha]_{g_{v \mapsto e}}^M = T$. It follows from clause (3.2) of the definition of a Tarski extension (definition 15) that $[\exists v\alpha]_g^M = T$. From this, we can infer that $[\exists v\alpha \vee \exists v\beta]_g^M = T$. Case 2: Suppose, on the other hand, that $[\beta]_{g_{v \mapsto e}}^M = T$. It follows from clause (3.2) of the definition of a Tarski extension (definition 15) that $[\exists v\beta]_g^M = T$. From this, we can infer that $[\exists v\alpha \vee \exists v\beta]_g^M = T$. Either way, $[\exists v\alpha \vee \exists v\beta]_g^M = T$. Thus, by the definition of satisfaction, $\langle M, g \rangle$ satisfies $\exists v\alpha \vee \exists v\beta$. In short, since $\langle M, g \rangle$ is an arbitrarily chosen structure variable assignment pair satisfying $\exists v(\alpha \vee \beta)$, we conclude that each structure variable assignment pair satisfying $\exists v(\alpha \vee \beta)$ satisfies the formula $\exists v\alpha \vee \exists v\beta$. We have established, then, the first clause in the definition of semantic equivalence (definition 26) for the pair of formulae $\exists v(\alpha \vee \beta)$ and $\exists v\alpha \vee \exists v\beta$.

We now have to establish the second clause of the definition of semantic equivalence (definition 26). We suppose, then, that $\langle M, g \rangle$ is an arbitrarily chosen structure variable assignment pair satisfying $\exists v\alpha \vee \exists v\beta$. We wish to show, of course, that it satisfies $\exists v(\alpha \vee \beta)$. From the supposition that $\langle M, g \rangle$ satisfies $\exists v\alpha \vee \exists v\beta$, we infer, by clause (2.2.2) of the definition of the Tarski extension (definition 15), either $[\exists v\alpha]_g^M = T$ or $[\exists v\beta]_g^M = T$. We now consider each of these two cases. Case 1: Suppose, on the one hand, that $[\exists v\alpha]_g^M = T$. It follows from its clause (3.2) that U_M has a member—call it e—such that $[\alpha]_{g_{v \mapsto e}}^M = T$. From this, we can infer that $[\alpha \vee \beta]_{g_{v \mapsto e}}^M = T$. By clause (3.2), we infer that $[\exists v(\alpha \vee \beta)]_g^M = T$. Case 2: Suppose, on the other hand, that $[\exists v\beta]_g^M = T$. It follows from clause (3.2) that U_M has a member—call it e—such that $[\beta]_{g_{v \mapsto e}}^M = T$. It follows further that $[\alpha \vee \beta]_{g_{v \mapsto e}}^M = T$. By clause (3.2) we infer that $[\exists v(\alpha \vee \beta)]_g^M = T$. Either way, $[\exists v(\alpha \vee \beta)]_g^M = T$. In short, since $\langle M, g \rangle$ is an arbitrarily chosen structure variable assignment pair satisfying $\exists v\alpha \vee \exists v\beta$, we conclude that each structure variable assignment

Classical Quantificational Logic

pair satisfying $\exists v\alpha \vee \exists v\beta$ satisfies the formula $\exists v(\alpha \vee \beta)$. We have established, then, the first clause in the definition of semantic equivalence (definition 26) for the pair of formulae $\exists v(\alpha \vee \beta)$ and $\exists v\alpha \vee \exists v\beta$.

Having established that each structure variable assignment that satisfies $\exists v(\alpha \vee \beta)$ satisfies $\exists v\alpha \vee \exists v\beta$ and that each structure variable assignment that satisfies $\exists v\alpha \vee \exists v\beta$ satisfies $\exists v(\alpha \vee \beta)$, we conclude by the definition of semantic equivalence that $\exists v\alpha \vee \exists v\beta$ and $\exists v(\alpha \vee \beta)$ are semantically equivalent.

We end this discussion of semantic equivalence with a very compact demonstration of the claim in (4). All that is required is the semantic equivalence of CPL that $\alpha \to \beta$ and $\neg \alpha \vee \beta$ and the semantic equivalences in CQL given in (1.1) and (3.2), which were just established.

$\exists v(\alpha \to \beta)$ semantically equivalent $\exists v(\neg \alpha \vee \beta)$ by CPL
semantically equivalent $\exists v \neg \alpha \vee \exists v\beta$ by (3.2)
semantically equivalent $\neg \forall v\alpha \vee \exists v\beta$ by (1.1)
semantically equivalent $\forall v\alpha \to \exists v\beta$ by CPL

We assume in this demonstration something that is evident but which we have not proved, namely, that semantic equivalence is transitive. Indeed, semantic equivalence is an equivalence relation, that is, a relation that is reflexive, symmetric, and transitive.

We now turn to the second relation, entailment, which a set of formulae bears to a single formula.

Definition 27 Entailment for CQL
A set of formulae *entails* a formula, or $\Gamma \models \alpha$, iff each structure variable assignment pair that satisfies the set Γ satisfies the formula α.

Here are two examples of the entailment relation.

$\exists x \forall y Rxy \models \forall y \exists x Rxy$

$\neg \exists x (Fx \wedge Gx) \models \forall x (Fx \to \neg Gx)$

When the set of formulae is a singleton set, it is customary to write just the formula itself.

We now turn to facts about entailment whose analogues are familiar to us from CPL and from CPDL.

Fact 7 Facts about entailment
Let α and β be formulae. Let Γ and Δ be sets of formulae.

(1)   If $\alpha \in \Gamma$, then $\Gamma \models \alpha$.
(2)   If $\Gamma \subseteq \Delta$ and $\Gamma \models \alpha$, then $\Delta \models \alpha$.

(3) $\Gamma \cup \{\alpha\} \models \beta$ iff $\Gamma \models \alpha \to \beta$.
(4) $\Gamma \models \alpha \wedge \beta$ iff $\Gamma \models \alpha$ and $\Gamma \models \beta$.

These facts and those that follow are all established in the same fashion as the analogous facts set out in section 2.2.3 of chapter 6. As before, readers are urged to review the arguments used to establish them.

Fact 8 Facts about entailment and tautologies
Let α be a formulae. Let Γ be a set of formulae.

(1) α is a tautology iff $\emptyset \models \alpha$.
(2) α is a tautology iff, for each Γ, $\Gamma \models \alpha$.

There are facts about the entailment relation for CQL that have no analogues among the facts pertaining to entailment either in CPD or in CPDL.

Fact 9 Important entailments for formulae of CQL

(1) $\forall v \alpha \vee \forall v \beta \models \forall v(\alpha \vee \beta)$;
(2) $\exists v(\alpha \wedge \beta) \models \exists v \alpha \wedge \exists v \beta$;
(3) $\exists v \forall w \alpha \models \forall w \exists v \alpha$

We shall establish the first and third, leaving the second to readers to establish.

To establish the first claim, we must demonstrate that every structure variable assignment pair satisfying $\forall v \alpha \vee \forall v \beta$ satisfies $\forall v(\alpha \vee \beta)$. Once this is accomplished, we can use the definition of entailment (definition 27) to conclude that $\forall v \alpha \vee \forall v \beta \models \forall v(\alpha \vee \beta)$. Thus, we suppose that $\langle M, g \rangle$ is an arbitrarily chosen structure variable assignment pair that satisfies a formula of the form $\forall v \alpha \vee \forall v \beta$. We wish to show that it satisfies $\forall v(\alpha \vee \beta)$. From the supposition, we infer by the definition of satisfaction (definition 20) that $[\forall v \alpha \vee \forall v \beta]_g^M = T$. Hence, by clause (2.2.2) of the definition of a Tarski extension (definition 15), we infer either $[\forall v \alpha]_g^M = T$ or $[\forall v \beta]_g^M = T$. We now consider each of these two cases. Case 1: Suppose, on the one hand, $[\forall v \alpha]_g^M = T$. Let e be an arbitrarily chosen member of U_M. It follows from clause its (3.1) that $[\alpha]_{g_{v \mapsto e}}^M = T$. From this, we use clause (2.2.2) of the same definition to infer that $[\alpha \vee \beta]_{g_{v \mapsto e}}^M = T$. Since e is arbitrarily chosen, we infer by the definition's clause (3.1) that $[\forall v(\alpha \vee \beta)]_g^M = T$. Case 2: Suppose, on the other hand, that $[\forall v \beta]_g^M = T$. Let e be an arbitrarily chosen member of U_M. It follows from clause (3.1) of the definition of a Tarski extension (definition 15) that $[\beta]_{g_{v \mapsto e}}^M = T$. From this, we can infer that $[\alpha \vee \beta]_{g_{v \mapsto e}}^M = T$. Since e is arbitrarily chosen, we infer by clause (3.1) that $[\forall v(\alpha \vee \beta)]_g^M = T$. Either way, then, $[\forall v(\alpha \vee \beta)]_g^M = T$. Therefore, by the definition of satisfaction (definition 20), $\langle M, g \rangle$ satisfies $\forall v(\alpha \vee \beta)$. In short, each structure variable assignment pair that satisfies $\forall v \alpha \vee \forall v \beta$ satisfies $\forall v(\alpha \vee \beta)$. By the definition of entailment (definition 27), we conclude that $\forall v \alpha \vee \forall v \beta \models \forall v(\alpha \vee \beta)$.

Classical Quantificational Logic

We leave the demonstration of the claim in (2) to readers and turn our attention to the claim in (3). We suppose that $\langle M, g \rangle$ is an arbitrarily chosen structure variable assignment pair that satisfies a formula of the form $\exists v \forall w \alpha$. It follows by the definition of satisfaction (definition 20) that $[\exists v \forall w \alpha]_g^M = T$. By clause (3.2) of the definition of a Tarski extension (definition 15), it follows that U_M has a member— call it e—such that $[\forall w \alpha]_{g_{v \mapsto e}}^M = T$. Next, let e' be an arbitrarily chosen member of U_M. By clause (3.1) of the same definition, it follows that $[\alpha]_{g_{v \mapsto e; w \mapsto e'}}^M = T$. By clause (3.2), $[\exists v \alpha]_{g_{w \mapsto e'}}^M = T$. Since e' is arbitrarily chosen, we infer by clause (3.1) that $[\forall w \exists v \alpha]_g^M = T$. Hence, $\langle M, g \rangle$ satisfies $\forall w \exists v \alpha$. In short, each structure variable pair that satisfies $\exists v \forall w \alpha$ satisfies $\forall w \exists v \alpha$. Therefore, by the definition of entailment (definition 27), $\exists v \forall w \alpha \models \forall w \exists v \alpha$.

We draw this section to a close by bringing to readers' attention three facts connecting entailment and unsatisfiability. Their analogues were established in chapter 6.

Fact 10 Facts about entailment and unsatisfiability
Let α be a formulae. Let Γ be a set of formulae.

(1) $\Gamma \models \alpha$ iff $\Gamma \cup \{\neg \alpha\}$ is unsatisfiable.
(2) Γ is unsatisfiable iff, for any α, $\Gamma \models \alpha$.
(3) Γ is unsatisfiable iff, for at least one α, a contradiction, $\Gamma \models \alpha$.

Exercises: Semantic properties and relations

1. For each of the following claims, indicate whether it is true or false. If it is true, explain why it is true; if it is false, provide a counterexample.

(a) Each unsatisfiable set contains a contradiction.
(b) No set of contingencies can form a satisfiable set.
(c) Each set of tautologies is satisfiable.
(d) No satisfiable set contains a tautology.
(e) Some satisfiable sets contain a tautology.
(f) α is a contingency iff $\{\alpha\}$ is satisfiable.
(g) A contingent formula's negation may be a tautology.
(h) $\Gamma \models \alpha \vee \beta$ iff $\Gamma \models \alpha$ or $\Gamma \models \beta$.
(i) $\{\alpha\}$ and $\{\beta\}$ are semantically equivalent iff $\alpha \leftrightarrow \beta$ is a tautology.
(j) If $\Delta \subseteq \Gamma$ and Γ is unsatisfiable, then Δ is unsatisfiable.
(k) If $\Delta \subseteq \Gamma$ and Δ is satisfiable, then Γ is satisfiable.
(l) If $\Gamma \models \alpha$, then $\alpha \in \Gamma$.
(m) If $\Delta \subseteq \Gamma$ and $\Gamma \models \alpha$, then $\Delta \models \alpha$.

2. Let v not be free in β. For which of the following pairs are the formulae semantically equivalent?

(a) $\forall v \beta$ and $\exists v \beta$ and β
(b) $\forall v (\alpha \wedge \beta)$ and $\forall v \alpha \wedge \beta$
(c) $\exists v (\alpha \wedge \beta)$ and $\exists v \alpha \wedge \beta$
(d) $\exists v (\alpha \vee \beta)$ and $\exists v \alpha \vee \beta$
(e) $\forall v (\alpha \vee \beta)$ and $\forall v \alpha \vee \beta$
(f) $\exists v (\beta \to \alpha)$ and $\beta \to \exists v \alpha$
(g) $\forall v (\beta \to \alpha)$ and $\beta \to \forall v \alpha$
(h) $\exists v (\alpha \to \beta)$ and $\forall v \alpha \to \beta$
(i) $\forall v (\alpha \to \beta)$ and $\exists v \alpha \to \beta$

3 Deduction

Deduction in CQL includes, in addition to the rules of deduction of CPDL, themselves the same as those of CPL, two pairs of rules, each corresponding to each of the two quantifiers. As one might expect, each pair of rules comprises an introduction rule and an elimination rule. We shall discuss each of the rules separately. The rules are formulated for natural deductions in the formula column format.

3.1 ∀ Elimination

Let us begin with the simplest inference of all. It is the rule that permits one to infer from a universal claim any of its substitution instances. In particular, if one is justified in asserting that every person ran, one is justified in asserting that Alan ran—provided Alan is in the universe of discourse. In other words, the following is a sound inference:

Every person ran.	$\forall x\, Px$
Alan ran.	Pa

The rule that renders such inferences legitimate, designated ∀ Elimination, is formulated as follows:

∀ Elimination

$\forall v \alpha$

$[c/v]\alpha$

In other words, this rule sanctions the inference from a universal claim to any of its substitution instances.

Here are two deductions illustrating the rule. The first deduction establishes the truth of the claim: if everyone admires himself, then Alan admires himself ($\forall x Rxx \rightarrow Raa$).

$\vdash \forall x Rxx \rightarrow Raa$

1	$\forall x Rxx$	supposition
2	Raa	from 1 by $\forall$ E
3	$\forall x Rxx \rightarrow Raa$	from 1–2 by $\rightarrow$ I

The second deduction establishes the truth of this claim: if everyone admires everyone, then Alan admires Bill ($\forall x \forall y Rxy \rightarrow Rab$).

$\vdash \forall x \forall y Rxy \rightarrow Rab$

1	$\forall x \forall y Rxy$	supposition
2	$\forall y Ray$	from 1 by $\forall$ E
3	Rab	from 2 by $\forall$ E
4	$\forall x \forall y Rxy \rightarrow Rab$	from 1–3 by $\rightarrow$ I

It is important to note that the rule of $\forall$ Elimination applies only to an entire formula, and not to proper subformula of a formula. If this restriction is not observed, it becomes possible to infer a false conclusion from a true premise. Thus, for example, one could infer from the premise that it is not the case that every person ran that Bill did not run.

It is not the case that every person ran.	$\neg \forall x Px$
*Bill did not run.	$\neg Pb$

But clearly, the premise could be true and the conclusion false. After all, the universe might comprise just two people, Bill and Carl, where Bill ran but Carl did not.

3.2 ∃ Introduction

Another intuitively clear rule of inference is the rule of ∃ Introduction. No one can deny that if one is warranted in asserting that Alan is human, one is warranted in asserting that someone is human. In other words, surely the following is a sound inference.

Alan is human. Ha
―――――――――   ―――
Someone is human. $\exists x Hx$

The rule of ∃ Introduction is formulated as follows.

∃ Introduction
$[c/v]\alpha$
―――
$\exists v\, \alpha$

The formulation of this rule differs in an important way from the previous rule. In the previous rule, the conclusion is a substitution instance of the premise. In this rule, the premise is a substitution instance of the conclusion.

This fact may strike one as odd. However, consideration of a few more simple inferences will make it clear why the rule is so formulated. To begin with, it is clear that the following inference is just as sound as the previous one.

Alan admires himself. Raa
―――――――――――   ―――
Someone admires himself. $\exists x Rxx$

However, just as sound is this inference.

Alan admires himself. Raa
―――――――――――   ―――
Alan admires someone. $\exists x Rax$

One might be tempted to formulate the previous rule in a different way, namely, one in which the formula of the conclusion is obtained from the formula of the premise by substituting a variable for a individual symbol. Adapting our notation for the substitution of an individual symbol for a variable, one might formulate the rule as follows.

∃ Introduction
α
―――
$\exists v\, [v/c]\alpha$

While the first two examples of the use of the rule of ∃ Introduction given previously conform to this reformulated rule, the third example does not, as readers should verify.

Like the rule of ∀ Elimination, the rule of ∃ Introduction applies only to an entire formula. Applying it to a proper subformula of an entire formula can give rise to invalid

Classical Quantificational Logic

inferences. Thus, for example, one could infer from the premise that Bill did not run that it is not the case that some one ran.

Bill did not run.	$\neg Pb$
*It is not the case that some person ran.	$\neg \exists x Px$

Again, consider a universe of just two people, Bill and Reed, where Reed ran but Bill did not run. Clearly, the premise is true, but the conclusion false.

3.3 ∀ Introduction

This rule is difficult to illustrate in isolation. However, it can be illustrated nicely in combination with the previous rule of ∀ Elimination. Consider the inference of the conclusion *everyone admires himself* ($\forall x Rxx$) from the single premise, *everyone admires everyone* ($\forall x \forall y Rxy$).

$\forall x \forall y Rxy \vdash \forall x Rxx$

1	$\forall x \forall y Rxy$	premise
2	$\forall y Ray$	from 1 by ∀ E
3	Raa	from 2 by ∀ E
4	$\forall x Rxx$	from 3 by ∀ I

The rule is formulated in the same way as the rule of ∃ Introduction.

∀ Introduction
$[c/v]\alpha$
$\forall v\, \alpha$

Thus, like the rule of ∃ Introduction, and unlike the rule of ∀ Elimination, the premise is a substitution instance of the conclusion.

The point of this rule is to allow one to infer certain truths, which are universal in two ways: they can be represented by a formula containing a universally quantified prefix with scope over the entire formula and the formula is true no matter how it is interpreted. Obviously, such truths should not be deducible by generalizing on a premise about a specific individual. In particular, one should not be able to use this rule to prove the conclusion *every person ran* ($\forall x Px$) from the single premise *Alan ran* (Pa).

Now, a very important question arises: which rule, if any, prevents such an inference? The answer is none. To block such a patently illegitimate inference, one must restrict the application of the rule of ∀ Introduction. The restriction is that the rule cannot be applied to any formula where the individual symbol giving rise to the universal generalization occurs in an undischarged supposition or a premise.

1	Pa	premise
*2	$\forall x Px$	from 1 by ∀ I

The use of the rule of ∀ Introduction is disallowed by the restriction just stated.

There is one other restriction that must be adopted. Consider the deduction of the conclusion *everyone admires everyone* ($\forall x \forall y Rxy$) from the premise *everyone admires himself* ($\forall x Rxx$).

1	$\forall x Rxx$	premise
2	Raa	from 1 by ∀ E
*3	$\forall y Ray$	from 2 by ∀ I
*4	$\forall x \forall y Rxy$	from 3 by ∀ I

Such an inference can be blocked by the restriction that the individual symbol with respect to which the universal assertion is being made does not occur in the resulting universal assertion.

Bearing in mind these two restrictions, we reformulate the rule of ∀ introduction as follows.

∀ Introduction

$$[c/v]\,\alpha$$

$$\forall v\,\alpha$$

c does not occur in α

c occurs in no premise or undischarged supposition

Recall again that this rule, like the preceding two, applies to only an entire formula.

3.4 ∃ Elimination

To understand this rule, let us begin with a legitimate use of it. Consider the claim: if someone admires himself, then someone admires someone ($\exists x Rxx \rightarrow \exists x \exists y Rxy$). Surely, this claim should be provable. But how? The mere supposition of the protasis $\exists x Rxx$ does not enable one to infer anything, for it does not have a form that fits with any of our rules of inference formulated so far. However, we do know the following: for some suitable choice of an individual symbol, say a, Raa is true. After all, if $\exists x Rxx$ is true, there must be something in the universe of discourse underpinning, as it were, the assertion $\exists x Rxx$. Let us call that thing a: that is, let us suppose Raa. Two applications of the rule of $\exists$ Introduction permit one to infer $\exists x \exists y Rxy$. Thus, for some suitable choice of a, $Raa \to \exists x \exists y Rxy$.

Let us formalize the reasoning as follows.

1		$\exists x Rxx$	supposition
2		Raa	supposition
3		$\exists y Ray$	from 2 by $\exists$ I
4		$\exists x \exists y Rxy$	from 3 by $\exists$ I
5		$Raa \to \exists x \exists y Rxy$	from 1–4 by $\to$ I
6		$\exists x \exists y Rxy$	from 1, 5 by $\exists$ E
7		$\exists x Rxx \to \exists x \exists y Rxy$	from 1–6 by $\to$ I

The rule used in line 6 of the proof requires two premises: one premise has the form of $\exists v \alpha$ and the other has the form $[c/v]\alpha \to \beta$, where the protasis of the second premise, namely $[c/v]\alpha$, is a substitution instance of the first premise, namely $\exists v \alpha$. The conclusion of the rule is identical with the apodosis of the second premise.

The rule, as it stands, permits invalid inferences. Two restrictions that must be introduced are the ones introduced for the rule of $\forall$ Introduction: namely, that c does not occur in α and that c does not occur in any undischarged supposition or any premise.

However, even with the addition of these two restrictions, fallacious inferences are still possible. Thus, for example, one can still deduce the conclusion *everyone admires someone* ($\forall x \exists y Rxy$) from the single premise *someone admires himself* ($\exists x Rxx$)—a clearly undesirable result.

1		$\exists x Rxx$	premise
2		Raa	supposition
3		$\exists y Ray$	from 2 by $\exists$ I
4		$Raa \to \exists y Ray$	from 1–3 by $\to$ I
5		$\exists y Ray$	from 1, 5 by $\exists$ E
6		$\forall x \exists y Rxy$	from 5 by $\forall$ I

To prevent such an inference, one must add a third restriction, namely, that c does not occur in β. The complete formulation of the rule of $\exists$ Elimination, then, is this:

$\exists$ Elimination
$\exists v\, \alpha,\ [c/v]\alpha \to \beta$
───────────────────
β
c does not occur in β
c does not occur in α
c occurs in no premise or undischarged supposition

We conclude this section with several examples of deductions. We preface these examples with a restatement of the deduction rules for the quantifiers.

	Introduction	Elimination
$\exists$	$[c/v]\alpha$ ─────── $\exists v\, \alpha$	$\exists v\, \alpha,\ [c/v]\alpha \to \beta$ ─────────────────── β c does not occur in β c does not occur in α c does not occur in any undischarged assumption
$\forall$	$[c/v]\alpha$ ─────── $\forall v\, \alpha$ c does not occur in α c does not occur in any undischarged assumption	$\forall v\, \alpha$ ─────── $[c/v]\alpha$

Several example deductions follow. The first two establish the fact that the universal quantifier distributes, as it were, across the $\wedge$.

$\forall x(Px \wedge Qx) \vdash \forall x Px \wedge \forall x Qx$.

1	$\forall x(Px \land Qx)$	premise
2	$Pa \land Qa$	from 1 by $\forall$ E
3	Pa	from 2 by $\land$ E
4	$\forall x Px$	from 3 by $\forall$ I
5	Qa	from 2 by $\land$ E
6	$\forall x Qx$	from 5 by $\forall$ I
7	$\forall x Px \land \forall x Qx$	from 6 by $\land$ I

$\forall x Px \land \forall x Qx \vdash \forall x(Px \land Qx)$.

1	$\forall x Px \land \forall x Qx$	premise
2	$\forall x Px$	from 1 by $\land$ E
3	$\forall x Qx$	from 1 by $\land$ E
4	Pa	from 2 by $\forall$ E
5	Qa	from 3 by $\forall$ E
6	$Pa \land Qa$	from 4, 5 by $\land$ I
7	$\forall x(Px \land Qx)$	from 6 by $\forall$ I

Each of the preceding deductions uses only one premise. The next deduction uses two.

$\{\forall x(Px \rightarrow Qx), \exists x(Px \land Tx)\} \vdash \exists x(Qx \land Tx)$.

1	$\forall x(Px \rightarrow Qx)$	premise
2	$\exists x(Px \land Tx)$	premise
3	$Pa \land Ta$	supposition
4	Pa	from 3 by $\land$ E
5	$Pa \rightarrow Qa$	from 1 by $\forall$ E
6	Qa	from 4, 5 by $\rightarrow$ E
7	Ta	from 3 by $\land$ E
8	$Qa \land Ta$	from 6, 7 by $\land$ I
9	$\exists x(Qx \land Tx)$	from 8 by $\exists$ I
10	$(Pa \land Ta) \rightarrow \exists x(Qx \land Tx)$	from 3–9 by $\rightarrow$ I
11	$\exists x(Qx \land Tx)$	from 2, 10 by $\exists$ E

The next deduction revisits a deduction in CPL that established the validity of the rule of disjunctive syllogism: $\{p \vee q, \neg q\} \vdash p$. It is here to remind readers that all of the deduction rules of CPL apply in CQL.

$\{\forall x(Px \vee Qx), \exists x \neg Px\} \vdash \exists x Qx$.

1	$\forall x(Px \vee Qx)$	premise
2	$\exists x \neg Px$	premise
3	$\neg Pa$	supposition
4	$Pa \vee Qa$	from 1 by $\forall$ E
5	Pa	supposition
6	$\neg Qa$	supposition
7	Pa	from 5 by Reit
8	$\neg Pa$	from 3 by Reit
9	Qa	from 6–8 by $\neg$ E
10	$Pa \rightarrow Qa$	from 5–9 by $\rightarrow$ I
11	Qa	supposition
12	Qa	from 11 by Reit
13	$Qa \rightarrow Qa$	from 11–12 by $\rightarrow$ I
14	Qa	from 4, 10, 13 by $\vee$ E
15	$\exists x Qx$	from 14 by $\exists$ I
16	$\neg Pa \rightarrow \exists x Qx$	from 3–15 by $\rightarrow$ I
17	$\exists x Qx$	from 2, 16 by $\exists$ E

The next two deductions establish the validity of an important logical equivalence involving negation and the quantifiers. The logical truth established by the first deduction is illustrated by the universal truth of the following claim: if it is not the case that something is square, then everything fails to be square.

It is not hard to see the truth of this illustration. Suppose it is true that it is not the case that something is square. Now pick something arbitrarily. Call it *a*. Suppose that *a* is

Classical Quantificational Logic

square. Well, then, something would be square, contrary to the supposition that it is not the case the something is square. Thus, a is not square. Since a was picked arbitrarily, we conclude that everything fails to be square. In short, then, if it is not the case that something is square, then everything fails to be square.

We restate the preceding informal proof formally here.

$\vdash \neg \exists x Px \rightarrow \forall x \neg Px$.

1	$\neg \exists x Px$	supposition
2	Pa	supposition
3	$\exists x Px$	from 2 by $\exists$ I
4	$\neg \exists x Px$	from 1 by Reit
5	$\neg Pa$	from 2–4 by $\neg$ I
6	$\forall x \neg Px$	from 5 by $\forall$ I
7	$\neg \exists x Px \rightarrow \forall x \neg Px$	from 1–6 by $\rightarrow$ I

The universal truth of the claim that if everything fails to be square, then it is not the case that something is square illustrates the logical truth established next. The truth of this illustration is as evident as the truth of the previous one. Moreover, its informal proof is also as clear.

Suppose that everything fails to be square. Suppose further that something is square. Let us call the thing that is square a. Thus, a is square. Well, our initial supposition is that everything fails to be square. So, it must be true that a fails to be square. This contradiction means that our second supposition—namely, that something is square—must be false. So, its negation is true. That is, it is not the case that something is square.

This informal deduction is formalized next. The preceding informal deduction has its counterparts in lines 1–2, 5–6, and 12–13 of the formal deduction; nothing in the informal reasoning serves as counterparts to lines 3–4 and lines 7–10. These lines are required by the complexity of the rule of $\exists$ Elimination. Recall that this complexity arises from safeguards built in to ensure that the rule does not give rise to invalid arguments.

$\vdash \forall x \neg Px \rightarrow \neg \exists x Px$.

1	$\forall x \neg Px$		supposition
2		$\exists x Px$	supposition
3		Pa	supposition
4		$\neg Pb$	supposition
5		Pa	from 3 by Reit
6		$\neg Pa$	from 1 by $\forall$ E
7		Pb	from 4–6 by $\neg$ I
8		$Pa \to Pb$	from 3–7 by $\to$ I
9		Pb	from 2, 8 by $\exists$ E
10		$\neg Pb$	from 1 by $\forall$ E
11		$\neg \exists x Px$	from 2–10 by $\neg$ I
12	$\forall x \neg Px \to \neg \exists x Px$		from 1–11 by $\to$ I

Exercises: Deductions

Establish the following.

1. $\vdash \forall x \forall y Rxy \to Raa$.
2. $\vdash \forall x \forall y Rxy \to \forall x Rxx$.
3. $\{\forall x((Px \land Qx) \to Fx), Qa \land \forall z Pz\} \vdash Pa \land Fa$.
4. $\exists x Px \vdash \forall x Qx \to \exists x (Px \land Qx)$.
5. $\{\exists z Rzz, \exists y \forall x Syx\} \vdash \exists y \exists z (Szy \to Ryy)$.
6. $\exists x (Px \land Qx) \vdash \exists x Px \land \exists x Qx$.
7. $\{\forall x \forall y (Rxy \leftrightarrow (Px \land \neg Py)), \exists x \exists y (Rxy \land Ryx)\} \vdash \exists x (Px \land \neg Px)$.
8. $\forall x (Px \to Qx) \vdash \exists x \neg Px \lor \exists x Qx$.
9. $\vdash \exists x \exists y Rxy \leftrightarrow \exists y \exists x Rxy$.
10. $\vdash \forall x \forall y Rxy \leftrightarrow \forall y \forall x Rxy$.
11. $\exists x \forall y Rxy \vdash \forall y \exists x Rxy$.
12. $\neg \exists x (Px \land Qx) \vdash \forall x (Px \to \neg Qx)$.

Classical Quantificational Logic 517

13. $\forall x(Px \to \neg Qx) \vdash \neg\exists x(Px \land Qx)$.
14. $\exists x(Px \lor Qx) \vdash \exists x Px \lor \exists x Qx$.
15. $\exists x Px \lor \exists x Qx \vdash \exists x(Px \lor Qx)$.

SOLUTIONS TO SOME OF THE EXERCISES

1 Notation

1. (a) Yes   (g) Yes
   (c) Yes   (i) No
   (e) No   (k) Yes

2. (a) The single quantifier prefix binds both occurrences of z.

   (c) The first quantifier prefix binds the only occurrence of x; the second quantifier prefix binds both occurrences of z; the only occurrence of y is free.

   (e) The first quantifier prefix binds the first and last occurrence of x; the second quantifier prefix binds the second and third occurrence of x.

   (g) The first quantifier prefix binds both occurrences of x; the second quantifier prefix binds the first two occurrences of y; the third quantifier prefix binds the first two occurrences of z; the last three occurrences of variables are free.

   (i) The first quantifier prefix binds the three occurrences of x; the second quantifier prefix binds the two occurrences of z; the third quantifier prefix binds the two occurrences of y.

   (k) The first quantifier prefix binds the first occurrence of y; the second quantifier prefix binds the second occurrence of x; all other variable occurrences are free.

   (m) The first quantifier prefix binds the two occurrences of z; the second quantifier prefix binds the first two occurrences of x; the third quantifier prefix binds the last three occurrences of x.

   (o) The first quantifier prefix binds the two occurrences of z; the second quantifier prefix binds the three occurrences of y; the third quantifier prefix binds the first three occurrences of x; the fourth quantifier prefix binds the last two occurrences of x.

3. (a) Raa (g) $(\exists z \forall x Ryx \land Ga) \land Raa$
   (c) Tax (i) $\exists y Ryx \land (Px \to Pa)$
   (e) $\exists x Syx$ (k) $\exists x Rxc \land \exists y Py$

4. (a) *Raa* and *Rab* are substitution instances of $\forall x\, Rax$. *Rac* is an unlisted substitution instance.

    (c) *Rab* and *Rbb* are substitution instances of $\exists x\, Rxb$. *Rcb* is an unlisted substitution instance.

    (e) $\exists y\, Rcy$ is a substitution instance of $\forall x \exists y\, Rxy$. The two other substitution instances are unlisted.

    (g) *Raa* and *Rbb* are substitution instances of $\forall z\, Rzz$. The third substitution instance is unlisted.

    (i) None of the substituion instances given are substituion instances of $\forall y\, Rcy$. There are, in fact, three.

2.1.1 Medieval valuations for CQL

1. (a) True   (f) False   (k) True
   (c) False   (h) False   (m) True
   (e) True   (j) False   (o) False

2. (b) True   (g) False   (l) True
   (d) True   (i) True   (n) True

2.1.2 Marcus valuations for CQL

1. (a) True   (f) False   (k) True
   (c) True   (h) True   (m) True
   (e) False   (j) False   (o) False

2. (a) True   (f) False   (k) True
   (c) True   (h) True   (m) True
   (e) False   (j) True   (o) False

2.1.3 Fregean valuations for CQL

1. (a) False   (f) True
   (c) True   (h) True
   (e) True   (j) True

2.1.4 Tarski extensions and Tarski classical valuations

1. All are variable assignments but (b).

Classical Quantificational Logic

4. (a) True   (f) True   (k) True
   (c) True   (h) False   (m) True
   (e) False   (j) True   (o) False

5. (a) True   (f) True   (k) True
   (c) False   (h) True   (m) True
   (e) True   (j) True   (o) False

12 Enrichments of Classical Quantificational Logic

1 Introduction

In spite of its name, classical quantificational logic permits one to express very little about quantity. It permits one to express universal statements and existential statements. That is to say, a universally quantified formula is true in all and only those structures in which its matrix is satisfied by everything in the structure's universe, and an existentially quantified formula is true in all and only those structures in which its matrix is satisfied by something in the structure's universe. This is not surprising; after all, the clauses of the definition are designed to ensure exactly this. However, one might well wonder whether there is a formula of CQL that is true if and only if its matrix is true in all and only those structures in which it is satisfied by, say, exactly one thing in the structure's universe. Consider the formula $\exists x(Px \wedge Ex)$. Its matrix $Px \wedge Ex$ is true of exactly one thing in any structure whose universe is the set of natural numbers and in which P denotes the set of prime numbers and E denotes the set of even numbers, since in such structures $Px \wedge Ex$ is satisfied by exactly one thing, the number 2. However, many are the structures in which this very same matrix is satisfied by more than one thing: for example, the structure whose universe is, again, the natural numbers but in which P denotes the set of nonprime numbers and E denotes the set of even numbers. In such a structure, an infinite number of things satisfy the matrix $Px \wedge Ex$. Therefore, the formula $\exists x(Px \wedge Ex)$ is not one that is true in all and only those structures in which its matrix is satisfied by exactly one thing.

In this chapter, we shall learn about various ways in which CQL can be enriched so that other statements about quantity can be expressed in logical notation. Thus, for example, we shall see how CQL can be enriched so that one obtains formulae that are true in all and only those structures in which its matrix is satisfied by exactly one thing. Indeed, all kinds of formulae expressing quantity that are not expressible in CQL are expressible in these enriched logics.

The simplest and best known enrichment of CQL that permits one to express a wide variety of statements pertaining to quantity is CQL with identity (CQLI). It results from the addition of a logical two-place relational symbol to express identity. In section 2, we shall explore some of what it can express and some of what it cannot express. Next, we shall turn to a much more substantial enrichment of CQL, which goes by the name of *generalized quantificational logic*. It was initially conceived of by Andrzej Mostowski (1913–1975) (Mostowski 1957) and further elaborated by Per Lindström (1936–2009) (Lindström 1966). In fact, generalized quantificational logic now constitutes an area of active research within logic (see Westerstahl 1989). Here, we shall explore only a small part of this area of logic, namely, the part that pertains to one-place and to two-place *monadic quantifiers*. We shall explore two-place monadic quantifiers, as they are commonly applied to the study of noun phrases in natural language; we shall explore one-place monadic quantifiers, as they present the general study of monadic quantification in the simplest setting. But first we turn to CQLI.

2 Classical Quantificational Logic with Identity

CQLI adds to CQL the logical constant for identity, to which we shall assign the symbol I. Now, the very same signature can serve as a signature both for CQL and for CQLI. The atomic formulae of CQLI are just like the formulae of CQL, except one additional kind of formula may be formed, namely formulae with the symbol for identity.

Definition 1 Atomic formulae (AF) of CQLI

Let $\langle \text{IS, RS, ad} \rangle$ be a signature. Let VR be a nonempty set of variables. Let TM be IS $\cup$ VR. α is an atomic formula of CQLI ($\alpha \in \text{AF}$) iff

(1) $\alpha = \Pi t_1 \ldots t_n$, for some $\Pi \in \text{RS}$ where $\text{ad}(\Pi) = n$ and for some n occurrences of terms from TM — $t_1, \ldots, t_n$, or

(2) $\alpha = I t_i t_j$, for some $t_i, t_j \in \text{TM}$.

Composite formulae are formed in the same way as in CQL. Moreover, scope, binding, bound and free variables, closed and open formulae, and substitution instances are defined as they are for CQL.

A structure for a CQLI signature is defined just as one is defined for a CQL signature. However, because CQLI has a logical symbol that CQL does not have, the definition of a Tarski extension for the formulae of CQL definition 15 of chapter 11 has an additional clause, namely, one for the atomic formulae containing the symbol I. We merely state the relevant clause, leaving readers to refer back to the definition of the Tarski extension for CQL, if necessary.

Definition 2 Tarski extension for CQLI

Let M, or $\langle U, i \rangle$, be a structure for a signature $\langle \text{IS, RS, ad} \rangle$. Let VR be a set of variables. Let TM be IS $\cup$ VR. And let g be a variable assignment from VR into U. Let t_i and t_j be members of TM. Then,

(1.3) $[I t_i t_j]_g^M = T$ iff $[t_i]_g^M = [t_j]_g^M$.

Of course, the definitions of the semantic concepts adopted for CQL are also adopted for CQLI without change. The addition of the logical constant symbol I considerably enriches the expressive capacity of the notation of CQL. It is now possible to find formulae, for each natural number n, that are true in all and only those structures in which their matrices are satisfied by at least n things from the structure's universe. Let us see how this might be done for the case where $n = 2$. One might think that, since the formula $\exists x\, Px$ is true in any structure in which one thing from the structure's universe satisfies the matrix Px, and since, similarly, the formula $\exists y\, Py$ is true in any structure in which one thing from the structure's universe satisfies the matrix Py, the composite formula $\exists x\, Px \land \exists y\, Py$ is true in all and only those structures in which at least two things from the structure's universe satisfy the matrices, one satisfies Px and another satisfies Py.

Indeed, it is true that the composite formula will be true in any structure in which at least two things from the structure's universe are in the set assigned to the one-place relational symbol P. Consider any structure in which P is assigned a two-element set, say the set $\{1, 2\}$. Px is satisfied in such a structure, since $[Px]^M_{x \mapsto 1} = T$. So, $[\exists x\, Px]^M = T$. Similarly, Py is satisfied in the same structure, since $[Py]^M_{y \mapsto 2} = T$. So $[\exists y\, Py]^M = T$. It follows, then, that $[\exists x\, Px \land \exists y\, Py]^M = T$. However, consider any structure in which P is assigned a singleton set, say the set $\{1\}$. Px is satisfied in such a structure, since $[Px]^M_{x \mapsto 1} = T$. So $[\exists x\, Px]^M = T$. Similarly, Py is satisfied in the same structure, since $[Py]^M_{y \mapsto 1} = T$. So $[\exists y\, Py]^M = T$. It follows, then, that $[\exists x\, Px \land \exists y\, Py]^M = T$. Thus, the formula fails to express the fact that at least two elements in a structure satisfy a certain property.

The reason the formula fails to be false in a structure where its matrix is satisfied by only one thing from the structure's universe is because distinct variables can be assigned the same value. The addition of a logical constant symbol for identity permits one to give a formula that prevents such an assignment:

$\exists x \exists y (Px \land Py \land \neg Ixy)$.

The truth of this formula in a structure requires that something satisfy the matrix Px and that something *else* satisfy the open matrix Py. Consider again a structure in which P is assigned the singleton set $\{1\}$. The matrix $Px \land Py \land \neg Ixy$ is not satisfied by anything in the structure's universe, for only one value is available in the structure to render Px and Py true, and under the assignment of that value to both x and y, the formula $\neg Ixy$ is false.

Now suppose one wants to say that at most one entity has some property expressed by the relational symbol P. This means that it is false that at least two things have the property. Hence, the negation of the formula $\exists x \exists y (Px \land Py \land \neg Ixy)$.

$\neg \exists x \exists y (Px \land Py \land \neg Ixy)$,

which is equivalent to the formula

$\forall x \forall y ((Px \land Py) \to Ixy)$.

Having seen how to express the fact that there are at least two things with property named by P, one might wonder how to express the fact that there are exactly two things with the property. Clearly, one way to express this is as the conjunction of an expression for there being at least two things with the property, an expression for which we have just seen, and an expression for there being at most two things with the property, that is, the negation of the expression for there being at least three things with the property. The latter can be expressed as follows: there are things named a, b, and c, each with the property named by P, and any two of these names name different things:

$$\neg\exists x\exists y\exists z\big((Px \wedge Py \wedge Pz) \wedge (\neg Ixy \wedge \neg Iyz \wedge \neg Ixz)\big).$$

This is equivalent to the formula

$$\forall x\forall y\forall z\big((Px \wedge Py \wedge Pz) \to (Ixy \vee Iyz \vee Ixz)\big).$$

CQLI is often augmented with symbols for functions, which permit the formulation of complex terms. A one-place function symbol combines with one occurrence of a term to yield a term; a two-place function symbol combines with two occurrences of terms to yield a term; and in general, an n place function symbol combines with n occurrences of terms to yield a term.

Let FS be the set of function symbols. Let FS be disjoint from IS, RS, VR, as well as the six logical constants of CQLI. In addition, one extends to adicity function ad to assign members of FS positive integers ($\mathbb{Z}^+$).

Definition 3 Terms of CQLI (syncategorematic version)
TM, the terms of CQLI, is the set defined as follows:

(1) CN $\cup$ VR $\subseteq$ TM;

(2) if f $\in$ FS, ad(f) = n and $t_1, \ldots, t_n$ are occurrences of members of TM, then f$t_1 \ldots t_n \in$ TM;

(3) nothing else is.

To get a better understanding of how the notation works, let us consider a few examples. To do so, let us stipulate that the numbers 0 through 6 are assigned to the first seven lowercase letters of the Roman alphabet and that the two-place functions of addition and multiplication are assigned to p and m, respectively. The customary arithmetic expression $2 \cdot 3$ is formulated as mcd. Notice that in the formal notation the function symbol for multiplication (m) is prefixed to the two terms to which it applies, whereas in customary arithmetic notation the function symbol for multiplication (here $\cdot$) is infixed between the two terms to which it applies. As a further example, the following two more complex customary terms $(x + y) \cdot z$ and $x \cdot (y + z)$ are rendered as $mpxyz$ and $mxpyz$, respectively. Were we to use the function notation introduced in chapter 2, these last two expressions would be recast as $m(p(x, y), z)$ and $m(x, p(y, z))$.

CUSTOMARY NOTATION	FORMAL NOTATION
0	a
1	b
2	c
3	d
4	e
5	f
6	g
$2 \cdot 3$	mcd
$2 \cdot x$	mcx
$x + 1$	pxb
$(2 \cdot x) + 1$	$pmcxb$
$x \cdot (y + z)$	$mxpyz$

Here are some examples of identity statements:

CUSTOMARY NOTATION	FORMAL NOTATION
$x = 2$	Ixc
$2 \cdot 3 = 6$	$Imcdg$
$(2 + x) \cdot 4 = y$	$Impcxey$

In spite of this enrichment, many important concepts cannot be expressed in CQLI. For example, one cannot express with respect to the usual structure for arithmetic either that there are infinitely many prime numbers or that two one-place relational symbols are true of the same number of numbers or that a one-place relational symbol holds of an odd number of numbers. Unfortunately, we shall not be able to set out the proofs of these assertions.

Exercises: Classical quantificational logic with identity

1. Write out in the notation of CQLI formulae that correspond to the customary expressions in arithmetic:

   (a) 1
   (b) $1 + 2$
   (c) $3 \cdot x$
   (d) $2 + x = 5$
   (e) $2 \cdot x = 4$
   (f) $1 + (2 \cdot x) = y$

2. Write out in the notation of CQLI formulae that express that the one-place relational symbol P is true of the following number of things:

   (a) more than one thing
   (b) at least three things
   (c) at most three things
   (g) fewer than three things
   (h) fewer than two things
   (i) less than one thing

(d) exactly three things   (j) no more than two things

(e) more than two things   (k) between two and four things

(f) more than three things (l) exactly three things

3. Write out in the notation of CQLI formulae that express that the universe has the following number of things:

(a) more than one thing   (g) fewer than three things

(b) at least three things   (h) fewer than two things

(c) at most three things   (i) less than one thing

(d) exactly three things   (j) no more than two things

(e) more than two things   (k) between two and four things

(f) more than three things (l) exactly three things

4. Let P be a relational symbol and f a functional symbol where $\text{ad}(P) = 1$ and $\text{ad}(f) = 2$. For each of the following formulae, find a structure whose universe is $\mathbb{N}$ and a variable assignment that satisfies the formula and one that does not.

(a) $\forall y f x y = x$ (b) $\exists x \forall y f x y = y$ (c) $\exists x (Px \wedge \forall y P f x y)$

5. Determine which of the following formulae are true and which are false

(a) $\forall x \exists y I p x y a$ (e) $\forall x \exists y I p x x y$ (i) $\forall x \exists y I m y y x$

(b) $\forall x \exists y I m x y b$ (f) $\exists y \forall x I p x x y$ (j) $\forall x \forall y I m x y m y x$

(c) $\forall x \exists y I p x y b$ (g) $\forall x \exists y I p y y x$ (k) $\forall x \exists y I m x y p x y$

(d) $\exists y \forall x I m x y a$ (h) $\exists y \forall x I p y y x$ (l) $\exists x \exists y I p x y m x y$

with respect to three structures whose universes are $\mathbb{N}$, $\mathbb{Z}$, and $\mathbb{Q}$, respectively, and where, in each of these structures, p denotes addition, m multiplication, a 0, and b 1.

6. Using the notation of CQLI and the symbols introduced following definition 3, write a formula that is

(a) true in $\mathbb{N}$ but false in $\mathbb{Z}^+$ (d) true in $\mathbb{Z}$ but false in $\mathbb{Z}^+$

(b) true in $\mathbb{N}$ but false in $\mathbb{Z}$ (e) true in $\mathbb{Z}$ but false in $\mathbb{N}$

(c) true in $\mathbb{Q}$ but false in $\mathbb{Z}$

3 Monadic Quantificational Logic

We now turn to the topic of generalized quantificational logic. The essential idea, as its name suggests, is to generalize the notion of a quantifier. This topic is broad and technically demanding. Fortunately, we shall be concerned with only two small parts of generalized quantificational logic, namely, the part that results from the use of one-place monadic quantifiers and the part that results from two-place monadic quantifiers. We shall denote the set of one-place monadic quantifiers as QC^1 and the set of two-place monadic quantifiers as QC^2. The quantificational logic that results from taking the set QC^1 as the quantificational connectives will be called *one-place monadic quantificational logic*, or MQL^1, and that which results from taking the set QC^2 will be called *two-place monadic quantificational logic*, or MQL^2.

The signatures for MQL^1 and MQL^2, like those for CQLI, are just like the signatures for CQL. The difference comes from having additional logical symbols. CQLI had just one additional logical symbol, the symbol denoting the identity relation. MQL^1 and MQL^2 have, instead of the two quantifiers of CQL and CQLI, an infinite number of quantifiers. The addition of quantifiers does not affect how atomic formulae are formed, nor how the composite formulae formed with the unary propositional connective and the binary propositional connectives are formed. The difference comes with the formation of composite formulae to which quantifiers are prefixed. In other words, the formation rules for MQL^1 and MQL^2 are just like those for CQL, except for the clauses involving quantifiers. In addition, the structures for MQL^1 and MQL^2 are just like the structures for CQL. Finally, the Tarski extensions for MQL^1 and MQL^2 are just like the Tarski extensions for CQL, except for the clauses involving the quantifiers. Readers shall be reminded of these differences as needed.

While the model theory of MQL^2 is what is apposite to the study of the meaning of English noun phrases, MQL^1 is easier to grasp and so provides an apt point of entry into the novelties of generalized quantificational logic in general and into those of MQL^2 in particular. Thus, we shall first acquaint ourselves with MQL^1 before turning to MQL^2.

3.1 One-Place Monadic Quantificational Logic

To understand the insight that led Andrzej Mostowski (1957) to the development of generalized quantificational logic, let us recall what satisfaction sets are and their use to define the truth of formulae whose scope is that of a universal or an existential quantifier prefix. Recall that the closed formula $\forall x Px$ is true in a structure if and only if Px is true no matter what member of the structure's universe is assigned to x. Put in terms of satisfaction sets, this amounts to saying that $\forall x Px$ is true in a structure if and only if the satisfaction set for Px is the structure's universe.

The closed formula $\exists x Px$, one might also remember, is true in a structure if and only if Px is true with respect to some or other member of the structure's universe assigned to x. This can be restated in terms of satisfaction sets as follows: $\exists x Px$ is true in a structure if and only if the satisfaction set for Px is a nonempty subset of the structure's universe. In other words,

(1.1) $[\forall x Px]^M = T$ iff $[Px]^M_x = U_M$;

(1.2) $[\exists x Px]^M = T$ iff $[Px]^M_x \neq \emptyset$.

The right side of each biconditional can be restated in terms of cardinalities. Let us begin with the second biconditional. Only the empty set has the cardinality of 0. Any nonempty set, then, has a cardinality greater than zero. So the formula $\exists x\, Px$ is true in any structure in which the cardinality of the satisfaction set for Px is greater than 0. Turning to the first biconditional, let us recall that one set exhausts a set precisely if the subtraction of the former from the latter yields the empty set. Thus, the formula $\forall x\, Px$ is true in a structure just in case the cardinality of the set that results from subtracting the satisfaction for Px from the structure's universe is 0. Here, then, is a statement in terms of the cardinalities of sets equivalent to the biconditionals in (1).

(2.1) $\quad [\forall x\, Px]^M = T \quad$ iff $\quad |U_M - [Px]_x^M| = 0;$

(2.2) $\quad [\exists x\, Px]^M = T \quad$ iff $\quad |[Px]_x^M| \neq 0.$

The essential idea of MQL^1 is to add to the set of quantifiers others that express the cardinality of the satisfaction set. To begin with, one may introduce to MQL^1 quantifiers corresponding to such English expressions as *at most n, fewer than n, at least n, more than n,* and *exactly n,* for any positive integer n. The symbols are these: for each $n \in \mathbb{Z}^+$, $\exists^1_{>n}$ (*there are more than n*), $\exists^1_{\geq n}$ (*there are at least n*), $\exists^1_{=n}$ (*there are exactly n*), $\exists^1_{<n}$ (*there are fewer than n*), and $\exists^1_{\leq n}$ (*there are at most n*). If these were the only quantifiers in MQL^1, then it would be no more expressive than CQLI, for CQLI permits one to express all of these cardinals, albeit with very, very cumbersome formulae. However, quantifiers can be introduced into MQL^1 that permit one to express cardinalities not expressible in CQLI. For example, one may introduce a quantifier requiring that a satisfaction set be infinite, for which we shall use the symbol $\exists^1_\infty$, as well as another requiring that the cardinality of a satisfaction set be greater than the cardinality of its complement, for which we shall use the symbol M^1. Either of these two quantifiers take us beyond the expressiveness of CQLI. Finally, we shall include three more quantifiers, the counterparts to the universal and the existential quantifiers of CQL, whose symbols are $\forall^1$ and $\exists^1$, and N^1, that require that the cardinality of a satisfaction set be 0. In short, the set of quantifiers of MQL^1 is QC^1, where

$$\text{QC}^1 = \{\forall^1, \exists^1, \text{N}^1, \exists^1_\infty, \text{M}^1\} \cup \bigcup_{n \in \mathbb{Z}^+} \{\exists^1_{>n}, \exists^1_{\geq n}, \exists^1_{=n}, \exists^1_{<n}, \exists^1_{\leq n}\}$$

As stated, a signature for MQL^1 is just like one for CQL and the atomic formulae for MQL^1 are formed just like the ones for CQL. Moreover, the composite formulae with propositional connectives are also formed in precisely the way the ones in CQL are formed. In fact, the only difference is that the set QC of CQL, which contains only two quantifiers, is replaced by the set QC^1, which has an infinite number of quantifiers. Thus, where the third clause in the definition of the formulae of CQL refers to QC, the third clause in the definition of the formulae of MQL^1 refers to QC^1. In other words, if a formula of CQL fails to be a formula of MQL^1, then it contains a quantifier from QC, and inversely, if a formula of MQL^1 fails to be a formula of CQL, then it contains a quantifier from QC^1. In fact, any formula of MQL^1 can be obtained from a formula in MQL^1 either by leaving the formula unchanged, if it contains no quantifiers, or if it does contain a quantifier, by replacing every occurrence of a quantifier from QC with one from QC^1.

Definition 4 Formulae of MQL^1
FM, the set of formulae of MQL^1, is defined as follows:

(1) $AF \subseteq FM$;

(2.1) if $\alpha \in FM$ and $* \in UC$, then $*\alpha \in FM$;

(2.2) if $\alpha, \beta \in FM \circ \in BC$, then $(\alpha \circ \beta) \in FM$;

(3) if $Q \in QC^1$, $v \in VR$ and $\alpha \in FM$, then $Qv\alpha \in FM$;

(4) nothing else is.

Let us now consider some examples of formulae from MQL^1 and some examples of nonformulae.

FORMULA   NONFORMULA
$\exists^1_{\leq 2} x\, Px$ $\exists x\, Px$
$\forall^1 x\, M^1 y\, Rxy$ $\forall x\, M^1 y\, Rxy$
$\exists^1 x\, (Px \wedge Rxy)$ $N^1 a\, \exists x\, Rxy$

The concepts of quantifier prefix and quantifier matrix, defined for CQL, apply to the formulae of MQL^1, as do the concepts of scope, binding, bound and free variables, closed and open formulae, and substitution instances.

As we stated, the structure for an MQL^1 signature is just like a structure for a CQL signature. However, the Tarski extension for MQL^1 differs from one for CQL, but only with respect to the clauses pertaining to quantifiers. While CQL has just two clauses pertaining to its quantifiers, MQL^1 has a countable infinity of such clauses pertaining to its quantifiers, arising from the schematic nature of clauses (3.6.1) through (3.6.5).

Definition 5 Tarski extension for MQL^1 (syncategorematic version)
Let M, or $\langle U, i \rangle$, be a structure for a signature $\langle IS, RS, ad \rangle$. Let VR be a set of variables. Let TM be $IS \cup VR$. And let g be a variable assignment from VR into U. Then $[\ ]_g^M$, a Tarski extension for MQL^1, is a function satisfying the following conditions.

(0) SYMBOLS

Let v be a member of VR, let c be a member of IS, and let Π be a member of RS.

(0.1) $[v]_g^M = g(v)$;

(0.2) $[c]_g^M = i(c)$;

(0.3) $[\Pi]_g^M = i(\Pi)$.

(1) ATOMIC FORMULAE

Let Π be a member of RS, let ad(Π) be 1, and let t be a member of TM.

(1.1) $[\Pi t]_g^M = T$ iff $[t]_g^M \in [\Pi]_g^M$.

Let Π be a member of RS, let ad(Π) be n (where $n > 1$), and let $t_1, \ldots, t_n$ be occurrences of members of TM.

(1.2) $[\Pi t_1 \ldots t_n]_g^M = T$ iff $\langle [t_1]_g^M, \ldots, [t_n]_g^M \rangle \in [\Pi]_g^M$.

(2) COMPOSITE FORMULAE

Let α and β be members of FM.

(2.1) $[\neg \alpha]_g^M = T$ iff $[\alpha]_g^M = F$;

(2.2.1) $[\alpha \wedge \beta]_g^M = T$ iff $[\alpha]_g^M = T$ and $[\beta]_g^M = T$;

(2.2.2) $[\alpha \vee \beta]_g^M = T$ iff either $[\alpha]_g^M = T$ or $[\beta]_g^M = T$;

(2.2.3) $[\alpha \to \beta]_g^M = T$ iff either $[\alpha]_g^M = F$ or $[\beta]_g^M = T$;

(2.2.4) $[\alpha \leftrightarrow \beta]_g^M = T$ iff $[\alpha]_g^M = [\beta]_g^M$.

(3) QUANTIFIED COMPOSITE FORMULAE

Let v be a member of VR and let α be a member of FM.

(3.1) $[\forall^1 v\, \alpha]_g^M = T$ iff $|U_M - [\alpha]_{g;v}^M| = 0$;

(3.2) $[\exists^1 v\, \alpha]_g^M = T$ iff $|[\alpha]_{g;v}^M| > 0$;

(3.3) $[N^1 v\, \alpha]_g^M = T$ iff $|[\alpha]_{g;v}^M| = 0$;

(3.4) $[M^1 v\, \alpha]_g^M = T$ iff $|[\alpha]_{g;v}^M| > |[U_M - \alpha]_{g;v}^M|$;

(3.5) $[\exists^1_\infty v\, \alpha]_g^M = T$ iff $|[\alpha]_{g;v}^M|$ is infinite;

(3.6.1) $[\exists^1_{>n} v\, \alpha]_g^M = T$ iff $|[\alpha]_{g;v}^M| > n$;

(3.6.2) $[\exists^1_{\geq n} v\, \alpha]_g^M = T$ iff $|[\alpha]_{g;v}^M| \geq n$;

(3.6.3) $[\exists^1_{<n} v\, \alpha]_g^M = T$ iff $|[\alpha]_{g;v}^M| < n$;

(3.6.4) $[\exists^1_{\leq n} v\, \alpha]_g^M = T$ iff $|[\alpha]_{g;v}^M| \leq n$;

(3.6.5) $[\exists^1_{=n} v\, \alpha]_g^M = T$ iff $|[\alpha]_{g;v}^M| = n$.

Enrichments of Classical Quantificational Logic   531

Let us see how to apply the Tarski extension for MQL^1 by considering the formula of MQL^1

(3) $\text{M}^1 x \,\exists^1 y \,(Rxy \wedge Py)$

and the structure M, whose the universe is the set $\{1, 2, 3\}$ and whose interpretation function assigns $\{\langle 1, 2\rangle, \langle 2, 1\rangle, \langle 3, 1\rangle, \langle 3, 3\rangle\}$ to R and assigns $\{2, 3\}$ to P. Let g be an arbitrarily chosen variable assignment.

By clause (3.4) of definition 5, we see that

$$[\text{M}^1 x \,\exists^1 y \,(Rxy \wedge Py)]_g^M = T \text{ if and only if } |[\exists^1 y \,(Rxy \wedge Py)]_{g;x}^M| >$$
$$|U_M - [\exists^1 y \,(Rxy \wedge Py)]_{g;x}^M|.$$

To ascertain whether the right-hand side of this biconditional is true, we determine the cardinality of the set

$$[\exists^1 y (Rxy \wedge Py)]_{g;x}^M,$$

which requires us to determine its membership. To ascertain its membership, we must answer the following questions:

(3.1) $[\exists^1 y(Rxy \wedge Py)]_{g_{x \mapsto 1}}^M = T$?

(3.2) $[\exists^1 y(Rxy \wedge Py)]_{g_{x \mapsto 2}}^M = T$?

(3.3) $[\exists^1 y(Rxy \wedge Py)]_{g_{x \mapsto 3}}^M = T$?

To determine which, if any, of these equalities hold, we turn to clause (3.2) of definition 5.

We begin with the equality in (3.1). It holds if and only if $|[Rxy \wedge Py]_{g_{x \mapsto 1};y}^M| > 0$. So, what is the membership of $[Rxy \wedge Py]_{g_{x \mapsto 1};y}^M$? To answer this last question, we raise three further questions:

(3.1.1) $[Rxy \wedge Py]_{g_{x \mapsto 1};y \mapsto 1}^M = T$?

(3.1.2) $[Rxy \wedge Py]_{g_{x \mapsto 1};y \mapsto 2}^M = T$?

(3.1.3) $[Rxy \wedge Py]_{g_{x \mapsto 1};y \mapsto 3}^M = T$?

To answer question (3.1.1), we must ascertain whether the open formula $Rxy \wedge Py$ is true with respect to M when x and y are both assigned the value 1. Clearly the answer is no, since $\langle 1, 1\rangle \notin [R]_{g_{x \mapsto 1};y \mapsto 1}^M$. Similarly, the answer to question (3.1.3) is no, as $\langle 1, 3\rangle \notin [R]_{g_{x \mapsto 1};y \mapsto 3}^M$. However, the answer to question (3.1.2) is yes, because $\langle 1, 2\rangle \in [R]_{g_{x \mapsto 1};y \mapsto 2}^M$. Thus, $[Rxy]_{g_{x \mapsto 1};y \mapsto 2}^M = T$. Moreover, since $2 \in [P]_{g_{x \mapsto 1};y \mapsto 2}^M$, $[Py]_{g_{x \mapsto 1};y \mapsto 2}^M = T$. This means, then, that $[Rxy \wedge Py]_{g_{x \mapsto 1};y \mapsto 2}^M = T$. So, the satisfaction set of $[Rxy \wedge Py]_{g_{x \mapsto 1};y}^M$ is $\{2\}$. Because the cardinality of this set is greater than 0, one concludes that $1 \in [\exists y(Rxy \wedge Py)]_{g;x}^M$.

Next, we determine whether equality (3.2) holds. We proceed in the same fashion as we just did, but addressing this new trio of questions:

(3.2.1) $[Rxy \land Py]^M_{g_{x \mapsto 2; y \mapsto 1}} = T$?

(3.2.2) $[Rxy \land Py]^M_{g_{x \mapsto 2; y \mapsto 2}} = T$?

(3.2.3) $[Rxy \land Py]^M_{g_{x \mapsto 2; y \mapsto 3}} = T$?

The answer in each case is no. After all, though $[Rxy]^M_{g_{x \mapsto 2; y}} = \{1\}$, $1 \notin [Py]^M_{g_{x \mapsto 2; y}}$. It follows, then, that $[Rxy \land Py]^M_{g_{x \mapsto 2; y}} = \emptyset$. Its cardinality is 0. Therefore, by clause (3.2), $2 \notin [\exists y(Rxy \land Py)]^M_{g; x}$.

Finally, we come to equality (3.3). We must determine the membership of $[Rxy]^M_{g_{x \mapsto 3; y}}$. So, we ask a fresh trio of questions:

(3.3.1) $[Rxy \land Py]^M_{g_{x \mapsto 3; y \mapsto 1}} = T$?

(3.3.2) $[Rxy \land Py]^M_{g_{x \mapsto 3; y \mapsto 2}} = T$?

(3.3.3) $[Rxy \land Py]^M_{g_{x \mapsto 3; y \mapsto 3}} = T$?

We observe that $1 \notin [P]^M_{g_{x \mapsto 3; y \mapsto 1}}$. So $[Py]^M_{g_{x \mapsto 3; y \mapsto 1}} = F$. Therefore the answer to question (3.3.1) is no. We also observe that $\langle 3, 2 \rangle \notin [R]^M_{g_{x \mapsto 3; y \mapsto 2}}$. So $[Rxy]^M_{g_{x \mapsto 3; y \mapsto 2}} = F$. And therefore the answer to question (3.3.2) is no. However, the answer to the third question is yes, since $[Rxy]^M_{g_{x \mapsto 3; y \mapsto 3}} = T$ and $[Py]^M_{g_{x \mapsto 3; y \mapsto 3}} = T$. Therefore, by clause (3.2), $3 \in [\exists y(Rxy \land Py)]^M_{g; x}$.

We have now ascertained the membership of the satisfaction set $[\exists y(Rxy \land Py)]^M_{g; x}$: its members are 1 and 3. The cardinality of this set is 2. The cardinality of its complement is 1. Thus, $|[\exists y(Rxy \land Py)]^M_{g; x}| > |U - [\exists y(Rxy \land Py)]^M_{g; x}|$. So, by clause (3.4) of definition 5, we conclude that $[M^1 x \exists^1 y(Rxy \land Py)]^M = T$.

Let us now state the definition of the categorematic version of the Tarski extension for MQL^1. To do so, we must first state which values are to be assigned to the quantifier prefixes of MQL^1.

Definition 6 Values for members of QC^1

For each structure M for a signature and for each $n \in \mathbb{Z}^+$,

$\forall^1 v \mapsto \{U_M\}$ $\exists^1_{>n} v \mapsto \{X \subseteq U_M : |X| > n\}$

$\exists^1 v \mapsto \text{Pow}(U_M) - \{\emptyset\}$ $\exists^1_{<n} v \mapsto \{X \subseteq U_M : |X| < n\}$

$N^1 v \mapsto \{\emptyset\}$ $\exists^1_{=n} v \mapsto \{X \subseteq U_M : |X| = n\}$

$M^1 v \mapsto \{X \subseteq U_M : |X| > |U_M - X|\}$ $\exists^1_{\geq n} v \mapsto \{X \subseteq U_M : |X| \geq n\}$

$\exists^1_\infty v \mapsto \{X \subseteq U_M : |X| \text{ is infinite}\}$ $\exists^1_{\leq n} v \mapsto \{X \subseteq U_M : |X| \leq n\}$

The foregoing defines, for every structure, a function from the quantifier prefixes of QC^1 into the power set of the structure's universe. We shall avail ourselves that such a function to give ourselves a general way to name the values to be associated with the quantificational connectives: o_Q is the value assigned to the the quantifier prefix Qv. With this notational convention, we state the categorematic definition of the Tarski extension for MQL^1.

Definition 7 Tarski extension for MQL^1 (categorematic version)
Let M, $\langle U, i \rangle$, be a structure for MQL^1. Let g be a variable assignment from VR into U. Then $[\ ]_g^M$ is a function satisfying the following conditions:

(0) SYMBOLS

Let v be a member of VR, let c be a member of IS, and let Π be a member of RS.

(0.1) $[v]_g^M = g(v)$;
(0.2) $[c]_g^M = i(c)$.

(1) ATOMIC FORMULAE

Let Π be a member of RS with $ad(\Pi) = 1$ and let t be a member TM.

(1.1) $[\Pi t]_g^M = T$ iff $[t]_g^M \in i(\Pi)$.

Let Π be a member of RS with $ad(\Pi) = n$ $(n > 1)$ and let $t_1, \ldots, t_n$ be occurrences of members of TM.

(1.2) $[\Pi t_1 \ldots t_n]_g^M = T$ iff $\langle [t_1]_g^M, \ldots, [t_n]_g^M \rangle \in i(\Pi)$.

(2) COMPOSITE FORMULAE

Let α and β be members of FM, $*$ be a member of UC, and $\circ$ be a member of BC.

(2.1) $[*\alpha]_g^M = o_*([\alpha]_g^M)$;
(2.2) $[\alpha \circ \beta]_g^M = o_\circ([\alpha]_g^M, [\beta]_g^M)$.

Let v be a member of VR, Q be a member of QC^1, and n be a positive integer.

(2.3) $[Qv\,\alpha]_g^M = T$ iff $[\alpha]_{g;v}^M \in o_Q$.

The following relations and properties, defined for CQL, have their analogues for MQL^1: the semantic relation of satisfaction between a structure and either a formula or a set of formulae; the properties of a formula of being a tautology, a contingency, or a contradiction; the properties of a formula and of a set of formulae being satisfiable or unsatisfiable;

the relation of semantic equivalence between a formula and a set of formulae; and the relation of entailment between a set of formulae and a formula. They are so familiar by now that there is no point in stating them again.

Exercises: MQL^1

For these exercises, assume that $IS = \{a, b, c\}$, $VR = \{x, y, z\}$, $RS_1 = \{F, G, H\}$, and $RS_2 = \{R, S, T\}$.

1. Which of the following are formulae of MQL^1?

(a) $(Fa \to (Gb \vee Rb))$
(b) $\exists^1_{=1} x\, Px$
(c) $\forall^1 x\, \exists^1_{=2} y\, (Rax \vee Sax)$
(d) $M^1 y\, (Ray \wedge N^1 x\, Fax)$
(e) $\forall^1 x\, (Hx \to Qx)$
(f) $M^1 y\, Hy$
(g) $M^1 x\, (Hx \wedge Gx)$
(h) $\forall x\, \exists^1_{>0} y\, Rxy$
(i) $\exists^1_{=3} x\, \exists y\, Rxy$
(j) $M^1 x\, \exists^1_{\geq 1} y\, Sxy$
(k) $M^1 z\, \exists^1 y\, (Gy \wedge Sxy)$
(l) $M^1 x\, N^1 y\, (Hx \vee Ryx)$

2. For each occurrence of each variable in each of the following formulae, state which quantifier prefix binds which variable and state which variables are free.

(a) $\exists^1 z\, Gz$
(b) $\forall^1 z\, Rzx$
(c) $\exists^1 x\, \forall^1 z\, Syz$
(d) $\exists^1 y\, Ga$
(e) $\exists^1 x\, (\forall^1 x\, Rxa \wedge Hx)$
(f) $\forall^1 x\, (Fx \to \exists^1 y\, Rxy)$
(g) $\forall^1 x\, (\forall^1 y\, \exists^1_{=2} z\, Ryz \leftrightarrow (Gy \wedge Tzx))$
(h) $(Tay \leftrightarrow (\exists^1 x\, Ryx \wedge Gy))$
(i) $\exists^1 x\, \exists^1 z\, \exists^1 y\, (Rzx \vee Syx)$
(j) $((\exists^1 z\, \forall^1 x\, Ryx \wedge Gz) \wedge Rzz)$
(k) $(Fz \vee \exists^1 y\, Gx) \to \forall^1 x\, Say$
(l) $\exists^1 x\, \exists^1 y\, (Rzx \vee \exists^1 z\, Syx)$
(m) $(\exists^1 z\, Fz \to \exists^1 x\, Gx) \wedge \forall^1 x\, (Gx \vee Sax)$
(n) $\forall^1 z\, \exists^1 x\, (Rzx \to Syx) \vee (\exists^1 y\, Fx \leftrightarrow Tay)$
(o) $\exists^1 z\, Fz \vee \forall^1 y\, \exists^1 x\, ((Tax \to Syx) \wedge \forall^1 x\, (Gx \vee Say))$

3. Determine the truth value of the following formulae of MQL^1 in the structure: $M = \langle U, i \rangle$, where $U = \{1, 2, 3, 4\}$ and

$i \quad a \mapsto 1 \quad F \mapsto \{1, 2, 3\} \quad R \mapsto \{\langle 1, 2\rangle, \langle 1, 3\rangle, \langle 2, 3\rangle, \langle 3, 3\rangle\}$
$\quad b \mapsto 2 \quad G \mapsto \{2, 3, 4\} \quad S \mapsto \{\langle 1, 2\rangle, \langle 1, 3\rangle, \langle 1, 4\rangle, \langle 1, 1\rangle\}$
$\quad c \mapsto 3 \quad H \mapsto \emptyset \phantom{\{2,3,4\}} \quad T \mapsto \emptyset$

(a) Fa
(b) $\exists^1_{=1} x\, Fx$
(c) $\exists^1_{>1} x\, (Fx \wedge Gx)$
(d) $\exists^1_{>2} x\, (Fx \wedge Gx)$
(g) $M^1 x\, (Gx \wedge Fx)$
(h) $\forall^1 x\, \exists^1_{>0} y\, Rxy$
(i) $\exists^1_{=3} x\, \exists^1_{>0} y\, Rxy$
(j) $M^1 x\, \exists^1_{\geq 1} y\, Sxy$

(e) $\forall^1 x\,(Fx \to Gx)$ (k) $\exists^1 y \mathrm{M}^1 z\,(Fz \wedge Syz)$
(f) $\mathrm{M}^1 y\, Fy$ (l) $\mathrm{M}^1 x\, \mathrm{N}^1 y\,(Gx \vee Ryx)$

4. Which of the formulae in the left-hand column are semantically equivalent to which of the formulae in the right-hand column?

(a) $\exists^1_{\leq 1} \alpha$ (c) $\exists^1 \alpha$
(b) $\exists^1_{\geq 1} \alpha$ (d) $\mathrm{N}^1 \alpha$

5. Which, if any, of the quantifiers $\forall^1, \exists^1, \mathrm{N}^1, \exists^1_i, \mathrm{E}^1, \mathrm{M}^1$ can be eliminated from QC^1 without loss of expressiveness?

6. Consider the following symbols: $\exists^1_{>0}, \exists^1_{\geq 0}, \exists^1_{=0}, \exists^1_{<0}, \exists^1_{\leq 0}$.

(a) Suppose they are added to QC^1. State the corresponding clauses of the Tarski extension for them.

(b) What, if any, advantages would there be in adding them to the quantifiers for QML^1? If there are no advantages, explain why.

3.2 Two-Place Monadic Quantificational Logic

Nine years after Andrzej Mostowski published his pioneering paper, Per Lindström (1966) carried Mostowski's ideas further. One-place monadic quantificational connectives express facts about the cardinality of a subset of the structure's universe. Lindström's idea was to compare the cardinalities of subsets of a structure's universe or of subsets of n tuples of its universe. Lindström's extension is very substantial and requires quite a bit of work to grasp. Fortunately, we do not require the full power of Lindström's elaboration of generalized quantificational model theory to treat the semantics of English quantified noun phrases. All we require is the model theory of what we dubbed *two-place monadic quantificational logic*, or MQL^2.

As stated earlier, a signature for MQL^2 is just like one for MQL^1 and its atomic formulae are formed in just the same way. Moreover, its composite formulae with propositional connectives are also formed in precisely the same way. The difference arises with respect to the quantifier symbols and with respect to the formation of formulae with quantifier prefixes.

Recall that the set of quantifers for MQL^2 is QC^2. And though we reuse the symbols for quantifiers used in MQL^1, we alter the superscript from 1 to 2 to remind ourselves that the symbols are distinct from those in MQL^1.

$$\mathrm{QC}^2 = \{\forall^2, \exists^2, \mathrm{N}^2, \exists^2_\infty, \mathrm{M}^2\} \cup \bigcup_{n \in \mathbb{Z}^+} \{\exists^2_{>n}, \exists^2_{\geq n}, \exists^2_{=n}, \exists^2_{<n}, \exists^2_{\leq n}\}$$

Moreover, they are given interpretations different from their counterparts in MQL^1.

MQL^2 can be presented in two different, but equivalent ways. Both presentations differ from MQL^1 and from each other in how formulae with quantifiers are formed. In MQL^1, a quantifier from QC^1 and a variable from VR combine with a single formula to form a more complex formula. In MQL^2, a quantifier from QC^2 and a variable from VR combine with two formulae to form a more complex formula. However, there are two different ways in which the two formula combine with the quantifier and the variable to form a more complex formula. Correlated with these two ways of forming formulae with quantifiers in MQL^2 are two different kinds of values assigned to quantifiers, though, when the formulae arrived at comprise the same formulae, quantifier, and variable, they are equivalent. We shall explore both presentations in each of the following two subsections. One presentation is easier to grasp, at least initially, while the second, which bears a closer resemblance to expressions in natural language, requires a little effort. We begin with the presentation that is easier to grasp.

3.2.1 Quantifiers as binary relations

In MQL^1, a quantifier from QC^1, a variable from VR, and a single formula combine in that order to yield a formula; however, in this first presentation of MQL^2, a quantifier from QC^2, a variable from VR, and a pair of formulae combine in that order to yield a formula. Thus, as we shall see from definition 8, while $\forall^1 x \, Px$ is a formula of MQL^1, $\forall^2 x \, Px$ is not a formula of MQL^2, for, even though the quantifier $\forall^2$ is a member of QC^2, $\forall^2 x$ must combine with two formulae, not with just one. Thus, $\forall^2 x \, Px \, Qx$ is a formula of MQL^2, whereas $\forall^2 x \, Px$ is not, for in the former case $\forall^2$ combines with two formulae, whereas in the latter case it combines with just one formula.

Definition 8 Formulae of MQL^2 (first presentation)
FM, the set of formulae of MQL^2, is defined as follows.

(1) AF ⊆ FM;

(2.1) If $\alpha \in$ FM and $* \in$ UC, then $*\alpha \in$ FM;

(2.2) If $\alpha, \beta \in FM$, $\circ \in BC$, then $(\alpha \circ \beta) \in$ FM;

(3) If $Q \in$ QC^2, $v \in$ VR and $\alpha, \beta \in$ FM, then $Qv \, \alpha \, \beta \in$ FM;

(4) Nothing else is.

Here are some contrasting examples:

FORMULA	NONFORMULA
$\exists^2 x \, Px \, Qx$	$\exists^2 x \, (Px \wedge Qx)$
$\forall^2 x \, Px \, \exists^2 y \, Qy \, Rxy$	$\forall^2 x \, \exists^2 y \, Rxy$
$N^2 x \, Px \, (Qx \wedge Rxy)$	$N^2 x \, Px \, Qx \, Rxy$
$N^2 x \, \exists^2 x \, Qx \, Px \, \forall^2 y \, Rxy \, Sx$	$N^2 x \, \exists^2 x \, Qx \to Px \, \forall^2 y \, Rxy \, Sx$

Enrichments of Classical Quantificational Logic

Again, the concepts of scope, binding, bound and free variables, closed and open formulae, and substitution instances, defined for CQL, for CQLI and for MQL^1, apply to the formulae of MQL^2, both as defined for the first presentation and for the second presentation.

As we stated, the structure for an MQL^2 signature is just like a structure for an MQL^1 signature, itself just like a structure for a CQL signature. The Tarski extension, however, for MQL^2 differs from one for MQL^1 formula, but only with respect to the clauses pertaining to quantifiers.

Definition 9 Tarski extension for MQL^2 (first presentation)

Let M, or $\langle U, i \rangle$, be a structure for a signature $\langle IS, RS, ad \rangle$. Let VR be a set of variables. Let TM be IS $\cup$ VR. And let g be a variable assignment from VR into U. Then, $[\ ]_g^M$, a Tarski extension for MQL^2 (first presentation), is a function satisfying the following conditions:

(0) SYMBOLS

Let v be a member of VR, c be a member of IS, and Π be a member of RS.

(0.1) $[v]_g^M = g(v)$;

(0.2) $[c]_g^M = i(c)$;

(0.3) $[\Pi]_g^M = i(\Pi)$.

(1) ATOMIC FORMULAE

Let Π be a member of RS, $ad(\Pi)$ be 1, and t be a member of TM.

(1.1) $[\Pi t]_g^M = T$ iff $[t]_g^M \in [\Pi]_g^M$.

Let Π be a member of RS, $ad(\Pi)$ be n (where $n > 1$), and $t_1, \ldots, t_n$ be occurrences of members of TM.

(1.2) $[\Pi t_1 \ldots t_n]_g^M = T$ iff $\langle [t_1]_g^M, \ldots, [t_n]_g^M \rangle \in [\Pi]_g^M$.

(2) COMPOSITE FORMULAE

Let α and β be members of FM.

(2.1) $[\neg \alpha]_g^M = T$ iff $[\alpha]_g^M = F$;
(2.2.1) $[\alpha \wedge \beta]_g^M = T$ iff $[\alpha]_g^M = T$ and $[\beta]_g^M = T$;
(2.2.2) $[\alpha \vee \beta]_g^M = T$ iff   either $[\alpha]_g^M = T$ or $[\beta]_g^M = T$;
(2.2.3) $[\alpha \to \beta]_g^M = T$ iff   either $[\alpha]_g^M = F$ or $[\beta]_g^M = T$;
(2.2.4) $[\alpha \leftrightarrow \beta]_g^M = T$ iff $[\alpha]_g^M = [\beta]_g^M$.

(3) QUANTIFIED COMPOSITE FORMULAE

Let v be a member of VR, Q be a member of QC^2, and n be a positive integer.

(3.1) $[\forall^2 v\, \alpha\, \beta]^M_g = T$ iff $|[\alpha]^M_{g;v} - [\beta]^M_{g;v}| = 0$;

(3.2) $[\exists^2 v\, \alpha\, \beta]^M_g = T$ iff $|[\alpha]^M_{g;v} \cap [\beta]^M_{g;v}| > 0$;

(3.3) $[N^2 v\, \alpha\, \beta]^M_g = T$ iff $|[\alpha]^M_{g;v} \cap [\beta]^M_{g;v}| = 0$;

(3.4) $[M^2 v\, \alpha\, \beta]^M_g = T$ iff $|[\alpha]^M_{g;v} \cap [\beta]^M_{g;v}| > |[\alpha]^M_{g;v} - [\beta]^M_{g;v}|$;

(3.5) $[\exists^2_\infty v\, \alpha\, \beta]^M_g = T$ iff $|[\alpha]^M_{g;v} \cap [\beta]^M_{g;v}|$ is infinite;

(3.6.1) $[\exists^2_{>n} v\, \alpha\, \beta]^M_g = T$ iff $|[\alpha]^M_{g;v} \cap [\beta]^M_{g;v}| > n$.

(3.6.2) $[\exists^2_{\geq n} v\, \alpha\, \beta]^M_g = T$ iff $|[\alpha]^M_{g;v} \cap [\beta]^M_{g;v}| \geq n$.

(3.6.3) $[\exists^2_{<n} v\, \alpha\, \beta]^M_g = T$ iff $|[\alpha]^M_{g;v} \cap [\beta]^M_{g;v}| < n$.

(3.6.4) $[\exists^2_{\leq n} v\, \alpha\, \beta]^M_g = T$ iff $|[\alpha]^M_{g;v} \cap [\beta]^M_{g;v}| \leq n$.

(3.6.5) $[\exists^2_{=n} v\, \alpha\, \beta]^M_g = T$ iff $|[\alpha]^M_{g;v} \cap [\beta]^M_{g;v}| = n$.

Let us now see how such clauses can be applied. Consider the formula

(4) $M^2 x\, Px\, \exists^2 y\, Qy\, Rxy$

and a structure M, whose the universe is the set $\{1, 2, 3, 4\}$ and whose interpretation function assigns $\{1, 2, 3\}$ to P, $\{1, 4\}$ to Q, and $\{\langle 1, 4\rangle, \langle 2, 1\rangle, \langle 4, 2\rangle\}$ to R. Let g be an arbitrarily chosen variable assignment.

By clause (3.4) of definition 9, we see that

$[M^2 x\, Px\, \exists^2 y\, Qy\, Rxy]^M_g = T$ if and only if $|[Px]^M_{g;x} \cap [\exists^2 y\, Qy\, Rxy]^M_{g;x}|$
$$> |[Px]^M_{g;x} - [\exists^2 y\, Qy\, Rxy]^M_{g;x}|.$$

To ascertain whether the right-hand side of this biconditional is true, we determine what members of the universe are in $[Px]^M_{g;x}$ and what members are in $[\exists^2 y\, Qy\, Rxy]^M_{g;x}$. Obviously, $[Px]^M_{g;x} = \{1, 2, 3\}$. However, some work is required to determine the membership of $[\exists^2 y\, Qy\, Rxy]^M_{g;x}$. In particular, we must answer the following questions.

(4.1) $[\exists^2 y\, Qy\, Rxy]^M_{g_{x \mapsto 1}} = T$?

(4.2) $[\exists^2 y\, Qy\, Rxy]^M_{g_{x \mapsto 2}} = T$?

(4.3) $[\exists^2 y\, Qy\, Rxy]^M_{g_{x \mapsto 3}} = T$?

(4.4) $[\exists^2 y\, Qy\, Rxy]^M_{g_{x \mapsto 4}} = T$?

Enrichments of Classical Quantificational Logic

To determine which, if any, of these equalities hold, we turn to clause (3.2). Now, equality (4.1) holds if and only if $|[Qy]^M_{g_{x \mapsto 1;y}} \cap [Rxy]^M_{g_{x \mapsto 1}}| \neq 0$. This, in turn, requires that one determine the membership of $[Qy]^M_{g_{x \mapsto 1;y}}$ and of $[Rxy]^M_{g_{x \mapsto 1;y}}$. Clearly, the former is the set $\{1, 4\}$. To determine the latter, we must now answer these four questions.

(4.1.1) $[Rxy]^M_{g_{x \mapsto 1; y \mapsto 1}} = T?$

(4.1.2) $[Rxy]^M_{g_{x \mapsto 1; y \mapsto 2}} = T?$

(4.1.3) $[Rxy]^M_{g_{x \mapsto 1; y \mapsto 3}} = T?$

(4.1.4) $[Rxy]^M_{g_{x \mapsto 1; y \mapsto 4}} = T?$

The only ordered pair in $[R]^M_g$ with 1 as a first coordinate is $\langle 1, 4 \rangle$. So, because only the fourth question has the answer yes, it follows that $[Rxy]^M_{g_{x \mapsto 1;y}} = \{4\}$. Moreover, because $[Qy]^M_{g_{x \mapsto 1;y}} = \{1, 4\}$, $[Rxy]^M_{g_{x \mapsto 1;y}} \cap [Qy]^M_{g_{x \mapsto 1;y}} = \{4\}$. Since the cardinality of this set is greater than 0, it follows by clause (3.2) of definition 9 that, the answer to question (4.1) is yes. Therefore, $1 \in [\exists^2 y \, Qy \, Rxy]^M_{g; x}$.

Next, to answer question (4.2), we must first answer following these additional questions.

(4.2.1) $[Rxy]^M_{g_{x \mapsto 2; y \mapsto 1}} = T?$

(4.2.2) $[Rxy]^M_{g_{x \mapsto 2; y \mapsto 2}} = T?$

(4.2.3) $[Rxy]^M_{g_{x \mapsto 2; y \mapsto 3}} = T?$

(4.2.4) $[Rxy]^M_{g_{x \mapsto 2; y \mapsto 4}} = T?$

$[R]^M_g$ has only one ordered pair with 2 as its first coordinate, namely, $\langle 2, 1 \rangle$. Thus, only the first question has the answer yes. As a result, $[Rxy]^M_{g_{x \mapsto 2;y}} = \{1\}$. Moreover, because $[Qy]^M_{g_{x \mapsto 2;y}} = \{1, 4\}$, $[Rxy]^M_{g_{x \mapsto 2;y}} \cap [Qy]^M_{g_{x \mapsto 2;y}} = \{1\}$. Its cardinality set is greater than 0. So, by clause (3.2) of definition 9, the answer to question (4.2) is yes. Therefore, $2 \in [\exists^2 y \, Qy \, Rxy]^M_{g; x}$.

The answer to question (4.3) is no, since the answer to each of the following questions is also no.

(4.3.1) $[Rxy]^M_{g_{x \mapsto 3; y \mapsto 1}} = T?$

(4.3.2) $[Rxy]^M_{g_{x \mapsto 3; y \mapsto 2}} = T?$

(4.3.3) $[Rxy]^M_{g_{x \mapsto 3; y \mapsto 3}} = T?$

(4.3.4) $[Rxy]^M_{g_{x \mapsto 3; y \mapsto 4}} = T?$

Therefore, $3 \notin [\exists^2 y \, Qy \, Rxy]^M_{g; x}$.

Finally, we turn to question (4.4), whose answer depends on a positive answer to at least one of the following questions.

(4.4.1) $[Rxy]^M_{g_{x \mapsto 4;\, y \mapsto 1}} = T?$

(4.4.2) $[Rxy]^M_{g_{x \mapsto 4;\, y \mapsto 2}} = T?$

(4.4.3) $[Rxy]^M_{g_{x \mapsto 4;\, y \mapsto 3}} = T?$

(4.4.4) $[Rxy]^M_{g_{x \mapsto 4;\, y \mapsto 4}} = T?$

$[R]^M_g$ has only one ordered pair with 4 as its first coordinate, namely, $\langle 4, 2 \rangle$. The second question, then, is the only question with an affirmative answer. Thus, $[Rxy]^M_{g_{x \mapsto 4;\, y}} = \{2\}$. However, because $[Qy]^M_{g_{x \mapsto 4;\, y}} = \{1, 4\}$, $[Rxy]^M_{g_{x \mapsto 4;\, y}} \cap [Qy]^M_{g_{x \mapsto 4;\, y}} = \emptyset$. Its cardinality set is not greater than 0. So, by clause (3.2) of definition 9, the answer to question (4.4) is no. Therefore, $4 \notin [\exists^2 y\, Qy\, Rxy]^M_{g;\, x}$.

We are, at last, in a position to state the membership of $[\exists^2 y\, Qy\, Rxy]^M_{g;\, x}$: it is $\{1, 2\}$. This, in turn, permits us to determine that, on the basis of clause (3.4), the truth value of formula (4) is true. Here are the relevant calculations:

$$
\begin{array}{rcl}
|[Px]^M_{g;\, x} \cap [\exists^2 y\, Qy\, Rxy]^M_{g;\, x}| & > & |[Px]^M_{g;\, x} - [\exists^2 y\, Qy\, Rxy]^M_{g;\, x}| \\
|\{1, 2, 3\} \cap \{1, 2\}| & > & |\{1, 2, 3\} - \{1, 2\}| \\
|\{1, 2\}| & > & |\{3\}| \\
2 & > & 1
\end{array}
$$

The model theoretic properties and relations known from CQL and said in section 3.1 to hold analogously for MQL^1 also hold analogously for MQL^2. We shall not dwell on these. Instead, we shall close this section with an exposition of other model theoretic properties that emerge from the fact that quantifier prefixes of MQL^2 denote binary relations on the power set of a structure's universe. In particular, we shall ask which kinds of properties of binary relations on sets hold for the values assigned to the various quantifier prefixes. In addition to the properties of binary relations—such as reflexivity, irreflexivity, symmetry, asymmetry, antisymmetry, transitivity, left comparability, and right comparability (chapter 2)—there are other properties pertinent to the model theory of MQL^2 and of interest in its adaptation to the semantics of natural language. They include the properties of *conservativity*, *monotonicity*, and *invariance under permutation*. We confine our attention to these properties, as they are properties that seem to apply also to various kinds of noun phrases.

The property of *conservativity* is thought to be a distinguishing property of the natural language expressions corresponding to quantifiers. To put it simply but not fully accurately, if a relation from one set to another is conservative, one can ascertain that it is merely by inspecting the members of the set. Consider, for example, the overlap relation. One can ascertain whether a set A overlaps another set by merely looking among the members of A to see whether any of them are members of B; there is no need to look at members beyond A.

Definition 10 Conservativity
Let D be a set. Let R be a binary relation over $\text{Pow}(D)$. R is conservative iff, for each $X, Y \in \text{Pow}(D)$, RXY iff $RX(X \cap Y)$.

Not every binary relation between sets is conservative. There are binary relations on the power set of a set where one set is related to another but the first set is not related to the intersection of the two sets, and where one set is related to its intersection with another set but the first set is not related to the second. Consider, for example, the relation of sets having the same cardinality. The set $\{1, 2, 3\}$ has the same cardinality as the set $\{2, 3, 4\}$; however, the set $\{1, 2, 3\}$ does not have the same cardinality as the set $\{2, 3\}$, which is the intersection of $\{1, 2, 3\}$ with $\{2, 3, 4\}$. Moreover, while the set $\{1, 2, 3\}$ is the same as the set $\{1, 2, 3\} \cap \{1, 2, 3, 4\}$, and hence they have the same cardinality, the set $\{1, 2, 3\}$ does not have the same cardinality as the set $\{1, 2, 3, 4\}$.

We turn to the property of *monotonicity*, of which there are four kinds for binary relations on sets. As we shall see in chapter 14, these properties seem to be pertinent to determining the distribution of a class of natural language expressions known as *negative polarity items*. Here are the general definitions of monotonicity.

Definition 11 Monotonic relations
Let R be a binary relation on a set; let $\leq$ be a partial order on the domain of R.

(1) R is monotonically decreasing on the left iff for each $x, y, z \in D_R$, if $x \leq y$ and Ryz, then Rxz;

(2) R is monotonically increasing on the left, iff for each $x, y, z \in D_R$, if $x \leq y$ and Rxz, then Ryz;

(3) R is monotonically decreasing on the right, iff for each $x, y, z \in D_R$, if $y \leq z$ and Rxz, then Rxy;

(4) R is monotonically increasing on the right iff for each $x, y, z \in D_R$, if $y \leq z$ and Rxy, then Rxz.

The definition of monotonicity is stated generally. When we apply it to the structures of MQL^2, the relations will be relations on the power set of a structure's universe and the partial order will be the subset relation. To illustrate, let us consider the disjointness relation $\perp$. It is monotonically decreasing both on its left and on its right, and it is monotonically increasing on neither its left or its right. These facts can be gleaned by considering an Euler diagram for disjointness. It should be evident that any subset of A is disjoint from B, as any subset of B is from A. At the same time, while a superset of A may be disjoint from B, not every superset is. Similarly, while a superset of B may be disjoint from A, not every superset is.

The third property is that of *invariance under permutation*. A permutation is a bijection whose domain and codomain are the same. A relation is invariant under permutation just

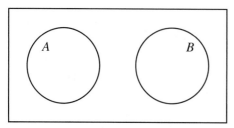

Figure 12.1
Euler diagram for $A \perp B$.

in case the relation is maintained even when the members of the domain on which the relation is defined are permuted. For example, consider the set comprising four members: a, b, c, and d. Consider the following permutation: $\pi(a) = b$, $\pi(b) = c$, $\pi(c) = d$, and $\pi(d) = a$. Clearly, $\{a, b\}$ is a subset of $\{a, b, c\}$. Equally, the set $\{\pi(a), \pi(b)\}$ is a subset of $\{\pi(a), \pi(b), \pi(c)\}$, since $\{\pi(a), \pi(b)\} = \{b, c\}$ and $\{\pi(a), \pi(b), \pi(c)\} = \{b, c, d\}$ and $\{b, c\} \subseteq \{b, c, d\}$.

Definition 12 Invariance under permutation
Let D be a set. Let R be a binary relation over $\text{Pow}(D)$. Let π be a permutation of D. R is invariant under π iff for each $X, Y \in \text{Pow}(D)$, RXY iff $R\pi(X)\pi(Y)$.

Exercises: Quantifiers as binary relations

For these exercises, assume that $\text{IS} = \{a, b, c\}$, $\text{VR} = \{x, y, z\}$, $\text{RS}_1 = \{F, G, H\}$, and $\text{RS}_2 = \{R, S, T\}$.

1. Which of the following are formulae of MQL^2 (presentation 1)?

   (a) Ha
   (b) $Fx \wedge Gb$
   (c) Rxy
   (d) $\exists^2_{\leq 3} x\, Fx$
   (e) $\exists^2_{\geq 3} z\, Fz\, Gz$
   (f) $N^2 z\, Raz \wedge Hz$
   (g) $\forall^2 x\, \exists^2_{>2} y\, Rxy$
   (h) $\exists^2_{<0} z\, Fz\, M^2 y\, Gy\, Syz$
   (i) $M^2 y\, \exists^2 z\, Fz\, Gy\, Tyz$
   (j) $M^2 z\, Fz\, \forall^2 x\, Gx\, Rxy$
   (k) $M^1 z\, Tyz\, \forall^2 x\, Hx\, Rxy$
   (l) $\exists^2_{>2} z\, Fz\, M^2 y\, Gy\, Sxy$

2. For each occurrence of each variable in each of the following formulae, state which quantifier prefix binds which variable and state which variables are free.

   (a) $\exists^2 z\, Fz\, Gz$
   (b) $M^2 z\, \exists^2 y\, Fa\, Txz\, Gy$
   (c) $M^2 z\, Szx\, (Fz \rightarrow Gz)$
   (d) $\exists^2_{=n} y\, Gx\, (Saz \wedge Tbz)$
   (e) $\forall^2 x\, Fx\, Ray$
   (f) $Fx \wedge \exists^2 y\, Tyx\, Gy$
   (g) $\exists^2 x\, Fx\, Rzx \vee \exists^2 x\, Syx\, Gy$
   (h) $\exists^2 x\, \forall^2 x\, Ryx\, Gz\, Rzx$
   (i) $\forall^2 z\, (N^2 y\, Fy\, Haz \rightarrow \exists^2 z\, Saz\, Gz)\, (Gx \vee Saz)$

3. Determine truth value of the formulae of MQL^2 listed in the structure: $M = \langle U, i \rangle$, where $U = \{1, 2, 3, 4, 5\}$ and

$i: \quad a \mapsto 1 \quad F \mapsto \{1, 3, 5\} \quad R \mapsto \{\langle 1, 2\rangle, \langle 3, 3\rangle, \langle 2, 2\rangle, \langle 5, 4\rangle, \langle 1, 3\rangle\}$
$ b \mapsto 2 \quad G \mapsto \{2, 3, 4\} \quad S \mapsto \{\langle 1, 2\rangle, \langle 1, 3\rangle, \langle 1, 4\rangle, \langle 1, 1\rangle\}$
$ c \mapsto 5 \quad H \mapsto \{1, 2, 3, 4\} \quad T \mapsto \emptyset$

(a) Fa
(b) Tbc
(c) Hc
(d) $\exists^2_{=1} x \, Fx \, Gx$
(e) $\exists^2_{=2} x \, Gx \, Hx$
(f) $\forall^2 y \, Gy \, Hy$
(g) $\exists^2_{>1} x \, Fx \, Ga$
(h) $\forall^2 y \, (Fy \wedge Gy) \, Hy$
(i) $N^2 z \, Raz \, Fz$
(j) $M^2 z \, Fz \, Hz$
(k) $\exists^2_{>1} x \, Fc \, Gc$
(l) $\forall^2 x \, Fx \, \exists^2_{\leq 1} y \, Gy \, Rxy$
(m) $\exists^2_{=1} y \, Fy \, \forall^2 z \, Hz \, Syz$
(n) $\forall^2 x \, Fx \, \exists^2_{\leq 1} y \, Gy \, Txy$
(o) $M^2 x \, Hx \, \exists^2 y \, Rxy \, (Fy \wedge Gy)$
(p) $\exists^2 y \, Fy \, M^2 z \, Syz \, Hz$

4. Consider the following binary relations on the power set of a set: subset, proper subset, superset, proper superset, overlap, disjointness, incomparability, and being most. Which of these binary relations has which of these properties: conservativity, invariance under permutation, left or right, increasing or decreasing monotonicity? In each case, justify your answer.

3.2.2 Quantifiers as a family of sets

We now come to the second presentation of MQL^2. Here, rather than forming a formula with a quantifier in one step, drawing on a pair of formulae, a quantifier, and a variable, we form a formula with a quantifier in two steps. First, a quantifier, a variable, and a formula combine to form an expression that is not a formula. We shall call it a *restrictor* and designate the set of such expressions as RT. Then, a formula with a quantifier prefix arises from putting together a restrictor with a formula. Thus, for example, the quantifier prefix $\forall^2 x$ and the formula Px form a restrictor $(\forall^2 x \, Px)$ and this restrictor and the formula Qx combine to form the formula $(\forall^2 x \, Px) \, Qx$. Notice that this formula of the second presentation of MQL^2 is not a formula of the first presentation, and the formula $\forall^2 x \, Px \, Qx$ of the first presentation is not one of the second presentation, though, as we shall see, they are true in the same structures.

Definition 13 Formulae of MQL^2 (second presentation)

FM, the set of formulae of generalized quantificational logic MQL^2, is defined as follows:

(1) $AF \subseteq FM$;
(2.1) If $\alpha \in FM$ and $\star \in UC$, then $\star\alpha \in FM$;
(2.2) If $\alpha, \beta \in FM$, $\circ \in BC$, then $(\alpha \circ \beta) \in FM$;
(3.1) If $Q \in QC^2$, $v \in VR$, and $\alpha \in FM$, then $(Qv\alpha) \in RT$;

(3.2) If $\alpha \in \text{RT}$ and $\beta \in FM$, then $\alpha\beta \in FM$;

(4) Nothing else is.

The following are examples of formulae from the second presentation of MQL^2 that are paired with their counterparts from the first presentation.

Formulae of MQL^2

PRESENTATION 1	PRESENTATION 2
$\exists^2 x\ Px\ Qx$	$(\exists^2 x\ Px)\ Qx$
$\forall^2 x\ Px\ \exists^2 y\ Qy\ Rxy$	$(\forall^2 x\ Px)\ (\exists^2 y\ Qy)\ Rxy$
$\text{N}^2 x\ Px\ (Qx \wedge Rxy)$	$(\text{N}^2 x\ Px)\ (Qx \wedge Rxy)$
$\text{N}^2 x\ Qx\ \exists^2 x\ Px\ \forall^2 y\ Rxy\ Sx$	$(\text{N}^2 x\ Qx)\ (\exists^2 x\ Px)\ (\forall^2 y\ Rxy)\ Sx$

In spite of the difference in the ways in which the formulae with quantifier prefixes are formed, the definition of scope as defined for the formulae of CQL, MQL^1, and MQL^2 (first presentation) applies to the formulae of MQL^2 (second presentation). The scope of an occurrence of a logical connective in a formula α is the subformula of α which contains the logical connectives occurrence but whose immediate subformulae do not. Consider, for example, the formulae of MQL^2 (second presentation), $(\text{N}^2 x\ Qx)\ (\exists^2 x\ Px)\ (\forall^2 y\ Rxy)\ Sx$. The scope of the quantifier prefix $\forall^2 y$ in this formula is $(\forall^2 y\ Rxy)\ Sx$, which is the smallest subformulae containing the only occurrence of $\forall^2 y$; it is not $(\forall^2 y\ Rxy)$, for though it contains $\forall^2 y$, it is not a formulae, rather it is a restrictor. Since the concept of scope as defined earlier can be applied here, so can the concepts of binding, bound and free variable, and closed and open formula, which were defined for CQL, MQL^1, and MQL^2 (first presentation).

We shall now define a categorematic Tarski extension for the second presentation of MQL^2. We begin by assigning values to the quantifier prefixes of MQL^2.

Definition 14 Values for members of QC^2

Let M be a structure for a signature and let n be a positive integer. Each quantifier is assigned a function from $\text{Pow}(U_M) \to \text{Pow}\big(\text{Pow}(U_M)\big)$ as follows.

$\forall^2 v:\quad X \mapsto \{Y \subseteq U_M\colon |X - Y| = 0\}$

$\exists^2 v:\quad X \mapsto \{Y \subseteq U_M\colon |X \cap Y| \neq 0\}$

$\text{N}^2 v:\quad X \mapsto \{Y \subseteq U_M\colon |X \cap Y| = 0\}$

$\text{M}^2 v:\quad X \mapsto \{Y \subseteq U_M\colon |X \cap Y| > |X - Y|\}$

$\exists^2_\infty v:\quad X \mapsto \{Y \subseteq U_M\colon |X \cap Y| \text{ is infinite}\}$

$\exists^2_{>n} v:\quad X \mapsto \{Y \subseteq U_M\colon X \cap Y > n\}$

$\exists^1_{<n} v:\quad X \mapsto \{Y \subseteq U_M\colon X \cap Y < n\}$

$\exists^1_{=n} v:\quad X \mapsto \{Y \subseteq U_M\colon X \cap Y = n\}$

Enrichments of Classical Quantificational Logic

$\exists^1_{\geq n} v:$ $\quad X \mapsto \{Y \subseteq U_M : X \cap Y \geq n\}$
$\exists^1_{\leq n} v:$ $\quad X \mapsto \{Y \subseteq U_M : X \cap Y \leq n\}$

Definition 15 Tarski extension for MQL^2 (second presentation)

Let M, $\langle U, i \rangle$, be a structure for MQL^2. Let g be a variable assignment from VR into U. Then [ ]$^{M, g}$ is a function from VR, IS, RS, and FM into {T, F} satisfying the following conditions.

(0) SYMBOLS

Let v be a member of VR, let c be a member of IS, and let Π be a member of RS.

(0.1) $[v]^M_g = g(v)$;

(0.2) $[c]^M_g = i(c)$;

(0.3) $[\Pi]^M_g = i(\Pi)$.

(1) ATOMIC FORMULAE

Let Π be a member of RS with $ad(\Pi) = 1$ and let t be a member TM.

(1.1) $[\Pi t]^M_g = T$ iff $[t]^M_g \in i(\Pi)$.

Let Π be a member of RS with $ad(\Pi) = n$ (n > 1) and let $t_1, \ldots, t_n$ be occurrences of members of TM.

(1.2) $[\Pi t_1 \ldots t_n]^M_g = T$ iff $\langle [t_1]^M_g, \ldots, [t_n]^M_g \rangle \in i(\Pi)$.

(2) COMPOSITE FORMULAE

Let α and β be members of FM.

(2.1) $[*\alpha]^M_g \;=\; o_*([\alpha]^M_g)$;

(2.2) $[\alpha \circ \beta]^M_g \;=\; o_\circ([\alpha]^M_g, [\beta]^M_g)$.

Let v be a member of VR, Q be a member of QC^2, and n be a positive integer.

(2.3.1) $[Qv\, \alpha]^M_g \;=\; o_Q([\alpha]^M_{g;v})$.

Let α be a member of RT and β a member of FM.

(2.3.2) $[\alpha\, \beta]^M_g = T$ iff $[\beta]^M_{g;v} \in [\alpha]^M_{g;v}$.

We conclude this discussion by recasting the properties of binary relations discussed in connection with the first presentation of MQL^2 in terms adapted to the second. To do so, we use the notion of a binary relation's associated function, im_R, defined in chapter 2, sec. 6.3.5. We begin with conservativity. It was defined for binary relations on a power set of a set as follows: RXY iff $RX(X \cap Y)$. Its counterpart for the second presentation of MQL^2 is this: $Y \in im_R(X)$ iff $X \cap Y \in im_R(X \cap Y)$. Next we turn to monotonicity. There are four kinds: right, left, increasing, and decreasing. Consider first a right monotonically increasing binary relation. Its definiens is this: if Rxy and $y \leq z$, then Rxz. Its counterpart for the second presentation of MQL^2 is this: if $y \in im_R(x)$ and $y \leq z$, then $z \in im_R(x)$. A left monotonically increasing relation is defined in this way: if Ryx and $y \leq z$, then Rzx. It too has a version for the second presentation of MQL^2: if $x \in im_R(y)$ and $y \leq z$, then $x \in im_R(z)$. And finally, there is permutation invariance. We leave it to the readers to write down the counterparts of relations invariant under permutation as well as relations that are right and left monotonically decreasing.

Exercises: Quantifers as families of sets

For these exercises, assume that IS $= \{a, b, c\}$, VR $= \{x, y, z\}$, RS$_1 = \{F, G, H\}$, and RS$_2 = \{R, S, T\}$.

1. Determine which of the following are formulae of MQL^2 (presentation 2)?

(a) Fb
(b) $Sab \wedge Hx$
(c) $\exists^2_{\leq 3} x\, Gx\, Hx$
(d) $(\exists^2_{\leq 3} x\, Gx)\, Hx$
(e) $\exists^2_{\geq 3} z\, Hz \vee Gz$
(f) $\forall^2 x\, (\exists^2_{>2} y\, Rxy)\, Fz$
(g) $(\exists^2_{<0} z\, Fz)\, (M^2 y\, Hy)\, Tyz$
(h) $(M^2 y\, (\exists^2 z\, Hz)\, Gy)\, Syz$
(i) $Hc \wedge N^2 z\, (\forall^2 x\, Fx)\, Txz$
(j) $(\exists^2 x\, Hx)\, (\forall^2_y Rxy)\, Fz$

2. For each occurrence of each variable in each of the following formulae, state which quantifier prefix binds which variable and state which variables are free.

(a) $\exists^2 z\, Fz\, Gz$
(b) $(M^2 z\, (\exists^2 y\, Fa)\, Txz)\, Gy$
(c) $(M^2 z\, Szx)\, (Fz \rightarrow Gz)$
(d) $(\exists^2_{\leq 3} y\, Gx)\, (Saz \wedge Tbz)$
(e) $(\forall^2 x\, Fx)\, Ray$
(f) $Fx \wedge (\exists^2 y\, Tyx)\, Gy$
(g) $(\exists^2 x\, Fx)\, Rzx \vee (\exists^2 x\, Syx)\, Gy$
(h) $(\exists^2 x\, (\forall^2 x\, Ryx)\, Gz)\, Rzx$
(i) $(\forall^2 z\, ((E^2 y\, Fy)\, Haz \rightarrow (\exists^2 z\, Saz)\, Gz))\, (Gx \vee Saz)$

3. Determine the truth value of the formulae of MQL^2 listed in the structure $M = \langle U, i \rangle$, where $U = \{1, 2, 3, 4, 5\}$ and

i $a \mapsto 1$ $F \mapsto \{1, 3, 5\}$ $R \mapsto \{\langle 1, 2\rangle, \langle 3, 3\rangle, \langle 2, 2\rangle, \langle 5, 4\rangle, \langle 1, 3\rangle\}$
    $b \mapsto 2$ $G \mapsto \{2, 3, 4\}$ $S \mapsto \{\langle 1, 2\rangle, \langle 1, 3\rangle, \langle 1, 4\rangle, \langle 1, 1\rangle\}$
    $c \mapsto 5$ $H \mapsto \{1, 2, 3, 4\}$ $T \mapsto \emptyset$

(a) Fa
(b) Rbc
(c) Hc
(d) $(\exists^2_{\geq 1} x\, Fx)\, Gx$
(e) $(\exists^2_{\geq 1} x\, Gx)\, Fx$
(i) $(\forall^2 z\, Raz)\, Fz$
(j) $(M^2 z\, Hz)\, Gz$
(k) $(\exists^2_{\geq 1} x\, Fc)\, Gc$
(l) $(\forall^2 x\, Fx)\, (\exists^2_{\geq 1} y\, Gy)\, Rxy$
(m) $(\forall^2 x\, Hx)\, (\exists^2_{\geq 1} y\, Fy)\, Sxy$

(f) $(\forall^2 y\, Hy)\, Fy$
(g) $(\exists^2_{\geq 1} x\, Fx)\, Ga$
(h) $\bigl(\forall^2 y\, (Fy \wedge Gy)\bigr)\, Hy$
(n) $(M^2 x\, Fx)\, (\exists^2_{\geq 1} y\, Hy)\, Txy$
(o) $(\exists^2_{=1} y\, Fy)\, (\forall^2 z\, Hz)\, Syz$
(p) $(\exists^2_{\geq 1} z\, Hz)\, (M^2 x\, Rxz)\, Sxy$

4. Write out the defining conditions on the function im_R that correspond to the defining conditions on binary relations for the following properties: (a) monotonically decreasing on the left, (b) monotonically decreasing on the right, (c) invariance under a permutation, (d) reflexivity, (e) irreflexivity, (f) symmetry, (g) asymmetry, (h) antisymmetry, and (i) transitivity.

4 Conclusion

The idea that the values assigned to quantifier prefixes are suitable as values for quantified noun phrases started to receive serious linguistic attention in the last fifth of the twentieth century. The earliest published contributions came from the logician Jon Barwise and the linguist Robin Cooper (Barwise and Cooper 1981). Work was also being done at that time by the linguist Edward Keenan, which saw publication a few years later (Keenan and Moss 1985; Keenan and Stavi 1986). Further work appeared shortly thereafter by the logician Dag Westerstahl (Westerstahl 1989). The most comprehensive presentation of generalized quantifiers and their application to natural language appears in Peters and Westerstahl (2011). A concise overview can be found in Keenan and Westerstahl (1997). Keenan (2018) has now written an accessible summary of his work on this and related topics.

SOLUTIONS TO SOME OF THE EXERCISES

2 Classical quantificational logic with identity

1. (a) *b*
   (c) *mdx*
   (e) *Imcxd*
   (f) *Ibmcxy*

2. (a) $\exists x \exists y (Px \wedge Py \wedge \neg Ixy)$
   (c) $\forall x \forall y \forall z \forall w \bigl((Px \wedge Py \wedge Pz \wedge Pw) \rightarrow (Ixy \wedge Ixz \wedge Ixw \wedge Iyz \wedge Iyw \wedge Izw)\bigr)$
   (j) $\forall x \forall y \forall z \bigl((Px \wedge Py \wedge Pz) \rightarrow (Ixy \wedge Ixz \wedge Iyz)\bigr)$

3. (a) $\exists x \exists y \neg Ixy$

4. (b) The formula is true, if f denotes addition, and it is false, if it denotes a function which adds two numbers and then adds 1.

5. (a) The formula is false with respect to $\mathbb{N}$ but true with respect to $\mathbb{Z}$ and $\mathbb{Q}$: the latter two have both positive and negative numbers, but the first does not.

(d) the formula is true with respect to all three sets: each number in each set multiplied by 0 is 0.

(g) the formula is true with respect to all three sets: each number in each set has a square.

(k) For each number, there is a number such that their product and sum is the same. This is false with respect to all three sets.

6. (a) $\exists x \exists y I p x y y$

   (c) $\exists x \exists y (\neg I x b \wedge \neg I y b \wedge I m x y b)$

   (d) $\forall x \exists y I x p y b$

3.1 MQL[1]

1. (a) no   (e) no
   (c) Yes   (g) Yes
   (i) no   k) Yes

2. (a) $\exists^1 z$ binds both occurrences of z.

   (c) $\exists^1 x$ binds only occurrence of x; $\forall^1 z$ binds both occurrences of z; the only occurrence of y is free.

   (e) $\exists^1 x$ binds first and last occurrences of x; $\forall^1 z$ binds second and third occurrences of x.

   (g) $\forall^1 x$ binds both occurrences of x; $\forall^1 y$ binds first two occurrences of y; $\exists^1_{=2} z$ binds first two occurrences of z; last occurrences of y, z, and x are free.

   (i) $\exists^1 x$ binds three occurrences of x; $\exists^1 z$ binds both z; $\exists^1 y$ binds both y.

   (k) $\exists^1 y$ binds first occurrence of y; $\forall^1 x$ binds last occurrence of x; free are the only occurrence of z, the first occurrence of x, and the last occurrence of y.

   (m) ($\exists^1 z$ binds both occurrences of z; $\exists^1 x$ binds first two occurrences of x; $\forall^1 x$ binds last three occurrences of x.

   (o) $\exists^1 z$ binds both occurrences of z; $\forall^1 y$ binds three occurrences of y; $\exists^1 x$ binds first, second, and third occurrences of x.

3. (a) True   (g) True
   (c) True   (i) True
   (e) False   (k) True

4. (a) and (d); (b) and (c).

5. Any two of the first three can be eliminated without loss of expressiveness.

6. (a) The clause for the $\exists^1_{>0}$ is this:

   $[\exists^1_{\geq 0} v \alpha]^M_g = T$ iff $|[\alpha]^M_{g;v}| > 0$.

   (b) There are no advantages.

3.2.1 Quantifiers as binary relations

1. (a) Yes   (g) No
   (c) Yes   (i) Yes
   (e) Yes   (k) No

2. (a) $\exists^2 z$ binds the three occurrences of z.

   (c) $M^2 z$ binds the four occurrences of z; the only occurrence of x is free.

   (e) $\forall^2 x$ binds both occurrences of x; the only occurrence of y is free.

   (g) $\exists^2 x$ binds first three occurrences of x; $\exists^2 x$ binds the next two occurrences of x; both occurrences of y and the only occurrence of z are free.

   (i) $\forall^2 z$ binds the first two and the last occurrence of z; $N^2 y$ binds first two occurrences of y; $\exists^2 z$ binds the third, fourth, and fifth occurrences of z; all occurrences of x.

3. (a) True   (i) False
   (c) False   (k) False
   (e) True   (m) True
   (g) False   (o) False

4. Partial answers:

   (a) The subset relation and the disjointness relation are conservative.

   (b) The subset relation is invariant under permutation.

   (c) The proper subset relation is left monotonic decreasing, but not right monotonic decreasing.

   (d) The overlap relation is not left monotonic decreasing, but it is left monotonic increasing.

3.2.2 Quantifiers as families of sets

1. (a) Yes   (f) No
   (c) No   (h) Yes
   (e) No   (j) Yes

3. (a) True   (i) False
   (c) False   (k) False
   (e) True   (m) True
   (g) False   (o) False

4. monotonically decreasing on the right:

(b) R is monotonically decreasing on the right iff, for each $x, y, z \in D_R$, if $x \leq y$ and $(y \in im_R(z)$, then $x \in im_R(z)$.

13 The Lambek Calculus and the Lambda Calculus

1 Introduction

In this chapter, we shall learn about two calculi: the *Lambek calculus* and the *Lambda calculus*. Before turning to the exposition of these two calculi, readers might find it useful to know what is meant by the word *calculus*[1] here. A calculus is a system for calculation, which means that it comprises a set of expressions and rules for transforming expressions in the set into other expressions in the same set. For this reason, the area of mathematics often called *analysis* by mathematicians is also called differential and integral *calculus* for it comprises a set of expressions and the rules of differentiation and integration by which one differentiates or integrates expressions in the set. The deductive part of CPL is also called the *propositional calculus*, for it too comprises a set of expressions, the formulae of CPL, and rules, the deduction rules, by which formulae are transformed into formulae. Indeed, we learned about one version of CPL known as the Gentzen sequent calculus, where the expressions transformed by the rules were not formulae, but sequents, which nonetheless themselves contain formulae.

The Lambek calculus, called a syntactic calculus by its formulator, the German born, Canadian mathematician Joachim (Jim) Lambek (1922–2014), has its origins in the work of Polish logician Kazimierz Ajdukiewicz (1935), who, inspired by ideas of Edmund Husserl (1900 bk. 5) and of Stanislaw Leśniewski, developed a calculus, a set of expressions and rules, for determining which strings of symbols in a logical notation are formulae and which are not. Though Ajdukiewicz alludes to the application of these ideas to the study of natural language, it is only with the work of the Austrian born, Israeli philosopher Yehoshua Bar-Hillel (1953) that its application to natural language is first seriously explored. Jim Lambek (1958) expanded the mathematical ideas of his predecessors into his syntactic calculus, pointing out the expanded calculus's application to the determination of which expressions in a natural language are correct expressions and which are not.

---

1. The word *calculus* in Latin means pebble. Pebbles were used by the Romans and other ancient people to do arithmetic calculation.

We shall also learn about the Lambda calculus. In the 1930s, starting with work in logic by the Austrian logician Kurt Gödel (1906–1978), a number of logicians turned their minds to trying to give a precise characterization of a computable function. Inspired by Gödel's work, the English logician, Alan Turing (1912–1954) arrived at another characterization, in which the functions defined were metaphorically called *machines* and now are called *Turing machines*. Still another characterization, due to Alonzo Church (1903–1995), an American logician, is given in terms of what he called the *Lambda calculus*.[2] The Lambda calculus comprises a set of expressions for functions, one key symbol of which is the lowercase of the Greek letter lambda (λ), and rules for transforming these expressions in expressions of the same set.

In this chapter, we shall first present the rudiments of the Lambek calculus, then the rudiments of the Lambda calculus. Finally, in preparation for the application of the Lambek and Lambda calculi to the study of natural language, we shall extend both and establish an isomorphism between them.

2 The Lambek Calculus

The Lambek calculus comprises a set of expressions, which we shall call *formulae,* and a set of rules whereby formulae are transformed into formulae, which we shall call *deduction rules*. The formulae of the Lambek calculus are obtained by forming complex expressions from a set of basic expressions and the three connectives of $\backslash$, $/$, and $\cdot$. We take our basic expressions to be the atomic formulae AF, the atomic formulae we used for CPL, though other sets could have served the same purpose. Here, then, is the formal definition of the set of all Lambek formulae, or LF.

Definition 1 Formulae of the Lambek calculus
Let AF be the atomic formula.

(1) $AF \subseteq LF$;
(2)   If $\alpha, \beta \in LF$, then $\alpha \backslash \beta, \alpha/\beta, \alpha \cdot \beta \in LF$;
(3)   Nothing else is in LF.

For example, if p, q, and r are atomic Lambek formulae, then not only are p, q, and r Lambek formulae, but so are $(p \backslash q)$, (q/r), $(r \cdot p)$, $((p \cdot q)/r)$, $((q/(q/r))$, $((p \backslash q)/(r \cdot p))$, and so on. We shall also adopt the usual convention of omitting the final pair of

---

2. Other characterizations of computable functions are one due to the Polish born, American mathematician, Emil Post (1897–1954), called *Post machines,* arrived at independently from the work of Alan Turing, as well as others due to Joachim Lambek (1922–2014), called metaphorically *abaci,* and Marvin Minsky (1927–2016), an American computer scientist, to mention but a few.

The Lambek Calculus and the Lambda Calculus

parentheses of a Lambek formula. Thus, instead of $(p\backslash q)$, $((p \cdot q)/r)$, and $((p\backslash q)/(r \cdot p))$, we shall often write $p\backslash q$, $(p \cdot q)/r$, and $(p\backslash q)/(r \cdot p)$.

We now turn to deduction in the Lambek calculus. We shall consider three presentations: the first is the presentation of formula deduction in the tree format, which is analogous to the presentation of formula natural deduction in the tree format; the second is the presentation of sequent deduction, also, in the tree format, which is analogous to the presentation of sequent natural deduction in the tree format; and the third is the presentation of Gentzen sequent deduction, again, in the tree format, which is analogous to the presentation of the Gentzen sequent calculus. Since each of the presentations of Lambek calculus given here uses the tree format, we shall usually omit this qualification hereafter.

An important question is whether these three presentations of the Lambek calculus are equivalent. We shall establish that they are. It should be noted that, in chapter 7, we could have asked the question of whether the five presentations, four natural deduction presentations and one the Gentzen sequent calculus presentation, of CPL are equivalent. To avoid burdening readers with too many questions in chapter 7, we assumed that readers would take it for granted that the five presentations are equivalent.

2.1 Formula Deduction

We begin with the definition of formula deduction (tree format) for the Lambek calculus.

Definition 2 Formula deduction for the Lambek calculus (tree format)
A deduction of a formula from a set of formulae, the premises of the deduction, is a tree of formulae each one of which is either a formula at the top of the tree, in which case it is taken from among the set of premises, or is obtained from formulae immediately above it in the formula tree by one of the rules specified below.

In analogy with natural deduction, we have six rules, an elimination and an introduction rule for each of the three basic connectives. Let us consider first the elimination and introduction rules for $\backslash$.

$\backslash$	Elimination	Introduction
	$\dfrac{\alpha \quad \alpha\backslash\beta}{\beta}$	$\dfrac{\begin{array}{c}[\,\alpha\,]\\ \vdots\\ \beta\end{array}}{\alpha\backslash\beta}$

This pair of rules is similar to the elimination and introduction rules for $\rightarrow$ given for natural deduction in the formula tree format, repeated here for ease of reference.

$\rightarrow$	Elimination	Introduction
	$\alpha, \alpha \rightarrow \beta$ $$\overline{}$$ β	$[\alpha]$ $\vdots$ β $$\overline{}$$ $\alpha \rightarrow \beta$

However, there is an important difference. We begin with the difference between the elimination rules. Formulae in the premises of the $\rightarrow$ elimination rule for formula natural deduction (tree format) are unordered with respect to one another. Indeed, the comma in the notation for the premise signals this fact. What this lack of order means is that the rule applies whether the formula tree terminating in α is to the left or the right of the formula tree terminating in $\alpha \rightarrow \beta$. In case of the $\backslash$ elimination rule, no comma occurs in the notation for the rule's premise. This signals that the order of the formula trees terminating in $\alpha \backslash \beta$ and in α makes a difference: if the formula tree terminating in $\alpha \backslash \beta$ occurs to the immediate right of the formula tree terminating in α, the rule applies; if it occurs to the immediate left, it does not.

A similar difference occurs in the case of the introduction rules. It makes no difference whether the formula α occurs to the left or the right of the formula $\alpha \rightarrow \beta$ in a deduction in CPL, but it does make a difference whether the formula α occurs to the left or the right of the formula $\alpha \backslash \beta$ in a deduction in the Lambek calculus. For the rule of $\backslash$ elimination to apply, the formula α must be the top, leftmost formula of the deduction terminating in β.

We turn next to the pair of rules for $/$. This pair of rules is just like the pair for $\backslash$, except that the order of the formulae relevant to the application of the rules is reversed. The $/$ elimination rule applies only if the formula tree terminating in β/α is to the left of the formula tree terminating in α. The $/$ introduction rule applies only if the formula α must be the top, rightmost formula of the deduction terminating in β.

$/$	Elimination	Introduction
	$\beta/\alpha \quad \alpha$ $$\overline{}$$ β	$\vdots \quad [\alpha]$ $\vdots \quad \vdots$ β $$\overline{}$$ β/α

The Lambek Calculus and the Lambda Calculus

Finally, we come to the pair of rules for $\cdot$.

$\cdot$	Elimination	Introduction
	$\begin{array}{c} \vdots \quad \alpha \cdot \beta \quad \vdots \\ \vdots \quad \rule{2cm}{0.4pt} \quad \vdots \\ \vdots \quad \alpha \quad \beta \quad \vdots \\ \vdots \quad \vdots \; \vdots \quad \vdots \\ \rule{3cm}{0.4pt} \\ \gamma \\ \rule{1cm}{0.4pt} \\ \gamma \end{array}$	$\begin{array}{c} \alpha \quad \beta \\ \rule{2cm}{0.4pt} \\ \alpha \cdot \beta \end{array}$

The $\cdot$ elimination rule is reminiscent of the $\vee$ elimination rule in natural deduction. Just as the rule of $\vee$ elimination in CPL is used to derive a formula of the form γ from a formula of the form $\alpha \vee \beta$, provided that there are formulae of the form $\alpha \to \gamma$ and $\beta \to \gamma$ available in the deduction, so the rule of $\cdot$ elimination in CPL is used to derive a formula of the form γ from a formula of the form $\alpha \cdot \beta$, provided that the formula γ is deducible both from the formula α and the formula β. Moreover, here too order is pertinent: if α is to the left of β in the formula $\alpha \cdot \beta$, then the subtree terminating in γ must have two top nodes, one labeled with α and another labelled with β with α occurring to the immediate left of β.

We now define formula deducibility in the Lambek calculus.

Definition 3 Formula deducibility for the Lambek calculus (tree format)

Let α be a formula in LF. Let Γ be a list of formulae taken from LF. Then, α is deducible in the formula tree format from Γ, or $\Gamma \vdash_{FD} \alpha$, iff there is a deduction (in the formula tree format) terminating with α such that (1) each formula at the top of the tree that is not enclosed in square brackets appears in Γ and (2) the order of appearance of the formulae unenclosed in square brackets at the top of the tree is the same as the order of the formulae listed in Γ.

Here is a number of theorems for the formula deduction presentation of the Lambek calculus.

THEOREMS

1.1 $\alpha \cdot (\beta \cdot \gamma) \vdash_{FD} (\alpha \cdot \beta) \cdot \gamma$
1.2 $(\alpha \cdot \beta) \cdot \gamma \vdash_{FD} \alpha \cdot (\beta \cdot \gamma)$
2.1 $\alpha \vdash_{FD} (\alpha \cdot \beta)/\beta$
2.2 $\beta \vdash_{FD} \alpha \backslash (\alpha \cdot \beta)$
3.1 $(\alpha/\beta) \cdot \beta \vdash_{FD} \alpha$
3.2 $\alpha \cdot (\alpha \backslash \beta) \vdash_{FD} \beta$

4.1 $\alpha \vdash_{FD} (\beta/\alpha)\backslash\beta$ 4.2 $\alpha \vdash_{FD} \beta/(\alpha\backslash\beta)$

5.1 $(\gamma/\beta)\cdot(\beta/\alpha) \vdash_{FD} \gamma/\alpha$ 5.2 $(\alpha\backslash\beta)\cdot(\beta\backslash\gamma) \vdash_{FD} \alpha\backslash\gamma$

6.1 $\gamma/\beta \vdash_{FD} (\gamma/\alpha)/(\beta/\alpha)$ 6.2 $\beta\backslash\gamma \vdash_{FD} (\alpha\backslash\gamma)\backslash(\alpha\backslash\beta)$

7.1 $(\alpha\backslash\beta)/\gamma \vdash_{FD} \alpha\backslash(\beta/\gamma)$ 7.2 $\alpha\backslash(\beta/\gamma) \vdash_{FD} (\alpha\backslash\beta)/\gamma$

8.1 $(\alpha/\beta)/\gamma \vdash_{FD} \alpha/(\gamma\cdot\beta)$ 8.2 $\alpha/(\gamma\cdot\beta) \vdash_{FD} (\alpha/\beta)/\gamma$

We now prove some of these theorems, leaving the proof of the others to readers as an exercise.

PROOF: Theorem 1.1

$$\cfrac{\cfrac{\cfrac{\cfrac{\alpha\cdot(\beta\cdot\gamma)}{\alpha}\quad \cfrac{\beta\cdot\gamma}{\beta}\ \cdot E}{\alpha\cdot\beta}\ \cdot I}{(\alpha\cdot\beta)\cdot\gamma}\ \cdot I}{}$$

PROOF: Theorem 4.1

$$\cfrac{\cfrac{[\beta/\alpha]\quad \alpha}{\beta}\ /E}{(\beta/\alpha)\backslash\beta}\ \backslash I$$

PROOF: Theorem 7.1

$$\cfrac{\cfrac{\cfrac{[\alpha]\quad \cfrac{(\alpha\backslash\beta)/\gamma\quad [\gamma]}{\alpha\backslash\beta}\ /E}{\beta}\ \backslash E}{\cfrac{\beta/\gamma}{\alpha\backslash(\beta/\gamma)}\ \backslash I}\ /I}{}$$

Exercises: Formula deductions

1. Using the formula deduction rules of this section, establish the theorems which have not already been established.

2. Use the same formula deduction rules to establish the following.

1.1 $\alpha\,\beta\,\gamma \vdash_{FD} (\alpha\cdot\beta)\cdot\gamma$ 1.2 $\alpha\,\beta\,\gamma \vdash_{FD} \alpha\cdot(\beta\cdot\gamma)$

2.1 $\alpha/\beta\,\beta \vdash_{FD} \alpha$ 2.2 $\alpha\,\alpha\backslash\beta \vdash_{FD} \beta$

3.1 $\gamma/\beta\,\beta/\alpha \vdash_{FD} \gamma/\alpha$ 3.2 $\alpha\backslash\beta\,\beta\backslash\gamma \vdash_{FD} \alpha\backslash\gamma$

2.2 Sequent Deduction

Let us turn to sequent deductions in the Lambek calculus. But before doing so, we must say something about the nature of the relation denoted by the turnstile. In chapter 7, the turnstile denoted a relation between a set of formulae on the one hand and a single formula on the other. We noted at the end of that chapter, however, that Gentzen thought of the relation denoted by the turnstile as a relation between a finite list of formulae, possible empty, and a single formula. It is this latter view that is adopted in the sequent deduction presentation of the Lambek calculus. The turnstile denotes a relation between a finite list of formulae, not empty, and a single formula.

Sequent deduction (tree format) in the Lambek calculus is defined as follows.

Definition 4 Sequent deduction for the Lambek calculus (tree format)

A deduction of a sequent is a tree of sequents each one of which either is an axiom or is obtained from sequents immediately above it in the sequent tree by one of the rules specified below.

Like the rules for sequent deduction (tree format) of CPL, which comprise an axiom and pairs of rules for each propositional connective, the rules for sequent deduction in the Lambek calculus comprise an axiom and pairs of rules for each connective.

Axiom

Axiom
$\Gamma \vdash_{SD} \alpha$
α is in Γ.

Next come the pair of rules for the elimination and introduction of $\backslash$.

$\backslash$	Elimination	Introduction
	$\Gamma \vdash_{SD} \alpha, \Delta \vdash_{SD} \alpha\backslash\beta$	$\alpha\ \Gamma \vdash_{SD} \beta$
	$\Gamma\ \Delta \vdash_{SD} \beta$	$\Gamma \vdash_{SD} \alpha\backslash\beta$

This pair of rules for the Lambek calculus is very similar to the pair of rules for the elimination and introduction or $\to$ for deductions in CPL in the sequent tree format.

$\to$	Elimination	Introduction
	$\dfrac{\Gamma \vdash_{SD} \alpha \,,\, \Delta \vdash_{SD} \alpha \to \beta}{\Gamma \cup \Delta \vdash_{SD} \beta}$	$\dfrac{\Gamma \cup \{\alpha\} \vdash_{SD} \beta}{\Gamma \vdash_{SD} \alpha \to \beta}$

We observe that the difference between the two pairs is ordered. First, what is to the left of the turnstile in the Lambek calculus is a list of formulae, whose order is that of the list, whereas what is to the left of the turnstile in CPL is a set of formulae, for which no order is defined. The order of the list is pertinent to the formulation of both rules for the \. In the case of the \ elimination rule, the left-hand side of the sequent in the conclusion of the rule comprises the list of formulae from which α is deducible followed by the list of formulae from which $\alpha \backslash \beta$ is deducible. In the case of the \ introduction rule, the left-hand side of the sequent in the conclusion of the rule is the same as the left-hand side of the sequent in the premise of the rule, except that the initial formula in the list has been removed.

It should be no surprise that the pair of rules for / are just like the pair for \, except that the points pertaining to order just mentioned are reversed. Readers should check that the comments about order do indeed apply, mutatis mutandis, to the pair of rules for \.

/	Elimination	Introduction
	$\dfrac{\Gamma \vdash_{SD} \alpha \,,\, \Delta \vdash_{SD} \beta/\alpha}{\Delta\,\Gamma \vdash_{SD} \beta}$	$\dfrac{\Gamma\,\alpha \vdash_{SD} \beta}{\Gamma \vdash_{SD} \beta/\alpha}$

Next we come to the pair of rules for · elimination and introduction.

·	Elimination	Introduction
	$\dfrac{\Delta \vdash_{SD} \alpha \cdot \beta,\, \Gamma\,\alpha\,\beta\,\Theta \vdash_{SD} \gamma}{\Gamma\,\Delta\,\Theta \vdash_{SD} \gamma}$	$\dfrac{\Gamma \vdash_{SD} \alpha,\, \Delta \vdash_{SD} \beta}{\Gamma\,\Delta \vdash_{SD} \alpha \cdot \beta}$

The Lambek Calculus and the Lambda Calculus

We add one further rule, for which there is no counterpart among the rules for the deductions in the formula tree format, namely, the the *cut* rule. It is so called, since it permits one to cut out, as it were, a formula that occurs both as the formula on the right side of a sequent's turnstile and in the list of formulae on the left side of a sequent's turnstile. This turns out to be a handy rule, which we shall use in some of the following proofs. As we shall see, this rule is dispensable: any sequent that can be derived with its aid can also be derived without it.

$$\begin{array}{|c|}\hline \text{Cut} \\ \hline \Delta \vdash_{SD} \alpha, \; \Gamma\alpha\Theta \vdash_{SD} \beta \\ \hline \Gamma\Delta\Theta \vdash_{SD} \beta \\ \hline \end{array}$$

In the preceding statement of the rules, we allow the variables ranging over a list of formulae from LF to be empty only if that assumption does not result in the list to the left of the turnstile being empty. Thus, the rule for · elimination permits either Δ or Θ or both to be empty; in all other cases, the lists are nonempty.

The analogs of the theorems for the formula deduction presentation of the Lambek calculus also holds for the sequent deduction presentation. Readers are asked to prove these theorems in the exercises.

THEOREMS

1.1 $\alpha\,\beta\,\gamma \vdash_{SD} (\alpha \cdot \beta) \cdot \gamma$ 1.2 $\alpha\,\beta\,\gamma \vdash_{SD} \alpha \cdot (\beta \cdot \gamma)$

2.1 $\alpha \vdash_{SD} (\alpha \cdot \beta)/\beta$ 2.2 $\beta \vdash_{SD} \alpha\backslash(\alpha \cdot \beta)$

3.1 $\alpha/\beta\,\beta \vdash_{SD} \alpha$ 3.2 $\alpha\,\alpha\backslash\beta \vdash_{SD} \beta$

4.1 $\alpha \vdash_{SD} (\beta/\alpha)\backslash\alpha$ 4.2 $\alpha \vdash_{SD} \beta/(\alpha\backslash\beta)$

5.1 $\gamma/\beta\,\beta/\alpha \vdash_{SD} \gamma/\alpha$ 5.2 $\alpha\backslash\beta\,\beta\backslash\gamma \vdash_{SD} \alpha\backslash\gamma$

6.1 $\gamma/\beta \vdash_{SD} (\gamma/\alpha)/(\beta/\alpha)$ 6.2 $\beta\backslash\gamma \vdash_{SD} (\alpha\backslash\gamma)\backslash(\alpha\backslash\beta)$

7.1 $(\alpha\backslash\beta)/\gamma \vdash_{SD} \alpha\backslash(\beta/\gamma)$ 7.2 $\alpha\backslash(\beta/\gamma) \vdash_{SD} (\alpha\backslash\beta)/\gamma$

8.1 $(\alpha/\beta)/\gamma \vdash_{SD} \alpha/(\gamma \cdot \beta)$ 8.2 $\alpha/(\gamma \cdot \beta) \vdash_{SD} (\alpha/\beta)/\gamma$

PROOF: Theorem 2.1

$$\cfrac{\cfrac{\alpha \vdash_{SD} \alpha \quad \beta \vdash_{SD} \beta}{\alpha\,\beta \vdash_{SD} \alpha \cdot \beta}\, \cdot I}{\alpha \vdash_{SD} (\alpha \cdot \beta)/\beta}\, /I$$

To be complete, the next proof requires that the proofs for theorems 3.1 and 4.1 be placed above the starting points of what is displayed.

PROOF: Theorem 5.1

$$\frac{\frac{\vdots}{\beta/\alpha\ \alpha \vdash_{SD} \beta}\ /E \quad \frac{\vdots}{\beta \vdash_{SD} (\gamma/\beta)\backslash\gamma}\ \backslash I}{\frac{(\beta/\alpha)\ \alpha \vdash_{SD} (\gamma/\beta)\backslash\gamma}{\frac{(\gamma/\beta)\ \beta/\alpha\ \alpha \vdash_{SD} \gamma}{\gamma/\beta\ \beta/\alpha \vdash_{SD} \gamma/\alpha}\ /I}\ \backslash E}\ cut$$

PROOF: Theorem 8.1

$$\frac{\gamma \cdot \beta \vdash_{SD} \gamma \cdot \beta}\ axiom \quad \frac{\frac{\frac{(\alpha/\beta)/\gamma \vdash_{SD} (\alpha/\beta)/\gamma}\ axiom}{(\alpha/\beta)/\gamma\ \gamma \vdash_{SD} \alpha/\beta}\ /E}{\frac{(\alpha/\beta)/\gamma\ \gamma\ \beta \vdash_{SD} \alpha}\ /E}$$
$$\frac{(\alpha/\beta)/\gamma\ \gamma \cdot \beta \vdash_{SD} \alpha}{(\alpha/\beta)/\gamma \vdash_{SD} \alpha/(\gamma \cdot \beta)}\ /I$$

Exercises: Sequent deductions

1. Using the sequent deduction rules, establish the theorems not established in this section.
2. Use the sequent deduction rules to establish the following.

    1.1 $\alpha \cdot (\beta \cdot \gamma) \vdash_{SD} (\alpha \cdot \beta) \cdot \gamma$ 1.2 $(\alpha \cdot \beta) \cdot \gamma \vdash_{SD} \alpha \cdot (\beta \cdot \gamma)$

    2.1 $(\alpha/\beta) \cdot \beta \vdash_{SD} \alpha$ 2.2 $\alpha \cdot (\alpha\backslash\beta) \vdash_{SD} \beta$

    3.1 $(\gamma/\beta) \cdot (\beta/\alpha) \vdash_{SD} \gamma/\alpha$ 3.2 $(\alpha\backslash\beta) \cdot (\beta\backslash\gamma) \vdash_{SD} \alpha\backslash\gamma$

2.2.1 Equivalence of formula deduction and sequent deduction

We now wish to show that the formula deduction presentation of the Lambek calculus is equivalent to the sequent deduction presentation. What we mean by equivalence is that the very same pairs comprising a finite list of formulae and a single formula are paired by $\vdash_{FD}$ as are paired by $\vdash_{SD}$: in other words, that, for each Θ, a finite list of formulae, and for each θ, a formula, $\Theta \vdash_{FD} \theta$ iff $\Theta \vdash_{SD} \theta$. We shall prove this in two long steps. The first long step is to show that, for each Θ, a finite list of formulae, and for each θ, a formula, if $\Theta \vdash_{FD} \theta$, then $\Theta \vdash_{SD} \theta$. The second is to show the converse: for each Θ, a finite list of formulae, and for each θ, a formula, if $\Theta \vdash_{SD} \theta$, then $\Theta \vdash_{FD} \theta$.

FIRST LONG STEP

To take the first long step, we must first introduce the concept of the depth of a tree. Each point where an expression, be it a formula or a sequent, occurs in a tree is a *node*. A *branch* in a tree is a sequence of all the nodes from a node in the tree, to one node immediately above it, and so forth, up to a top node. Thus, a tree has, from its bottom node, precisely as many branches as top nodes. The number of nodes in a branch is the branch's length. The *depth* of a tree is the length of its longest branch.[3]

The concept of the depth of a tree permits us to put into one group all trees of depth 1, into another all trees of depth 2, and in general, in its own group all trees of depth n, a positive integer.

Next, we recall that $\Theta \vdash_{FD} \theta$ holds iff there is a formula deduction tree whose top formulae, from left to right, are the formulae listed in Θ and whose bottom formula is θ. Now, should we show that, for each instance of $\Theta \vdash_{FD} \theta$ that is underpinned by a formula deduction tree of depth 1, there is a sequent deduction whose bottom sequent is $\Theta \vdash_{SD} \theta$, and that, for each instance of $\Theta \vdash_{FD} \theta$ that is underpinned by a formula deduction tree of depth 2, there is a sequent deduction whose bottom sequent is $\Theta \vdash_{SD} \theta$, and that, for each instance of $\Theta \vdash_{FD} \theta$ that is underpinned by a formula deduction tree of depth 3, there is a sequent deduction whose bottom sequent is $\Theta \vdash_{SD} \theta$, and indeed that, for each instance of $\Theta \vdash_{FD} \theta$ that is underpinned by a formula deduction tree of depth n, there is a sequent deduction whose bottom sequent is $\Theta \vdash_{SD} \theta$, then it follows that, for each Θ, a finite list of formulae, and for each θ, a formula, if $\Theta \vdash_{FD} \theta$, then $\Theta \vdash_{SD} \theta$.

It is not possible to write out a proof of what we want by writing out a proof starting from the case of formula deduction trees of depth 1, then the case of formula deduction trees of depth 2, and so on, since we shall never come to the end. What we can, and must, do is to apply a kind of proof that uses the principle known as the *principle of mathematical induction*. To apply this principle here means that we first show that, for each instance of $\Theta \vdash_{FD} \theta$ that is underpinned by formula deduction of depth 1, we have a sequent deduction of $\Theta \vdash_{SD} \theta$. This part of the proof is known as this proof's *base case*. Next, we show that, if for each instance of $\Theta \vdash_{FD} \theta$ that is underpinned by formula deduction of depth n or less, we have a sequent deduction of $\Theta \vdash_{SD} \theta$, then for each instance of $\Delta \vdash_{FD} \beta$ that is underpinned by formula deduction of depth $n + 1$ or less, we have a sequent deduction of $\Delta \vdash_{SD} \beta$. This part of the proof is known as this proof's *induction case*. Once the base case and the induction case have been completed, we shall invoke the principle of mathematical induction to conclude that, for each Θ, a finite list of formulae, and for each θ, a formula, if $\Theta \vdash_{FD} \theta$, then $\Theta \vdash_{SD} \theta$.

---

3. On this definition, a tree may have more than one longest branch. Of course, if there are two longest branches, they have the same length.

BASE CASE

$\Theta \vdash_{FD} \theta$ results from a formula deduction of depth 1. Such a formula deduction comprises a single formula, call it α. Thus, $\Theta \vdash_{FD} \theta$ is just $\alpha \vdash_{FD} \alpha$. However, $\alpha \vdash_{SD} \alpha$ is an axiom of sequent deduction. We have thereby proved the base case.

INDUCTION CASE

INDUCTION HYPOTHESIS

For each $\Theta \vdash_{FD} \theta$ that results from a formula deduction of depth n or less (where $n > 1$), there is a sequent deduction of $\Theta \vdash_{SD} \theta$.

Now consider any formula deduction of depth $n + 1$. This deduction must result from the application of one of the six rules of formula deduction to two formula deductions, one of depth n, the other of depth no greater than n. By the induction hypothesis, each of these two deductions results in pair of a formula list and a single formula for which there is a sequent deduction. We then use these two sequent deductions to deduce the sequent corresponding to the pair of formula list and single formula established by the formula deduction of depth $n + 1$. Since we must show this to be so no matter which of the six formula deduction rules are used in the last step of the formula deduction of depth $n + 1$, we have six cases to consider. We now consider each of these cases in turn.

CASE: \ Elimination

Suppose that $\Theta \vdash_{FD} \theta$ results from a formula deduction the last applied rule of which is \ Elimination. Then the underlying formula deduction has the following form:

$$\begin{array}{cc} \Gamma & \Delta \\ \vdots & \vdots \\ \alpha & \alpha\backslash\beta \\ \hline & \beta \end{array}$$

where Θ is $\Gamma \, \Delta$ and θ is β. By the induction hypothesis, we know that there are sequent deductions of $\Gamma \vdash_{SD} \alpha$ and of $\Delta \vdash_{SD} \alpha\backslash\beta$. We can use these two sequent deductions to obtain a sequent deduction of $\Gamma \, \Delta \vdash_{SD} \beta$, as follows.

$$\cfrac{\cfrac{\vdots}{\Gamma \vdash_{SD} \alpha} \, IH \quad \cfrac{\vdots}{\Delta \vdash_{SD} \alpha\backslash\beta} \, IH}{\Gamma \, \Delta \vdash_{SD} \beta} \, \backslash E$$

CASE: \ Introduction

Suppose that $\Theta \vdash_{FD} \theta$ results from a formula deduction the last applied rule of which is \ Introduction. Then the underlying formula deduction has this form:

The Lambek Calculus and the Lambda Calculus

$$
\frac{\begin{array}{cc} [\alpha] & \Delta \\ \vdots & \vdots \\ \beta \end{array}}{\alpha\backslash\beta}
$$

where Θ is Δ and θ is $\alpha\backslash\beta$. By the induction hypothesis, we know that there is a sequent deduction of $\alpha\ \Delta \vdash_{SD} \beta$. We can use this sequent deduction to obtain a sequent deduction of $\Delta \vdash_{SD} \alpha\backslash\beta$, as follows.

$$
\frac{\dfrac{\vdots}{\alpha\ \Delta \vdash_{SD} \beta}\ IH}{\Delta \vdash_{SD} \alpha\backslash\beta}\ \backslash I
$$

CASE: / Elimination

This case will be solved as an exercise.

CASE: / Introduction

This case will be solved as an exercise.

CASE: · Elimination

Suppose that $\Theta \vdash_{FD} \theta$ results from a formula deduction the last applied rule of which is · Elimination. Then the underlying formula deduction is this:

$$
\begin{array}{ccc} \Gamma & \alpha\cdot\beta & \Delta \\ \vdots & \overline{} & \vdots \\ \vdots & \alpha\quad\beta & \vdots \\ \vdots & \vdots\quad\vdots & \vdots \\ \hline & \gamma & \\ & \overline{\gamma} & \end{array}
$$

where Θ is $\Gamma\ \alpha\cdot\beta\ \Delta$ and θ is γ. By the induction hypothesis, we know that there is a sequent deduction, one of $\Gamma\ \alpha\ \beta\ \Delta \vdash_{SD} \gamma$. We can use this sequent deduction to obtain a sequent deduction of $\Gamma\ \alpha\cdot\beta\ \Delta \vdash_{SD} \gamma$, as follows.

$$
\frac{\dfrac{}{\alpha\cdot\beta \vdash_{SD} \alpha\cdot\beta}\ axiom \quad \dfrac{\vdots}{\Gamma\ \alpha\ \beta\ \Delta \vdash_{SD} \gamma}\ IH}{\Gamma\ \alpha\cdot\beta\ \Delta \vdash_{SD} \gamma}\ \cdot E
$$

CASE: · Introduction

Suppose that $\Theta \vdash_{FD} \theta$ results from a formula deduction the last applied rule of which is · Introduction. Then the form of the underlying formula deduction is this:

$$
\begin{array}{cc}
\Gamma & \Delta \\
\vdots & \vdots \\
\alpha & \beta \\
\hline
\multicolumn{2}{c}{\alpha \cdot \beta}
\end{array}
$$

where Θ is $\Gamma \, \Delta$ and θ is $\alpha \cdot \beta$. By the induction hypothesis, we know that there are two sequent deductions, one of $\Gamma \vdash_{SD} \alpha$ and another of $\Delta \vdash_{SD} \beta$. We can use these sequent deductions to obtain a sequent deduction of $\Gamma \, \Delta \vdash_{SD} \alpha \cdot \beta$, as follows.

$$
\cfrac{\cfrac{\vdots}{\Gamma \vdash_{SD} \alpha} \, IH \quad \cfrac{\vdots}{\Delta \vdash_{SD} \beta} \, IH}{\Gamma \, \Delta \vdash_{SD} \alpha \cdot \beta} \, \cdot I
$$

This completes the proof of the induction case, which itself comprises six subcases.

We now invoke the principle of mathematical induction to conclude that, for each Θ, a finite list of formulae, and for each θ, a formula, if $\Theta \vdash_{FD} \theta$, then $\Theta \vdash_{SD} \theta$.

SECOND LONG STEP

The second step is to show that for each Γ, a finite list of formulae, and for each α, a formula, if $\Gamma \vdash_{SD} \alpha$, then $\Gamma \vdash_{FD} \alpha$. The way to do this is to show that, besides the fact that $\alpha \vdash_{FD} \alpha$ holds, the analog of each of the deduction rules for sequent deduction holds for formula deductions; that it to say, each of the following holds.

DERIVED RULES

$$\cfrac{\Gamma \vdash_{FD} \alpha \quad \Delta \vdash_{FD} \alpha\backslash\beta}{\Gamma \, \Delta \vdash_{FD} \beta} \, DR \, \backslash E \qquad \cfrac{\alpha \, \Gamma \vdash_{FD} \beta}{\Gamma \vdash_{FD} \alpha\backslash\beta} \, DR \, \backslash I$$

$$\cfrac{\Gamma \vdash_{FD} \beta/\alpha \quad \Delta \vdash_{FD} \alpha}{\Gamma \, \Delta \vdash_{FD} \beta} \, DR \, /E \qquad \cfrac{\Gamma \, \alpha \vdash_{FD} \beta}{\Gamma \vdash_{FD} \beta/\alpha} \, DR \, /I$$

$$\cfrac{\Delta \vdash_{FD} \alpha \cdot \beta \quad \Gamma \, \alpha \, \beta \, \Theta \vdash_{FD} \gamma}{\Gamma \, \Delta \, \Theta \vdash_{FD} \gamma} \, DR \cdot E \qquad \cfrac{\Gamma \vdash_{FD} \alpha \quad \Delta \vdash_{FD} \beta}{\Gamma \, \Delta \vdash_{FD} \alpha \cdot \beta} \, DR \cdot I$$

We already saw that because each formula comprises a formula deduction of the formula from itself, $\alpha \vdash_{FD} \alpha$ holds. We now turn to the derived rules.

The Lambek Calculus and the Lambda Calculus

PROOF: $DR \setminus E$

Suppose that $\Gamma \vdash_{FD} \alpha$ and that $\Delta \vdash_{FD} \beta$. By the definition of $\vdash_{FD}$, there are two formula deductions, one with the list of formulae in Γ as its top formulae and with α as its bottom formula, the other with the list of formulae in Δ as its top formulae and with $\alpha \setminus \beta$ as its bottom formula. These two formula deductions may be used to arrive at the follow formula deduction, whose top formulae comprise the list $\Gamma\ \Delta$ and whose bottom formula is β.

$$\begin{array}{cc} \Gamma & \Delta \\ \vdots & \vdots \\ \alpha & \beta \setminus \alpha \\ \hline & \beta \end{array}$$

PROOF: $DR \setminus I$

This will be proved as an exercise.

PROOF: DR / E

This will be proved as an exercise.

PROOF: DR / I

Suppose that $\Gamma\ \alpha \vdash_{FD} \beta$. By the definition of $\vdash_{FD}$, there is a formula deduction the list of whose top formula from left to right is $\Gamma\ \alpha$ and whose bottom formula is β. This formula deduction may be extended to one the list of whose top formula from left to right is Γ and whose bottom formula is β/α, as shown here.

$$\begin{array}{cc} \Gamma & [\alpha] \\ \vdots & \vdots \\ & \beta \\ \hline & \beta/\alpha \end{array}$$

PROOF: $DR \cdot E$

This will be proved as an exercise.

PROOF: $DR \cdot I$

This will be proved as an exercise.

Once readers supply a proof for this last derived rule, they will have shown that, for each Γ, a finite list of formulae, and for each α, a formula, if $\Gamma \vdash_{SD} \alpha$, then $\Gamma \vdash_{FD} \alpha$.

Exercises: Equivalence of formula deduction and sequent deduction

1. Show that for each positive integer n, there is a formula deduction tree of depth n.
2. Supply the proofs signaled as exercises in this section.

2.3 Gentzen Sequent Deduction

At last we come to Gentzen sequent deduction for the Lambek calculus.

Definition 5 Gentzen sequent deduction for the Lambek calculus (tree format)
A deduction of a sequent is a tree of sequents each one of which either is an axiom or is obtained from sequents immediately above it in the sequent tree by one of the rules specified below.

As we recall from chapter 7, the Gentzen sequent calculus uses the same axiom and introduction rules as the deduction rules for the sequent tree format, but it replaces the elimination rules with another set of introduction rules. What are called introduction rules in the sequent tree format are relabeled as right introduction rules in the Gentzen sequent calculus and the new set of introduction rules are called left introduction rules. An analogous change is made here with respect to the Lambek calculus.

Here, then, is the axiom for the Gentzen sequent deduction presentation of the Lambek calculus.

Axiom
$\Gamma \vdash_{GD} \alpha$
α is in Γ.

Here are the introduction rules for the Gentzen sequent deduction of the Lambek calculus.

\	Left	Right
	$\Gamma \vdash_{GD} \alpha,\ \Delta\ \beta\ \Theta \vdash_{GD} \gamma$	$\alpha\ \Gamma \vdash_{GD} \beta$
	$\Delta\ \Gamma\ \alpha\backslash\beta\ \Theta \vdash_{GD} \gamma$	$\Gamma \vdash_{GD} \alpha\backslash\beta$

The Lambek Calculus and the Lambda Calculus

/	Left	Right
	$\Gamma \vdash_{GD} \alpha,\ \Delta\ \beta\ \Theta \vdash_{GD} \gamma$	$\Gamma\ \alpha \vdash_{GD} \beta$
	$\Delta\ \beta/\alpha\ \Gamma\ \Theta \vdash_{GD} \gamma$	$\Gamma \vdash_{GD} \beta/\alpha$

·	Left	Right
	$\Gamma\ \alpha\ \beta\ \Delta \vdash_{GD} \gamma$	$\Gamma \vdash_{GD} \alpha,\ \Delta \vdash_{GD} \beta$
	$\Gamma\ \alpha \cdot \beta\ \Delta \vdash_{GD} \gamma$	$\Gamma\ \Delta \vdash_{GD} \alpha \cdot \beta$

Cut
$\Delta \vdash_{GD} \alpha,\ \Gamma \alpha \Theta \vdash_{GD} \beta$
$\Gamma \Delta \Theta \vdash_{GD} \beta$

The analogs of the theorems for the formula deduction of the Lambek calculus also holds for the Gentzen sequent deduction. Readers are asked to prove these theorems in the exercises.

THEOREMS

1.1 $\alpha \cdot (\beta \cdot \gamma) \vdash_{GD} (\alpha \cdot \beta) \cdot \gamma$
1.2 $(\alpha \cdot \beta) \cdot \gamma \vdash_{GD} \alpha \cdot (\beta \cdot \gamma)$

2.1 $\alpha \vdash_{GD} (\alpha \cdot \beta)/\beta$
2.2 $\beta \vdash_{GD} \alpha\backslash(\alpha \cdot \beta)$

3.1 $\alpha/\beta\ \beta \vdash_{GD} \alpha$
3.2 $\alpha\ \alpha\backslash\beta \vdash_{GD} \beta$

4.1 $\alpha \vdash_{GD} (\beta/\alpha)\backslash\alpha$
4.2 $\alpha \vdash_{GD} \beta/(\alpha\backslash\beta)$

5.1 $\gamma/\beta\ \beta/\alpha \vdash_{GD} \gamma/\alpha$
5.2 $\alpha\backslash\beta\ \beta\backslash\gamma \vdash_{GD} \alpha\backslash\gamma$

6.1 $\gamma/\beta \vdash_{GD} (\gamma/\alpha)/(\beta/\alpha)$
6.2 $\beta\backslash\gamma \vdash_{GD} (\alpha\backslash\gamma)\backslash(\alpha\backslash\beta)$

7.1 $(\alpha\backslash\beta)/\gamma \vdash_{GD} \alpha\backslash(\beta/\gamma)$
7.2 $\alpha\backslash(\beta/\gamma) \vdash_{GD} (\alpha\backslash\beta)/\gamma$

8.1 $(\alpha/\beta)/\gamma \vdash_{GD} \alpha/(\gamma \cdot \beta)$
8.2 $\alpha/(\gamma \cdot \beta) \vdash_{GD} (\alpha/\beta)/\gamma$

PROOF: Theorem 3.1

$$\dfrac{\dfrac{}{\alpha \vdash_{GD} \alpha}\ axiom \quad \dfrac{}{\beta \vdash_{GD} \beta}\ axiom}{(\alpha/\beta)\ \beta \vdash \alpha}\ /L$$

PROOF: Theorem 6.1

$$\cfrac{\cfrac{\cfrac{\cfrac{\overline{\gamma \vdash_{GD} \gamma}\ axiom \quad \overline{\beta \vdash_{GD} \beta}\ axiom}{\gamma/\beta\ \beta \vdash \gamma}\ /L \quad \overline{\alpha \vdash_{GD} \alpha}\ axiom}{\gamma/\beta\ \beta/\alpha\ \alpha \vdash \gamma}\ /L}{\gamma/\beta\ \beta/\alpha \vdash \gamma/\alpha}\ /R}{\gamma/\beta \vdash (\gamma/\alpha)/\beta/\alpha}\ /R$$

Exercises: Gentzen deductions

1. Using the Gentzen deduction rules, establish the theorems not established in this section.
2. Use the Gentzen deduction rules to establish the following.

1.1 $\alpha \cdot (\beta \cdot \gamma) \vdash_{GD} (\alpha \cdot \beta) \cdot \gamma$ 1.2 $(\alpha \cdot \beta) \cdot \gamma \vdash_{GD} \alpha \cdot (\beta \cdot \gamma)$

2.1 $(\alpha/\beta) \cdot \beta \vdash_{GD} \alpha$ 2.2 $\alpha \cdot (\alpha\backslash\beta) \vdash_{GD} \beta$

3.1 $(\gamma/\beta) \cdot (\beta/\alpha) \vdash_{GD} \gamma/\alpha$ 3.2 $(\alpha\backslash\beta) \cdot (\beta\backslash\gamma) \vdash_{GD} \alpha\backslash\gamma$

2.3.1 Equivalence of sequent deduction and Gentzen deduction

As we are about to see, the sequent deduction presentation of the Lambek calculus is equivalent to the Gentzen deduction presentation: that is, for each Θ, a finite list of formulae, and for each θ, a formula, $\Theta \vdash_{SD} \theta$ iff $\Theta \vdash_{GD} \theta$.

As with the proof of the equivalence of the formula deduction presentation and the sequent deduction presentation of the Lambek calculus, we proceed in two steps. The first step is to show that, for each Θ, a finite list of formulae, and for each θ, a formula, if $\Theta \vdash_{SD} \theta$, then $\Theta \vdash_{GD} \theta$. The second is to show the converse: for each Θ, a finite list of formulae, and for each θ, a formula, if $\Theta \vdash_{GD} \theta$, then $\Theta \vdash_{SD} \theta$. Since both presentations share half of their rules, these two steps will be shorter than their earlier counter parts.

FIRST STEP

PROOF: $DR \backslash E$

$$\cfrac{\Delta \vdash_{GD} \alpha\backslash\beta\ H \quad \cfrac{\Gamma \vdash_{GD} \alpha\ H \quad \cfrac{\overline{\alpha \vdash_{GD} \alpha}\ axiom \quad \overline{\beta \vdash_{GD} \beta}\ axiom}{\alpha\ \alpha\backslash\beta \vdash \beta}\ \backslash L}{\Gamma\ \alpha\backslash\beta \vdash \beta}\ cut}{\Gamma\ \Delta \vdash_{GD} \beta}\ cut$$

PROOF: DR /E

Exercise.

The Lambek Calculus and the Lambda Calculus

PROOF: $DR \cdot E$

$$\dfrac{\Delta \vdash_{GD} \alpha \cdot \beta \quad H \quad \dfrac{\dfrac{\Gamma\,\alpha\,\beta\,\Theta \vdash_{GD} \gamma}{\Gamma\,\alpha \cdot \beta\,\Theta \vdash_{GD} \gamma}\ \cdot L \quad H}{}}{\Gamma\,\Delta\,\Theta \vdash_{GD} \gamma}\ cut$$

SECOND STEP

The second step requires two rules that show that a finite list and a formula that is the $\cdot$ conjunction, as it were, of the formula are equivalent. The rule labeled $\cdot LI$ is a generalization of $\cdot L$ and the rule labeled $\cdot LE$ is a derived rule that results in a generalized version of $\cdot E$ on the left side of the turnstile:

$$\dfrac{\Gamma\,\alpha_1\,\alpha_2 \ldots \alpha_n\,\Delta \vdash_{GD} \beta}{\Gamma\,(\ldots(\alpha_1 \cdot \alpha_2)\ldots) \cdot \alpha_n\,\Delta \vdash_{GD} \beta}\ DR \cdot LI$$

$$\dfrac{\Gamma\,(\ldots(\alpha_1 \cdot \alpha_2)\ldots) \cdot \alpha_n\,\Delta \vdash_{GD} \beta}{\Gamma\,\alpha_1\,\alpha_2 \ldots \alpha_n\,\Delta \vdash_{GD} \beta}\ DR \cdot LE$$

where both Γ and Δ may be empty lists.

These rules must be proved using the principle of mathematical induction. The base case considers an instance with just two formulae, α_1 and α_2.[4] The induction case hypothesizes that the rule holds for n formulae, α_1 through α_n, and shows that the rule also holds for $n+1$ formulae, α_1 through α_{n+1}.

PROOF: $DR \cdot LI$

BASE CASE

$$\dfrac{\dfrac{\overline{\alpha_1 \vdash_{GD} \alpha_1}\ axiom \quad \overline{\alpha_2 \vdash_{GD} \alpha_2}\ axiom}{\alpha_1\,\alpha_2 \vdash_{GD} \alpha_1 \cdot \alpha}\ \cdot R \quad \dfrac{\vdots}{\Gamma\,\alpha_1 \cdot \alpha_2\,\Delta \vdash_{GD} \beta}\ H}{\Gamma\,\alpha_1\,\alpha_2\,\Delta \vdash_{GD} \beta}\ cut$$

INDUCTION CASE

$$\dfrac{\dfrac{\overline{\alpha_1 \ldots \alpha_n \vdash_{GD} (\ldots(\alpha_1 \cdot \alpha_2)\ldots) \cdot \alpha_n}\ H \quad \overline{\alpha_{n+1} \vdash_{GD} \alpha_{n+1}}\ axiom}{\alpha_1 \ldots \alpha_{n+1} \vdash_{GD} (\ldots((\alpha_1 \cdot \alpha_2)\ldots) \cdot \alpha_n) \cdot \alpha_{n+1}}\ \cdot R \quad \dfrac{\vdots}{\Gamma\,\alpha_1 \ldots \alpha_{n+1}\,\Delta \vdash_{GD} \beta}\ H}{\Gamma\,\alpha_1 \ldots \alpha_{n+1}\,\Delta \vdash_{GD} \beta}\ cut$$

PROOF: $DR \cdot LE$
Exercise.

---

4. Those familiar with proofs using the principle of mathematical induction know that one may take as the base case an instance of just one formula, α_1.

PROOF: $DR \setminus L$

$$\cfrac{\Delta \vdash_{GD} \alpha\setminus\beta \quad H \quad \cfrac{\cfrac{\Gamma \vdash_{GD} \alpha}{\Gamma \alpha\setminus\beta \vdash \beta} \; H \quad \cfrac{\overline{\alpha \vdash_{GD} \alpha} \; axiom \quad \overline{\beta \vdash_{GD} \beta} \; axiom}{\alpha \; \alpha\setminus\beta \vdash \beta} \setminus L}{\Gamma \alpha\setminus\beta \vdash \beta} \; cut}{\cfrac{\gamma/\beta \; \beta/\alpha \vdash \gamma/\alpha}{\Delta \; \Gamma \vdash_{GD} \beta} \; /R} \; cut$$

PROOF: DR / L

Exercise.

PROOF: $DR \cdot L$

$$\cfrac{\Delta \vdash_{GD} \alpha\cdot\beta \quad H \quad \cfrac{\overline{\Gamma \; \alpha \; \beta \; \Theta \vdash_{GD} \gamma} \; H}{\Gamma \; \alpha\cdot\beta \; \Theta \vdash_{GD} \gamma} \cdot L}{\Gamma \; \Delta \; \Theta \vdash_{GD} \gamma} \; cut$$

Exercises: Equivalence of sequent deduction and Gentzen deduction

1. Prove, without availing yourself of the equivalences discussed, that the formula deduction presentation of the Lambek calculus is equivalent to the Gentzen deduction presentation: that is, for each Θ, a finite list of formulae, and for each θ, a formula, $\Theta \vdash_{FD} \theta$ iff $\Theta \vdash_{GD} \theta$.

2.4 Cut Elimination

We now turn to the cut elimination theorem for the Lambek calculus. The cut elimination theorem, as its name suggests, asserts that the cut rule can be eliminated. In other words, for any deduction from axioms alone of a sequent in the Lambek calculus in which it is applied, there is a corresponding deduction from axioms of the very same sequent in which the cut rule is not applied at all. Or, to put it another way, every sequent theorem of the Lambek calculus can be proved without any application of the cut rule.

The proof proceeds by induction on the degree of a cut. The degree of a cut is the degree of the constituents of the rule. Recall the form of the cut rule.

$$\cfrac{\Gamma \vdash \alpha \quad \Delta \; \alpha \; \Theta \vdash \beta}{\Delta \; \Gamma \; \Theta \vdash \beta} \; cut$$

The constituents of an application of the cut rule are Γ, Δ, Θ, α, and β. The degree of a formula is the number of appearances of a connective in the formula and the degree of a list is the sum of the degrees of the formulae in the list. The Lambek calculus has,

The Lambek Calculus and the Lambda Calculus

besides the cut rule, six rules and one axiom. The six rules comprise three R rules, one for each connective, and three L rules, again one for each connective. The proof is to show that, for any application of the cut rule in a deduction, either the cut rule can be eliminated altogether or it can be replaced by other applications of the cut rule of lesser degree. Moreover, the resulting deduction has no greater depth and no greater number of nodes. This means that, by repeated elimination of applications of the cut rule or by repeated reductions of the degree of any application of the cut rule, any proof applying the cut rule can be turned into one that does not apply it at all; this can be done in a finite number of eliminations or reductions. In total, there are fourteen cases to consider. There are seven cases to consider for the left-hand premise of the cut rule and seven cases for the right-hand premise. The seven cases comprise the case when the premise in question is an axiom, the three cases where the premise, is the last step in a subdeduction resulting from an application of an L rule and the three cases where the premise is the last step in a subdeduction resulting from an application of an R rule.

We shall start with the two cases in which one of the premises is an axiom.

CASE 1.1

We begin by instantiating the left premise as the axiom $\gamma \vdash \gamma$.[5] This means that γ is substituting for α as well as for Γ. The result is the following:

$$\frac{\gamma \vdash \gamma \quad \Delta \, \gamma \, \Theta \vdash \beta}{\Delta \, \gamma \, \Theta \vdash \beta} \; cut$$

The conclusion of this application of the cut rule is identical with its right-hand premise. Thus, this application of the cut rule is entirely superfluous: any deduction in which such a configuration is found can be shortened by excising both the left- and right-hand sides.

CASE 1.2

Next, we instantiate the right premise as the axiom $\gamma \vdash \gamma$. This means that Δ and Θ are the empty list and that α and β are γ. One thereby obtains the following instantiation.

$$\frac{\Gamma \vdash \gamma \quad \gamma \vdash \gamma}{\Gamma \vdash \gamma} \; cut$$

This time the conclusion of the application of the cut rule is identical with its left-hand premise. Again, the application of the cut rule is entirely superfluous.

We now turn to the cases where one of the premises results from one of the rules. We shall consider initially the cases where the L rules apply, considering first the three cases

---

5. The two premises are unordered with respect to one another so that, technically speaking, there are no left- and right-hand premises. However, we shall persist in distinguishing the two premises as left and right, thinking of the particular expression of the cut rule stated earlier.

where their application results in the left-hand premise and then the three cases where their application results in the right-hand premise.

CASE 2.1: Application of the L rules

CASE 2.1.1: Left-hand premise

We have three subcases to consider, corresponding to applications of / L rule, \ L rule, and the · L rule, which yield the left-hand premise.

SUBCASE: / L rule (left-hand premise)

Let the left-hand premise, which is of the form $\Gamma \vdash \alpha$, result from an application of the / L rule. Then its list has a formula such as δ/γ. Furthermore, because the left-hand premise results from an application of / L, it itself is immediately dominated by two premises, one of the form $\Psi \vdash \gamma$ and the other of the form $\Upsilon \, \delta \, \Lambda \vdash \alpha$. These premises, in turn, require Γ in the left-hand premise to have the form $\Upsilon \, \delta/\gamma \, \Psi \, \Lambda$. Thus, the left-hand premise has the form $\Upsilon \, \delta/\gamma \, \Psi \, \Lambda \vdash \alpha$. All this implies that the application of the cut rule that we are considering appears in the following configuration.

$$\frac{\dfrac{\Psi \vdash \gamma \quad \Upsilon \, \delta \, \Lambda \vdash \alpha}{\Upsilon \, \delta/\gamma \, \Psi \, \Lambda \vdash \alpha} /L \quad \Delta \, \alpha \, \Theta \vdash \beta}{\Delta \, \Upsilon \, \delta/\gamma \, \Psi \, \Lambda \, \Theta \vdash \beta} \text{ cut}$$

We can now rearrange this portion of any longer deduction in which the previous configuration appears into the same deduction, but replacing the previous portion by the following.

$$\frac{\Psi \vdash \gamma \quad \dfrac{\Upsilon \, \delta \, \Lambda \vdash \alpha \quad \Delta \, \alpha \, \Theta \vdash \beta}{\Delta \, \Upsilon \, \delta \, \Lambda \, \Theta \vdash \beta} \text{ cut}}{\Delta \, \Upsilon \, \delta/\gamma \, \Psi \, \Lambda \, \Theta \vdash \beta} /L$$

REMARKS

1. Each deduction tree has the same depth.

2. Each deduction tree has the same number of nodes.

3. Each deduction tree has exactly the same number of applications of the cut rule.

4. The application of cut rule in the portion of the second deduction is higher in its deduction tree than the application of the cut rule in the portion of the first deduction tree.

5. Moreover, the degree of the application of the cut rule in the portion of the first deduction tree is greater than the degree of the application of the cut rule in the portion of the second deduction tree. Since $d(\delta/\gamma) = d(\delta) + d(\gamma) + 1$, $d(\Delta \, \Theta \, \Psi \, \Upsilon \, \Lambda \, \alpha \, \beta \, \delta/\gamma) > d(\Delta \, \Theta \, \Upsilon \, \Lambda \, \alpha \, \beta \, \delta)$.

SUBCASE: \ L rule (left-hand premise)

Let $\Gamma \vdash \alpha$ result from an application of the \ L rule to yield the left-hand premise. (This is left as an exercise.)

SUBCASE: · L rule (left-hand premise)

Let the left-hand premise, which is of the form $\Gamma \vdash \alpha$, result from an application of the · L rule. Then its list has a formula such as $\delta \cdot \gamma$. Γ, then, has the form $\Upsilon \, \delta \cdot \gamma \, \Lambda$. So, the left-hand premise is $\Upsilon \, \delta \cdot \gamma \, \Lambda \vdash \alpha$. Now this sequent results from an application of · L. For this rule to apply, the left-hand premise must be immediately dominated by a premise of the form $\Upsilon \, \delta \cdot \gamma \, \Lambda \vdash \alpha$. Putting all this together, one obtains the following instantiation of the cut rule whose application we are considering.

$$\frac{\dfrac{\Upsilon \, \delta \, \gamma \, \Lambda \vdash \alpha}{\Upsilon \, \delta \cdot \gamma \, \Lambda \vdash \alpha} \cdot L \qquad \Delta \, \alpha \, \Theta \vdash \beta}{\Delta \, \Upsilon \, \delta \cdot \gamma \, \Lambda \, \Theta \vdash \beta} \, cut$$

We can now rearrange this portion of any longer deduction in which the preceding portion appears into the same deduction, but replacing the preceding portion by the following:

$$\frac{\dfrac{\Upsilon \, \delta \, \gamma \, \Lambda \vdash \alpha \qquad \Delta \, \alpha \, \Theta \vdash \beta}{\Delta \, \Upsilon \, \delta \, \gamma \, \Lambda \, \Theta \vdash \beta} \, cut}{\Delta \, \Upsilon \, \delta \cdot \gamma \, \Lambda \, \Theta \vdash \alpha} \cdot L$$

REMARKS

1. Each deduction tree has the same depth.
2. Each deduction tree has the same number of nodes.
3. Each deduction tree has exactly the same number of applications of the cut rule.
4. The application of cut rule in the portion of the second deduction is higher in its deduction tree than the application of the cut rule in the portion of the first deduction tree.
5. Moreover, the degree of the application of the cut rule in the portion of the first deduction tree is greater than the degree of the application of the cut rule in the portion of the second deduction tree. Since $d(\delta \cdot \gamma) = d(\delta) + d(\gamma) + 1$, $d(\Delta \, \Theta \, \Psi \, \Upsilon \, \Lambda \, \alpha \, \beta \, \delta \cdot \gamma) > d(\Delta \, \Theta \, \Upsilon \, \Lambda \, \alpha \, \beta \, \delta \, \gamma)$.

CASE 2.1.2: Right-hand premise

Again, we have three subcases to consider, corresponding to applications of / L rule, \ L rule, and the · L rule, this time to the right-hand premise.

SUBCASE: / L rule (right-hand premise)

Let the right-hand premise, which is of the form $\Delta\ \alpha\ \Theta \vdash \beta$, result from an application of the / L rule. So α is a formula such as δ/γ. Since δ/γ results from an application of / L rule, Θ must have the form of $\Xi\ \Lambda$, where Λ may be empty, but Ξ is not. The right-hand premise, then, has the form $\Upsilon\ \delta/\gamma\ \Xi\ \Lambda \vdash \beta$ and it is immediately dominated by two sequents, one of the form $\Xi \vdash \gamma$ and the other of the form $\Upsilon\ \delta\ \Lambda \vdash \beta$. In addition, the left-hand premise has the form $\Psi \vdash \delta/\gamma$. This, in turn, requires that it result from an application of the / R rule and that it be immediately dominated by a sequent of the form $\Psi\ \gamma \vdash \delta$. Putting this all together, one obtains the following instantiation of the application of the cut rule.

$$\dfrac{\dfrac{\Psi\ \gamma \vdash \delta}{\Psi \vdash \delta/\gamma}\ /R \qquad \dfrac{\Upsilon\ \delta\ \Lambda \vdash \beta \qquad \Xi \vdash \gamma}{\Upsilon\ \delta/\gamma\ \Xi\ \Lambda \vdash \beta}\ /L}{\Upsilon\ \Psi\ \Xi\ \Lambda \vdash \beta}\ cut$$

Any deduction that includes such a configuration can be rearranged into a deduction where the relevant portion is reconfigured as follows.

$$\dfrac{\Xi \vdash \gamma \qquad \dfrac{\Psi\ \gamma \vdash \delta \qquad \Upsilon\ \delta\ \Lambda \vdash \beta}{\Upsilon\ \Psi\ \gamma\ \Lambda \vdash \beta}\ cut}{\Upsilon\ \Psi\ \Xi\ \Lambda \vdash \beta}\ cut$$

REMARKS

1. Each deduction tree has the same depth.

2. The second deduction tree has one less node than the first.

3. Although the portion of the second deduction tree shown has two applications of the cut rule, while the portion of the first has only one, each of the applications in the portion of the second deduction tree has a lesser degree than does the single application of the cut rule in the portion shown of the first deduction tree. Since $d(\delta/\gamma) = d(\delta) + d(\gamma) + 1$, $d(\Xi\ \Psi\ \Upsilon\ \Lambda\ \beta\ \delta/\gamma) > d(\Psi\ \Upsilon\ \Lambda\ \beta\ \delta\ \gamma)$ and $d(\Xi\ \Psi\ \Upsilon\ \Lambda\ \beta\ \delta/\gamma) > d(\Xi\ \Psi\ \Upsilon\ \Lambda\ \beta\ \gamma)$.

SUBCASE: \ L rule (right-hand premise)

Let $\Delta\ \alpha\ \Theta \vdash \beta$ result from an application of the \ L rule to the right-hand premise. (This is left as an exercise.)

SUBCASE: · L rule (right-hand premise)

Let $\Delta\ \alpha\ \Theta \vdash \beta$ result from an application of the · L rule. So α is a formula such as $\delta \cdot \gamma$. Since the right-hand premise results from an application of the · L rule, it has the form $\Upsilon\ \delta \cdot \gamma\ \Lambda \vdash \beta$ and it is immediately dominated by a sequent of the form $\Upsilon\ \delta\ \gamma\ \Lambda \vdash \beta$. At the same time, the left-hand premise has the form $\Gamma \vdash \delta \cdot \gamma$. So, it has resulted from an

The Lambek Calculus and the Lambda Calculus

application of the · R rule. Thus, Γ has two sub lists, Ξ and ϒ, and the left-hand premise is immediately dominated by two sequents of the form Ξ ⊢ δ and Ψ ⊢ γ. The following is, then, the instantiation of the application of the cut rule we are considering.

$$\dfrac{\dfrac{\Xi \vdash \delta \quad \Psi \vdash \gamma}{\Xi\,\Psi \vdash \delta \cdot \gamma}\,\cdot R \quad \dfrac{\Upsilon\,\delta\,\gamma\,\Lambda \vdash \beta}{\Upsilon\,\delta\cdot\gamma\,\Lambda \vdash \beta}\,\cdot L}{\Upsilon\,\Xi\,\Psi\,\Lambda \vdash \beta}\,\text{cut}$$

The preceding configuration within any deduction can be rearranged in the following way to yield a fresh deduction:

$$\dfrac{\Psi \vdash \gamma \quad \dfrac{\Xi \vdash \delta \quad \Upsilon\,\delta\,\gamma\,\Lambda \vdash \beta}{\Upsilon\,\Xi\,\gamma\,\Lambda \vdash \beta}\,\text{cut}}{\Upsilon\,\Xi\,\Psi\,\Lambda \vdash \beta}\,\text{cut}$$

REMARKS

1. Each deduction tree has the same depth.
2. The second deduction tree has one less node than the first.
3. Although the portion of the second deduction tree shown has two applications of the cut rule, while the portion of the first has only one, each of the applications in the portion of the second deduction tree shown has a lesser degree than does the single application of the cut rule in the portion shown of the first deduction tree. Since $d(\delta \cdot \gamma) = d(\delta) + d(\gamma) + 1$, $d(\Xi\,\Psi\,\Upsilon\,\Lambda\,\beta\,\delta\cdot\gamma) > d(\Xi\,\Upsilon\,\Lambda\,\beta\,\delta\,\gamma)$ and $d(\Xi\,\Psi\,\Upsilon\,\Lambda\,\beta\,\delta\cdot\gamma) > d(\Xi\,\Psi\,\Upsilon\,\Lambda\,\beta\,\gamma)$.

CASE 2.2: Application of the R rules

We now turn to the deductions with applications of the R rules. Again, we start with those cases where the left-hand premise of the cut rule results from an application of an R rule and then we turn to those cases where the right-hand premise results from such applications.

CASE 2.2.1: Left-hand premise

We have three subcases to consider, corresponding to applications of / R rule, \ R rule, and the · R rule.

SUBCASE: / R rule (left-hand premise)

Let the left-hand premise, which is of the form Γ ⊢ α, result from an application of the / R rule. So α is a formula such as δ/γ. The list of the right-hand premise, then, contains the formula δ/γ. So, the premise itself has the form Δ δ/γ Θ ⊢ β. But then this sequent will have resulted from an application of the / L rule. This implies that it is immediately dominated by two premises, one of the form Ξ ⊢ γ and the other of the form ϒ δ Λ ⊢ β. The application of the cut rule under consideration, then, is instantiated as follows.

$$\frac{\Psi\,\gamma\vdash\delta}{\Psi\vdash\delta/\gamma}\,/R \qquad \frac{\Xi\vdash\gamma \qquad \Upsilon\,\delta\,\Lambda\vdash\beta}{\Upsilon\,\delta/\gamma\,\Xi\,\Lambda\vdash\beta}\,/L$$
$$\frac{}{\Upsilon\,\Psi\,\Xi\,\Lambda\vdash\beta}\,cut$$

This arrangement can be reconfigured as follows.

$$\frac{\Xi\vdash\gamma \qquad \dfrac{\Psi\,\gamma\vdash\delta \quad \Upsilon\,\delta\,\Lambda\vdash\beta}{\Upsilon\,\Psi\,\gamma\,\Lambda\vdash\beta}\,cut}{\Upsilon\,\Psi\,\Xi\,\Lambda\vdash\beta}\,cut$$

REMARKS

1. Each deduction tree has the same depth.

2. The second deduction tree has one less node than the first.

3. Although the portion of the second deduction tree shown has two applications of the cut rule, while the portion of the first has only one, each of the applications in the portion of the second deduction tree shown has a lesser degree than does the single application of the cut rule in the portion shown of the first deduction tree. Since $d(\delta/\gamma) = d(\delta) + d(\gamma) + 1$, $d(\Xi\,\Psi\,\Upsilon\,\Lambda\,\beta\,\delta/\gamma) > d(\Xi\,\Psi\,\delta\,\gamma)$ and $d(\Xi\,\Psi\,\Upsilon\,\Lambda\,\beta\,\delta/\gamma) > d(\Xi\,\Psi\,\Upsilon\,\Lambda\,\beta\,\delta)$.

SUBCASE: \ R rule (left-hand premise)

Let the left-hand premise result from an application of the \ R rule. (This is left as an exercise.)

SUBCASE: · R rule (left-hand premise)

Let the left-hand premise result from an application of the rule · R. Thus, the left-hand premise has the form $\Xi\,\Psi\vdash\delta\cdot\gamma$ and it is immediately dominated by two premises, $\Xi\vdash\delta$ and $\Psi\vdash\gamma$. At the same time, the right-hand premise has the form $\Upsilon\,\delta\cdot\gamma\,\Lambda\vdash\beta$. It results from an application of the · L rule and so it is immediately dominated by the sequent $\Upsilon\,\delta\,\gamma\,\Lambda\vdash\beta$. The application of the cut rule under consideration instantiates to the following.

$$\frac{\dfrac{\Xi\vdash\delta \quad \Psi\vdash\gamma}{\Xi\,\Psi\vdash\delta\cdot\gamma}\,\cdot R \qquad \dfrac{\Upsilon\,\delta\,\gamma\,\Lambda\vdash\beta}{\Upsilon\,\delta\cdot\gamma\,\Lambda\vdash\beta}\,\cdot L}{\Upsilon\,\Xi\,\Psi\,\Lambda\vdash\beta}\,cut$$

The preceding configuration within any deduction can be rearranged in the following way to yield a fresh deduction:

$$\frac{\Psi\vdash\gamma \qquad \dfrac{\Xi\vdash\delta \quad \Upsilon\,\delta\,\gamma\,\Lambda\vdash\beta}{\Upsilon\,\Xi\,\gamma\,\Lambda\vdash\beta}\,cut}{\Upsilon\,\Xi\,\Psi\,\Lambda\vdash\beta}\,cut$$

The Lambek Calculus and the Lambda Calculus

REMARKS

1. Each deduction tree has the same depth.

2. The second deduction tree has one less node.

3. Although the portion of the second deduction tree shown has two applications of the cut rule, while the portion of the first has only one, each of the applications in the portion of the second deduction tree shown has a lesser degree than does the single application of the cut rule in the portion shown of the first deduction tree. Since $d(\delta \cdot \gamma) = d(\delta) + d(\gamma) + 1$, $d(\Xi \Psi \Upsilon \wedge \beta \delta \cdot \gamma) > d(\Xi U \wedge \beta \delta \gamma)$ and $d(\Xi \Psi \Upsilon \wedge \beta \delta \cdot \gamma) > d(\Xi \Psi \Upsilon \wedge \beta \gamma)$.

CASE 2.2.2: Right-hand premise

Again, we have three subcases to consider corresponding to the applications of / R rule, \ R rule, and the · R rule.

SUBCASE: / R rule (right-hand premise)

Let the right-hand premise result from an application of the / R rule. Then an application of the cut rule leads to the following.

$$\cfrac{\Psi \vdash \alpha \quad \cfrac{\Upsilon \, \alpha \wedge \gamma \vdash \delta}{\Upsilon \, \alpha \wedge \vdash \delta/\gamma} \, /R}{\Upsilon \, \Psi \wedge \vdash \delta/\gamma} \, cut$$

This arrangement can be reconfigured as follows.

$$\cfrac{\cfrac{\Psi \vdash \alpha \quad \Upsilon \, \alpha \wedge \gamma \vdash \delta}{\Upsilon \, \Psi \wedge \gamma \vdash \delta} \, cut}{\Upsilon \, \Psi \wedge \vdash \delta/\gamma} \, /R$$

REMARKS

1. Each deduction tree has the same depth.

2. Each deduction tree has the same number of nodes.

3. Each deduction tree has exactly the same number of applications of the cut rule.

4. The application of cut rule in the portion of the second deduction is higher in its deduction tree than the application of the cut rule in the portion of the first deduction tree.

5. Moreover, the degree of the application of the cut rule in the portion of the first deduction tree is greater than the degree of the application of the cut rule in the portion of the second deduction. Since $d(\delta/\gamma) = d(\gamma) + d(\delta) + 1$, $d(\Psi \Upsilon \wedge \alpha \, \delta/\gamma) > d(\Psi \Upsilon \wedge \delta \gamma)$.

SUBCASE: \ R rule (right-hand premise)

Let right-hand premise result from the rule \ R. (This is left as an exercise.)

SUBCASE: · R rule (right-hand premise)

Let right-hand premise result from the rule $\cdot R$.

$$\cfrac{\Psi \vdash \alpha \quad \cfrac{\Upsilon\, \alpha\, \Lambda \vdash \delta \quad \Xi \vdash \gamma}{\Upsilon\, \alpha\, \Lambda\, \Xi \vdash \delta \cdot \gamma}\cdot R}{\Upsilon\, \Psi\, \Lambda\, \Xi \vdash \delta \cdot \gamma}\text{cut}$$

This arrangement can be reconfigured as follows.

$$\cfrac{\cfrac{\Psi \vdash \alpha \quad \Upsilon\, \alpha\, \Lambda \vdash \delta}{\Upsilon\, \Psi\, \Lambda \vdash \delta}\text{cut} \quad \Xi \vdash \gamma}{\Upsilon\, \Psi\, \Lambda\, \Xi \vdash \delta \cdot \gamma}\cdot R$$

REMARKS

1. Each deduction tree has the same depth.
2. Each deduction tree has the same number of nodes.
3. Each deduction tree has exactly the same number of applications of the cut rule.
4. The application of cut rule in the portion of the second deduction is higher in its deduction tree than the application of the cut rule in the portion of the first deduction tree.
5. Moreover, the degree of the application of the cut rule in the portion of the first deduction tree is greater than the degree of the application of the cut rule in the portion of the second deduction tree. Since $d(\delta \cdot \gamma) = d(\gamma) + d(\delta) + 1$, $d(\Xi\, \Psi\, \Upsilon\, \Lambda\, \alpha\, \delta \cdot \gamma) > d(\Psi\, \Upsilon\, \Lambda\, \alpha\, \delta)$.

We have now completed the proof of the cut elimination theorem for the Lambek calculus. We have shown that, for any proof in the sequent calculus that uses the cut rule, there is a proof of the same sequent without any use of the cut rule. We established this result using induction by showing that, for any configuration in a proof of a sequent in which the cut rule appears, the configuration can be replaced with another configuration in which use of the cut rule is moved up the proof tree or in which the cut rule is replaced by an application of the cut rule of a smaller degree. Since each proof is of finite depth, any application of a cut rule can ultimately be eliminated.

Exercises: Cut elimination

1. Complete the proofs omitted in this section.

3 The Lambda Calculus

We now come to the Lambda calculus. Since it is a calculus, it comprises a set of expressions and rules for transforming expressions in the set into other expressions in the same set. The expressions are expressions for functions and the rules transform expressions for functions into expressions for functions. In section 3.1, we shall see how the expressions

The Lambek Calculus and the Lambda Calculus

are formed; in section 3.2, how the expressions are expressions for functions; and in section 3.3, how to transform these expressions into other expressions.

3.1 Notation of the Lambda Calculus

There are various versions of the notation for the Lambda calculus. We shall be concerned just with two, one known as the *single typed,* or *monotyped,* Lambda calculus[6] and the other as the *simply typed* Lambda calculus. Though we shall be mainly concerned with the latter, we begin with the former, simpler version, as a propaedeutic to the latter.

3.1.1 Notation: Monotyped

The two basic categories of symbols in the Lambda calculus are constants, or CN, and variables, or VR. We are already familiar with both these categories of symbols from CQL, where variables were distinguished from individual and relational symbols, which are CQL's counterpart to the constants of the Lambda calculus. And just as VR is disjoint from IS and RS in CQL, so VR is disjoint from CN in the Lambda calculus.

We begin our discussion of the notation of the Lambda calculus with the notation of the monotyped version.

Definition 6 Formation of monotyped terms (syncategorematic version)

(1.1) $CN \subseteq TM$;

(1.2) $VR \subseteq TM$;

(2.1)   If $\sigma \in TM$ and $\tau \in TM$, then $(\sigma \tau) \in TM$;

(2.2)   If $v \in VR$ and $\tau \in TM$, then $(\lambda v.\tau) \in TM$;

(3)   Nothing else is in TM.

The two clauses in (1) of definition 6 require that the constants and the variables be terms. These terms are *atomic terms*. The two clauses in (2) enlarge the set of terms to include composite terms. The composite terms are formed in two ways: the first way (2.1) concatenates any two terms and encloses the result with a pair of parentheses, and the second way (2.2) introduces the lambda operator (λ) followed by a variable, a period, and a term and encloses the result with a pair of parentheses. By analogy with the terminology in chapter 11, it is useful to divide a term of the form $\lambda v.\tau$ into its first part λv, which comprises the lambda operator and the variable and which we shall call the *(lambda) prefix,* and its second part τ, which comprises the remainder of the term and which we call the *(lambda) matrix.*

As we shall see in section 3.2, each clause in (2) introduces a way to express an operation: the latter does so by introducing a complex symbol comprising the symbol λ, the Greek letter called lambda, and a variable; the former does so through simple juxtaposition.

---

6. Another customary, but potentially misleading, name is the *untyped* Lambda calculus.

This practice is familiar from secondary school algebra, where multiplication of two variables or of a parameter and a variable or of a numeral and a variable is expressed by simple juxtaposition. For example, one may write xy for $x \cdot y$, or ay for $a \cdot y$ or $2y$ for $2 \cdot y$.

We shall adopt here conventions for constants and variables of the Lambda calculus analogous to those we adopted in chapter 11 for CQL. We shall use for constants lowercase letters from the beginning of the Roman alphabet, such as a, b, and c; and if a large number of constants is required, we shall use the letter c, subscripted with positive integers. Similarly, we shall use for variables lowercase letters from the end of the Roman alphabet, such as w, x, y, and z; and again, if a large number of variables are required, we shall use the letter v, subscripted with positive integers. In addition, for terms, atomic or composite, we shall use the letters t, s, and p. Should a larger number of terms be needed, we shall subscript t with positive integers.

Before turning to the abbreviatory conventions governing parentheses to be adopted hereafter, let us pause to see how we display in a tree diagram, of the kind we first met in chapter 1, the reasoning whereby we establish whether an expression is indeed a term in the monotyped Lambda calculus.

Suppose that a and b are constants and that x and y are variables. By clause (1.1) a and b are terms and by clause (1.2) x and y are also terms. Since a and y are both terms, by clause (2.1), (ay) is a term; similarly, since b and x are terms, (xb) is a term, also by clause (2.1). Since x and y are variables and (ay) and (xb) are terms, then by clause (2.2) both $(\lambda x.(xb))$ and $(\lambda y.(ay))$ are terms. And since these last two expressions are terms, then by clause (2.1) $((\lambda x.(xb))(\lambda y.(ay)))$ is a term. This reasoning is displayed in the following diagram.

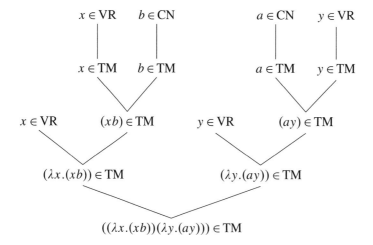

Just as we simplified such diagrams in the case of CPL, CPDL, and CQL, we simplify the term formation diagrams, first by restricting our attention to terms, thereby omitting the nodes labeled with CN and VR, and second, as all the remaining expressions are terms, we omit the indication that they are members of TM. The result is that the preceding diagram simplifies to the following.

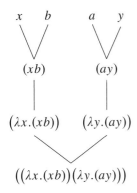

Now we turn to abbreviatory conventions governing parentheses. There are three. The first, already familiar to us from the conventions governing the notations for CPL, CPDL, and CQL, is to omit the outer most parentheses. Thus, for example, the formula in PL, $((p \wedge q) \rightarrow r)$, is abbreviated as $(p \wedge q) \rightarrow r$. Similarly, the terms $(\sigma\tau)$ and $(\lambda v.\tau)$ are abbreviated as $\sigma\tau$ and $\lambda v.\tau$, respectively. This convention applies only to a term, not to any subterm of a term.

We now turn to the second convention. Without any specific guidance, the arithmetic expression $4 + 2 \cdot 3$ is ambiguous: either it could denote 10, which results from carrying out the operation of multiplication before carrying out the operation of addition, or it could denote 18, which results from carrying out the operation of addition before carrying out the operation of multiplication. The use of parentheses makes unambiguous which operation is performed first. Thus, the expression $4 + (2 \cdot 3)$ unambiguously denotes 10, while the expression $(4 + 2) \cdot 3$ unambiguously denotes 18. A convention one can adopt is to give the multiplication sign priority over the addition sign. Under this convention, the expression without parentheses, $4 + 2 \cdot 3$, unambiguously denotes 10. To denote 18, one then must resort to parentheses and write the expression $(4 + 2) \cdot 3$.

The second convention is of this kind. The two clauses in (2) of the definition of the set of terms (definition 6) introduce notation for two operations, the application operation and the abstraction operation. The convention is that the application operation has priority over the abstraction operation. More exactly, if the term to the right of a lambda prefix itself results from clause (2.1), then the parentheses introduced by the rule are omitted. Thus, $\lambda v.\sigma\tau$ abbreviates, by the second convention, $\lambda v.(\sigma\tau)$, which itself abbreviates as, by the first convention, $(\lambda v.(\sigma\tau))$. $\lambda v.\sigma\tau$ is to be distinguished from $(\lambda v.\sigma)\tau$ in the same way as $4 + 2 \cdot 3$ is from $(4+2) \cdot 3$.

The third convention pertains to expressions of application. We shall assume that application associates to the left. Thus, the term $(\rho\sigma)\tau$, which, by the first convention, abbreviates as $((\rho\sigma)\tau)$, is abbreviated further by the second convention as $\rho\sigma\tau$. This convention does not compromise the distinction between the terms $(\rho\sigma)\tau$ and $\rho(\sigma\tau)$. Only the first is subject to this convention. The third convention also applies to subterms of a term.

To shore up our understanding of these conventions, let us consider the term $(\lambda x.((yx)(\lambda x.(zx))))$. By the first convention, one eliminates the outer parentheses, obtaining $\lambda x.((yx)(\lambda x.(zx)))$. By the second convention, one eliminates the outer parentheses enclosing the subterm $((yx)(\lambda x.(zx)))$, obtaining $\lambda x.(yx)(\lambda x.(zx))$. The same convention permits the elimination of the parentheses surrounding the subterm (zx), yielding $\lambda x.(yx)(\lambda x.zx)$. Finally, by the third convention, one eliminates the parentheses surrounding the subterm (yx), yielding $\lambda x.yx(\lambda x.zx)$.

Just as in CQL, the scope of an occurrence of a quantifier and the scope of the prefix in which it occurs is the same, so in the Lambda calculus, the scope of an occurrence of the lambda operator symbol and the scope of the prefix in which it occurs is the same. Moreover, the definition of the scope of an occurrence of a lambda prefix is completely analogous to the definition of the scope of an occurrence of a quantifier prefix.

Definition 7 Scope of a lambda prefix

The scope of an occurrence of a lambda prefix in a term τ is the subterm of τ, which contains the lambda prefix's occurrence but whose immediate subterms do not.

We now define what it is for an occurrence of a lambda prefix to bind a particular occurrence of a variable. For example, the scope of the first lambda operator in the term $(\lambda x.((yx)(\lambda x.(zx))))$ is the entire formula, whereas that of the second lambda operator is $(\lambda x.(zx))$.

Definition 8 Binding

Let τ be in TM. Let λv be a lambda prefix in τ and let u be a variable in τ. (An occurrence of) the lambda prefix λv binds (an occurrence of) the variable u iff

(1) $v = u$ (i.e., v and u are the same variable);
(2) (The occurrence of) u is within the scope of (the occurrence of) the lambda prefix λv; and
(3) No other occurrence of the lambda prefix λv has the relevant occcurrence of u within its scope and is itself within the scope of the relevant occurrence of the lambda prefix λv.

Evidently, then, the lambda prefix λx binds both occurrences of the variable x in the term $\lambda x.x$, which, by the first convention, abbreviates $(\lambda x.x)$. The lambda prefix λy

The Lambek Calculus and the Lambda Calculus

binds both occurrences of the variable y in the term $\lambda y.xy$, which abbreviates $(\lambda y.(yx))$. But, the lambda prefix λy does not bind both occurrences of the variable y in the term $(\lambda y.x)y$, a conventional abbreviation of $((\lambda y.x)y)$. Finally, let us consider again the term $\lambda x.yx(\lambda x.zx)$, an abbreviation of $(\lambda x.((yx)(\lambda x.(zx))))$, as we explained earlier. In this term, the second occurrence of the lambda prefix λx binds the last two occurrences of the variable x in the term, while the first occurrence of the lambda prefix binds only the first two occurrences of the variable x.

We now distinguish, as we did in CQL, between bound variables, variables bound by some prefix, here a lambda prefix, and free variables, variables bound by no lambda prefix.

Definition 9 Bound variable
(An occurrence of) a variable v is bound in a term τ iff (an occurrence of) a lambda prefix in the term τ binds it.

Definition 10 Free variable
(An occurrence of) a variable v is free in a term τ iff no (occurrence of a) lambda prefix in the term τ binds it.

Thus, in the term $\lambda x.yx(\lambda x.zx)$, each occurrence of the variable x is bound and neither the occurrence of y nor the occurrence of z is bound. No single occurrence of a variable may be both bound and free, though the variable itself may be. For example, the variable y is both free and bound in the term in $\lambda x.xy(\lambda y.yz)$, while each of its occurrences is one or the other and not both.

The foregoing distinction between variables that are bound and variables that are free is used to distinguish two kinds of terms: those that are closed and those that are open. An open term is a term with at least one free variable occurring in it.

Definition 11 Open term
A term is open iff it has at least one (occurrence of a) free variable.

A closed term is one in which no free variables occur.

Definition 12 Closed term
A term is closed iff it has no (occurrences of a) free variable.

It is convenient to define a function that applies to a term and retrieves any variable with at least one free occurrence within the term.

Definition 13 Free variables in a term

(1) $\mathrm{Fvr}(v) = \{v\}$, for each $v \in \mathrm{VR}$;
(2) $\mathrm{Fvr}(c) = \emptyset$, for each $c \in \mathrm{CN}$;
(3) $\mathrm{Fvr}(\sigma\tau) = \mathrm{Fvr}(\sigma) \cup \mathrm{Fvr}(\tau)$;
(4) $\mathrm{Fvr}(\lambda v.\tau) = \mathrm{Fvr}(\tau) - \{v\}$.

Open and closed terms could also be defined in terms of the function that determines, for any terms, what its free variables are. A term τ is closed just in case $\mathrm{Fvr}(\tau)$ is empty and it is open otherwise.

In chapter 11, section 1, we were introduced to the syntactic operation of substitution. It was defined as an operation that mapped terms to terms and formulae to formulae by replacing each occurrence of a free variable with an occurrence of an individual symbol. Here it will be defined as an operation that maps terms to terms by replacing each occurrence of a free variable with an occurrence of a term. The term that substitutes for the free variable may be any term, a constant or a variable, or indeed a composite term.

Definition 14 Substitution of a term for a free variable
Let ρ, σ, and τ be terms and let u, v, and w be variables.

(1.1) $[\tau/v]\sigma = \sigma$, if $\sigma \in \mathrm{CN}$;
(1.2) $[\tau/v]\sigma = \sigma$, if $\sigma \in \mathrm{VR}$ and $\sigma \neq v$;
(1.3) $[\tau/v]\sigma = \tau$, if $\sigma \in \mathrm{VR}$ and $\sigma = v$;

(2.1) $[\tau/v](\rho\sigma) = [\tau/v]\rho[\tau/v]\sigma$;
(2.2.1) $[\tau/v](\lambda w.\rho) = \lambda w.\rho$, if $w = v$;
(2.2.2) $[\tau/v](\lambda w.\rho) = \lambda w.[\tau/v]\rho$, if $w \neq v$.

The substitution operation substitutes a term for a free variable in a term. Should the term in which the substitution is to take place have no free variables, then the term to which the substitution operation applies is unaltered. Thus, as clause (1.1) makes clear, if the term in which the substitution is to take place is a constant, then the substitution operation leaves the constant unaltered. Should the term contain a free occurrence of a variable, but the variable is not the same as the substituens, then again the term to which the substitution operation applies is left unaltered. As clause (1.2) makes clear, a term that is just a variable, and therefore free, is unaltered by a substitution operation if the substituens is a distinct variable. Should the term in which the substituion is to take place be just a variable and should it be identical with the substituens, then the substitution takes place. Clause (1.3) states this.

Next, if a composite term comprises two terms, the substitution operation applied to the composite term is just the result of the operation applied to its two immediate subterms. If the composite term comprises a lambda prefix and a term, whether the substitution operation applied to the composite term alters the term depends on whether the term in question has at least one occurrence of the substituens and its occurrence is free. And, it is free just in case the substituens is distinct from the variable in the lambda prefix.

We illustrate these various cases next, the expression above indicates the substitution to take place and the expression into which the substitution takes place and the expression below is the result of the substitution.

CLAUSE

(1.1)	$[a/x]b$	$[y/x]b$	$[\lambda z.az/x]b$
	b	b	b
(1.2)	$[a/x]y$	$[z/x]y$	$[\lambda z.az/x]y$
	y	y	y
(1.3)	$[a/x]x$	$[z/x]y$	$[\lambda z.az/x]x$
	a	y	$\lambda z.az$
(2.1)	$[a/x](xy)$	$[z/x](xy)$	$[\lambda z.az/x](xy)$
	ay	zy	$(\lambda z.az)y$
(2.2.1)	$[a/x](\lambda z.xz)$	$[y/x](\lambda z.xz)$	$[\lambda w.yw/x](\lambda z.xz)$
	$\lambda z.az$	$\lambda z.yz$	$\lambda z.(\lambda w.yw)z$
(2.2.2)	$[a/x](\lambda x.xz)$	$[y/x](\lambda x.xz)$	$[\lambda w.yw/x](\lambda x.xz)$
	$\lambda x.xz$	$\lambda x.xz$	$\lambda x.xz$

Before turning to exploring some of the details of substitution, we remind readers of the two neologisms introduced in chapter 11 in connection with substitution: namely, the word *substituens,* or τ in the notation of $[\tau/v]$, the term replacing the various occurrences of a free variable, v, and the word *substituendum,* or v in the notation of $[\tau/v]$, the free variable whose various free occurrences τ replaces.

When substitution was defined for expressions of CQL, a substituendum is a variable and the substituens is an individual symbol. However, whereas in CQL only individual symbols served as substituens, in the Lambda calculus any term, atomic or composite, closed or open, may serve as a substituens. Since substitution in CQL replaces a free variable with only an individual symbol, such a replacement never disturbs the binding relations between an occurrence of a quantifier prefix and the occurrences of the variable it binds. Because substitution in a term of the Lambda calculus permits open terms to substitute for the free occurrences of a variable, the relationship between the occurrence of a quantifier prefix and the free occurrences of a variable may be disturbed.

Let us see a few examples of how this relationship can be disturbed. Consider the substitution of the variable x for the free variable y in the term $\lambda x.xy$. The result is $\lambda x.xx$: that is, $[x/y]\lambda x.xy = \lambda x.xx$. The term into which the substitution is to take place has one lambda prefix λx, one bound occurrence of x, and one free occurrence of y. The term that results has one lambda prefix λx, two bound occurrences of x, and no free occurrence of any variable. This substitution contrasts with the substitution of the variable z for the free variable y in the term $\lambda x.xy$. The result is $\lambda x.xz$: that is, $[z/y]\lambda x.xy = \lambda x.xz$. The one lambda prefix λx binds the same variables in the same positions as it did before the substitution, and the position of the only free occurrence of a variable in $\lambda x.xy$ is the position of the only free occurrence of a variable in $\lambda x.xz$, though the variable has changed. Here is another example of a disturbance of the binding relationship. The substituens y in the term $\lambda z.yz$ is free before substitution and occurs as bound after substitution for v in the term $\lambda y.vy$, for $[\lambda z.yz/v]\lambda y.vr = \lambda y.(\lambda z.yz)r$.

Such a disturbance does not arise if the substituens is a closed term. If it is an open term, the substituendum position is not within the scope of a lambda prefix whose variable is among the free variables in the substituens. For example, no harm arises in substituting the closed term b for y in $\lambda x.xy$. Moreover, no harm arises in substituting the closed term $\lambda z.zc$ for x in $\lambda y.xy$. Nor does any harm arise in substituting an open term, say $\lambda z.zv$ for x in $\lambda x.xy$, since the free variable v in $\lambda z.zv$ is distinct from the variable x, which occurs in the lambda prefix of the term containing the substituendum.

We now define substitution in such a way that the relationship between the occurrences of lambda prefixes and the variable occurrences they bind will not be disturbed.

Definition 15 Free for a term to substitute for a variable
Let σ, τ be terms and let u and v be variables. Then, σ is free to substitute for v in τ iff one of the following holds:

(1.1) $\tau \in \mathrm{CN}$;
(1.2) $\tau \in \mathrm{VR}$;
(2.1) τ has the form $(\pi \rho)$ and σ is free to substitute for v in both π and ρ;
(2.2) τ has the form $\lambda u.\pi$ and σ is free to substitute for v in π and if $v \in \mathrm{Fvr}(\pi)$, then $u \notin \mathrm{Fvr}(\sigma)$.

Thus, x is not free to substitute for y in $\lambda x.xy$, though a, a constant, clearly is. In fact, any closed term is free to substitute for y in $\lambda x.xy$. Moreover, any variable distinct from x, say z, is free to substitute for y in $\lambda x.xy$, as is any term, provided the substituens does not have x among its free variables.

Exercises: Monotyped Lambda calculus

1. For each sequence of symbols, determine whether it is a term of the Lambda calculus, a term by convention only or not a term at all. If it is a term by convention, state the relevant convention. If it is not a term, explain why it is not.

(a) ab
(b) (xy)
(c) $(z.\lambda z)$
(d) $\lambda xy.xyz$
(e) $\big(\lambda x.(\lambda y.(\lambda z.a))\big)$
(f) $abcd$
(i) $\big(\lambda a.(ay)\big)$
(j) $\big(\lambda z.((\lambda x.x)(\lambda y.((xy)z)))\big)$
(k) $\big(z(\lambda z.x)\big)$
(l) $\lambda z.(\lambda x.x)(\lambda y.xyz)$
(m) $\lambda x.\lambda x.x$
(n) $\lambda x.(ax)$

(g) $\lambda z.(\lambda x.x)$ (o) $\lambda x.x(\lambda y.y)$
(h) $\lambda z.y\lambda x.x$ (p) $\lambda x.x(\lambda y.yxx)x$

2. For each variable occurrence in each term, state which lambda prefix occurrence binds which variable occurrence and state which variable occurrences are free.

(a) $y\lambda x.z$ (d) $\lambda x.(\lambda x.y)zx$
(b) $\lambda x.\lambda y.xyz$ (e) $\lambda x.(\lambda y.yz)(xy)$
(c) $(\lambda y.xz(\lambda z.ya))(\lambda x.wy(\lambda z.z))$ (f) $(z(\lambda z.x))$

3. In each case, determine which term results from the application of the substitution function.

(a) $[x/y]y$ (d) $[\lambda x.xy/y]y$
(b) $[\lambda z.wz/v]\lambda y.vx$ (e) $[\lambda x.xy/x]\lambda z.(\lambda x.x)(\lambda y.xyz)$
(c) $[\lambda x.yx/x]\lambda z.zx$ (f) $[\lambda y.ayx/z](z(\lambda z.x))$

4. State which substituenda, listed in the first column, are free to substitute for the substituens, specified in the second column, in the term, specified in the same row as the substituens but in the third column.

SUBSTITUENDA	SUBSTITUENS	TERM
a	z	az
x	x	az
y	y	$\lambda y.ay$
z	z	$\lambda x.az$
$\lambda x.bx$	z	$\lambda y.ayz$
$\lambda y.ay$	x	$\lambda x.\lambda y.yx$
$\lambda z.xy$	w	$\lambda x.yw(\lambda y.xy)$
$\lambda x.\lambda y.xy$	y	$\lambda x.yw(\lambda y.xy)$
$\lambda w.wy$		
$\lambda x.xy$		
$\lambda x.xw(\lambda y.by)$		

5. Using the definition for Fvr (definition 13) as a guide, provide an explicit definition of the set of variables occurring in a term, using the notation Ovr. Also write an explicit definition for a function Bvr that identifies variables with at least one bound occurrence in a term.

3.1.2 Notation: Simply typed

We now turn to the simply typed Lambda calculus. In the monotyped Lambda calculus, there is but one type, TM; in the simply typed Lambda calculus, there is a (countably) infinite number of types. Our first task is to define this infinite set of types. To define this set of types, we shall avail ourselves of an index set. Readers may recall that we were able to partition the set of relational symbols, introduced in chapter 9, into a countably infinite number of subsets, where the one-place relational symbols are in one set, the two-place relational symbols in a second set, and in general, the n-place relational symbols are in an nth set. Indeed, what we did was to index the family of sets using the positive integers: an n-place relational symbol is a member of the set RS_n. The set of relational symbols, RS, then is just $\bigcup_{i \in \mathbb{Z}^+} RS_i$. In a similar fashion, we can define the set of all nonlogical symbols for CPDL as follows: let NS_0 be IS and, for each $i \in \mathbb{Z}^+$, let NS_i be RS_i. NS, the set of nonlogical symbols, is then defined as $\bigcup_{n \in \mathbb{N}} NS_n$. In this way, the natural numbers are used as indices whereby a countably infinite set of categories, or types, is defined and into which categories, or types, the nonlogical constants (the individual symbols and the relational symbols) of CQL, can be put.

In the same way, we shall define a countably infinite set of indices and use them to define a countably infinite set of categories, or types, by which atomic terms of the simply typed Lambda calculus can be categorized, or typed. Not surprisingly, the set of indices, called types, is defined recursively. The set of types, named Typ, includes two atomic types, e and t, as well as the composite types obtained from them, stated in the following definition.

Definition 16 Set of simple types

Let e and t be distinct entities.

(1) $e, t \in \text{Typ}$;
(2)   If $x, y \in \text{Typ}$, then so is (x/y);
(3)   Nothing else is in Typ.

As with the notation of CQL, we assume that the set of constants and the set of variables are disjoint from one another. In addition, as we did with the relational symbols of CPDL, we shall segregate both the constants and the variables into disjoint subsets, labeling distinct subsets with distinct types, just as we labeled the disjoint subsets of the set of relational symbols of CPDL with distinct positive integers. Thus, if x and y are distinct types, then CN_x, the *constants of type x*, and CN_y, the *constants of type y*, are disjoint from one another, as are the *variables of type x*, or VR_x, and the *variables of type y*, or VR_y. The set of all constants, or CN, then is just $\bigcup_{x \in \text{Typ}} CN_x$ and the set of all variables, or VR, is $\bigcup_{x \in \text{Typ}} VR_x$. We shall refer to the members of either VR or CN as *atomic terms*.

We now wish to extend the atomic terms to include *composite terms*, just as we extended the atomic formulae, or propositional variables of CPL, to include composite formulae. We shall do this by extending the atomic terms of each type.

The Lambek Calculus and the Lambda Calculus

Definition 17 Formation of simply typed terms (syncategorematic version)
Let x and y be in Typ.

(1) ATOMIC TERMS

(1.1) $CN_x \subseteq TM_x$;

(1.2) $VR_x \subseteq TM_x$;

(2) COMPOSITE TERMS

(2.1) If $\sigma \in TM_{x/y}$ and $\tau \in TM_y$, then $(\sigma\tau) \in TM_x$;

(2.2) If $\upsilon \in VR_y$ and $\tau \in TM_x$, then $(\lambda\upsilon.\tau) \in TM_{x/y}$.

Just as we defined CN to be $\bigcup_{x \in Typ} CN_x$ and VR to be $\bigcup_{x \in Typ} VR_x$, we define TM to be $\bigcup_{x \in Typ} TM_x$.

Let us see how these formation rules work. Suppose that $a, b \in CN_e$, $x \in VR_e$, and $y \in VR_{t/e}$. By clause (1.1) $a, b \in TM_e$ and by clause (1.2) $y \in TM_{t/e}$. Since $y \in TM_{t/e}$ and $a \in TM_e$, it follows by clause (2.1) that $ya \in TM_t$. Since $x \in VR_e$ and $ya \in TM_t$, clause (2.2) shows us that $\lambda x.ya \in TM_{t/e}$. Finally, clause (2.1) establishes that $(\lambda x.ya)b \in TM_t$, since $\lambda x.ya \in TM_{t/e}$ and $b \in TM_e$. As we have come to expect, such reasoning can be recapitulated in a tree diagram:

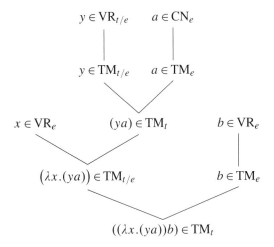

Here too we can simplify the diagram, though not as much as we were able to with the diagrams for the monotyped terms. The reason is that whether two terms can form a term depends on which category each belongs to. For example, there is no term formed from two terms, one taken from $TM_{t/e}$ and one taken from $TM_{e/t}$. Indeed, no two terms taken from the very same TM_x, regardless of what x is, cannot be put together to form a term. Moreover, even if one term is taken from $TM_{t/e}$ and another from TM_e, they form a term only if the second one is put after the first one. It follows, then, that in simplifying the diagrams, we cannot omit information about the type of the category. We can, however, confine our attention to terms and we can compress the labels so that label mentions only the term and its type.

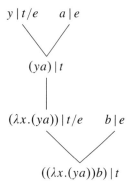

Attentive reader will have noticed that the assignment of types in the term tree conform to the rules of / Elimination and / Introduction of the Lambek calculus. We shall discuss this in detail in section 4.

We adopt the same abbreviatory conventions for the simply typed Lambda calculus as we did for the monotyped Lambda calculus. The various notions of the monotyped Lambda calculus—namely, of scope, of binding, of bound and free variables, of open and closed terms, of substitution, and of being free to substitute for a variable in a term—all apply mutatis mutandis to the simply typed Lambda calculus. Two definitions, nonetheless, deserve comment. Binding is a relation between the occurrence of a lambda prefix in a term and the occurrence of a free variable in the same term. In a Lambda Calculus with more than one type, the variable in the lambda prefix must be of the same type as the variable occurrence that the lambda prefix binds. However, it turns out that the definition as stated for the monotyped Lambda calculus applies here without any change, for the first clause of its definiens, which requires the variable occurrence in the lambda prefix occurrence and the variable occurrence that it binds be occurrences of the same variable, and so, ipso facto, the variable occurrences of variables of the same type.

In contrast, the definition of substitution requires the explicit mention of types. If substitution is to yield a term, then the substituens must have the same type as the free variable for which it substitutes. Here is the revised definition of substitution containing the emendation.

Definition 18 Substitution of a term for a free variable (simply typed)
Let ρ, σ and τ be terms and let $u, v,$ and w be variables. Let term τ and variable v be of the same type.

The Lambek Calculus and the Lambda Calculus

(1.1) $[\tau/v]\sigma = \sigma$, if $\sigma \in \text{CN}$;

(1.2) $[\tau/v]\sigma = \sigma$, if $\sigma \in \text{VR}$ and $\sigma \neq v$;

(1.3) $[\tau/v]\sigma = \tau$, if $\sigma \in \text{VR}$ and $\sigma = v$;

(2.1) $[\tau/v](\rho\sigma) = [\tau/v]\rho[\tau/v]\sigma$;

(2.2.1) $[\tau/v](\lambda w.\rho) = \lambda w.\rho$, if $w = v$;

(2.2.2) $[\tau/v](\lambda w.\rho) = \lambda w.[\tau/v]\rho$, if $w \neq v$.

Since the definition of being free to substitute for a variable explicitly uses the notation for substitution, its definition for the simply typed Lambda calculus also requires the explicit mention of types.

Definition 19 Free for a term to substitute for a variable

Let σ, τ be terms and let u and v be variables. Let term σ and variable v be of the same type. Then, σ is free to substitute for v in τ iff one of the following holds:

(1.1) $\tau \in \text{CN}$;

(1.2) $\tau \in \text{VR}$;

(2.1) τ has the form $(\pi\rho)$ and σ is free to substitute for v in both π and ρ;

(2.2) τ has the form $\lambda u.\pi$ and σ is free to substitute for v in π and if $v \in \text{Fvr}(\pi)$, then $u \notin \text{Fvr}(\sigma)$.

Exercises: Simply typed Lambda calculus

1. Let a, b, c, and d be terms and let x, y, and z be variables, where c and x are of type e, d, and y are of type t/e, b and z are of type $(t/e)/(t/e)$, and a of type $t/(t/e)$. Determine which of the following are well-formed terms, taking into account the abbreviations introduced in this section. If a term is well formed, restore its unabbreviated form, and for each well-formed term, determine its type.

(a) cd (f) zyc (k) $(\lambda x.yc)c$

(b) dc (g) $z(yc)$ (l) $a(zd)$

(c) zy (h) ay (m) $(\lambda x.xy)a$

(d) da (i) $a(zd)$ (n) $(\lambda z.dx)z$

(e) $\lambda x.dx$ (j) $\lambda y.ad$ (o) $\lambda x.adx$

2. Using the same type assignment as used in the previous example, determine the type of p and the type of q as they occur within the following terms. Assume that each term is of type t. In the event that either p or q admits of more than one type, indicate what they are.

(a) *pdx* (d) *q(byp)*
(b) *p(dx)* (e) *p(qd)*
(c) *pad* (f) *q(pbd)*

3.2 Semantics: Functional Structures

In this section, we shall show how the terms of the simply typed Lambda calculus denote functions. We do this by defining a structure for the constants of the simply typed Lambda calculus. However, before presenting the definition, we shall illustrate the formulation of the definition by restating the definition for a structure for a CPDL signature.

Recall that the point of a structure is to assign values to each of the nonlogical symbols used in the notation of CPDL. To do this, one requires a way to characterize the nonlogical symbols. To this end, we defined a CPDL signature, which is a trio, comprising the set of individual symbols, the set of relational symbols, and the adicity function that assigns to each relational symbol its adicity. Just now, we saw an alternative characterization of the set of nonlogical symbols for CPDL. Using this new characterization of the notation of CPDL, let us recharacterize what a structure for CPDL is. Recall that a structure for CPDL is the ordered pair $\langle U, i \rangle$, where i is the interpretation function and U is a nonempty set. i's domain is the set NS, which was defined to be $\mathrm{IS} \cup \mathrm{RS}$, and its codomain comprises various sets built from U. However, not any function from the domain of symbols to the codomain is an interpretation function. Certain constraints must be respected. In particular, each member of IS, that is, of NS_0, is assigned a member of U and each member of RS_n, that is, of NS_n, is assigned a set of n-tuples drawn from U.

By exhaustively segregating the members of the codomain of an interpretation function into a family of sets, one can index the members of the family in such a way that the constraint on the function becomes easy to state. Here is how: let D_0 be U, let D_1 be $\mathrm{Pow}(U)$, and in general, for each $n > 1$, let D_n be $\mathrm{Pow}(U^n)$. Finally, let D be $\bigcup_{i \in \mathbb{N}} D_i$. The codomain of an interpretation function becomes simply D.

We then define a structure for CPDL as follows.

Let NS, or $\bigcup_{n \in \mathbb{N}} \mathrm{NS}_n$, be the set of nonlogical symbols for CPDL.
$\langle U, i \rangle$ is a structure for NS iff

(1) U is a nonempty set;
(2) i is a function from NS into D such that, each $n \in \mathbb{N}$, if $\sigma \in \mathrm{RS}_n$, then $i(\sigma) \in D_n$.

As with CPDL, we start with a nonempty set, called U. We let U be D_e and we let D_t be the set with just the truth values T and F. We use the other types to extend our names for various subsets of the codomain of the interpretation function as follows: $D_{x/y} = D_x^{D_y}$, that is, $\mathrm{Fnc}(D_y, D_x)$, or the set of all functions from D_y into D_x. We let $D = \bigcup_{x \in \mathrm{Typ}} D_x$. An

The Lambek Calculus and the Lambda Calculus

interpretation function for the simply typed Lambda calculus, then, is any function from CN into D respecting the types.

Definition 20 Structure for *CN* for the simply typed Lambda calculus
Let *CN* be the set of constants of the simply typed Lambda Calculus.
$\langle U, i \rangle$ is a structure for CN iff

(1) U is a nonempty set;
(2) i is a function from CN into D such that, for each $x \in Typ$, if $\tau \in CN_x$ then $i(\tau) \in D_x$.

To assign a value to terms that have free variables, one requires a variable assignment. In the simply typed Lambda calculus, variables have types and they must be assigned values from the codomain of the corresponding type.

Definition 21 Variable assignment for the simply typed Lambda calculus
A variable assignment is any function g from VR into U such that if $v \in VR_x$ then $g(v) \in D_x$.

Equipped with a structure and a variable assignment, we can now proceed to assign values to composite terms, based on the values assigned to the atomic terms, in exactly a way analogous to the way used for CQL.

Definition 22 Extension for the simply typed Lambda calculus
Let M, or $\langle U, i \rangle$, be a structure for CN. Let g be a variable assignment.

(1.1) If $\tau \in CN$, then $[\tau]_g^M = i(\tau)$;
(1.2) If $\tau \in VR$, then $[\tau]_g^M = g(\tau)$;
(2.1) If $\sigma \in TM_{x/y}$ and $\tau \in TM_y$, then $[\sigma\tau]_g^M = [\sigma]_g^M([\tau]_g^M)$;
(2.2) If $v \in VR_y$ and $\tau \in TM_x$, then $[\lambda v.\tau]_g^M = f$, where f is a function from D_y into D_x such that $f(a) = [\tau]_{g_{v \mapsto a}}^M$, for each a in D_y.

We note that the expression in (2.1) expresses the application of a function to a value and its interpretation is the application of the function to the value: $[\sigma]_g^M$ is the function, $[\tau]_g^M$ is a value to which the function applies, and $[\sigma\tau]_g^M$ is the result of application of the function to the value. The expression in (2.2) expresses the abstraction of a function from the function denoted by its matrix: the function is the one that arises by permitting the values assigned to the variable v in $[\tau]_{g_{v \mapsto a}}^M$, assigning any other variables free in τ the values assigned to them by the variable assignment g.

By the definition of a structure for CN (definition 20) and by the definition of a variable assignment (definition 21), we know that each atomic term is assigned a value suited to its type. But what about composite terms? Could it be that $\tau \in TM_x$ but $[\tau]_g^M \notin D_x$; that is, could it be, for example, that $a \in TM_{(t/e)/e}$ but $[a]_g^M \notin D_{(t/e)/e}$? The answer is that such a situation does not arise. And it is established by a proof using the principle of mathematical induction.

Fact 1 Type soundness
For each $\tau \in \text{TM}_x$, for each structure M and for each variable assignment g, $[\tau]_g^M \in D_x$.

PROOF
Let τ be a term. Let M, or $\langle U, i \rangle$, be a structure and g a variable assignment.

BASE CASE

Suppose that τ is an atomic term. Then, either it is in CN_x, for some $x \in \text{Typ}$, or it is in VR_x, for some $x \in \text{Typ}$. Suppose, on the one hand, that $\tau \in \text{CN}_x$. Then, by clause (2) of the definition of a structure (definition 20), $i(\tau) \in D_x$. And by clause (1.1) of the definition of an extension (definition 22), $[\tau]_g^M \in D_x$. Suppose, on the other hand, that $\tau \in \text{VR}_x$. Then, by the definition of a variable assignment (definition 21), $g(\tau) \in D_x$. And by clause (1.2) of the definition of an extension (definition 22), $[\tau]_g^M \in D_x$. Thus, whether the atomic term τ is a constant or a variable, $[\tau]_g^M \in D_x$.

INDUCTION CASE

Suppose that τ, a member of TM_x, is a composite term. Then, it arises either from clause (2.1) or from clause (2.2) of the definition of the formation of terms (definition 17). In the former case, there is $y \in \text{Typ}$ such that $\sigma \in \text{TM}_{x/y}$, $\pi \in \text{TM}_y$, and $\sigma\pi \in \text{TM}_x$; in the latter case, there are $y, z \in \text{Typ}$ such that $x = z/y$, $\upsilon \in \text{VR}_y$, $\rho \in \text{TM}_z$ and $\lambda\upsilon.\rho \in \text{TM}_{z/y}$.

INDUCTION HYPOTHESIS
Let $[\sigma]_g^M$ be in $D_{x/y}$, $[\pi]_g^M$ in D_y, $[\upsilon]_g^M$ in D_y, and $[\rho]_g^M$ in D_x.

CASE 1

Suppose that τ, a member of TM_x, has the form $\sigma\pi$. Then, by the induction hypothesis, $[\sigma]_g^M \in D_{x/y}$ and $[\pi]_g^M \in D_y$. Now $D_{x/y} = D_x^{D_y}$, which just is $\text{Fnc}(D_y, D_x)$, or the set of all functions from D_y into D_x. Thus, $[\sigma]_g^M([\pi]_g^M) \in D_x$. But, by clause (2.1) of the definition of extension (definition 22), $[\sigma\pi]_g^M = [\sigma]_g^M([\pi]_g^M)$. Therefore, $[\sigma\pi]_g^M \in D_x$.

CASE 2

Suppose that τ, a member of TM_x, has the form $\lambda\upsilon.\rho$. Then, by the induction hypothesis, $[\upsilon]_g^M \in D_y$ and $[\rho]_g^M \in D_z$. By clause (2.2) of definition 22, $[\lambda\upsilon.\rho]_g^M = f$, where f is a function from D_y into D_z, that is, $f \in \text{Fnc}(D_y, D_z)$, or $D_z^{D_y}$, which is just $D_{z/y}$. So $[\lambda\upsilon.\rho]_g^M \in D_{z/y}$. But $x = z/y$. Therefore, $[\lambda\upsilon.\rho]_g^M \in D_x$.

Thus, in either case, if $\tau \in \text{TM}_x$, then $[\tau]_g^M \in D_x$. Therefore, by the principle of mathematical induction, we conclude that for each $\tau \in \text{TM}_x$, for each structure M and for each variable assignment g, $[\tau]_g^M \in D_x$.

The Lambek Calculus and the Lambda Calculus

Definition 23 Equivalence of terms

Let σ and τ be terms of the simply typed Lambda calculus. Then, $\sigma \equiv \tau$, or σ and τ are equivalent, iff, for each structure M and for each variable assignment g, $[\sigma]_g^M = [\tau]_g^M$.

3.3 Deduction

We have introduced the notation of the Lambda calculus, and we have shown that it is used to express functions. The latter is not surprising, since the point of devising the notation is to have a systematic way to express functions. Two questions arise: are there different terms that are terms for the same function? If so, how can one establish that two terms for the same function are indeed terms denoting the same function?

To understand better what the import of these questions is, let us consider two familiar, analogous problems, one that we learned to tackle in primary school, another that we learned to tackle in secondary school. Let us begin with the one from primary school.

Consider the set of what one might call the elementary arithmetic expressions. It comprises the set of Indic numerals (IN) as well as all the terms that can be formed from them using the symbols for addition and multiplication. We call the set of all such expressions EA and we define it as follows.

DEFINITION: Elementary arithmetic expressions

(1) $IN \subseteq EA$;
(2.1) If $\sigma, \tau \in EA$, then $(\sigma + \tau) \in EA$;
(2.2) If $\sigma, \tau \in EA$, then $(\sigma \cdot \tau) \in EA$;
(3)   Nothing else is in EA.

Expressions of EA include: 0, 1, $1298 + 3422$, $155 \cdot (18 + 34)$ and $(542 + 174) \cdot 352$. Clearly, each natural number has many different expressions for it. Indeed, there are infinitely many expressions in EA for each natural number. One might wonder how does one determine whether two members of EA are expressions for the same natural number? The answer is to use the rules we have learned in primary school for addition and multiplication to convert one expression to another. The usual strategy is to convert a more complex expression into a simpler one, with a view to converting the two expressions into the simplest expression, an Indic numeral. Two complex expressions are expressions for the same natural number just in case the two complex expressions can be converted, in a finite number of steps, into the same Indic numeral, using the rules for addition and multiplication. By way of illustration, consider the following three expressions of EA: $\big(((3 \cdot 5) + (7 \cdot 11)) \cdot (3 + 7)\big)$, $(1 + 3) \cdot \big((6 + 4) \cdot (15 + 5)\big)$, and $\big((13 \cdot 17) \cdot 2\big) + \big((5 \cdot (5 \cdot 3)) \cdot (3 \cdot 2)\big)$.

$$\big(((3 \cdot 5)+(7 \cdot 11)) \cdot (3+7)\big)$$
$$\big((15+(7 \cdot 11)) \cdot (3+7)\big)$$
$$\big((15+77) \cdot (3+7)\big)$$
$$\big(92 \cdot (3+7)\big)$$
$$(92 \cdot 10)$$
$$920$$

$$(1+3) \cdot \big((6+4) \cdot (15+5)\big)$$
$$4 \cdot \big((6+4) \cdot (15+5)\big)$$
$$4 \cdot \big(10 \cdot (15+8)\big)$$
$$4 \cdot (10 \cdot 23)$$
$$4 \cdot 230$$
$$920$$

$$\big((13 \cdot 17) \cdot 2\big) + \big((5 \cdot (5 \cdot 3)) \cdot (3 \cdot 2)\big)$$
$$(221 \cdot 2) + \big((5 \cdot (5 \cdot 3)) \cdot (3 \cdot 2)\big)$$
$$442 + \big((5 \cdot (5 \cdot 3)) \cdot (3 \cdot 2)\big)$$
$$442 + \big((5 \cdot 15) \cdot (3 \cdot 2)\big)$$
$$442 + \big(75 \cdot (3 \cdot 2)\big)$$
$$442 + (75 \cdot 6)$$
$$442 + 450$$
$$892$$

We see from these examples that the first two expressions from EA, $\big(((3 \cdot 5)+(7 \cdot 11)) \cdot (3+7)\big)$ and $(1+3) \cdot \big((6+4) \cdot (15+5)\big)$, denote the same natural number, whereas the third expression, $\big((13 \cdot 17) \cdot 2\big) + \big((5 \cdot (5 \cdot 3)) \cdot (3 \cdot 2)\big)$, denotes a different one.

In a similar fashion, one can define the set of all polynomials in, say, the variable x and ask which of these polynomial expressions are expressions of the same function. Two polynomial expressions are expressions of the same function just in case each can be converted into the other, using the various rules of elementary algebra. A typical strategy is to convert the two polynomials into the same polynomial. For example, the polynomials $x^2 - 5 + x + 3$ and $(x+2) \cdot (x-1)$ can be both be converted into the polynomial $x^2 + x - 2$, as a result of which we know that they are polynomials denoting the same function.

The Lambda calculus, as we saw in section 3.2, is a general notation for expressing functions. When, one might wonder, are two such expressions expressions of the same function? In particular, are there rules whereby each term can be converted into the other, as a result of which we know that the two terms denote the same function? The answer is that there are such rules. The rules, as presented here, are framed in terms of the conversion relation, denoted by $\Rightarrow$.[7] In particular, we are interested in defining a conversion relation between pairs of terms. We do so by defining a fundamental conversion relation and then enlarging it recursively. The fundamental conversion relation comprises three kinds of conversions: *alpha* conversion, *beta* conversion, and *eta* conversion.

Recall that the Lambda calculus has the lambda prefix that binds variables. CQL also has quantifier prefixes that bind variables. In chapter 11, we learned that pairs of formulae of CQL, such as $\forall x P x$ and $\forall y P y$ as well as such pairs as $\forall z R z x$ and $\forall y R y x$, are logically equivalent. Indeed, we said that they are mere alphabetical variants one of the other. The Lambda calculus also has terms that are alphabetical variants. We could say, after the fashion of the characterization of alphabetically variant formulae of CQL, that terms are alphabetical variants of one another just in case, first, they have been formed in the same way, second, their corresponding occurrences of constants are occurrences of the same constant, and, third, their corresponding free variable occurrences are occurrences of the same variable. In other words, two terms are alphabetical variants if they differ only with respect to the variables bound by their corresponding lambda prefixes: where one has a lambda prefix λu binding the occurrences of the variable u, the other has a lambda prefix λv binding the occurrences of the variable v in positions corresponding to the occurrences of u bound by the occurrence of λu. The terms $\lambda x.ax$ and $\lambda y, ay$, for example, are alphabetical variants, whereas the terms $\lambda x.ax$ and $\lambda y, by$ are not, nor are the terms $\lambda x.xa$ and $\lambda y, ay$.

Alpha conversion relates any term to a distinct term that is an alphabetical variant of it. Alpha conversion stipulates that, for each term τ and for variables u and v, if $u \notin \text{Fvr}(\tau)$ and u is free to substitute for v in τ, then $\lambda v.\tau \Rightarrow \lambda u.[u/v]\tau$. We now reformulate in a form suited for deductions as follows.

ALPHA (α) CONVERSION

Alpha (α) conversion
$\lambda v.\tau \Rightarrow \lambda u.[u/v]\tau$
u is free to substitute for v in τ.
$u \notin \text{Fvr}(\tau)$.

---

7. This symbol is used here to denote a different relation from the relation it was used to denote in chapter 3.

Notice that what is written above the line, unlike what is written below, is not an expression of the form of $\Rightarrow$ flanked by two Lambda terms, but is a statement of the conditions for what is written below the line, which is used in a deduction as an axiom.

Let us see whether a pair of terms, $\lambda v.\tau$ and $\lambda u.[u/v]\tau$, related by alpha conversion, are indeed alphabetical variants, as described previously. Clearly, the tree diagrams for the two terms are exactly alike, since the latter term arises from the former by substituting for the free occurrences of v in τ occurrences of u. Moreover, the two conditions on alpha conversion guarantee that the number of free occurrences of v in τ is the same as the number of free occurrences of u in $[u/v]\tau$. Let us see how. Suppose, on the one hand, that $[u/v]\tau$ has more free occurrences of u in $[u/v]\tau$ than τ has of free occurrences of v. Then τ must have free occurrences of u. But the second condition (1) of alpha conversion rules this out. For example, $\lambda y.yy$ and $\lambda x.xy$ are not alphabetical variants, as the number of free occurrences of y in yy is greater than the number of free occurrences of x in xy. At the same time, neither of $\lambda y.yy$ and $\lambda x.xy$ can arise from the other by alpha conversion. $\lambda x.xy$ cannot arise from $\lambda y.yy$ by alpha conversion, since $\lambda x.[x/y]yy$ yields $\lambda x.xx$, which is distinct from the term $\lambda x.xy$.[8] And $\lambda y.yy$ cannot arise from $\lambda x.xy$ by alpha conversion, for if it did, y would have to substitute for x in xy, contrary to the second condition, which requires that y not be free in xy. In contrast, $\lambda x.xy$ and $\lambda z.zy$ are alphabetical variants and each can arise from the other by alpha conversion.

Suppose, on the other hand, that τ has more free occurrences of v than $[u/v]\tau$ has of free occurrences of u in $[u/v]\tau$. Then τ must have free occurrences of v whose corresponding occurrences of u in $[u/v]\tau$ are not free. This means that, in the substitution of u for v in τ, an occurrence of u becomes bound. Consider, for example, $\lambda y.\lambda y.yy$ and $\lambda x.\lambda y.xy$. The former is not an alphabetical variant of the latter, as $\lambda y.yy$ has fewer free occurrences of x than $\lambda y.xy$. However, $\lambda y.\lambda y.yy$ does not arise from $\lambda x.\lambda y.xy$ by alpha conversion, for it would require y to substitute for x in $\lambda y.xy$, contrary to the first condition, which requires that y be free to substitute for x in $\lambda y.xy$. In contrast, $\lambda z\lambda y.zy$ is an alphabetical variant of $\lambda x\lambda y.xy$ and $\lambda z\lambda y.zy$ can be obtained from $\lambda x\lambda y.xy$ by alpha conversion, since z is not free in xy and z is free to substitute for x in xy.

The next rule is that of beta conversion. Readers are no doubt familiar with the practice learned in secondary school of substituting a numeral for a variable in a polynomial: For example, one substitutes the numeral 5 into the polynomial $2x^2 + 3x - 4$. The result is: $2 \cdot 5^2 + 3 \cdot 5 - 4$, which equals 51. Now, customary mathematical practice permits such polynomials to serve both to denote a function and to denote a value. The latter is possible, if in the context of discussion the variable has been stipulated to have a value. Thus, should

---

8. Of course, $\lambda x.xx$ is an alphabetical variant of $\lambda y.yy$.

The Lambek Calculus and the Lambda Calculus

somewhere in the context it be stipulated that x takes on the value of 5, then $2x^2 + 3x - 4$ denotes, not the function, but the number 51. By prefixing the lambda prefix λx to a polynomial, any free occurrence of the variable x in the polynomial becomes bound, indeed bound by the prefixed λx, which thereby shields it from assuming any value from the context. Thus, while $2x^2 + 3x - 4$ may denote either a function or a number, $(\lambda x.(2x^2 + 3x - 4))$ denotes only the function. To show that one wishes to apply this function to a number, say 5, one writes $((\lambda x.(2x^2 + 3x - 4))5)$. Beta conversion says that the term $((\lambda x.(2x^2 + 3x - 4))5)$ converts to the term $(2 \cdot 5^2 + 3 \cdot 5 - 4)$.

Let us look at the details of the rule of beta conversion. So long as either the substituens has no variables or the term into which the substitution takes place has no lambda prefix, no unwanted consequences arise. However, in the Lambda calculus, any term may be a substituens and the term into which the substitution is to take place may have lambda prefixes. This opens up the possibility that an open term, once it has substituted for a free variable in a second term, becomes a closed term. Consider $(\lambda x.\lambda y.xy)(\lambda z.zy)$. Notice that the third occurrence of the variable y is free. Hence, the composite term is open. However, beta conversion yields the term $\lambda y.(\lambda z.zy)y$, which has no free variables and so is a closed term. It is therefore necessary to restrict the rule of beta conversion to cases where the substituendum is free to substitute for the variable in the term into which the substitution is to take place. Beta conversion states that, for all terms σ and τ and for each variable v, if σ is free to substitute for v in τ, then $(\lambda v.\tau)\sigma \Rightarrow [\sigma/v]\tau$. This too can be reformulated in a form suited for deductions.

BETA (β) CONVERSION

Beta (β) conversion
$(\lambda v.\tau)\sigma \Rightarrow [\sigma/v]\tau$
σ is free to substitute for v in τ.

Again, we have an axiom, where a condition on the form of the axiom is written below it.

Finally, we come to eta conversion. The equivalence it asserts for terms of the Lambda calculus is the counterpart of the equivalence of two formulae of CQL where one formula, say α, is the quantifier matrix of the other, $\exists v\alpha$, and the quantifier prefix of the latter, $\exists v$, binds no variable in its matrix, α: for example, $\exists x(Pa \wedge Ryc)$ and $Pa \wedge Ryc$.

Eta conversion says that, for each term τ and for each variable v, if $v \notin \text{Fvr}(\tau)$, then $\lambda v.\tau v \Rightarrow \tau$. Here it is stated as an axiom for a deduction.

ETA (η) CONVERSION

Eta (η) conversion

$v \notin \text{Fvr}(\tau)$

$\lambda v.\tau v \Rightarrow \tau$

As before, we have an axiom, where a condition on the form of the axiom is written below it.

Suppose that q denotes the squaring function on the natural numbers. If we take the universe to be natural numbers, then q has type e/e: apply it to a natural number, say 3, and one obtains the natural number 9. Let x be a variable of type e. qx is therefore a term of type e, though it denotes nothing until x, the free variable, is assigned a value. From the term qx, we can form the term $\lambda x.qx$. This term is of type e/e. Since the Indic numerals are constants of type e, let us compare the terms q and $\lambda x.qx$. $q0$ is 0, $q1$ is 1, $q2$ is 4, and so on $(\lambda x.qx)0$, by beta conversion, converts to $q0$, which is 0; $(\lambda x.qx)1$ to $q1$, which is 1; $(\lambda x.qx)2$ converts to $q2$, which is 4. In other words, the two lambda expressions are expressions for the same function, the squaring function on the set of natural numbers. We therefore wish the term $\lambda x.qx$ to convert to the term q. However, $\lambda x.(px)x$ should not eta convert to px. For example, let p be a function of type $(e/e)/e$ that when applied to a natural number, say 5, returns a function, which when applied to a number adds 5 to it. For example, $(p5)2$ is 7 and $(p5)6$ is 11, while $(p3)2$ is 5, $(p3)4$ is 7. Now, $(\lambda x.(px)x)2$ is 4, $(\lambda x.(px)x)3$ is 6, and in general, $(\lambda x.(px)x)n$ is $2n$. However, $p2$ is not a natural number; rather, it is a function that adds 2 to a natural number.

The foregoing conversions are also spoken of as reductions. The idea is that the term to the left of $\Rightarrow$ reduces, or simplifies, to the term to the right. This, obviously, is not the case for alpha conversion. However, alpha conversion preserves the structure of the term and therefore does not render the right-hand term any more complex. Now eta conversion certainly does amount to a simplification, or reduction, for the term to the right of $\Rightarrow$ is a proper subterm of the term to the left. Beta conversion also relates a more complex term to a simpler one in the simply typed Lambda calculus, though it may not in the monotyped Lambda calculus. As readers can easily check, beta conversion applied to the term $(\lambda x.xx)(\lambda y.yyy)$ of the monotyped Lambda calculus yields a more complex term. Thus, in the simply typed Lambda calculus, the second term of the conversion relation is either as simple as or simpler than the first term.

We do not wish to confine conversion relation only to the terms directly related by alpha, beta, and eta conversion. In fact, nothing in the conversion rules permit us to assert $\lambda x.x \Rightarrow$

$\lambda x.x$; in particular, this expression is not an instance of alpha conversion. However, just as in classical logic, where each formula is deducible from itself, we wish that each term in the Lambda calculus be a conversion of itself. In other words, we wish for the relation denoted by $\Rightarrow$ to be reflexive.

REFLEXIVITY OF CONVERSION

> **Reflexivity of conversion**
>
> $$\overline{\tau \Rightarrow \tau}$$

This too is an axiom. There are no conditions on its form.

At the same time, we wish for the relation to be transitive. In other words, if one term is related by conversion to a second and the second to a third, we wish that the first be related by conversion to the third. Thus, for example, $\lambda x.((\lambda y.y)x)a \Rightarrow (\lambda y.y)a$ and $(\lambda y.y)a \Rightarrow a$. But again, nothing in the rules permit us to assert $\lambda x.((\lambda y.y)x)a \Rightarrow a$.

TRANSITIVITY OF CONVERSION

> **Transitivity of conversion**
>
> $$\frac{\sigma \Rightarrow \tau, \tau \Rightarrow \rho}{\sigma \Rightarrow \rho}$$

Finally, we wish that the conversion relation be *congruent,* or consistent, with application and abstraction. Thus, as the terms $(\lambda y.xy)a$ and $(\lambda z.zy)b$ convert to xa and bx, respectively, so we wish that the composite term $((\lambda y.xy))a)((\lambda z.zy)b)$ convert to the composite term $(xa)(by)$: in other words, $((\lambda y.(xy))a)((\lambda z.zy)b) \Rightarrow (xa)(by)$. This is the point of the next rule: if $\sigma \Rightarrow \tau$ and $\rho \Rightarrow \pi$, then $\sigma\rho \Rightarrow \tau\pi$.

APPLICATION CONGRUENCE OF CONVERSION

> **Application congruence of conversion**
>
> $$\frac{\sigma \Rightarrow \tau, \rho \Rightarrow \pi}{\sigma\rho \Rightarrow \tau\pi}$$

Similarly, we expect conversion to be congruent, or consistent, with abstraction. Thus, if $(\lambda y.zxy)a$ converts to zxa, that is, $(\lambda y.zxy)a \Rightarrow zxa$, then we wish that $(\lambda x.(\lambda y.zxy)a)$ convert to $\lambda x.zxa$, that is, $(\lambda x.(\lambda y.zxy)a) \Rightarrow \lambda x.zxa$. More generally, if $\sigma \Rightarrow \tau$, then $\lambda v.\sigma \Rightarrow \lambda v.\tau$.

ABSTRACTION CONGRUENCE OF CONVERSION

Abstraction congruence of conversion
$\sigma \Rightarrow \tau$
$\lambda v.\sigma \Rightarrow \lambda v.\tau$

We now define the conversion relation as the transitive, reflexive closure of the relations of alpha, beta, and eta conversions, which is congruent with respect to application and abstraction. To restate the definition formally, we first define the notion of a *conversion formula*, which is any expression of the form $\sigma \Rightarrow \tau$, where σ and τ are Lambda terms.

Definition 25 Conversion relation (tree format)
Let σ and τ be Lambda terms. $\sigma \Rightarrow \tau$ holds iff there is a tree of conversion formulae each one of which either is a formula at the top of the tree, in which case it is an instance of one of the axioms specified above, or is a conversion formula obtained from conversion formulae immediately above it in the formula tree by one of the rules specified above.

Exercises: Deduction

1. Define the relation of alphabetical variant for the formula of CQL.

2. Assume that each of the following expressions are all terms. For each term in the first column, which of the other two terms in the same row are its alphabetical variants?

ax	ay	ax
$\lambda x.x$	$\lambda x.x$	$\lambda y.y$
$\lambda x.b$	$\lambda y.b$	$\lambda x.a$
$\lambda x.ax$	$\lambda v.av$	$\lambda v.va$
$\lambda x.xy$	$\lambda z.zv$	$\lambda z.zy$
$\lambda x.\lambda y.xy$	$\lambda v.\lambda w.vw$	$\lambda y.\lambda x.yx$
$\lambda x.xc(\lambda y.ay)$	$\lambda y.yc(\lambda y.ay)$	$\lambda v.vc(\lambda z.az)$

3. Assume that each of the following expressions are all terms. Simplify them as much as possible using alpha, beta, and eta conversions. It may be useful to restore parentheses.

(a) $\lambda x.xa$
(b) $(\lambda x.bx)a$
(c) $\lambda z.lzy$
(d) $(\lambda x.cxy)a$
(e) $(\lambda x.\lambda y.cxy)(ab)$
(f) $(\lambda x.\lambda x.cx)ab$
(g) $(\lambda x.cxy)y$
(h) $(\lambda z.\lambda z.ca)(ab)$
(i) $(\lambda x.cxx)z$
(j) $(\lambda x.c(dx))x$
(k) $(\lambda x.xz)(\lambda v.va)$
(l) $(\lambda x.\lambda y.axy)y$
(m) $(\lambda z.\lambda x.zxx)(ca)b$
(n) $\lambda z.yz$
(o) $(\lambda x.\lambda y.cyx)ab$
(p) $\lambda x.ax$
(q) $(\lambda x.\lambda y.ayx)y$
(r) $(\lambda z.\lambda y.\lambda x.a(zx)(zy))(cab)$
(s) $(\lambda w.wx)(\lambda y.\lambda x.cxy)$
(t) $(\lambda z.za)((\lambda v.\lambda x.vxx)(\lambda y.\lambda w.gway))$
(u) $(\lambda x.x)(\lambda z.za)$
(v) $(\lambda x.\lambda y.xcy)ab$
(w) $(\lambda x.xa)(\lambda y.yb)$
(x) $ab(\lambda z.(\lambda v.\lambda x.v(\lambda y.dyx))(\lambda w.wz))$

4 The Lambek Typed Lambda Calculus

In section 3, we observed that the assignment of types in the term tree for terms of the simply typed Lambda calculus conform to the rules of / elimination and / introduction of the Lambek calculus. In fact, the types required to categorize all the terms of the simply typed Lambda calculus are precisely the formulae of Lambek calculus obtained by taking two expressions, here *e* and *t,* as atomic formulae and generating from them the composite formulae whose only connective is /. In addition, as readers can easily see by inspecting the type subscripts used in clause (2) of the definition of simply typed terms (definition 17), the assignment of types to composite terms conforms to the rules of / elimination and / introduction.

As we shall now see, one can expand the terms of the Lambda calculus in such a way that the terms in the expanded set comprise all and only terms categorized by not just one, but all three connectives of the Lambda calculus and that the corresponding formation rules reflect the rules of elimination and introduction for all three connectives. We shall call this expanded version of the Lambda calculus the *Lambek typed Lambda calculus.*

In the two versions of the Lambda calculus presented, we have expressions for functions and expressions for arguments, the latter expressions being possibly expressions also for functions. Both versions follow the predominant mathematical wont of writing an expression showing the application of a function to an argument by putting the function expression to the left and the argument expression on the right. Suppose that *s* denotes the successor function on the natural numbers. If we take the universe to be natural numbers, then s has type e/e: apply it to a natural number, say 3, and one obtains the natural number 4: that is, $s3$ equals 4. However, it sometimes happens that one writes the function expression on the right and the argument expression of the left. This is the custom, for example, in the notation for exponentiation. In the expression x^2, the argument expression is the variable x and the function expression is ^{2}. If we wish to reflect this custom in the Lambda calculus, then we assign the symbol introduced for the squaring function, not the type e/e, but the type $e\backslash e$, and we write for example, not $q3$ as we did in section 3.3, but $3q$ instead. The Lambek typed Lambda calculus allows for function expressions whose argument expressions are written to their right and for function expressions whose argument expressions are written to their left. A function expression whose argument expression is written to its right has the familiar type α/β, where β is the type of the argument expression and α is the type of the composite expression obtained from the function expression and the argument expression taken in that order. A function expression whose argument expression is written to its left has the new type $\beta\backslash\alpha$, where β is the type of the argument expression and α is the type of the composite expression obtained from the argument expression and the function expression taken in that order.

The Lambek typed Lambda calculus also has expressions for pairs. Given two expressions σ and τ, we can form an expression for the pair of the entities denoted by σ and τ, namely, the expression $\text{pr}(\sigma, \tau)$. Inversely, given an expression for a pair of entities, we have expressions for the its first and second coordinates. Should σ be an expression for a pair, then $\text{pj}_1(\sigma)$ is an expression for the first coordinate of the pair and $\text{pj}_2(\sigma)$ for the second. The type of the complex expression is $\alpha \cdot \beta$, where α is the type of the first coordinate and β is the type of the second coordinate. Thus, $\text{pr}(3, 7)$ is an expression that denotes the ordered pair $\langle 3, 7 \rangle$ and $\text{pr}(q,s)$ is the ordered pair comprising the squaring function and the successor function.

The types used for the Lambek typed Lambda calculus, which we shall call *Lambek types*, are defined as follows.

Definition 26 Set of Lambek types

Let e and t be distinct entities.

(1) $e, t \in \text{Typ}$;
(2) If $x, y \in \text{Typ}$, then so is (x/y), $(x\backslash y)$, and $(x \cdot y)$;
(3) Nothing else is in Typ.

As with the simple types of the Lambda calculus, the symbols in CN_x are *constants of type x* and the symbols in VR_x are *variables of type x*. The set of all constants, or CN, is $\bigcup_{x \in \text{Typ}} \text{CN}_x$ and the set of all variables, or VR, is $\bigcup_{x \in \text{Typ}} \text{VR}_x$. As we did with the

The Lambek Calculus and the Lambda Calculus

simply typed Lambda calculus, we refer to VR and CN as *atomic terms*. We extend the atomic terms to include composite terms, just as we extended the atomic terms of the simply typed Lambda calculus to its composite terms.

Definition 27 Formation of Lambek typed terms (syncategorematic version)
Let x and y be in Typ.

(1) ATOMIC TERMS

(1.1) $\quad CN_x \subseteq TM_x$;

(1.2) $\quad VR_x \subseteq TM_x$;

(2) COMPOSITE TERMS

(2.1.1) $\quad$ If $\sigma \in TM_{x/y}$ and $\tau \in TM_y$, then $(\sigma\tau) \in TM_x$;

(2.1.2) $\quad$ If $v \in VR_y$ and $\tau \in TM_x$, then $(\lambda v.\tau) \in TM_{x/y}$.

(2.2.1) $\quad$ If $\sigma \in TM_{y\backslash x}$ and $\tau \in TM_y$, then $(\sigma\tau) \in TM_x$;

(2.2.2) $\quad$ If $v \in VR_y$ and $\tau \in TM_x$, then $(\lambda v.\tau) \in TM_{y\backslash x}$.

(2.3.1) $\quad$ If $\sigma \in TM_{x \cdot y}$, then $pj_1(\sigma) \in TM_x$ and $pj_2(\sigma) \in TM_y$;

(2.3.2) $\quad$ If $\sigma \in TM_x$ and $\tau \in TM_y$, then $pr(\sigma, \tau) \in TM_{x \cdot y}$.

The set of terms for this expanded calculus is TM, or $\bigcup_{x \in Typ} TM_x$.

Punctilious readers may be wondering why, having justified the adoption of types of the form $\alpha\backslash\beta$ by the occasional mathematical practice of writing an expression for a function to the right of the expression for its argument, we have not reflected this practice in the formation of Lambek typed Lambda terms by formulating clause (2.2.1) as follows—if $\sigma \in TM_{y\backslash x}$ and $\tau \in TM_y$, then $(\tau\sigma) \in TM_x$—thereby having expressions where the function expression occurs to the right of the expression for its argument. The reason is that having Lambda terms of both types renders reading the Lambda notation more difficult, and this without any compensating gain in insight. One might then wonder why types of the form $\alpha\backslash\beta$ are introduced at all? The answer is that such types are required for the syntactic analysis of natural language expressions in a type logical grammar, a topic of chapter 14.

We now define a structure for the Lambek typed Lambda calculus. To do so, we need to pair the various types of terms with various types of values. As with the simply typed Lambda calculus, we start with a nonempty set, called U and we let U be D_e and let D_t be the set with just the truth values, T and F. As before, we use the other types to extend our names for various subsets of the codomain of the interpretation function as follows: $D_{x/y} = Fnc(D_y, D_x)$ (or, $D_x^{D_y}$), $D_{y\backslash x} = Fnc(D_y, D_x)$, and $D_{x \cdot y} = D_x \times D_y$. We let $D = \bigcup_{x \in Typ} D_x$. An interpretation function for the Lambek typed Lambda calculus, then, is any function from CN into D respecting the types.

Definition 28 Structure for CN for the Lambek typed Lambda calculus

Let CN be the set of constants of the Lambek typed Lambda Calculus.
$\langle U, i \rangle$ is a structure for CN iff

(1) U is a nonempty set;

(2) i is a function from CN into D such that, for each $x \in Typ$, if $\tau \in \text{CN}_x$ then $i(\tau) \in D_x$.

To assign a value to terms that have free variables, one requires a variable assignment.

Definition 29 Variable assignment

A variable assignment is any function g from VR into U such that, if $v \in \text{VR}_x$, then $g(v) \in D_x$.

Equipped with a structure and a variable assignment, we can now proceed to assign values to composite terms, based on the values assigned to the atomic terms, in an exactly analogous way to the way used for the simply typed Lambda calculus.

Definition 30 Extension for the Lambek typed Lambda calculus

Let M, or $\langle U, i \rangle$, be a structure for CN. Let g be a variable assignment.

(1)   ATOMIC TERMS

(1.1)   If $\tau \in \text{CN}$, then $[\tau]_g^M = i(\tau)$;

(1.2)   If $\tau \in \text{VR}$, then $[\tau]_g^M = g(\tau)$;

(2)   COMPOSITE TERMS

(2.1.1)   If $\sigma \in \text{TM}_{x/y}$ and $\tau \in \text{TM}_y$, then $[\sigma\tau]_g^M = [\sigma]_g^M([\tau]_g^M)$;

(2.1.2)   If $v \in \text{VR}_y$ and $\tau \in \text{TM}_x$, then $[\lambda v.\tau]_g^M = f$, where f is a function from D_y into D_x such that $f(a) = [\tau]_{g_{v \mapsto a}}^M$, for each a in D_y;

(2.2.1)   If $\sigma \in \text{TM}_{x \backslash y}$ and $\tau \in \text{TM}_y$, then $[\tau\sigma]_g^M = [\tau]_g^M([\sigma]_g^M)$;

(2.2.2)   If $v \in \text{VR}_y$ and $\tau \in \text{TM}_x$, then $[\lambda v.\tau]_g^M = f$, where f is a function from D_y into D_x such that $f(a) = [\tau]_{g_{v \mapsto a}}^M$, for each a in D_y;

(2.3.1)   If $\sigma \in \text{TM}_{x \cdot y}$, then $[\text{pj}_1(\sigma)]_g^M = \pi_2^1[\sigma]_g^M$ and $[\text{pj}_2(\sigma)]_g^M = \pi_2^2[\sigma]_g^M$;

(2.3.2)   If $\sigma \in \text{TM}_x$ and $\tau \in \text{TM}_y$, then $[\text{pr}(\sigma, \tau)]_g^M = \langle [\sigma]_g^M, [\tau]_g^M \rangle$.

(π_2^1 and π_2^2 are functions that apply to ordered pairs and yield the first coordinate and the second coordinate, respectively.)

Next, we show how deductions are done when Lambek types are paired with terms from the expanded Lambda calculus. The following are the six rules of the Lambek calculus where each schematic formula is pairs with a Lambda term. We start with the rules for /

The Lambek Calculus and the Lambda Calculus

and \ elimination. Corresponding to the rule of / elimination is the formation of a Lambda term expressing function application, while corresponding to the rule of / introduction is the formation of a Lambda term expression function abstraction.

/	Elimination	Introduction
	$x/y \mapsto f \quad y \mapsto a$ ─────────── $x \mapsto fa$	$\vdots \quad [\,x \mapsto v\,]$ $\vdots$ $y \mapsto f$ ─────────── $y/x \mapsto \lambda v.f$ v fresh

(To say that a variable is *fresh* is to say that it does not occur in any term above the term in question in the deduction.)

The rules for \ elimination and introduction are similar. The rule of \ elimination corresponds to the formation of a Lambda term expressing function application and the rule of \ introduction to the formation of a Lambda term expression function abstraction. (Again, we bring to readers' attention that the Lambda terms resulting from the / and \ elimination rules are schematically the same, as are the Lambda terms resulting from the introduction rules.)

\	Elimination	Introduction
	$y \mapsto a \quad y\backslash x \mapsto f$ ─────────── $x \mapsto fa$	$[\,x \mapsto v\,] \quad \vdots$ $\vdots \quad \vdots$ $y \mapsto f$ ─────────── $x\backslash y \mapsto \lambda v.f$

We come to the last pair of rules, those for · elimination and introduction. Here appear Lambda terms that are proper to the expanded Lambda calculus. The rule of · elimination corresponds to the appearance of two terms, each an immediate subterm of the term associated with the schematic type with the ·. The rule of · introduction corresponds to the appearance of a single term whose immediate subterms of the terms associated with the types to the left and the right of the ·.

	Elimination	Introduction
	$\vdots \quad x \cdot y \mapsto a \quad \vdots$ $\vdots \quad \rule{4cm}{0.4pt} \quad \vdots$ $\vdots \quad x \mapsto \text{pj}_1(a) \quad y \mapsto \text{pj}_2(a) \quad \vdots$ $\vdots \quad \vdots \quad \vdots \quad \vdots$ $\rule{6cm}{0.4pt}$ $z \mapsto b$ $\rule{2cm}{0.4pt}$ $z \mapsto b$	$x \mapsto a \quad y \mapsto b$ $\rule{3cm}{0.4pt}$ $x \cdot y \mapsto \text{pr}(a,b)$

We shall illustrate Lambek calculus deductions with these terms presently. But first we set out the axioms and rules whereby the Lambda terms proper to the Lambek typed Lambda calculus are converted one into another. Terms proper to the expanded calculus are formed using the symbols pr, pj_1, and pj_2. The first is used to denote the pairing function, the function that forms a pair from one or two things. For example, should a and b be symbols in CN_e, denoting say 3 and 7, respectively, then $\text{pr}(a, b)$ is an expression in $TM_{e \cdot e}$ and denotes $\langle 3, 7 \rangle$. pj_1 and pj_2 denote two *projection* functions, the first function applies to an ordered pair yielding its first coordinate and the second, applied to the same ordered pair, yields its second coordinate. The first projection function, then, applied to $\langle 3, 7 \rangle$ yields 3, and the second projection function applied to the same ordered pair yields 7. In general, if one forms a pair and then takes the first projection of the pair formed, one gets the first element of the pair formed, and if one takes the second project of the pair, one gets the second element of the pair. These facts about the functions mean that composite terms either of the form $\text{pj}_1(\text{pr}(\sigma, \tau))$ or of the form $\text{pj}_2(\text{pr}(\sigma, \tau))$ may be converted, respectively, to the simpler terms σ and τ.

FIRST PROJECTION OF CONVERSION

First projection of conversion
$\rule{3cm}{0.4pt}$ $\text{pj}_1(\text{pr}(\sigma, \tau)) \Rightarrow \sigma$

SECOND PROJECTION OF CONVERSION

Second projection of conversion
$$\overline{\mathrm{pj}_2(\mathrm{pr}(\sigma, \tau)) \Rightarrow \tau}$$

In addition, it is evident that there is no difference between a pair and what results from forming a pair from the first and second projections of the initial pair.

PAIRING OF CONVERSION

Pairing of conversion
$$\overline{\mathrm{pr}(\mathrm{pj}_1(\sigma), \mathrm{pj}_2(\sigma)) \Rightarrow \sigma}$$

FIRST PROJECTION CONGRUENCE OF CONVERSION

First projection congruence of conversion
$$\sigma \Rightarrow \tau$$
$$\overline{\mathrm{pj}_1(\sigma) \Rightarrow \mathrm{pj}_1(\tau)}$$

SECOND PROJECTION CONGRUENCE OF CONVERSION

Second projection congruence of conversion
$$\sigma \Rightarrow \tau$$
$$\overline{\mathrm{pj}_2(\sigma) \Rightarrow \mathrm{pj}_2(\tau)}$$

PAIRING CONGRUENCE OF CONVERSION

Pairing congruence of conversion
$\sigma \Rightarrow \tau, \rho \Rightarrow \pi$
$\mathrm{pr}(\sigma, \rho) \Rightarrow \mathrm{pr}(\tau, \pi)$

Let us now see the examples promised earlier of deductions in the Lambek typed Lambda calculus. We shall use the tree format to display the deductions. Each node of the deduction tree comprises a pair of expressions flanking the symbol $\mapsto$. The expression to the left of the symbol is a formula of the Lambek calculus and the expression to its right is a term from the Lambda calculus. Each rule applied is, in fact, a pair of rules, one from the Lambek calculus applying to the formula or formulae of a pair or of pairs of expressions, the other from the Lambda calculus, applying to the term or terms of a pair or of pairs of expressions.

Next we will present a deduction corresponding to each of the following three theorems of the Lambek calculus.

Theorem 1.1 $\alpha \cdot (\beta \cdot \gamma) \vdash_{FD} (\alpha \cdot \beta) \cdot \gamma$

Theorem 3.1 $(\alpha/\beta) \cdot \beta \vdash_{FD} \alpha$

Theorem 5.1 $(\gamma/\beta) \cdot (\beta/\alpha) \vdash_{FD} \gamma/\alpha$

However, each formula is paired with a Lambda term. This means that the deductions are carried out with the deduction rules of this section, and not the deduction rules for formula deduction in the Lambek calculus. This also means that the deducibility relation is not relative to the Lambek calculus formula deduction rules, but relative to the deduction rules of the Lambek typed Lambda calculus. We therefore symbolize the deducibility relation here as $\vdash_{LL}$, rather than as $\vdash_{FD}$.

In our first deduction, which establishes the counterpart of Th. 1.1, the Lambek calculus formula $\alpha \cdot (\beta \cdot \gamma)$ is paired with the Lambda term a.

TO SHOW

$\alpha \cdot (\beta \cdot \gamma) \mapsto a \vdash_{LL} (\alpha \cdot \beta) \cdot \gamma \mapsto \mathrm{pr}(\mathrm{pr}(\mathrm{pj}_1(a), \mathrm{pj}_1(\mathrm{pj}_2(a))), \mathrm{pj}_2(\mathrm{pj}_2(a)))$

PROOF

$$\cfrac{\cfrac{\alpha \mapsto \mathrm{pj}_1(a) \qquad \cfrac{\cfrac{\alpha \cdot (\beta \cdot \gamma) \mapsto a}{\beta \cdot \gamma \mapsto \mathrm{pj}_2(a)} \cdot E}{\beta \mapsto \mathrm{pj}_1(\mathrm{pj}_2(a))} \quad \gamma \mapsto \mathrm{pj}_2(\mathrm{pj}_2(a))}{\alpha \cdot \beta \mapsto \mathrm{pr}(\mathrm{pj}_1(a), \mathrm{pj}_1(\mathrm{pj}_2(a)))} \cdot I}{(\alpha \cdot \beta) \cdot \gamma \mapsto \mathrm{pr}(\mathrm{pr}(\mathrm{pj}_1(a), \mathrm{pj}_1(\mathrm{pj}_2(a))), \mathrm{pj}_2(\mathrm{pj}_2(a)))} \cdot I$$

The Lambek Calculus and the Lambda Calculus

It is worth noting that we could have equally paired the Lambek formula $\alpha \cdot (\beta \cdot \gamma)$ with the Lambda term $\mathrm{pr}(a, \mathrm{pr}(b,c))$, by assuming that the Lambek formula α is paired with the Lambda term a, the Lambek formula β with the Lambda term b, and the Lambek formula γ with the Lambda term c. Doing so, would yield the following proof, where the tree for the proof is just the tree for the previous proof, except that the Lambda terms have been changed.

TO SHOW

$\alpha \cdot (\beta \cdot \gamma) \mapsto \mathrm{pr}(a,\mathrm{pr}(b,c)) \vdash_{LL} (\alpha \cdot \beta) \cdot \gamma \mapsto \mathrm{pr}(\mathrm{pr}(a,b), c)$:

PROOF

$$\cfrac{\cfrac{\cfrac{\alpha \cdot (\beta \cdot \gamma) \mapsto \mathrm{pr}(a,\mathrm{pr}(b,c))}{\alpha \mapsto a \quad \cfrac{\beta \cdot \gamma \mapsto \mathrm{pr}(b,c)}{\beta \mapsto b} \cdot E \quad \gamma \mapsto c}}{\alpha \cdot \beta \mapsto \mathrm{pr}(a,b)} \cdot I}{(\alpha \cdot \beta) \cdot \gamma \mapsto \mathrm{pr}(\mathrm{pr}(a,b),c)} \cdot I$$

We now turn to a pair of proofs in the Lambek typed Lambda calculus corresponding to a deduction of theorem 3.1.

TO SHOW

$(\alpha/\beta) \cdot \beta \mapsto a \vdash_{LL} \alpha \mapsto \mathrm{pj}_1(a)\mathrm{pj}_2(a)$:

PROOF

$$\cfrac{\cfrac{(\alpha/\beta) \cdot \beta \mapsto a}{\alpha/\beta \mapsto \mathrm{pj}_1(a) \quad \beta \mapsto \mathrm{pj}_2(a)} \cdot E}{\alpha \mapsto \mathrm{pj}_1(a)\mathrm{pj}_2(a)} /E$$

TO SHOW

$(\alpha/\beta) \cdot \beta \mapsto \mathrm{pr}(a,b) \vdash_{LL} \alpha \mapsto ab$:

PROOF

$$\cfrac{\cfrac{(\alpha/\beta) \cdot \beta \mapsto \mathrm{pr}(a,b)}{\alpha/\beta \mapsto a \quad \beta \mapsto b} \cdot E}{\alpha \mapsto ab} /E$$

We conclude our illustrative deductions with two corresponding to a deduction of theorem 5.1.

TO SHOW

$(\gamma/\beta) \cdot (\beta/\alpha) \mapsto a \vdash_{LL} \gamma/\alpha \mapsto \lambda v.\mathrm{pj}_1(a)(\mathrm{pj}_2(a)v)$:

PROOF

$$\frac{\dfrac{(\gamma/\beta)\cdot(\beta/\alpha)\mapsto a}{\gamma/\beta\mapsto\mathrm{pj}_1(a)\quad\beta/\alpha\mapsto\mathrm{pj}_2(a)}\cdot E\quad[\alpha\mapsto v]_1}{\dfrac{\beta\mapsto\mathrm{pj}_2(a)v}{\dfrac{\gamma\mapsto\mathrm{pj}_1(a)(\mathrm{pj}_2(a))v}{\gamma/\alpha\mapsto\lambda v.\mathrm{pj}_1(a)(\mathrm{pj}_2(a))v}/I_1}}$$

TO SHOW

$(\alpha\backslash\beta)/\gamma\mapsto a\vdash_{LL}\alpha\backslash(\beta/\gamma)\mapsto\lambda w.\lambda v.avw$:

PROOF

$$\frac{[\alpha\mapsto w]_2\quad\dfrac{\dfrac{(\alpha\backslash\beta)/\gamma\mapsto a\quad[\gamma\mapsto v]_1}{\alpha\backslash\beta\mapsto av}/E_1}{\dfrac{\beta\mapsto avw}{\dfrac{\beta/\gamma\mapsto\lambda v.avw}{\alpha\backslash(\beta/\gamma)\mapsto\lambda w.\lambda v.avw}\backslash I_2}/I_1}\backslash E}$$

Exercises: Lambek typed Lambda calculus

1. Assign Lambda terms to each formula and carry out the relevant deduction.

   (a) $\alpha\vdash_{FD}(\alpha\cdot\beta)/\beta$

   (b) $\alpha\vdash_{FD}(\beta/\alpha)\backslash\beta$

   (c) $\gamma/\beta\vdash_{FD}(\gamma/\alpha)/(\beta/\alpha)$

   (d) $(\alpha/\beta)/\gamma\vdash_{FD}\alpha/(\gamma\cdot\beta)$

SOLUTIONS TO SOME OF THE EXERCISES

3.1.1 Monotyped Lambda calculus

1. (a) Abbreviation   (i) Neither
   (c) Neither   (k) A term
   (e) A term   (m) Neither
   (g) Abbreviation   (o) Abbreviation

2. (a) $y\lambda x.z$ (d) $\lambda x.(\lambda x.y)zx$
   (c) $(\lambda y.xz(\lambda z.ya))(\lambda x.wy(\lambda z.z))$ (f) $(z(\lambda z.x))$

The Lambek Calculus and the Lambda Calculus   613

3. (a) $[x/y]y$ (d) $[\lambda x.xy/y]y$
   (e) $[\lambda x.xy/x]\lambda z.(\lambda x.x)(\lambda y.xyz)$
   (f) $[\lambda y.ayx/z](z(\lambda z.x))$

4. All terms are free to substitute for z in az;
   All terms are free to substitute for y in $\lambda y.ay$;
   All terms but y, $\lambda w.wy$, and $\lambda x.xy$ are free to substitute for z in $\lambda y.ayz$;
   All terms but $\lambda x.xw(\lambda y.by)$ are free to substitute for w in $\lambda x.yw(\lambda y.xy)$.

5. Definition of Ovr

   (1) $\mathrm{Ovr}(v) = \{v\}$, for each $v \in \mathrm{VR}$;
   (2) $\mathrm{Ovr}(c) = \emptyset$, for each $c \in \mathrm{CN}$;
   (3) $\mathrm{Ovr}(\sigma\tau) = \mathrm{Ovr}(s) \cup Ovr(t)$;
   (4) $\mathrm{Ovr}(\lambda v.t) = \mathrm{Ovr}(t) \cup \{v\}$.

3.1.2 Simply typed Lambda calculus

1. (a) cd (f) zyc (k) $(\lambda x.yc)c$
   (c) zy (h) ay (m) $(\lambda x.xy)a$
   (e) $\lambda x.dx$ (j) $\lambda y.ad$ (o) $\lambda x.adx$

2. (a) $p \mid (t/e)/(t/e)$ (d) $q \mid t \to t; p \mid e$
   (c) $p \mid (t/(t/e))/(t/(t/e))$ (f) $p \mid (x/(t/e))/((t/e)/(t/e));$
                                                      $q \mid t/x$

14 Noun Phrases in English

1 Introduction

In chapter 10, we began our investigation of how the truth values of minimal English clauses are determined by the constituent structure of the clauses and values assigned to their minimal constituents. We would like to expand our investigation of the English clause to a wider range of clauses so as to show how the logical tools set out in the preceding four chapters can be applied to shed light on the central question of natural language semantics, namely, the question of how the meanings of constituent expressions contribute to the meaning of the expression of which they are constituents.

A natural next step is to investigate what we shall call *simple clauses*, which are like minimal clauses, except that the noun phrase contains at most one noun, one determiner, and one adjective. Afterward, we shall venture still further by considering noun phrases that contain modifying prepositional phrases or restrictive relative clauses. Finally, we shall revisit coordination. Previously, the coordination we investigated was that of coordinated declarative clauses. However, English permits the coordination of phrases too. We shall briefly investigate them as well.

2 Simple Noun Phrases in English

Noun phrases can be quite complex. Indeed, as we saw in chapter 3, section 3.3.3, a noun phrase may contain another noun phrase, and in some cases, there is no limit to this iteration. Here, we shall confine our attention for the most part to simple noun phrases, which we stipulate to be noun phrase containing at most two nouns, one subordinate to the other. In order to arrive at a characterization of the syntax of simple English noun phrases, let us begin by looking carefully at the core of any English noun phrase, the noun.

2.1 English Nouns

It is useful, to begin with, to recognize that English nouns fall into four classes: pronouns, proper nouns, count nouns, and mass nouns. Count nouns and mass nouns constitute what, in grammar reaching back to the Middle Ages, are known as *common nouns*. A common noun was thought to apply to more than one individual, thereby being common to the two or more individuals to which it applies. In contrast, *proper* nouns were thought to apply to just one individual and hence to be proper to the single individual to which it applies. Pronouns were so called because they were thought to stand in for (*pro-*) other nouns. The terms *count* and *mass* are recent additions to the technical vocabulary of traditional grammar. Count nouns were thought to apply to things that can be counted, whereas mass nouns were thought to apply to things that can only be measured. As we shall see, like many of the definitions of technical terms from traditional grammar, the characteristics of the definition, while applying to instances that readily come to mind, do not generalize.

2.1.1 Proper nouns

Proper nouns that fell within the ambit of chapter 10 were simple, personal first names. This was done to ease exposition by minimizing complexity. It is now time to get a more complete overview of English proper nouns. To begin with, proper nouns are not always single words, rather they range from single nouns, for example, *Montreal, Bratislava, Kigali, Pune;* to strings of words that have the form of a noun phrase, for example, *the Northwest Territories, The Dream of the Red Chamber, the Children's Crusade, the Age of Reason, the International Monetary Fund;* and even to simple sentences, for example, *Who Is Afraid of Virginia Woolf,* the title of a novel. There are, in fact, millions of English proper nouns: after all, each English counting numeral is a proper noun.

(1) *Two* is a prime number, but *five million three hundred forty-two thousand, eight hundred ninety-five* is not.

A proper noun, as its name is intended to suggest, denotes a unique thing. However, as is well known, for many proper nouns, so-called forenames such as *Alice, Burton, Carl*, and so on, do not denote unique individuals. Many, people share such names. Some people share both a forename and a surname: *Michael Smith,* for example, is common to many unrelated people. Still, though the same proper noun names more than one person, the presumption is that, on any occasion of use, it applies uniquely. At the same time, some proper nouns denote nothing at all. These are names of fictitious entities: fictitious persons (*Sherlock Holmes, John Doe*), fictitious places (*Xanadu, Shangrila*, or *Mount Meru*), and fictitious events (*Armageddon*). (For further details pertaining to the semantics of proper nouns, see Larson and Segal 1995, chap. 5.)

Proper nouns that comprise more than one word may exclude any determiner, require a determiner, or, rarely, permit the determiner to be optional. If a proper noun admits a determiner, the determiner is the definite article.

Though personal first names in English exclude being preceded by a determiner, many English proper nouns require the definite article (*The Hague, the Maghreb, the Crimea, the Kremlin, the Vatican, The Iliad, the Vedas, the Koran, the Himalayas*). Some include modifiers: *the Forbidden City, the Great Salt Lake, the Black Forest, the Grand Canyon, the Great Plains, the Rocky Mountains*). And while personal first names in English occur only in the singular and exclude the definite article, many English proper nouns occur only in the plural and require the definite article: *the Great Lakes, the Pyrennes, the Seychelles, the Andes,* to name a few.

A good first-order generalization, then, is this: proper nouns in English do not tolerate the free alternation between singular and plural nor do they admit being immediately preceded by either a possessive noun phrase or determiners other than the definite article.

As is well known, many English proper nouns also appear as common nouns, having a different but predicable shift in meaning. First, a proper noun for someone may also serve as a common noun for people with that name. Second, a proper noun for a company, an artist, or a composer may serve as a common noun for products produced by the company, the artist, or the composer, respectively. Third, a proper noun for something or someone famous may serve in addition as a common noun for other things having similar qualities.

(2.1) Each Dan at the wedding had a sarcastic remark to make.
Each person named *Dan* at the wedding had a sarcastic remark to make.

(2.2) No one can afford to buy a Rembrandt.
No one can afford to buy a painting by Rembrandt.

(2.3) Bill's wife is no Florence Nightingale.
Bill's wife does not have the qualities of Florence Nightingale.

(For further details, see Algeo 1973; Bauer 1983, chap. 3.2.3; Payne and Huddleston 2002, sec. 20.)

Exercises: Proper nouns

1. We observed that English proper nouns can also be used as common nouns. Find five more examples of the second and third usages discussed, preferably attested, and provide paraphrases of each.
2. Show that the following two sentences are not synonymous. Explain why they are not.
   (1) Bill thinks that he is Picasso.
   (2) Bill thinks that he is a Picasso.
3. Find five examples of English proper nouns used as verbs, preferably attested, and provide paraphases of each.
4. We saw in chapter 12 that CQL can be enriched by adding to its set of logic constants the symbol *I*, a binary relational symbol, to denote the identity relation on a structure's

universe. Suppose that one adds to its set of logical constants instead the symbol E, a unary relational symbol, to denote a structure's universe. Using the unary relational symbol E and using the individual symbol c as a translation of the proper noun *Santa Claus,* write out formulae for each of the following English sentences.

(1) Everything exists.
(2) Nothing exists.
(3) Santa Claus does not exist.

Do the formulae adequately render their corresponding English sentences? In each case, explain your answer.

2.1.2 Pronouns

Pronouns, which, unlike proper nouns, show a productive singular plural contrast, and which, unlike common nouns, do not admit determiners, are fairly diverse, encompassing eight subcategories.

TYPE OF PRONOUN	EXAMPLES
Quantificational pronouns	*someone, somebody, something, everyone, everybody, everything, anyone, anybody, anything*
Interrogative pronouns	*what, who*
Relative pronouns	*which, who*
Demonstrative pronouns	*this, that*
Personal pronouns	*I, we, you, he, she, it, they*
Possessive pronouns	*mine, our, your, his, her, its, their*
Reflexive pronouns	*myself, yourself, himself, herself, themselves*
Reciprocal pronouns	*each other, one another*

While the first two subcategories of pronouns, namely quantificational and interrogative pronouns, do not evince the two kinds of context dependence discussed in chapter 4, the others do. Moreover, while the reciprocal pronouns are only cotext dependent and the first- and second-person singular personal, possessive, and reflexive pronouns are only setting dependent, the remaining pronouns are liable to both forms of contextual dependence.

However, as was known already to Apollonius Dyscolus (ca. second century CE), pronouns, in particular third-person personal pronouns, can be used both endophorically and exophorically. Though we discussed the distinction between exophoric and endophoric usages in chapter 4, we said nothing about the assignment of values to such expressions. The topic is complex. Here we shall confine ourselves to brief answers to these two questions: how are the values of third-person personal pronouns determined, when used exophorically? And how are their values determined, when used endophorically, get their values?

Willard Quine (1960, sec. 28) noticed that third-person personal pronouns, used exophorically, behave in a way similar to the way free variables in a formula of CQL, or indeed expressions of English, behave. Consider the following three expressions.

(3.1) Px
(3.2) x is prime.
(3.3) It is prime.

Even if a structure assigns P a value, say the set of prime numbers, the formula Px can not be assigned a value unless x is assigned a value. However, once assigned a value by a variable assignment, the truth value of the formula is determined. It is common in writing in English on mathematical topics for variables to be used, as exemplified by the quasi-English sentence in (3.2). Here too the truth value of the expression is determined only once a value is assigned to the variable. (Readers may wish to review section 2 of chapter 11, where this point was first made.) Quine's point is that the same situation obtains for the personal pronoun *it*. Once one finds a suitable value from the context, the truth or falsity of sentence (3.3) is determined.

English pronouns have gender, while variables do not. But that just means that the gender of the third-person personal pronoun puts a restriction on the possible values the pronoun could be assigned.

The situation is more complex than Quine's brief discussion suggests. To apply the insight, one must make special provisions to distinguish the circumstances of utterance from the circumstances of evaluation. This distinction and its application to pronouns and other endophoric expressions in natural language were pioneered and developed by David Kaplan, David Lewis, and Robert Stalnaker, using ideas taken from modal logic.[1]

What about third-person personal pronouns used endophorically. As we saw in section 3.1 of chapter 4, these pronous and other proforms have their values determined by cotext. The question is how. Traditional grammarians thought of pronouns as words that stand for nouns. Early transformational linguists formalized this idea in terms of transformations. The idea is that a sentence, such as the one in (4.1), which has the proform *he* whose antecedent is *Bill,* is analyzed as having a surface structure, corresponding to the expression in (4.1), and a deep structure, corresponding to the expression in (4.2), and the two are related by the transformational rule of *pronominalization,* whereby the second occurrence of *Bill* in (4.2) is replaced by *he*.

(4.1) Bill thinks that he is smart.
(4.2) Bill thinks that Bill is smart.
(5.1) Alice put on her coat and Bill put on his coat.
(5.2) Alice put on Alice's coat and Bill put on Bill's coat.

---

1. See section 1 of chapter 4. For further discussion, see Larson and Segal (1995, chap. 6).

However, this account does not work for pronouns with quantified noun phrases as antecedents:

(6.1) Each woman thinks that she is brillant.
(6.2) Each woman thinks that each woman is brillant.

Here, as Quine (1960, sec. 28) observed, one turns to logic for help. Let the pronoun *she* be assigned a value that varies as the values of its antecedent noun phrase varies, just as the values of variables vary with the quantifier matrix that binds it. So widespread is this view of how to treat such cases that linguists no longer speak of a pronoun having an antecedent, but they say instead that a pronoun is bound by its antecedent, extending this talk even to cases where the antecedent is a proper noun.

2.1.3 Common nouns

As stated at the beginning of section 2.1, common nouns have come to be divided into count nouns and mass nouns. There are clear morpho syntactic criteria by which to distinguish them. (See Jespersen 1924, 198–200, where the distinction between the two kinds of nouns is made, and Bloomfield 1933, where the patterns are set out.) We shall illustrate these patterns with the minimal pair of *advice* and *suggestion* noticed by Carl Lee Baker (1989, 8–12).

Here are eight criteria, all of which were known to Bloomfield (1933). First, count nouns have both singular and plural forms; mass nouns typically having only a singular form, do not.

(7.1) Bill heeded a suggestion/suggestions by Alice.
(7.2) Bill heeded advice/*advices by Alice.

A count noun, and not a mass noun, may serve as the antecedent for the pronouns *one* and *another* (205).

(8.1) Alice made a suggestion. Bill made one as well.
(8.2) *Alice gave advice. Bill gave one as well.

The indefinite article (*a*) and the determiners *either* and *neither* are used with singular count nouns and not with either mass nouns or plural count nouns (206).

(9.1) Bill heeded a suggestion by Alice.
(9.2) *Bill heeded an advice by Alice.

Cardinal adjectives for numbers greater than one as well as quasi-cardinal adjectives such as *a few, few, several, many* are used only with plural count nouns; whereas *more, all,* and *enough* are used with mass nouns and plural count nouns (ibid.).

(10.1) Alice made more suggestions/*suggestion to Bill.
(10.2) Alice gave more advice/*advices to Bill.

Plural count nouns and mass nouns may occur with no determiner, whereas a singular count noun requires a determiner (252).

(11.1) Alice made suggestions/*suggestion to Bill.
(11.2) Alice gave advice/*advices to Bill.

Mass nouns, but not count nouns, are preceded by *less, little, a little,* and *much* the counterparts of *fewer, few, a few,* and *many,* respectively (206).[2]

(12.1) Alice made few suggestions/*suggestion to Bill.
(12.2) Alice made few advices/*advice to Bill.
(12.3) Alice gave little advice/*advices to Bill.

All this is summarized in the following table.

DISTRIBUTIONAL PROPERTIES

DISTRIBUTIONAL PROPERTIES	MASS NOUN	COUNT NOUN
Exhibits singular/plural contrast	−	+
Antecedent for *another, one*	−	+
Modifiable by indefinite article, *either,* and *neither*	−	sg+, pl−
Modifiable by *all, enough, more*	+	sg−, pl+
Modifiable by cardinal numerals other than *one*	−	sg−, pl+
Tolerates having no determiner	−	sg−, pl+
Modifiable by *few, a few, many, several*	−	sg−, pl+
Modifiable by *less, little, a little, much*	+	−

As we noted, the contrasting terms *count noun* and *mass noun* are misleading. While it is true that count nouns do indeed apply only to things that can be counted, it is not true that mass nouns apply to things that cannot be. To be sure, many mass nouns apply to things that cannot be counted, though they can be measured:

NONCOUNTABLE MASS NOUNS

bacon, beef, bleach, bronze, broth, butter, calcium, cement, cereal, chalk, champagne, charcoal, cheese, chiffon, clay, copper, coral, corn, cotton, cream, curry, denim, dew, diesel, dirt, filth, foam, garlic, granite, gravel, grease, honey, ink, ivory, ivy, jade, jam, linen, liquor, liquorice, manure, mould, mustard, oxygen, paint, parsley, plaster, pollen, porcelain, pork, powder, rhubarb, rice, salt, satin, sherry, silk, soap, soup, spaghetti, steam, succotash, sulphur, sweat, syrup, tinsel, toast, tobacco, veal, velvet, wax, wool

---

2. The contrast between *less* and *fewer* has eroded for many North American speakers of English. It is not unusual to hear speakers say *there are less forks than knives on the table,* instead of there are fewer forks than knives on the table.

However, many nouns that evince the same morpho syntactic properties as these nouns denote countable things.

MASS NOUN	COUNT NOUN (near synonym)	MASS NOUN	COUNT NOUN (near synonym)
advice	suggestions	jewelry	jewels
ammunition	bullets	knowledge	beliefs
artillery	cannons	laundry	dirty clothes
bedding	sheets	laughter	laughs
carpeting	carpets	livestock	farm animals
change	coins	luggage	suitcases
clothing	clothes	machinery	machines
company	guests	mail	letters
crockery	pans	news	tidings
cutlery	knives	pasta	noodles
damage	injuries	pottery	pots
dishware	plates	property	possessions
drapery	drapes	silverware	spoons
evidence	clues	spaghetti	noodles
foliage	leaves	stuff	things
footwear	shoes	toiletry	toiletries
furniture	chairs	traffic	vehicles
glassware	glasses	underwear	undergarments
hardware	tools	weaponry	weapons
infantry	foot soldiers	wildlife	animals

Many English common nouns appear to satisfy the criteria of both categories (Bloomfield 1933, chap. 16.1). Often, however, there is an evident difference in construal that correlates with which criteria the word satisfies. Some mass nouns, when used as count nouns, denote kinds: for example, *breads* denote kinds of bread, *cheeses* denote kinds of cheese, *wheats* kinds of wheat, and *virtues* kinds of virtue (see Quirk et al. 1985, I.53; Payne and Huddleston 2002, 336). Others, when used as count nouns, denote standard units: for example, *cakes* denote standard units of cake, as opposed to slices of cake, *pizzas* a standard unit of pizza, as opposed to slices of pizza, *hamburger* a standard unit of hamburger, that is, a hamburger paddy (see Quirk et al. 1985, I.53; Payne and Huddleston 2002, 336). Notice that *coffees, teas,* and *beers* may denote either kinds or servings. Still other mass nouns, when used as count nouns, denote instances: for example, *details* denote instances of detail, *discrepancies* instances of discrepancy, *lights* instances (sources) of light,[3] *efforts* instances where effort is exercised, *actions* instances of action, *thoughts*

---

3. Notice that the mass noun *darkness* has no plural counterpart.

instances of thought (cf. *ideas*), *errors* instances of error (cf. *mistakes*), and *shortages* instances of shortage (see Quirk et al. 1985, I.53; Payne and Huddleston 2002, 337). Finally, some mass nouns, when used as count nouns, may denote sources of various kinds: for example, *fears* denote things that give rise to fear, or perhaps instances of fear, *embarassments* things that give rise to embarrassment, *surprises* things that give rise to surprise, *wonders* things that give rise to wonder, *delights* things that give rise to delight, and so on. Inversely, it is well known that many count nouns, satisfying the distributional criteria for mass nouns, are then construed as denoting a subset of the parts of items in their denotation as a count noun. Just which parts are included vary from word to word and from occasion of use to occasion of use.

COUNT NOUN	DENOTATION OF ITS MASS VERSION
turnip	the edible parts of turnips
potato	the edible parts of potatoes
apple	the edible parts of apples
carrot	the edible parts of carrots
duck	the edible parts of ducks
turkey	the edible parts of turkeys
chicken	the edible parts of chickens
lamb	the edible parts of lamb
crab	the edible parts of crabs
oak	the usable parts of oak trees
birch	the usable parts of birch trees
maple	the usable parts of maple trees
pine	the usable parts of pine trees
rabbit	the usable fur of rabbits
	the edible parts of rabbits

Moreover, as noted in descriptive grammars and demonstrated in psycholinguistic studies going back to Clark and Clark (1979), common nouns usually used as count nouns can be used, on the fly, as it were, as mass nouns.

(13.1)   The termite was living on a diet of book.
         (Payne and Huddleston 2002, p. 337, ex. 14 i)

(13.2)   There was cat all over the driveway.
         (ex. 14 ii)

(13.3)   Bill got a lot of house for $100,000.

(13.4)   How much floor did you lay today?

These four classes of English noun are easily distinguished on the basis of two criteria: first, whether the noun in question occurs equally freely in the singular and in the plural, and second, whether the noun in question tolerates a variety of determiners. On the one hand, proper nouns and pronouns do not tolerate determiners, though admittedly the definite article occurs in some proper nouns, while mass nouns and count nouns do. On the other hand, pronouns and count nouns evince the alternation between singular and plural, whereas proper nouns and mass nouns do not evince such an alternation.

DISTRIBUTIONAL PROPERTIES OF NOUNS	Occurs with a determiner	Admits the contrast of singular and plural
Proper noun	−	−
Pronoun	−	+
Mass noun	+	−
Count noun	+	+

Exercises: Common nouns

1. Here are three lists of words. For each list, state in what way the words in the list are exceptional with respect to the criteria set out in this section, find at least five similar words and explain how you think the exceptionality of these words should be treated.

(a) *hair, rock, rope*

(b) *antelope, deer, swine*

(c) *brains, dues, effects, goods*

2.2 Adjectives

We discussed English adjectives briefly in chapter 10. There, we pointed out that English adjectives may be used both predicatively and attributively. While many adjectives may be used either way, some are used exclusively predicatively and others exclusively attributively. Indeed, some languages, such as Slave, an indigenous North American language of the Dene (Athabaskan) language family, require all adjectives to be predicative, that is, to occur as a complement to a copular verb (Rice 1989, chap. 21; cited in Baker 2003, 194); others, such as Vata and Gbadi, West African languages, require that all adjectives occur attributively (Koopman 1984, 64–66; cited in Baker 2003, 206); still other languages, such as Russian, impose special morphology on the adjective, depending on whether it is used attributively, having a so-called short form, or predicatively, having a so-called long form (Baker 2003, 206).

Here we shall turn our attention to English attributive adjectives. When one thinks of attributive adjectives, one usually thinks of adjectives that may also occur predicatively. We shall call such adjectives *predictive* attributive adjectives. Predictive attributive adjectives are not, however, the only attributive adjectives. There are also *cardinal* adjectives and *thematic* adjectives.[4] A cardinal adjective is one that says something about the size of the denotation of the noun it modifies; in other words, they are the cardinal numerals (*one, two, three,* etc.) used as adjectives. A thematic adjective is one that restricts the set denoted by the noun it modifies by dint of a thematic relation the members of the set bear to other things.[5] These adjectives are typically obtained from nouns by the addition of a suitable suffix.

PARAPHRASAL PROPERTIES OF THEMATIC ADJECTIVES		
THEMATIC RELATION	PHRASE	PARAPHRASE
AGENT	presidential lies	lies told by a president
PATIENT	mental stimulation	stimulation of the mind
BENEFICIARY	avian sanctuaries	sanctuaries for birds
INSTRUMENT	solar generators	generators using the sun
LOCATION	marine life	life in the sea
MATERIAL	molecular chains	chains made out of molecules
POSSESSOR	musical comedies	comedy which have music
POSSESSEE	reptilian scales	scales had by reptiles
CAUSE	malarial mosquitoes	mosquitoes causing malaria
EFFECT	thermal stress	stress caused by heat

Notice that the phrase and the paraphrase observe similar restrictions on the relata of the thematic relation. For example, a lie requires an animate agent. Hence the oddity both of the phrase *reptilian lies* and of its paraphrase *lies by reptiles*.

When they modify a noun, the resulting constituent is often susceptible of a number of construals.

(14) atmospheric testing:
CONSTRUAL 1: testing of the atmosphere (patient)
CONSTRUAL 2: testing in the atmosphere (location)
CONSTRUAL 3: testing by the atmosphere (instrument)

Let us turn to the patterns whereby these three classes of attributive adjectives distinguish themselves from one another. First, as already mentioned, predictive adjectives may occur as complements to copular verbs. Cardinal and thematic adjectives either do not so occur at all or do so with less ease or with a shift in construal.

(15.1) The expensive sofa
The sofa is expensive.

---

4. The basic patterns were first identified in Levi (1978). Many of the examples pertaining to thematic adjectives are drawn from her work.

5. What are called *thematic roles* include agent, patient, beneficiary, instrument, and location. Since these adjectives are construed with such roles, or relations, we call them *thematic adjectives*.

(15.2) These two beliefs
        *These beliefs are two.
(15.3) The solar panel
        *The panel is solar.
        The panel is a solar one.

Second, coordinators may coordinate predictive adjectives, but may not coordinate thematic adjectives. Cardinal adjectives are only coordinated by the coordinator *or*.

(16.1) a rich and surly tourist
(16.2) those six or seven tourists
        *those six and seven tourists
(16.3) *solar but lunar module

Third, however, an adjective from one class may not coordinate with an adjective from another.

(17.1) *which five and governmental subsidies
(17.2) *those handsome and two friends
(17.3) *each departmental and large meeting

Fourth, though predictive adjectives may occur one after another, cardinal adjectives may not, and thematic adjectives do so only exceptionally.

(18.1) a short, ugly dog
        an obnoxious old man
(18.2) *six, seven stones
(18.3) *a dental, malarial infection

Fifth, cardinal adjectives precede predictive ones and predictive ones precede thematic ones.

(19.1) thirteen, expensive pencils
        *expensive, thirteen pencils
(19.2) large, malarial mosquito
        *malarial, large mosquito
(19.3) arrogant, criminal lawyer
        *criminal, arrogant lawyer
(19.4) eight, logical fallacies
        *logical, eight fallacies

(19.5) three, large, ugly reptilian scales
*large, ugly, reptilian, three scales
*reptilian, three, large, ugly scales
*three, reptilian, large, ugly scales

Finally, many predictive adjectives have acceptable comparative and superlative forms and admit modification by words of degree, such as *quite, rather, so,* and *very;* neither cardinal adjectives nor thematic adjectives have either comparative or superlative forms and do not admit modification by degree words, unless they shift their construal.

(20.1) rich, richer, richest
expensive, more expensive, most expensive

(20.2) five, *fiver, *fivest
five, *more five, *most five

(20.3) macular, *macularer, *macularest
macular, *more macular, *most macular

(21.1) very richer

(21.2) *so three

(21.3) *rather malarial

The foregoing criteria are summarized in this table.

	CARDINAL	PREDICTIVE	THEMATIC
LINEAR ORDER	1	2	3
FOLLOW A COPULAR VERB	No	Yes	No
COORDINATION WITHIN	No	Yes	No
COORDINATION ACROSS	No	No	No
ITERATION	No	Yes	No
COMPARATIVE/SUPERLATIVE	No	Many	No
DEGREE WORD MODIFICATION	No	Many	No

We stated that though many predictive adjectives have comparative and superlative forms, many do not. Those that do also admit of modification by degree words, and those that do not do not. The former predictive adjectives are known as *gradable* adjectives. Typical examples are *tall, short, good, bad, erroneous, accurate, beautiful, ugly, expensive,* and *gaudy*. The latter are known as *nongradable* adjectives. *Alive, dead, foreign,* and *pregnant* are examples of these adjectives.

Finally, there are seven other adjectives that behave like cardinal adjectives. They are: *few, a few, little, a little, many, much,* and *several*. They sound stilted when placed after a copular verb. They must precede predictive and thematic adjectives. They may not iterate or coordinate with themselves or cardinal adjectives. The first two, like cardinal adjectives, do not have comparative and superlative forms and they do admit modification by degree words; the last four do have comparative and superlative forms and do admit modification by degree words.

(22.1) many, more, most
(22.2) much, more, most
(22.3) few, fewer, fewest
(22.4) little, less, least

2.3 Determiners

As we noted in chapter 3, English noun phrases sometimes include words such as the definite article (*the*) and the indefinite article (*a*). These words form a substitution class with the demonstrative adjectives (*this* and *that*) and the interrogative adjectives (*which* and *what*). The substitution class, called *determiners* (Dt), also include such words as the indefinite article, *all, any, each, every, no,* and *some*. These are called *quantificational* determiners, because of their evident similarity to the quantifiers of CQL.

Determiners have three properties distinguishing them within noun phrases. First, though they do not occur in every noun phrase, if one does occur in a noun phrase, it occurs initially.

(23.1)   We enjoyed [NP that very tasty dish].
(23.2)   *We enjoyed [NP very tasty that dish].

Second, determiners do not iterate with one another.

(24)   *that a car   *a that car
       *this which tie   *which this tie
       *the each election   *each the election
       *what the friends   *the what friends
       *which what lawyer   *what which lawyer
       *what some guard   *some what guard
       *some these cars   *these some cars
       *no which contrivance   *which no contrivance
       *any no essay   *no any essay

And third, while some do coordinate with one another,

(25.1)   Each and every person must attend.
(25.2)   You must do this and that exercise.

they do not do so freely.

(26.1)   *Every and each person must attend.
(26.2)   *You must do this and some exercise.

Exercises: Determiners

1. In what way, if at all, does the fact that such English expressions as *What the hell!* and *That a boy!* are acceptable bear on the claim that English determiners do not iterate? Provide evidence to support your answer to this question.

2. Cardinal and quasi-cardinal adjectives may occur initially in a noun phrase. Does this fact warrant treating them as determiners? Provide evidence to support your answer to this question.

3 Putting Things Together

The Lambek typed Lambda calculus can be used to provide a grammar for the expressions of a natural language. Formulae, or types, of the Lambek calculus are assigned to the basic expressions of a natural language. These formulae, or types, are the syntactic categories of the basic expressions. The deduction rules are used to assign formulae, or types, to the composite expressions. In other words, they are the syntactic rules whereby the syntactic categories of composite expressions are obtained from the syntactic categories of their immediate subexpressions. At the same time, a basic expression is assigned a Lambda term of the same type as the expression and a composite expression is assigned a Lambda term of the same type as the composite expression that arises from the Lambda terms assigned to the composite expression's immediate subexpressions. The Lambek typed Lambda calculus used to treat the syntax and semantics of a natural language is often known as a *type logical grammar*. We shall call such a grammar a *Lambek typed grammar*.

In the remainder of this chapter, we shall introduce a Lambek typed grammar for English by showing how the Lambek typed Lambda calculus can be applied to a range of patterns in English. For comparison, we shall also show how the same range of patterns can be handled by an enriched constituency grammar supplemented with a transformational rule. To simplify the exposition and to facilitate the comparison, we shall alter a few of assumptions adopted earlier in chapters 3, 8, and 10.

Notice that a Lambek typed grammar does not assign values to the expressions of the language for which it is a grammar, rather it assigns Lambda terms. These Lambda terms may be assigned values through a structure for the Lambda terms. In this way, a structure for the Lambda terms of a Lambek typed grammar is indirectly a structure for the expressions of the language for which the Lambek typed grammar is a grammar. The interposition of Lambda terms between the expressions of natural language and their values in a structure provides for an elegant and compact formulation of the semantic rules corresponding to the Lambek calculus rules, though, as we shall see in section 3.6, it also has its drawbacks.

In section 3.2 of chapter 10, we formulated semantic rules, one for the constituency formation rule that forms a clause and one for each of the constituency formation rules that forms a phrase. These rules stated the value of an expression in terms of the values assigned to the expression's immediate constituents. However, the values to be assigned, though mathematically equivalent, are nonetheless different. In particular, a phrase is assigned, not a set, but its mathematically equivalent characteristic function, and a noun phrase comprising a proper noun is assigned, not a singleton set, but its member. At the same time, we add a category, commonly used by many syntacticians, known as a *determiner phrase*. We redesignate all proper nouns as determiner phrases (DP) and redesignate complements previously designated as noun phrases also as determiner phrases.[6] We shall also modify a few of the constituency formation rules found in section 3.1 of chapter 3. We shall indicate these changes when the first occasion to use one presents itself. These changes permit us to use Lambda terms in the formulation of the corresponding semantic rules, thereby making the rules easier to state and comparison with the rules of Lambek typed grammar easier to see.

The expressions of English to be investigated here are not only simple clauses but also clauses whose noun phrases include as modifiers not just adjectives but also prepositional phrases and relative clauses. We shall also revisit coordination, first discussed in chapter 8, adverting to coordination of nonclausal constituents. We end the chapter by seeing how some of the patterns pertaining to complements discussed in chapter 10 and requiring an extension of constituency grammar with enriched categories can be handled in Lambek typed grammar.

3.1 Simple Clauses with Quantified Noun Phrases

We begin our discussion by showing how Lambek typed grammar can be applied to minimal clauses whose verbs are either intransitive, transitive, or ditransitive, as illustrated by the following sentences. We shall also show how the corresponding constituency formation rules and their accompanying semantic rules can be recast in the deductive format used in Lambek typed grammars.

(27.1) Alice slept.
(27.2) Alice greeted Bill.
(27.3) Alice showed Bill Fido.

A Lambek typed grammar follows the customary practice of assigning to clauses truth values and proper nouns members of a structure's universe. This means that a clause has type t and is assigned a Lambda constant of type t, and hence assigned a truth value when the Lambda terms are interpreted in a structure, and all proper nouns are of type e and are

---

6. We shall continued to use *noun phrase* in our description of English expressions and use DP only in connection with the enriched constituency grammar analysis of this chapter.

assigned constants of type *e,* and hence assigned an individual in the universe of a structure for Lambda terms. Sentence (27.1) comprises a proper noun and an intransitive verb. Since the intransitive verb *slept* occurs to the right of the proper noun *Alice,* the intransitive verb must be assigned the type $e\backslash t$. In our altered constituency grammar, the proper noun *Alice* is assigned the category DP, the intransitive verb *slept* is still assigned the category V:⟨⟩, or VP, and a clause is still assigned S. Since we have changed the category NP to DP, the clause formation rule, NP VP → S changes to the rule, DP VP → S. To make clear the close analogy between the clause formation rule and the elimination rule of the Lambek calculus, we recast the clause formation rule in the form of an elimination rule.

$$\frac{\text{DP} \quad \text{VP}}{\text{S}} \text{S1} \qquad\qquad \frac{e \quad e\backslash t}{t} \backslash \text{E}$$

We now apply these rules to sentence (25.1).

$$\frac{\begin{array}{cc} \textit{Alice} & \textit{slept} \\ \text{DP} & \text{VP} \end{array}}{\text{S}} \text{S1} \qquad\qquad \frac{\begin{array}{cc} \textit{Alice} & \textit{slept} \\ e & e\backslash t \end{array}}{t} \backslash \text{E}$$

We shall call *derivations* not only constituency trees of constituency grammar recast in the format of a deduction but also deductions in the Lambek calculus insofar as they are applied to the syntactic analysis of natural language expressions, as just illustrated.

Using Lambda terms, we can state the syntactico-semantic version of DP, the clause formation, in a fashion as compact as that of the Lambek typed grammar. A proper noun, such as *Alice,* is assigned a member of a structure's universe. Let *a* be a Lambda constant of type *e*. An intransitive verb, such as *slept,* denotes a subset of a structure's universe. Let it be assigned a Lambda constant for the characteristic function of such as set. Call it *s*. The term *a* can be assigned to the proper noun, regardless of whether it is categorized as DP or as of type *e*. Similarly, the term *s* can be assigned to an intransitive verb, regardless of whether it is categorized as VP or as of type $e\backslash t$. We use these terms of the Lambda calculus to write down the syntactico-semantic version of the DP rule and the corresponding rule in a Lambek typed grammar.

$$\frac{\text{DP} \mapsto a \quad \text{VP} \mapsto s}{\text{S} \mapsto sa} \text{DP} \qquad\qquad \frac{e \mapsto a \quad e\backslash t \mapsto s}{t \mapsto sa} \backslash \text{E}$$

As readers might recall from the paragraph following the definition of the formation of Lambek typed terms (definition 27) chapter 13, the Lambda term of type $e\backslash t$ is written to the left of the term of type *e* when the two terms are combined into a composite term.[7]

---

7. The assignment of the type $e\backslash t$ to an English verb phrase reflects a fact about English: sometimes argument expressions occur to the left of the function expressions. Though as we noticed in chapter 13, this is also true of informal mathematical notation, it is avoided in the notation of the Lambda calculus, as to do so brings about needless notational complexity.

Next comes sentence (27.2). Its analysis in constituency grammar is just the way it was before, with one further alteration: what was previously specified as a noun phrase in a specification of a word's complement is now specified as a determiner phrase. Turning now to the Lambek typed analysis, we ask: what is the type of the transitive verb *greeted*? Since *Bill*, a proper noun, has type e and *greeted Bill*, the verb phrase, has type $e\backslash t$, *greeted* has the type $(e\backslash t)/e$. Here then are the derivations.[8]

$$
\begin{array}{ccc}
 & \text{greeted} & \text{Bill} \\
\text{Alice} & \text{V:}\langle\text{DP}\rangle & \text{DP} \\ \hline
\text{DP} & \multicolumn{2}{c}{\text{VP}} \\ \hline
\multicolumn{3}{c}{\text{S}}
\end{array}
\qquad
\begin{array}{ccc}
 & \text{greeted} & \text{Bill} \\
\text{Alice} & (e\backslash t)/e & e \\ \hline
e & \multicolumn{2}{c}{e\backslash t} \\ \hline
\multicolumn{3}{c}{t}
\end{array}
$$

The values associated with the expressions *Alice, Bill,* and *greeted* are those of the values of Lambda calculus terms of type e for the first two and of type $(e\backslash t)/e$ for the last. Let those terms be a, b, and g, respectively, which we take to be terms of the appropriate type in the Lambda calculus.

$$
\begin{array}{ccc}
 & \text{greeted} & \text{Bill} \\
\text{Alice} & \text{V:}\langle\text{DP}\rangle \mapsto g & \text{DP} \mapsto b \\ \hline
\text{DP} \mapsto a & \multicolumn{2}{c}{\text{VP} \mapsto gb} \\ \hline
\multicolumn{3}{c}{\text{S} \mapsto gba}
\end{array}
\qquad
\begin{array}{ccc}
 & \text{greeted} & \text{Bill} \\
\text{Alice} & (e\backslash t)/e \mapsto g & e \mapsto b \\ \hline
e \mapsto a & \multicolumn{2}{c}{e\backslash t \mapsto gb} \\ \hline
\multicolumn{3}{c}{t \mapsto gba}
\end{array}
$$

(Recall that gba is an abbreviation of the Lambda term $((gb)a)$.)

The very same values assigned in the Lambek typed derivation provide appropriate values for the constituency grammar derivation. To see this, let us pause to consider what the function denoted by g is. Let G be the set of ordered pairs formed from members of a structure's universe. Corresponding to this set is a function that assigns to each member of the universe x the set of those things in the universe that G pairs with x. The function denoted by g assigns to each member of the universe x, the characteristic function of the set of members of the universe that G pairs with x. This is just a restatement in terms of the Lambda calculus of the semantic rule stated in section 3.3.1 of chapter 10 for the formation of verb phrases from a transitive verb and its complement noun phrase. We can easily see the equivalence of the rule for the formation of a verb phrase from a transitive verb and its complement in a constituency grammar and the rule of \E where the terms are $(e\backslash t)/e$ and e, in that order.

$$
\begin{array}{cc}
\text{V:}\langle\text{DP}\rangle \mapsto g & \text{DP} \mapsto b \\ \hline
\multicolumn{2}{c}{\text{VP} \mapsto gb}
\end{array}
\qquad
\begin{array}{cc}
(e\backslash t)/e \mapsto g & e \mapsto b \\ \hline
\multicolumn{2}{c}{e\backslash t \mapsto gb}
\end{array}
$$

---

8. The labels for the rules applied in the derivations are omitted, lest the derivations become too cluttered with notation. In each case, it is evident which rule has been applied.

We now come to the last example, sentence (27.3). Following the reasoning used thus far, the type of *showed Bill Fido,* the verb phrase, is $e\backslash t$ and the type of *Bill* and *Fido* is e. There are two prima facie plausible type assignments for *showed*: $((e\backslash t)/e)/e$ and $(e\backslash t)/(e\cdot e)$. Within the context of our exposition here, the assignments are empirically equivalent. We choose the latter to emphasize the parallel with the constituency grammar analysis.

$$\frac{e \mapsto a \quad \dfrac{(e\backslash t)/(e\cdot e) \mapsto s \quad \dfrac{e \mapsto b \quad e \mapsto f}{e\cdot e \mapsto \mathrm{pr}(b,f)}}{e\backslash t \mapsto s(\mathrm{pr}(b,f))}}{t \mapsto (s(\mathrm{pr}(b,f)))a}$$

(Alice) (showed) (Bill) (Fido)

(Recall that pr is an expression in the Lambda calculus that denotes the pairing function. For example, $\mathrm{pr}(3,7)$ is an expression that denotes the ordered pair $\langle 3,7\rangle$.)

The constituency derivation tree is just like the Lambek typed derivation tree, except that the two proper nouns, *Bill* and *Fido,* do not form a constituent. The values assigned are the same.

$$\frac{\mathrm{DP}\mapsto a \quad \dfrac{V{:}\langle \mathrm{DP},\mathrm{DP}\rangle \mapsto s \quad \mathrm{DP}\mapsto b \quad \mathrm{DP}\mapsto f}{V{:}\langle\,\rangle \mapsto s(\mathrm{pr}(b,f))}}{S \mapsto (s(\mathrm{pr}(b,f)))a}$$

(Alice) (showed) (Bill) (Fido)

Again, it is easy to see the equivalence of the rule for the formation of a verb phrase from a ditransitive verb and its two complements in a constituency grammar and the rule of \ elimination where the terms are $(e\backslash t)/(e\cdot e)$ and $e\cdot e$, in that order.

$$\frac{V{:}\langle \mathrm{DP},\mathrm{DP}\rangle \mapsto s \quad \mathrm{DP}\mapsto b \quad \mathrm{DP}\mapsto f}{V{:}\langle\,\rangle \mapsto s(\mathrm{pr}(b,f))}$$

$$\frac{(e\backslash t)/(e\cdot e) \mapsto s \quad e\cdot e \mapsto \mathrm{pr}(b,f)}{e\backslash t \mapsto s(\mathrm{pr}(b,f))}$$

We now venture beyond English minimal clauses and explore how to treat clauses whose subject noun phrases are quantified noun phrases. There are four kinds of quantificational English noun phrases, depending on whether the head noun is singular or plural, count or noncount. We shall confined ourselves to singular quantified noun phrases with count nouns.

Right at the inception of the development of CQL, the great Italian mathematician, Giuseppe Peano (1858–1932) noted that sentences, such as the in (28), have paraphrases by sentences, such as those in (29), that are themselves readily put into logical notation, such as the formulae in (30).

(28.1)   Each boy sleeps.
(28.2)   Some girl sleeps.
(29.1)   For each *x*, if *x* is a boy, then *x* sleeps.
(29.2)   There is some *x* such that *x* is a girl and *x* sleeps.
(30.1) $\forall x (Bx \to Sx)$
(30.2) $\exists x (Gx \wedge Sx)$

While each of the sentences in (28) are monoclausal, comprising a subject noun phrase and a verb phrase with a single intransitive verb, the sentences in (29), paraphrases of the sentences in (28), are biclausal. Similarly, the formulae in (30), which render the sentences in (28) into the notation of CQL, are composite formulae, whereas formula (31.2), which renders the monoclausal sentence in (31.1), is an atomic formula.

(31.1)   Alice slept.
(31.2)   *Sa*.

This shows that the model theory of CQL cannot be adapted to the constituency of simple English clauses with quantified noun phrases. However, the model theory of the second presentation of MQL^2, set out in section 3.2.2 of chapter 12, provides values suited to the syntactic structure of such clauses. In fact, simple quantified noun phrases, such as those in (28), have the same syntactic structure as the restrictors of the second presentation of MQL^2.

Let us state the constituency formation rule for constituents comprising a quantificational determiner and a singular common noun, or a minimal quantified noun phrase. The relevant constituency formation rule given in chapter 3 is restated first and its reformulation using DP in lieu of NP is stated second.

NP2   Dt N_c → NP
DP   Dt N_c → DP

These rules can be recast as follows.

$$\frac{DP \quad N_c}{NP} NP2 \qquad \frac{DP \quad N_c}{DP} DP$$

Now a quantificational determiner is assigned a suitable function from the functions assigned to the quantifiers of the second presentation of MQL^2. For example, if the determiner is *each*, we assign it $o_{\forall^2}$, and if it is *some* or the indefinite article (*a*), we assign it $o_{\exists^2_{\geq 1}}$. A common noun is assigned a subset of the universe. The resulting determiner phrase is assigned a set of subsets of the universe, namely the family of sets that results

Noun Phrases in English

from applying the function assigned to the determiner to the set assigned to the common noun. To use the Lambda calculus, all this has to be recast in terms of functions. We shall use the symbols $\forall$ and $\exists$ as the Lambda constants for the functions $o_{\forall^2}$ and $o_{\exists^2_{\geq 1}}$, adapted to serve in structures for the Lambda calculus. Such functions are of the type $(t/(e\backslash t))/(e\backslash t)$, since they are functions from characteristic functions for sets, which have type $e\backslash t$, into characteristic functions for sets of sets, which have type $t/(e\backslash t)$. Letting o is the Lambda constant for functions of this type, we state the syntactico-semantic versions of the constituency formation rule DP and its corresponding Lambek typed grammar rule.

$$\frac{\text{Dt} \mapsto o \quad N_c \mapsto b}{\text{DP} \mapsto ob}\text{ DP} \qquad \frac{(t/(e\backslash t))/(e\backslash t) \mapsto o \quad e\backslash t \mapsto b}{t/(e\backslash t) \mapsto ob}\ \backslash\text{E}$$

We first apply the constituency formation rules with their semantic pairs to sentence (28.1), letting b denote the characteristic function for the set of boys and s the characteristic function for the set of sleepers.

$$\frac{\dfrac{\overset{each}{\text{Dt} \mapsto \forall} \quad \overset{boy}{N_c \mapsto b}}{\text{DP} \mapsto \forall b} \quad \overset{sleeps}{\text{VP} \mapsto s}}{\text{S} \mapsto \forall bs}$$

To analyze the sentences in (28) using Lambek typed grammar, we must assign a type to common nouns. Since they denote subsets of a structure's universe, they must have either the type $e\backslash t$ or t/e. For the time being, we shall assign them the former type. As readers are asked to show in the exercises, neither of these choices is empirically correct. We shall show how this empirical inadequacy can be addressed in section 3.5.

$$\frac{\dfrac{\overset{each}{(t/(e\backslash t))/(e\backslash t) \mapsto \forall} \quad \overset{boy}{e\backslash t \mapsto b}}{t/(e\backslash t) \mapsto \forall b} \quad \overset{sleeps}{e\backslash t \mapsto s}}{t \mapsto \forall bs}$$

We close this section by bringing to readers' attention a contrast between the constituency grammar and the Lambek typed grammar set out here. As we observed in chapter 3, section 3.1, a constituency grammar assigns the same syntactic category, namely NP, both to an expression which comprises a single proper noun and to an expression comprising a determiner followed by a common noun. The modified constituency grammar adopted in this chapter also assigns both kinds of expressions the same syntactic category, namely DP. In contrast, a Lambek typed grammar assigns to an expression comprising a single proper noun the type e and to an expression comprising a quantificational determiner followed by a common noun the type $t/(e \backslash t)$. Yet both grammars assign to an expression comprising just a proper noun an individual from a structure's universe and to an expression comprising a quantificational determiner followed by a common noun a family of subsets of a structure's universe, or equivalently, its counterpart using characteristic functions.

Exercises: Simple clauses with quantified noun phrases

1. Recast the constituency trees in section 3.3 of chapter 10 as derivations.
2. Find evidence to show that common nouns cannot have either the type $e\backslash t$ or the type t/e.

3.2 Adjectives Again

We saw in section 2.2 that there are various subcategories of attributive adjectives, the main categories being cardinal, predictive, and thematic. In general, little, if any, attention has been given to how thematic adjectives might be treated semantically. Most attention has been devoted to cardinal adjectives and to predictive adjectives. Since we shall discuss cardinal adjectives when we revisit quantified noun phrases (section 3.5), we shall discuss only predicating adjectives here.

Predictive adjectives do not form a uniform semantic class; rather, they comprise three principal classes, two of which restrict the denotations of the nouns they modify and one of which excludes the denotation of the nouns they modify. Restrictive predicating adjectives that, when occurring attributively, can be paraphrased by a pair of coordinated restrictive relative clauses are called *intersective* (predictive) adjectives. Color adjectives furnish ready examples.

(32.1)   A pink elephant is in the cage.
(32.2)   Something which is pink and which is an elephant is in the cage.

Another criterion for such adjectives is that they give rise to judgments of entailment of the sort that follows. The first sentence is judged to entail the second and the third.

(33.0)   This is a pink elephant.
(33.1)   This is pink.
(33.2)   This is an elephant.

The second kind of restrictive predictive adjective does not admit the kind of paraphrase illustrated in (32): sentence (34.2) is not a paraphrase of the one in (34.1).

(34.1)   A small elephant is in the cage.
(34.2)   Something which is small and which is an elephant is in the cage.

Noun Phrases in English

However, such adjectives, called *subsective* adjectives, are judged to have entailments that contrast with those for intersective adjectives. In particular, though sentence (35.0) is judged to entail sentence (35.2), it is not judged to entail the one in (35.1).

(35.0)   This is a small elephant.

(35.1)   This is small.

(35.2)   This is an elephant.

In general, what is judged to be a small elephant does need not be judged as small.

Finally, we come to what one might call *exclusive* adjectives. Like subsective adjectives, and unlike intersective adjectives, the noun phrases in which they occur are judged not to be paraphrased properly by a noun phrase with a pair of coordinated restrictive relative clauses, one for the adjective and one for the noun.

(36.1)   A plastic flower is in the vase.

(36.2)   Something which is plastic and which is a flower is in the vase.

Sentence (36.2) is not judged to be a correct paraphrase of the one in (36.1). These adjectives are judged to have entailments that contrast both with intersective and with subsective adjectives. In particular, such adjectives give rise to noun phrases whose denotations are disjoint from the denotations of the nouns they modify. For example, sentence (37.0) is judged to entail sentence (37.1) and is judged to entail the falsity of sentence (37.2).

(37.0)   This is a plastic flower.

(37.1)   This is plastic.

(37.2)   This is a flower.

In a constituency grammar, predictive adjectives are assigned the same category regardless of whether they occur as predicates, that is, as complements to copular verbs, or they occur as attributes, that is as modifiers of a common noun. They are therefore assigned the same value in each position.

The constituency formation rule NP3 in chapter 3, restated first, forms a noun phrase from a determiner, an adjective phrase, and a common noun. We shall form the same constituent, now called a determiner phrase, using the constituency formation rule DP and another constituency formation rule N3, given second.

NP3   Dt AP N_c → NP

N3   AP N_c → N_c

Using the derivational format, we can easily see that the constituents formed by NP3 are also formed by a combination of N3 and DP.

$$\frac{\text{Dt} \quad \text{AP} \quad \text{N}_c}{\text{NP}} \text{NP3} \qquad \frac{\text{Dt} \quad \frac{\text{AP} \quad \text{N}_c}{\text{N}_c} \text{N3}}{\text{DP}} \text{DP}$$

Since intersective predictive adjectives are indeed intersective, we use the Lambda constant $\cap_{e \backslash t}$ to state the syntactico-semantic version of the constituency formation rule N3.

$$\frac{\text{AP} \mapsto a \quad \text{N}_c \mapsto b}{\text{N}_c \mapsto \cap_{e \backslash t} ab} \text{N3}$$

What is the function denoted by the Lambda constant $\cap_{e \backslash t}$? $\cap$ is a set theoretic operation that for each pair of sets, yields their intersection. Next, consider some fixed set A. We can define a function that assigns to each set its intersection with A. More formally, let f be a function that assigns $X \cap A$ to set X. $\cap_{e \backslash t}$ denotes the counterpart of this function that assigns to the characteristic function of X the characteristic function of $X \cap A$. It is important to stress that $\cap_{e \backslash t}$ is a term of the Lambda calculus, which in turn denotes the function from characteristic functions to characteristic functions just described, whereas $\cap$ is the symbol for the operation of set intersection. (The use of a subscript that is a type with $\cap$ is intended to remind readers that the symbol is a Lambda term and is distinct from the set theoretic operation of intersection to which it bears a systematic relation. The motivation for this Lambda notation will be explained in section 3.6.)

What happens in a Lambek typed grammar? If one follows the long-standing view that the copular verb denotes the identity relation, then the copular verb has the type $(e \backslash t)/(e \backslash t)$ and the predictive adjective in predicative position has the type $e \backslash t$. However, the very same adjective in attributive position must be assigned the type $(e \backslash t)/(e \backslash t)$, since it modifies a common noun, which has the type $e \backslash t$ to yield an expression of the same type. This means that a predictive adjective in predicative position has the type $e \backslash t$ and in attributive position the type $(e \backslash t)/(e \backslash t)$. This means that the Lambda terms assigned in these positions must not only be different but they must have different values. Suppose, for example, that a predictive adjective in predicative position is assigned the Lambda term. The term must be of type $e \backslash t$ and therefore is a characteristic function for a subset of a structure's universe. However, the predictive adjective in attributive position must be assigned a Lambda term for a function from (charactistic functions of) subsets of a structure's universe to (characteristic functions of) the universe's subsets. Put in terms of the Lambda calculus, if the Lambda term assigned to the adjective in predicative position is a, then the Lambda term assigned to it in attributive position must be $\cap_{e \backslash t} a$; the former term denotes the characteristic function of a set, and the latter term denotes a function from characteristic functions to characteristic functions as we just described.

The derivation in the Lambek typed grammar corresponding to the one in the constituency grammar is therefore this:

$$\frac{(e \backslash t)/(e \backslash t) \mapsto \cap_{e \backslash t} a \quad e \backslash t \mapsto b}{e \backslash t \mapsto \cap_{e \backslash t} ab}$$

3.3 Prepositional Phrases

In chapter 10, we saw that a preposition that occurs as a complement of the verb *to be* denotes a relation other than the identity relation (section 3.3.2), whereas should it occur as a complement of other verbs, it denotes the identity relation (section 3.3.3).

(38.1)   Dan is on the bus.
(38.2)   Dan relies on Beverly.

But prepositional phrases are not always complements; they also occur as modifiers, modifying either a common noun or a verb.

(39.1)   A man is in Calgary.
(39.2)   A man in Calgary sleeps.
(39.3)   A man sleeps in Calgary.

In chapter 3, constituency formation rule NP4, given first, was used to analyze sentences such as the one in (39.2); however here, we shall use N4, given second, together with the DP constituency formation rule given previously.

NP4   Dt N_c PP $\to$ NP
N4 N_c PP $\to$ N_c

Again, we use the derivational format to show that the constituents formed by NP4 are also formed by a combination of N4 and DP.

$$\frac{\text{Dt} \quad N_c \quad \text{PP}}{\text{NP}} \text{NP4} \qquad \frac{\text{Dt} \quad \dfrac{\dfrac{N_c \quad \text{PP}}{N_c}\text{N4}}{\text{DP}}}{\text{DP}}$$

Next, we turn to the question of what values to assign to the various constituents. Here, the prepositional phrases are not complements, but modifiers. The customary view is that the denotation of a constituent that is a modifier restricts the denotation of the constituent that it modifies. Though we shall see exceptions, the cases we shall be examining here are all cases of restriction corresponding to set intersection. We shall use the Lambda constant $\cap_{e \backslash t}$, introduced earlier, and obtain the following syntactico-semantic version of N4.

$$\frac{N_c \mapsto a \quad \text{PP} \mapsto b}{N_c \mapsto \cap_{e \backslash t} ba} \text{N4}$$

These rules have two desirable consequences. First, these constituency formation rules reflect an ambiguity that arises when two or more prepositional phrases modify a common noun.

(40.0)   A book on the table near the lamp
(40.1)   A [NP [N_c [N_c book [PP on the table]] [PP near the lamp]]]
(40.2)   A [NP [N_c book [PP on the [NP [N_c table [PP near the lamp]]]]]]

Second, these rules reflect the judgment of speakers of English that each of the sentences in (39.2) and (39.3) entails the other and each entails sentence (39.1). Showing this is left as an exercise.

We turn, now, to how sentences (39.2) and (39.3) are handled by Lambek typed grammar. (We shall discuss sentence (39.1) afterward.) Let us consider first prepositional phrases that modify common nouns, such as the one in (39.2). Common nouns have been assigned the type $e\backslash t$. Since *man* and *man in Calgary* form intersubstitutable expressions, they must have the same type. Therefore, *in Calgary* must have the type $(e\backslash t)\backslash(e\backslash t)$. Moreover, since the preposition *in* may immediately precede a proper noun, *in* has the type $((e\backslash t)\backslash(e\backslash t))/e$.

$$\begin{array}{c c c}
& in & Calgary \\
man & ((e\backslash t)\backslash(e\backslash t))/e \mapsto i & e \mapsto c \\
\hline
e\backslash t \mapsto m & \multicolumn{2}{c}{(e\backslash t)\backslash(e\backslash t) \mapsto ic} \\
\hline
\multicolumn{3}{c}{e\backslash t \mapsto icm}
\end{array}$$

As the readers are asked to show in the exercises, this rule reflects the ambiguity noted with respect to the noun phrase in (40.0).

What type is assigned to prepositional phrases that are modifiers of intransitive verbs, as in sentence (39.3)? Intransitive verbs are assigned the type $e\backslash t$. Since *sleeps* and *sleeps in Calgary,* when occurring to the right of a proper noun, form a clause, they must have the same type. Therefore, *in Calgary* must have the type $(e\backslash t)\backslash(e\backslash t)$. Moreover, since the preposition *in* may immediately precede a proper noun, *in* has the type $((e\backslash t)\backslash(e\backslash t))/e$.

$$\begin{array}{c c c}
& in & Calgary \\
sleeps & ((e\backslash t)\backslash(e\backslash t))/e \mapsto j & e \mapsto c \\
\hline
e\backslash t \mapsto s & \multicolumn{2}{c}{(e\backslash t)\backslash(e\backslash t) \mapsto jc} \\
\hline
\multicolumn{3}{c}{e\backslash t \mapsto jcs}
\end{array}$$

However, should *in Calgary* follow the verb *to be,* as it does in (38.1), we run into a conflict. Again, if one follows the long-standing view that the copular verb denotes the identity relation, then the copular verb has the type $(e\backslash t)/(e\backslash t)$ and the prepositional phrase in predicative position must have the type $e\backslash t$. However, the very same prepositional phrase after a common noun or an intransitive verb, each of which has the type $e\backslash t$, must be assigned the type $(e\backslash t)/(e\backslash t)$. As with predictive adjectives, prepositions must be assigned different types in different positions, where the difference in types precludes the same Lambda term from being used in all the different positions.

Exercises: Prepositional phrases

1. Show that the expression *book on the table near the lamp* is amphibolous in a Lambek typed grammar, that is, that it has two syntactic derivations. Assume for the sake of the exercise that *the* has the type $e/(t/e)$.

2. Consider the sentences in (39.2) and (39.3).

(a) Explain how the constituency grammar adopted here renders these sentences equivalent, given that the prepositional phrase *in Calgary* denotes the set of things in Calgary.

(b) Show that the Lambek typed grammar where common nouns have typed $e \backslash t$ also renders these sentences equivalent.

(c) Show that the Lambek typed grammar where common nouns have typed t/e fails to render these sentences equivalent. equivalent.

3.4 Restrictive Relative Clauses

We shall investigate the syntax and semantics of relative clauses. As we remarked in section 3.4.3 of chapter 3, typically a relative clause begins with either a relative pronoun or a prepositional phrase that has a relative pronoun as an immediate constituent. Moreover, the initial relative pronoun or prepositional phrase corresponds to a noun phrase that could serve either as the subject or as the complement of the verb or to a prepositional phrase that could serve either as a complement to, or a modifier of, the verb. We indicate this correspondence in the sentences that follow.

(41.1) A man [RC [DP who] __ bought a yacht] was found dead in the marina.
(41.2) A guest [RC [PP to whom] Don gave the key __] is in the lobby.
(41.3) The woman [RC [DP whose dog] Alice fed __] is waiting for Colleen.
(41.4) The country [RC [PP in which] the president declared martial law __] is suffering from food shortages.

This observation that a relative clause is a clause with a missing element suggests, as noted by Willard Quine (1960, sec. 22 and 23), that a restrictive relative clause corresponds to a formula of CQL which has one free variable in it.

Not all relative clauses have relative pronouns. Some begin with the word *that* instead of starting with a relative pronoun or a phrase one of whose immediate constituents is a relative pronoun.

(42.1) A man [RC that __ bought a yacht] was found dead in the marina.
(42.2) A guest [RC that Don gave the key to __] is in the lobby.

(42.3) The country [RC that the president declared martial law in __] is suffering from food shortages.

Others have no special word to signal the beginning of the clause.

(43.1) A guest [RC Don gave the key to __] is in the lobby.
(43.2) The country [RC the president declared martial law in __] is suffering from food shortages.

Finally, the verb of the relative clause need not be finite, it may be infinitival.

(44) A book [RC for Don to give __ to Carol] is on the table.

Relative clauses may be either appositive or restrictive. These two relative clauses are distinguished in written English by the use of commas: an appositive relative clause is put between commas, as in (45.1), a restrictive relative clause is not, as in (45.2). This convention of punctuation reflects the fact that, when a relative clause is used appositively, the clause is uttered with a special intonation in which the voice drops. This intonation is not used when the relative clause is used restrictively.

(45.0) Each book [RC which Beverly bought __] has a red dust jacket.
(45.1) Each book, which Beverly bought, has a red dust jacket.
(45.2) Each book which Beverly bought has a red dust jacket.

A sentence with a relative clause, such as the one in (45.0), has different truth conditions. Consider the following circumstances. There are five books that are being talked about. Beverly had bought precisely three of them. The three that Beverly bought have red dust jackets; the other two, which she did not buy, do not. The sentences in (45) may be true or false. It is true that each book Beverly bought has a red dust jacket, but it is false that she bought each book. On the restrictive use of the relative clause, it *restricts* the denotation of *book* to the three Beverly bought and the sentence says of those three that they have red dust jackets. On the appositive use of the relative clause, rather than restricting the denotation of *book,* the sentence says that each of the five books has a red dust jacket and that Beverly had bought all five.

Let us now consider restrictive relative finite clauses with relative pronouns. Here are three examples.

(46.1) a dog which __ slept
(46.2) a city which Beverly likes __
(46.3) a toy which Alice gave __ to Bill.

We begin with the example in (46.1). In light of the type assignments made earlier to the indefinite article (*a*) to common nouns, such as *dog,* and the intransitive verb, *slept,* the relative pronoun, *which,* must be assigned the type, $((e\backslash t)\backslash(e\backslash t))/(e\backslash t)$. Since the denotation of *dog which slept* is the set of things that are dogs and which slept, the value

Noun Phrases in English   643

assigned to *which* is the function whose term in the Lambda calculus is $\cap_{e\backslash t}$, which denotes the function corresponding to set intersection and which we have now encountered twice in connection with modification.

$$
\cfrac{
  \cfrac{a}{(t/(e\backslash t))/(e\backslash t) \mapsto \exists}
  \quad
  \cfrac{
    \cfrac{dog}{e\backslash t \mapsto d}
    \quad
    \cfrac{
      \cfrac{which}{((e\backslash t)\backslash(e\backslash t))/(e\backslash t) \mapsto \cap_{e\backslash t}}
      \quad
      \cfrac{slept}{e\backslash t \mapsto s}
    }{(e\backslash t)\backslash(e\backslash t) \mapsto \cap_{e\backslash t} s}
  }{e\backslash t \mapsto \cap_{e\backslash t} s d}
}{t/(e\backslash t) \mapsto \exists(\cap_{e\backslash t} s d)}
$$

We turn to the example in (46.2). Under the type assignments made to its various words, there is no type that can be assigned to the entire expression. One possibility is to assign *likes*, not the type $(e\backslash t)/e$, but the mathematically equivalent type $e\backslash(e\backslash t)$.

$$
\cfrac{
  \cfrac{city}{e\backslash t \mapsto c}
  \quad
  \cfrac{
    \cfrac{which}{((e\backslash t)\backslash(e\backslash t))/(e\backslash t) \mapsto \cap_{e\backslash t}}
    \quad
    \cfrac{
      \cfrac{Beverly}{e \mapsto b}
      \quad
      \cfrac{likes}{e\backslash(e\backslash t) \mapsto l}
    }{e\backslash t \mapsto bl}
  }{(e\backslash t)\backslash(e\backslash t) \mapsto \cap_{e\backslash t}(lb)}
}{e\backslash t \mapsto \cap_{e\backslash t}(lb)c}
$$

Though the type assignments determine Lambda terms whose interpretation guarantees that the expression denotes (the characteristic function for) the set of things that are cities and which Beverly likes, it does so at the expense of assigning *likes* in (46.2) the type $e\backslash(e\backslash t)$, while *likes* in the sentence *Beverly likes Calgary* is assigned the type $(e\backslash t)/e$. Because of the different type assignments, it is possible for sentences like the following to be jointly satisfiable,

(47.1)   a city which Beverly likes is Calgary.
(47.2)   Beverly does not like Calgary.

for *like* can be assigned one relation in the first sentence and another in the second.

Fortunately, there is an alternative. It is possible to assign *likes* the usual type assigned to transitive verbs, namely, $(e\backslash t)/e$. To do so, though, we must avail ourselves of one of the rules of which we have not availed ourselves so far, the rule of / Introduction.

$$
\cfrac{
  \cfrac{city}{e\backslash t \mapsto c}
  \quad
  \cfrac{
    \cfrac{which}{((e\backslash t)\backslash(e\backslash t))/(e\backslash t) \mapsto \cap_{e\backslash t}}
    \quad
    \cfrac{
      \cfrac{Beverly}{e \mapsto b}
      \quad
      \cfrac{
        \cfrac{likes}{(e\backslash t)/e \mapsto l} \quad \cfrac{}{e \mapsto x}
      }{e\backslash t \mapsto lx}
    }{\cfrac{t \mapsto lxb}{e\backslash t \mapsto \lambda x.lxb}}
  }{(e\backslash t)\backslash(e\backslash t) \mapsto \cap_{e\backslash t}(\lambda x.lxb)}
}{t/e \mapsto \cap_{e\backslash t}(\lambda x.lxb)c}
$$

However, even here, it has been necessary to assign the relative pronoun *which* two different types, depending on whether the noun phrase containing the relative pronoun is to be connected with what would be the subject of the relative clause or the object. Moreover, the types we have do not permit us to handle relative clauses where the relative pronoun gap is not at the edge of the clause, as for example in (46.3).

There are a number of ways to address this problem. We turn to one due to Michael Moortgat (1988, 1996). He introduces a new connective to the expanded Lambek calculus, together with the following introduction rule.

$\uparrow$	Introduction
	$\vdots \quad [x \mapsto v] \quad \vdots$
	$\vdots \quad \quad \vdots$
	$y \mapsto f$
	$y \uparrow x \mapsto \lambda v.f$
	v fresh

The semantic value associated with expressions of the type $x \uparrow y$ is $\text{Fnc}(D_y, D_x)$. In other words, $D_{x \uparrow y} = D_x^{D_y}$, which is the same as $D_{x/y}$ and $D_{y \backslash x}$. The relative pronouns *which, who,* and *whom* are assigned the type $\big((t/e)\backslash(t/e)\big)/(t \uparrow e)$. Using this new type assignment, one arrives at the following derivation for the example in (46.1).

$$
\begin{array}{c}
\\
\\
\\
\\
dog \\
e\backslash t \mapsto d
\end{array}
\begin{array}{c}
\\
\\
which \\
\big((e\backslash t)\backslash(e\backslash t)\big)/(t \uparrow e) \mapsto \cap_{e\backslash t} \\
\hline
(e\backslash t)\backslash(e\backslash t) \mapsto \cap_{e\backslash t}(\lambda x.sx)
\end{array}
\begin{array}{c}
slept \\
\dfrac{e \mapsto x \quad e\backslash t \mapsto s}{t \mapsto sx} \\
t \uparrow e \mapsto \lambda x.sx
\end{array}
$$
$$e\backslash t \mapsto \cap_{e\backslash t}(\lambda x.sx)d$$

We note that eta conversion permits the last Lambda term to be reduced to $\cap_{e\backslash t} sd$.[9] Readers are encouraged to carry out the derivations for the other two sentences in (46).

We noted in chapter 3, section 3.4.3, that relative clauses are just one of a number of patterns exhibiting discontiguity. We alluded to a number of syntactic approaches to discontiguity and sketched how it can be handled by a constituency grammar supplemented with transformational rules. A relative clause, such as the one in (47.1), repeated in (48.0),

---

9. Though this might look like a case of beta conversion, it is not. Look carefully at the types of x and d.

Noun Phrases in English

is analyzed as having a deep structure, shown in (48.1) and a surface structure, shown in (48.2), the latter arising from the former by a transformation rule that moves, so to speak, the noun phrase containing the relative pronoun *which* from its object position to a position at the beginning of the clause containing it.

(48.0) A city which Beverly likes ___ is Calgary.

(48.1) A city [S Beverly [VP likes [DP which]]] is Calgary.

(48.2) A city [RC [DP,i which] [S Beverly [VP likes [DP,i t]]]] is Calgary.

To analyze sentence (48.0), we require a constituency formation rule for modification by a restrictive relative clause. To simplify the discussion, we shall take the category of relative clause to be primitive. We adopt the constituency formation rule N5: N_c RC → N_c. We now state its syntactico-semantic version, using the familiar Lambda term for intersection.

$$\frac{N_c \mapsto a \quad RC \mapsto b}{N_c \mapsto \cap_{e\backslash t} ba} \, N5$$

Next is the derivation for the expression *city which Beverly likes* in sentence (48.0). It illustrates the point made by Quine (1960) that the part of the relative clause that excludes the relative pronoun, here *Beverly likes,* corresponds to an open formula with just one free variable, here the Lambda term lxb, instead of a formula.

$$\frac{\dfrac{city}{N_c \mapsto c} \quad \dfrac{\dfrac{which}{DP,i \mapsto \lambda x} \quad \dfrac{\dfrac{Beverly}{DP \mapsto b} \quad \dfrac{likes}{\dfrac{V:\langle DP\rangle \mapsto l \quad DP,i \mapsto x}{VP \mapsto lx}}}{S \mapsto lxb}}{RC \mapsto \lambda x.lxb}}{N_c \mapsto \cap_{e\backslash t}(\lambda x.lxb)c}$$

As we observed in section 3.4.3 of chapter 3, some discontiguous constituents are, as it were, local, confined to a clause or a phrase; others are unbounded, the two parts being separated by what seems to be an unbounded number of clauses. It was soon recognized that whether an unbounded dislocation is allowed depends in part on the nature of the constituents intervening between the gap and the dislocated constituent. The first person to investigate this in a systematic fashion was John R. (Haj) Ross (1967). Ross identified constituents that would block, as it were, unbounded dislocations, dubbing them *islands*. Though we shall illustrate some islands with unbounded dislocations involving relative clauses, these islands apply to other forms of discontiguity as well.

In the pair of sentences in (47), the dislocated constituent and the gap are separated by a clause. However, in the first sentence the clause is a complement to a verb, in the second the clause is a clause in apposition to a noun.

(49.1) Carl read a book on a topic [PP on which] Dan said [S that Beverly wrote a paper __].
(49.2) *Carl read a book on a topic [PP on which] Dan repeated the rumor [S that Beverly wrote a paper __].

Another contrast depends on whether the gap occurs in an uncoordinated clause or only one of a pair of coordinated clauses.

(50.1) Colleen knows the person [DP whom] Bill greeted __ .
(50.2) *Colleen knows the person [DP whom] Alice saw Carl and Bill greeted __ .

Stating constituency formation rules that respect such islands is a central empirical challenge to grammatical theories of English. Indeed, it constitutes an area of research unto itself. Needless to say, we shall not pursue the problem further here.

3.5 Quantified Noun Phrases Again

We now return to the treatment of quantified noun phrases. Earlier we saw how to treat quantified noun phrases that occur in subject position both with an enriched constituency grammar and with a Lambek typed grammar. Though we used a sentence with an intransitive verb to illustrate the analyses, sentences where the verbs are either transitive or ditransitive could have been used as well to illustrate the point, provided that the complements of the verbs are proper nouns. However, should a position other than the subject position contain a quantified noun phrase, a problem arises.

Consider sentence (51), whose direct object is a quantified noun phrase.

(51) Alice greeted each boy.

Whereas a constituency grammar permits a syntactic derivation of the verb phrase, indeed the entire sentence, the semantic rules associated with the constituency formation rules do not assign any value to the verb phrase, as shown here.

$$
\begin{array}{c c}
 & \textit{each} \quad\quad \textit{boy} \\
\textit{greeted} & \text{Dt} \mapsto \forall \quad N_c \mapsto b \\
\underline{\text{V:}\langle\text{DP}\rangle \mapsto g} & \underline{\text{DP} \mapsto \forall b} \\
\multicolumn{2}{c}{\text{VP} \mapsto ?}
\end{array}
$$

To see why this is so, recall what the function g is. Let G be the set of ordered pairs formed from members of a structure's universe. Corresponding to this set is a function that assigns to each member of the universe x the set of those things in the universe that G pairs with x. The function denoted by g assigns to each member of the universe x the characteristic function of the set of members of the universe that G pairs with x. Thus, the domain of g is a structure's universe. However, $\forall b$ is not a subset of a structure's universe, or more accurately, a characteristic function for a subset of a structure's universe, rather it is a family of its subsets, or more accurately, a characteristic function for a set of characteristic functions for its subsets. Thus, the function denoted by g is undefined for the value assigned to the quantified noun phrase *each boy*.

A Lambek typed grammar also does not permit a syntactic derivation of the verb phrase, let alone the clause containing the verb phrase. If there is no syntactic derivation, there can be no assignment of values either.

$$\frac{greeted \quad \dfrac{each \quad boy}{\dfrac{(t/(e\backslash t))/(e\backslash t) \mapsto \forall \quad e\backslash t \mapsto b}{t/(e\backslash t) \mapsto \forall b}}}{e\backslash t \mapsto ?}$$

$$(e\backslash t)/e \mapsto g$$

The problem of quantified noun phrases in nonsubject position can be handled both in Lambek typed grammar and in transformational grammar. We shall present one solution for each. We start with Lambek typed grammar and show how Michael Moortgat (1990) proposed a variation of his idea of how to handle restrictive relative clauses to handle the problem posed by quantified noun phrases in nonsubject position. He introduces still another connective to the expanded Lambek calculus, whose introduction rule is this:

⇑	Introduction
	⋮ ⋮ ⋮
	⋮ $y \Uparrow x \mapsto f$ ⋮
	⋮ ─────── ⋮
	⋮ $y \mapsto v$ ⋮
	⋮ ─────── ⋮
	⋮ ⋮ ⋮
	───────
	$x \mapsto h$
	───────
	$x \mapsto \lambda v.h$
	(v fresh)

$D_{x \Uparrow y} = D_y^{\left(D_y^{D_x}\right)}$, that is, $\text{Fnc}(\text{Fnc}(D_x, D_y), D_y)$, where the determiners of quantified noun phrases are assigned values from $D_{e \Uparrow t}$, precisely the type that corresponds to the quantifiers of the second presentation of MQL[2].

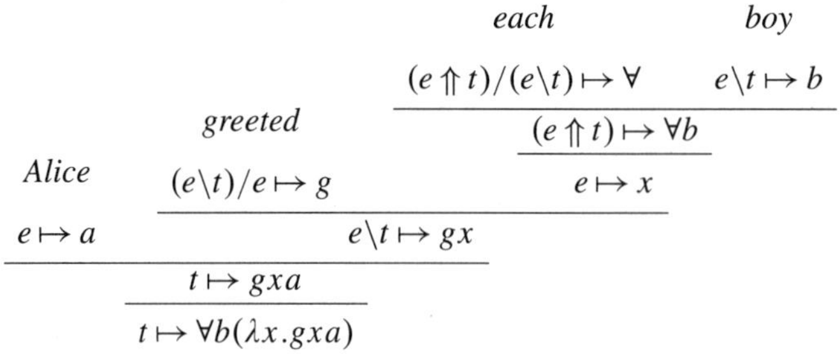

Treatment here can be applied to single clauses with more than one quantified noun phrase, as readers will see.

One way to handle the problem posed by clauses with quantified noun phrases in non-subject positions in a transformational grammar, as observed and developed by Robert May (1977), is to suppose that, in addition to deep structure and surface structure, there is what he called *logical form*. Moreover, just a transformational rule of wh-movement moves, as it were, a constituent containing a wh word to a clause initial position, so a rule of quantifier raising, or QR, moves a quantified noun phrase to an initial position in its clause. For example, sentence (51), repeated as (52.0), has, on this account, a deep structure and a surface structure corresponding to the analysis in (52.1) and a logical form corresponding to the analysis in (52.2).

(52.0) Alice greeted each boy.

(52.1) [S [DP Alice] [VP [V:⟨DP⟩ greeted] [DP each boy]]].

(52.2) [DP,i each boy] [S Alice greeted [DP,i t]].

The assignment of values is done with respect to the sentence's logical form.

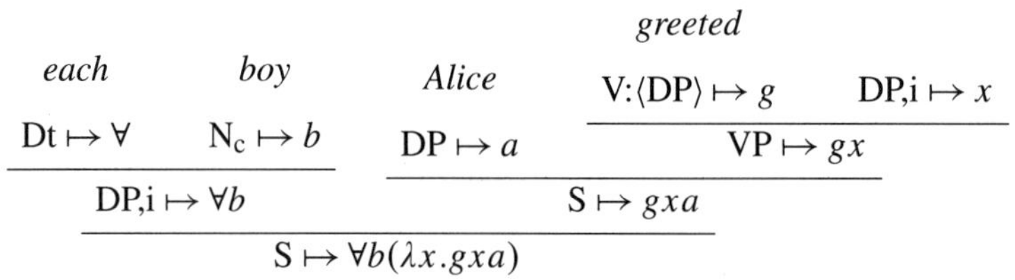

These two ways to treat the syntax and semantics of English quantified noun phrases raise two questions: first, which words are to be treated as quantificational determiners? and second, how well do these treatments accord with how speakers judge such sentences? We shall begin with the second question. To avoid distracting complexities, we shall confine the words to be regarded as quantificational determiners to a proper subset of those discussed earlier, namely, those that have a use with singular count nouns, namely, *each* and *every,* assigning them the value $o_{\forall 2}$, symbolized in our notation of the Lambda calculus as $\forall$, the indefinite article (*a*) and *some,* assigning them the value $o_{\exists^2_{\geq 1}}$, symbolized in our notation of the Lambda calculus as $\exists$, and *no*, assigning it the value o_{N^2}, symbolized in our notation of the Lambda Calculus as N.

We now turn to four open problems pertaining to the syntax and semantics of quantified noun phrases. The first two pertain to sentences in which more than one quantified noun phrase occurs, the third pertains to sentences in which occur not only a quantified noun phrase but also the adverb *not* and the fourth pertains to what value is to be assigned to cardinal and quasi-cardinal adjectives. However, before turning to these problems, we say something about the sort of evidence used to assess the empirical adequacy of the semantics of quantified noun phrases.

3.5.1 Scope judgments

The principal kind of judgment used to investigate the semantics of quantified noun phrases are what are often called *scope judgments*. Before explaining what such judgments are, let us be clear about what the term *scope* means. The point that we are about to make, though subtle, is important.

In earlier chapters, we were introduced to the notion of the scope of a logical constant. This notion is a technical one, defined for a formally defined notation. In logic, the scope of an occurrence of a logical constant is the smallest subformula containing the occurrence of the logical constant. Quantifiers and the negation symbol of CQL are logical constants and therefore their occurrences in a formula have scope. In the Lambda calculus, the scope of an occurrence of a constant term is the smallest subterm containing the constant term's occurrence. Scope, then, is a purely syntactic notion. It has semantic consequences. The values assigned to two formulae that are alike except that the relative scopes of two logical constants may well be different. Thus, for example, we know that in CQL the following pair of formulae may very well have different values assigned to them in the very same structure: $\forall x \exists y Rxy$ and $\exists y \forall x Rxy$. This difference arises from the difference in the order, and hence scope, of the quantifier prefixes. To determine the truth of the formula $\forall x \exists y Rxy$ in a structure, one first chooses a value for x and then hunts for a suitable value for y so that Rxy is true with respect to those choices and one proceeds in that way for each possible assignment of a value to x. It might turn out that, for different choices of values for x, one must choose different values for y so that Rxy is true for each choice of a value for x. In this way, it is said that the choice of value for y might depend on the choice of value for x. But in the case of the formula $\exists y \forall x Rxy$, the choice of a value for y does not depend on choice of any value for x. It should be noted that not every transposition of quantifier prefixes in a formula results in inequivalent formulae: $\forall x \forall y Rxy$ is equivalent to $\forall y \forall x Rxy$, just as $\exists x \exists y Rxy$ is equivalent to $\exists y \exists x Rxy$.

By analogy with formal notation, scope can be defined for expressions of a natural language, once the expressions are given a syntactic analysis. There is no universally accepted definition of scope for constituency grammars and their enrichments, though the various definitions used bear a close resemblance to the definition scope for formal notation. For constituency grammar and its transformational enrichment set out here, we adopt the following definition: the scope of an occurrence of a quantified noun phrase is the constituent of which the quantified noun phrase is an immediate constituent. We then say that an occurrence of constituent A falls within the scope of an occurrence of constituent B just in case the latter occurrence is a constituent of the scope of the former occurrence. For example, in sentence (53.1), the quantified noun phrase *some guest* falls within the scope of the quantified noun phrase *each host,* since the scope of the latter is the entire sentence and the former constituent is an immediate constituent of the entire sentence, whereas in sentence (53.2), the quantified noun phrase *each host* falls within the scope of the quantified noun phrase *some guest.*

(53.1) Each host thinks some guest is tired.
[S [NP Each host] thinks [S [NP some guest] is tired]].

(53.2) Some guest greeted each host.
[NP Some guest] [VP greeted [NP each host]].

The converse holds for the quantified noun phrases in the second sentence.

We observe that, just as the transposition of quantifier prefixes in a formula of CQL may result in inequivalent formulae, so the transposition of quantified noun phrases in a natural language sentence may result in sentences judged to be inequivalent, as are the sentences in (53).

Besides its use as a technical term for syntax, the word *scope* is also used to describe speaker judgments regarding how sentences with more than one quantified noun phrase are construed. Consider the English sentence in (54).

(54) Each investigator believes some tourist is a spy.

CONSTRUAL 1
Each investigator believes some tourist is a spy, where different investigators may have different tourists in mind.

CONSTRUAL 2
There is a tourist that each investigator believes to be a spy.

Contemporary scholars, linguists, and philosophers alike, say that the sentence has two construals, one where the quantified noun phrase *each investigator* has scope of the quantified noun phrase *a tourist,* which happens to correspond to the scope relation as defined with respect to the constituency of the sentence, and one in which the quantified noun phrase *a tourist* has scope of the quantified noun phrase *each investigator,* which does not correspond to the constituency of the sentence. The idea that is the basis for this descriptive usage is that, should we symbolize sentence (54) in some notation of logic, the quantifier prefix corresponding to *each investigator* would have scope over the quantifier prefix corresponding to *some tourist,* on the first construal, whereas the converse relation describes the second construal. This description then reflects the fact that, on the first construal, the choice of tourist might depend on the choice of investigator, whereas on the second construal, there is no such dependence.

Such judgments of choice dependence, described in terms of which quantified noun phrase falls within the scope of which quantified noun phrase, regardless of the constituency of an expression's surface structure, is the principal source of data for the assessment of the syntax and semantics of quantified noun phrases.

3.5.2 The scope of quantified noun phrases

What, then, are the facts pertaining to sentences with more than one occurrence of quantified noun phrases? We shall confine our discussion to sentences comprising a single clause with more than one quantified noun phrase. We shall first consider cases where no quantified noun phrase is a constituent of any other quantified noun phrase and then cases where one quantified noun phrase is a constituent of another.

QUANTIFIED NOUN PHRASES: NO ONE A CONSTITUENT OF THE OTHERS

We begin by considering sentence (55). Even though it comprises a single clause, like the biclausal sentence in (54), it is liable to two construals.

(55)   Each pilot inspected some airplane.

CONSTRUAL 1: ∀ ∃
For each pilot, there is an airplane that he or she inspected.

CONSTRUAL 2: ∃ ∀
There is an airplane that each pilot inspected.

In other words, though syntactically the quantified noun phrase *some airplane* falls within the scope of the quantified noun phrase *each pilot,* at least in terms of the sentence's surface structure, the sentence is liable to two construals, the first where the former quantified noun phrase is described as falling within the scope of the latter, annotated with ∀ ∃, and the second where the latter is described as falling within the scope of the former, annotated with ∃ ∀.

In a transformational grammar with the rule of QR, the existence of these two construals is treated as a matter of amphiboly, where the very same string of sounds accommodates two constituent structures, not with respect to the so-called surface structure, but with respect to its logical form. In other words, associated with sentence (55) are two triples of constituent structure, which are the same with respect to deep structure and surface structure but different with respect to logical form. The two logical forms for sentence (55) and their corresponding Lambda terms are shown in (56).

(56.1)   LOGICAL FORM 1
[S [DP,i some airplane] [S [DP each pilot] inspected [DP,i t]]]
$\exists a (\lambda y. \forall p (\lambda x. i y x))$

(56.2) LOGICAL FORM 2
$\big[$S [DP,j each pilot] $\big[$S [DP,i some airplane] [S [DP,j t] inspected [DP,i t]]$\big]\big]$
$\forall p(\lambda x.\exists a(\lambda y.iyx))$

Lambek typed grammar also furnishes two analyses of sentence (55), whose associated Lambda terms are the same.

$$
\frac{\dfrac{\dfrac{each}{(e\Uparrow t)/(e\backslash t) \mapsto \forall} \quad \dfrac{pilot}{e\backslash t \mapsto p}}{e \Uparrow t \mapsto \forall p} \quad \dfrac{inspected}{(e\backslash t)/e \mapsto i} \quad \dfrac{\dfrac{some}{(e\Uparrow t)/(e\backslash t) \mapsto \exists} \quad \dfrac{airplane}{e\backslash t \mapsto a}}{\dfrac{e\Uparrow t \mapsto \exists a}{e \mapsto y}}}{\dfrac{\dfrac{\dfrac{e \mapsto x \qquad \dfrac{e\backslash t \mapsto iy}{t \mapsto iyx}}{t \mapsto \forall p(\lambda x.iyx)}}{t \mapsto \exists a(\lambda y.\forall p(\lambda x.iyx))}}{}}
$$

$$
\frac{\dfrac{\dfrac{each}{(e\Uparrow t)/(e\backslash t) \mapsto \forall} \quad \dfrac{pilot}{e\backslash t \mapsto p}}{\dfrac{e \Uparrow t \mapsto \forall p}{e \mapsto x}} \quad \dfrac{inspected}{(e\backslash t)/e \mapsto i} \quad \dfrac{\dfrac{some}{(e\Uparrow t)/(e\backslash t) \mapsto \exists} \quad \dfrac{airplane}{e\backslash t \mapsto a}}{\dfrac{e\Uparrow t \mapsto \exists a}{e \mapsto y}}}{\dfrac{\dfrac{e\backslash t \mapsto iy}{\dfrac{t \mapsto iyx}{t \mapsto \exists a(\lambda y.iyx)}}}{t \mapsto \forall p(\lambda x.\exists a(\lambda y.iyx))}}
$$

In case of transformational grammar, these two construals are reflected in the scopal relations of the quantified noun phrases, not in the sentence's surface structure, but in its logical form. In the Lambek typed grammar, the construals are reflected in the order of the discharge of the assumptions introduced by the quantified expressions, which, in turn, is reflected in the accompanying Lambda terms by the scope of the subterms corresponding to the natural language quantified expressions.

In a single-clause sentence with two quantified noun phrases, neither of which is a constituent of the other, a consequence of both transformational grammar and the Lambek typed grammar treatments is that each such sentence is liable to two distinct syntactic analyses, which, for a suitable choice of quantifier noun phrases, means that the sentence has two distinguishable construals. In fact, more generally, for n quantified noun phrases in a single-clause sentence no two occurrences of which are such that one is a constituent of the other, the sentence has $n!$, that is, $n \cdot (n-1) \cdot \ldots \cdot 1$, syntactic analyses, and with perhaps as many distinguishable construals. However, this is not always the case. Exceptions may arise from the choice of verb, from the choice of quantificational determiner and from difference in syntactic structure.

We begin with the pair of verbs, *to grow out of* and *to grow into* that express converse relations. We wish to compare the range of construals that minimal sentences having these verbs and the same quantified noun phrases admit. The first sentence in (57) is liable to two construals, the first consistent with common sense beliefs and the second inconsistent. The second sentence is just like the first, except that the quantificational determiners have been transposed. It, however, is liable to only one construal. Moreover, the available construal is not the one that is consistent with common sense beliefs, but rather the one that is not.

(57.1)   Each oak tree grew out of some acorn.

CONSTRUAL 1: ∀ ∃

For each oak tree, there is some acorn out of which it grew.

CONSTRUAL 2: ∃ ∀

There is an acorn out of which each oak tree grew.

(57.2)   Some oak tree grew out of each acorn.

CONSTRUAL 1: ∀ ∃ (unavailable)

For each oak tree, there is some acorn out of which it grew.

CONSTRUAL 2: ∃ ∀

There is an oak tree which grew out of each acorn.

The same disparity obtains when the verb *to grow out of* is replaced by its converse *to grow into*.

(58.1)   Each acorn grew into some oak tree.

CONSTRUAL 1: ∀ ∃

For each acorn, there is some oak tree into which it grew.

CONSTRUAL 2: ∃ ∀

There is an oak tree into which each acorn grew.

(58.2)   Some acorn grew into each oak tree.

CONSTRUAL 1: ∀ ∃ (unavailable)

For each oak tree, there is some acorn into which it grew.

CONSTRUAL 2: ∃ ∀

There is an oak tree into which each acorn grew.

This important observation is reported by Ray Jackendoff (1983, 207), who ascribes it to Jeffrey Gruber (1965).

Another kind of exception arises from the presence of the quantificational determiner *no*. Quantified noun phrases that have *no* as a determiner often fail to evince both scopal construals.

(59.1) Each girl greeted no boy.

> CONSTRUAL 1: ∀ N
> For each girl, there is no boy she greeted.
>
> CONSTRUAL 2: N ∀
> There is no boy whom each girl greeted.

(59.2) No boy was greeted by each girl.

> CONSTRUAL 1: ∀ N (unavailable)
> For each girl, there is no boy she greeted.
>
> CONSTRUAL 2: N ∀
> There is no boy whom each girl greeted.

Finally, verbs with double complements also do not evince both scope construals.[10]

(60.1) Alice told each lie to a boy.

> CONSTRUAL 1: ∀ ∃
> For each lie, there is a boy to whom Alice told it.
>
> CONSTRUAL 2: ∃ ∀
> There is a boy to whom Alice told each lie.

(60.2) Alice told a boy each lie.

> CONSTRUAL 1: ∃ ∀
> There is a boy to whom Alice told each lie.
>
> CONSTRUAL 2: ∀ ∃ (unavailable)
> For each lie, there is a boy to whom Alice told it.

QUANTIFIED NOUN PHRASES: ONE A CONSTITUENT OF ANOTHER

So far we have been considering monoclausal sentences in which no quantified noun phrase is a constituent of another. Let us now consider monoclausal sentences in which one quantified noun phrase is a constituent of another.

(61) Each pupil in some class slept.

Both Lambek typed grammar and a transformation grammar with quantifier raising give rise to two derivations. Let us examine the Lambek typed derivation first. There are two derivations that are the same up to the point of the formation of the subject noun phrase, as shown here.

---

10. See Larson (1990, sec. 3.1), where he credits the initial observation to an unpublished manuscript by Patricia Schneider-Zioga and to personal communication from David Lebeaux. For a systematic treatment, see Bruening (2001).

Noun Phrases in English

$$
\cfrac{
  \text{each} \qquad \cfrac{
    \text{pupil} \qquad \cfrac{
      \text{in} \qquad \cfrac{
        \text{some} \qquad \text{class} \\
        (e \Uparrow t)/(e\backslash t) \mapsto \exists \quad e\backslash t \mapsto c
      }{
        \cfrac{e \Uparrow t \mapsto \exists c}{e \mapsto y}
      }
    }{
      \cfrac{((e\backslash t)\backslash(e\backslash t))/e \mapsto l}{(e\backslash t)\backslash(e\backslash t) \mapsto ly}
    }
  }{
    e\backslash t \mapsto p \qquad e\backslash t \mapsto lyp
  }
}{
  \cfrac{(e \Uparrow t)/(e\backslash t) \mapsto \forall \qquad e\backslash t \mapsto lyp}{
    \cfrac{e \Uparrow t \mapsto \forall(lyp)}{e \mapsto x}
  }
}
$$

Once the next step is taken, the derivation may proceed in either of two ways, depending on which assumption is discharged first. In the first continuation of the derivation, it is the last assumption introduced, the one that corresponds to the *each pupil,* which is discharged first, whereas in the second continuation, it is the first assumption introduced, the one that corresponds to *some class,* which is discharged first.

each pupil in some class   *slept*

$$
\cfrac{
  \cfrac{
    \cfrac{e \mapsto x \qquad e\backslash t \mapsto s}{t \mapsto sx}
  }{t \mapsto \forall(lyp)(\lambda x.sx)}
}{t \mapsto \exists c(\lambda y.\forall(lyp)(\lambda x.sx))}
$$

each pupil in some class   *slept*

$$
\cfrac{
  \cfrac{
    \cfrac{e \mapsto x \qquad e\backslash t \mapsto s}{t \mapsto sx}
  }{t \mapsto \exists c(\lambda y.sx)}
}{t \mapsto \forall(lyp)(\lambda x.\exists c(\lambda y.sx))}
$$

Look carefully at the Lambda terms at the end of each derivation. In the first case, there are no occurrences of free variables, and in the second case, the variable *y,* introduced in connection with the quantified noun phrase *some class,* has a free occurrence. The first derivation yields a closed Lambda term, which means that any structure for Lambda terms assigns a truth value to it, while the second yields an open Lambda term, which means that, without a variable assignment, no structure for Lambda terms assigns a truth value to it.

A similar result obtains when the sentence is analyzed by a transformational grammar with quantifier raising, at least when the rule is formulated as we have done. The surface structure constituency of sentence (61) is given in (62.1), while the two constituent structures of the logical forms are given in (62.2) and (62.3). The logical form in (62.2) arises from quantifier raising applying first to the subject noun phrase *each pupil in some class,* then to the noun phrase *some class* in its dislocated position. The logical form in

(62.3) arises from quantifier raising applying first to the noun phrase *some class* within the noun phrase *each pupil in some class*, still in the subject position, then applying to what remains in the subject noun phrase, [DP each pupil [PP in [DP,i t]]. In the first logical form, the dislocated determiner phrase *some class,* which carries the index i, has within its scope the gap with which it is coindexed; however, it does not in the second logical form.

(62.1) SURFACE STRUCTURE
[S [DP [DP each pupil] [PP in [DP some class]]] slept].

(62.2) LOGICAL FORM 1
[S [DP,i some class] [S [DP,j each pupil in [DP,i t]] [S [DP,j t] [VP slept]]]]
$\exists c(\lambda y.\forall (lyp)(\lambda x.sx))$

(62.3) LOGICAL FORM 2
[S [DP,j each pupil [PP in [DP,i t]]] [S [DP,i some class] [S [DP,j] [VP slept]]]]
$\forall (lyp)(\lambda x.\exists c(\lambda y.sx))$

In other words, the transformational analysis and the Lambek typed analysis arrive at precisely the same results: each assigns only one closed Lambda term to sentence (61), which corresponds to the construal where the second quantified noun phrase, which is a proper constituent of the first, is construed as having the first within its scope, that is, the construal to the effect that there is some class in which each pupil slept. Construals such as these are often called the *inverse linked* ones. This is because the left to right order of the quantified noun phrases in the surface structure receives a construal in which the order is, as it were, *inverted*.

In recent work, Robert May and Alan Bale (2006), developing earlier work by Robert May (1977), point out that the pattern pertaining to the possible scope construals of quantified noun phrases within the same clause and none a constituent of the other is similar to the possible scopes construals of quantified noun phrases where one is the constituent of the other.

3.5.3 Quantified noun phrases and *not*

We have followed linguistic custom and extended the use of the word *scope* so as to describe different construals to which sentences containing quantified noun phrases are susceptible. This custom is extended to other words that are the natural language counterparts of other logical constants, for example, the English adverb *not,* which, in the presence of quantified noun phrases, gives rise to different construals. Consider the following minimal pair of monoclausal sentences, both containing the adverb *not* and a quantified noun phrase, the first of which admits only one construal, the second two.

(63.1) Some guest did not sleep.
CONSTRUAL 1: $\exists \neg$
There is some guest who did not sleep.

(63.2) Each guest did not sleep.
CONSTRUAL 1: ∀ ¬
Each guest is such that he or she did not sleep.
CONSTRUAL 2: ¬ ∀
Not every guest slept.

Such different construals are well established (see Horn 1989, chap. 4.3 and 7.3). Moreover, such different construals arise with quantified noun phrases whose determiner is *no*, as pointed out by Horn with respect to an advertising slogan from the 1960s:

(64) Everybody doesn't like something; but nobody doesn't like Sara Lee.

Further complications arise in cases where the adverb *not* occurs in a clause with two quantified noun phrases. Particularly challenging here is to distinguish just what the various construals are.

(65) Each pilot did not inspect some airplane.

This sentence accommodates six orders for the two quantified noun phrases and the one adverb.

(66.1) CONSTRUAL 1: ¬ ∀ ∃
There is a man who admires no women.

(66.2) CONSTRUAL 2: ¬ ∃ ∀
There is no woman whom each man admires.

(66.3) CONSTRUAL 3: ∀ ¬ ∃
Each man admires no women.

(66.4) CONSTRUAL 4: ∃ ¬ ∀
There is a woman whom some man does not admire.

(66.5) CONSTRUAL 5: ∀ ∃ ¬
Each man is such that there is a woman whom he does not admire.

(66.6) CONSTRUAL 6: ∃ ∀ ¬
There is a woman whom no man admires.

They, in turn, give rise to four logically distinct construals.

Another complication, as noted in chapter 8 (section 7.2), is that the adverb *not* may occur at the beginning of a finite clause, provided it is followed by a noun phrase whose determiner is a universal one.

(67.1) Not every bird flies.
(67.2) *Not some bird flies.

Moreover, when the adverb *not* does occur initially, rather than in its usual position of being to the immediate right of an auxiliary verb, there are fewer construals.

(68.1) Not every bird flies.

CONSTRUAL: ¬ ∀

It is not the case that every bird flies.

(68.2) Not every man admires some woman.

CONSTRUAL: ¬ ∀ ∃

There is a man who admires no women.

These complications, which have yet to receive a satisfactory treatment, will not be pursued further here.

3.5.4 The quantificational determiners of English

Words of quantity include not only determiners called quantificational but also adjectives, which we previously called *cardinal adjectives* and *quasi-cardinal adjectives*. With the advent of the application of generalized quantifiers to the study of the semantics of natural language, many linguists took to treating these adjectives as denoting generalized quantifiers. One fact about cardinal and quasi-cardinal adjectives is that, with the exception of the cardinal numeral *one,* they must modify plural count nouns.

One important feature of plural noun phrases is that they are often liable to two construals: a collective one and a distributive one. It has been long recognized that, in English, plural noun phrases in subject position give rise to so-called collective and distributive construals. An example of this is found in the sentence given in (69), which is, in fact, true on both a distributive and a collective construal.

(69) Whitehead and Russell wrote a book.

CONSTRUAL 1: collective

Whitehead and Russell wrote a book *together*.

CONSTRUAL 2: distributive

Whitehead and Russell *each* wrote a book.

It is true on the collective construal, since *Principia Mathematica* was written as a collaborative effort of Whitehead and Russell. This construal can be forced by the use of the adverb *together*. The sentence is also true on the distributive construal, since Russell wrote at least one book on his own, for example, *An Inquiry into Meaning and Truth*, and Whitehead also wrote a book on his own, for example, *A Treatise on Universal Algebra*. This construal can be enforced by the use of the adverb *each*.

It is important to stress that collective and distributive construals are not confined to sentences with plural subject noun phrases and verb phrases with verbs denoting actions that can, but need not be, undertaken collaboratively. Sentence (70) has both collective

Noun Phrases in English

and distributive construals but the verb does not denote an action that can be undertaken collaboratively.

(70)   These two suitcases weigh 50 kilograms.

   CONSTRUAL 1: collective
   These two suitcases weigh 50 kilograms *together*.

   CONSTRUAL 2: distributive
   These suitcases weigh 50 kilograms *each*.

Though we shall not show it here, a plural noun phrase in any complement position is liable to collective and distributive construals, depending on the choice of word to which the noun phrase is a complement. (See Gillon 1999 for a survey of the data.) Finally, we note that many plural noun phrases are liable to construals intermediate, as it were, between collective and distributive construals. (See Gillon 1987 for discussion.)

The point is that the distributive construal of plural noun phrases exhibits construals reminiscent of those construals described in terms of scope judgments. For example, on the distributive construals of sentence (69), the choice of book depends on the choice of man. This dependence is evinced in the following paraphrases of sentence (69), where the relevant men are Whitehead and Russell.

(71.1)   These men wrote a book.
(71.2)   These two men wrote a book.
(71.3)   Two men wrote a book.

The first sentence is devoid of any quantified noun phrase, yet each of the sentences evince the same possible dependence of choice of book upon the choice of man.

Finally, recall sentence (55), which has two construals, one on which the subject noun phrase is construed as having scope over the object noun phrase, the other on which the object noun phrase is construed as having scope over the subject noun phrase. In the case of the sentences in (72), there is no construal of the object noun phrase having scope over the subject noun phrase.

(72.1)   Two men wrote two books.
(72.2)   Two books were written by two men.

In particular, we do not construe the first sentence, for example, as one in which, for each choice of book, there is a choice of two men, distinct pairs of men for each choice of book; whereas, we do construe it in such a way that for each choice of man there is a choice of two books, distinct pairs of books for each choice of man. In other words, *two books* distributes with respect to *two men,* but *two men* does not distribute with respect to *two books.* The opposite holds for the second sentence. The sentence permits the construal that one book has been written by two men and the other has been written by two other men;

however, the sentence does not permit the construal that one man wrote two books and the other man wrote two other books.

These facts suggest that the choice dependence evinced by noun phrases with cardinal and quasi-cardinal adjectives may not arise from these adjectives having the values of generalized quantifiers but may arise instead from the distributivity to which plural noun phrases are liable and that the value of cardinal and quasi-cardinal adjectives may be simply the imposition of a cardinality on the denotation of the noun phrase.

Exercises: The quantificational determiners of English

1. For each of the following sentences, write out the logical forms that QR assigns to it. Provide the derivations that type logical grammar assigns it. Make sure that each is accompanied by a Lambda term.

(a) No pilot slept.

(b) Each host greeted some guest.

(c) Each oak tree grew out of some acorn.

(d) Some oak tree grew out of each acorn.

(e) Some host greeted no guest.

(f) No guest was greeted by each host.

(g) Bill carved each figurine from a stick.

(h) Bill carved each stick into some figurine.

(i) Bill carved some figurine from each stick.

(j) Bill carved some stick into each figurine.

2. Briefly describe the circumstances in which the adverb *not* may occur initially in a clause.

3.6 Nonclausal Coordination

In chapter 8, we investigated the coordination of clauses in English in great detail. However, coordination in English is not confined to clauses: almost any pair of constituents of the same category may be coordinated in English. However, as we shall see, it is also possible to coordinate constituents of different categories. Indeed, it is possible to coordinate two expressions one of which is not even a constituent. In this section, we shall review the basic patterns.[11] We shall then consider various ways in which some of the patterns pertaining to coordination have been analyzed, both by constituency grammar and by a Lambek typed grammar.

---

11. Interested readers should consult the thorough presentation of the patterns found in Quirk et al. (1985, chap. 13), as well as in Huddleston, Payne, and Peterson (2002).

Noun Phrases in English

BASIC PATTERNS

The best known pattern is this: except for coordinators themselves and determiners, two constituents of the same category may be coordinated. Moreover, the resulting clause is often well paraphrased by a corresponding pair of coordinated clauses. We shall call this pattern *homogenous constituent coordination*. Here are a few examples.

(73.1)   Carla [VP hit the ball] [CNJ but] [VP did not run to first base].
PARAPHRASE
Carla hit the ball but Carla did not run to first base.

(73.2)   Adam [V met] [CNJ and] [V hugged] Beverly.
PARAPHRASE
Adam met Beverly and Adam hugged Beverly.

(73.3)   My friend seemed [AP rather tired] [CNJ and] [AP somewhat peevish].
PARAPHRASE
My friend seemed rather tired and my friend seemed somewhat peevish.

(73.4)   [NP A man in a jacket a box] [CNJ or] [NP a woman in a dress] left the store.
PARAPHRASE
A man in a jacket left the store or a woman in a dress left the store.

(73.5)   [NP [NP Alice's] [CNJ or] [NP Bill's] house] burned down.
PARAPHRASE
Alice's house burned down or Bill's house burned down.

(73.6)   Bill remained [PP in the house] [CNJ and] [PP on the telephone]
PARAPHRASE
Bill remained in the house and Bill remained on the telephone.

(73.7)   Bill remained [P in] [CNJ or] [P near] the house.
PARAPHRASE
Bill remained in the house or Bill remained near the house.

(73.8)   Bill walked [Adv quietly] [CNJ and] [Adv deliberately].
PARAPHRASE
Bill walked quietly and Bill walked deliberately].

However, while being constituents of the same category is close to a sufficient condition for coordination, it is not a necessary one. One pattern involves the coordination of prepositional phrases with adverbial phrases or the coordination of prepositional phrases with temporal noun phrases.

(74.1)   The enemy attacked *very quickly* and *with great force*.

(74.2)   Bill works *Sunday afternoons* or *on weekdays*.

Another pattern of the coordination of constituents of different categories arise from appended coordination, a form of ellipsis discussed in chapter 4. This occurs when a nonclausal constituent, often introduced by a coordinator such as *and* or *but,* is appended to a clause.

(75.1)   [NP Alice] has been charged with perjury, and [NP her secretary] __ too.
  PARAPHRASE
  Alice has been charged with perjury, and her secretary had been charged with perjury too.

(75.2)   The judge found [NP Beverly] guilty, but not __ [NP Fred].
  PARAPHRASE
  The judge found Beverly guilty, but the judge did not find Fred guilty.

(75.3)   The speaker lectured about the periodic table, but __ only briefly.
  PARAPHRASE
  The speaker lectured about the periodic table, but the speaker lectured about the periodic table only briefly.

(75.4)   Fred goes to the cinema, but __ seldom with his friends.
  PARAPHRASE
  Fred goes to the cinema, but Fred seldom goes to the cinema with his friends.

In some cases, the appended constituent corresponds to a constituent in the preceding clause. In that case, what the appended constituent conveys is what would be conveyed by a clause just like the preceding clause, except that the appended phrase replaces its counterpart in the preceding clause. This is seen in the first two sentences in (75). In other cases, the preceding clause may contain no counterpart to the appended constituent. In that case, what is conveyed is what is conveyed by the same clause with the appended constituent. This is exemplified in the last two sentences in (75).

A third pattern, known as gapping, arises when a series of phrases that do not themselves form a constituent is coordinated with a clause. Gapping, a form of ellipsis discussed in chapter 4, occurs, roughly, under the following circumstances: an independent clause is followed by an expression that, though not itself a constituent, comprises two constituents, neither of which is a constituent of the other; the first of these latter two constituents, corresponds to the initial constituent in the preceding clause and the second to the clause's final constituent; the point of ellipsis, or gap, is the point between the two constituents that follow the clause; and the expression between the initial and final constituent of the clause is the antecedent for the gap.

(76.1)   On Monday <u>Alice had been</u> in Paris and on Tuesday __ in Bonn.
  PARAPHRASE
  On Monday Alice had been in Paris and on Tuesday Alice had been in Bonn.

(76.2) Bill came to <u>Fiji in 1967</u> and Evan __ the following year.
PARAPHRASE
Bill came to Fiji in 1967 and Evan came to Fiji the following year.

(76.3) Max had not <u>finished</u> the assignment, nor (had) Jill __ hers.
PARAPHRASE
Max had not finished the assignment, nor (had) Jill finished hers.

In the first sentence of (76), for example, the prepositional phrases, *on Tuesday* and *in Bonn*, do not form a constituent; however, the noun phrase *on Tuesday* corresponds to the first phrase in the preceding clause, *on Monday*, and the phrase *in Bonn* to the last phrase in the preceding clause, *in Berlin*.

Not only may a series of phrases be coordinated with a clause, a series of phrases may be coordinated with a second series of phrases, as we see in the sentences that follow.

(77.1) The mother gave [NP a cookie] [PP to one of her children] [PP on Wednesday] [CNJ and] [NP a piece of cake] [PP to the other] [PP on Thursday].

(77.2) Colleen painted [NP the bedroom] [AP blue] [CNJ but] [NP the kitchen] [AP purple].

In the first sentence of (77), for example, the pair of phrases *a cookie* and *to one of her children* is coordinated with the pair of phrases *a piece of cake* and *to the other*.

The last pattern, sometimes called *delayed right constituent coordination* and known in transformational linguistics as *right node raising,* occurs where two expressions neither of which need be constituents are coordinated and followed by a constituent which, if taken with the two coordinated nonconstituents, would make them constituents; it is as if one common constituent has been factored out, as it were, of the coordinated expressions, thereby rendering them nonconstituents. In the first sentence of (74), neither the expression *Dan may accept* nor the expression *Bill will certainly reject* are constituents, yet they are coordinated by *but*. Moreover, as we see from the paraphrase, these expressions, when supplemented by the expression *the management's new proposal,* each form a constituent.

(78.1) [S Dan may accept __] but [S Bill will certainly reject __] [NP the management's new proposal]].
PARAPHRASE
Dan may accept the management's new proposal but Bill will certainly reject the management's new proposal.

(78.2) [S I enjoyed __] but [S everyone else seemed to find fault with __] [NP her new novel]].

PARAPHRASE
I enjoyed her new novel but everyone else seemed to find fault with her new novel.

(78.3) Alice [VP knew of __] but [VP never mentioned __] [NP Bill's other work]].
PARAPHRASE
Alice knew of Bill's other work but Alice never mentioned Bill's other work.

SOME ANALYSES

We have just reviewed four patterns pertaining to coordination: homogenous constituent coordination, appended coordination, gapping, and delayed right constituent coordination. As readers will have noticed, each example of coordination is paired with a paraphrase comprising a pair of sentences coordinated with the same coordinator. In early transformational grammar, the analysis was evident: posit a transformational rule to connect the constituency analysis of the paraphrasing sentence with the coordinated clauses, the paraphrased sentence's so-called deep structure, with the constituency analysis of the paraphrased sentence, the paraphrase sentence's so-called surface structure. The pattern of homogeneous constituent coordination was treated by a rule called *conjunction reduction,* now generally abandonned. Appended coordination, gapping and delayed right constituent coordination were treated by the transformational rules of *stripping, gapping,* and *right node raising,* respectively.

Since this is a book on semantics, we shall not divert our attention to the various transformational treatments of these patterns. Rather, we shall confine our attention to semantic analyses of just two patterns, homogeneous constituent coordination and delayed right constituent coordination, based on the apparent syntactic structure of the two patterns.

We begin with the analysis of the pattern of homogenous constituent coordination provided by a Lambek typed grammar. To enhance the readability of the notation in the exposition to follow, we adopt the following abbreviation for types. Any type expression of the form $x \backslash x / x$ is an abbreviation for a type expression of the form $(x \backslash x)/x$. For example, $t \backslash t / t$ is an abbreviation for $(t \backslash t)/t$. Similarly, any derivation of the form on the left abbreviates a derivation of the form on the right:

$$\frac{x \quad x\backslash x/x \quad x}{x} \qquad \frac{x \quad \frac{(x\backslash x)/x \quad x}{x\backslash x}}{x}$$

We first present derivations for the three sentences in (79), without values,

(79.1) Don jogged and Carol swam.
(79.2) Don jogged and swam.
(79.3) Don accompanied and hosted Carol.

and then we shall discuss the derivations and their associated values.

```
Don      jogged              Carol    swam
 e        e\t       and        e       e\t
 ─────────────      t\t/t     ─────────────
       t                            t
       ─────────────────────────────────
                       t
```

```
             jogged          and              swam
  Don         e\t      (e\t)\(e\t)/(e\t)      e\t
              ──────────────────────────────────
   e                        e\t
  ──────────────────────────────────────
                      t
```

```
          accompanied    and      hosted
            (e\t)/e     x\x/x    (e\t)/e     Carol
  Don       ──────────────────────────         e
              (e\t)/e
   e         ──────────────────────────────────
                         e\t
             ──────────────────────────────────────
                              t
```

(where, in the last case, x is $(e\backslash t)/e$).

If we look at the derivations closely, we see that the coordinator *and* is assigned a different type in each derivation. In the first derivation, the coordinator coordinates two expressions of type t. This means that *and* in that context must have type $t\backslash t/t$. In the second sentence, the coordinator coordinates two expressions of type $e\backslash t$, so it has the type $(e\backslash t)\backslash(e\backslash t)/(e\backslash t)$. And in the third derivation, it has type $((e\backslash t)/e)\backslash((e\backslash t)/e)/((e\backslash t)/e)$, since it coordinates expressions of type $(e\backslash t)/e$. In fact, in general, as the type of the coordinate expressions vary, so does the type of the coordinator, even though it appears to be the very same word. In other words, a coordinator is assigned as many different types as there are types for the expressions which it coordinates. Moreover, different types correspond to different values, the very same word is assigned as many different values as it has types of expressions it coordinates.

To incorporate this into a Lambek typed grammar, we must first define a set of subset of the types, which we shall call *Boolean types*.

Definition 1 Set of Boolean types

(1) $t \in$ Btp;

(2) if $x \in$ Typ and $y \in$ Btp, then so are $x\backslash y$ and y/x;

(3) Nothing else is in Btp.

In fact, since we are interested in coordination, we shall confine our attention to a proper subset of the Boolean types, the set we call the *binary Boolean types*. These are types of the form $x\backslash x/x$, where x is a Boolean type.

Definition 2 Set of binary Boolean types
$z \in$ Bbt iff, for some $x \in$ Btp, $z = x \backslash x / x$.

Next we define two sets of Boolean constant terms, one set for those to be assigned to *and*, depending on its type, and another set for those to be assigned to *or*, depending on its type. We shall designate the former set of constant terms as CN_I and the latter as CN_U.

The family CN_I comprises as many constant terms as there are binary Boolean types. It is therefore convenient to use the binary Boolean types to distinguish the various intersective constant terms. If x is a binary Boolean type, then $\cap_{(x \backslash y) \backslash (x \backslash y)/(x \backslash y)}$ and $\cap_{(y/x) \backslash (y/x)/(y/x)}$ are intersective constant terms. While the indexation is mnemonically convenient, it is graphically cumbersome. So, we abbreviate $\cap_{(x \backslash y) \backslash (x \backslash y)/(x \backslash y)}$ as $\cap_{x \backslash y}$ and $\cap_{(y/x) \backslash (y/x)/(y/x)}$ as $\cap_{y/x}$.

Finally, we must state what values are to be assigned to the terms in CN_I. We stipulate that $o_\wedge$ be assigned to $\cap_t$, an abbreviation of $\cap_{t \backslash t / t}$. We shall treat all the other intersective constant terms as abbreviations. We state the definition as follows.

Definition 3 Values for CN_I
(1) $\cap_t$ denotes $o_\wedge$;
(2) If $\sigma, \tau \in TM_x$, $\cap_y \in CN_I$, and $v \in VR_x$, then $\sigma \cap_{x \backslash y} \tau = \lambda v. \sigma v \cap_y \tau v$ and $\sigma \cap_{y/x} \tau = \lambda v. v \sigma \cap_y v \tau$.

Using the Lambek typed grammar, we analyze the first two sentences in (75). To do so, we assign the expressions *jogged* and *swam* the Lambda constants j and s, each of type $e \backslash t$, and the expressions *Don* and *Carol* the Lambda constants d and c, each of type e.

$$
\begin{array}{cccccc}
\text{Don} & \text{jogged} & & \text{Carol} & \text{swam} \\
e \mapsto d & e \backslash t \mapsto j & \text{and} & e \mapsto c & e \backslash t \mapsto s \\
\hline
t \mapsto jd & & t \backslash t / t \mapsto \cap_t & & t \mapsto sc \\
\hline
& & t \mapsto jd \cap_t sc & &
\end{array}
$$

$$
\begin{array}{cccc}
& \text{jogged} & \text{and} & \text{swam} \\
& e \backslash t \mapsto j & (e \backslash t) \backslash (e \backslash t)/(e \backslash t) \mapsto \cap_{e \backslash t} & e \backslash t \mapsto s \\
\cline{2-4}
\text{Don} & & e \backslash t \mapsto j \cap_{e \backslash t} s & \\
e \mapsto d & & e \backslash t \mapsto \lambda v. jv \cap_t sv & \\
\hline
& t \mapsto (\lambda v. jv \cap_t sv) d & & \\
& e \backslash t \mapsto jd \cap_t sd & &
\end{array}
$$

To analyze the last sentence in (79), we assign the expressions *accompanied* and *hosted* the Lambda constants a and h, each of type $(e \backslash t)/e$. (Note that x in the following derivation stands for the type $(e \backslash t)/e$.)

$$
\begin{array}{c}
\begin{array}{ccc}
\textit{accompanied} & \textit{and} & \textit{hosted} \\
(e\backslash t)/e \mapsto a & x\backslash x/x \mapsto \cap_x & (e\backslash t)/e \mapsto h
\end{array}
\end{array}
$$

(derivation with Don $e \mapsto d$, Carol $e \mapsto c$)

$$
\begin{array}{c}
(e\backslash t)/e \mapsto a \cap_x h \\ \hline
(e\backslash t)/e \mapsto \lambda v.av \cap_{e\backslash t} hv \\ \hline
e\backslash t \mapsto (\lambda v.av \cap_{e\backslash t} hv)c \\ \hline
e\backslash t \mapsto ac \cap_{e\backslash t} hc \\ \hline
e\backslash t \mapsto \lambda w.acw \cap_t hcw \\ \hline
t \mapsto (\lambda w.acw \cap_t hcw)d \\ \hline
t \mapsto acd \cap_t hcd
\end{array}
$$

As we remarked previously, a coordinator is assigned as many different types as the types of the constituents it coordinates and, accordingly, is assigned as many different values. But this apparent ambiguity is a fact about the notation we have chosen, not about the language we are analyzing. Nothing in the patterns described suggest there is any such ambiguity.[12]

It is possible to achieve the same coverage of homogenous constituent coordination as is achieved by a Lambek typed grammar, but without treating the coordinators as ambiguous. To do so in an enriched constituency grammar,[13] we require, to begin with, constituency formation rules for the various categories of constituents that may be coordinated: $X \, C_c \, X \rightarrow X$, where X is a lexical category, a phrasal category, or S. And we assign $\cap$, intersection, to $and|C_c$, and $\cup$, union, to $or|c_c$. We stress that here the symbols $\cap$ and $\cup$ are not terms of the Lambda calculus, but the usual symbols for intersection and union. We illustrate this analysis with derivations for the last two sentences in (79), where *jogged* and *swam* are assigned J and S, subsets of the universe, respectively, and *accompanied* and *hosted* A and H, sets of ordered pairs of members of the universe, *Don* the singleton set {1}, and *Carol* the singleton set {2}.

$$
\begin{array}{c}
\begin{array}{ccc}
& \textit{jogged} & \textit{and} & \textit{swam} \\
\textit{Carol} & V:\langle\rangle \mapsto J & C_c \mapsto \cap & V:\langle\rangle \mapsto S
\end{array} \\
\text{NP} \mapsto \{2\} \quad\quad V:\langle\rangle \mapsto J \cap S \\ \hline
S \mapsto T \text{ (iff } \{2\} \subseteq J \cap S)
\end{array}
$$

$$
\begin{array}{c}
\begin{array}{ccc}
\textit{accompanied} & \textit{or} & \textit{hosted} \\
V:\langle NP \rangle \mapsto A & C_c \mapsto \cup & V:\langle NP \rangle \mapsto H
\end{array} \quad \textit{Carol} \\
\textit{Don} \quad\quad V:\langle NP \rangle \mapsto A \cup H \quad\quad\quad \text{NP} \mapsto \{2\} \\
\text{NP} \mapsto \{1\} \quad\quad V:\langle\rangle \mapsto \{x : \langle x, 2 \rangle \in A \cup H\} \\ \hline
S \mapsto T \text{ (iff } \{1\} \subseteq \{x : \langle x, 2 \rangle \in A \cup H\})
\end{array}
$$

---

12. This point was made to me a number of years ago by Ed Keenan.

13. This is essentially a constituency grammar adaptation of the treatment of this pattern in Keenan and Faltz (1985).

Now one problem, which is easily handled, is this. The semantic values assigned to declarative clauses are the truth values T and F. But it makes no sense to speak of the union or intersection of a pair of truth values. It appears, therefore, that this assignment of values to *and* and *or* does not permit an assignment of truth values to coordinated declarative clauses, such as sentence (79.1). However, it is easy to find sets that behave with respect to union, intersection, and complementation just as T and F behave with respect to $o_\vee$, $o_\wedge$, and $o_\neg$. They are the sets $\{\emptyset\}$ and $\emptyset$, and as readers can see from inspecting the following tables, the functions are isomorphic.

$\wedge$	T	F
T	T	F
F	F	F

$\cap$	$\{\emptyset\}$	$\emptyset$
$\{\emptyset\}$	$\{\emptyset\}$	$\emptyset$
$\emptyset$	$\emptyset$	$\emptyset$

$\vee$	T	F
T	T	T
F	T	F

$\cup$	$\{\emptyset\}$	$\emptyset$
$\{\emptyset\}$	$\{\emptyset\}$	$\{\emptyset\}$
$\emptyset$	$\{\emptyset\}$	$\emptyset$

$\neg$	T	F
	F	T

$-$	$\{\emptyset\}$	$\emptyset$
	$\emptyset$	$\{\emptyset\}$

(where $-$ is complementation over the set $\{\emptyset, \{\emptyset\}\}$.)

While many instances of the patterns in (73) are successfully treated by conjunction reduction, important problems remained. It was quickly realized in the early days of transformational grammar that conjunction reduction failed to handle a variety of cases of homogeneous constituent coordination. We now detail these.

To begin with, coordinated constituents that have a quantified noun phrase either as a subject or as a complement do not always have the requisite paraphase, as the pair of sentences in (80.2) shows.

(80.1)   Each attendee smokes and drinks.
          Each attendee smokes and each attendee drinks.

(80.2)   Some attendee smokes and drinks.
          Some attendee smokes and some attendee drinks.

The second sentence in (80.2) is true even if the smokers and the drinkers form disjoint sets; the first sentence is not true with respect to those circumstances. Another kind of sentence that lacks the requisite paraphrase are those that have coordinated noun phrases serving as the antecedents of reciprocal pronouns. Indeed, the required paraphrases of such sentences are judged unacceptable.

(81.1)   Alice and Alexis admire each other.
         NONPARAPHRASE
       *Alice admires each other and Alexis admires each other.
         PARAPHRASE
         Alice admires Alexis and Alexis admires Alice.
(81.2)   Bill introduced Jules and Jim to each other.
         NONPARAPHRASE
       *Bill introduced Jules to each other and Bill introduced Jim to each other.
         PARAPHRASE
         Bill introduced Jules to Jim and Bill introduced Jim to Jules.

Sentences in which reciprocal polyadic words are predicated of coordinated noun phrases also lack the requisite paraphrase.

(82.1)   Alice and Alexis are friends.
         NONPARAPHRASE
         Alice is a friend and Alexis is a friend.
         PARAPHRASE
         Alice is a friend of Alexis and Alexis is a friend of Alice.
(82.2)   Audrey and Alexis are alike.
         NONPARAPHRASE
       *Audrey is alike and Alexis is alike.
         PARAPHRASE
         Audrey is like Alexis and Alexis is like Audrey

Finally, the conjunction reduction rule entails that noun phrases coordinated by *and* have only distributive construals. But as we saw earlier, such coordinated noun phrases often also have collective construals.

These patterns, except the one illustrated by sentence (80.2), also pose a problem both for the Lambek typed grammar analysis and the enriched constituency grammar analysis of homogenous constituent coordination.

We close this treatment of nonclausal coordination by showing how a Lambek typed grammar can nicely handle the pattern of delayed right constituent coordination right, as evinced in the two sentences in (83).

(83.1)   Bill likes and Carol dislikes Atlanta.
(83.2)   Alexie showed Colleen Banff and Don Fresno.

In the first derivation that follows, x below stands for the type t/e. In order to fit the entire derivation on the page, we break it up into two parts.

$$
\begin{array}{c}
\begin{array}{cc}
\begin{array}{c} \text{likes} \\ \text{Bill} \quad (e\backslash t)/e \mapsto l \quad e \mapsto v \\ \hline e \mapsto b \quad \quad e\backslash t \mapsto lv \\ \hline t \mapsto lvb \\ \hline t/e \mapsto \lambda v.lvb \end{array} & \begin{array}{c} \text{and} \\ \\ x\backslash x/x \mapsto \cap_x \end{array}
\end{array}
\quad
\begin{array}{c} \text{dislikes} \\ \text{Carol} \quad (e\backslash t)/e \mapsto d \quad e \mapsto w \\ \hline e \mapsto c \quad \quad e\backslash t \mapsto dw \\ \hline t \mapsto lvc \\ \hline t/e \mapsto \lambda w.dwc \end{array}
\\[2em]
\hline t/e \mapsto \lambda v.lvb \cap_x \lambda w.dwc \\
\hline t/e \mapsto \lambda u.(\lambda v.lvb)u \cap_t (\lambda w.dwc)u \\
\hline t/e \mapsto \lambda u.lub \cap_t duc
\end{array}
$$

$$
\begin{array}{c}
\begin{array}{cc} \text{Bill likes and Carol dislikes} & \text{Atlanta} \\ t/e \mapsto \lambda u.lub \cap_t duc & e \mapsto a \end{array} \\
\hline t \mapsto (\lambda w.lub \cap_t duc)a \\
\hline e \mapsto lab \cap_t dac
\end{array}
$$

Exercises: Nonclausal coordination

1. Find an example where the quantified noun phrase is not a subject noun phrase and fails to give rise to an equivalent paraphrase.

2. Identify which of the following types are Boolean types and, of the Boolean types, which are binary Boolean types. Recall that $x\backslash x/x$ abbreviates $(x\backslash x)/x$.

(a) t
(b) e
(c) $t\backslash e$
(d) $t\backslash t/t$
(e) $e\backslash t$

(f) $e\backslash e/e$
(g) $(e\backslash e)\backslash t$
(h) $(t/e)\backslash (t/e)/(t/e)$
(i) $(t/t)\backslash e$
(j) $(e\backslash t)\backslash (e\backslash t)$

(k) $(e\backslash t)\backslash (t\backslash e)$
(l) $t\backslash e/t$
(m) $e/(e/(e\backslash t))$
(n) $(t/e)\backslash (t\backslash e)/(t/e)$
(o) $e\backslash (e\backslash (e\backslash t))$

4 Conclusion

In this chapter, we have expanded both our empirical and theoretical horizons. In chapter 10, we confined our attention to minimal clauses, declarative clauses whose noun phrases are all proper nouns and whose verb phrases contain the verb and its complements. In this chapter, we considered first simple clauses, clauses with noun phrases having at most one determiner and one adjective. We then ventured further to consider simple clauses whose nouns are modified either by prepositional phrases or by restrictive relative clauses. And we ended the chapter by considering monoclausal declarative sentences with various coordinated phrasal constituents. Nonetheless, we avoided much detail. While we described the various kinds of English nouns, we confined our theoretical treatment to singular count nouns. Although reported on the different kinds of English determiners, we only gave a theoretical treatment for quantificational determiners. In addition, we choose to assign values only to intersective, predictive adjectives, though we had noted the existence of various other kinds. And while distinguishing two kinds of relative clauses, we analyzed only restrictive ones.

In chapter 3, we introduced the notion of a constituency grammar. We showed that, while immediate constituency analysis brought to light many patterns in English syntax, constituency grammar failed to do justice to many of the patterns. In chapter 10, we set out to address two of the problems brought to light in chapter 3, namely the projection problem and the subcategorization problem, as well as a new problem, namely the problem of defining a structure of an English lexicon. However, the new patterns, pertaining to modification, homogeneous constituent coordination, and different ways of construing choice dependence in clauses with quantified noun phrases, required further modifications of constituency grammar. At the same time, these new patterns afforded us the opportunity to see how Lambek typed grammars, also known as type logical grammars, can be applied to the study of the same patterns. We saw that each kind of grammar has advantages and disadvantages.

SOLUTIONS TO SOME OF THE EXERCISES

2.1.1 Proper nouns

2. We can show that the two sentences are not synonymous by finding a circumstance of evaluation where one is true and the other is false. Consider the situation in which Bill knows that he is not the artists, Picasso, but he believes that he has the same artist genius as Picasso. In this situation, the first sentence of the pair is false and the second is true.

(1) Bill thinks that he is Picasso.

(2) Bill thinks that he is a Picasso.

4. *Everything exists*:

This sentence can be rendered as $\forall x Ex$. The formula is a tautology and the sentence is judged true, no matter what.

Nothing exists:

This sentence can be rendered as $\neg \exists x Ex$. Depending on the circumstance of evaluation, the sentence may be true or false. The formula, however, is a contradiction, since the definition of a structure requires that its universe be nonempty. Should this requirement be omitted, then the formula is a contingency.

Santa Claus does not exist:

This sentence is true, though it could have been false. Yet the formula *Ec* is a tautology, since a structure requires that its universe is nonempty and its interpretation function assign a value from the universe for each individual symbol.

2.1.3   Common nouns

1. These words satisfy the criteria both for count nouns and for mass nouns.
2. These words satisfy the criteria for count nouns, except that the plural suffix *-s* cannot be added to them.
3. In omitting the suffix *-s* from these words, the result either has a different meaning or it is not an English word.

2.3   Determiners

1. English expressions such as *What the hell!* and *That a boy!* are idioms. Each word fails to be intersubstitutable with other words of the same word class.

3.1   Simple clauses with quantified noun phrases

2. It cannot be that proper nouns are assigned the type e and common nouns either the type $e \backslash t$ or the type t/e, since neither *Alice person* nor *person Alice* are expressions of English, let alone expressions which can be judged either true or false.

3.3   Prepositional phrases

2.(a) In a constituency grammar, the sentences are equivalent, since the intersection of set of things sleeping with the intersection of the set of men with the set of things in Calgary is the same set as the intersection of the set of men with the intersection of the set of things sleeping and the set of things in Calgary.

2.(c) Notice that the preposition *in* is assigned the type $((t/e) \backslash (t/e)/e$ when it forms a prepositional phrase modifying a common noun and is assigned the type $((e\backslash t) \backslash (e\backslash t)/e$ when modifying a verb phrase. These types are different. Even though their domains of interpretation, namely $\text{Fnc}(D_e, \text{Fnc}(D_{e/t}, D_{e/t}))$ and $\text{Fnc}(D_e, \text{Fnc}(D_{e\backslash t}, D_{e\backslash t}))$, are the same, since $D{e/t} = D{e \backslash t}$, the values assigned to the occurrence of the preposition *in* when modifying a verb phrase need not be the same as the value assigned to it when modifying a noun phrase. Thus, the sentences in (39.2) and (39.3) do not entail one another, contrary to speaker judgments.

3.6   Nonclausal coordination

2.(a)   Btp: a, d, e, g, h, j, n, o
Bbt: d, h

15 Conclusion

1 Introduction

We have now arrived at the end of the book. It is therefore time to review what in the domain of natural language semantics has been covered and what has not been. It is also the place to point readers to other approaches to natural language semantics that, like this book, take the insights of logic as their point of departure.

2 Retrospective

There are many, many aspects of natural language that one can study. One fundamental aspect is the patterns, or regularities, in the expressions of a natural language. Few classical civilizations studied natural language, and of the few that did, only two, the classical civilizations of ancient Greece and of ancient India, went beyond the simple identification of words and produced grammars of their respective languages. The earlier, and the more comprehensive, of the two grammars appeared in classical India, where the grammar formulated by Pāṇini had all the attributes of a modern generative grammar: it identified all the minimal expressions of the language, formulated a finite set of rules whereby all and only the correct expressions of the language are formed, and set out how the understanding of complex expressions depends on the understanding of the basic expressions.

The recognition that the understanding of complex expressions arises from an understanding of the minimal expressions making them up invites the question of how a speaker of a language understands its minimal expressions, a question that classical Indian thinkers raised and for which still no really satisfactory answer has been found. Indeed, even the question of how a speaker of a language understands such simple expressions as its proper nouns remains a question without a satisfactory answer. In contrast, the question of how a speaker understands a complex expression has received, over the past half century, much attention, above all by linguists and philosophers availing themselves of ideas either taken directly, or adapted, from logic. This book provided an introduction to how contemporary linguists and philosophers address this question, looking almost exclusively at the patterns found in the expressions of English. Chapters are devoted either to the exposition of logic or to the description of the patterns of English and to showing how the ideas of logic can be applied to the patterns described.

The first chapter, Language, Linguistics, Semantics: An Introduction, explained what the central question of natural language semantics is and, with respect to an example, showed how this question can be raised and how the ideas of logic can be applied to answer it. After we situated natural language semantics within linguistic theory and situated linguistic theory with respect to psychology and mathematics, two fields with which linguistic theory has deep connections, we showed how an infinite set of expressions, called SL, can be defined recursively in terms of a finite set of minimal expressions L, or {A, B, C, D}, and how values assigned to the minimal expressions determine the values of the complex expressions formed from them. To mitigate the impression that the example is contrived, readers were asked in the exercises to apply what is illustrated in the example to the notation for positive integers found in various cultures from around the world as well as to the expressions for positive integers found in various natural languages around the world.

To go beyond expressions for positive integers in a natural language to expressions more typical of language, chapter 3 (Basic English Grammar) introduced the notion of a constituency grammar, a formalization of the kind of informal grammar that has its origins in Pāṇini's treatment of Classical Sanskrit and of the notion of immediate constituency analysis developed by American structuralist linguists. In chapter 3, constituency formation rules were set out only for English, though such formation rules have been used with great success to analyze the patterns in expressions found among the world's languages. We also saw in chapter 3 that English exhibits many patterns that the formation rules of constituency grammar do not naturally characterize, patterns, again, found in natural languages throughout the world. Nonetheless, the constituency formation rules do correctly characterize many of the patterns of coordination and subordination in English, patterns again that are found in many of the other languages of the world.

However, it was not until chapter 8 (English Connectors) that constituency formation rules were paired with valuation rules. While to state the constituency formation rules for the coordination and subordination of declarative clauses is not difficult, once the patterns have been identified, to pair them with valuation rules required further substantial preparation. On the one hand, we needed to learn about the pertinent tools from logic. These were elaborated in chapter 6 (Classical Propositional Logic: Notation and Semantics). On the other hand, we needed to acquire knowledge about assessing a new form of data, speakers' truth value judgments. In determining constituency formation rules, as we did in chapter 3, we relied on acceptability judgments; however, to determine valuation rules, we rely on speakers' judgments of the truth values of declarative clauses. Assessing speakers' truth value judgments requires some care, for speakers' truth value judgment of a declarative clause is shaped by their understanding of the clause part of which understanding is invariant from context to context and part of which varies from context to context. Therefore, before reaching chapter 8, we learned about the ways the context in which an expression is uttered helps to shape how a speaker understands the expression. In chapter 4 (Language and Context), we saw that English is rich in words whose understanding depends crucially on the interlocutors having an understanding of the context in which the expression is being uttered. And in chapter 5 (Language and Belief), we saw how an interlocutor's understanding of an expression is also modulated by his or her beliefs about the subject matter of what is being said.

Equipped with knowledge of constituency formation rules, the semantics of CPL and the way in which context helps to shape how speakers understand expressions of their language, we investigated how compound declarative clauses are formed from simple declarative clauses through coordination and subordination and formulated constituency formation rules and valuation rules for such clauses. In the course of chapter 8, we saw much evidence to support the customarily adopted hypothesis that the English coordinators *and* and *or*, when coordinating declarative clauses, behave like the binary propositional connectives of $\wedge$ and $\vee$, and that the English subordinator *if*, when subordinating one declarative clause to another, behaves like the binary propositional connective $\rightarrow$.

In addition, however, we learned an important lesson. Although the values assigned to the coordinators *and* and *or* could be taken off the shelf of CPL, as it were, we could not take the value for any subordinator off its shelf. In particular, since the English coordinators *and* and *or* may take a pair of clauses to form a clause, just as the binary connectives of CPL take a pair of formulae to form a formula, the English coordinators *and* and *or* are assigned $o_\wedge$ and $o_\vee$, respectively. However, the English subordinator *if*, when subordinating one indicative clause to another, cannot be assigned the truth function $o_\rightarrow$. The reason is that the truth function $o_\rightarrow$ is adapted to the connective $\rightarrow$, which takes pairs of formulae to form a formula, whereas the subordinator *if* takes a clause to form a subordinate clause, which itself cannot be judged either true or false, and must be subordinated to an independent clause in order to yield an expression that can be judged either true or false. In short, whereas CPL has the counterpart to coordinated clauses, it has no counterpart to subordinated clauses. However, there is a function, mathematically equivalent to $o_\rightarrow$, namely o_{if}, which is adapted to the syntax of subordination.

Independent English clauses ultimately comprise clauses none of whose proper constituents is a clause. So we next turned our attention to the least complex of such clauses, which we called minimal clauses. The minimal constituents of minimal clauses are verbs and proper nouns. Anticipating the question of what kinds of values are to be assigned to proper nouns and verbs of various kinds, we turned first to the study of a part of CQL, dubbed here CPDL, whose notation includes propositional connectives as well as individual and relational symbols, but excludes variables and quantificational symbols, for clearly individual symbols are the logical analog of proper nouns and relational symbols of verbs. One important point that emerged in chapter 9 (Classical Predicate Logic) is that interpretation functions, which are the functions that assign values to individual and relational symbols, are confined to those functions that respect the classification of the symbols as either individual symbols or as relational symbols of a specified adicity. In other words, the kinds of values that can be legitimately assigned to a nonlogical symbol depends on what kind of symbol it is. It was pointed out at the start of chapter 10 (Grammatical Predicates and Minimal Clauses in English) that the basic categories of constituency grammar provide no basis for an analogous determination. For example, simply knowing that a word is a verb tells us nothing about what sets of things it is true of: it could be true of individuals, if it is an intransitive verb; of pairs of individuals, if it is a transitive verb; and of triples of individuals, if it is a ditransitive verb. However, we saw that a simple enrichment of constituency grammar provided a solution not only to the subcategorization problem and the projection problem, discussed in chapter 3, but also to the failure of the simple categories of unenriched constituency grammar to constrain the assignment of values to the minimal expressions.

Chapter 10 also illustrated a lesson learned in chapter 8: the tools of logic often cannot be adopted off the logical shelf, but they must be adapted to the syntax of the patterns being investigated. First, we saw that the classification of the basic expressions of English is more complex than the classification of the basic symbols of CPDL. An interpretation function to be used for the basic expressions of English minimal clauses must be adapted then to the more complex classification of the basic expressions of English. Second, we saw that the formation of the minimal clauses of English is more complex than the formation of the atomic closed formula of CPDL, for clauses in English all contain a verb phrase, something for which there is no counterpart in the notation of CPDL. This means that a valuation function, extending the interpretation function, has to take into account the additional structure of English minimal clauses. Just as the existence of subordinate clauses in English meant that the value assigned to the subordinator *if* could not be $o_\rightarrow$, but rather o_{if}, so the existence of the verb phrase in English introduced a complication in the definition of the valuation function that does not arise in CPDL.

Finally, in chapter 14 (Noun Phrases in English), we expanded our horizons both empirically and theoretically. To broaden our horizons empirically, we expanded our investigation of independent English clauses to include those that, like minimal clauses, have a single verb, but unlike minimal clauses, may also have noun phrases, such as quantified noun phrases, more complex than those comprising just single proper nouns. Indeed, we looked in some detail at English quantified noun phrases. To analyze such English noun phrases, we first learned about CQL, set out in detail in chapter 11 (Classical Quantificational Logic). However, CQL affords little help in seeing how to assign values to quantified noun phrases. However, an extension of CQL, called *generalized quantifier theory,* is of great value in this regard. In chapter 12 (Enrichments of Classical Quantificational Logic), we set out just enough of generalized quantifier theory as is necessary to its application to the semantics of quantified noun phrases.

Conclusion

We also availed ourselves of our expanded empirical domain to broaden our theoretical horizons. We had learned in chapter 1 (section 2.2.1) that there are two ways used by linguists to characterize recursively the expressions of natural language: one way is to use some extension of constituency grammar and another is to use some extension of categorial grammar. However, up to chapter 14, no mention was made of approaches based on some descendant of categorial grammar. In fact, the complement lists introduced into constituency grammar in chapter 10 amounts to an incorporation of one of the central ideas of categorial grammar.[1]

The particular descendant of categorial grammar we developed is called here the Lambek typed Lambda calculus, frequently called *type logical grammar*. This calculus comprises a pair of calculi, the Lambek calculus, set out in the first half of chapter 13, and the Lambda calculus, set out in the second half. Indeed, as explained in chapter 13, these two calculi are isomorphic, a mathematical result known as the Curry Howard isomorphism. This theorem has the pleasant consequence that once an expression is assigned a category in the Lambek calculus, it is ipso facto assigned a term in the Lambda calculus, whose interpretation in a model provides a value for the expression. In other words, once one has the syntactic label for an expression, one knows what kind of semantic value it can be assigned. Even semanticists who do not use the Lambek calculus as a syntax, but use some form of constituency grammar instead, nonetheless use the Lambda calculus as notation to refer to the values of natural language expressions. Indeed, so widely used is the Lambda calculus that many semanticists labor under the mistaken assumption that, unless an expression is assigned a Lambda term, its semantics has not been specified.

3 What Has Not Been Covered

Of course, much has been left out. To begin with, almost nothing has been said about languages other than English, though much of what applies to English applies to many, many other languages. Moreover, as already stressed repeatedly throughout the earlier chapters, much about the semantics of English has either not been mentioned at all or merely passed over in a most cursory fashion. For example, except for in the survey of traditional English grammar, adverbs were not even mentioned.

In chapter 8 (English Connectors), we distinguished English connectors into three classes: coordinators (*and, or*, and *but*), subordinators (e.g., *if, because*) and conjuncts (e.g., *however, moreover*). English has only three coordinators, it has a score of subordinators and it has dozens of conjuncts. Yet, our attention was limited to the two of the three coordinators, *and* and *or,* saying almost nothing about *but*. We also ignored most of

---

1. Though many linguists do not realize it, when they formulate rules involving argument structure or subcategorization frames, they are, in effect, using categorial grammar.

the subordinators, confining our attention to *if,* in the text, and *unless,* in the exercises. And finally we had nothing to say about conjuncts, the largest class of English connectors. Our choice, of course, was determined by where the insights of logic are thought to shed light.

When we discussed the meanings of prepositions, both in chapter 10 (Grammatical Predicates and Minimal Clauses in English) and in chapter 14 (Noun Phrases in English), we limited our inquiry to their use to signal the complements of verbs (e.g., *of* in *to approve of*), where their values are that of the identity relation, and to their use to form propositional phrases that either modify nouns (e.g., *from Jiayi,* as in *a man from Jiayi*) or are predicated of a subject noun phrase (as in *the man is from Jiayi*). Their use to modify the meaning of verbs (e.g., *to call up* or *to call over*) or to form prepositional phrases that function like adverbs (e.g., *with haste;* cf. *quickly*), though signaled, was not pursued.

In chapter 10, and to a greater extent in chapter 14, we discussed English adjectives, distinguishing them into cardinal, predictive, and thematic. However, we did not go beyond the topics of cardinal adjectives, assigned values assigned to generalized quantifiers, and of a subset of predictive adjectives known as intersective adjectives. We discussed neither thematic adjectives nor nonintersecting predictive adjectives.

As we also saw in chapter 14, English nouns comprise subcategories: proper nouns and personal pronouns, on the one hand, and common nouns, on the other, which themselves can be distinguished into mass nouns and count nouns. While we sketched the properties that distinguish these subcategories of noun, our attention was confined to singular proper nouns that denote individual things, typically people, and to singular count nouns. We said nothing about the many proper nouns that denote events and nonexistent objects, nor about count nouns denoting abstract objects, nor about plural count nouns, nor about mass nouns. There is an extensive literature on all of these topics, and an especially large one on the last two. Moreover, we gave a most cursory description of pronouns, partly in chapter 4 and partly in chapter 14, confining our attention to their general context dependent properties.

In chapter 14, we looked in some detail at quantified noun phrases, showing how the values assigned to generalized quantifiers can be suitably assigned to them. We also looked at the variation in range of construals to which clauses containing more than one quantified noun phrase are liable, a variation described in terms of the notion of scope taken from CQL. However, we did not discuss the challenges presented by indefinite noun phrases, which have patterns that distinguish them from quantified noun phrases. Indeed, the patterns exhibited by indefinite noun phrases have led some researchers to treat them not as generalized quantifiers but as something more akin to a restricted variable.

Verbs were extensively discussed in chapter 10. However, attention was focussed on the kinds of complements they require. Verbs in English, as well as those in many other languages, have properties such as mood, tense, and aspect. English has several moods, including the indicative, the subjunctive, the imperative, and the interrogative. In our discussion of clausal coordination in chapter 8, we distinguished the indicative mood from the imperative and subjunctive moods. The latter two were not pursued, since to do so would take us beyond the confines of CQL and generalized quantifier theory. Two other properties of English verbs, as well as of verbs of many other languages, are aspect and grammatical tense. Though we did not discuss either, we did explain in chapter 4 that aspect pertains to whether or not the action denoted by the verb is to be regarded as complete or incomplete. In the same chapter, we delved into grammatical tense, pointing out its essentially contextual nature, namely to convey a temporal order between the moment of speech and the time of the event or activity denoted by the verb. There is now an extensive literature on mood, aspect, and grammatical tense.

Context dependence, a pervasive property of natural language expressions, was surveyed in chapter 4. There, we distinguished between exophora, the phenomenon in which the full understanding of an expression requires an understanding of the setting in which it is used, and endophora, the phenomenon in which the full understanding of an expression requires an understanding of the cotext surrounding the expression used. We further distinguished endophora into endophora resulting from the use of proforms and endophora resulting from ellipsis. Ellipsis, we explained, arises from some apparent defect in an expression, where a suitable expression in the cotext fills in, so to speak, the defect. But we did not describe under what conditions a constituent may serve as an antecedent for either a proform or a point of ellipsis. Nor did we discuss, except in the most general way in chapter 14, how values are assigned either to endophors or to exophors. There is, unsurprisingly, a very extensive literature on these topics. Finally, in chapter 4, we alluded to, but did not discuss, a pervasive phenomenon of natural languages, namely the phenomenon in which a nonclausal constituent conveys what a clause conveys.

To help us address in a rigorous fashion the central question of natural language semantics, we turned to the discipline part of whose concern is first to define sets of expressions and then to assign values to their complex expressions based on values assigned to the expressions making them up—logic. Logic is a very large subject. At the core of modern logic is what is called here CQL, which appeared at the end of the nineteenth century and blossomed into a field of study in its own right during the first third of the twentieth. Since then, the field has expanded rapidly. It now includes such topics as defeasibility logic, dynamic logic, free logic, fuzzy logic, intuitionistic logic, many-valued logic, modal logic, partial logic, relevance logic, just to mention a few.[2] All of these areas of logic are thought to have some application to the study of natural language. This book has concentrated on CQL and its applications and adaptations to natural language semantics, which were covered in four chapters. Though we did venture beyond the confines of CQL to introduce generalized quantifier theory, the Lambek calculus and the Lambda calculus, we did

---

2. One can get an idea of just how large the field is by considering the fact that the *Handbook of Philosophical Logic,* edited by Dov Gabbay and Franz Guenthner, which aims to give an overview of each of the main subfields of logic, now runs to fourteen volumes.

not touch on modal logic, a very rich area of logic with applications to such phenomena as exophora, mood, aspect, grammatical tense, among many other topics. An excellent introduction to modal logic and its application to the study of natural language semantics is found in the second volume of the very lucidly written book by Gamut (1991). It spans four chapters taking up three hundred pages. The presentation is both thorough and clear.

4 Related Approaches

There are many approaches to natural language semantics that draw on the insights of logic. Readers are strongly encouraged to familiarize themselves with them. To aid in that task, let us indicate briefly some resources to which readers can turn.

One very important and general lesson it is hoped that the reader has drawn is that the part of natural language semantics addressed in this book, namely the part that addresses the question of how an understanding of complex expressions arises from an understanding of their parts, presupposes a recursive characterization of a natural language's set of expressions: in other words, it presupposes a syntax for the language. As we explained in chapter 1, there are two lineages, as it were, of such recursive characterizations, one based on categorial grammar and another based on constituency grammar. Though there is no mathematical reason for it, those based on categorial grammar assume that an expression only has the structure worn on its sleeve, so to speak, while those based on constituency grammar assume that an expression has both an overt and a covert structure. This contrast was illustrated in chapter 14 by the two treatments of quantified noun phrases.

Transformational grammar began as a theory of syntax, unpaired with any semantic theory. One person who sought to supplement transformational syntax with a semantic theory based on logic was Barbara Partee. Indeed, her idea was to try to bring together Chomsky's transformational syntax and Montague's semantics.[3] The most widely adopted version of this marriage is that set out in a textbook by Irene Heim and Angelika Kratzer (1998), both students of Barbara Partee. Once readers adjust to their somewhat idiosyncratic notation for the Lambda calculus, it too can be read quickly. Two more textbooks that presuppose some form of movement are one by Gennaro Chierchia and Sally McConnell-Ginet (1990; second edition, 2000) and one by Richard Larson and Gabriel Segal (1995). The first draws from ideas familiar from Richard Montague, the latter from ideas sketched by the philosopher Donald Davidson.

Another approach based on constituency grammar, but not using transformations, is *discourse representation theory*, developed by Hans Kamp, a student of Richard Montague's. The fundamental idea is that, while the apparent syntax is that of a

---

3. There are several introductory textbooks to the ideas of Richard Montague. An early introduction is one by David Dowty, Robert Wall, and Stanley Peters (1981). A more recent introductory work is by Ronnie Cann (1993). A third introduction is provided in the book by Gamut (1991).

constituency grammar, there is a second level of representation, known as discourse representations. The original idea was motivated by observations regarding pronouns whose antecedents are indefinite noun phrases.[4] However, at the same time, it took up a treatment of tense and aspect. A very limited introduction is found in Gamut (1991, 2: 7.4). A more well rounded but brief introduction is given by Hans Kamp and Uwe Reyle (2011), whereas van Eijck and Kamp (1997) provide a technically advanced introduction. A comprehensive introduction can be found in a textbook by the same authors (Kamp and Reyle 1993).

A competing approach to much of the same phenomena that seeks to eschew a second level of representation is *dynamic Montague grammar*. The key logical ideas come from dynamic predicate logic, a branch of logic. The approach was pioneered by Jeroen Groenendijk and Martin Stokhof (1991). Paul Dekker (2011) provides a brief introduction.

A very nice and thorough introduction to type logical grammar is Bob Carpenter's *Type Logical Semantics* (1997). Michael Moortgat (1997), whose ideas were used in chapter 14, gives a very compact and advanced introduction to type logical grammars. Related approaches are set out by Mark Steedman (2000), Glyn Morrill (2011), and Pauline Jacobson (2014).

---

4. Similar proposals with respect to the same data were independently arrived at by Irene Heim (1983) and Pieter Seuren (1985).

References

Ajdukiewicz, Kazimierz. 1935. Die syntaktische Konnexität. *Studia Logica* 1: 1–27. English translation: McCall, Storrs, ed. 1967. *Polish Logic.* Oxford: Oxford University Press. 207–231.

Algeo, John. 1973. *On Defining the Proper Name.* Gainesville: University of Florida Press.

Allan, Keith. 1986. *Linguistic Meaning.* London: Routledge & Kegan Paul.

Allerton, David J. 1975. Deletion and Proform Reduction. *Journal of Linguistics* 11 (2): 213–237

Allerton, David J. 1982. *Valency and the English Verb.* New York: Academic Press.

Anderson, Stephen R., and Edward L. Keenan. 1985. Deixis. In *Language Typology and Syntactic Description*, edited by Timothy Shopen, 259–308. 3 vols. Cambridge: Cambridge University Press. Second edition, 2007.

Bach, Emmon. 1964. Subcategories in Transformational Grammars. In *Proceedings of the Ninth International Congress of Linguists*, edited by H. Lunt, 672–678. The Hague: Mouton.

Bach, Emmon, and Robert T. Harms. 1968. *Universals in Linguistic Theory.* New York: Holt, Rinehart and Winston.

Baker, Carl Lee. 1989. *English Syntax.* Cambridge, MA: MIT Press. Second edition, 1995.

Baker, Mark C. 2003. *Lexical Categories: Verbs, Nouns, and Adjectives.* Cambridge: Cambridge University Press.

Bally, Charles, and Albert Sechehaye, in collaboration with Albert Riedlinger, eds. 1972. *Cours de Linguistique Générale: Ferdinand de Saussure.* Paris: Payot. Critical ed., Tullio de Mauro. 1982. *Bibliothèque scientifique Payothèque.*

Bar-Hillel, Yehoshua. 1953. A Quasi-arithmetical Notation for Syntactic Description. *Language* 29: 47–58. Reprinted in Bar-Hillel, Yehoshua, ed. 1964. *Language and Information.* Reading, MA; Addison-Wesley. 61–74.

Bar-Hillel, Yehoshua. 1954. Indexical Expressions. *Mind* 63: 359–379.

Bar-Hillel, Yehoshua, ed. 1964. *Language and Information.* Reading, MA: Addison-Wesley.

Bar-Hillel, Yehoshua. 1971. *Pragmatics of Natural Language.* Dordrecht, the Netherlands: D. Reidel.

Bar-Lev, Zev, and Arthur Palacas. 1980. Semantic Command over Pragmatic Priority. *Lingua* 51: 467–490.

Bartsch, Renate, and Theo Vennemann. 1972. *Semantic Structures.* Frankfurt: Athenäum.

Barwise, Jon, and Robin Cooper. 1981. Generalized Quantifers and Natural Language. *Linguistics and Philosophy* 4 (2): 159–219.

Bauer, Laurie. 1983. *English Word-formation.* Cambridge Textbooks in Linguistics. Cambridge: Cambridge University Press.

Bäuerle, Rainer, Christoph Schwarze, and Arnim von Stechow, eds. 1983. *Meaning, Use and Interpretation of Language.* Berlin: De Gruyter.

Blakemore, C., and G. F. Cooper. 1970. Development of the Brain Depends on Visual Environment. *Nature* 228: 447–448.

Blakemore, Diane. 1992. *Understanding Utterances.* Vol. 6 of *Blackwell Textbooks in Linguistics.* London: Longman.

Bloch, Bernard. 1946. Studies in Colloquial Japanese II: Syntax. *Language* 22 (3): 200–248.

Bloomfield, Leonard. 1933. *Language*. New York: Holt.

Bresnan, Joan. 1978. A Realistic Transformational Grammar. In *Linguistic Theory and Psychological Reality*, edited by Morris Halle, Joan Bresnan, and George A. Miller, 1–59. Cambridge, MA: MIT Press.

Bresnan, Joan, ed. 1982. *The Mental Representation of Grammatical Relations*. Cambridge, MA: MIT Press.

Bronkhorst, Johannes. 1998. Les Éléments Linguistiques Porteurs de Sens dans la Tradition Grammaticale du Sanskrit. *Histoire Épistémologie Langage* 20 (1): 29–38.

Bruening, Benjamin. 2001. QR Obeys Superiority: ACD and Frozen Scope. *Linguistic Inquiry* 32: 233–273.

Bühler, Karl. 1934. *Sprachtheorie: Die Darstellungsfunktion der Sprache*. Jena, Germany: Fisher.

Bunt, Harry, and Arthur van Horck, eds. 1996. *Discontinuous Constituency*. Vol. 6 of *Natural Language Processing*. Berlin: De Gruyter.

Burks, Arthur W. 1949. Icon, Index and Symbol. *Philosophy and Phenomenological Research* 9: 673–689.

Bursill-Hall, G. L, ed. and trans. 1972. *Thomas of Erfurt: Grammatica Speculativa*. Vol. 1 of *The Classics of Linguistics*. London: Longman.

Buszkowski, Wojciech, Witold Marciszewski, and Johan van Benthem, eds. 1988. *Categorial Grammar*. Amsterdam, the Netherlands: John Benjamins.

Cann, Ronnie. 1993. *Formal Semantics: An Introduction*. Cambridge Textbooks in Linguistics. Cambridge: Cambridge University Press.

Carpenter, Bob. 1997. *Type Logical Semantics*. Cambridge, MA: MIT Press.

Chao, Yuen Ren. 1968. *A Grammar of Spoken Chinese*. Berkeley, CA: University of California Press.

Chierchia, Gennaro, and Sally McConnell-Ginet. 1990. *Meaning and Grammar*. Cambridge, MA: MIT Press. Second edition, 2000.

Chomsky, Noam. 1956. Three Models for the Description of Language. *IRE Transactions of Information Theory* 2(3): 113–124.

Chomsky, Noam. 1957. *Syntactic Structures*. Janua Linguarum Series Minor 4. The Hague, the Netherlands: Mouton.

Chomsky, Noam. 1958. A Transformational Approach to Syntax. In *Proceedings of the Third Texas Conference on Problems of Linguistics Analysis of English*, edited by Archibald A. Hill, 1962. 124–158. Reprinted in Fodor and Katz, eds., 1964.

Chomsky, Noam. 1959a. Review of *Verbal Behaviour* by B. F. Skinner. *Language* 35: 26–58.

Chomsky, Noam. 1959b. On Certain Formal Properties of Grammars. *Information and Control* 2(2): 137–167.

Chomsky, Noam. (1960) 1962. Explanatory Models in Linguistics. In *Logic, Methodology, and Philosophy of Science: Proceedings of the 1960 International Congress,* edited by Ernst Nagel, Patrick Suppes, and Alfred Tarski, 528–550. Stanford, CA: Stanford University Press.

Chomsky, Noam. 1963. Formal Properties of Grammars. In *Handbook of Mathematical Psychology*, edited by Robert D. Luce, Robert Bush, and Eugene Galanter, 323–418. New York: Wiley.

Chomsky, Noam. 1965. *Aspects of the Theory of Syntax*. Cambridge, MA: MIT Press.

Chomsky, Noam. 1967. Recent Contributions to the Theory of Innate Ideas. *Synthèse* 17: 2–11.

Chomsky, Noam. 1970. Remarks on Nominalization. In *Readings in English Transformational Grammar*, edited by Roderick A. Jacobs and Peter Rosenbaum, 184–221. Waltham, MA: Ginn.

Chomsky, Noam. 1976. Conditions on Rules of Grammar. *Linguistic Analysis* 4: 303–351.

Chomsky, Noam. 1981. *Lectures on Government and Binding*. Dordrecht, the Netherlands: Foris.

Chomsky, Noam. 1991. Some Notes on Economy of Derivation and Representation. In *Principles and Parameters in Comparative Grammar*, edited by Robert Freidin, 417–454. Cambridge, MA: MIT Press.

Clark, Eve V., and Hubert H. Clark. 1979. When Nouns Surface as Verbs. *Language* 55: 767–811.

Cohen, L. Jonathan. 1971. The Logical Particles of Natural Language. In *Pragmatics of Natural Language*, edited by Y. Bar-Hillel, 1971. 50–68.

Cole, Peter, ed. 1978. *Pragmatics*. Vol. 9 of *Syntax and Semantics*. New York: Academic Press.

Cole, Peter, and Jerry L. Morgan, eds. 1975. *Speech Acts*. Vol. 3 of *Syntax and Semantics*. New York: Academic Press.

Comte-Sponville, André. 1995. *Petit Traité des Grandes Vertus*. Paris: Presses Universitaires de France.

Cook, Eung-Do, and Keren D. Rice, eds. 1989. *Athapaskan Linguistics: Current Perspectives on a Language Family*. Berlin: Mouton De Gruyter.

Copi, Irving. 1953. *Introduction to Logic*. New York: Macmillan. Sixth edition, 1982.

Cresswell, Max J. 1973. *Logics and Languages*. London: Methuen.

Cruse, D. A. 1986. *Lexical Semantics*. Cambridge Textbooks in Linguistics. Cambridge: Cambridge University Press.

Curtiss, Susan. 1988. Abnormal Language Acquisition and the Modularity of Language. In *Linguistics: The Cambridge Survey*, edited by Frederick J. Newmeyer, 2:96–116. Cambridge: Cambridge University Press.

Davidson, Donald. 1967. The Logical Form of Action Sentences. In Rescher ed. 1966. 81–95. Reprinted in Davis and Gillon eds., 2004. Chap. 36.

Davidson, Donald, and Gilbert Harman, eds. 1972. *Semantics of Natural Language*. Dordrecht, the Netherlands: D. Reidel.

Davis, Steven, and Brendan S. Gillon, eds. 2004. *Semantics: A Reader*. Oxford: Oxford University Press.

Davy, Humphry. 1812. *Elements of Chemical Philosophy*. Philadelphia, PA: Bradford and Inskeep.

Dekker, Paul. 2011. Dynamic Semantics. In Maienborn, von Heusinger, and Portner, eds., 2011. Chap. 38.

Donnellan, Keith. 1966. Reference and Definite Descriptions. *The Philosophical Review* 75: 281–304.

Dowty, David. 1979. *Word Meaning and Montague Grammar*. Dordrecht, the Netherlands: D. Reidel.

Dowty, David. 1991. Thematic Proto-roles and Argument Selection. *Language* 67 (3): 547–619.

Dowty, David, Robert E. Wall, and Stanley Peters. 1981. *Introduction to Montague Semantics*. Dordrecht, the Netherlands: D. Reidel.

Ellman, Jeffrey L., E. A. Bates, M. H. Johnson, Annette Karmiloff-Smith, D. Parisi, and K. Plunkett. 1996. *Rethinking Innateness: A Connectionist Perspective on Development*. Cambridge, MA: MIT Press.

Erfurt, Thomas of. n.d. *Grammatica Speculativa*. English translation in Bursill–Hall, G. L., ed. and trans. 1972. *Thomas of Erfurt: Grammatica Speculativa*. Vol. 1 of *The Classics of Linguistics*. London: Longman.

Everaert, Martin, and Henk van Riemsdijk, eds. 2006. *The Blackwell Companion to Syntax*. Malden, MA: Blackwell.

Fillmore, Charles. 1965. *Indirect Object Constructions in English and the Ordering of Transformations*. The Hague, the Netherlands: Mouton and Co. In *Monographs on Linguistic Analysis*: no. 1.

Fillmore, Charles. 1971. *Santa Cruz Lectures on Deixis*. Stanford, CA: Center for the Study of Language and Information. Lecture notes no. 65. 1997.

Fillmore, Charles. 1986. Pragmatically Controlled Zero Anaphora. *Proceedings of the Annual Meeting of the Berkeley Linguistics Society* 12: 95–107.

Fillmore, Charles J., and D. Terence Langendoen, eds. 1971. *Studies in Linguistic Semantics*. New York: Holt, Rinehart and Winston.

Findlay, John Niemeyer, trans. 1970. *Logical Investigations*. London: Routledge and Kegan Paul. English Translation of *Logische Untersuchungen,* Edmund Husserl. 1900.

Fodor, Jerry A. 1971. *The Language of Thought*. New York: Crowell.

Fodor, Jerry A., ed. 1981. The Present Status of the Innateness Controversy. *In RePresentations: Philosophical Essays on the Foundations of Cognitive Science*, 257–316. Cambridge, MA: MIT Press.

Fodor, Jerry A., and Jerrold J. Katz, eds. 1964. *The Structure of Language: Readings in the Philosophy of Language*. Englewood Cliffs, NJ: Prentice-Hall.

Frege, Gottlob. 1892. Über Sinn und Bedeutung. *Zeitschrift für Philosophie und Philosophische Kritik* 100: 25–50. English translation in Geach and Black, trans., 1950. 56–78.

Frei, Henri. 1944. Systèmes de Déictiques. *Acta Linguistica* 4: 111–129.

Freiden, Robert, ed. 1991. *Principles and Parameters in Comparative Grammar*. Cambridge, MA: MIT Press.

French, Peter, Theodore E. Uehling, and Howard K. Wettstein, eds. 1979. *Contemporary Perspectives in the Philosophy of Language*. Minneapolis: University of Minnesota Press.

Frisby, John. 1980. *Seeing: Illusion, Brain and Mind*. Oxford: Oxford University Press.

Frisch, Karl von. 1950. *Bees: Their Vision, Chemical Senses and Language*. Ithaca, NY: Cornell University Press.

Frisch, Karl von. 1974. Decoding the Language of Bees. *Science* 185: 663–668.

Gabbay, Dov M., and Franz Guenthner, eds. 1989. *Handbook of Philosophical Logic*. 14 vols. Dordrecht, the Netherlands: Kluwer Academic Publishers. Second edition, 2001.

Gamut, L. T. F. 1991. *Logic, Language and Meaning*. 2 vols. Chicago: University of Chicago Press.

Gazdar. Gerald. 1979. *Pragmatics: Implicature, Presupposition and Logical Form*. New York: Academic Press.

Gazdar, Gerald. 1982. Phrase Structure Grammar. In *The Nature of Syntactic Representations*, edited by Pauline Jacobson and Geoffrey Pullum, 131–186. Dordrecht, the Netherlands: D. Reidel.

Gazdar, Gerald, Ewan Klein, Geoffrey Pullum, and Ivan Sag. 1985. *Generalized Phrase Structure Grammar*. Cambridge, MA: Harvard University Press.

Geach, Peter Thomas. 1950. Russell's Theory of Descriptions. *Analysis* 10: 84–88.

Geach, Peter Thomas. 1962. *Reference and Generality: An Examination of Some Medieval and Modern Theories*. Ithaca, NY: Cornell University Press. Third edition, 1980.

Geach, Peter Thomas, and Max Black, trans. 1950. *Translations from the Philosophical Writings of Gottlob Frege*. Oxford: Blackwell. Third edition, 1980.

Gentzen, Gerhard. 1934. Untersuchungen über das Logische Schliessen. *Mathematische Zeitschrift* 39: 176–210, 405–431.

Gillon, Brendan S. 1987. The Readings of Plural Noun Phrases in English. *Linguistics and Philosophy* 10: 199–219.

Gillon, Brendan S. 1999. Collectivity and Distributivity Internal to English Noun Phrases. *Language Sciences* 18 (1–2): 443–468

Green, Georgia M. 1974. *Semantics and Syntactic Regularity*. Bloomington: Indiana University Press.

Grice, H. Paul. 1975. Logic and Conversation. In Cole and Morgan, eds. 1975. 41–57.

Grice, H. Paul. 1989. *Studies in the Way of Words*. Cambridge, MA: Harvard University Press.

Groenenkijk, J. A. G, Theo Janssen, and Martin Stokhof, eds. 1981. *Formal Methods in the Study of Language*. Amsterdam, the Netherlands: Mathematisch Centrum, Universiteit van Amsterdam (Mathematical Center Tracts: v. 135–136).

Groenendijk, Jeroen, and Martin Stokhof. 1991. Dynamic Predicate Logic: Towards a Compositional, Non-Representative Semantics of Discourse. *Linguistics and Philosophy* 13 (1): 39–100

Gruber, Jeffery S. 1965. "Studies in Lexical Relations." Unpublished PhD diss., Massachusetts Institute of Technology.

Hacking, Ian. 1983. *Representing and Intervening: Introductory Topics in the Philosophy of Natural Science*. Cambridge: Cambridge University Press.

Harris, Zellig. 1946. From Morpheme to Utterance. *Language* 22: 61–183.

Harris, Zellig. 1951. *Methods in Structural Linguistics*. Chicago: University of Chicago Press.

Heim, Irene. 1983. File Change Semantics and the Familiarity Theory of Definiteness. In Bäuerle, Schwarze, and von Stechow, eds. 1983. 164–1898. Reprinted in Portner and Partee, eds. 2002. Chap. 9.

Heim, Irene, and Angelika Kratzer. 1998. *Semantics in Generative Grammar*. Oxford: Blackwell.

Hill, Archibald A., ed. 1962. *Proceedings of the Third Texas Conference on Problems of Linguistic Analysis in English*. Presented 9–12 of May 1958. Austin: University of Texas Press.

Hintikka, Kaarlo Jaakko Juhani, Julius Matthew Emil Moravcsik, and Patrick Suppes, eds. 1973. *Approaches to Natural Language: Proceedings of the 1970 Stanford Workshop on Grammar and Semantics*. Dordrecht, the Netherlands: D. Reidel.

References

Hockett, Charles Francis. 1954. Two Models of Grammatical Description. *Word* 10: 210–231.

Hockett, Charles Francis. 1958. *A Course in Modern Linguistics*. New York: Macmillan.

Horn, Laurence. 1989. *A Natural History of Negation*. Chicago: University of Chicago Press.

Howe, Herbert M. 1965. A Root of van Helmont's Tree. *ISIS* 56 (4): 408–419.

Huddleston, Rodney. 2002. The Clause: Complements. In Huddleston and Pullum, eds. 2002. 213–322.

Huddleston, Rodney, John Payne, and Peter Peterson. 2002. Coordination and Supplementation. In Huddleston and Pullum, eds. 2002. 1273–1362.

Huddleston, Rodney, and Geoffrey K. Pullum, eds. 2002. *The Cambridge Grammar of the English Language*. Cambridge: Cambridge University Press.

Husserl, Edmund. 1900. *Logische Untersuchungen*. Halle, Germany: N. Niemeyer. Second edition, 1913. English translation in Findlay, John Niemeyer, trans. 1970. *Logical Investigations*. 2 vols.

Ifrah, Georges. 1994. *Histoire Universelle des Chiffres: L' Intelligence des Hommes Racontée par les Nombres et le Calcul*. 2 vols. Paris: Robert Laffont.

Ihde, Aaron John. 1964. The Development of Modern Chemistry. New York: Harper and Row.

Jackendoff, Ray. 1983. *Semantics and Cognition*. Vol. 9 of *Current Studies in Linguistics*. Cambridge, MA: MIT Press.

Jacobs, Roderick A., and Peter Rosenbaum, eds. 1970. *Readings in English Transformational Grammar*. Waltham, MA: Ginn.

Jacobson, Pauline. 2014. *Compositional Semantics: An Introduction to the Syntax/Semantics Interface*. Oxford: Oxford University Press.

Jacobson, Pauline, and Geoffrey Pullum. 1982. *The Nature of Syntactic Representations*. Dordrecht, the Netherlands: D. Reidel.

Jaśkowski, Stanisław. 1934. On the Rules of Suppositions in Formal Logic. *Studia Logica* 1: 5–32.

Jennings, Ray E. 1994. *The Genealogy of Disjunction*. Oxford: Oxford University Press.

Jesperson, Otto. 1924. *Philosophy of Grammar*. London: G. Allen & Unwin.

Kamp, Hans. 1981. A Theory of Truth and Semantic Representation. In *Formal Methods in the Study of Language*, edited by Groenendijk, et al. 1981. 277–322.

Kamp, Hans, and Uwe Reyle. 1993. *From Discourse to Logic: Introduction to Model Theoretic Semantics of Natural Language, Formal Logic, and Discourse Representation Theory*. Dordrecht, the Netherlands: Kluwer. In *Studies in Linguistics and Philosophy* 42.

Kamp, Hans, and Uwe Reyle. 2011. Discourse Representation Theory. In Maienborn, von Heusinger, and Portner, eds. 2011. Chap. 37.

Kaplan, David. 1978. Dthat. *Syntax and Semantics* 9, edited by Peter Cole, 221–243. New York: Academic Press.

Kaplan, David. 1979. On the Logic of Demonstratives. *Journal of Philosophical Logic* 8: 81–98.

Keenan, Edward, ed. 1975. *Formal Semantics of Natural Language*. Cambridge: Cambridge University Press.

Keenan, Edward L. 2018. *Eliminating the Universe: Logical Properties of Natural Language*. Singapore: World Scientific.

Keenan, Edward L., and Leonard M. Faltz. 1985. *Boolean Semantics for Natural Language*. Vol. 23 of *Synthèse Language Library*. Dordrecht, the Netherlands: D. Reidel.

Keenan, Edward L., and Larry Moss. 1985. Generalized Quantifers and the Expressive Power of Natural Language. In van Benthem and ter Meulen, eds. 1985. 73–124.

Keenan, Edward L., and Jonathan Stavi. 1986. A Semantic Characterization of Natural Language Determiners. *Linguistics and Philosophy* 9 (3): 253–326.

Keenan, Edward L., and Dag Westerstahl. 1997. Generalized Quantifiers in Linguistics and Logic. In van Benthem and ter Meulen, eds. 1997. 835–893.

Kielhorn, F., ed. 1880. *The Vyâkarana-Mahâbhâshya of Patanjali*. 3 vols. Poona, India: Bhandarkar Oriental Research Institute. Third edition, revised and annotated by K.V. Abhyankar, 1962. Fourth edition, 1985.

Klein, Ewan. 1980. A Semantics for Positive and Comparative Adjectives. *Linguistics and Philosophy* 4(1): 1–45.

Klibansky, Raymond, ed. 1968. *Contemporary Philosophy*. Florence: La Nuova Italian Editrice.

Klima, Edward S. 1965. "Studies in Diachronic Syntax." Unpublished PhD diss., Harvard University.

Koopman, Hilda. 1984. *The Syntax of Verbs: From Verb Movement Rules in the Kru Languages to Universal Grammar*. Dordrecht, the Netherlands: Foris.

Kripke, Saul. 1979. Speaker's Reference and Semantic Reference. In French, Uehling, and Wettstein, eds. 1979. 6–27.

Lakoff, Robin. 1971. If's, And's and But's about Conjunction. In Fillmore and Langendoen, eds. 1971. 3–114.

Lambek, Joachim. 1958. The Mathematics of Sentence Structure. *American Mathematical Monthly* 65: 154–170. Reprinted in Buszkowski, Marciszewski, and van Benthem, eds. 1988. 57–84.

Lane, Harlan L. 1976. *The Wild Boy of Aveyron*. Cambridge, MA: Harvard University Press.

Larson, Richard K. 1990. Double Objects Revisited: Reply to Jackendoff. *Linguistic Inquiry* 21 (4): 589–632.

Larson, Richard, and Gabriel Segal. 1995. *Knowledge of Meaning*. Cambridge, MA: MIT Press.

Leech, Geoffrey. 1974. *Semantics*. Harmondsworth, UK: Penguin.

Leech, Geoffrey N. 1970. *Towards a Semantic Description of English*. Indiana University Studies in the History and Theory of Linguistics. Bloomington: Indiana University Press.

Lemmon, Edward. 1966. Sentences, Statements and Propositions. In Williams and Montefiore, eds. 1966. 87–107.

Lenneberg, Eric H. 1967. *The Biological Foundations of Language*. New York: Wiley.

Levi, Judith N. 1978. *The Syntax and Semantics of Complex Nominals*. New York: Academic Press

Levinson, Stephen C. 1983. *Pragmatics*. Cambridge: Cambridge University Press (Cambridge Textbooks in Linguistics).

Levinson, Stephen C. 2000. *Presumptive Meanings: The Theory of Generalized Conversational Implicature*. Cambridge, MA: MIT Press (Language, Speech, and Communication series).

Lewis, David. 1970. General Semantics. *Synthése* 22: 18–67. Reprinted in *Semantics of Natural Language*, edited by Davidson and Harman, 1972. 169–218. Reprinted in Davis and Gillon, eds. 2004. Chap 11.

Lewis, David. 1975. Language Games. In Keenan, ed. 1975. 3–15. Reprinted in Davis and Gillon, eds. 2004. Chap. 30.

Lindström, Per. 1966. First Order Predicate Logic with Generalized Quantifiers. *Theoria* 32: 186–195.

Lloyd, Geoffrey Ernest Richard. 1973. *Greek Science After Aristotle*. London: Chatto and Windus.

Luce, Robert D., Robert Bush, and Eugene Galanter, eds. 1963. *Handbook of Mathematical Psychology*. New York: Wiley.

Lunt, Horace G., ed. 1964. *Proceedings of the Ninth International Congress of Linguists*. Presented in Cambridge, MA, 27–31 of August, 1962. The Hague, the Netherlands: Mouton and Co.

Lyons, John. 1968. *Introduction to Theoretical Linguistics*. Cambridge: Cambridge University Press. Chap. 7, Sec. 6.

Lyons, John. 1977. *Semantics*. 2 vols. Cambridge: Cambridge University Press.

Lyons, John. 1995. *Linguistic Semantics: An Introduction*. Cambridge: Cambridge University Press.

Maienborn, Claudia, Klaus von Heusinger, and Paul Portner, eds. 2011. *Semantics: An International Handbook of Natural Language Meaning*. Berlin: De Gruyter Mouton.

Marcus, Ruth Barcan. 1961. *Modalities: Philosophical Essays*. Oxford: Oxford University Press.

Martin, Robert M. 1992. *There Are Two Errors in the the Title of This Book*. Peterborough, Canada: Broadview Press. Second edition, 2002.

May, Robert C. 1977. "The Grammar of Quantification." Unpublished PhD diss., Massachusetts Institute of Technology.

May, Robert C., and Alan Bale. 2006. Inverse Linking. In Everaert and van Riemsdijk, eds. 2006. 639–667.

McCall, Storrs, ed. 1967. *Polish Logic*. Oxford: Oxford University Press.

McCawley, James D. 1968. The Role of Semantics in a Grammar. In Bach and Harms, eds. 1968. 124–168.

McCawley, James. 1981. *Everything That Linguists Have Always Wanted to Know About Logic but Were Ashamed to Ask*. Chicago: University of Chicago Press. Second edition, 1993.

Michael, Ian. 1970. *English Grammatical Categories: And the Tradition to 1800*. Cambridge: Cambridge University Press.

Mill, John Stuart. (1843) 1881. *A System of Logic, Ratiocinative and Inductive, Being a Connected View of the Principles of Evidence and the Methods of Scientific Investigation*. Eighth edition. New York, n.p.

Montague, Richard. 1968. Pragmatics. In *Contemporary Philosophy, a Survey*, edited by Raymond Klibansky, 102–122. Florence, La Nuova Italia Editrice. Reprinted in Thomason 1974, 148–187.

Montague, Richard. 1970a. English as a Formal Language. In *Linguaggi nella società e nella tecnica*, edited by Bruno Visentini, et al., 189–224. Milan: Edizioni di Communità. Reprinted in Thomason 1974, 188–221.

Montague, Richard. 1970b. Universal Grammar. *Theoria* 36: 373–398. Reprinted in Thomason 1974, 222–246.

Montague, Richard. 1970c. Pragmatics and Intensional Logic. *Synthèse* 22: 68–94. Reprinted in Thomason 1974, 119–147.

Montague, Richard. 1973. The Proper Treatment of Quantification in Ordinary English. In *Approaches to Natural Language: Proceedings of the 1970 Stanford Workshop on Grammar and Semantics*, edited by Kaarlo Jaakko Juhani Hintikka, Julius Matthew Emil Moravcsik, and Patrick Suppes, 221–242. Dordrecht, the Netherlands: D. Reidel. Reprinted in Thomason 1974, 247–270.

Moortgat, Michael. 1988. *Categorial Investigations*. Dordrecht, the Netherlands: Foris.

Moortgat, Michael. 1990. The Quantification Calculus: Questions of Axiomatization. In Deliverable R1.2A of DYNANA: Dynamic Interpretation of Natural Language. ESPRIT Basic Research Action BR 3175. Centre for Cognitive Science, Edinburgh.

Moortgat, Michael. 1996. Generalized Quantification and Discontinuous Type Constructors. In Bunt and van Horck, eds. 1996. 181–207.

Moortgat, Michael. 1997. Categorial Type Logics. In van Benthem and ter Meulen, eds. 1997. 95–180.

Morrill, Glyn V. 2011. *Categorial Grammar: Logical Syntax, Semantics, and Processing*. New York: Oxford University Press.

Mostowski, Andrzej. 1957. On a Generalization of Quantifiers. *Fundamenta Mathematicae* 44 (1): 12–36.

Munitz, Milton Karl, and Peter Unger. 1974. *Semantics and Philosophy*. New York: New York University Press.

Nagel, Ernst, Patrick Suppes, and Alfred Tarski, eds. 1962. *Logic, Methodology, and Philosophy of Science: Proceedings of International Congress for Logic, Methodology, and Philosophy of Science*. Stanford, CA: Stanford University Press.

Newmeyer, Frederick J., ed. 1988. *Linguistics: The Cambridge Survey*. 4 vols. Cambridge: Cambridge University Press.

Nida, Eugene A. 1948. The Analysis of Immediate Constituents. *Language* 24: 168–177.

Nunberg, Geoffrey. 1993. Indexicality and Deixis. *Linguistics and Philosophy* 16 (1): 1–43.

Parsons, Terence. 1990. *Events in the Semantics of English*. Cambridge, MA: MIT Press.

Payne, John, and Rodney Huddleston. 2002. Nouns and Noun Phrases. In Huddleston and Pullum, eds. 2002. 323–524.

Pelletier, Francis Jeffry. 1999. A Brief History of Natural Deduction. *History and Philosophy of Logic* 20 (1): 1–31.

Perlmutter, David, ed. 1983. *Studies in Relational Grammar*. Chicago: University of Chicago Press.

Peters, P. Stanley, and Robert W. Ritchie. 1973. On the Generative Power of Transformation Grammars. *Information Sciences* 6: 49–83.

Peters, Stanley, and Dag Westerstahl. 2006. *Quantifiers in Language and Logic*. Oxford: Oxford University Press.

Pollard, Carl, and Ivan Sag. 1994. *Head Driven Phrase Structure Grammar*. Chicago: The University of Chicago Press.

Portner, Paul, and Barbara Hall Partee, eds. 2002. *Formal Semantics: The Essential Readings*. Oxford: Blackwell.

Prawitz, Dag. 1965. *Natural Deduction: A Proof-Theoretical Study*. Stockholm, Sweden: Almqvist & Wicksell. In *Acta Universitatis Stockholmiensis, Stockholm Studies in Philosophy*. Vol. 3.

Prior, Arthur. 1957. *Time and Modality*. Oxford: Oxford University Press.

Pullum, Geoffrey, and Rodney Huddleston. 2002. Adjectives and Adverbs. In Huddleston and Pullum, eds. 2002. 525–596.

Putnam, Hilary. 1967. The "Innateness Hypothesis" and Explanatory Models in Linguistics. *Synthèse* 17:12–22.

Quine, Willard. 1960. *Word and Object*. Cambridge, MA: MIT Press.

Quirk, Randolph, Sidney Greenbaum, Geoffrey Leech, and Jan Svartik. 1985. *A Comprehensive Grammar of the English Language*. London: Longman Group.

Reichenbach, Hans. 1947. *Elements of Symbolic Logic*. New York: Macmillan.

Rescher, Nicholas, ed. 1966. *Logic of Decision and Action*. Pittsburgh, PA: University of Pittsburgh Press.

Rice, Keren D. 1989. A Grammar of Slave. In Cook and Rice, eds. 1989. Chap. 21.

Rice, Sally. 1988. Unlikely Lexical Entries. In *Proceedings of the Annual Meeting of the Berkeley Linguistics Society* 14: 202–212.

Robins, Robert H. 1966. The Development of the Word Class System of the European Grammatical Tradition. *Foundations of Language* 2 (1): 3–19.

Robinson, Cyril Edward. 1948. *Hellas: A Short History of Ancient Greece*. New York: Pantheon Books.

Ross, John Robert. 1967. "Constraints on Variables in Syntax." Unpublished PhD diss., Massachusetts Institute of Technology.

Rosten, L. 1968. *The Joys of Yiddish*. New York: McGraw–Hill.

Russell, Bertrand. 1940. *An Inquiry into Meaning and Truth*. New York: W. W. Norton.

Ryle, Gilbert. 1949. *The Concept of Mind*. London: Hutchinson.

Sag, Ivan, Thomas Wasow, and Emily M. Bender. 1999. *Syntactic Theory: A Formal Introduction*. Stanford, CA: Center for the Study of Language and Information. In CSLI Lecture Notes: no. 152. Second edition, 2003.

Schachter, Paul. 1962. Review of Lees 1963. *International Journal of American Linguistics* 28 (2): 134–145.

Schmerling, Susan F. 1975. Asymmetric Conjunction and Rules of Conversation. In Cole and Morgan, eds. 1975. 211–232

Seuren, Pieter A. M. 1985. *Discourse Semantics*. Oxford: Blackwell.

Shaw, Harry. 1963. *Punctuate It Right*. New York: Harper and Row.

Shopen, Timothy, ed. 1985. *Language Typology and Syntactic Description*. 3 vols. Cambridge: Cambridge University Press. Second edition, 2007.

Skinner, Burrhus Frederic. 1957. *Verbal Behavior*. New York: Appleton–Century–Crofts.

Soames, Scott. 1989. Presuppositions. In Gabbay and Guenthner, eds. 1989. Chap. 9.

Sognnaes, Reider F. 1957. Tooth Decay. *Scientific American* 197 (6): 109–117.

Staal, J. F. 1965. Context-Sensitive Rules in Pāṇini. *Foundations of Language* 1: 63–72. Reprinted in *Universals: Studies in Indian Logic and Linguistics*, edited by J. F. Staal, 171–180. Chicago: University of Chicago Press.

Staal, J. F. 1969. Sanskrit Philosophy of Language. In *Linguistics in South Asia*, edited by Thomas A. Sebeok, 499–531. Vol. 5 of *Current Trends in Linguistics*. The Hague, the Netherlands: Mouton.

Staal, J. F., ed. 1988. *Universals: Studies in Indian Logic and Linguistics*. Chicago: University of Chicago Press.

Stalnaker, Robert. 1970. Pragmatics. *Synthèse* 22 (1): 272–289. Reprinted in Davidson and Harman, eds. 1972. 380–397.

Stalnaker, Robert. 1974. Pragmatics and Presupposition. In Munitz and Unger, eds. 1974. 141–177.

Stalnaker, Robert. 1978. Assertion. In *Syntax and Semantics* 9, edited by Peter Cole, 315–332. New York: Academic Press.

Steedman, Mark. 2000. *The Syntactic Process*. Cambridge, MA: MIT Press.

References

Stirling, Leslie, and Rodney Huddleston. 2002. Deixis and Anaphora. In Huddleston and Pullum, eds. 2002. 1449–1564.

Strawson, Peter F. 1950. On Referring. *Mind* 59: 320–344.

Suppes, Patrick. 1957. *Introduction to Logic*. New York: D. Van Nostrand Company: The University Series in Undergraduate Mathematics.

Sweet, Henry. 1913. *Collected Papers of Henry Sweet*. Oxford: The Clarendon Press.

Tarski, Alfred. 1932. The Concept of Truth in Formalized Languages. Reprinted in Woodger, trans. 1956. 152–278.

Tarski, Alfred. 1941. *Introduction to Logic and to the Methodology of Deductive Sciences*. New York: Oxford University Press. Third edition, 1965.

Tarski, Alfred. 1944. The Semantic Conception of Truth and the Foundations of Semantics. *Philosophy and Phenomenological Research* 4: 341–376.

Thomason, Richmond H., ed. 1974. *Formal Philosophy: Selected Papers of Richard Montague*. New Haven, CT: Yale University Press.

Thompson, D'Arcy Wentworth. 1917. *On Growth and Form*. Cambridge: Cambridge University Press. Second edition, 1942.

Trow, Charles E. 1905. *The Old Shipmasters of Salem*. New York: G. P. Putnam's Sons.

van Benthem, Johan, and Alice ter Meulen, eds. 1985. *Generalized Quantifiers in Natural Language*. Vol. 4 of *Groningen-Amsterdam Studies in Semantics*. Dordrecht, the Netherlands: Foris.

van Benthem, Johan, and Alice ter Meulen, eds. 1997. *Handbook of Logic and Language*. Cambridge, MA: MIT Press.

van Eijck, Jan, and Hans Kamp. 1997. Representing Discourse in Context. In van Benthem and ter Meulen, eds. 1997. Chap. 3.

Visentini, Bruno, and Camillo Olivetti, eds. 1968. *Linguaggi nella Società e nella Tecnica*. Milan, Italy: Edizioni di Communità. 1970.

Watson, John Broadus. 1925. *Behaviorism*. New York: W. W. Norton.

Wells, Rulon S. 1947. Immediate Constituents. *Language* 23: 81–117.

Westerstahl, Dag. 1989. Quantifiers in Formal and Natural Languages. In Gabbay and Guenthner, eds. 1989. 1–131. Second edition, 223–338.

Williams, Bernard, and Alan Montefiore, eds. 1966. *British Analytical Philosophy*. New York: Humanities Press.

Woodger, J. H., trans. 1956. *Logic, Semantics, Metamathematics*. Oxford: Oxford University Press. Second edition, edited and with an introduction by John Corcoran. Indianapolis: Hackett Publishing.

Wundt, Wilhelm. 1904. *Völkerpsychologie: Eine Untersuchung der Entweklungsgasetze von Sprache, Mythus und Sitte*. 10 vols. Leipzig: W. Engelmann.

Zwicky, Arnold. 1978. Arguing for Constituents. *Chicago Linguistic Society*: 503–512. Papers from the 14th regional meeting.

List of Symbols

A	Adjective	124
ad	Adicity	125
Adv	Adverb	369
A-ENTAILS	Answerhood entailment	340
AF	Atomic formulae, set of (CPL)	249
AF	Atomic formulae, set of (CPDL)	378
AF	Atomic formulae, set of (CQL)	457
AF	Atomic formulae, set of (CQLI)	522
AP	Adjectival Phrase	125
Av	Auxiliary verb	369
BC	Basic categories, set of	426
BC	Binary connectives, set of	247
BF	Basic formulae, set of	249
BX	Basic expressions, set of	136
CF	Composite formulae, set of	249
Chr	Characteristic functions, set of	97
C_c	Coordinator	326
C_i	Category	136
C_s	Subordinator	326
CN	Constants, set of (monotyped Lambda calculus)	579
CN	Constants, set of (simply typed Lambda calculus)	589
CN_x	Constants of type x, set of (simply typed Lambda calculus)	589
CT	Categories, set of	136
CS	Constituents, set of	137
Dg	Degree	125
DP	Determine phrase	631
Dt	Determiner	124
EA	Elementary arithmetic expressions	595
ENTAILS	Entailment for English	331

FM	Formulae, set of (CPL)	245
FM	Formulae, set of (CPDL)	379
FM	Formulae, set of (CQL)	458
FM	Formulae, set of (MQL^1)	529
FM	Formulae, set of (MQL^2)	536
FR	Formation rules, set of	136
FRs	Suffixation formation rule	18
FRp	Prefixation formation rule	20
FS	Function symbols of CQLI, set of	524
Fvr	Free variables, set of (formulae of CQL)	461–462
Fvr	Free variables, set of (terms of Lambda calculus)	583
I	Identity symbol	522
iff	if and only if	55
im_R	image set function	99
IN	Idealized counting numerals, set of	32
IN	Indic numerals, set of	595
IS	Individual symbols, set of)	377
LC	Lexical categories, set of	426
LX	Lexicon	136
$\mathbb{N}$	Natural numbers, set of	53
N_c	Common noun	123
N_p	Proper noun	123
NP	Noun phrase	124
P	Preposition	125
P_-	One-place relational symbol	377
PC	Propositional connectives, set of	244
PC	Phrasal categories, set of	426
pj_1	First projection	608
pj_2	Second projection	609
PP	Prepositional phrase	125–126
Pow	Power set	70
pr	Pairing	610
PV	Propositional variables, set of	244
QC	Quantificational connectives, set of (CQL)	457
QC^1	Quantificational connectives, set of (MQL^1)	528
QC^2	Quantificational connectives, set of (MQL^2)	535
QR	Quantifier raising	648
$\mathbb{R}$	Real numbers, set of	99n7
RS	Relational symbols, set of	377
RS$_n$	n-place relational symbols, set of	378

List of Symbols

S	Clause	127
TM	Terms, set of (CQLI)	522
TM	Terms, set of (monotyped Lambda calculus)	579
TM	Terms, set of (simply typed Lambda calculus)	589
TM_x	Terms of type x, set of (simply typed Lambda calculus)	589
Typ	Types, set of	588
U	Universe	384
UC	Unary connectives, set of	247
V	Universal set	54
V_i	Intransive verb	126
VR	Variables, set of (CQL)	457
VR	Variables, set of (CQLI)	522
VR	Variables, set of (monotyped Lambda calculus)	579
VR	Variables, set of (simply typed Lambda calculus)	589
VR_x	Variables of type x, set of (simply typed Lambda calculus)	589
VRh	Valuation rule	29
VRi	Valuation rule	22
VRj	Valuation rule	24
VRk	Valuation rule	25
V_s	Verb taking a clausal complement	127
V_t	Transive verb	126
VP	Verb phrase	128
WN	Welsh counting numerals, set of	32
$\mathbb{Z}$	Integers, set of	53
$\mathbb{Z}^+$	Positive integers, set of	53
$\mathbb{Z}^-$	Negative integers, set of	53
$\mathbb{Z}_n$	Positive integers up to and including n, set of	53
ZN	Chinese counting numerals, set of	32
{}	List notation	52
$\{x :\}$	Abstraction notation	52
$\in$	Set membership	52
$\emptyset$	Empty set	54
$||$	Cardinality	54
$\subseteq$	Subset relation	55
$\subset$	Proper subset relation	59
$\perp$	Disjointness relation	60
$\cup$	Union operation	63
$\cap$	Intersection operation	64
—	Difference operation	65
—	Complementation operation	65

Symbol	Description	Page
⟨⟩	Sequence	67
×	Cartesian product operation	68
⋃	Generalized union operation	72
⋂	Generalized intersection operation	72
⨉	Generalized Cartesian product operation	73
→	Function	91
→	Synthesis rule	132
→	Material implication	242
→ E	→ Elimination (formula column)	280–282
→ E	→ Elimination (formula tree)	300
→ E	→ Elimination (sequent column)	307–308
→ E	→ Elimination (sequent tree)	313–314
→ I	→ Introduction (formula column)	287–289
→ I	→ Introduction (formula tree)	300
→ I	→ Introduction (sequent column)	308
→ I	→ Introduction (sequent tree)	313–314
→ L	→ Left Introduction	317–318
→ R	→ Right Introduction	317–318
⇒	Analysis rule	132
⇒	Conversion (Lambda calculus)	597
↦	Assignment	92
$f(\ )$	Function	94
$f_{x \mapsto y}$	Near variant function	96
$<f, g>$	Product of functions	98
·	Concatenation	98
·	Conjunction symbol	244
·	Connective (Lambek calculus)	552
· E	· Elimination (formula tree)	553
· E	· Elimination (sequent tree)	558
· I	· Introduction (formula tree)	553
· I	· Introduction (sequent tree)	558
· L	· Left Introduction	567
· R	· Right Introduction	567
X^*	Strings over a set	98
*	Unacceptability	8, 123
|	Lexical entry	131
¬	Negation symbol	241, 244
¬ E	¬ Elimination (formula column)	289–291
¬ E	¬ Elimination (formula tree)	301

List of Symbols

$\neg$ E	$\neg$ Elimination (sequent column)	308
$\neg$ E	$\neg$ Elimination (sequent tree)	314–315
$\neg$ I	$\neg$ Introduction (formula column)	289–291
$\neg$ I	$\neg$ Introduction (formula tree)	301
$\neg$ I	$\neg$ Introduction (sequent column)	309
$\neg$ I	$\neg$ Introduction (sequent tree)	314–315
$\neg$ L	$\neg$ Left Introduction	319
$\neg$ R	$\neg$ Right Introduction	319
$\wedge$	Conjunction symbol	242, 244
$\wedge$ E	$\wedge$ Elimination (formula column)	278–279
$\wedge$ E	$\wedge$ Elimination (formula tree)	297–298
$\wedge$ E	$\wedge$ Elimination (sequent column)	304–305
$\wedge$ E	$\wedge$ Elimination (sequent tree)	310–311
$\wedge$ I	$\wedge$ Introduction (formula column)	279–280
$\wedge$ I	$\wedge$ Introduction (formula tree)	297–298
$\wedge$ I	$\wedge$ Introduction (sequent column)	304–305
$\wedge$ I	$\wedge$ Introduction (sequent tree)	310–311
$\wedge$ L	$\wedge$ Left Introduction	316
$\wedge$ R	$\wedge$ Right Introduction	316
$\vee$	Disjunction symbol	242, 244
$\vee$ E	$\vee$ Elimination (formula column)	282–283
$\vee$ E	$\vee$ Elimination (formula tree)	298–299
$\vee$ E	$\vee$ Elimination (sequent column)	305
$\vee$ E	$\vee$ Elimination (sequent tree)	311–313
$\vee$ I	$\vee$ Introduction (formula column)	283–284
$\vee$ I	$\vee$ Introduction (formula tree)	298–299
$\vee$ I	$\vee$ Introduction (sequent column)	306
$\vee$ I	$\vee$ Introduction (sequent tree)	310–311
$\vee$ L	$\vee$ Left Introduction	318–319
$\vee$ R	$\vee$ Right Introduction	318–319
$\leftrightarrow$	Material equivalence symbol	242–243, 244
$\leftrightarrow$ E	$\leftrightarrow$ Elimination (formula column)	284–287
$\leftrightarrow$ E	$\leftrightarrow$ Elimination (formula tree)	299–300
$\leftrightarrow$ E	$\leftrightarrow$ Elimination (sequent column)	306
$\leftrightarrow$ E	$\leftrightarrow$ Elimination (sequent tree)	313
$\leftrightarrow$ I	$\leftrightarrow$ Introduction (formula column)	284–287
$\leftrightarrow$ I	$\leftrightarrow$ Introduction (formula tree)	299–300
$\leftrightarrow$ I	$\leftrightarrow$ Introduction (sequent column)	307
$\leftrightarrow$ I	$\leftrightarrow$ Introduction (sequent tree)	313

Symbol	Description	Page
↔ L	↔ Left Introduction	317
↔ R	↔ Right Introduction	317
∼	Negation symbol	244
&	Conjunction symbol	244
⊃	Material implication symbol	244
≡	Material equivalence symbol	244
⊤	Truth symbol	244
⊥	Falsity symbol	244
|	Sheffer stroke	244
v_a	Extension of a bivalent truth value assignment	259–260
$o_\neg$	Truth function for negation symbol	262
$o_\wedge$	Truth function for conjunction symbol	263
$o_\vee$	Truth function for disjunction symbol	263
$o_\rightarrow$	Truth function for material implication symbol	263
$o_\leftrightarrow$	Truth function for material equivalence symbol	263
+	Exclusive *or*	347
⟨IS, RS, ad⟩	Signature	377
⟨U, i⟩	Structure for a signature	384
∃	Existential quantifier symbol (CQL)	457–458
∀	Universal quantifier symbol (CQL)	457–458
$\forall^1$	Universal quantifier symbol (MQL^1)	528
$\exists^1$	Existential quantifier symbol (MQL^1)	528
N^1	"No" Quantifier symbol (MQL^1)	528
$\exists^1_\infty$	"Infinitely many" Quantifier symbol (MQL^1)	528
M^1	"Most" Quantifier symbol (MQL^1)	528
$\exists^1_{>n}$	"Exists more than n" Quantifier symbol (MQL^1)	528
$\exists^1_{\geq n}$	"Exists at least n" Quantifier symbol (MQL^1)	528
$\exists^1_{=n}$	"Exists exactly n" Quantifier symbol (MQL^1)	528
$\exists^1_{<n}$	"Exists fewer than n" Quantifier symbol (MQL^1)	528
$\exists^1_{\leq n}$	"Exists at most n" Quantifier symbol (MQL^1)	528
$\forall^2$	Universal quantifier symbol (MQL^2)	535
$\exists^2$	Existential quantifier symbol (MQL^2)	535
N^2	"No" Quantifier symbol (MQL^2)	535
M^2	"Most" Quantifier symbol (MQL^2)	535
$\exists^2_{>n}$	"Exists more than n" Quantifier symbol (MQL^2)	535
$\exists^2_{\geq n}$	"Exists at least n" Quantifier symbol (MQL^2)	535
$\exists^2_{=n}$	"Exists exactly n" Quantifier symbol (MQL^2)	535
$\exists^2_{<n}$	"Exists fewer than n" Quantifier symbol (MQL^2)	535
$\exists^2_{\leq n}$	"Exists at most n" Quantifier symbol (MQL^2)	535

List of Symbols

$\models$	Entailment (CPL)	269–270
$\models$	Entailment (CPDL)	392
$\models$	Entailment (CQL)	503
/	Connective (Lambek calculus)	552
/ E	/ Elimination (formula tree)	554
/ E	/ Elimination (sequent tree)	558
/ I	/ Introduction (formula tree)	554
/ I	/ Introduction (sequent tree)	558
/ L	/ Left Introduction	567
/ R	/ Right Introduction	567
\	Connective (Lambek calculus)	552
\ E	\ Elimination (formula tree)	553
\ E	\ Elimination (sequent tree)	557
\ I	\ Introduction (formula tree)	553
\ I	\ Introductiontion (sequent tree)	557
\ L	\ Left Introduction	566
\ R	\ Right Introduction	566
λ	Lambda operator	579
$\vdash$	Deducibility (CPL, Gentzen sequent calculus)	316
$\vdash$	Deducibility (CPL, Natural deduction)	292, 302
$\vdash_{FD}$	Deducibility (Lambek calculus, formula tree)	555
$\vdash_{SD}$	Deducibility (Lambek calculus, sequent tree)	557
$\vdash_{GC}$	Deducibility (Lambek calculus, Gentzen sequent)	566
$\uparrow$		644
$\uparrow$ I	$\uparrow$ Introduction	644
$\Uparrow$		647
$\Uparrow$ I	$\Uparrow$ Introduction	647

CONSTITUENCY RULES

AP1	Adjectival phrase: rule 1	132
AP2	Adjectival phrase: rule 2	132
D1	Dependent clause: rule 1	325
DP	Determiner phrase rule	634
N3	Noun: rule 3	637
N4	Noun: rule 4	639
NP1	Noun phrase: rule 1	132
NP2	Noun phrase: rule 2	132
NP3	Noun phrase: rule 3	132

NP4	Noun phrase: rule 4	132
PP1	Prepositional phrase: rule 1	132
S1	Clause: rule 1	132
S2	Clause: rule 2	325
S3	Clause: rule 3	325
S4	Clause: rule 4	325
S5	Clause: rule 5	369
VP1	Verb phrase: rule 1	132
VP2	Verb phrase: rule 2	132
VP3	Verb phrase: rule 3	132
VP4	Verb phrase: rule 4	132
VP5	Verb phrase: rule 5	132
VP6	Verb phrase: rule 6	369
VP7	Verb phrase: rule 7	369

Index

Abstraction notation, 52. *See also* List notation
Abstract noun, 107
Acceptability, 154
Action, 107, 112
Active voice, 113, 412
Addressee, 180. *See also* Bystander; Speaker
Adicity, 377
Adjectival clause, 119
Adjective, 107, 115–116, 152, 185, 413–417, 421, 624–628, 636–638. *See also* Attributive adjective; Cardinal adjective; Comparative adjective; Demonstrative adjective; Exclusive adjective; Gradable adjective; Intersective adjective; Polyadic adjective; Polyvalent adjective; Predicative adjective; Pronominal adjective; Quasi cardinal adjective; Subsective adjective; Superlative adjective; Thematic adjective
Adjective clause. *See* Adjectival clause
Adjective phrase, 152, 398, 413, 421, 425, 637
Adverb, 107, 116, 349, 364, 366
  not, 356–357, 367–371
Adverbial clause, 119
Agreement (grammar), 143, 146
  number (*see* Number agreement)
Agreement (method). *See* Method of agreement
Ajdukiewicz, Kazimierz, 21, 551
Ambiguity, 233, 639, 667
  lexical (*see* Lexical ambiguity)
  structural (*see* Structural ambiguity)
  vs context dependence, 177, 201–204
  vs implicature, 219–220
American structuralist linguist, 11, 21, 106, 122
Amphiboly. *See* Structural ambiguity
Analysis tree, 135–136
  categorematic (logic), 248
  immediate constituency (*see* Immediate constituency analysis)
  syncategorematic (logic), 246–247
  See also Synthesis tree

Anaphora, 188. *See also* Endophora; Proform; Pronoun
Ancillary deduction, 277, 278, 282, 287–289, 290–291, 297, 301
Ancillary premiss. *See* Supposition
Anomaly. *See* Semantic anomaly
Answerhood conditions, 339–341, 350–351. *See also* Compliance conditions; Truth conditions
Antecedent (clause). *See* Protasis
Antecedent (gap, pronoun), 107, 110, 189–195, 196–200. *See also* Endophora; Gap; Pronoun
Antisymmetry (binary relation), 81
Anvaya. *See* Method of agreement
Apodosis, 229n3, 343, 354, 359. *See also* Protasis
Apollonius Dyscolus, 106, 618
Appended coordination, 196, 198, 207. *See also* Ellipsis; Gapping
Arc. *See* Directed edge
Argument (function). *See* Pre-image
Argument (reasoning), 17, 237–239
  invalid, 237, 239
  valid, 237–239
Aristarchus of Samos, 42
Aristarchus of Samothrace, 106
Aristotle, 3, 17, 28, 36, 45, 222, 238, 289, 375, 429
Arithmetic, 13
Arrow. *See* Directed edge
Article, 106, 117–118
  definite, 118, 190, 617, 628
  indefinite, 118, 434, 620, 628
Aspect, 112, 184
  continuous (*see* Progressive aspect)
  imperfect (*see* Progressive aspect)
  perfect (*see* Perfect aspect)
  progressive (*see* progressive aspect)
  See also Perfect aspect; Progressive aspect
Assignment, 93
  variable (*see* Variable assignment)
  See also Function
Association (rule of), 92–93

Aṣṭādhyāyī, 9–11, 21, 27, 122, 163n7
Astronomy, 39–42
Asymmetry (binary relation), 81
Atomic formula
  CPDL, 375, 378–380
  CPL, 249
  CQL, 457–458
  CQLI, 522
  Lambek calculus, 552
  MQL^1, 528
  MQL^2, 535
Atomic term
  monotyped Lambda calculus, 579
  simply typed Lambda calculus, 588
Attribute adjective, 413, 624–625, 636
Austinian conditional, 363–364, 366
Autonomy, 13–16
Auxiliary verb, 113
Axiom
  Gentzen sequent calculus, 315–316, 320
  Lambek calculus, 557, 566
  sequent natural deduction, 303, 309–310
  simply typed Lambda calculus, 597–601

Bacon, Roger, 28
Bar-Hillel, Yehoshua, 21, 179, 551
Bartsch, Renate, 27
Base case, 561. *See also* Mathematical induction
Basic category, 422, 426
Basic formula
  CPL, 249
  CPDL, 380
Bees, 5
Behavior, 3–6, 9
Behaviorism, 3, 6
Belief, 155–156, 181, 209, 212, 217. *See also* World knowledge
Belonging to (a set). *See* Membership
Beta conversion, 598–599
Biconditional, 55n1, 284–285
Bijection, 94, 97
Binary Boolean type, 665–666
Binary connective, 244, 247
Binary relation, 75–91
  associated function, 98–99
  empty (*see* Empty binary relation)
  identity (*see* Identity relation)
  universal (*see* Universal binary relation)
Binary relation from a set to a set, 86–90
  property (*see* Left monogamy; Left totality; Right monogamy; Right totality)
Binary relation on a set, 76–91
  property (*see* Antisymmetry; Asymmetry; Connectedness; Intransitivity; Irreflexivity; Reflexivity; Symmetry; Transitivity)

Binding
  CQL, 459–461
  CQLI, 522
  monotyped Lambda calculus, 582
  MQL^1, 529
  MQL^2, 537
  simply typed Lambda calculus, 590
Bipartite directed graph, 86–90, 91–92
Bipartite graph, 86–91, 92. *See also* Directed graph
Bivalent truth value assignment, 254–256
Bloch, Bernard, 11, 21, 122
Bloomfield, Leonard, 3, 6, 11, 28, 122, 149, 620
Boolean type, 665
Botany, 41
Bound variable
  CQL, 461
  CQLI, 522
  monotyped Lambda calculus, 583
  MQL^1, 529
  MQL^2, 537
  simply typed Lambda calculus, 590
Brahe, Tycho, 39
Branch, 561
Bühler, Karl, 179
Burks, Arthur, 179
Bystander, 180. *See also* Addressee; Speaker

Calculus
  Lambda (*see* Lambda calculus)
  Lambek (*see* Lambek calculus)
Cancelability, 218, 337
Capacity, 4–6, 13
  linguistic (*see* Linguistic capacity)
  grammatical (*see* Grammatical competence)
Cardinal adjective, 620, 625–628, 658, 660
Cardinality, 54
Cardinal numeral. *See* Cardinal adjective
Cartesian product, 68–69
  generalized (*see* Generalized Cartesian product)
Case, 108–109. *See also* Nominative case; Objective case; Possessive case
Categorial grammar, 21, 426, 677, 680
Category, 106–107, 123–129. *See also* Lexical category; Phrasal category
Character, 178–179, 209. *See also* Content
Characteristic function, 96–97
Chemistry, 37
Chinese counting numeral, 32–33
Chomsky, Noam, 6, 8, 21, 161, 165
Church, Alonzo, 552
Circumstances of evaluation, 196, 201–203, 330. *See also* Circumstances of utterance
Circumstances of utterance, 178, 201. *See also* Circumstances of evaluation; Exophora
Classical predicate logic (CPDL), 375–394
Classical propositional logic (CPL), 237–273, 275–320

Index

Classical quantificational logic (CQL), 457–516
Classical valuation
  CPDL, 385–386
  CPL, 256–261, 262–264
  CQL, 465–466
    cylindric algebra, 493–496
    Fregean, 473–476
    Marcus, 470–472
    medieval, 466–469
    Tarski, 477–486 (*see also* Tarski extension)
Clause, 118–120, 322–326, 339–341, 350–351
  adjective (*see* Adjectival clause)
  noun (*see* Nominal clause)
  principal (*see* Main clause)
  See also Adjectival clause; Adverbial clause; Comparative clause; Compound clause; Conditional clause; Coordinate clause; Declarative clause; Gerundial phrase; Imperative clause; Independent clause; Infinitival phrase; Interrogative clause; Main clause; Minimal clause; Nominal clause; Participial clause; Relative clause; Sentence; Simple clause; Subordinate clause; Superordinate clause
Clause formation rule, 127, 128, 132, 325, 345, 369, 426, 436, 631
Clause valuation rule, 436
Cleft sentence, 138
Closed formula, 461
  CQL, 461
  MQL^1, 529
  MQL^2, 537
  monotyped Lambda calculus, 583–584
  simply typed Lambda calculus, 590
Closed term
  monotyped Lambda calculus, 583–584
  simply typed Lambda calculus, 590
Codomain, 75–76
  binary relation on a set, 86
  function, 90
  See also Domain; Range
Collective noun, 108
Command. *See* Imperative clause
Common noun, 107, 616–617, 620–623, 634–635
Communication, 5, 209. *See also* Intraspecial communication
Commutativity, 331, 334–337, 344, 348–349
Comparative adjective, 115
Comparative clause, 120
Competence, 6–8, 13
  grammatical (*see* Grammatical competence)
  See also Performance
Complement, 151, 399–402, 404, 406, 414–415, 419, 420, 423–424
  objective (*see* Objective complement)
  optional (*see* Optional complement)
  permutation (*see* Complement permutation)
  polyadicity (*see* Polyadic complement)
  polyvalance (*see* Polyvalent complement)
Complementation, 65
Complementizer, 402
Complement list, 422–424, 426, 427
Complement permutation, 408, 441–444
Compliance conditions, 339–341, 350–351. *See also* Answerhood conditions; Truth conditions
Composite formula
  CPDL, 379
  CPL, 349
  CQL, 462
  CQLI, 522
  MQL^1, 528, 530
  MQL^2, 535
    first presentation, 537–538
    second presentation, 545
Composite term
  monotyped Lambda calculus, 579
  simply typed Lambda calculus, 593
Compositionality, 21–22
Compound clause, 321–322
Compound noun, 108
Concatenation, 98
Conclusion, 237–239. *See also* Premise
Concomitance. *See* Method of agreement
Concrete noun, 107
Conditional clause, 354, 358, 361. *See also* Austinian conditional; Counterfactual conditional; Subjunctive conditional
Congruence
  abstraction, 602
  application, 601
  first projection, 609
  pairing, 610
  second projection, 609
Conjunct, 322
Conjunction, 106–107, 117, 121, 242
  coordinating, 117 (*see also* Coordinator)
  subordinating, 117, 121 (*see also* Subordinator)
Conjunction reduction, 664, 668–669
Connectedness (binary relation), 83
Connectionism, 8
Connective, 244, 376, 381. *See also* Binary connective; Main logical connective; Propositional connective; Quantificational connective; Unary connective
Connector, 322–326. *See also* Conjunct; Coordinator; Subordinator
Consequent. *See* Apodosis
Conservativity, 540, 541
Consistency, 293. *See also* Inconsistency
Constant, 376
  monotyped Lambda calculus, 579–580
  simply typed Lambda calculus, 588
  See also Individual symbol; Variable

Constant function, 94
Constituency, 11–12, 21, 130, 136. *See also* Constituent
Constituency analysis rule, 124
Constituency formation rule, 136–137, 398, 431, 634–635, 637, 639, 645–646, 667
Constituency grammar, 21, 137, 631–633, 635, 637, 646, 649–650, 667
   enriched (*see* Enriched constituency grammar)
Constituency synthesis rule, 131–132, 136. *See also* Clause formation rule
Constituency valuation rule, 429–430. *See also* Clause valuation rule
Constituent, 11–12, 122, 137, 398–399, 427. *See also* Discontiguous constituent; Immediate constituent
Content, 179. *See also* Character
Context, 175–177
   dependence, 177, 188
   See also Cotext; Setting
Contingency
   CPL, 266
   CPDL, 390–391
   CQL, 498
   MQL^1, 533
   See also Contradiction; Tautology
Continuous aspect. *See* Progressive aspect
Contradiction
   CPL, 266
   CPDL, 390–391
   CQL, 498
   MQL^1, 533
   See also Contingency; Tautology
Conversational maxim, 211
Conversational record, 228
Conversion
   alpha, 597–598
   beta, 597, 598–599
   eta, 598–599
   reflexivity, 601
   transitivity, 601
Coordinate clause, 324–325, 328, 329–330, 661–664
Coordinate (sequence), 67
Coordination, 117, 324, 660
   appended, 662
   asyndetic, 328
   delayed right constituent, 663, 669
   homogenous constituent coordination, 661, 664, 668–669
   nonclausal, 660
   syndetic, 328
   See also Coordinator
Coordinator, 117, 322–325, 358, 626, 665, 667
   and, 328, 330–343
   but, 328
   or, 344–353 (*see also* Exclusive *or*)

Copula. *See* Copular verb
Copular complement ellipsis, 196–198
Copular verb, 109, 111, 402, 406, 431–432, 625, 638, 640
Cotext, 177, 188–189, 209
Counterfactual conditional, 36–37, 363–363
Counting numeral, 28–33
   Chinese (*see* Chinese counting numeral)
   idealized (*see* Idealized counting numeral)
   Welsh (*see* Welsh counting numeral)
Count noun, 616, 620–624
Cow pox, 42–43
CPDL. *See* Classical predicate logic
CPL. *See* Classical propositional logic
CQL. *See* Classical quantificational logic
Cresswell, Max, 27, 179
Critical period, 4–5, 8–9
Curry Howard isomorphism, 677
Cut elimination theorem, 570–578
Cut rule, 559, 567
Cut theorem, 302
Cylindrical algebra, 493–496

Dative shift. *See* Complement permutation
Davidson, Donald, 184
Davy, Humphry, 37–38
Declarative clause, 321, 329–330, 402
Deducibility
   CPL, 292, 302, 303, 310
Deduction
   Gentzen sequent calculus (*see* Gentzen sequent calculus deduction)
   Lambda calculus, 595–602
   natural (*see* Natural deduction)
Deduction rule
   cut (*see* Cut rule)
   contraction, 320
   elimination (*see* Elimination rule)
   introduction (*see* Introduction rule)
   left introduction (*see* Left introduction rule)
   permutation, 320
   right introduction (*see* Right introduction rule)
   reiteration, 291
   weakening, 320
Deep structure, 163. *See also* Logical form; Surface structure
Defeasibility. *See* Cancelability
Definite article, 118, 190
Definite noun phrase, 193
Degree of cut, 570–571
Deixis, 180, 202. *See also* Exophora
Demonstration. *See* Deixis
Demonstrative adjective, 185
Demonstrative pronoun, 110, 184–185, 193, 618
DeMorgan's Law, 65
Denial of the antecedent, 239
Deprivation experiment, 4–5, 9

Index

Depth, 571
Derivability (implicature), 218
De Saussure, Ferdinand, 2
DesCartes, René
Determiner, 616–617, 624, 628
  demonstrative, 628
  interrogative, 628
  quantificational, 628, 634
Diachronic, 2. *See also* Synchronic
Diagram, 19, 23
  bipartite directed graph (*see* Bipartite directed graph)
  directed graph (*see* Directed graph)
  Euler (*see* Euler diagram)
  Venn (*see* Venn diagram)
  See also Tree
Difference (set), 65
Dionysius of Thrax, 106
Directed edge, 77
Directed graph, 77–85. *See also* Bipartite graph
Direct object, 111
Direct product. *See* Cartesian product
Discharge (of a supposition)
  formula natural deduction, 287, 300–301
Discontiguity. *See* Discontiguous constituent
Discontiguous constituent, 158–160, 645
Discontinuity. *See* Discontiguous constituent
Discourse representation theory
Disjointness relation, 60–61
Disjunction, 242
Disjunctive syllogism, 352
Dislocated constituent, 158, 645
Ditransitive verb, 376, 630, 633
Domain, 75–76
  binary relation on a set, 86
  function, 90
  See also Codomain; Range
Donatus, 106
Doubleton set, 54
Dowty, David, 184
Dynamic Montague grammar, 681

Easy movement. *See* Tough movement
Elementary arithmetic expression, 595–596
Element of. *See* Membership
Elimination rule, 276, 557–559
  formula column format, 278–291, 506–507, 510–512
    $\wedge$, 278–279
    $\vee$, 282–283
    $\rightarrow$, 280–282
    $\leftrightarrow$, 285
    $\neg$, 289–290
    $\forall$, 506–507
    $\exists$, 510–512
  formula tree format
    $\wedge$, 297–298
    $\vee$, 298–299
    $\leftrightarrow$, 299–300
    $\rightarrow$, 300
    $\neg$, 301
    $\backslash$, 553
    $/$, 554
    $\cdot$, 555
  sequents in a column
    $\wedge$ 304–305
    $\leftrightarrow$, 306
    $\neg$, 308
    $\rightarrow$, 307–308
    $\vee$, 305
  sequents in a tree
    $\wedge$, 310–311
    $\vee$, 311–313
    $\leftrightarrow$, 313
    $\rightarrow$, 313–314
    $\neg$, 314–315
    $\backslash$, 557
    $/$, 558
    $\cdot$, 558
  See also Introduction rule
Ellipsis, 188–189, 196–199. *See also* Appended coordination; Copular complement ellipsis; Gapping; Interrogative ellipsis; Nominal ellipsis; Verb phrase ellipsis
Empty binary relation, 77
Empty set, 54
Endophora, 189–190, 209, 618–619. *See also* Antecedent; Gap; Pronoun
Endowment, innate. *See* Innate endowment
Enriched constituency grammar, 425–426
Entailment
  CPL, 269–270
  CPDL, 392
  CQL, 503–505
  MQL^1, 533
  natural language, 210–211, 218–219, 228–229, 231, 330, 340–341 (*see also* Implicature; Presupposition)
Equivalence
  material, 243–244
  semantic (*see* Semantic equivalence)
Eratosthenes of Cyrene, 40
Eta conversion, 644
Ethology, 3–5
Eudoxus of Cnidus, 39
Euler, Leonhard, 55
Euler diagram, 55–59
Exclusion. *See* Method of difference
Exclusive adjective, 537
Exclusive *or*, 346–348
Existence (function), 93. *See also* Left totality
Exophora, 177–180, 209, 232, 618–619
Experiment, 3–5, 9, 37–38, 43–46
  deprivation (*see* Deprivation experiment)
  See also Observation

Extension (function), 95
Extension (logic)
  simply typed Lambda calculus, 593
  See also Classical valuation; Tarski extension
Extrapolation. *See* Observation
Extraposition, 159

Fallacy, 239
Falsity, 329–330. *See also* Truth
Family (of sets), 70
Fillmore, Charles, 179
Finite verb, 114
First order predicate logic. *See* Classical quantificational logic
First person pronoun, 110, 180
Fixed action pattern, 4
Flouting (a maxim), 213. *See also* Observing (a maxim)
Fodor, Jerry, 2, 28
Formation rule
  clause (*see* Clause formation rule)
  constituency (*see* Constituency formation rule)
  prefixation, 20
  suffixation, 18
Formula. *See also* Atomic formula; Basic formula; Closed formula; Composite formula; Open formula
  CPDL, 378–380
    definition (categorematic), 379
  CPL
    definition (categorematic), 247–249
    definition (syncategorematic), 245–247
  CQL, 458
    definition (categorematic), 458
  CQLI, 522
  Lambek calculus, 552–553
  MQL^1, 529
  MQL^2
    first presentation, 536
    second presentation, 543–544
Formula natural deduction, 276–277
  column format
    CPL, 277–293
    CQL, 506–516
  tree format
    CPL, 297–302
    Lambek Calculus, 553–556
  See also Elimination rule; Introduction rule
Free to substitute
  monotyped Lambda calculus, 586
  simply typed Lambda calculus, 590
Free variable
  CQL, 461–462
  monotyped Lambda calculus, 583–584

MQL^1, 529
MQL^2, 537, 544
  simply typed Lambda calculus, 590
Frege, Gottlob, 10, 28, 179
Frei, Henri, 179
Fresh variable, 607
Function, 90–93, 98–99. *See also* Bijection; Characteristic function; Constant function; Extension (function); Injection; Near variant function; Partial function; Product function; Restriction; Surjection
Functional structure. *See* Structure: simply typed Lambda calculus
Future tense, 112, 183

Gödel, Kurt, 552
Galileo, 11
Gap, 158, 190, 334. *See also* Antecedent; Endophora; Pronoun
Gapping, 196, 207–208, 323–324. *See also* Appended coordination; Ellipsis
Geach, Peter, 194
Gender (grammatical), 108
Generalized Cartesian product, 72–74. *See also* Cartesian product
Generalized direct product. *See* Generalized Cartesian product
Generalized intersection, 72. *See also* Intersection
Generalized phrase structure grammar (GPSG), 163
Generalized quantificational logic, 526–547
Generalized union, 72. *See also* Union
Generative grammar, 9
Genie, 9
Gentzen, Gerhard, 297, 303, 320
Gentzen sequent calculus, 276, 315–317, 566–568
Gentzen sequent calculus deduction
  CPL, 276, 315–319, 320
  Lambek calculus, 566–568
  See also Left introduction rule; Right introduction rule
Gerund, 115, 120
Gerundial clause. *See* Gerundial phrase
Gerundial phrase, 402
Gradable adjective, 627
Grammar, 1–2, 9–10, 26–27, 105–106. *See also* Categorial grammar; Constituency grammar; Enriched constituency grammar; Generative grammar; Traditional grammar; Transformational grammar
Grammatical competence, 6–13. *See also* Idealization
Grammatical form, 226
Grammatical tense, 182–183
Graph
  bipartite (*see* Bipartite graph)
  directed (*see* Directed graph)
  See Binary relation
Greek, classical, 1–2

Index

Greek numeral, 31
Grice, Herbert Paul, 210, 217
Gricean maxim, 212
  cooperation, 213
  manner, 213, 217
  quality, 213–214
  quantity, 213, 215–216, 334–336
  relevance, 213, 216, 333
Grossetete, Robert, 42

Harris, Zellig, 11, 21, 122
Head (of a constituent), 398–399
Head driven phrase structure grammar (HPSG), 163
Helping verb. *See* Auxiliary verb
Historical linguistics, 2
Hockett, Charles, 11, 21, 122, 141–142
Husserl, Edmund, 21, 28, 179, 551
Hypothesis. *See* Theoretical hypothesis

Idealization, 13
Idealized counting numeral, 32
Identity relation, 77–78, 432–434, 521, 523, 638–640
I-language. *See* Grammatical competence
Image (function), 94
Immediate constituency analysis, 11–12, 121–122
Immediate constituent, 122, 128
Immediate subformula
  CPL, 249–251
  CPDL, 380–381
Imperative clause, 339–341, 350–352
Imperative mood, 113, 339–340, 361
Implicature, 210–212, 217–220, 228–231, 337–338, 346, 353, 371
Inaccessibility (formula natural deduction), 287
Inclusive *or*, 346–347
Inconsistency, 293. *See also* Consistency
Indefinite pronoun (*one*), 189
Independent clause, 322, 329
Indexation (of sets), 72–74
Indexicality. *See* Deixis
Indigenous languages, 2
Indirect object, 111
Indirect question, 160
Individual symbol, 376–378, 457. *See also* Relational symbol; Structure (for a signature)
Indo-Aryan, 9
Indo-European language, 9, 108
Induction case, 561. *See also* Mathematical induction
Infinitival clause. *See* Infinitival phrase
Infinitival phrase, 104, 402
Infinitive, 114, 402
Injection, 94
Innate endowment, 3–9
  language, 3–6 (*see also* Grammatical competence; Linguistic capacity)
Instinct, 3

Integer, 53
  negative, 53
  positive, 53
Interdeducibility, 293
Interjection, 107, 118
Interlocutor, 209. *See also* Addressee; Speaker
International Phonetic Alphabet (IPA), 105
Interpretation function, 383–384, 467
Interrogative clause, 340–343, 350–351, 402
Interrogative ellipsis, 197
Interrogative mood, 339–341, 361
Interrogative pronoun, 110, 618
Intersection (of sets), 63–64
Intersective adjective, 636, 638
Intersubstitutability, 122
Intransitive verb, 110, 401, 405–406, 429, 631, 640
Intransitivity (binary relation), 83
Intraspecial communication, 5
Introduction rule, 276, 557–559
  formula column format, 278–291, 506–507, 510–512
  $\wedge$, 279–280
  $\vee$, 283–284
  $\leftrightarrow$, 285–286
  $\rightarrow$, 287–289
  $\neg$, 290–291
  $\exists$, 507–509
  $\forall$, 509–510
  formula tree format
  $\wedge$, 297–298
  $\vee$, 298–299
  $\leftrightarrow$, 299–300
  $\rightarrow$, 300
  $\neg$, 301
  \, 553
  /, 554
  ·, 555
  sequents in a column
  $\wedge$, 304–305
  $\leftrightarrow$, 306
  $\neg$, 308
  $\rightarrow$, 307–308
  $\vee$, 305
  sequents in a tree
  $\wedge$, 310–311
  $\vee$, 311–313
  $\leftrightarrow$, 313
  $\rightarrow$, 313–314
  $\neg$, 314–315
  \, 557
  /, 558
  ·, 558
  See also Elimination rule
Invalid argument. *See* Argument: invalid
IPA. *See* International Phonetic Alphabet
Irreflexivity (binary relation), 79
Island, 645
Iteration. *See* Recursion

Jaśkowski, Stanislaw, 276, 277
Jenner, Edward, 42–43
Jespersen, Otto, 620
Joint method, 43–44. *See also* Mill's methods

Kamp, Hans, 184
Kaplan, David, 179, 619
Keenan, Edward L., 547, 667n12
Kepler, Johannes, 39

Labeled analysis tree diagram. *See* Analysis tree
Labeled bracketing, 129, 134
Labeled synthesis tree diagram. *See* Synthesis tree
Lambda calculus
  Lambek typed, 603–612
  monotyped, 579–587
  simply typed, 588–603
  untyped (*see* Lambda calculus: monotyped)
Lambda matrix, 579
Lambda prefix, 579
Lambda term, 629–630, 638
  closed, 655–656
  open, 655
Lambek, Joachim (Jim), 21, 551
Lambek calculus, 21, 34, 320, 552–578
Lambek type, 604
Lambek typed grammar, 629–631, 635, 647, 652, 654
Lambek typed lambda calculus, 603–612
Language acquisition, 5–9
Language acquisition device, 6
Language faculty. *See* Language acquisition device
Language learning. *See* Language acquisition
Latin, 1–2
Law of excluded middle, 367
Law of noncontradiction, 368
Left embedding, 164–165
Left introduction rule (Gentzen sequent calculus), 316–319
  ∧, 316
  ↔, 317
  →, 317–318
  ∨, 318–319
  ¬, 319
Left monogamy (binary relation), 88–90, 92–93
Left totality (binary relation), 86–87, 90, 92–93
Leibniz, Gottfried, 28, 55
Lemmon, Edward John, 179
Length (of sequence), 67
Leśniewski, Stanislaw, 551
Lewis, David, 27, 179, 619
Lexical ambiguity, 219
Lexical category, 106–107, 423, 425, 436–437
Lexical entry, 136, 427
Lexical Functional Grammar (LFG), 161
Lexicon, 26, 130, 423, 426, 427–428
Linguistic behavior, 5
Linguistic capacity, 6–9, 209

Linguistics, 1–3, 35
  historical, 2
Linking verb. *See* Copular verb
Linnaeus, Carl, 36
Linström, Per, 535
Lippershey, Hans, 201
List notation, 52. *See also* Abstraction notation
Literal. *See* Basic formula
Locke, John, 28
Logic
  classical propositional logic (*see* Classical propositional logic)
  classical predicate logic (*see* Classical predicate logic)
  classical quantificational logic (*see* Classical quantificational logic)
  defeasibility, 679
  dynamic, 679
  free, 679
  fuzzy, 679
  intuitionistic, 320
  linear, 320
  many-valued, 679
  modal, 679
  partial, 679
  predicate (*see* Classical quantificational logic)
  propositional (*see* Classical propositional logic)
  relevance, 320
  substructural, 320
Logical form, 226, 648, 651–652, 655–656. *See also* Deep structure; Surface structure
Long distance dependency, 159
Loop (directed graph), 79
Lorenz, Konrad, 3n2
Lyons, John, 179

Mahābhāṣya (Great Commentary), 10
Main clause, 118
Main logical connective, 252, 381
Map. *See* Function
Mapping. *See* Function
Mass noun, 620–624
Material implication, 242–244
Mathematical induction, 561
Matrix (binary relation), 76–77
Maxim. *See* Conversational maxim; Gricean maxim
Membership (in a set), 52
Mendeleev, Dimitri, 36
Metalinguistic negation, 371
Method of agreement, 42–43. *See also* Mill's methods
Method of difference, 43. *See also* Mill's methods
Method of exclusion. *See* Method of difference
Middle voice, 412
Mill, John Stuart, 42–44
Mill's methods, 42–44
Minimal clause, 397, 398–399, 426, 429–431, 630
Minimal pair, 44

Index

Minoan numeral, 30
Minsky, Marvin, 552n2
Model. *See* Structure (for a signature)
Model theory, 18, 21–22, 27
Modification. *See* Modifier
Modifier, 115–117, 399
Modus ponens, 355, 363
Modus tollens, 239, 355, 363
Monadic Quantificational Logic
   one place monadic quantificational logic (MQL^1), 527–535
   two place monadic quantificational logic (MQL^2), 535–546
   first presentation, 536–543
   second presentation, 543–546
Monogamy (binary relation). *See* Left monogamy; Right monogamy
Monotonicity, 546
  decreasing, 546
  increasing, 546
  left, 546
  right, 546
Montague, Richard, 27, 179
Mood, 113. *See also* Imperative mood; Interrogative mood; Subjunctive mood
Morphological tense, 182
Morphology, 14–15
Mostowski, Andrzej, 522
Movement (transformational grammar), 160, 648. *See also* Transformational rule

Natural deduction, 276
  formula (*see* Formula natural deduction)
  sequent (*see* Sequent natural deduction)
  See also Gentzen sequent calculus deduction
Natural number, 53
Near variant function, 96
Negation, 241, 244
Negative integer. *See* Integer: negative
Nida, Eugene, 11, 21, 122
Node, 77, 86
Nominal clause, 118
Nominal ellipsis, 198
Nominative case, 109
Nonconventionality, 217–218
Nondetachability, 218
Nonfinite verb. *See* Gerund; Infinitive; Participle
Noun, 107–110, 420, 422. *See also* Abstract noun; Collective noun; Common noun; Compound noun; Concrete noun; Count noun; Mass noun; Pronoun; Proper noun
Noun clause. *See* Nominal clause
Noun phrase, 124, 425. *See also* Plural noun phrase; Quantified noun phrase
Null binary relation. *See* Empty binary relation
Null set. *See* Empty set
Number. *See* Integer; Natural number; Real number

Number (grammar)
  singular, 108
  plural, 108
  See also Number agreement
Number agreement, 146–149, 199
Numeral, 28–33. *See also* Counting numeral; Greek numeral; Minoan numeral; Roman numeral; Tally numeral

Object, 109, 115. *See also* Direct object; Indirect object
Objective case, 109
Objective complement, 109, 115
Observation
  common sense, 35–42
  regularity (*see* Observational regularity)
  See also Experiment; Theoretical hypothesis
Observational regularity, 35–44
Observing (a maxim), 213. *See also* Flouting
Open formula
  CQL, 461
  CQLI, 522
  MQL^1, 529
  MQL^2, 537
Open term
  monotyped Lambda calculus, 583–584
  simply typed Lambda calculus, 590
Optional complement, 410, 418, 445
Order
  partial (*see* Partial order)
  strict (*see* Strict order)
Ordered pair, 67, 75, 77. *See also* Sequence
Origin, 180. *See also* Spatial origin; Temporal origin

Pāṇini, 9–11, 21, 27, 122, 130, 163n7
Partial function, 94–95
Partial order, 84–85, 541
Participial clause. *See* Participial phrase
Participial phrase, 120
Participle, 106, 114–115, 120
Parts of speech, 105–106, 117–118, 121
Passive voice, 113, 412, 450
Pasteur, Louis, 43
Past tense, 112, 183
Patañjali, 10, 27–28
Pattern. *See* Regularity
Peirce, Charles Sanders, 179
Perfect aspect, 112, 183
Performance, 13. *See also* Competence
Person (grammar), 110, 112, 178
Personal pronoun, 110, 121, 178, 189–195, 618–620. *See also* First person pronoun; Second person pronoun; Third person pronoun
Philology, 1
Phone, 105, 136

Phonology, 14–15, 105, 136
Phrasal category, 152, 422
Phrase. *See* Adjective phrase; Gerundial phrase; Infinitival phrase; Noun phrase; Participial phrase; Prepositional phrase; Verb phrase
Phrase formation rule schema, 425–426, 437
  permuting complement, 443
  polyadic complement, 447
  polyvalent complement, 440
Phrase valuation rule schema, 437
  permuting complement, 443
  polyadic complement, 448
  polyvalent complement, 441–442
Planets, 38–39
Plato, 39
Plural, English, 7
Plural noun phrase, 658
  collective construal, 658–659
  distributive construal, 658–659
Polyadic adjective, 415–417
  contextual, 417
  reciprocal, 415
Polyadic complement, 410–412, 415–416, 418, 445–449, 450–451
Polyadic preposition, 418
Polyadic verb
  causative, 412
  contextual, 411–412
  indefinite, 411
  middle, 412
  reciprocal, 410–411
  reflexive, 410
Polynomial expression, 596
Polyvalent adjective, 415
Polyvalent complement, 406–410, 415–417, 439–441, 441–444
Polyvalent verb, 406–410
Positive integer. *See* Integer: positive
Possessive case, 109
Possessive pronoun, 618
Possessive s, 138
Post, Emil, 552n2
Post machine, 552n2
Poverty of the stimulus, 6–8
Power set, 70, 96–97
PP preposing, 159
Prawitz, Dag, 297
Predicate (grammar)
Predicate (logic). *See* Relational symbol
Predicate nominative, 109
Predicative adjective, 413, 624–627, 636–638
Pre-image (function), 94
Premise, 237–239. *See also* Conclusion
Preposition, 107, 109, 111, 116, 121, 417–419, 422, 432, 639–640
  complementless (*see* polyadic preposition)
Prepositional phrase, 117, 425, 433, 641, 661
  modifier, 639–640

Present tense, 112, 183
Presupposition, 222–224, 228–230, 335, 371
  trigger, 224–225, 227
  See also Entailment; Implicature
Principal clause. *See* Main clause
Prior, Arthur, 184
Priscian, 106
Product function, 97–98
Proform, 188–190, 194–195, 399. *See also* Endophora; Pronoun
Progressive aspect, 112
Projection problem, 152–153, 439, 454
Pronominal adjective, 115–116
Pronominalization, 619
Pronoun, 107, 110, 189–195. *See also* Demonstrative pronoun; Indefinite pronoun (*one*); Interrogative pronoun; Personal pronoun; Reciprocal pronoun; Reflexive pronoun; Relative pronoun
Proper name. *See* Proper noun
Proper noun, 107, 110, 616–617, 630–631, 635–636
Proper subformula
  CPL, 250–251
  CPDL, 380
Proper subset, 59–60. *See also* Subset
Propositional connective, 244–245, 247, 376
Propositional logic. *See* Classical propositional logic
Propositional variable, 239, 244–245
Protasis, 229n3, 343, 354, 359. *See also* Apodosis
Pseudocleft sentence, 139, 165, 399–400
Psychology, 3–9
Putnam, Hilary, 8

Quality, 107, 115, 121
Quantificational connective
  CQL, 457
    existential, 457, 466–469, 470–472, 473–476, 477–486, 489–492, 493–496, 506–512
    universal, 457, 466–469, 470–472, 473–476, 477–486, 489–492, 493–496, 506–512
  MQL^1, 528–529
  MQL^2, 535
Quantificational determiner, 628. *See also* Quantified noun phrase
Quantificational pronoun, 618
Quantified noun phrase, 634–636
  nonsubject position, 646–648
  scope, 649–656
  scope judgments, 649–651
Quantifier. *See* Quantificational connective; Quantificational determiner
Quantifier matrix
  CQL, 459
  MQL^1, 529
Quantifier prefix
  CQL, 459
  MQL^1, 529
Quantifier raising (QR), 648, 655–656

Index

Quasi cardinal adjective, 620, 658, 660. *See also* Cardinal adjective
Quine, Willard van Ormine, 194, 619, 641

Range, 93. *See also* Codomain
Real number, 99n7
Reciprocal pronoun, 110, 191, 618
Recursion, 12–13, 18–26, 144–145
Recursion theory, 18
Recursive rule. *See* Recursion
Redi, Francesco, 43
Reduction. *See* Conversion
Reflexive pronoun, 110
Reflexivity (binary relation), 78–79
Regularity. *See* Observational regularity; Statistical regularity
Regularization, 13
Reichenbach, Hans, 179
Relation, 75
  binary (*see* Binary relation from a set to a set; Binary relation on a set)
Relational grammar, 161
Relational symbol, 376–378, 457. *See also* Individual symbol; Structure (for a signature)
Relative clause, 641–646
  appositive, 642
  restrictive, 642, 645
Relative pronoun, 110, 160, 618
Response, 3. *See also* Stimulus
Restriction (of a function), 95
Rewriting system, 21
Right introduction rule (Gentzen sequent calculus)
  $\wedge$, 316
  $\leftrightarrow$, 317
  $\rightarrow$, 317–318
  $\vee$, 318–319
  $\neg$, 319
Right monogamy (binary relation), 89
Right totality (binary relation), 88
Role, 180–181
Roman numeral, 30–31
Ross, John R. (Haj), 645
Rotation of the earth, 41–42
Russell, Bertrand, 179

Sanskrit, 1–2, 9–10, 27, 108, 122
Satisfaction
  CPL, 265
  CPDL, 389–390
  CQL, 497
  MQL^1, 533
Satisfaction set
  CQL, 489–490
  MQL^1, 527–528
Satisfiability
  CPL, 267
  CPDL, 391
  CQL, 498–499
  MQL^1, 533
  See also Unsatisfiability
Scope
  CPL, 252
  CPDL, 381
  CQL, 459
  judgment (*see* Quantified noun phrase: scope judgments)
  Lambda calculus, 582
  monotyped Lambda calculus, 582
  MQL^1, 527
  MQL^2, 537
  simply typed Lambda calculus, 590
  See also Quantified noun phrase
Secondary predication, 407
Second person, 110, 180
Second person pronoun, 110, 181
Segment. *See* Phone
Segmentation, 122
Semantic anomaly, 16
Semantic equivalence
  CPL, 268–269
  CPDL, 392–393
  CQL, 499–503
  MQL^1, 534
Semantics, 18, 27
Semi-Thue system, 21
Sentence, 106. *See also* Clause
Sequence, 66–67. *See also* Ordered pair
Sequent, 276–277. *See also* Gentzen sequent calculus; Sequent natural deduction
Sequent natural deduction
  column format
    CPL, 303–309
  tree format
    CPL, 309–315
  Lambek calculus, 557–560
  See also Elimination rule; Introduction rule
Set, 51
  finite, 54
  infinite, 54
  null (*see* Empty set)
  See also Doubleton set; Empty set; Singleton set; Universal set
Set builder notation. *See* Abstraction notation
Setting, 177, 209, 232–233
  modification (thereof), 186
Signature
  CPDL, 377
  CQL, 457–458
  CQLI, 522
  MQL^1, 527
  MQL^2, 527
    first presentation, 528
    second presentation, 535
Simple clause, 615, 630–636
Simplified analysis tree, 135

Simplified synthesis tree, 133–134
Singleton set, 54
Single valuedness. *See* Left monogamy
Size (of set). *See* Cardinality
Skinner, Burrhus Frederick, 6
Sluicing. *See* Interrogative ellipsis
Small clause, 407
Small pox, 42–43
Soundness
 type, 594
Spatial origin, 180
Speaker, 180. *See also* Addressee; Bystander
Stalnaker, Robert, 179, 619
Statistical regularity, 35n12
Stimulus, 3–4. *See also* Response
Stimulus, poverty of. *See* Poverty of the stimulus
Stoics, 17, 28
Striate cortex, 5, 3
Striate neuron, 5
Strict order, 84
Stripping. *See* Appended coordination
Structural ambiguity, 141–143, 219
Structuralist linguist. *See* American structuralist linguist
Structure (English lexicon), 427–428
Structure (for a signature), 384–385
 CPDL, 384–385
 CQL, 465
 CQLI, 522
 MQL^1, 527
 MQL^2, 527
 simply typed Lambda calculus (*see* Functional structure)
Subcategorization, 150–152
Subcategorization problem, 425, 454
Subformula. *See* Immediate subformula; Proper subformula
Subject, 109
Subjunctive conditional, 361
Subjunctive mood, 113. *See also* Counterfactual conditional
Subordinate clause, 118. *See also* Counterfactual conditional; Subjunctive conditional
Subordinating conjunction. *See* Subordinator
Subordination. *See* Subordinator
Subordinator, 117, 322–325, 327, 358–360, 362–366
 adjunctive, 365–366
 disjunctive, 365–366
 if, 353–364, 366
Subsective adjective, 637
Subset, 55, 57. *See also* Proper subset
Substituendum
 CQL, 462
 monotyped Lambda calculus, 585
Substituens
 CQL, 462

monotyped Lambda calculus, 586
simply typed Lambda calculus, 590
Substitution
 CQL, 462
 monotyped Lambda calculus, 584
 simply typed Lambda calculus, 590
Substitution instance
 CQL, 463
Superlative adjective, 115
Superordinate clause, 118
Suppes, Patrick, 303
Supposition, 278, 287
Surface structure, 163. *See also* Deep structure; Logical form
Surjection, 94
Sweet, Henry, 149
Symmetry (binary relation), 80
Syncategoremata, 17
Synchronic, 2. *See also* Diachronic
Syntax, 14–17
Synthesis tree, 132–134
 categorematic (logic), 248
 immediate constituency (*see* Immediate constituency analysis)
 syncategorematic (logic), 246
 See also Analysis tree

Table. *See* Matrix
Tally numeral, 29
Tarski, Alfred, 27, 166, 483
Tarski extension
 CQL, 484, 490–491
 CQLI, 522
 MQL^1, 533
 MQL^2
  first presentation, 537
  second presentation, 545
Tautology
 CPL, 266
 CPDL, 390
 CQL, 498
 MQL^1, 533
 See also Contingency; Contradiction
Techne Grammatike, 106
Temporal adverb, 183
Temporal order, 182
Temporal origin, 180
Tense, 112–113, 182–183. *See also* Future tense; Past tense; Present tense
Term
 CQL, 457–459
 CQLI, 522
 monotyped Lambda calculus, 579
 MQL^1, 529
 MQL^2, 537
 See also Atomic term; Closed term; Composite term; Open term

Index

Thematic adjective, 625–628
Theorem, 293
Theoretical hypothesis, 25, 39–42. *See also* Observation
Third person, 110, 121, 180, 189
Third person pronoun, 110, 181
Thomas of Erfurt, 28
Tinbergen, Nikolaas, 3n2
Topicalization, 159
Totality. *See* Left totality; Right totality
Tough movement, 160
Trace, 163
Traditional grammar, 105–106
Transformational grammar, 161, 648, 651–652, 655–656, 664, 668
Transformational rule, 161, 163, 619, 645, 664. *See also* Conjunction reduction; Dative shift; Quantifier raising; Tough movement; wh movement
Transitive verb, 111, 401, 431
Transitivity (binary relation), 82
Tree, 246–248, 458. *See also* Analysis tree; Synthesis tree
Trigger, 4
Truth, 329–330. *See also* Falsity
Truth conditions, 330, 339, 343. *See also* Answerhood conditions; Compliance conditions
Truth function, 262–264
Truth value, 254, 321, 329–330
Truth value assignment, 254–256
 bivalent (*see* Bivalent truth value assignment)
Turing, Alan, 552
Turing machine, 552
Type
 simple, 588
 See also Binary Boolean type; Lambek type
Type assignment, 633, 643
Type logical grammar. *See* Lambek typed grammar

Unary connective, 244, 247
Union (of sets), 63
Uniqueness. *See* Left monogamy
Universal binary relation, 77
Universal grammar. *See* Language acquisition device
Universal set, 54
Universal statement, 35–39
Universe, 383–384. *See also* Structure (for a signature)
Universe of discourse. *See* Universal set
Unsatisfiability
 CPL, 267–268
 CPDL, 391
 CQL, 499
 MQL^1, 533
 See also Satisfiability

Valid argument. *See* Argument: valid
Valuation, 254–257, 465–486, 490–491, 493–496
 bivalent, 255–256
 classical (*see* Classical valuation)
Valuation rule
 clause (*see* Clause valuation rule)
 constituency (*see* Constituency valuation rule)
 prefixation, 22, 24
 suffixation, 25
Value (function). *See* Image
Variable, 52–53, 457. *See also* Bound variable; Constant; Free variable; Fresh variable; Propositional variable
Variable assignment, 466, 477–478, 482–489, 493–496
 CQL, 466, 477–478, 478–486, 493–496
 simply typed Lambda calculus, 593
Venn, John, 55
Venn diagram, 55–59
Vennemann, Theo, 27
Verb, 107, 109, 110, 112, 399–414. *See also* Auxiliary verb; Copular verb; Ditransitive verb; Finite verb; Intransitive verb; Nonfinite verb; Polyadic verb; Polyvalent verb; Transitive verb
Verbal Behavior, 6
Verb phrase, 126–127, 399–414, 424, 431, 432, 435–436
Verb phrase ellipsis, 197–198
Vervet monkey, 5
Voice, 113, 412. *See also* Active voice; Middle voice; Passive voice
von Frisch, Kurt, 3n2, 5
VP preposing, 159
Vyatireka. *See* Method of difference

Waggle dance, 5
Watson, John Broadus, 3
Wells, Rulon, 11, 21, 122
Welsh counting numeral, 32
wh movement, 160
Wild child (of Aveyron), 9
William of Ockham, 28
Wittgenstein, Ludwig, 28
World knowledge, 16. *See also* Belief
Wrapping, 158–159
Wundt, Wilhelm, 3, 28